Electric Arc Furnace: Methods to Decrease Energy Consumption

Alberto N. Conejo

Electric Arc Furnace: Methods to Decrease Energy Consumption

 Springer

Alberto N. Conejo
School of Metallurgical and Ecological
Engineering
University of Science and Technology
Beijing (USTB)
Beijing, China

ISBN 978-981-97-4052-9   ISBN 978-981-97-4053-6 (eBook)
https://doi.org/10.1007/978-981-97-4053-6

This Springer imprint is published by the registered company Springer Nature Singapore Pte Ltd.
The registered company address is: 152 Beach Road, #21-01/04 Gateway East, Singapore 189721,
Singapore

If disposing of this product, please recycle the paper.

This book is dedicated to the memory of my mother Ma. Cruz Nava-Romero (†) and to my family: Carmen, Alberto and Roberto

Preface

Steel is the material that has had the largest impact on the development of civilizations since the beginning of the Iron Age in human history. Without steel, the Industrial Revolution would not have happened and our daily lives would be closer to the Middle Ages. In 2023, steel continues its supremacy as the largest metal produced with 1888.2 Mton, followed in a second distant place by aluminum with 70.5 Mton. The production of steel will increase in the following decades due to the need of industrialization of developing countries/regions (China, India, Latin America, Africa) and to improve their economies.

Steel consumption per capita is fully coupled with the gross domestic product (GDP). China has been in the first two decades of the twenty-first century a strong driving force of the world economy, with a rise in steel consumption per capita from 123 kg in 2001 to 645 kg in 2022. As the top steel producer worldwide with a share of more than 50%, any shift in the steel industry in China has a big influence on the whole steel market. The second steel producer worldwide is India. India comes from the bottom in terms of steel production per capita, with 26.8 kg in 2001, growing threefold in the following two decades, reaching 81.1 kg in 2022. It is expected that by 2030, its production per capita will increase to 160 kg. Currently, the world's average steel consumption per capita is 230 kg. Since steel is critical to urban development and creation of national infrastructure, it will keep growing. It is estimated that by 2050, its growth will be 30%, reaching about 2.5 billion tons of crude steel.

However, the dark side of steel production has been a dramatic increase in emissions of carbon dioxide (CO_2). For many centuries, steelmaking has relied on fossil fuels as a source of energy and for the reduction of iron ores. Steelmaking is energy intensive because of the need to reach high temperatures to melt the iron ores and separate the gangue content. The technology to produce steel is divided into two stages, the first step is ironmaking, where oxygen in the iron oxides is removed using carbon, obtaining liquid iron with a high concentration of carbon which requires a second step, steelmaking, to remove the excess carbon, using oxygen. As a result, the integrated route produces 2 tons of CO_2 per ton of liquid iron and puts the steel industry as the largest producer of CO_2 among the industrial sector. This result is not because the steel industry is the largest producer of CO_2 per ton of metal but because is the metal with the largest annual production. The contribution from the steel industry to global emissions of CO_2 is about 7%. The final effect of these emissions is global warming and potential threat of annihilation of humankind. Today, there is no time to look back at the reasons why actions were not taken 50 years ago when the global average temperature rise on earth compared to pre-industrial era was 0.2 degrees and CO_2 emissions were growing exponentially, because today the global average temperature rise ranges from 1.2 to 1.4 degrees and global warming is a phenomenon that is not only affecting our daily lives but a serious threat to our existence. It is urgent to focus on the decarbonization of the steel industry. It is the huge generation of CO_2 emissions the main reason why the blast furnace has reached its peak as the dominant process in the integrated route to produce steel. It is not because of a low thermal efficiency or low productivity. It is because its "DNA" is made of carbon. In the last two decades, the steel industry started to lay out the foundations of a sustainable steel industry which in principle means the goal of net-zero carbon emissions. New breakthrough technologies are under development; many of them replacing carbon by hydrogen. However, its full commercialization will take at least two more decades. In this scenario, the electric arc furnace (EAF) has attracted most of the attention because it can help to solve the problem.

The EAF has a history of more than one century and has become a symbol of steel scrap melting. In its entire existence, the main raw material for the EAF has been steel scrap. Unfortunately, its high dependence on steel scrap was also the reason why the EAF was considered unsuitable for about six to seven decades to produce high-quality steels. The development of the minimill in the 1970s and 1980s and the high flexibility of the EAF in terms of raw materials demonstrated that it is capable to produce high-quality steels. The fact that the EAF can use 100% steel scrap is today one of its major strengths because it allows to consume a lower amount of energy and produce lower CO_2 emissions. In 2021, the World Steel Association (WSA) reported that "Every metric ton of scrap used for steel production avoids the emission of 1.5 tons of carbon dioxide and also avoids the consumption of 1.4 tons of iron ore, 740 kilograms of coal and 120 kilograms of limestone." According to BigMint.com, in 2022, the total consumption of steel scrap was about 598 Mton, consequently contributing to a decrease in the production of 897 Mton CO_2 per year.

Worldwide, the steel industry has progressively shifted from the integrated plant to the minimill; however, this change offers different challenges based on the availability of electric energy, availability and cost of high-quality scrap, as well as reserves of natural gas to produce direct reduced iron (DRI). Countries that in the past had a large production of steel through the BF-BOF, for example, China and Japan, have announced decisions to shift to the EAF. China is perhaps the country with the biggest challenges due to several reasons; it is the largest producer of CO_2 emissions, 90% of its steel production is dominated by the integrated route, a large number of integrated plants are relatively new, it has poor scrap availability, it has poor reserves of natural gas, the suppression of fossil fuels involves labor displacement of 5 million workers, 57% electricity is produced by coal fired power plants (using about 48% of the total coal), about half of existing EAF's are small and inefficient, etc. Despite all of these obstacles, China is rapidly installing new EAFs with the best available technologies, a factor that guarantees success in their goal to achieve carbon neutrality by 2060. In 2023, the first DRI plant in the world using Coke Oven Gas (COG), instead of natural gas, started operations at HBIS. Nippon Steel also decided to shift from the blast furnace to the electric arc furnace, marking 2023 as the peak and subsequent decline for the integrated route, opening the way to the supremacy by the EAF; however, the growth rate by the EAF in the next decades will depend on its capacity to overcome several challenges.

The EAF has a large number of limitations in comparison with the BOF, for example: (1) use of an expensive type of energy (electric energy), (2) very poor stirring conditions which results in lower decarburization rates, (3) residual elements in steel scrap, (3) cost of scrap can be higher than iron ore, (4) dependence on DRI to produce higher quality steels, (5) higher heat losses, (6) lower metallic yield (slag leaves the furnace losing iron and heat). If the EAF process overcomes these limitations, it will be able to fully overcome the BF-BOF route and become the dominant process for steelmaking.

The decrease in energy consumption is a common subject to all industrial processes, but it is of special interest for the EAF because the EAF is not only an energy-intensive process but also consumes an expensive form of energy; electric energy. The high energy intensity of the EAF results in higher production of CO_2 due to the current dependence on fossil fuels. Energy consumption is the most important variable in the production costs after raw materials, with a cost in the order of 15–20%. Due to the continuous degradation in the quality of raw materials, the gangue content will keep increasing, which results in a higher demand in energy consumption. The replacement of electric energy by chemical energy has been done in the past using carbon; however, the need to fully replace carbon by other fuels, such as hydrogen, will demand higher inputs of electric energy. This book discusses in detail 15 methods to decrease energy consumption in the EAF. Decreasing energy consumption requires an integral approach which means that all methods should be understood and optimized. Chapter 1 provides a summarized description of these methods. Slag foaming in particular is considered one of the most important methods because of its wide benefits not only in terms of energy consumption but also on obtaining higher melting rates. If these methods are applied and optimized, it will result not only in a decrease in the total energy consumed but also in a re-design of the EAF as we know it today. Its main features have remained the same since it was invented more than a century ago. The EAF limitations previously described can be eliminated with new designs that provide better stirring conditions and higher decarburization rates; in the end, the ultimate goal is to reach the sustainable production of steel. The development of an intelligent EAF is based on the development of process control algorithms which provide higher levels of automation. The number of workers directly in operation will continue to decrease but the number of skilled workers developing mathematical models will be in higher demand.

I want to acknowledge the support provided by many people at the former steel plant SICARTSA, currently ArcelorMittal Lazaro Cardenas (AMLC) in Mexico. My collaboration with them for almost 20 years was a fundamental learning experience.

I also acknowledge the support from USTB for a seed grant that started the writing of this book.

The knowledge available on the subject of this book is huge and impossible to handle by one single individual; however, the author hopes it provides a useful and solid knowledge base. I ask the readers for their help in order to improve a subsequent version that can include missing information, mistakes and questions that clarify the current contents of this book.

Beijing, P.R. China   Alberto N. Conejo
September 2024

Contents

About the Author

Alberto N. Conejo (Cuitzeo Mexico, September 25, 1959) is a Mexican professor who works at the University of Science and Technology Beijing (USTB) at the School of Metallurgical and Ecological Engineering since September 2018. Before this, he was a professor of ironmaking and steelmaking in Mexico for 30 years. He also worked for several metallurgical plants from 1982 until 1988 in several positions; welding supervisor at TEISA, a producer of turbines, head of the Heat Treatment Department at ENCO, a producer of gears for the auto industry and manager of Research and Development at NKS, a company producer of large castings and forged products. He did his Ph.D. in metallurgy at Colorado School of Mines in Golden CO, USA. His Ph.D. thesis was related to gas-solid reactions and the production of iron carbide from iron oxides.

For more than 15 years, he had a strong collaboration with the steel industry in Mexico, in particular with ArcelorMittal Lazaro Cardenas. He provided more than 1200 man-hours of technical training on ironmaking and steelmaking to union workers, supervisors and process engineers.

He has been a visiting professor at Tohoku University, Japan, with Prof. Shin-Ya Kitamura, for three months from July to September 2009; ArcelorMittal laboratories in Aviles Spain for six months from February to July 2011; School of Metallurgical and Ecological Engineering at the University of Science and Technology Beijing

(USTB), P.R. China, with Prof. Lifeng Zhang, for one year from August 2014 to July 2015; Indian Institute of Technology Kanpur, India, with Prof. Dipak Mazumdar, for six months, December 2017–May 2018.

He has received several national and international awards: Charles W. Briggs Award 2002, Iron and Steel Society, USA. Best Paper Award on EAF steelmaking; Michoacán State Award: Technology Award, 2005. Due to his work on EAF slag foaming. National Award (2nd place): Technology and Science Award granted by the Mexican Steel Producers Association (Canacero). 2010/2011. Due to his work on EAF modeling. In October 2021, he won the Chinese Government Friendship Award. This is a national award, granted due to prominent contributions to China or other outstanding accomplishments in teaching and academic studies.

He is an editorial member of the journal *Metallurgical Research and Technology* (previously known as *Revue de Metallurgie*), since January 2017, he has been an editorial member of the *International Journal of Minerals, Metallurgy and Materials*, Beijing, China, since July 2019, he has been a reviewer of more than 12 international journals.

He has published more than 90 technical papers in journals, with an index from Scopus, h = 22. He has supervised or co-supervised 20 Master and 3 Ph.D. students.

In 2018, he founded FeMRI (Ferrous Metallurgy Research Institute; www.femri.org), a company provider of consulting, training and research services for the steel industry worldwide.

Chapter 1
Introduction

1.1 Steel Production Through the EAF

1.1.1 Worldwide Production of Steel

Figure 1.1 shows the global trend on steel production worldwide by country in the last 50 years until 2020 [1]. Until about 1990, Russia, USA and Japan were the main steel producing countries. In 1993 China overtook USA as the main producer of steel worldwide.

Figure 1.2 shows the total amount of steel produced worldwide based on data from the World Steel Association [2]. The production of steel was very low before 1920s. After WWII the production of steel experienced its first large expansion until the 1970s. During the 1970s until the end of the twentieth century the production of steel remained almost stagnant. Starting the twenty-first century China has been the driving force for a second stage of a larger production of steel.

Figure 1.3 is a schematic comparison on steel production by process. By process, the production of steel was dominated by the Open Hearth Furnace (OHF) from 1900 until 1960 and then by the BOF process from 1960 until today. It can be observed a very low share by the EAF process before World War II (WWII). The slow growth of the EAF process can be attributed to the following reasons:

- Availability of steel scrap in large amounts and at low price. Initially the growth of the EAF process mainly occurred in developed in countries that generated steel scrap
- Its initial market was mainly the production of low-quality steels due to the presence of residual elements in steel scrap
- Depends on an expensive source of energy; electric energy. Its development depends on new technologies which decrease the price of electrical energy

A. N. Conejo, *Electric Arc Furnace: Methods to Decrease Energy Consumption*, https://doi.org/10.1007/978-981-97-4053-6_1

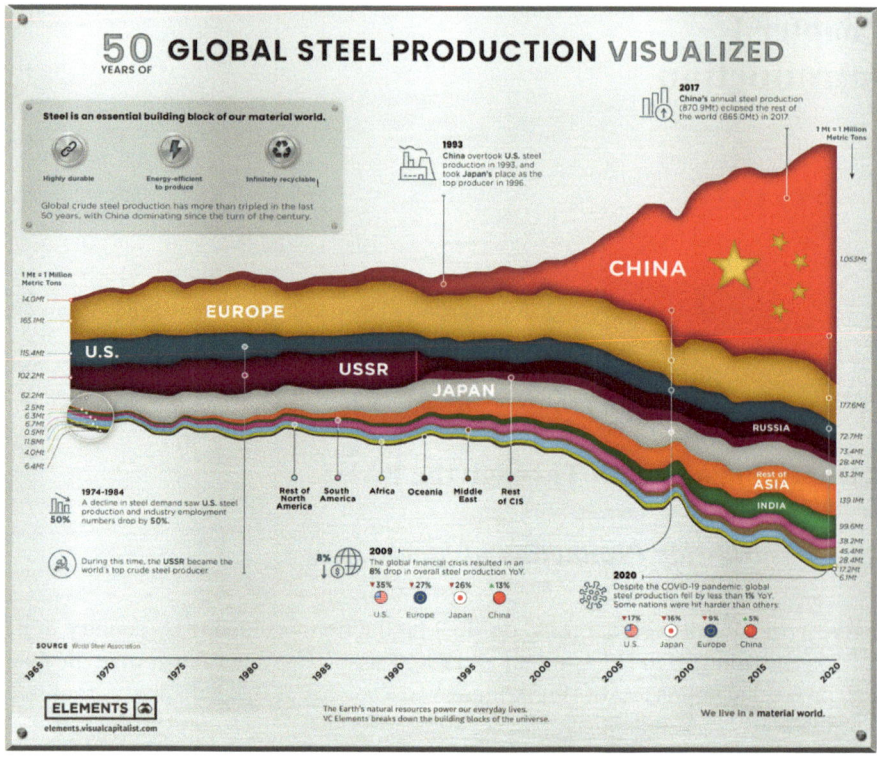

Fig. 1.1 Steel production worldwide in the last 50 years, by country. After [1]

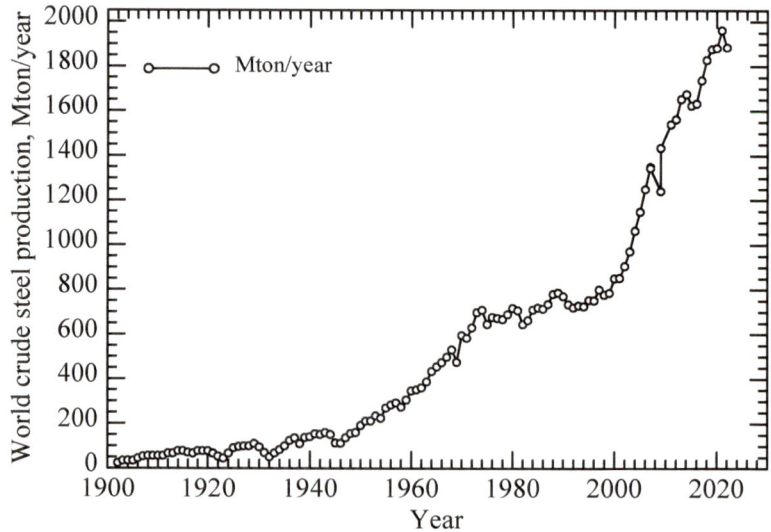

Fig. 1.2 Worldwide production of steel from 1900 to 2022. After [2]

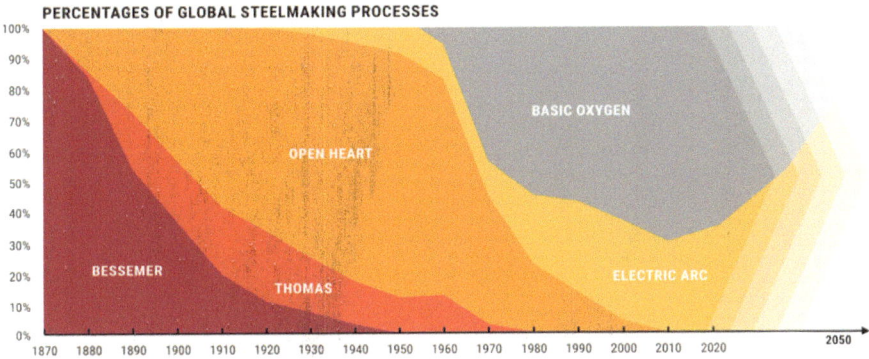

PERCENTAGES OF GLOBAL STEELMAKING PROCESSES

Fig. 1.3 Steel production worldwide in the last 50 years, by country. After [3]

- Advantages of the OHF and BOF over the EAF, such as energy recycling with the regenerative furnace and very large decarburization rates in the BOF, furthermore, both furnaces use a cheaper source of energy; chemical energy.

One of the reasons to develop the Electric Arc Furnace (EAF) was to melt steel scrap which became available due to the first massive production of steel through the Bessemer process and the subsequent end of life of steel products. The fuel for the Bessemer process is the heat generated by oxidation of elements contained in the hot metal, mainly carbon and silicon. Steel scrap contains a low concentration of carbon, at least one order of magnitude lower than hot metal and doesn't produce the chemical energy required for scrap melting. Using electrical energy instead of chemical energy solved the problem to melt high scrap ratios, up to 100%. However, a metallic charge entirely of steel scrap contains residual elements that cannot be removed under steelmaking conditions, therefore, the EAF was originally constrained to the production of low-quality steels such as long products; rebar and steel wire.

Figure 1.4 shows the growth of the EAF process worldwide from 1930 to 2022. In 1913 the total production of steel by the EAF reaches 0.75 Mton, 50% from Germany and 20% from the USA. The typical furnace capacity was 5 ton. During the first 30 years of the twentieth century the production of steel by the EAF process was less than 2.5%. After WW-II EAF steelmaking increased faster reaching about 10% by 1960. This was as a consequence of the need to produce steel in a more flexible way, with a broader range of raw materials, lower capital investment and faster production rates. A heat in the OHF process took about 12 h or even up to 24 h [4], in comparison with 3–4 h for the EAF in its early years. Steel production in the EAF kept growing steadily for the next 40 years, from 29 Mton and a share of 8.4% in 1960 to 283 Mton and a share of 33.3% in the year 2000. A contribution to the growth rate of the EAF was the birth of the minimill concept in the USA in the early 1980s. In spite that the production of steel by the EAF kept increasing after 2000, its percentage decreased due to an acceleration in the production of steel in China by

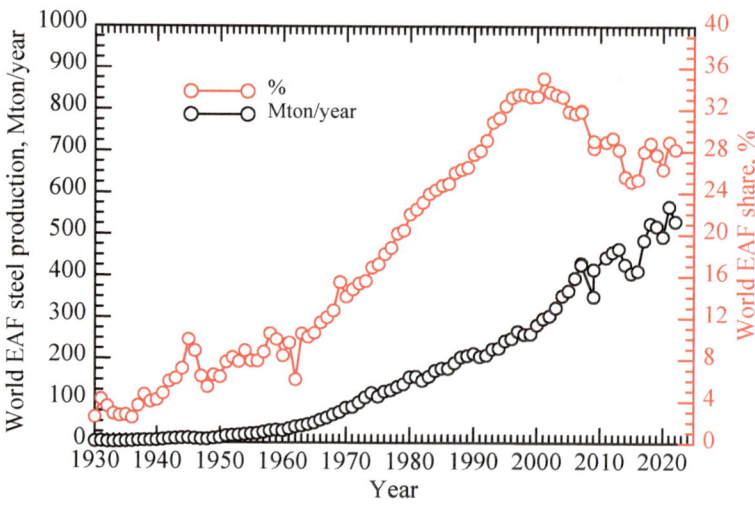

Fig. 1.4 Production of steel worldwide by the EAF process from 1900 to 2022. After [5]

the BF/BOF processes. Steel production increased to 531 MTon in 2022, equivalent to 28.2%.

1.1.2 Share of EAF Steel Output in the USA

The modern EAF process was developed in France but it was the USA who extensively adopted this method. In 1901 the United States Steel Corporation (USS) was founded when J. P. Morgan (1833–1913) buys out Carnegie Steel, Federal Steel, and National Steel. At one time, it was the largest steelmaker and largest corporation in the world. Bethlehem Steel Corporation is co-founded in 1904 by Charles M. Schwab (1862–1939), who had recently resigned from U.S. Steel, and the American industrialist, Joseph Wharton (1826–1919), founder of the Wharton School of Business, which then became the second largest steel producer in the United States.

In 1906 Héroult installed the first EAF at the Sanderson-Halcomb company in Syracuse New York, with a capacity of 3 ton. The second EAF was built by the Firth Stirling company at McKeessport, a suburb of Pittsburgh PA in 1908. Then in 1909 a plant of 13.5 ton was built by the Illinois steel company in South Chicago and in 1910 the American steel and wire company (USS subsidiary) built another plant with a similar capacity in Worcester Mass [6]. The USS in 1910 acquired the rights to commercialize the EAF process.

Figure 1.5 summarizes the production and share of the EAF in the USA from 1900 to 2022, taken from different sources; Rogers data [7] are based on Hogan's reports and the US geological survey (USGS) [8]. Up to 1915 the share of the EAF in the US was almost negligible (less than 0.2%) but grew seven-fold from 1915 to 1920

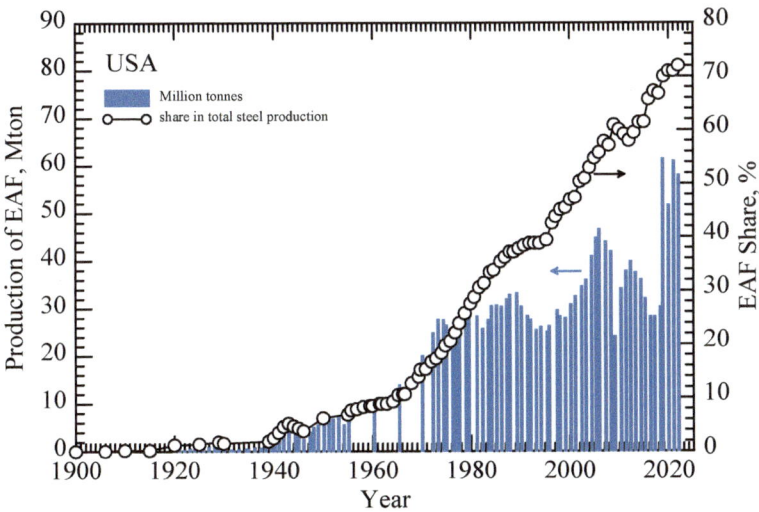

Fig. 1.5 Production of steel in the USA by the EAF process from 1900 to 2022. After [7, 8]

to reach 0.5 Mton/year. This result doesn't reflect the large expansion of the EAF because in spite of its large number, their capacity was small, from 1 to 20 ton. The number of EAF's from 1910 to 1929 increased from 10 to 650 with a corresponding increment in steel production from 52,975 ton to 1.01 Mton, equivalent to 0.2% and 1.74%, respectively [9]. In 1921, Pring [6] mentions new plants, one with three 25-ton EAF's and a production capacity of 12,000 ton/month and another with ten-30 ton EAF's. The growth from 1940 to 1950 was from 1.7 Mton to 6.03 Mton, representing 2.5% and 6.2%, respectively, a fast growth influenced by the end of WW-II.

The first EAF conference was organized by the American Institute of Mining, Metallurgical and Petroleum Engineers (AIME), chaired by Charles W. Briggs in 1943. During the decade from 1950 to 1960 the share of EAF steel production remained with minor changes, from 6.2 to 8.4%. However, the share of the EAF kept increasing from the 1960s until today as a result of the shutdown of OHF's and higher production share of steel by the BOF which caused a larger availability of scrap at lower prices because the demand of scrap by the BOF is lower than that of the OHF's. Herty [10] suggested an optimum range for the scrap in the OHF from 65 to 72%, a much larger value than that for scrap in the BOF which is usually below 20%.

The production of steel in the USA by the EAF increased from 1960 to 2022 as follows: 1960, 8.37 Mton (8.4%), in 1970, 19.9 Mton (15.2%), in 1980, 31.1 Mton (27.9%), in 1995, 38 Mton (40.4%), in 2002, 46 Mton (50.4%) and 59 Mton (72%) in 2022. These values clearly show the USA made the decision to produce steel primarily by the EAF process. The last OHF in the USA was closed in 1991 and almost a decade later, in 2002, the EAF became the main steelmaking reactor.

The evolution of the EAF industry in the USA gives a good reference of this process in the twentieth century. There are many sources of information that allow to study the EAF steel industry in the USA, for example: the book by Rogers [7] gives a detailed analysis of this steel industry in the USA from 1870 to 2001, Israilevich [11] compares the minimill and the integrated plants from 1958 to 1982 in terms of productivity, and both Stubbles [12] and Simcoe [13] provided a detailed history of the minimills in the USA. The concept of the minimill involves a steel plant with scrap-based EAF and continuous casting, with a small production capacity in the order of 0.1–0.3 Mton/year. According to Stubbles the first steel plant with these characteristics was Lasco (currently Gerdau Whitby), founded by Gerald Heffernan in 1964. In 1968 Heffernan built North Star steel in St. Paul Minnesota, it was the first minimill in the USA, with a production capacity of 0.2 Mton/year.

Nucor, today the largest company based on the EAF process, started operations using the EAF in 1969 in Darlington South Carolina, then a second minimill in the USA had a production capacity of 0.2 Mton/year. This company had its origins in 1905 as the Oldsmobile car manufacturer, then renamed Nuclear Corporation of America with the unsuccessful idea to provide nuclear services and later on acquired Vulcraft, a Joist manufacturer. Ken Iverson was the CEO of Vulcraft and the person who consolidated the success of the minimill industry from 1965 to 1990. Under his leadership Nucor implemented the latest technological developments into the EAF. In the decade from 1960 to 1970 steel production by the EAF doubled.

In 1969 Willy Korf built Georgetown steel. At this time the Midrex process had just passed the pilot plant trials at Oregon steel and Korf bought the rights for its commercialization. He built MIDREX plants both in Georgetown USA and Hamburg Germany in 1971 making him the pioneer using DRI in the EAF. In 1974 Korf bought MIDREX.

Chaparral steel (currently Gerdau Midlothian) was founded by Heffernan in 1975 with a projected annual capacity of 0.2 Mton. Its capacity grew to 1 Mton by 1984 and more than 2 Mton in 1998. This plant was managed by G. Forward and became a role model of an efficient minimill, supported on employee training, incentives, safety, recycling of waste products and incorporation of new technologies into the EAF.

A marking point in the evolution of the minimill was the decision of Iverson to apply a new continuous casting technology in 1988 at Nucor Crawfordsville; thin slab casting. This technology was a key factor for the minimills to expand and compete in much larger volumes, similar to those of the integrated plants, furthermore, produce high-quality steels and enter the flat-products market previously a monopoly of the integrated steel plants. Thin casting made a drastic simplification of the production line from 800 to 250 m. This plant casted slabs with a thickness of 51 mm, producing 2 Mton/year with an investment of 1 billion USD. From this point forwards it was not a minimill anymore but the name has remained. Nucor is not only the largest company with minimills but also a leader in terms of development and application of the most recent technologies. They developed a technology (CASTRIP) based on the concept of near-net shape casting invented in 1865 by Bessemer. The technology was developed in collaboration with BlueScope (Australia) and IHI (Japan) capable to directly cast thin strips with a thickness from 1.5 to 2 mm. After almost 20 years of research work, in 2002 Nucor Crawfordsville IN started operations using this technology and later on a second plant Nucor Blytheville AK in 2009. In addition to this, Nucor also built one of the largest DRI plants in the world in Louisiana expanding the options of raw materials for the EAF.

1.1.3 EAF Steel Industry in China

China is by far the largest producer of steel in the world. This should not be a surprise. In the history of the Chinese steel industry by Song [14], he cites the work by Hartwell who studied the production of iron and steel during the northern Song dynasty (960–1126 CE) and concluded that the scale of total production that China achieved was obtained by Europe until 700 hundred years later.

The first integrated plant was Hanyang Iron Works built in 1890. The production of steel in China from 1929 to 2022 is shown in Fig. 1.6 [5, 14–17]. Anshan iron works was built by Japan during the invasion years in Anshan-Liaoning in 1916. For the next 30 years Anshan produced almost 90% of all the steel in China using the OHF. In 1943 Anshan produced 1.3 Mton of pig iron and 0.9 Mton of steel [14, 15]. When the P.R. China was founded in 1949 the production of steel was only 0.158 Mton [18]. The first five-year plan (1953–1957) included China's industrial development as its primary goal, followed by the Great Leap Forward campaign during the second five-year plan (1958–1963), which was stopped in 1961. Anshan remained the most important steel complex until the early 1970s [16]. From 1966 to 1968 produced 5–6 Mton, about 50% of the annual production. The period from 1950 to 1993 can be separated in three stages [18]; from 1950 to 57 is the recovery period, marked by the technical support from URSS engineers in the period from 1949 to 1960 and one of the largest exchanges of foreign experts in history [19]; from 1958 to 1976 there is an overall increase but also large fluctuations and the period from 1977 to 1993 is the first stage of rapid growth in comparison with the previous 77 years.

Steel production by the EAF was resumed in 1953. According to Dong et al. [20] by 1978 there was a total of 982 BF's and steelmaking was dominated by the Open Hearth Process, there was 276 converters producing 3.03 Mton and 1678 EAF's producing 4.1 Mton of steel. The share of continuous casting was less than 10%.

Den Xiaoping (in office, 1982–1987) as the architect of modern China, after a visit to Nippon steel in Japan, founded Baosteel in 1978 and provided the foundations for the modern steel industry. The cost to build phase I and phase II of Baosteel had a cost of 30 billion Yuan (4.2 billion USD) [20]. In ten years, from 1978 to 1988, the steel production increased from 24 to 58 Mton. With the reform and opening up, the steel industry was allowed to retain profits for future expansions and pay bonuses. The last open-hearth in China was shut down in 2001, it had been in operation since 1960. It can be observed that steel production by the integrated route has dominated the production of steel during the last century. The exponential growth started at the dawn of the twenty-first century, increasing from 128 Mton in 2000 to 1094 Mton

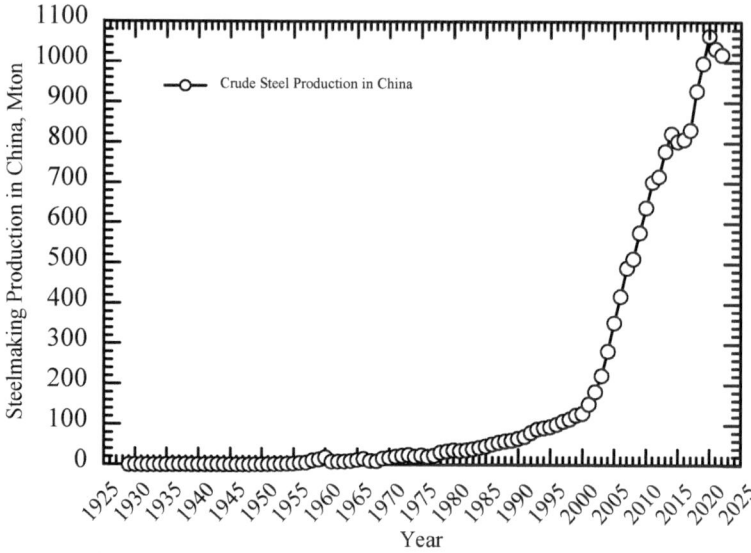

Fig. 1.6 Steel production in China from 1929 to 2022. After [5, 14–17]

in 2020. The stage between 1978 and 2000 is called sometimes a stage of capacity construction and from 2000 until 2013 a stage of high-speed development. The large demand of steel in China during the last 20 years is due to the large urbanization ratio. The urbanization ratio was 19% in 1978 and increased to 65% in 2022 [14].

The achievements in terms of steel production in China in the last 20 years are unique in the history of humankind. Before 1939 the steel consumption per capita was in the range 0.8–1.2 kg, decreased to even lower values at the end of their civil war, below 0.4 kg/year by 1949. In 1951 the steel consumption per capita in the USA was 530 kg, in this year the values in Canada, England and Mexico were 316, 240 and 31 kg respectively [15]. By 1978 it increased to 33 kg [21]. The values from 1980, 1990 and 2000 were 43.5, 64.4 and 100.2 kg, respectively [22]. The rapid growth in steel production since 2000 increased the consumption per capita of finished steel products to 691.3 kg in 2020, second largest in the world after South Korea [5].

The production of steel by the EAF process has played a minor role in China, as shown in detail in Fig. 1.7. Before 1990 the EAF share reached values in the order of 25% but as China kept increasing its share by the BF-BOF, the share of the EAF gradually decreased. In 2003 the share of steel through the EAF was 17.6%. In 2019, in spite of a higher production of steel, the share of the EAF in China decreased to 10.4% [23] due to the large steel output by the BF-BOF. Even with this modest share, China is currently the largest producer of steel through the EAF process, with an annual production similar to the whole North-America.

In China most of the steel companies are State Owned Enterprises (SOE) at the central and provincial level. Data reported in 2019 illustrate the share of SOE; out of 342 BOF's, 68 were central SOE, 209 provincial SOE and 65 belonged to private

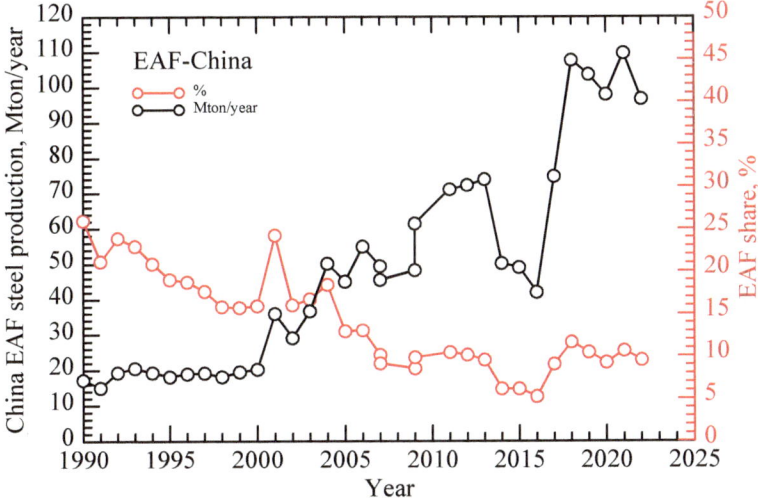

Fig. 1.7 Steel production in China by the EAF process from 1990 to 2022

companies [24]. In 2022, from the top 10 largest steel companies in the world, 6 are Chinese; Baowu (131 Mton), Ansteel (55 Mton), Shagang (41.4 Mton), HBIS (41.0 Mton), Jianlong (36 Mton) and Shougang (33 Mton). Shagang and Jianlong are private companies. Shagang's subsidiary, Zhangjiagang Hongchang is the largest steel complex in the world with a capacity of 30 Mton. In order to make the Chinese steel industry no only big but also strong, mergers were needed, for example the Hebei Iron and Steel (HBIS) group was formed in 2008 merging Xuanhua Iron and Steel (founded in 1918) and Tangsteel (founded in 1942), the group has 13 steel plants in Hebei province with a steel production capacity of about 57 Mton mainly through the BF/BOF. The capacity of the EAF in this group is about 9% [25].

Li et al. [26] illustrated in Fig. 1.8 the provincial steel production capacity in Mton in 2020 based on the Global Energy Monitor [25]. Hebei, Jiangsu and Shandong have the largest capacity with 299, 119 and 83 Mton per year, respectively.

Figure 1.9 illustrates a map based on data reported by the Global Energy Monitor [25], indicating the EAF steel capacity on a provincial level in China. It can be observed that Jiangsu has the largest installed capacity.

In 2023 the China Iron and Steel Industry Association (CISA) made a report on the situation and perspectives of the EAF process in China [28]. This section summarizes some of their results. In 1993 the EAF share was 23% but due to a much larger production of steel by the integrated route, its share has been decreasing; 17.6% in 2003 and 10.4% in 2019. For most part of the twenty-first century the EAF share has been around 10%. The reasons that explain this result are the following:

1. China was a non-developed country and lacked of steel scrap (importing steel scrap makes it more expensive than hot metal). In 2022 China consumed 263

Fig. 1.8 Steel capacity in China in 2020. After [26]

Mton of steel scrap, 80% consumed by the steel industry. The expected availability of steel scrap for 2030 is about 360 and 410 Mton for 2040. The scrap generated is estimated in the order of 10–15 Mton/year in the following three decades, however many organizational changes in scrap procurement need to be made to guarantee this goal.

2. China doesn't have large reserves of natural gas to produce DRI using the conventional gas-based technologies (imported DRI can be more expensive than hot metal)
3. The electric power networks were poor in the past
4. China didn't produce electrodes in the past
5. The environmental laws were not strict in the past. The environmental product declaration (EPD) platform was created recently in 2022, allowing consumers to recognize environmentally friendly companies. Carbon tax policies are being developed.
6. Lack of national engineering companies in the past that specialized in the construction of DRI-EAF plants, especially EAF plants above 100 ton/heat with high levels of automation.

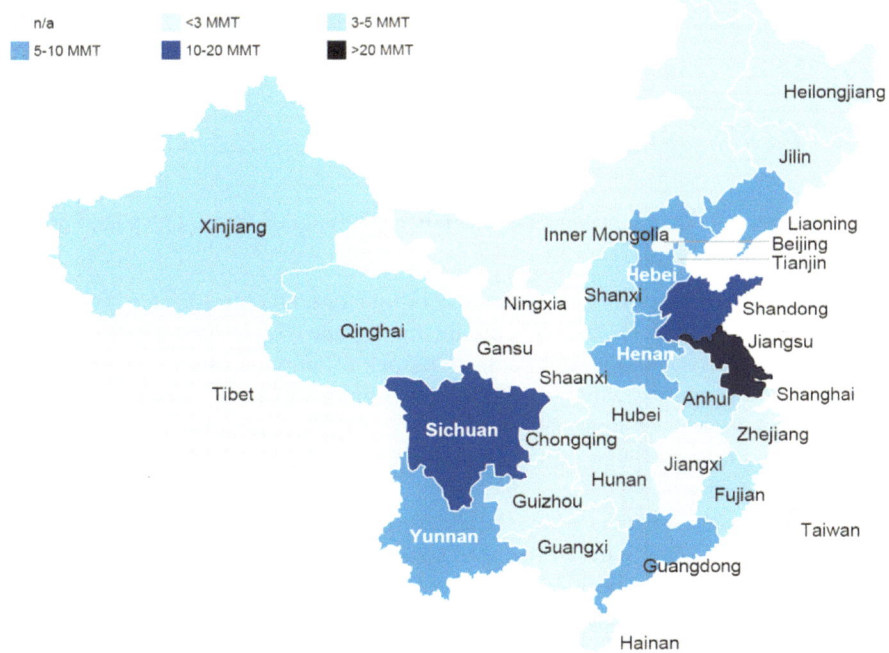

Fig. 1.9   EAF installed capacity in China in 2019. After [25, 27]

7. The CISA report [28] also indicated that scrap melting in the BOF is cheaper compared to the EAF, about 4–7 USD/ton and 40–50 USD/ton, respectively. The BOF in China commonly employs from 10 to 20% scrap.

Without the basic raw materials employed in the EAF, a poor electric power supply and expensive electrodes, it was not possible to consider the EAF as an alternative route to produce steel. In the last five years things have changed and there is no doubt that China will expand its EAF industry, as will be explained below.

A survey by CISA indicated that at the end of 2022 there was a total of 370 EAF's with a capacity of 190 Mton. Another source reported a higher number, indicating a total of 360 EAF's in 2012 [29] and in 2017 the number of new EAF's would be 100. The geographic distribution of the EAF's indicates two main zones, east and central south China, with 34% and 33% capacity, respectively. Six provinces; Guangdong, Jiangsu, Guangxi, Sichuan, Fujian and Hubei had a capacity of 54% of the total EAF capacity in China.

In developed countries, such as USA, Japan and Germany, the cost of steel by the EAF process is about 10–25% lower in comparison with the integrated plant because of a larger availability of steel scrap and efficient recycling systems. In China, due to the shortage of steel scrap there is an opposite situation, steel made by the EAF can be more expensive than the conventional route, for example in 2018 the price difference was about 100 USD/ton but this gap keeps decreasing over time, by 2021 the price difference was about 25–40 USD/ton.

1.1.4 EAF Steel Industry in India

India is the second largest producer of steel through the EAF, this country has several particularities [30].

- It has huge reserves of iron ores, about 29 billion tons but only 13% are high-grade iron ore with > 65% Fe and 47% are medium grade (62–65%Fe). In addition to this, the iron ore contains high concentrations of alumina (pellet contains 2.4–3%) and phosphorous which results in higher amounts of slag. The BF produces 375–420 kg/ton in comparison with Chinese BF that produce 350–375 kg/ton. The extra amount of slag increases the requirements of energy and therefore a higher coke rate
- Most of the steel is produced by electric furnaces, as shown in Fig. 1.10, in 2019 there were 56 EAF's with an installed capacity of 42 Mton and 1174 induction furnaces (IF) with an installed capacity of 49 Mton. In 2019 the production was as follows; BOF, 49 Mton; IF, 33 Mton and EAF, 28 Mton, a total of about 110 Mton.
- The induction furnace predominates over the EAF due to several reasons; lower investment, higher metallic yield, capable to decarburize low carbon DRI. However, plant capacity is limited to less than 1 Mton per year, it has limitations

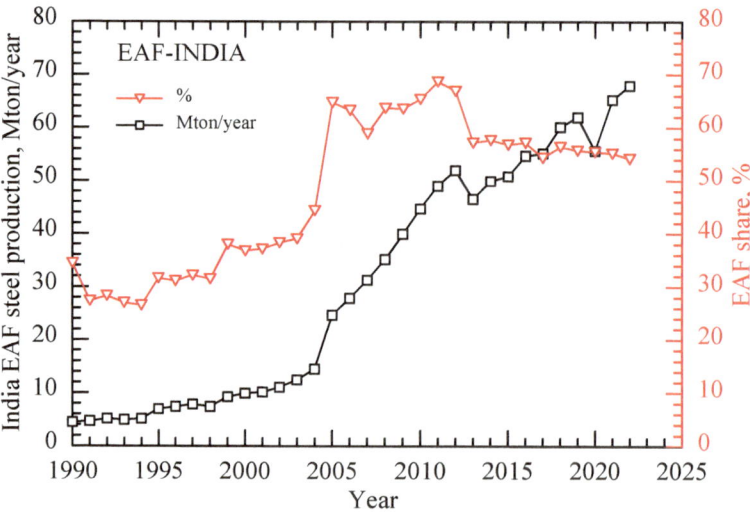

Fig. 1.10 Steel production in India by the EAF process, from 1990 to 2022. Ref. [32]

Table 1.1 Steel consumption per capita and total steel production in India

Year	1991	2001	2006	2010	2016	2022	2030 [30]
kg per capita	14.3	26.8	39.6	55	65	81.1	(160)
Mton/year	17	27	51	68	95	110	(250)

on the metallic charge because it doesn't employ oxygen injection and water consumption is high (10 ton water/h per ton of furnace capacity). The largest IF's have a capacity of 50 ton.

- India is the largest producer of DRI. In 2019 it produced 37 Mton DRI with 318 furnaces. DRI is extensively produced using coal as the main reducing agent which produces a large amount of CO_2 and low carbon DRI (1–3%C for gas-based processes and 0.2–0.25%C for coal-based processes). About 80% of DRI is produced using solid carbon (29 out of 37 tons) The total emissions of CO_2 considering DRI + IF is in the order of 3 ton CO_2/ton steel [31].
- Due to the previous conditions, the consumption of energy by the Indian steel industry is quite high, about 27.3 GJ/ton for the integrated plants and for EAF steelmaking about 600 kWh/ton with cold scrap, which is much higher than the world average of 416 kWh/ton.

The steel consumption per capita in India is still low compared with the world average which ranges from 221 to 233 kg per capita in the last 5 years from 2018 to 2022 but their global steel production has a large growth rate in the last decade, from 68 Mton in 2010 to 110 Mton in 2022, as shown in Table 1.1 based on data from the WSA.

Currently the Indian steel industry consumes high amounts of energy and produces significant amounts of CO_2. It is expected a replacement of DRI coal-based processes for DRI-gas based processes and also more EAF's replacing induction furnaces [30].

1.1.5 Share of EAF Steel Output in Other Countries

Currently, if China is excluded, the Electric Arc Furnace (EAF) steelmaking has the same relevance as the Basic Oxygen Process (BOP or BOF); in this scenario the volume of steel produced by the EAF process is close to 50%. The production of steel in Japan, Germany, South Korea, Iran and Mexico is briefly discussed below, with emphasis on the EAF.

Japan: The steelmaking industry in Japan is based on the integrated route, about 70% of steel was produced by the BF-BOF route until 2017 and from 2017 to 2019 it increased to 75%, consequently the share of the EAF decreased from 30 to 25%, below the world average. Japan was the largest producer of steel until 1995 and second only to China from 1996 to 2017. In 2018 it was relegated by India to the third place with a production in the order of 102 Mton, about 5.6% of the world total. Japan had a huge increment in production of steel from 1948 until 1973 when its production raised from almost 0 to more than 120 Mton per year, employing up to 150,000 workers [33]. For more than two decades after that, Japan led the steel industry due to many technological improvements. In the 1990s, Japan had six integrated steelmakers but then it was reduced to three major players, Nippon steel, JFE steel and Kobe steel. Kawabata [34] criticized the slow response by the big steel producers to shift to the EAF process, however Japan has similar problems to China with the lack of iron ores and natural gas to produce DRI which allows the production of high quality steels. Emi [35] indicates a long history of the EAF in Japan, with the first 1.5 ton EAF built in 1916 at Daido steel, then the number of EAF's increased to 7 in 1927 but with very low capacity, less than 10 ton, by 1934 the number increased to 104 EAF's with capacities below 20 ton, then in 1951 the EAF production increased to 2.39 Mton with 406 EAF's with a capacity below 40 ton. In 1952 Chubu Kokan installed a jumbo Lectromelt EAF with a capacity of 250 ton with a low-capacity transformer, 40 MVA. In 1962, 12 EAF's had a capacity above 50 ton. From 1968 to 1973, Kobe steel-built steelworks with EAF's ranging from 70 to 120 ton. The development of the ladle furnace in Japan in the late 1970s gave the EAF a boost in productivity. Japan also adopted new technologies such as twin-shell EAF in 1985, DC-EAF in 1990, scrap pre-heating since 1992 and the largest EAF in the world, the GigaHP Transformer-Jumbo DC-EAF at Tokyo steel in 2010 with 450 ton of nominal capacity.

Germany: Germany is the largest producer of steel in Europe. Before WWII, in 1938, Germany produced 22.7 Mton of steel and decreased to 19.1 Mton in 1940. After WWII the Allies imposed a limit of about 25% of the previous maximum capacity, set at about 5.8 Mton/year. The steel industry of the German Federal Republic recovered at a fast rate, reaching 15.8 Mton in 1952, 34.1 Mton in 1960 and 43.7 Mton in 1972. The story of this growth is explained in detail by Brandi [36]. The restrictions on steel production were relaxed and eventually lifted when the European Coal and Steel Community (ECSC) was created in 1952.

Germany is an example of a society devoted to technology adaptation and development, which explains its rapid growth not only in steel production but also as a technology supplier. Arens and Worrell [37] explained the decrease in energy consumption in the German steel industry from 8.10 GJ/ton 1958 to 6.10 GJ/ton in 2012 due to fast adaptation of six different technologies. In a previous study, Arens et al. [38] evaluated the total Specific Energy Consumption (SEC) primary energy in Germany by the EAF from 1994 to 2007, indicating a roughly constant value in the order of 6.2 GJ/ton ($\pm$ 2%) with a small tendency to decrease, about 0.3%/year, attributed to the increase in share of steel production by the EAF.

In the period from 1970 to 1990 the annual production of steel was on average 40 Mton [39]. The reunification of east and west Germany in 1990 didn't have a strong impact on steel production because of the limited production in east Germany, for example, in 1954 the GFR produced 19.7 Mton and the GDR 2.6 Mton [40], this gap kept widening over time. Production of steel by the integrated route remained

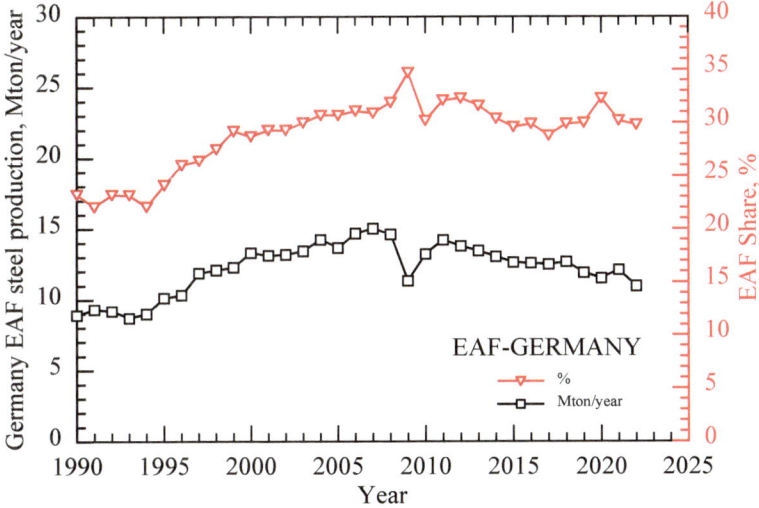

Fig. 1.11   Share of steel production in Germany. After [38, 41]

almost constant, about 30 Mton/year but the share of the EAF increased from about 6.5 Mton/year in 1980 to 13 Mton/year in 2000, about 30%, as shown in Fig. 1.11.

Thyssen-Krupp was formed after the merge of both companies in 1999, creating the largest steel company in Germany, based on the BF-BOF processes. It has a production capacity of 13 Mton/year. Duisburg has the largest concentration of steel plants in Germany. Hüttenwerke Krupp Mannesmann (HKM) in Duisburg is also an integrated plant with a capacity of 6 Mton/year. Salzgitter is another integrated plant with a capacity of 5.2 Mton/year located in Salzgitter.

The largest steel company based on the EAF is Badische Stahlwerke (BSW) located in Kehl with a production capacity of 2.5 Mton, with two AC-EAF's of 102 ton. Its website indicates a workforce of 1300 employees and a production of 2 Mton/year, which would make it one of the most productive EAF steel plants in the world. Riva operates two minimills near Berlin with capacities of 1.8 Mton/year in Brandenburg, using 2 EAF's of 150 ton and another one in Hennigsdorfer with a capacity 1.0 Mton/year.

The number of blast furnaces decreased from 104 in 1970 to 22 in 2000. The number of EAF's also decreased from 71 in 1980 to 29 in 2000 [41] but steel production increased due to higher productivity. Production of steel started to decrease since 2019 before the Ukraine-Russia war in February 2022, but due to higher prices of electricity and natural gas it was also a factor that affected steel production in the last two years.

South-Korea: South-Korea produced 65.8 Mton of steel in 2022, 68.5% by the BOF and 31.5% by the EAF, indicating a dominant role from the integrated route. Their steel consumption per capita is by far the highest in the world with 1081 kg in 2021 much higher than the world average of 233 kg per capita. Their high consumption rate is related with a strong manufacturing industry that export automobiles, machinery and other steel products.

POSCO is South Korea's largest integrated steel company, with a production, in 2018, of 42 Mton per year, followed by Hyundai with 21 Mton/year and Dongkuk with 3.7 Mton/year. POSCO started operations in 1973 as a state-owned enterprise with an initial capacity of 1 Mton/year. In 2000 was fully privatized. It is considered one of the most productive steel companies in the world. Lieberman and Kang [42] compared its productivity with Nippon steel and US Steel in terms of value added per worker hour from 1973 to 2003; POSCO's productivity increased from 1164 to 21,586 (1980 Yen per hour), equivalent to an annual growth rate of 8.9%, in comparison to Nippon steel and USS with 4.6% and 0.3%, respectively. POSCO's first CEO from 1970 to 1993, Park Tae-Joon is credited with the expression "You can import coal and machines, but you cannot import talent" which emphasized the role of talent in the success of their organization. In 1986 he founded POSTECH, Korea's first science and technology research-oriented university. Kim [43] describes in detail the success of the Korean economy, which resulted from a combination of government policies focused on education to form skilled workers (industrial warriors) and industrialization strategies as well as strong worker's unions. The result today is a country with large and strong family-owned corporations (chaebols) like Samsung, Hyundai, etc. Hyundai and Dongkuk are EAF steel producers in contrast to POSCO, however POSCO like most steel plants all over the world has started changes to shift into the EAF.

Iran: The first blast furnace in Iran was built in 1960 with technical support from the Soviet Union and a second one completed in 1982. In the 1990s Iran built DRI reduction plants using their own technology Ghaem and ZamZam DRI processes [44]. Ghaem was developed by Esfahan steel company. Iran steel production kept increasing from less than 2 Mton in 1990 to more than 30 Mton in 2022. Figure 1.12 shows the share of production of steel in the EAF in Iran from 1990 to 2022. Production of steel in Iran has several characteristics: high reserves of iron ores and natural gas, most of the steel is produced by the EAF (more than 90%), it is the world second largest producer of DRI (number one through gas-based processes), it developed its own DRI technology (PERED) and most of the steel industry are state owned enterprises.

Mobarakeh Steel (MSC) is the largest steel company with a production of steel of 10.2 Mton in 2022. The steel shop has 8 EAF's with nominal capacities from 180 to 200 tons, charged with 15% scrap and the rest is DRI. Esfahan Steel Company (ESCO) is the largest integrated steel plant with a capacity of 3.6 Mton/year.

The production of DRI in Iran in 2022 was 32.9 Mton, three times higher compared to 10 years before. The Persian Reduction process (PERED) was developed in Iran and started commercialization in 2017. The production from 2018 to 2022 has been in the order from 2.5 to 3 Mton/year in 4 plants built in Iran. This technology is similar to other gas-based processes but has developed several design improvements, such as elimination of the cluster breaker and according to Mohsenzadeh et al. [45] has better performance in natural gas consumption (295 vs. 280 Nm^3/ton) and water

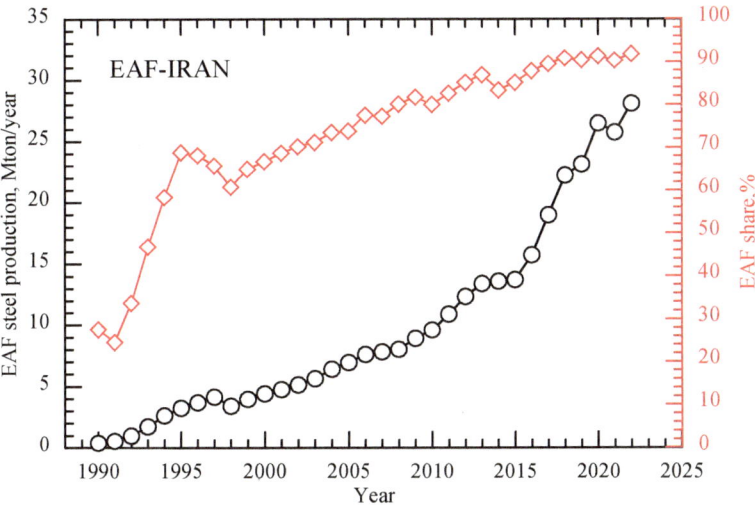

Fig. 1.12 Production of steel in Iran. After [32]

consumption (1.2 vs. 1.0 Nm^3/ton). Water consumption is an important issue. The Iranian steel industry consumes 56 Nm^3/ton much higher than the benchmarking established by China at 12.3 Nm^3/ton [46].

Türkiye: Türkiye is the largest producer of steel in the Middle East/North Africa (MENA) region and has a similar level of production to Germany, the largest steel producer in Europe. Steel consumption per capita was 100 kg in 1986. It increased to 310–390 kg in the last 5 years.

Turkish Anatolia is referred as the cradle of the iron age, attributed to the Hittite empire around 1300 BC. A full review of the Turkish steel industry from its foundation as republic in 1923 until 2008 was given by Günay [47]. In 1976 the production of steel was only 1.5 Mton. The production capacity reached 4.2 Mton in 1980. Before the 1980s most facilities were based on the integrated route: MKEK 1928; Kardemir 1937, Erdemir 1965; Icdas 1970; Isdemir 1977.

In 1990 steel production reached 9.44 Mton followed by a slow growth during this decade reaching 14.3 Mton in 2000. In the first decade of the 2000s the growth rate was higher, increasing to 29.1 Mton in 2010. Steel production in the last decade has remained at about 35 Mton/year, on average.

The 1980s defined the future orientation of the Turkish steel industry. Privatization in the 1980s and new investments enhanced the EAF production capacity, reaching a EAF-share close to 50% in 1990, as shown in Fig. 1.13. The share of the EAF in the last 15 years has been around 70%. The main raw materials for the EAF are steel scrap and DRI, however due to the lack of abundant reserves of natural gas in Türkiye there is no local production of DRI. In 2004 produced only 3% of the natural gas consumed. Therefore, steel scrap is the main raw material. This situation makes Türkiye the largest importer of steel scrap in the world. Günay [47] identified

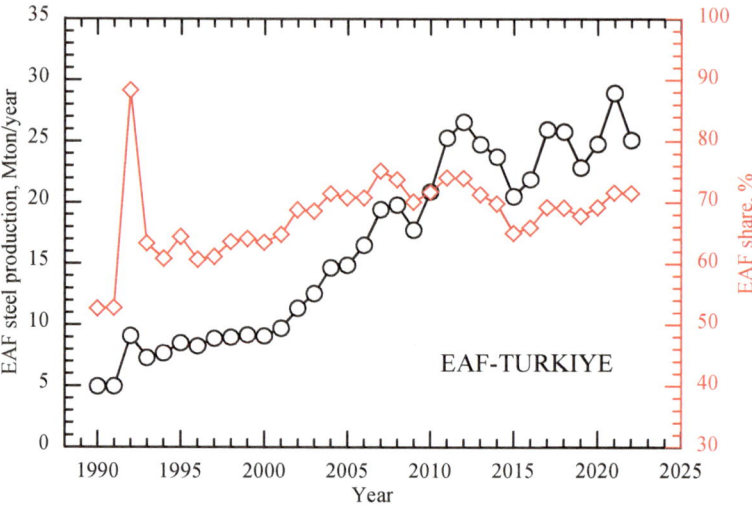

Fig. 1.13 Turkish production of steel. After [5]

as strengths of the Turkish steel industry its proximity to the European market which reduces transportation costs but also identified some weaknesses; Only 15% of the steel produced is for flat products (which have a higher added value), high electricity prices (106 USD/MWh in 2005), insufficient steel scrap and high macro-economic instability.

Russia: As the former Soviet Union (USSR), since 1970 it became the largest producer of steel in the world, reaching its peak in 1987 with 163 Mton [48]. Ukraine was the second largest producer. After the collapse of the USSR on December 1991, Russia remained on the top 4th until 2009, displaced to the 5th place by India in 2010. The steel industry was privatized within a decade after Russia was formed. All the new owners were Russian people. The transition is explained by Fortescue [49].

The Russian steel industry is characterized by the following elements [50–52]; (1) low-cost raw iron ores with one of the largest reserves of iron ores in the world but also low iron content (2) low cost of natural gas, the price from 2008 to 2012 in the EU ranged from 348 to 560 USD/1000 m^3, compared with their domestic prices which ranged from 84 to 115 USD/1000 m^3, (3) low cost of electricity, (4) outdated equipment, (5) cheaper but also excessive labor/ton of steel produced, (6) production is well above consumption, allowing to be a net exporter of steel, (7) one of the top producers of CO_2 in the world.

The main process to produce steel is the BF-BOF (+ OHF). Russia and Ukraine are the two countries which still use the very obsolete OHF process. In 2010 the share of the OHF in Russia was 10% and kept decreasing to 1.9% in 2022. The share of the EAF in 2004 was 15% and increased to 33% in 2022 due to new investments. In the period from 2005 to 2012 Russian steelmakers invested on average about 8 billion USD per year [52].

The decrease in the share of the OHF and increase in the share for the EAF allowed a decrease in the production of CO_2 by the Russian steel industry, considered in 2007 the 4th emitter after China and the USA, and similar to India [53]. In the period from 1990 to 2009 the share of the EAF increased from 14.9 to 27.2%, in the same period the emissions from the steel industry decreased from 1.98 to 1.64 ton CO_2/ton steel [54]

The largest Russian producers of steel in 2004 were [50]; Evraz with 14 Mton, followed by MMK with 11.5 Mton, Severstal with 10.2 Mton, NMLK with 9.1 Mton and Mechel with 4.8 Mton. In 2010 [51] Severstal was number one with 18.2 Mton, followed by Evraz with 16.3 Mton, NMLK with 11.9 Mton, MMK with 11.4 Mton, Mechel with 6.1 Mton and Metalloinvest with 6.1 Mton. Before the Russia-Ukraine war, the production of steel in Russia reached a maximum of 77 Mton in 2021, then decreased to 71.5 Mton in 2022. In 2022 the largest producer was NLMK with 16 Mton, followed by Evraz, MMK and Severstal with 12.5, 11.69 and 10.69 Mton, respectively. The steel consumption per capita in 2004 was 180 kg, then from 2018 to 2022 increased to 280–300 kg.

Figure 1.14 shows steel production by the EAF process from 1992 until 2022. It can be shown that before 2006 the EAF share was less than 15%, then increased to values in the range from 30 to 35%.

Mexico: The first integrated plant, FUMOSA, was built in Monterrey in 1900 with one blast furnace and one OHF, with an initial capacity of 0.1 Mton/year, three further revampings increased its capacity to 1.5 Mton/year in 1977 [55]. Both AHMSA and HyLSA were founded in 1942 in northern Mexico. AHMSA as an integrated plant and HyLSA as a mini-mill. Both AHMSA and HyLSA produced flat products. TAMSA started operations in 1952 producing seamless steel pipes for the oil industry. The

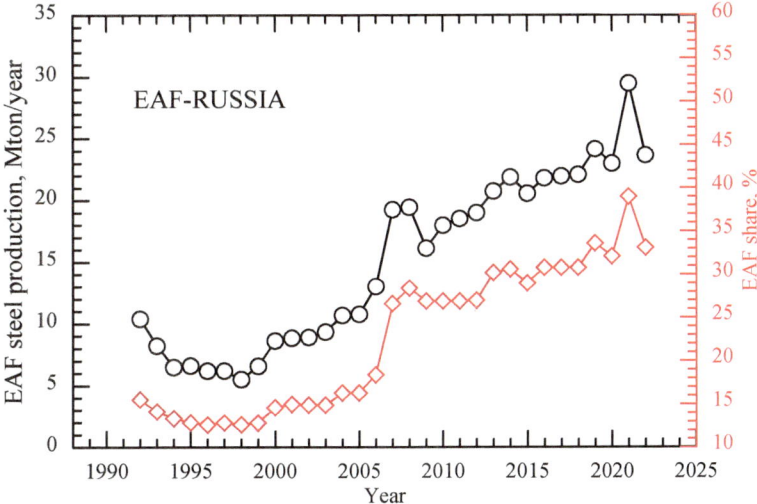

Fig. 1.14 Production of steel in the USSR and Russia

minimill HyLSA Puebla started operations in 1969 using DRI. In the late 1950s the annual steel production reached 1 Mton. SICARTSA-I, an integrated, state-owned company, started operations in 1976 with a production capacity of 1 Mton. The birth and start-up of SICARTSA's operations was not easy, as explained in detail by Avila [56]. In the period from 1950 to 1980 steel production grew six-fold, reaching 7.15 Mton in 1980. However, the 1980s was a decade of stagnation. The inflation rate increased to about 20% in the early 1970s but sky rocketed for about 10 years since 1982 with a peak of 141% in 1987. This period was marked by nationalization of steel plants and huge financial support to private plants. Unable to pay huge debts the government took control of FUMOSA in 1977 and AHMSA in 1978, forming Sidermex the same year as a control entity. HyLSA was also hit by the crisis of 1982 but the Mexican government absorbed 55% of their debt [56]. FUMOSA was closed in 1986. The privatization of AHMSA and SICARTSA (I, II) in 1991 represented a huge financial lose for Mexico; the companies were sold for less than 10% of its real value, furthermore, the government had to pay more than 6000 MUSD in debts [57]. In 1982 there were 1155 state owned enterprises (SOE), by 1988 there were only 618 and 252 in 1994 [58]. A national research center, IMIS, was formed in 1975 with impressive pilot scale facilities but was closed when the SOE were sold. Privatization involved massive layoffs and modernization which aided to increase productivity. The new company names became Ternium (HyLSA), ArcelorMittal (SICARTSA) and Tenaris (TAMSA). SICARTSA-II was sold in 1991 to Villacero but later on in 2006 was sold to ArcelorMittal. Techint bought TAMSA in 1993 and HyLSA in 2005. Tenova, the technology branch of Techint commercializes the HyL technology to produce DRI.

Only two companies have blast furnaces, AHMSA and ArcelorMittal. In Mexico the major part of steel is produced by the EAF using scrap and Direct Reduced Iron (DRI). ArcelorMittal has 4 EAF's with 220 ton of nominal capacity and use 100% DRI. TAMSA was the first company to acquire a license to produce (DRI) in 1968 by the HyL process. This technology was developed by a team led by Juan Celada, an electric engineer, in the late 1950s. It was the first successful commercial technology to produce DRI injecting a reducing gas. The first version was called HyL-I. The second version was never disclosed. In the 1970s the version HyL III was developed. A DRI plant with the HyL III process started operations at ArcelorMittal in 1988.

DeAcero has two plants, the first one started operations in Saltillo in 1985 and the second one in 1998 in Celaya, with expansions in 2006 and 2013. Their production capacity is 4.5 Mton/year. These plants employ 100% steel scrap and produce long products.

SIMEC group or ICH originated with the purchase of CH in 1991, a small minimill specialized in the production of alloyed steels. From 2001 to 2008 bought three more plants in Guadalajara, Apizaco and San Luis Potosi. In 2020 produced 1.5 Mton of steel in Mexico. The success of this company is in part due to the growth of the automotive industry in Mexico. Worldwide, in 2002 Mexico was the 7th largest producer of vehicles and 4th largest producer of auto parts. The number of units produced increased from 1.5 million units 2004 to 4.1 million units in 2019.

TYASA has two EAF's, located in a town close to Orizaba, started operations in 1993 with an EAF of 50 ton and expanded its capacity in 2013 with a Quantum EAF to reach a production capacity of 1.2 Mton/year.

Gerdau acquired Sidertul and Corsa and has a production capacity of 1.5 Mton/year using the EAF.

The production of steel steadily increased from the 1950s until 1980, then remained stagnant during the 1980s. Privatization and new investments in technology slowly increased the output of steel to 20 Mton in 2018, decreasing to 18.2 Mton in 2022. Steel production experienced two crises, in 2001 and 2009. The crisis from 2001 was due to recession in the USA caused by the dot-com bubble and attack on the twin towers. The great recession in the USA in 2008 also strongly affected the production of steel in 2009. Since USA is the biggest customer (72% of Mexican steel exports in 2022), its economy is fully tied to the USA economy.

Figure 1.15 shows the production of steel in Mexico since 1960. It can be observed that Mexico has a large share of steel production by the EAF, increasing from 50% in 1990 to 85% in 2022. The production of steel by the Mexican steel industry is below its demand. In the decade from 2000 to 2010 the difference ranged from 4 to 7 Mton/year but this gap has increased in the last 10 years. In 2022 the difference was 10 Mton. This result is a consequence of a higher steel consumption per capita. In the last 10 years the steel consumption per capita has been close to 200 kg; in 2022 it was 194.8 kg, below the world average of 221.8 kg for the same year.

As will be discussed in detail in this book, the steel industry is highly energy intensive. In 2016 the industrial sector consumed 444 TWh of energy, the largest fraction, 16% was consumed by the steel industry [59].

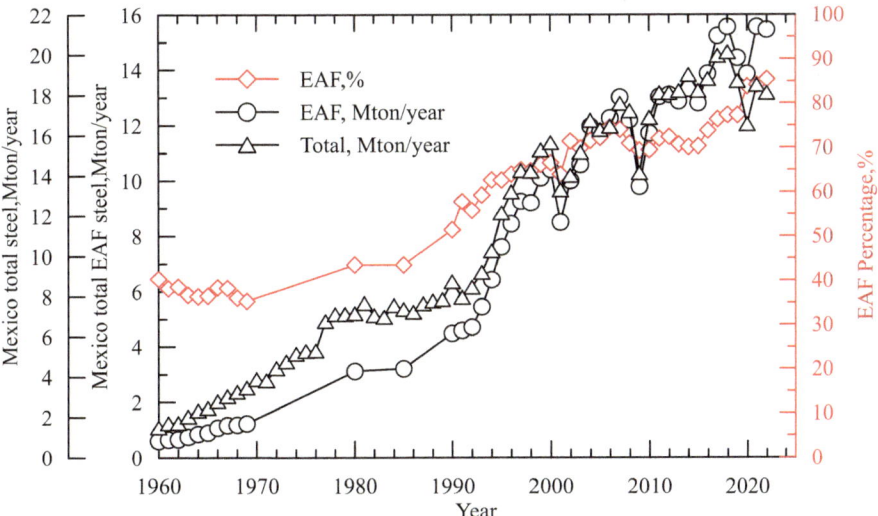

Fig. 1.15 Production of steel in Mexico

Table 1.2 Countries with a larger share by the EAF process in 2021

Country	Total, Mton	EAF, Mton	%EAF
China	1032.8	109.5	10.6
United States	85.5	59.2	69.2
Turkey	40.4	28.9	71.6
Iran	28.5	25.7	90.3
Italy	24.4	20.5	84.0
Mexico	18.5	15.6	84.1
Spain	14.2	9.7	68.3

In the last decade the following five countries have a share of production of steel by the EAF more than 50%; USA, Turkey, Iran, Italy and Mexico. In 2021, according with the World Steel Association (WSA) these countries produced more than 70% of the steel by the EAF process, as shown in Table 1.2.

1.1.6 Number of EAF's Worldwide

The total number of EAF's operating in the world is not accurately known but some reports have given an estimated value: In 1995, Dollé [60] suggested a number above 1200 EAF's. Ouvradou [61] in 2006 reported a number above 1400 EAF's all over the world, 10% in the European Union. Odenthal [62] in 2017 reported that the total number of EAF's in the world in operation was about 1100. Approximately 980 AC-EAF's and 100 DC-EAF's.

China has the largest number of EAF's in the world, about 370, more than one third of the world total, however the average furnace capacity is only 68 tons, lower in comparison with the countries shown in Table 1.3 [28]. Currently only 20% of the EAF's in China have a capacity ≥ 100 ton, the rest, 80%, is below 100 ton. 40% of the EAF's have a capacity in the range from 50 to 70 ton.

In the USA, 85 new EAF's were built in the period from 1945 to 1960 with a capacity from 12 to 175 ton capacity and transformer capacity from 5 to 36 MVA [63]. According with Bosley et al. [64], in 1983 the USA had 209 EAF's, about half with a capacity lower than 50 ton and about the other half with a capacity from 50 to 200 ton. Worrell et al. [65] reported a total of 226 EAF's in 1997 in the USA in 122

Table 1.3 Distribution of the average EAF capacity in different countries in 2020. Ref. [28]. India Ref. [30]

Country	USA	Türkiye	South Korea	Germany	Japan	China	India
Number EAF's	110	37	34	18	59	370	56
Capacity, ton	113	112	100	100	84	68	42

minimills owned by 85 companies. The age of the EAF's ranged from 0 to 74 years with an average of 24 years. The number of EAF's decreased to 173 in 2014 [66].

1.2 EAF Technological Progress

1.2.1 Advantages of the EAF Process

The anticipated takeover by the EAF process as the dominant route for steelmaking is based on the following arguments:

1. Lower production of CO_2 (2000 kg CO_2/ton for the integrated route versus 400 kg CO_2/ton for the scrap-based EAF).
2. Lower capital cost in comparison the BF-BOF (10 times lower [61])
3. Higher flexibility in the nominal production capacity (EAF nominal capacity from 20 to 300 ton)
4. Higher flexibility in raw materials (hot metal, scrap, DRI). It can automatically adapt to changes in the price market for raw materials
5. Lower prices of scrap in comparison with hot metal (subject to market control)
6. Energy recycling (sensible heat in the off-gas can be used to preheat the scrap)
7. Lower production of dust (half [61])
8. Four times lower production of slag [67] (Integrated plant 350–450 kg/ton versus 100 kg/ton for the scrap-based EAF). The total from the integrated plant correspond to 250–300 kg/ton from the BF and 100–150 kg/ton from the BOF. For DRI-EAF operations the production of slag is from 150 to 250 kg/ton
9. Higher productivity (lower labor force [61])
10. Implementation of a large number of new technologies
11. It employs expensive energy but can be minimized using hot metal and a number of methods to decrease energy consumption.

1.2.2 Capital Investment

Christmas [68] reported the capital investment for an integrated plant in the order of 2–3 billion USD in comparison with 200–300 million for a minimill, 10 times lower. Similar values had been reported by Ouvradou [61] in 2006, shown in Table 1.4. The capital cost for the scrap-EAF route is about 3 times lower. In addition to this, the installation costs are also lower by a factor of 1:7 and overheads cost 3 times lower for the EAF. He summarized that for a production of 1 Mton, the total investment by the EAF is 10 times lower in comparison with the integrated route.

In the late 1980s Bosley et al. [64] estimated the cost of the pre-heating system for the Consteel process around 3–5 million USD.

Table 1.4 Typical capital investment costs for different steelmaking routes (€/ton per year). Ref. [61]

		Steel shop	
Expenses	Integrated, €/ton	Scrap, €/ton	DRI, €/ton
Agglomeration, coke and BF	350[a]–450[b,c]		
DRI			100–150[b,e]
Steel shop and C casting	150[a]–200[b,d]	150[b,d]	150[b,d]
Total	450[a]–650[b,d]	150[b,d]	250–300[b,d]

Notes
[a] Existing site
[b] New site
[c] Per ton of pig iron
[d] Per ton of steel
[e] Per ton of DRI or HBI

1.2.3 EAF Performance

An analysis of the EAF steel industry in the USA in 1987 [64] reviewed the status of the EAF industry, new technologies available and its implementation as well as the advantages and challenges of the EAF in the context of the USA. In that decade the BF-BOF route still dominated the production of steel because it was cheaper, however the scrap-EAF route was rapidly growing due to lower capital investment, about 20% lower, higher size flexibility, lower labor intensity, lower environmental impact and similar operational costs if the price of scrap was at 90 USD/ton. The average EAF had a nominal capacity of 80 ton, a transformer of 35.5 MVA and a shell diameter of 5.37 m. At that time there was poor application of new technologies such as continuous scrap pre-heating, slag foaming, oxy-fuel burners, bottom gas injection, EBT, automation (temperature and chemical analysis, predictive computer models, etc.). The production of low carbon steels (0.03–0.05%C) and low nitrogen steels (40–50 ppm N) was only possible in the BOF. Electricity was an expensive energy, as it still is. There was no use of DRI due to its higher cost.

BSW in Germany reported in 1986 the advantages to apply the latest technologies to double the productivity and get much lower energy consumption, as shown in Table 1.5. This table summarizes improvements with an EAF of 66 ton and transformer of 40/48 MVA equipped with four 1.5 MW oxy-fuel burners, scrap preheated in a bucket to 420°C and a ladle furnace with a transformer of 10/12 MVA.

Table 1.5 Performance with state-of-the-art EAF. After [64]

	Tap-to-tap time, min	electric energy, kWh/ton	Electrodes, kg/ton	Yield, %
1979	130	504	6.0	84.4
1985	60	363 + 9 (LMF)	2.7 + 0.04 (LMF)	90.0

In order for the EAF share to grow, is necessary abundant availability of steel scrap and electricity. These authors concluded that both components were abundant in the USA. In 1985, out of 120 Mton of obsolete scrap generated, 34 Mton were sold domestically and 9 Mton exported, the rest was accumulated to form about 720 Mton of recoverable scrap. They made an observation that low quality scrap can increase the production costs, citing as an example that 1% of non-metallics in scrap lead to; 1.5% iron losses, 1% extra lime, extra 0.18 kg/ton of electrodes and 15 kWh/ton of extra power. A major challenge to produce high quality steels with 100% scrap was the residual elements, in particular copper. They reported that all the available methods to remove copper (such as vacuum treatment, hydrometallurgy or selective melting) were not economically attractive. Hot metal was mentioned in this report as a possible scrap substitute but most probably due to limitations on oxygen injection in the EAF its use was not common until two decades later. In regard with electric power the authors indicated that the steel industry consumed 137.4 TWh in 1984 and even if the EAF capacity was doubled it would represent less than 1% of the US generating capacity, indicating large availability of electric energy. Only three out of seven of the new technologies discussed in their report have succeeded today; DC-EAF, scrap pre-heating and continuous scrap charging. The limitations exposed for the other options were the following:

1. Thermal plasma EAF: A thermal plasma EAF does not depend on expensive electrodes to heat the metallic charge but in order to have adequate power, it should be operated at high voltage and long arc length which result in severe heat losses from the plasma torch. The torch life is also short.
2. Coreless induction melting (IF): This is an option to replace small EAF's of 20 ton or less. The optimum efficiency for IF's is from 10 to 20 ton. For a higher capacity, the energy consumption increases from 10 to 20%.
3. Energy Optimization Furnace (EOF): This furnace is a BOF that can be theoretically charged with 100% scrap with high levels of post-combustion using coal and oxygen. The scrap is placed above the converter where it is pre-heated. Most commonly the metallic charge is hot metal and 40–60% scrap.
4. K-ES process: This process was developed in the late 1980s. Its main attraction is the large introduction of chemical energy by injection of coal/oxygen through the bottom and large post-combustion of the CO produced.

1.2.4 Forecasts of EAF Global Steel Production

Forecasts made in the last 20 years anticipated that sometime between 2010 and 2016 the EAF would be the main route to produce steel [61, 69]. As of 2023 this has not happened because this change depends on the production of steel in China.

All countries with a high share in steel production by the EAF process have access to steel scrap (domestic and imported) and a strong electric power network. Production of DRI, one of the main scrap alternatives requires natural gas. These three conditions have been a limitation for the massive growth of EAF steelmaking in China. However, on the virtual address to the UN general assembly on September 2020 by president Xi Jinping, indicating that China would reach the peak on CO_2 emissions before 2030 and achieve carbon neutrality before 2060, it was the signal for major changes in the Chinese steel industry. Currently there are at least two projects for DRI plants based on new technologies which do not depend on natural gas.

According with S&P global [70], based on industry sources, in April 2022 it was announced that China would install new EAF plants in 10 provinces, mainly in Hebei province, increasing China's share to 15–20% by 2025. Even more recently, Shanghai Metals Market [71] in February 2023, based on information from the Ministry of industry and information technology from July 2022, indicated that in 2025 China would produce more than 180 Mton of steel by the EAF, a share of more than 15%, and by 2030 the share would be increased to 20%, with 82 new EAF's and new steel capacity of almost 100 Mton. Zhang et al. [72] in 2016 forecasted two scenarios for China in 2050 which suggest a share of 35% if hot metal is still employed in the EAF and 45% if hot metal is not employed. The China Iron and Steel Association (CISA) recent forecast from 2023 suggest a share of 30% in 2035, and 39% in 2050. The total production of steel by 2050 is expected to be in the order of 2.5 billion tons [73]. The US Energy Information Administration (US-EIA) indicates that China's BF-BOF fleet is relatively new, about 15 years on average, its life span ranges from 40 to 50 years, however after about 25 years the BF requires major refractory relining with a cost about half of a new BF, therefore, if the country decides to stop using BF's the transition to the EAF can be accelerated, which is the reason they forecast a 50% share of steel production through the EAF by 2050 [74]. In all of these scenarios the production of steel by the BF-BOF will keep decreasing.

1.3 Challenges for the EAF

There is almost full agreement that the decline of the blast furnace has started due to environmental issues, which implies that few or even zero blast furnaces will be built in the future. The OECD [75] made a report that still considers the construction of new blast furnaces in Asia but most probably will be to replace old facilities. This scenario opens the possibility for a dominant role by the EAF process in 2050. Assuming the EAF continues its growth on steel production, there are several challenges that need to be taken into account; availability of iron ore, availability of steel scrap, availability of green electricity and large improvements to decrease energy consumption. The first three subjects are briefly discussed in this section and the fourth one, which is the subject of the whole book, is summarized in the next section.

1.3.1 Reserves of Iron Ore Quality for DRI

It is expected a high rise in the production of DRI in the next three decades, from about 100 Mton to 400 Mton by 2050 [76], in order to replace steel scrap in steelmaking operations. The Chinese iron and steel association forecast [28] estimate a production of DRI in China of 3 Mton for 2025, 9 Mton for 2030, 40 Mton for 2050 and 50 Mton for 2060. Currently one plant started operations in 2023 with an annual capacity of 0.6 Mton/year but three more plants are under construction. These targets depend on the availability of iron ores with enough quality. Barrington [76] argues that the challenge is not the amount but the quality of iron ores. Currently a high-quality iron ore can be defined in terms of a high iron content (> 67%), low acid gangue (< 3.5%) and a low phosphorous concentration (< 0.015%P_2O_5). In practice, the depletion of high-quality iron ores is a big issue, the only guarantee is that iron ores regarded unsuitable for the production of DRI will be used more and more. Barrington also mentions that in practice the iron content decreased from 63.9 to 61.9% and the acid gangue increased from 5.11 to 7.08% in the last 20 years.

The use of low quality DRI will need removal of the gangue content through melting. The technological route suggested include direct reduction with hydrogen which drastically decreases CO_2 emissions in the reduction process, further melting in an electric smelting furnace to produce hot metal and final steelmaking in a BOF process [77, 78]. An estimated value of 400 kg CO_2/ton has been suggested using this route [77].

1.3.2 Availability, Cost and Residual Elements in Steel Scrap

The challenges on steel scrap are availability, price and residual elements. Steel scrap was cheap for the entire twentieth century compared with the soaring prices it experienced in the last 20 years. At the time Willy Korf built Georgetown steel in the USA in 1969, the price of steel scrap was 30 USD/ton [12]. From 1960 to 2000 the price of steel scrap fluctuated around 100 ± 50 USD/ton [79]. During this time the price had spikes with high prices but on average the values remained in the range indicated, however when the production of steel in China started to raise exponentially, the price of many raw materials went very high. The scrap prices raised to 250 and for a few months in 2008 to more than 500 USD/ton. The prices have been highly unstable between 200 and 400 USD/ton. The price of raw materials shows a large variability on a monthly basis. At some times the iron ore is cheaper than the price of steel scrap but at other times this situation is reversed [80]. One solution is to apply optimization models to define the mixture of metallics which produce the steel composition at the lowest cost.

One of the main quality problems in steel scrap is the presence of residual elements, which was the main limitation to produce high-quality steels in the EAF for more than 60 years. In spite that steel scrap is almost 100% recyclable, the accumulation of residual elements is a major issue. The possibility to increase the share of steel scrap in steelmaking will increase if the sorting systems are improved and dilution practices are applied using alternate iron units.

1.3.3 Availability of Large Amounts of Green Electricity

Worldwide the production of electricity has largely increased, about 110 times in 20 years, with a total of 1.95 billion kW, this increment is almost half from wind power and the rest solar power. China is also moving fast in the direction of producing green electricity. According with the CISA report [28] at the end of 2022 the installed capacity of wind power was 365 million kW and 393 million kW from solar power, which corresponds to 14.3% and 15.3% respectively of the installed capacity. The expected growth of new energy is 40% by 2030 and 50% by 2035.

The transition from a BF/BOF to DRI/Scrap/EAF dominated process will be slowly. Palone et al. [81] suggested a transition in 4 steps which would cause a decrease from 8.5 Mton CO_2/year to 0.68 Mton CO_2/year when the transition process is completed. In the final stage DRI is produced with green hydrogen and the metallic charge is completed with 10% scrap. The success of this transition depends on the cost and availability of green hydrogen and the availability of both scrap and DRI. They estimate that 1.42 GW of electrolyser will be required and the cost of the transition would be almost 3 billion euros.

1.4 Energy Consumption in the EAF

In the previous sections of this chapter the importance and growth of the EAF worldwide has been discussed. There is no doubt that in the near future the EAF will dominate the production of steel worldwide. However, the challenges previously discussed should be properly addressed. This book is concerned with energy consumption in the EAF. The EAF uses a source of expensive energy, electric energy. For the EAF to become the dominant process in a shorter period of time requires to increase its thermal efficiency and decrease energy consumption.

The study of energy consumption in the EAF requires a holistic approach. In addition to metallurgical engineering tools, it is also important to involve electric engineering, statistical engineering, computer science, physical and mathematical modelling, as well as economics. In order to decrease energy consumption is important to collect information from the thermal, chemical and fluid flow states of the EAF using multiple sensors of liquid steel, liquid slag, off-gases, equipment, etc. and feed that information into control algorithms to anticipate changes in the EAF operation that promote higher melting rates with minimum energy losses.

1.4.1 Methods to Decrease Energy Consumption

The major cost to produce steel is the metallic charge but in a second place is the cost of the electric energy, about 15% [82]. Energy consumption in the EAF is also another major challenge because it employs not only an expensive source of energy (electric energy) but also because, currently, most of the electric energy produced is based on fossil fuels. The purpose of this book is to discuss in detail 15 methods to decrease energy consumption in the EAF which involve a comprehensive optimization of current metallurgical practices to increase the reactor's thermal efficiency. A summarized version of these methods is given below.

(1) *Optimization models of mass and energy*

The nature of the raw materials, its physical and chemical composition, initial temperature, amount and gangue content define the theoretical energy requirements which can be predicted by thermodynamics. This value cannot be decreased and is the limit imposed by nature, therefore the critical aspect is a wise selection of the raw materials. In general, high quality raw materials will consume lower energy because of a higher iron content, less gangue content, higher density, etc., but also those raw materials are more expensive. Optimization tools can be used to define the best choice of raw materials considering the target in steel chemical composition, actual price of raw materials, availability of raw materials, and then define the best choice that yields the lowest production cost of steel. The optimization models include mass and energy balances. The accuracy in the prediction of energy consumption depends on the method to estimate the thermal efficiency. As the thermal efficiency decreases, the energy requirements on energy consumption increase. Chapter 2 provides the basis of mass and energy balances and Chap. 3 discusses the methods to predict energy consumption, which includes the optimization models.

(2) *Oxygen injection*

Oxygen injection plays multiple roles in steelmaking; decarburization of liquid steel, is a source of chemical energy and also is a source to increase the poor stirring conditions in the EAF.

The specific oxygen consumption in the EAF has been increased in the last decade in order to provide additional chemical energy and replace electric energy, which is more expensive. Today is even possible to reach zero consumption of electric energy using hot metal. However, an optimum oxygen injection practice should take into account multiple factors, not only its contribution to decrease electric energy, otherwise its overall effects can be detrimental to the final steel quality and metallic yield. Chapter 4 describes the fundamental aspects of oxygen injection and the need to keep a C-O balance that avoids excessive formation of FeO.

(3) *Oxy-fuel burners*

The current technology for oxygen injection involves the use of coherent jets, which can be operated in different modes; oxy-fuel burners, oxygen jets, co-injection of carbon particles and co-injection of other particles. Oxy-fuel burners are employed in scrap-based operations to pre-heat the scrap charged which also accelerates its melting rate and then contributes to decrease energy consumption. An optimum burner operation has to define the power profile as a function of time because its thermal efficiency decreases as the scrap is pre-heated and to avoid excess scrap oxidation. Chapter 5 describes burner operation and its contribution to the decrease in electric energy consumption.

(4) *Post combustion*

Steelmaking is an oxidizing process, with the C–O reaction as the most important reaction in liquid steel. Its product is CO and as more carbon is oxidized more CO is generated. This CO is useful in the removal of nitrogen and also serves to enhance the poor stirring conditions in the EAF, furthermore, CO has huge potential chemical energy and its further oxidation into CO_2 releases heat that can be used to pre-heat scrap charged into the EAF, decreasing the consumption of electric energy. Burners operated in oxygen injection mode promote the post-combustion reaction. The optimum number, location and oxygen flow rates should be defined in order to ensure a maximum residence time of the gases inside the EAF chamber and a higher heat transfer rate from the off-gas to the solid scrap. Chapter 6 provides a detailed description of post-combustion and its benefits on energy consumption.

(5) *Slag foaming*

From all the methods to decrease energy consumption in the EAF, slag foaming is one of the most important methods because of multiple benefits in all steelmaking operations. Consequently, Chap. 7 is the largest chapter in this book. Foaming phenomena has been investigated for centuries in different fields but only in the last three decades in EAF steelmaking. The concept is quite simple because only involves the formation of a gas phase, CO, and formation of a slag with physical properties that extend the residence time of the bubbles, achieving full electric arc coverage. In practice, there are many variables that need to be considered simultaneously to optimize slag foaming. This chapter provides not only the fundamentals on foaming but also a description of the methods to measure slag foaming and a detailed analysis of models developed to predict slag foaming.

(6) *Stirring*

One of the main current limitations of the EAF is its poor stirring conditions which translate into low decarburization rates, if we compare with the alternate process, the BOF. In a conventional EAF liquid steel remains almost static because the forces that promote the motion of steel are localized forces; electromagnetic forces, oxygen jets and subsequent CO formation and bubble escape from the zone around the injection jets. Even if porous plugs are installed for bottom gas injection, the height/ diameter ratio is too low that makes gas injection inefficient, if the gas injection layout is not properly designed. Chapter 8 reviews the contribution of all the forces that promote the motion of liquid steel with emphasis on forced convection due to bottom gas injection and electromagnetic stirring devices. Improving the stirring conditions increases the melting rate and therefore can also decrease energy consumption.

(7) Scrap quality and preheating

Up to now, scrap is the main raw material in EAF steelmaking, therefore all aspects about scrap have an impact on the EAF process and energy consumption in particular; its quality, its physical and chemical composition, its initial temperature, its size and density, its cost, its concentration of residual elements and its availability. One of the most recent and largest technological changes in the EAF is scrap pre-heating technologies. Most of these technologies recover energy from the off-gases, promoting an overall higher thermal efficiency and a large decrease in the overall energy consumption. Chapter 9 analyses the physical and chemical properties of steel scrap as well as a detailed description of the technologies that have been developed for scrap pre-heating.

(8) Hot DRI

Current DRI technologies have the capability to transport and charge hot DRI into the EAF, especially for new plants built close to or above the EAF steel shop. Hot DRI, like scrap pre-heating require less amount of energy to complete the heating and melting processes, therefore saving electric energy. Chapter 10 describes the production of hot DRI and the savings in energy consumption as a function of DRI temperature and chemical composition.

(9) Hot metal

Hot metal is an ideal raw material for the BOF because its large concentration of carbon and silicon which allows an operation with only chemical energy. In the last decade due to increasing pressure to decrease electric energy consumption, the share of hot metal in the EAF has also been increased, converting the EAF into a BOF, decreasing electric energy consumption to zero when hot metal is about 70%. Since the EAF has not the same oxygen injection capacity or the geometry of the BOF, hot metal is charged in lower proportions which largely depend on the oxygen injection capacity and capacity of off-gas extraction system. Chapter 10 provides a discussion on the requirements on oxygen injection when the metallic charge includes hot metal and its benefits on energy consumption.

(10) Hot heel

The remaining mass of liquid steel after tapping is the widely applied hot-heel practice in EAF steelmaking which promotes a higher melting rate and early slag foaming for the next heat, reducing energy consumption. The mass of hot heel has been defined on empirical basis; therefore, its value varies for different steel shops. Chapter 11 summarizes the effect of hot heel on energy consumption and values employed by different steel shops.

(11) *Tapping control*

The tapping temperature depends on the type of steel, tapping hole conditions and superheat required to deliver liquid steel to the next process. In old furnaces with a tapping spout the temperature losses during tapping had to be compensated with higher tapping temperatures but today most modern furnaces employ Eccentric Bottom Tapping (EBT) which significantly decreases tapping time and the stream of steel is more uniform. The superheat should be as small as possible in order to decrease oxygen solubility and energy consumption.

(12) *Energy recovery*

The thermal efficiency of the EAF is low, about 50%, which indicates that only about half of all the supplied heat is employed to produce liquid steel and the rest, also about half of all the heat supplied is lost as sensible heat in the off-gas, sensible heat in the slag, sensible heat in water cooling and other heat radiation losses. Much of the sensible heat in the off-gas can be recovered for scrap pre-heating but also has other potential applications. In addition to heat recovery from the off-gas, several technologies are under development to recover the sensible heat in liquid slag. Chapter 11 makes a detailed comparison of different technologies.

(13) *Water cooling*

All modern EAF's employ water cooling in the EAF shell and EAF roof. The introduction of this method allowed an EAF operation with long arcs which yield a higher electric power. In the 1960s it was recognized the importance to increase the power transformer to increase the melting rate but the lack of an efficient operation with foamy slags restricted the operation to short arcs. Today, the foamy slag practice is common but still the heat radiated to the walls can be significant under flat bath conditions. In order to decrease heat losses to the water-cooled panels and increase its life it is important to promote a thick slag layer with a high melting point on the surface of the water-cooled panels. Chapter 12 reviews design aspects and compares heat losses as a function of the slag layer thickness adhered to the walls of the water-cooled panels.

(14) *Optimization of electric parameters*

The supply of electric energy from the transformer at the sub-station to the metallic charge requires optimization of a large number of variables and obtain an ideal smooth delivery of electric power. To reach this goal it is important to employ the proper instrumentation and an efficient electrode-regulation system which yields a higher active power. Chapter 13 provides some fundamentals on the electric power system and key aspects to decrease electric disturbances, such as flicker and harmonics.

(15) *EAF design and automation*

Design and automation are two different aspects of EAF technology. The evolution in design is reviewed since the first concept of EAF was proposed by Siemens in the late nineteenth century including an evaluation of the current design limitations that

Table 1.6 Methods to decrease energy consumption and summary of energy savings, according to Worrell et al. [65] in 1999

Technology measure	Energy savings		Share of production measure applied, %
	GJ/ton	kWh/ton	
Improved process control [83]	0.11	30	90
Flue gas monitoring and control [84]	0.05	15	50
UHP transformers improved efficiency c	0.06	17	40
Bottom stirring and gas injection [85]	0.07	20	11
Foamy slag practice [65]	0.07	19	35
Oxy-fuel burners [86]	0.14	39	25
Post-Combustion [87, 88]	0.23	64	–
Eccentric bottom tapping [89]	0.05	15	52
DC-EAF [65]	0.32	89	5
Scrap pre-heating (consteel) [90]	0.22	61	20
Scrap pre-heating (shaft) [65]	0.43	119	20
Twin shell DC with scrap pre-heating [91]	0.07	19	10
Preventive maintenance [65]	0.24	67	100
Energy monitoring and management system [65]	0.06	17	100

affect the performance of this process and its future development. The chapter ends with a review on the role of automation on EAF productivity, including the role of technical training to engineers in charge of the operation.

Worrell et al. [65] in 1999 summarized savings in energy consumption using different methods, as shown in Table 1.6. This information is provided only as a reference because those results were obtained when many of these methods were in its early stage of development and sometimes is only one reference that was used to define the average on energy savings. For a more detailed discussion the reader should review the information provided in the corresponding chapters in this book.

1.4.2 Energy Consumption and the Environment

One of the main reasons to reduce energy consumption is not just to improve the performance of the EAF but to contribute to reduce CO_2 emissions. The steel industry is the largest consumer of energy in the world among industrial sectors, with 22%, equivalent to 8% of the total energy used in 2019, according with the International Energy Agency [92]. Energy consumption in the steel industry is large in spite of a 60% reduction in specific energy consumption from 1970 to 2017, from 50 to 20

GJ/ton, respectively, according with the World Steel Association [93]. According to Pfeifer et al. [94], introduction of the latest innovative technologies have decreased electric energy consumption from 630 to 350 kWh/ton. In spite that using the BAT and energy consumption has almost reached the theoretical limits, it is still possible to further decrease by 10–20% if outdated equipment is replaced with BAT [95].

There is an intimate relationship between energy consumption in the steel industry and the environment [96], as more steel is produced, more energy is required and more CO_2 is produced because steelmaking is currently based on energy derived from fossil fuels.

Considering the forecast from the World Steel Association to produce 2.5 billion tons of steel in 2050, the challenge is to decrease the production of CO_2 by 60%, from its current average of 1.8–2 ton CO_2/ton in integrated steel plants to 0.5 ton CO_2/ton of liquid steel in order to reach the global emission target [96, 97]. Chapter 15 provides a discussion on energy consumption and CO_2 emissions, including a brief review of all the methods employed to decrease and control CO_2.

1.4.3 Radical Improvements to Existing EAF Technologies

The EAF has experienced large transformations in the last 123 years. However, its design offers a number of limitations, for example:

- Uses expensive energy (electric energy): Solutions to this problem include optimization of thermal and electric efficiency, and replacement of electric energy with chemical energy.
- Extremely poor stirring conditions, promoted by a geometry that has remained almost unchanged, characterized by a low H/D ratio. This in turn causes a lower decarburization rate, lower melting rates, lower metallic yield, etc.
- Low thermal efficiency, in the order of 50%
- Cannot produce high quality steels if the raw materials, especially high-quality steel scrap, are not available at competitive prices.

The methods to decrease energy consumption described in his book describe in detail the fundamentals which help to define the optimum conditions to improve each method, shares information on the impact of each method on energy consumption and defines the remaining challenges to keep improving the EAF process.

References

1. Conte N (2021) Visualizing 50 years of global steel production. https://www.visualcapitalist.com/visualizing-50-years-of-global-steel-production/. Accessed 3 May 2023
2. Besson E (2023) World steel in figures. World Steel Association. https://worldsteel.org/steel-topics/statistics/world-steel-in-figures/. Accessed 7 Oct 2023

3. Luxmet (2021) Rising trends of EAF steelmaking. https://www.luxmet.fi/2021/02/18/rising-trends-of-eaf-steelmaking/. Accessed 6 Oct 2023
4. Trubetskov KM, Gurskii GL (1964) Increasing the productivity of open-hearth furnaces. Metallurgist 8:547–550
5. WSA (2023) Steel and raw materials. World Steel Association. https://worldsteel.org/wp-content/uploads/Fact-sheet-raw-materials-2023.pdf
6. Pring JN (1921) The electric furnace. Longmans, Green and Co., New York, USA
7. Rogers RP (2009) An economic history of the American steel industry. Routledge Taylor & Francis. https://doi.org/10.4324/9780203881033
8. USGS-NMIC (n.d.) Iron and steel statistics and information. Annual Report Iron Steel. https://www.usgs.gov/centers/national-minerals-information-center/iron-and-steel-statistics-and-information
9. Moore WE (1931) Twenty year advance in electric arc furnaces for the production of iron and steel. Trans Electrochem Soc 60:165. https://doi.org/10.1149/1.3497861
10. Herty CH (1927) Burnt lime and raw limestone in the basic open-hearth process. Ind Eng Chem 19:592–594. https://doi.org/10.1021/ie50209a025
11. Israilevich P (1986) The birth of the competitive market in the steel industry. Federal Reserve Bank of Cleveland, working paper no. 86-06
12. Stubbles JR (2009) The minimill story. Metall Mater Trans B Process Metall Mater Process Sci 40:134–144. https://doi.org/10.1007/s11663-008-9216-9
13. Simcoe CR (2018) The history of metals in America. ASM International
14. Song L (2023) China: steel industry. In: Encyclopedia of mineral and energy policy. Springer, Berlin Germany, pp 142–52
15. ErSelçuk M (1956) The iron and steel industry in China. Econ Geogr 32:347–371
16. Field RM (1970) Industrial production in communist China: 1957–1968. China Q: 46–64. https://doi.org/10.1017/s0305741000049742
17. CIA (1979) China: the steel industry in the 1970s and 1980s
18. Feng L (1994) China's steel industry. Resour Policy 20:219–234
19. Alitto GS, Shen Z (2002) A historical examination of the issue of soviet experts in China: basic situation and policy changes. Russ Hist Russe 29:377–400
20. Dong H, Liu Y, Wang L, Li X, Tian Z, Huang Y et al (2019) Roadmap of China steel industry in the past 70 years. Ironmak Steelmak 46:922–927. https://doi.org/10.1080/03019233.2019.1692888
21. Song L (2017) China: steel industry. In: Tiess G, Majumder T, Cameron P (eds) Encyclopedia of mineral energy policy. Springer, Berlin, pp 1–10. https://doi.org/10.1007/978-3-642-40871-7
22. Garnaut R, Golley J, Song L (2010) China: the next twenty years of reform and development. Australian National University Press
23. World Steel Association (2019) World steel in figures. https://www.worldsteel.org/dms/internetDocumentList/bookshop/2015/World-Steel-in-Figures-2015/document/World%20Steel%20in%20Figures%202015.pdf
24. Brandt L, Jiang F, Luo Y, Su Y (2019) Ownership and productivity in vertically-integrated firms: evidence from the Chinese steel industry. Economics Department, University of Toronto
25. Nace T (2023) Global energy monitor. In: Global steel plant tracker
26. Li Z, Andersson FNG, Nilsson LJ, Åhman M (2023) Steel decarbonization in China—a top-down optimization model for exploring the first steps. J Clean Prod 384:135550. https://doi.org/10.1016/j.jclepro.2022.135550
27. Chalabyan A, Li Y, Vercammen S, Zhou J, Tang R, Zhao V (2019) How should steelmakers adapt at the dawn of the EAF mini-mill era in China? In: McKinsey report
28. CISA (2023) Research report on the development of EAF process in China: current situation, potential of carbon reduction and development prospect. China Iron Steel Industrial Association. https://mp.weixin.qq.com/s/PdPO2oLX2_-Knu9j7oA1eQ. Accessed 14 Sept 2023

29. Ai L, He C (2016) The present situation and development trend of EAF in China. Ind Heat 45:75–80 (in Chinese)
30. Shanmugam SP, Nurni VN, Manjini S, Chandra S, Holappa LEK (2021) Challenges and outlines of steelmaking toward the year 2030 and beyond—Indian perspective. Metals 11. https://doi.org/10.3390/met11101654
31. Bhardwaj N, Seethamraju S, Bandyopadhyay S (2023) Decarbonisation options for rotary kiln-induction furnace process of crude steel. Production 103:607–612. https://doi.org/10.3303/CET 23103102
32. Kirschen M (2021) Visualization of slag data for efficient monitoring and improvement of steelmaking slag operation in electric arc furnaces, with a focus on MgO saturation. Metals 11:1–10. https://doi.org/10.3390/met11010017
33. Hasegawa H (1996) The steel industry in Japan: a comparison with Britain. Routledge Taylor & Francis
34. Kawabata N (2023) Evaluating the technology path of Japanese steelmakers in green steel competition. Japanese Polit Econ 49:231–252. https://doi.org/10.1080/2329194X.2023.225 8162
35. Emi T (2015) Steelmaking technology for the last 100 years: Toward highly efficient mass production systems for high quality steels. ISIJ Int 55:36–66. https://doi.org/10.2355/isijinter national.55.36
36. Brandi HT (1974) The development of the German steel industry during the past 25 years. Tetsu-To-Hagane 60:1179–1191
37. Arens M, Worrell E (2014) Diffusion of energy efficient technologies in the German steel industry and their impact on energy consumption. Energy 73:968–977. https://doi.org/10.1016/j.energy.2014.06.112
38. Arens M, Worrell E, Schleich J (2012) Energy intensity development of the German iron and steel industry between 1991 and 2007. Energy 45:786–797. https://doi.org/10.1016/j.energy.2012.07.012
39. Mannsbart W, Schlomann B (1992) The iron and steel industry in Germany (W.) between 1974 and 1990 with special regard to the Saarland
40. CIA (1955) The iron and steel industry of East Germany
41. Schleich J (2007) Determinants of structural change and innovation in the German steel industry—an empirical investigation. Int J Public Pol 2:109–123. https://doi.org/10.1504/IJPP.2007.012278
42. Lieberman MB, Kang J (2008) How to measure company productivity using value-added: a focus on Pohang steel (POSCO). Asia Pacific J Manag 25:209–224
43. Kim H-A (2020) Korean skilled workers: toward a labor aristocracy. University of Washington Press
44. Motlagh M (2003) Expansion of DRI–EAF based steel industry in Iran. Steel Times Int 27:16–17
45. Mohsenzadeh FM, Payab H, Abdoli MA, Abedi Z (2018) An environmental study on Persian direct reduction (PERED®) technology: comparing capital cost and energy saving with MIDREX® technology. Ekoloji 27:959–967
46. Ghazinoory S, Fatemi M, Adab A (2022) Iranian steel value chain: advantageous but unsustainable. Clean Technol Environ Policy: 2099–2115. https://doi.org/10.1007/s10098-022-023 00-6
47. Günay K (2008) The competitiveness of the Turkish iron and steel industry in the process of membership to the European Union. Ph.D. thesis. Isik University
48. Baggins B (2000) Soviet industrial output: 1940–1985. Slav Res Cent Libr: 1. https://www.marxists.org/history/ussr/government/economics/statistics/ind-out.htm. Accessed 13 Dec 2023
49. Fortescue S (2009) The Russian steel industry, 1990–2009. Eurasian Geogr Econ 50:252–274
50. Iperti L (2005) Do Russian steel producers really have a competitive advantage? Metall Ital 97:58–62
51. Berger (2012) Overview of Russian steel. Workshop, p 49

52. Shatokha V (2017) Post-Soviet issues and sustainability of iron and steel industry in Eastern Europe. Trans Inst Min Metall Sect C Miner Process Extr Metall 126:62–69. https://doi.org/10.1080/03719553.2016.1251750
53. Kundak M, Lazić L, Črnko J (2009) CO_2 emissions in the steel industry. Metalurgija 48:193–197
54. Shevelev LN (2010) A review of greenhouse gas emissions in the Russian iron and steel industry. Steel Times Int 34:33
55. Correa-Villanueva JL (1986) La liquidación de Fundidora Monterrey y la reconversión industrial. Cuad Políticos: 41–56
56. Ávila-Juárez JÓ (2011) Acero. Nacionalismo y neoliberalismo en México. Historia de la Siderúrgica Lázaro Cárdenas-Las Truchas, S.A. Universidad Autónoma de Querétaro
57. Sacristán-Roy E (2006) Las privatizaciones en México. Econ UNAM 3: 54–64
58. Kehoe TJ, Meza F (2013) Crecimiento rápido seguido de estancamiento: México (1950–2010). Trimest Econ 80:237–280
59. Sandoval E, Franco R (2021) Beneficios socioambientales derivados de la eficiencia energética en el sector industrial mexicano. Econ Teoría y Práctica 29:89–108
60. Dollé G (1995) L'évolution du four électrique à arc. Rev Metall 92:1177–1186
61. Ouvradou C (2006) The electric furnace situation and European perspectives. Rev Métallurgie: 218–225
62. Odenthal HJ, Kemminger A, Krause F, Vogl N (2017) A holistic CFD approach for standard and shaft-type electric arc furnaces. AISTech Iron Steel Technol Conf Proc 1:1101–1114
63. Ciotti JA (1971) A new era in melting. JOM 23:30–35
64. Bosley J, Clark J, Dancy T, Fruehan R, McIntyre E (2000) Techno-economic assessment of electric steelmaking through the year. Pittsburgh, PA, USA
65. Worrell E, Martin N, Price L (1999) Energy efficiency and carbon dioxide emissions reduction opportunities in the U. S. Iron and Steel Sector. NBNL report
66. Thekdi A, Nimbalkar S, Keiser J, Storey J (2015) Preliminary results from electric arc furnace off-gas enthalpy modeling. In: Iron steel technology conference on exposition. Oak Ridge National Laboratory (ORNL), Oak Ridge, USA, p 15
67. Conejo AN (2014) Steel slags: characterization and alternatives of recycling. CONAC, Monterrey NL Mexico, 23–26 Mar 2014. AIST Mexico, pp 1–12
68. Christmas I (2012) Changing economics of steel. Ironmak Steelmak 39: 258–262. https://doi.org/10.1179/0301923312Z.00000000063
69. Bell S, Davis B, Javaid A, Essadiqi E (2006) Final report on refining technologies of steel
70. S&P (2022) China's EAF capacity growth gathers pace in 2022 as steel sector tracks decarbonization goals. https://www.spglobal.com/commodityinsights/en/market-insights/latest-news/energy-transition/042922-chinas-eaf-capacity-growth-gathers-pace-in-2022-as-steel-sector-tracks-decarbonization-goals
71. SMM (2023) Nearly 100 Million MT EAF capacity will be put into operation. Shanghai Met Mark. https://news.metal.com/newscontent/102104950/nearly-100-million-mt-eaf-capacity-will-be-put-into-operation/. Accessed 24 May 2023
72. Zhang Q, Hasanbeigi A, Price L, Lu H, Arens M (2016) A bottom-up energy efficiency improvement roadmap for China's iron and steel industry up to 2050
73. Holappa L (2020) A general vision for reduction of energy consumption and CO_2 emissions from the steel industry. Metals 10:1–20. https://doi.org/10.3390/met10091117
74. US-EIA (2022) IEO2021 issues in focus: energy implications of potential iron and steel-sector decarbonization pathways
75. OECD (2022) Latest developments in steelmaking capacity
76. Barrington C (2022) The iron ore challenge for direct reduction on road to carbon-neutral steelmaking. Direct MIDREX: 3–7
77. Cavaliere P, Perrone A, Silvello A, Stagnoli P, Duarte P (2022) Integration of open slag bath furnace with direct reduction reactors for new-generation steelmaking. Metals 12:203. https://doi.org/10.3390/met12020203

78. Nicholas S, Basirat S (2022) Solving iron ore quality issues for low-carbon steel. IEFA: 14. https://ieefa.org/resources/solving-iron-ore-quality-issues-low-carbon-steel. Accessed 25 Sept 2022

79. Risser R, Hoffman M (2011) Scrap steel cost affects reinforcing steel prices. https://www. concreteconstruction.net/how-to/materials/scrap-steel-cost-affects-reinforcing-steel-prices_o. Accessed 5 Oct 2023

80. Nagatomi Y (2014) Demand side perspectives: cooperation and competition in the extractive industries (EI) sector. PECC: 30–47. https://www.pecc.org/pecc/208-publications/602-cooper ation-and-competition-in-the-extractive-industries-sector-perspectives-from-demand-and-sup ply-sides. Accessed 5 Oct 2023

81. Palone O, Barberi G, Di Gruttola F, Gagliardi GG, Cedola L, Borello D (2022) Assessment of a multistep revamping methodology for cleaner steel production. J Clean Prod 381. https://doi. org/10.1016/j.jclepro.2022.135146

82. Torres R, Aguilar S, Conejo AN (2000) Evolution of refractory performance and metallurgical practices at IMEXSA. In: 58th electric furnaces conference, 12–15 Nov 2000, Orlando, Florida, pp 415–423

83. Staib WE, Bliss NG (1995) Neural network control system for electric arc furnaces. Metall Plant Technol Int 18:58–61

84. Stockmeyer R, Heinen K-H, Veuhoff H, Siegert H (1990) Saving electrical energy arc furnace with a new off-gas exhaust control. Stahl Eisen 110:113–116

85. Schade RJ (1991) Bottom stirring in an electric arc furnace. CMP, Pittsburgh, PA, USA

86. Jones JAT (1996) New steel melting technologies: part III, application of oxygen lancing in the EAF. Iron Steelmak 23:41–44

87. Kleimt B, Köhle S (1997) Power consumption of electric arc furnaces with post-combustion. Metall Plant Technol Int 3: 56–57

88. Gregory DS, Ferguson DK, Slootman E, Viraize F, Luckhoff J (1996) Results of ALARC-PC post-combustion at cascade steel rolling mills. Iron Steelmak 23:49–54

89. CMP (1992) Electric arc furnace efficiency. Center for Materials Production. Report 92-10. Pittsburgh, PA, USA

90. Herin H, Busbee T (1996) The consteel process in operation at Florida steel. Iron Steelmak 23:43–46

91. Macauley D, Smailer RM (1997) Engineering fundamentals for a least cost/flexible steel-making. In: 25th advanced technology symposium new melting technology, St. Petersburg Beach, FL

92. International Energy Agency (2020) Energy technology perspectives

93. World Steel Association (2017) World steel position paper. Steel's contribution to a low carbon future and climate resilient societies

94. Pfeifer H, Kirschen M (2002) Thermodynamic analysis of EAF energy efficiency and comparison with statistical model of electric energy demand. In: 7th European electric conference, 26–29 May 2002, Venice, Italy

95. Tam C (2009) Energy technology transitions for industry. OECD/IEA. Int Energy Agency 2009:1–326. https://www.iea.org/reports/energy-technology-transitions-for-industry

96. Conejo AN, Birat JP, Dutta A (2019) A review of the current environmental challenges of the steel industry and its value chain. J Environ Manage 259. Available online Dec 2019

97. Holappa L (2017) A general approach to the reduction of CO_2 emissions from the steel industry. In: 2nd ISIJ-VDEh-Jernkontoret Jt. symposium, 12–13 June, Stocjolm Sweden, pp 61–72

Chapter 2
Energy Balance

2.1 Introduction

The first concept that should be clear when trying to understand energy consumption is the first law of thermodynamics, also called the law of conservation of energy which states that "energy can neither be created nor destroyed; rather, it can only be transformed or transferred from one form to another". This law was first demonstrated by Émilie du Châtelet in 1730. Is a balance of the various forms of energy. One simplified version is expressed as follows

$$\Delta U = Q - W$$

where; ΔU is the change in the thermal energy of the system, Q is the net heat transfer into the system and W is the net work done by the system.

Alternatively;

$$energy\ input = energy\ output + heat\ losses$$

Mass and energy balances are always interrelated, they cannot be separated. The "Handbook on material and energy balance calculations in materials processing" by Morris et al. [1] is a complete reference to learn about mass and energy balances in metallurgy.

In the EAF energy is consumed for heating and melting all the additions. Depending on their physical and chemical composition each material requires a fixed amount of energy that depends on the final temperature of the molten bath. This value is the theoretical energy consumption. Its real value can be higher but not lower. This concept is sometimes misunderstood by some management people in steel plants who demand to decrease energy consumption in order to decrease production costs. The law of conservation of energy in simple words indicates that if we add cheaper raw materials containing impurities with high melting point, the

A. N. Conejo, *Electric Arc Furnace: Methods to Decrease Energy Consumption*,
https://doi.org/10.1007/978-981-97-4053-6_2

theoretical energy requirements will be increased. This value becomes a constant when the mass, temperature and chemical composition of the charged materials is defined:

$$\text{Theoretical energy consumption} = \text{Electrical energy}$$
$$+ \text{Chemical energy} = \text{Constant}$$

The most common unit to measure electric energy consumption in the EAF is in kWh. Watt is a unit of power, which multiplied by time yields energy units. Usually, only the amount of electric energy is recorded during a heat and also usually is the value given as a reference on energy consumption, which is wrong because that is not the total energy consumption. To obtain the total energy consumed, the chemical energy should be calculated, however this is not a common practice. Prakash et al. [2] reported the total energy consumption for different mixtures of metallics, indicating a consumption of electric energy of 480 kWh/ton for scrap and a total energy consumption of 600 kWh/ton. Using 70% DRI the electric energy consumption increased to 500 kWh/ton and the total energy consumption was 700 kWh/ton. In modern practices it is intended to decrease the consumption of electric energy by increasing chemical energy in order to decrease production costs. Madias et al. [3] classified into four ranges of electric energy consumption the EAF's depending on the metallic charge: (1) < 300 kWh/ton; EAF's charged with more than 20% hot metal, (2) 300–400 kwh/ton; EAF's charged with pig iron and high quality scrap, (3) 400–450 kWh/ton; EAF's charged with intermediate quality scrap and hot DRI, and (4) > 450 kWh/ton; EAF's charged with cold DRI or low quality scrap.

The development of energy balances is necessary to define strategies to decrease energy consumption. The following are some of its applications:

- Optimum selection of the charge mixture
- Optimum C–O balance
- Quantify the effect of pre-heating temperature.

There are two types of energy balances, static and dynamic. The static model is easier and faster to develop but is mainly valid for the range of the variables investigated and cannot be extrapolated under different conditions. A dynamic model (also called mechanistic, analytic or phenomenological) is more complex and computationally expensive to develop but can be extrapolated for values outside the range of the variables investigated.

2.2 Components of an Energy Balance

2.2.1 Energy Inputs

The energy sources for the EAF are electric and chemical energy, as shown below.

i. Electric energy
ii. Chemical energy

    a. Chemical energy due to oxygen injection (oxidation reactions)
    b. Chemical energy due to carbon from electrodes
    c. Chemical energy due to organics from scrap
    d. Chemical energy from burners
    e. Chemical energy from post-combustion.

Electric energy: Steel scrap in contrast to hot metal, has a much lower carbon concentration and therefore cannot produce enough chemical energy to melt the metallic charge. The EAF is therefore a suitable reactor to melt steel scrap. Electric energy is still today the main source of energy in conventional EAF's. Electric energy is usually considered a clean energy, however, 62.8% of the electric energy in the world is produced by fossil fuels [4]; 36.4% from coal and 23.3% from natural gas. China which produces almost 30% of the total electricity in the world has the largest consumption of coal for the production of electricity.

Electric energy is supplied from generation plants at high voltages, 100–500 kV. At the steel plant, step-down transformers decrease the voltage to 34.5 kV and finally at the steel shop another transformer decrease the voltage to 100–1000 V and high currents from 40 to 60 kA. The final arc voltage is controlled by the arc length, approx. 10 V/cm.

Chemical energy due to oxygen injection: Oxygen injection is the primary source of chemical energy in the EAF. This is due to the exothermic nature of oxidation reactions. On average 70% of oxygen is consumed by decarburization of the molten metal and the remaining 30% with other alloying elements [5, 6]. The energy input from oxygen injection (kWh/m^3 O_2) depends on the chemical composition and temperature of liquid steel as well as the specific value of oxygen injected. As a reference, the energy input from oxygen injection ranges from 100 to 250 kwh/ton [7].

Chemical energy due to carbon from electrodes: The electrodes as a source of carbon is not desired however it is unavoidable to some extent. Two forms of electrode consumption provide carbon; tip sublimation and side oxidation. Bowman [8, 9] reported expressions indicating that tip consumption is proportional to the square of the electrical current and side oxidation is proportional to the electrode oxidizing surface.

$$E_T \text{ (kg/ton)} = A \frac{I^2}{P} \frac{t_{on}}{t_{tap}} = A \frac{I^2 t_{on}}{W}$$

$$E_S \text{ (kg/ton)} = 3B \frac{\pi \overline{D} L_{ox}}{P}$$

$$E_{total} = E_T + E_S$$

where: E_T and E_S are tip and side oxidation in kg/ton, respectively, I is the electric current in kA, $\overline{D}$ is the average electrode diameter in m, L_{ox} is the oxidizing length of the electrode, W are the tons of liquid steel, t_{on} is the power on time in hr and t_{tap} is the tap-to-tap time in hr. P is the productivity in tons per heat time. A and B are constants, as shown in Table 2.1.

The average diameter can be estimated as follows: $\overline{D} = (D_E + D_T)/2 = (D_E + 0.7D_E)/2$. The oxidizing length, can also be estimated as follows: $L_{ox}(m) = 2.8\sqrt[3]{W}/100$. About 40% is side oxidation and 50% tip oxidation. Electrode consumption in the range from 1.3 to 2.7 kg/ton releases from 12.4 to 24.8 kwh/ton [7]. The heat of formation of CO_2 is about 9.1 kWh/kg C. The oxygen sources for electrode oxidation can be injected oxygen, air infiltration and CO_2 from post-combustion. The oxidation reactions are accelerated due to the high temperature of the surface of electrodes.

Chemical energy due to organics from scrap: Scrap can contain from 0.25 to 1% of oils, plastics and paints which upon combustion releases chemical energy from 25 to 100 kWh/ton [7]. The heat released when some organics are burned is intense and observed in the form of large flames however this source of heat has negative side effects because during combustion polychlorinated dibenzodioxins (PCDD), for short dioxins, and dibenzofurans (PCDF), for short furans, can be formed in concentrations between 0.02 and 9.2 ngTE/m^3 [10], where TE is toxic equivalent. Dioxins and furans are extremely toxic. The low concentrations makes sampling and analysis a complex task [11, 12]. Prüm et al. reported that applying activated HOK lignite before the bag filter house decreased levels from 1.4 to less than 0.1 ngTE/m^3 and comply with EU regulations [10]. Even though dioxins are emitted in relatively low concentrations, they are toxic bio-accumulative compounds [13] and are also one reason to consider EAF dust as a hazardous material because they condense as liquids in the dust collecting system.

Chemical energy from burners and post-combustion: Burners and post-combustion are two important sources of chemical energy in scrap-based operations. They are currently employed to decrease electric energy consumption.

Chapter 4 discusses in detail chemical energy from oxygen injection. Chapters 5 and 6 explains in detail burners and post-combustion.

Table 2.1 Values of parameters A and B, required to estimate electrode consumption [8, 9]

	A, kg/kA^2 h		B, kg/m^{2}h
AC, arc voltage > 250 V	0.013 × 3	Ladle furnaces	3.0–4.0
Av, tip angle below 30°	0.010 × 3	Closed furnace, < 4 Nm^{3}O$_2$/ton	3.0–4.0
AC, tip angle over 40°	0.016 × 3	5–15 Nm^{3}O$_2$/ton	5.0–6.0
DC	0.0124	15–45 Nm^{3}O$_2$/ton	6.0–8.0
Ladle furnace	0.01–0.03	25–45 Nm^{3}O$_2$/ton + post comb	8.0–10.0

2.2.2 *Energy Outputs*

    i. Liquid steel
   ii. Liquid slag
  iii. Off-gas and dust
  iv. Water cooling
   v. Heat losses.

Liquid steel: The computation of the theoretical value of the energy required to produce one ton of any material is based on a simple equation that computes the amount of heat at a given temperature.

$$\int_{\Delta H^0_{T_1}}^{\Delta H^0_{T_2}} d(\Delta H^0) = \int_{T_1}^{T_2} \Delta C_p dT$$

$$\Delta H^0_{T_2} = \Delta H^0_{T_1} + \int_{T_1}^{T_2} \left(\sum v_p C_{p,\,products} - \sum v_r C_{p,\,reactants} \right) dT$$

where: ΔH^0 represents heat or enthalpy, $\Delta H^0_{T_i}$ is the amount of heat at temperature T_i, C_p is the heat capacity, v_p and v_r represent the stoichiometric coefficients for products or reactants, respectively.

The energy required to produce one ton of liquid steel at 1537 °C is 337 kWh. This value is for melting pure iron. Additional heat or thermal energy is required for superheating liquid steel and melting non-ferrous materials in the metallic charge. The highest proportion, 76% of the total energy, is required to reach the melting point and the rest is the energy for melting and superheat to 1600 °C.

Liquid slag: The volume of slag produced is defined from the following sources; (i) amount of impurities in the metallic charge, (ii) amount of fluxes added and (iii) oxidation of elements. Scrap and hot metal-based operations produce less slag compared with the use of DRI. The refining capacity of slags increases with its mass, but on the other hand, by increasing the mass of slag also decreases the metallic yield and energy losses. The blast furnace produces 250–300 kg/ton, the BOF produces 100–150 kg/ton and the EAF produces about 100 kg/ton using scrap and 100–150 kg/ton using DRI. In other words, production of ironmaking slags ranges from 25 to 30% and steelmaking slags ranges from 10 to 15% [14]. Therefore, it is to be expected higher energy losses in the slag using DRI in comparison with scrap. Energy losses in the slag in scrap-based operations has been reported in the range from 30 to 50 kWh/tonls [7].

The sensible heat in liquid slag at 1593 °C is 480 kWh/ton slag and increases 0.742 kWh/ton slag for every °C above that temperature [7].

Energy in the off-gas and dust: The rate of production of gases during steelmaking in the EAF is variable during a heat. Increases by increasing the power input, by increasing the oxidation of fossil fuels and by increasing air infiltration.

Energy in the off-gas in scrap-based operations has been reported in the range 80–120 kWh/ton, [5, 15] representing about 10–20% of the input energy.

Energy losses by water cooling: Energy losses in modern EAF's increased due to application of water-cooling systems, however it helped to solve the problem of a long arc operation. Energy losses due to water cooling has been reported in the range 60–100 kWh/ton [7]. Power losses by water cooling include shell, roof, slag door and the fourth hole. The energy in water cooling is defined by the following expression:

$$\Delta H_{wcp} = \left(\sum \dot{m} C_{pw} \Delta T \right) time_{tap\text{-}to\text{-}tap}$$

where: $\dot{m}$ is the water flow rate in water panels of different sections of the EAF, C_{pw} is the heat capacity of water (4180 J/kg °C), ΔT is the temperature gradient between the outlet and inlet temperatures, and $time_{tap\text{-}to\text{-}tap}$ is the tap-to-tap time.

The energy losses from water cooled panels depend on the water flow rate, the construction material, cooling system (water or spray cooling), the thickness of slag layer, energy radiated to the walls and tap-to-tap time.

The energy radiated to the walls increases by increasing the electric power supply especially if the electric arcs are not covered by foamy slag. Simon reported that the structure of the slag layer is formed by alternate layers of slag and metal [16]. Opfermann and Riediger [17] reported that the maximum energy losses occur during the flat bath period, about 1.7 kWh/ton per minute, however they didn't specify the arc coverage. During power-off time they suggest energy losses about 0.5 kWh/ton per minute, equivalent to 7 kWh/ton assuming a total power-off time of 11 min, which includes tapping.

Water cooling is reviewed in detail in Chap. 15.

2.2.3 Energy Losses

Energy losses define the thermal efficiency. In principle, the EAF's thermal efficiency is a ratio between the energy in liquid steel to the energy supplied:

$$\eta_{th} = \frac{Q_{steel}}{Q_{input}}$$

This definition would define energy losses as any energy outlet other than that in liquid steel. This is not the conventional definition for heat losses as explained in the following section. Energy losses is one of the most complex terms in the calculation of an energy balance. It is usually applied as a lump value to make possible the energy balance by which the energy inlets are equal to the energy outlets. EAF energy losses are influenced by many factors such as [18]:

1. Radiation from the electric arc to the water-cooled panels, which depend on:

   1.1 Foaming conditions
   1.2 Slag/steel temperature
   1.3 Furnace power (furnace transformer) and furnace size
   1.4 Tap to tap time

2. Energy losses with slag, which depend on:

   2.1 Amount and slag chemical composition
   2.2 Energy recovery from liquid slag

3. Energy losses in the off-gas, which depend on:

   3.1 Suction power by the fume extraction system
   3.2 Volume and chemical composition of the off-gas
   3.3 Energy losses due to infiltrated air
   3.4 Energy recovery from the off-gas

4. Energy losses due to scrap charging

   4.1 Number of scrap charges
   4.2 Duration of scrap charges

5. Energy losses due to delays

In order to better understand the term energy losses is necessary to look in detail the general energy balance equation:

$$\text{energy input} = \text{energy output} + \text{heat losses}$$

Energy input has two components; electric (ΔH_e^i) and chemical energy (ΔH_{ch}^i).

$$\text{energy input} = \left(\Delta H_e^i\right) + \left(\Delta H_{ch}^i\right)$$

The chemical energy results from the following chemical reactions:

($\Delta H_{ch,1}^i$) Oxidation reactions with elements dissolved in liquid steel.
($\Delta H_{ch,2}^i$) Oxidation of carbon from the electrodes.
($\Delta H_{ch,3}^i$) Combustion of organic matter in scrap.
($\Delta H_{ch,4}^i$) Combustion of fuel by burners.
($\Delta H_{ch,5}^i$) CO post-combustion.

The energy outputs correspond to the sensible heat in the streams of materials leaving the furnace, endothermic reactions, radiation and convection from different surfaces and energy losses. A total of about 17 terms:

(ΔH_{st}^o)   Sensible heat in liquid steel.

(ΔH^o_{sl})	Sensible heat in liquid slag.
(ΔH^o_{off})	Sensible heat in the off-gas.
(ΔH^o_{dust})	Sensible heat in the dust.
(ΔH^o_{wc})	Sensible heat in water cooling.
$(\Delta H^0_{endo,1})$	Endothermic heat due to reduction of FeO in the slag.
$(\Delta H^o_{endo,2})$	Endothermic slag's heat of formation.
(ΔH^o_{w1})	Endothermic heat to remove humidity in the charged materials.
(ΔH^o_{w2})	Endothermic heat to evaporate water cooling of electrodes.
(ΔH^o_{foam})	Heat losses due to arc radiation as a function of the foaming conditions.
$(\Delta H^o_{power\text{-}off})$	Heat due to radiation and convection from the molten bath during power-off.
(ΔH^o_{sch})	Heat due to radiation during scrap charging.
(ΔH^o_{sh})	Heat transfer from furnace shell to the surroundings.
(ΔH^o_{sd})	Heat transfer from slag door to the surroundings.
(ΔH^o_{tap})	Heat transfer from the stream of liquid steel during tapping.
(ΔH^o_{hh})	Sensible heat to keep temperature of hot heel.
(ΔH^o_{e})	Electric losses.

Taking into account the previous terms, a more detailed description of the heat balance is the following:

$$\Delta H^i_e + \left(\Delta H^i_{ch,1} + \Delta H^i_{ch,2} + \Delta H^i_{ch,3} + \Delta H^i_{ch,4} + \Delta H^i_{ch,5}\right)$$
$$= \Delta H^o_{st} + \Delta H^o_{sl} + \Delta H^o_{off} + \Delta H^o_{dust}$$
$$+ \Delta H^o_{wc} + \left(\Delta H^o_{endo,1} + \Delta H^o_{endo,2}\right) + \Delta H^o_{w1}$$
$$+ \Delta H^o_{w2} + \Delta H^o_{foam} + \Delta H^o_{power\text{-}off}$$
$$+ \Delta H^o_{sch} + \Delta H^o_{sh} + \Delta H^o_{sd} + \Delta H^o_{tap} + \Delta H^o_{hh} + \Delta H^o_e$$

If heat losses are all the energy outputs except the sensible heat of liquid steel:

$$\text{heat losses} = \text{energy input} - \text{sensible heat in liquid steel}$$
$$\text{heat losses} = \text{energy input} - \Delta H^o_{st}$$

The common idea about heat losses is that those terms that are usually more difficult to measure or to estimate and cannot be included in the heat balance are altogether lumped as the heat losses term and therefore result in a very subjective value. Taking into consideration two aspects; first, that it is possible to measure or to estimate the sensible heat in the slag and the sensible heat in the off-gas and, second, the potential for heat recovery from both sources, heat losses can be defined as follows:

$$\text{heat losses} = \text{energy input} - \left(\Delta H^o_{st} + \Delta H^o_{sl} + \Delta H^o_{off}\right)$$
$$\text{heat losses} = \Delta H^o_{endo,1} + \Delta H^o_{endo,2} + \Delta H^o_{dust}$$
$$+ \Delta H^o_{wc} + \Delta H^o_e \Delta H^o_{w1} + \Delta H^o_{w2}$$

$$+ \Delta H^o_{foam} + \Delta H^o_{power-off} + \Delta H^o_{sch}$$
$$+ \Delta H^o_{sh} + \Delta H^o_{sd} + \Delta H^o_{tap} + \Delta H^o_{hh}$$

Heat losses from the previous equation assumes that only the sensible heat in liquid steel, slag and off-gas is known leaving the rest of the terms as global heat losses, which include all endothermic reactions, arc radiation, delays, scrap charging, furnace shell heat transfer, tapping, energy in the hot heel and electric losses. Water cooling due to its relatively low temperature in comparison with liquid steel, liquid slag and the off-gas is not a source of exergy and its value should be added as part of the heat losses term, however this term can be easily measured and is usually not reported in the heat losses term.

The problem to have an accurate value of heat losses is not only the large number of terms involved which would require a large amount of experimental measurements but also that some parameters are difficult to measure or to estimate. Consequently, the term heat losses is commonly defined in a very loose way. Real heat losses should correspond to heat that leaves in the effluent streams or lost to the atmosphere and is not recovered back into the system. In practice, there is ambiguity in its definition. Heat losses in the EAF should be carefully reported. For example, Diaconu et al. [19] computed an important number of energy losses, reporting that the primary heat losses in a 6.7 ton EAF were due to water cooling (33%), followed by roof opening during scrap charging (29%) and the off-gas (18%), however since many details were not reported, these results are questionable because of inconsistencies in the values reported.

Table 2.2 summarizes a number of heat balances, showing large differences, primarily due to large differences in raw materials and operation condition but very useful to define a characteristic range.

In order to develop an energy balance with higher accuracy it is important to consider the need to start with reliable information. This type of information requires equipment properly calibrated, sensors to monitor the properties indicated, etc. The following list gives an idea of some of the requirements to obtain reliable information.

$\Delta H^i_{ch,1}$ Oxidation of elements in metallic charge $\rightarrow$ accurate amount of each component in the metallic charge and its chemical composition.

(ΔH^o_{sl}) Sensible heat in liquid slag $\rightarrow$ Accurate amount of slag produced and average chemical composition.

(ΔH^o_{off}) Sensible heat in the off-gas $\rightarrow$ Accurate chemical composition of the off-gas, temperature and volume.

$(\Delta H^i_{ch,4})$ Combustion of fuel by burners $\rightarrow$ Accurate value on burners thermal efficiency.

$(\Delta H^i_{ch,5})$ CO post-combustion $\rightarrow$ Accurate value on post-combustion efficiency.

(ΔH^o_{dust}) Sensible heat in the dust $\rightarrow$ Accurate amount of dust produced and chemical composition.

(ΔH^o_{wc}) Heat losses in water cooled panels $\rightarrow$ Inlet and outlet temperature as well as water flow rate for each panel.

Table 2.2 Different energy balances and heat losses reported

		ΔH_{st}^o, %	ΔH_{off}^o, %	ΔH_{sl}^o, %	ΔH_{wc}^o, %	ΔH_e^o, %	Other heat losses, %
1986	Pearce [20]	52.7	19.5	12.2	16.2	1.1	–
1998	Hornby [21]	54	36	7	6	–	1
2002	Pujadas [22]	46	27	7	–	–	–
2002	Kirschen [23]	50.8	10.5	4.0	20.0	1.6	4.2
2002	Kirschen [23]	46.7	15.5	4.6	22.9	4.0	6.3
2005	Pfeifer [6]	53.1	20.7	3.8	13.4	10.0	4.6
2007	Cardenas [15]	46.7	11.0	11.3	–	–	31
2007	Prakash [2]	48.8	10.6	4.9	20.0	11.6	4.0
-	Apfel [nd]	58.5	18.4	11.4	8.1	2.9	–
2009	Kirschen [24]	45–60	15–35	5–10	10–20	–	–
2012	Logar [25]	51.4	15.9	5.5	15.0	4.6	7.6
2014	Pesamosca [26]	54	(20)	11.0	(11)	2.0	–
2015	Thekdi [27]	52	16.7	3.1	7.9	1.7	(0.4)
2016	Yang [28]	53.7	18.9	5.8	3.2	-	-
2020	Diaconu [19]	44	18	7.9	33	2.93	29
2023	Liu [29]	53.9	(25)		8.8	3.21	

$(\Delta H_{ch,2}^i)$ Oxidation of C from electrodes → how much C from electrodes is oxidized?

$(\Delta H_{ch,3}^i)$ Combustion of organic matter in scrap → how much organic matter in the scrap?

$\Delta H_{ch,4}^i$ Energy from burners → what is the thermal burner's thermal efficiency?

$\Delta H_{ch,5}^i$ energy from post-combustion → what fraction of O_2 reacts with CO?

(ΔH_{foam}^o) Heat losses due to arc radiation → Coverage of electric arcs (foaming height) and electric power.

$(\Delta H_{endo,1}^o)$ Reduction of FeO in the slag → how much FeO is reduced?

$(\Delta H_{endo,2}^o)$ Slag's heat of formation → Accurate thermochemical database and slag amount.

(ΔH_{dust}^o) Quantity, temperature and chemical composition of EAF dust per heat.

(ΔH_{w1}^o) Heat to remove humidity → Humidity in charged materials.

(ΔH_{w2}^o) Heat to evaporate water cooling of electrodes → accurate total volume for water cooling, for each water-cooled panel.

$(\Delta H_{power-off}^o)$ Heat losses due to delays → Time due to delays, surface temperature of slag and furnace parts. Part of this heat is transferred to the water-cooled panels.

(ΔH_{sch}^o) Heat losses due to radiation during scrap charging → Charging time per scrap basket, number of scrap baskets, shell surface area and temperature, roof surface area and temperature.

(ΔH_{sh}^o) Heat losses due to heat transfer from slag door to the surroundings → Shell surface area and temperature.

(ΔH_{sd}^o) Heat losses through slag door → Area of slag door, volume of air ingress, time opened during a heat.

(ΔH_{hh}^o) Heat losses to keep temperature of hot heel → volume of hot heel and initial temperature.

(ΔH_{tap}^o) Heat losses during tapping → Surface and temperature of liquid steel stream, tapping time.

(ΔH_e^o) Electric losses by different components of the primary and secondary circuit.

Due to the large amount of information it is anticipated an error in any heat balance, its magnitude will depend on the quality of information available from the previous list.

The mass of slag can be estimated from a mass balance for CaO or MgO added in the fluxes, requiring knowledge of the slag and dust chemical composition as well as the amount of additions containing CaO or MgO, and additionally the amount of dust, according with the following expression:

$$W_{flux}X_{CaO}^{flux} + W_{DRI}X_{CaO}^{DRI} = W_{slag}X_{CaO}^{slag} + W_{dust}X_{CaO}^{dust}$$

In addition to oxidation reactions and gangue content in the feed materials, the real slag volume should consider refractory wear. Arzpeyama et al. [30] estimated a value of 3 kg/ton which is added to the previous calculation.

Energy losses depending on furnace capacity: Sims [31] in 1963 suggested the following expression to compute heat losses as a function of furnace capacity:

$$Q = 4.52 \times C_{furnace}^{-0.352}$$

where: Q are the heat losses in kWh/min ton and Ci is the furnace capacity in tons.

Energy losses by conduction and radiation from the crucible: Heat conduction is defined by Fourier's heat law of conduction.

$$\dot{Q} = k\frac{(T_2 - T_2)}{d}$$

where: $\dot{Q}$ is the flux of heat transfer, k is the thermal conductivity, d is the average thickness of the refractory lining, T_2 is the inner temperature, T_1 is the outer temperature.

Thermal or infrared radiation from the outer steel shell is defined by the Stefan-Boltzmann Law.

$$\frac{\dot{Q}}{A} = \sigma\left(T_2^4 - T_1^4\right)$$

where: $\dot{Q}$ is the flux of heat transfer, σ is the Stefan-Boltzmann constant (5.6 $\times$ 10^{-8} W/m^{2}K^4), T_2 is the surface temperature, T_1 is ambient temperature.

The energy losses from the crucible are much smaller in comparison with other heat losses and are therefore usually neglected in the overall energy balances.

Summarizing on the reliability of mass and heat balances, the accuracy between predicted and experimental data on an energy balance depends on [32]:

- Accuracy of mass, chemical composition and temperatures of materials included in the energy balance
- Accuracy to integrate all the chemical reactions taking place
- Accuracy of thermochemical database (heat of reaction, formation and dissolution)
- Assumptions: reaction mechanisms (type and physical state of chemical species).

Frequently is necessary to apply empirical approaches or make assumptions to define the parameters needed for an energy balance. For example, the amount of hot heel and its temperature are usually unknown. A typical value is 10% of the weight of the heat but can be as high as 50% for the Consteel EAF. Adams et al. [33] suggested an efficiency factor of energy conversion from chemical energy of oxidation reactions to the steel melt in the range from 0.7 to 0.8. Abraham and Chen [34] indicated that one of the most challenging aspects to develop an energy balance is to define the fraction of oxygen consumed with all the elements dissolved in liquid steel. To solve the problem, they assigned the following order of reactions:

- Combustion of natural gas
- Oxidation of Al, Si, Mn and Fe
- Oxidation of C–CO
- Oxidation of C–CO$_2$.

In addition to that, they also considered the following assumptions:

- 3% of the metallic charge is lost due to oxidation
- The slag's melting point was 1573 K
- The specific heat capacity of liquid slag is equal to liquid lime
- The total weight of the slag is twice the weight of the fluxes.

2.2.4 EAF Thermal and Electric Efficiency

Thermal efficiency (η_{th}) can be defined as the ratio of the energy in liquid steel to the total input energy.

$$\eta_{th} = \frac{Q_{steel}}{Q_{input}}$$

Pfeifer et al. [5] provided an alternate definition based on the ratio of energy in both liquid steel and slag to the input energy:

$$\eta_{th}' = \frac{Q_{steel} + Q_{slag}}{Q_{input}}$$

This definition, η_{th}', gives higher values in comparison with η_{th}. Table 2.3 shows a summary of reported thermal efficiencies in EAF's compiled by Pfeifer et al. indicating a range from 50 to 65%. The EAF low thermal efficiency increases the total minimum energy requirements from 350 kWh/ton for pure iron to 560–680 kWh/ton using 100% scrap [35].

Kirschen et al. [24] carried out an analysis of the thermal efficiency in 70 EAF's, as shown in Fig. 2.1. It can be observed the thermal efficiency ranges in general from 40 to 60%. As the thermal efficiency decreases due to higher heat losses, the total energy requirements increase.

Abuluwefa [36] reported a low thermal efficiency of 48% for an EAF with a nominal capacity of 100 ton charged with both scrap and DRI. Most of the energy employed was electric energy, 86%. After meltdown, the thermal efficiency largely depends on the foaming conditions. Ameling [37] described the thermal efficiency in the range from 36 to 100%, depending on the slag foaming conditions.

The instantaneous thermal efficiency results from the instantaneous heat transfer by conduction, convection and radiation among all the materials and phases present in the EAF, this value changes during a heat. In scrap-based operations, the instantaneous thermal efficiency is close to 100% at the initial period of melting because the arc radiation is used to melt the scrap and the off-gas partially preheats the scrap or leaves at low temperatures but once the arc is not shielded by solid scrap, the temperature of the off gas is higher and the electric arc radiates energy to the walls.

Electric efficiency: Electric efficiency depends on the effective transfer of heat radiated from the arc to the solid or liquid steel. During the boring and meltdown period electric efficiency is high, about 88–92%, because the scrap shields the arc [7]. Electric efficiency can be defined as the ratio between the electric arc power (P_{arc}) to the active power (P_A), as follows [38, 39]:

Table 2.3   Thermal efficiency of EAF (η_{th}') after ref. [5]

Year	Q_{input}	Q_{Elec}	Q_{s+s}	η_{th}, %	References
1982	789	541	503	63.8	Fett et al. Stahl und Eisen, 102, 1982, 461–465
2001	773	497	442	57.2	
1989	704	487	443	59.2	Brod et al., Stahl und Eisen, 109, 1989, 229–238
2001	758	477	449	59.2	Kirschen et al. 59th EAF Conf., 2001, 737–748
2002	807	510	401	49.7	
1990	729	427	429	58.8	Gripenberg et al. Iron and Steel Engineer, 1990,33–37
1998	680	400	440	64.7	Ehle et al. Steel World, 3 (2001) 2, 24–32
1999	810	393	434	53.5	Ehle et al. Steel World, 3 (2001) 2, 24–32

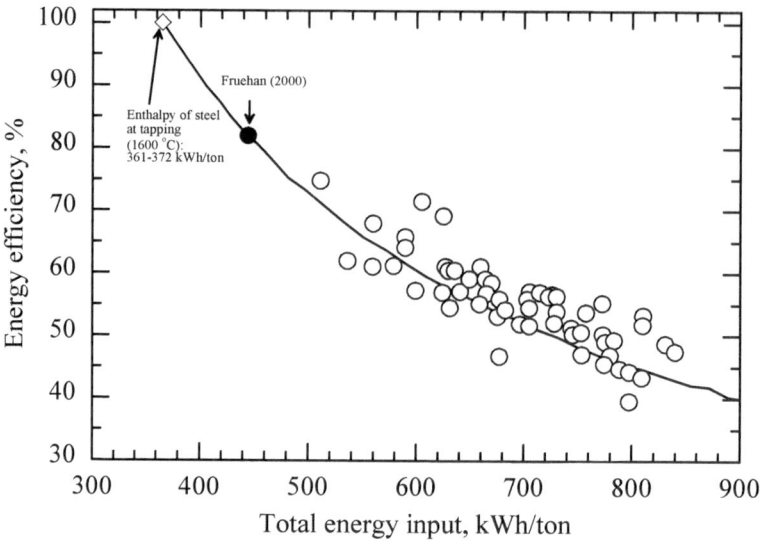

Fig. 2.1 EAF thermal efficiency, after Kirschen et al. [24]

$$\eta_{el} = \frac{P_{arc}}{P_A} = \frac{3\left(\sqrt{U^2 - I^2X^2} - IR\right)I}{3\left(\sqrt{U^2 - I^2X^2} - IR\right)I + 3I^2R} = \frac{3\left(\sqrt{U^2 - I^2X^2} - IR\right)I}{\sqrt{3}UIcos\emptyset}$$

where: U is voltage, R is electric resistance and cos∅ is the power factor.

Kirschen et al. [40] suggest that the electric efficiency is the product of two parts; an efficiency of electric energy transfer to the high current system which ranges from 0.88 to 0.92 and an efficiency of energy conversion from the electric arc which ranges from 0.36 to 0.93, depending on the foaming conditions. The higher value of energy conversion from the electric arc to the molten bath can be obtained with a long arc operation and good foaming conditions (0.88–0.92).

2.2.5 Sankey Diagrams

The metallic charge defines the energy balance. The main raw materials with a higher purity in iron units will result in a lower energy consumption and vice versa. Energy balances are specific to the chemical composition and temperature of the raw materials employed as well as a given set of chemical reactions and heat transfer efficiencies. A review of several cases can give an idea of the magnitude of the total energy inputs and outputs and how this energy is distributed. The graphic representation of an energy balance is given in the form of Sankey diagrams [41].

Table 2.4 EAF energy balance, kWh/ton [35]

Input		Output	
Electric energy	340–500	Liquid steel	350–380
Oxygen–carbon reactions	100–250	Liquid slag	30–50
Burners	25–80	Water cooling	60–100
Scrap grease and others	40–80	Off-gas	140–180
Electrodes	15–25	Recharges and electrical losses	20–80

Pujadas et al. [22] reported an energy balance for a 150 ton AC-EAF with a nominal diameter of 6.7 m, 90 MVA transformer, average active power of 78 MW, charged with 100% scrap, specific oxygen injection of 52 Nm^3/ton and 37.1% FeO, among some of the process conditions. On this basis, the total energy input was 775 kWh/ton, approximately 60% was chemical energy and 40% electric energy. Energy in liquid steel was 46%, in liquid slag 7% and 27% in the off-gas.

Jones et al. [35] reviewed several EAF energy balances from the late 1990s, considering scrap-based operations. For a total energy input of about 560–690 kWh/ton, the input and output energy values are defined in Table 2.4.

Yang et al. [28] reported an energy balance using scrap + 50% hot metal. The electric energy consumption was 23% and 77% was due to chemical energy (24% sensible heat in the hot metal and 34% due to exothermic reactions). For EAF's with intensive post-combustion the energy in the off-gas is about 35–38% [42].

2.3 Static Energy Balances

The theoretical energy consumption is computed using only thermodynamics, which depends only on the initial and final states. A model based only on thermodynamics is a static model.

In the energy balance equation; energy input = energy output + heat losses, the energy input comprises electric and chemical energy and the energy output correspond to the sensible heat in liquid steel, liquid slag, off-gases and energy losses. Energy transfer due to water cooling is part of the energy losses. Energy inputs correspond to the heat required for heating, melting and superheat of the feed materials. The chemical energy is due to oxidation of dissolved elements in liquid steel, reduction of iron oxide in the slag, heats of dissolution of solid carbon and combustion reactions. The main heat losses are due to the heat extracted by the water-cooling systems.

1. The energy required for heating (includes melting and superheat) of individual components in the metallic charge, fluxes, carbon and any other feed material into the EAF is computed with the following expression:

$$\frac{kWh}{ton} = \int_{T_1}^{T_2} \sum C_p dT + \Delta H_{transf}$$

where: kWh/ton is the specific energy consumption, C_p is the heat capacity of each material, T_1 and T_2 are the initial and final temperatures, and ΔH_{transf} is the heat of transformation (phase change during heating or melting). Increasing the melting point increases the energy requirements. Carbon plays a key role during steelmaking operations, the melting point of steel scrap is usually higher than that of DRI and pig iron due to differences in carbon concentration, typically about 0.4%, 1% and 4%, respectively. Increasing the carbon concentration decreases the melting point from pure iron, from 1535 to 1153 °C at the eutectic point, corresponding to 4.5%C.

2. Chemical energy for the following chemical reaction is computed using Kirchhoff's law of enthalpy:

$$aA + bB = mM + nN$$

$$\frac{kWh}{ton} = \int_{T_1}^{T_2} \left[\left(mC_{p,M} + nC_{p,N} \right) - \left(aC_{p,A} + bC_{p,B} \right) \right]$$

3. Sensible heat in liquid steel, slag, off-gas and water cooling can be calculated with the following equation:

$$\frac{kWh}{ton} = \int_{T_1}^{T_2} \sum C_p dT$$

Kirschen et al. [40] used a modified approach. Instead of using mass balances they considered a balance on the rate of mass inputs and outputs which allows to transform the derivatives as a function of time, essentially computing kWH/ton per minute. In the end, they also faced the problem to deal with heat losses or efficiency factors to adjust the heat balance. They reported that in order to reproduce the experimental values they had to adjust locally each one of the efficiency factors, suggesting the following ranges:

$$0.6 < \text{Electric efficiency} < 0.8$$
$$0.7 < \text{Chemical efficiency} < 0.8$$
$$\text{Burner efficiency} < 0.6$$
$$\text{Post-combustion efficiency} < 0.4$$

Using these correction factors these authors studied different cases. One case consisted of an 80-ton EAF charged with 90% DRI + 10% Scrap, the composition of DRI was 93% metallization (8.3% FeO), 2% C, 2.4% gangue content (1% CaO, 1.4% SiO_2), 57 kg fluxes/ton, injecting 14 kg C/ton and 32 Nm^3 O_2/ton. The amount of slag produced was 11.9%. Their energy balance calculations reported 311 kWh/ton of chemical energy, 582 kWh/ton of electric energy and a total energy demand of 893 kWh/ton.

2.3.1 Variables Affecting Energy Consumption

The total energy consumption in the EAF is not only affected by the amount and type of materials charged into the furnace. A list of the variables that affect energy consumption is given below:

- Quality of raw materials:
  - Steel scrap: density, organic material, residual elements, etc.
  - DRI: Chemical composition (metallization, carbon and gangue concentration) and physical properties (particle size distribution, fraction of fines)
  - Fluxes: amount and type of fluxes
- Metallurgical practice
  - Volume and flow rate of oxygen injected
  - Temperature of liquid steel during melting and tapping
  - Slag foaming practice
  - Amount of slag
  - Scrap charges: number and duration
  - DRI feeding rate
- Process time
  - Tap-to-tap time
  - Power-on time
  - Power-off time
- Heat losses
  - Water cooled panels: flow rate and water temperature, maintenance and slag coating
  - Off-gas extraction system: extraction rate of the off-gases and system's maintenance
  - Slag door opening: Amount of infiltrated air. Heat losses due to air ingress can be in the order of 5 kWh/ton. Closed slag door operation and good maintenance minimize air infiltration
  - Recycling sensible heat in the off-gas and slag
  - Energy radiated from the electric arcs
  - Refractory; refractory life (refractory thickness), corrosion resistance to the slag, erosion rate.

Table 2.5 Theoretical energy requirements to produce steel from 100% scrap with different amounts of slag

Temp., °C	Steel, kwh/ton	Steel + 5% slag, kwh/ton	Steel + 8% slag, kwh/ton	Slag, kwh/ton
1520	368	390	411	508
1594	386	413	428	534
1648	398	426	443	553

- Electric power profile

  - Transformer capacity and electric ancillary equipment for arc stability
  - Electrode regulation
  - Melting rate: Depends not only on the transformer capacity.

The theoretical requirements of energy to melt one ton of scrap is summarized in Table 2.5 as a function of superheat and the amount of slag.

2.3.2 Energy Consumed to Melt One Ton of Scrap

This section is a detailed calculation to compute the energy consumed to melt one ton of steel scrap. In order to simplify the energy requirements for scrap melting it is assumed it is pure iron. The heat capacities of iron as a function of temperature (Fe-α, Fe-β, Fe-γ, Fe-δ, Fe-l) and heat of transformation are shown in Tables 2.6 and 2.7, respectively.

Table 2.6 Heat capacity of iron as a function of temperature

	°C	°K	C_P, cal/°K mol
Fe-α	25–760	298–1033	$3.0 + 7.58 \times 10^{-3} T + 0.6 \times 10^5 T^{-2}$
Fe-β	760–910	1033–1183	11.13
Fe-γ	910–1400	1183–1673	$5.80 + 2.0 \times 10^{-3} T$
Fe-δ	1400–1537	1673–1810	$6.74 + 1.64 \times 10^{-3} T$
Fe-l	1537–2700	1810–2973	$9.77 + 0.40 \times 10^{-3} T$

Source O. Kubaschewski and C. Alcock; Metallurgical Thermochemistry, 5th edition, 1989

Table 2.7 Heat of transformation

Phase change	ΔH^0_{transf}, cal/mol
$\alpha \rightarrow \beta$	326
$\beta \rightarrow \gamma$	215
$\gamma \rightarrow \delta$	165
$\delta \rightarrow l$	3670

The amount of heat to increase the temperature of from T_1 to T_2 is given by the following general expression:

$$\Delta H^0_{T_2} = \Delta H^0_{T_1} + \int_{T_1}^{T_2} \sum C_P \, dT$$

If T_2 is 1600 °C (1873 K):

$$\Delta H^0_{1873°K} = \int_{298}^{1033} C_{P,Fe\alpha} \, dT + \Delta H^0_{transf(\alpha \to \beta)} + \int_{1033}^{1183} C_{P,Fe\beta} \, dT$$

$$+ \Delta H^0_{transf(\beta \to \gamma)} + \int_{1183}^{1673} C_{P,Fe\gamma} \, dT + \Delta H^0_{transf(\gamma \to \delta)}$$

$$+ \int_{1673}^{1810} C_{P,Fe\delta} \, dT + \Delta H^0_{transf(\delta \to l)} + \int_{1810}^{1873} C_{P,Fel} \, dT$$

Replacing the values of heat capacities:

$$\Delta H^0_{1873°K} = \int_{298}^{1033} \left(3.0 + 7.58 \times 10^{-3} \, T + 0.6 \times 10^5 \, T^{-2} \right) dT$$

$$+ \Delta H^0_{transf(\alpha \to \beta)} + \int_{1033}^{1183} (11.13) \, dT + \Delta H^0_{transf(\beta \to \gamma)}$$

$$+ \int_{1183}^{1673} \left(5.80 + 2.0 \times 10^{-3} \, T \right) dT + \Delta H^0_{transf(\gamma \to \delta)}$$

$$+ \int_{1673}^{1810} \left(6.74 + 1.64 \times 10^{-3} \, T \right) dT + \Delta H^0_{transf(\delta \to l)}$$

$$+ \int_{1810}^{1873} \left(9.77 + 0.40 \times 10^{-3} \, T \right) dT$$

$$\Delta H^0_{1873°K} = 3.0(1033 - 298) + \frac{7.58 \times 10^{-3}}{2} \left(1033^2 - 298^2 \right)$$

$$- 0.6 \times 10^5 \left(\frac{1}{1033} - \frac{1}{298} \right) + 326$$

$$+ 11.13(1183 \text{ - } 1033) + 215 + 5.80(1673 - 1183)$$

$$+ \frac{2.0 \times 10^{-3}}{2} \left(1673^2 - 1183^2\right) + 165$$

$$+ 6.74(1810 - 1673) + \frac{1.64 \times 10^{-3}}{2} \left(1810^2 - 1673^2\right) + 3670$$

$$+ 9.77(1873 - 1810) + \frac{0.40 \times 10^{-3}}{2} \left(1873^2 - 1810^2\right)$$

$$\Delta H^0_{1873°K} = 5993 + 326 + 1669.6 + 215 + 4241.4$$

$$+ 165 + 1314.6 + 3670 + 661.91 = 18,256.9 \text{ cal/mol}$$

$$\Delta H^0_{1873°K} = 18,256.9 \frac{\text{cal}}{\text{gmol}} \times \frac{1 \text{ gmol Fe}}{55.854 \text{ g}}$$

$$\times \frac{1000 \text{ g}}{\text{kg}} = 326,868 \frac{\text{cal}}{\text{kg Fe}°}$$

$$= 326.868 \frac{\text{kcal}}{\text{kg scrap}} \times \frac{1000 \text{ kg}}{\text{ton}} \times \frac{\text{kWh}}{860 \text{ kcal}}$$

$$= 380 \frac{\text{kWh}}{\text{ton scrap}} = 1.368 \frac{\text{GJ}}{\text{ton scrap}}$$

From the total amount of energy required to produce a ton of liquid steel from steel scrap, about 76% is employed to pre-heat the charged materials, 20% for melting and 4% to reach the final molten steel temperature (1600 °C in this case) as illustrated in Fig. 2.2. This information indicates that pre-heating consumes the largest amount of energy.

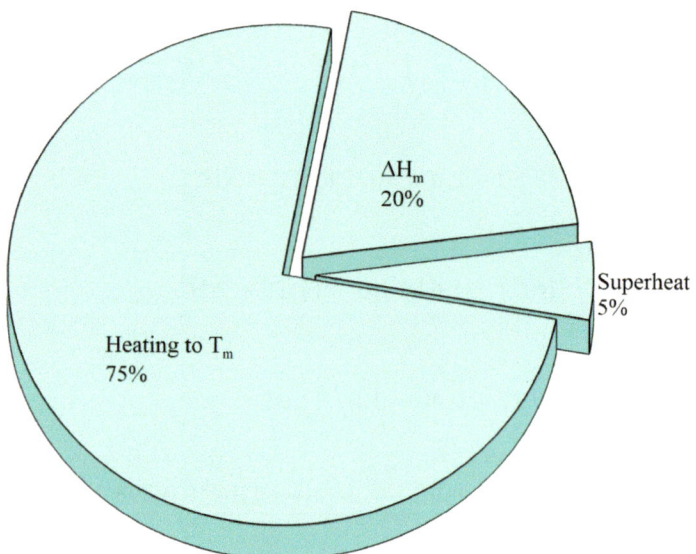

Fig. 2.2 Energy requirements to melt one ton of scrap up to 1600 °C

2.3.3 Energy Consumed to Melt One Ton of DRI

Direct Reduced Iron (DRI) in addition to metallic iron, also contains unreduced oxides of high melting point. The energy requirements to melt DRI are defined considering the specific energy demand for each component and its mass fraction. In addition to metallic iron, DRI also contains the following chemical species: Fe_3C, C, FeO, SiO_2, CaO, Al_2O_3 and MgO. Silica has three polymorphs; Quartz is stable up to 867 °C, tridymite is stable from 867 to 1470 °C and cristobalite from 1470 °C to the melting point of silica. Table 2.8 shows the heat capacities and Table 2.9 the heats of transformation (includes melting).

The following is a detailed calculation on the energy demand to bring all the previous chemical species up to 1600 °C (1873 K):
SiO_2:

$$\Delta H^0_{1873°K} = \int_{298}^{523} C_{P,SiO_2-\alpha} \, dT + \Delta H^0_{transf(\alpha \to \beta)} + \int_{523}^{1873} C_{P,SiO_2-\beta} \, dT$$

Table 2.8 Heat capacities

Phase	°C	K	C_P, cal/K mol	Source
α-cristobalite	25–250	298–523	$4.28 + 21.06 \times 10^{-3} \, T$	1
β-cristobalite	250–1722	523–1995	$17.39 + 0.31 \times 10^{-3} \, T - 9.90 \times 10^5 \, T^{-2}$	1
C	25–2027	298–2300	$4.10 + 1.02 \times 10^{-3} \, T - 2.10 \times 10^5 \, T^{-2}$	2
CaO	25–1600	298–1800	$11.67 + 1.08 \times 10^{-3} \, T - 1.56 \times 10^5 \, T^{-2}$	2
MgO	25–1827	298–2100	$10.18 + 1.74 \times 10^{-3} \, T - 1.48 \times 10^5 \, T^{-2}$	2
Al_2O_3	25–1600	298–1800	$27.43 + 3.06 \times 10^{-3} \, T - 8.47 \times 10^5 \, T^{-2}$	2
$Fe_{0.95}O$	25–1377	298–1650	$11.66 + 2.00 \times 10^{-3} \, T - 0.67 \times 10^5 \, T^{-2}$	2
$Fe_{0.95}O$	1377–1600	1650–1873	16.30	2

[a] O. Kubaschewski and C. Alcock; Metallurgical Thermochemistry, 5th edition, 1989
[b] Handbook on Material and Energy Balance calculations in Metallurgical Process Gordon H. Geiger Revised edition, 1993

Table 2.9 Heats of transformation

Phase	Transf	T, K	ΔH^0_{transf}, cal/mol
SiO_2	$\alpha \to \beta$	523	200
SiO_2	$\beta \to l$	1986	3100
FeO	$s \to l$	1650	7490

[a] J. Elliot & M. Gleiser; Thermochemistry for steelmaking, vol I, 1960
[b] Handbook on Material and Energy Balance calculations in Metallurgical Process. G. H. Geiger. Revised edition, 1993

$$\Delta H^0_{1873°K} = \int_{298}^{523} \left(4.28 + 21.06 \times 10^{-3}\, T\right)\, dT + \Delta H^0_{transf(\alpha \to \beta)}$$

$$+ \int_{523}^{1873} \left(17.39 + 0.31 \times 10^{-3}\, T - 9.90 \times 10^5\, T^{-2}\right)\, dT$$

$$\Delta H^0_{1873°K} = 4.28(523 - 298) + \frac{21.06 \times 10^{-3}}{2}\left(523^2 - 298^2\right)$$

$$+ 200 + 17.39(1873 - 523) + \frac{0.31 \times 10^{-3}}{2}\left(1873^2 - 523^2\right)$$

$$+ 9.90 \times 10^5 \left(\frac{1}{1873} - \frac{1}{523}\right)$$

$$\Delta H^0_{1873°K} = (963 + 1945) + 200$$

$$+ (23476.5 + 501 - 1364) = 25,721.5 \text{ cal/mol}$$

$$\Delta H^0_{1873°K} = 25,721.5 \frac{\text{cal}}{\text{gmol}} \times \frac{1 \text{ gmol SiO}_2}{60.0843 \text{ g}}$$

$$\times \frac{1000 \text{ g}}{\text{kg}} = 428,090.2 \frac{\text{cal}}{\text{kg SiO}_2}$$

$$= 428 \frac{\text{kcal}}{\text{kg SiO}_2} \times \frac{1000 \text{ kg}}{\text{ton}} \times \frac{\text{kwh}}{860 \text{ kcal}} = 497.8 \frac{\text{kwh}}{\text{ton SiO}_2}$$

C:

$$\Delta H^0_{1873°K} = \int_{298}^{1873} C_{P,C}\, dT$$

$$\Delta H^0_{1873°K} = \int_{298}^{1873} \left(4.10 + 1.02 \times 10^{-3}\, T - 2.10 \times 10^5\, T^{-2}\right)\, dT$$

$$\Delta H^0_{1873°K} = 4.10(1873 - 298) + \frac{1.02 \times 10^{-3}}{2}\left(1873^2 - 298^2\right)$$

$$+ 2.10 \times 10^5 \left(\frac{1}{1873} - \frac{1}{298}\right)$$

$$\Delta H^0_{1873°K} = (6457.5 + 1743.85575) - (592.57) = 7608.77 \text{ cal/mol}$$

$$\Delta H^0_{1873°K} = 7608.77 \frac{\text{cal}}{\text{gmol}} \times \frac{1 \text{ gmol C}}{12.0 \text{ g}} \times \frac{1000 \text{ g}}{\text{kg}} = 634,064 \frac{\text{cal}}{\text{kgC}}$$

$$= 634 \frac{\text{kcal}}{\text{kg C}} \times \frac{1000 \text{ kg}}{\text{ton}} \times \frac{\text{kwh}}{860 \text{ kcal}} = 737 \frac{\text{kwh}}{\text{ton C}}$$

CaO:

$$\Delta H^0_{1873°K} = \int\limits_{298}^{1873} C_{P,CaO} \, dT$$

$$\Delta H^0_{1873°K} = \int\limits_{298}^{1873} \left(11.67 + 1.08 \times 10^{-3}T - 1.56 \times 10^5 T^{-2}\right) dT$$

$$\Delta H^0_{1873°K} = 11.67(1873 - 298)$$

$$+ \frac{1.08 \times 10^{-3}}{2}\left(1873^2 - 298^2\right) + 1.56 \times 10^5 \left(\frac{1}{1873} - \frac{1}{298}\right)$$

$$\Delta H^0_{1873°K} = (18380.25 + 1846.4355) - (440.20)$$

$$= 19,786.5 \, \frac{cal}{mol \, CaO}$$

$$\Delta H^0_{1873°K} = 19,768 \, \frac{cal}{gmol} \times \frac{1 \, gmol \, CaO}{56.07 \, g} \times \frac{1000 \, g}{kg}$$

$$= 352,888 \, \frac{cal}{kg \, CaO}$$

$$= 352.8 \frac{kcal}{kg \, CaO} \times \frac{1000 \, kg}{ton} \times \frac{kwh}{860 \, kcal} = 410 \frac{kwh}{ton \, CaO}$$

MgO:

$$\Delta H^0_{1873°K} = \int\limits_{298}^{1873} C_{P,MgO} \, dT$$

$$\Delta H^0_{1873°K} = \int\limits_{298}^{1873} \left(10.18 + 1.74 \times 10^{-3} \, T - 1.48 \times 10^5 \, T^{-2}\right) dT$$

$$\Delta H^0_{1873°K} = 10.18(1873 - 298) + \frac{1.74 \times 10^{-3}}{2}\left(1873^2 - 298^2\right)$$

$$+ 1.48 \times 10^5 \left(\frac{1}{1873} - \frac{1}{298}\right)$$

$$\Delta H^0_{1873°K} = (16,033.5 + 2974.81275) - (417.62)$$

$$= 18,590 \, cal/mol$$

$$\Delta H^0_{1873°K} = 18,590 \, \frac{cal}{gmol} \times \frac{1 \, gmol \, MgO}{40.302 \, g} \times \frac{1000 \, g}{kg}$$

$$= 461,284 \, \frac{cal}{kg \, MgO}$$

$$= 461 \frac{kcal}{kg \, MgO} \times \frac{1000 \, kg}{ton} \times \frac{kWh}{860 \, kcal} = 536 \frac{kWh}{ton \, MgO}$$

Al_2O_3:

$$\Delta H^0_{1873^\circ K} = \int_{298}^{1873} C_{P,Al_2O_3} \, dT$$

$$\Delta H^0_{1873^\circ K} = \int_{298}^{1873} \left(27.43 + 3.06 \times 10^{-3} \, T - 8.47 \times 10^5 \, T^{-2}\right) dT$$

$$\Delta H^0_{1873^\circ K} = 27.43(1873 - 298) + \frac{3.06 \times 10^{-3}}{2}(1873^2 - 298^2)$$
$$+ 8.47 \times 10^5 \left(\frac{1}{1873} - \frac{1}{298}\right)$$

$$\Delta H^0_{1873^\circ K} = (43,202.25 + 5231.56725) - (2390.066)$$
$$= 46,043 \text{ cal/mol}$$

$$\Delta H^0_{1873^\circ K} = 46,043 \, \frac{\text{cal}}{\text{gmol}} \times \frac{1 \text{ gmol Al}_2\text{O}_3}{101.933 \text{ g}}$$
$$\times \frac{1000 \text{ g}}{\text{kg}} = 451,706 \, \frac{\text{cal}}{\text{kg Al}_2\text{O}_3}$$
$$= 451.7 \, \frac{\text{kcal}}{\text{kg Al}_2\text{O}_3} \times \frac{1000 \text{ kg}}{\text{ton}} \times \frac{\text{kWh}}{860 \text{ kcal}}$$
$$= 525.23 \, \frac{\text{kWh}}{\text{ton Al}_2\text{O}_3}$$

$Fe_{0.95}O$:

$$\Delta H^0_{1873^\circ K} = \int_{298}^{1650} C_{P,Fe_{0.95}O} \, dT + \Delta H^0_{\text{transf}(s \to l)} + \int_{1650}^{1873} C_{P,Fe_{0.95}O} \, dT$$

$$\Delta H^0_{1873^\circ K} = \int_{298}^{1650} \left(11.66 + 2.00 \times 10^{-3} \, T - 0.67 \times 10^5 \, T^{-2}\right) dT$$

$$+ \Delta H^0_{\text{transf}(s \to l)} + \int_{1650}^{1873} (16.30) \, dT$$

$$\Delta H^0_{1873^\circ K} = 11.66(1650 - 298) + \frac{2.00 \times 10^{-3}}{2}(1650^2 - 298^2)$$
$$+ 0.67 \times 10^5 \left(\frac{1}{1650} - \frac{1}{298}\right)$$
$$+ 7490 + 16.30(1873 - 1650)$$

$$\Delta H^0_{1873^\circ K} = (15,764.32 + 2633.696 - 184.22)$$

$$+ 7490 + (3634.9) = 29,522.9 \text{ cal/mol}$$

$$\Delta H^0_{1873°K} = 29,522.9 \ \frac{\text{cal}}{\text{gmol}} \times \frac{1 \text{ gmol Fe}_{0.95}\text{O}}{71.837 \text{ g}} \times \frac{1000 \text{ g}}{\text{kg}}$$

$$= 410,970 \ \frac{\text{cal}}{\text{kg Fe}_{0.95}\text{O}}$$

$$= 410.97 \frac{\text{kcal}}{\text{kg Fe}_{0.95}\text{O}} \times \frac{1000 \text{ kg}}{\text{ton}} \times \frac{\text{kWh}}{860 \text{ kcal}}$$

$$= 477.9 \frac{\text{kWh}}{\text{ton Fe}_{0.95}\text{O}}$$

Based on the previous calculations, Table 2.10 summarizes the specific energy consumption for the different components in DRI. In the literature there are different values, all having very small differences, due to different values for the heat capacities and heats of transformation, for example Fruehan et al. [43] reported a value of 368 kWh/ton for scrap melting up to 1600 °C, a difference of about 3% with respect to the present calculation.

Heat of formation of slag: Due to the high melting temperature of oxides, at 1600 °C only the melting temperature for pure FeO is reached but not for the other pure oxides. Obviously at 1600 °C the slag is formed and composed of all those oxides because, on one side, its real chemical composition is not in the form of pure oxides but complex compounds that have lower melting points and in addition to this, some high melting point oxides are melted at the high temperatures of the arc zone. A detailed analysis is out of the scope in this section, in order to estimate the heat of formation of the slag, as a first approximation it is assumed only liquid FeO is formed then due to the presence of carbon, its reduction, which is highly endothermic, should also be accounted for. With this approximation, the heat of formation of slag is defined by the reduction of FeO by dissolved carbon:

$$(\text{FeO}) + \underline{\text{C}} = \underline{\text{Fe}} + \text{CO}_{(g)}$$

The heat of reaction at 1600 °C was estimated as 651 kWh/ton FeO [44].

The previous values can be employed to get a first estimate of the energy requirements to melt one ton of DRI. As an example, consider a DRI with the following chemical composition:

Table 2.10 Energy requirements to raise the temperature of DRI components to 1600 °C

	SiO_2	CaO	MgO	Al_2O_3	$Fe_{0.95}O$	C	Fe
cal/kg	428,090	352,888	461,284	451,706	410,970	634,064	326,868
MJ/ton	1791	1476	1930	1889	1719	2652	1367
kWh/ton	497.8	410	536	525.23	477.9	737	380

	Fe$_t$	Fe°	C	SiO$_2$	CaO	Al$_2$O$_3$	MgO	FeO
X$_i$ (wt%)	90.8	85.8	2.48	2.41	0.61	0.87	0.69	6.43

$$\%FeO_{DRI} = (\%Fe_t - \%Fe°)\frac{FeO}{Fe} = (90.8 - 85.8)\frac{72}{56} = 6.43$$

	SiO$_2$	CaO	MgO	Al$_2$O$_3$	Fe$_{0.95}$O	C	Fe
Cal/kg	428,090	352,888	461,284	451,706	410,970	634,064	326,868
MJ/ton	1791	1476	1930	1889	1719	2652	1367
kWh/ton	497.8	410	536	525.23	477.9	737	380

$$\Delta H_{DRI}^{1600}\left(\frac{kWh}{ton\ DRI}\right) = \left(X_{Fe°} \times \Delta H_{Fe}°\right) + \left(X_C \times \Delta H_C°\right) + \left(X_{SiO_2} \times \Delta H_{SiO_2}°\right)$$
$$+ \left(X_{CaO} \times \Delta H_{CaO}°\right)$$
$$+ \left(X_{Al_2O_3} \times \Delta H_{Al_2O_3}°\right) + \left(X_{MgO} \times \Delta H_{MgO}°\right)$$
$$+ \left(X_{Fe_{0.95}O} \times \Delta H_{Fe_{0.95}O}°\right) + \left(X_{Slag} \times \Delta H_{Slag}°\right)$$
$$\Delta H_{DRI}^{1600}\left(\frac{kWh}{ton\ DRI}\right) = (0.858 \times 380) + (0.0248 \times 737)$$
$$+ (0.0241 \times 497) + (0.0061 \times 410)$$
$$+ (0.0087 \times 525) + (0.0069 \times 536)$$
$$+ (0.0643 \times 477) + (0.0643 \times 651)$$
$$= 326 + 18.27 + 11.97 + 2.5 + 4.56$$
$$+ 3.71 + 30.67 + 41.85 = 439.6\ kWh/ton\ DRI$$

In order to define the energy consumption on terms of per ton of steel, the following experimental conversion factor is employed:

$$1\ ton\ steel = 1.16\ ton\ DRI$$
$$\Delta H_{DRI}^{1600} = 1.16 \times 439.6 = 509\ \frac{kWh}{ton\ steel} = 1.83\ \frac{GJ}{ton\ steel}$$

2.3.4 Chemical Energy Sources

As mentioned before, the chemical energy derives from the following sources: oxidation reactions with elements contained in the metallic charge, combustion from burners, post-combustion, burning of organics in the scrap, electrode oxidation, dissolution of chemical species in liquid steel or liquid slag, calcination of limestone, etc. The analysis of chemical reactions is quite complex because involves not only the number of chemical reactions but the definition of the fraction of oxygen that reacts in a given chemical reaction, the accuracy to define the state of the reacting phases present, etc. The following list is an example of the chemical reactions to be considered in order to estimate the amount of chemical energy:

1. Heat released due to oxidation of Si: $Si + O_2 \rightarrow SiO_2$
2. Heat released due to oxidation of C to CO: $C + \frac{1}{2}O_2 \rightarrow CO$
3. Heat released due to oxidation of Fe: $Fe + \frac{1}{2}O_2 \rightarrow FeO$
4. Heat absorbed due to the reduction of FeO by carbon dissolved in liquid steel: $FeO + \underline{C} \rightarrow Fe + CO$
5. Reduction of FeO by silicon in scrap: $2FeO + \underline{Si} \rightarrow 2Fe + SiO_2$
6. Heat released by oxyfuel burners: $CH_4 + 2O_2 \rightarrow CO_2 + 2H_2O$
7. Heat released due to post-combustion of CO: $CO + \frac{1}{2}O_2 \rightarrow CO_2$
8. Heat released by organic matter in the scrap: $C_xH_y + (x + y/4)\,O_2 \rightarrow xCO_2 + \frac{y}{2}H_2O$
9. Heat of dissolution of FeO contained in DRI: $FeO_{DRI} \rightarrow \underline{Fe} + \frac{1}{2}O_2$
10. Heat of dissolution of Fe_3C in DRI: $Fe_3C \rightarrow 3\underline{Fe} + C$
11. Heat of formation of liquid slag.

In order to define the fraction of oxygen in each of the previous reactions, Abraham and Chen [34] developed an empirical approach, which was adjusted comparing model predictions and experimental data, defining the following priority in the consumption of oxygen:

- Combustion reactions by oxy-fuel burners
- Oxidation of dissolved elements in steel; Al, Si, Mn and Fe
- Oxidation of carbon dissolved in liquid steel to CO
- Post-combustion from CO to CO_2.

The heat of reaction can be estimated in several ways. One way is to let the reactions happen at room temperature and then raise the temperature of the products to the molten bath temperature. Another way is to raise the reactants to the molten bath temperature and then let the chemical reactions occur at that temperature. Figure 2.3 shows the second option.

The thermodynamic databases to carry out the calculations are shown in Tables 2.11, 2.12 and 2.13. Table 2.11 shows the heat capacities, Table 2.12 defines the heats of transformation and melting, Table 2.13 defines the heats of reactions at room temperature.

The heat of reaction at any temperature is computed using Kirchoff's law:

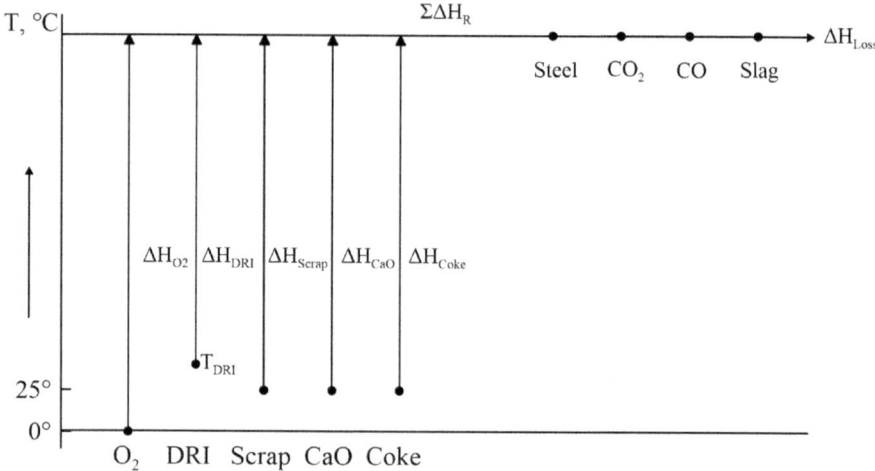

Fig. 2.3 Thermodynamic analysis of the heat of reaction

$$\Delta H_{R,T}^o = \Delta H_{R,298}^o + \int_{298}^{T} \Delta C_p dT$$

The following computation of heats of reactions is based on the work of Cardenas et al. [15, 44, 45].

(1) $FeO_{DRI} + \underline{C} \rightarrow Fe + CO$

$$\Delta H_R^o = \Delta H_{R,298}^o + \int_{298}^{1000} \Delta C_p\, dT + \int_{1000}^{1033} \Delta C_p\, dT$$

$$+ \int_{1033}^{1179} \Delta C_p\, dT + \int_{1179}^{1644} \Delta C_p\, dT + \int_{1644}^{1674} \Delta C_p\, dT$$

$$+ \int_{1674}^{1700} \Delta C_p\, dT + \int_{1700}^{1803} \Delta C_p\, dT$$

$$+ \int_{1803}^{T} \Delta C_p\, dT + \Delta H_{Fe,\alpha-\beta} + \Delta H_{Fe,\beta-\gamma}$$

$$+ \Delta H_{Fe,\gamma-\delta} + \Delta H_{Fe,\delta-l}^m + \Delta H_{FeO}^m + \Delta H_C^d$$

where:

Table 2.11 Heat capacities

Specie	State	C_p, cal/K mol	Range, K
Fe	c, α	$3.37 + 7.10 \times 10^{-3}T + 0.43 \times 10^5 T^{-2}$	298–1033[a]
	c, β	10.40	1033–1179[a]
	c, γ	$4.85 + 3.00 \times 10^{-3}T$	1179–1674[a]
	c, δ	10.30	1674–1803[a]
	l	10.00	1803–1900[a]
FeO	c	$-4.307 + 7.315 \times 10^{-3}\,T - 605,473.23\,T^{-2} + 358.7237\,T^{-0.5}$	298–1644[b]
	l	16.3	1644–2000[b]
Fe$_3$C	c, α	$19.64 + 20.00 \times 10^{-3}T$	298–463[b]
	c, β	$25.62 + 3.00 \times 10^{-3}T$	463–1500[b]
	l	29.0	1500–2000[b]
SiO$_2$	c_{crist}, α	$19.960 - 586,845.12\,T^{-2} - 89.553\,T^{-0.5} + 6,693,886\,T^{-3}$	298–1996[b]
Al$_2$O$_3$	c	$-3.760 + 7.157 \times 10^{-3}T + 1,193,897\,T^{-2} + 1689.376\,T^{-0.5} - 26,794\,T^{-1}$	298–1200[b]
	c	$-188.22 + 15.745 \times 10^{-3}\,T + 63,053,626\,T^{-2} + 159,86.07\,T^{-0.5} - 364,609.8\,T^{-1}$	1200–2327[b]
CaO	c	$14.051 - 274,174.47\,T^{-2} - 32.003\,T^{-0.5} + 24,612,521\,T^{-3}$	298–2845[b]
MgO	c	$14.605 - 148,459.37\,T^{-2} - 283.173\,T^{-0.5} + 1,396,895.8\,T^{-3}$	298–2845[b]
S	c_{rom}	$4.126 + 4.781 \times 10^{-3}\,T\,36.217\,T^{-1}$	298–368[b]
	c_{mon}	$4.189 + 4.7002 \times 10^{-3}T$	368–388[b]
	l	$-12.637 + 51.649 \times 10^{-3}T$	388–432[b]
	l	$13.461 - 8.765 \times 10^{-3}T$	432–500[b]
	l	$13.503 - 8.843 \times 10^{-3}T$	500–600[b]
	l	$10.062 - 3.107 \times 10^{-3}T$	600–800[b]
	l	7.565	800–1500[b]
C	c	$14.622 - 1.916 \times 10^{-3}\,T + 197,315.9\,T^{-2} - 244.74\,T^{-0.5} - 1,578,579\,T^{-3}$	298–1000[b]
	c	$5.922 + 0.2291 \times 10^{-3}\,T - 689,151.12\,T^{-2} - 9.486\,T^{-0.5}$	1000–6000[b]
CO	g	$21.69 - 1.49 \times 10^{-3}\,T - 368,368\,T^{-2} + 7663.3163\,T^{-1} - 618.846\,T^{-0.5}$	298–1700[b]
	g	$10.536 - 2,780,374.386\,T^{-2} + 4993.763\,T^{-1} - 164.263\,T^{-0.5}$	1700–6000[b]

(continued)

Table 2.11 (continued)

Specie	State	C_p, cal/K mol	Range, K
Si	c	$5.451 + 0.925 \times 10^{-3}\,T - 84367.12\,T^{-2}$	298–1685[b]
	c	6.5	1685–2500[b]
Mn	c, α	$5.7002 + 3.379 \times 10^{-3}\,T - 37,536.83\,T^{-2}$	298–980[b]
	c, β	$8.183 + 0.904 \times 10^{-3}\,T - 82,037.66\,T^{-2}$	980–1361[b]
	c, γ	$7.5979 + 1.985 \times 10^{-3}\,T$	1361–1412[b]
	c, δ	$8.0221 + 1.9759 \times 10^{-3}\,T$	1412–1519[b]
	l	11.00	1519–3000[b]
O_2	g	$6.435 + 4.058 \times 10^{-3}\,T + 54,811\,T^{-2} - 1.617 \times 10^{-6}\,T^2 - 18.92\,T^{-0.5}$	298–1000[b]
	g	$21.434 - 0.3459 \times 10^{-3}\,T - 4,465,269\,T^{-2} - 986.26^{-0.5} + 22,897.69\,T^{-1}$	1000–4000[b]

Table 2.12 Heat of transformation and melting

Reaction	ΔH_{Trans}, cal/mol	ΔH_{Fus}, cal/mol	T, K
$Fe_{c,\alpha} \to Fe_{c,\beta}$	410	–	1033[a]
$Fe_{c,\beta} \to Fe_{c,\gamma}$	210	–	1179[a]
$Fe_{c,\gamma} \to Fe_{c,\delta}$	110	–	1674[a]
$Fe_{c,\delta} \to Fe_l$	–	3700	1803[a]
$FeO_c \to FeO_l$	–	7490	1644[b]
$Fe_3C_{c,\alpha} \to Fe_3C_{c,\beta}$	180	–	463[b]
$Fe_3C_{c,\beta} \to Fe_3C_l$	–	12,300	1500[b]
$S_{c,rom} \to S_{c,mon}$	95.690	–	368.3[b]
$S_{c,mc} \to S_l$	–	411.328	388.4[b]
$Si_c \to Si_l$	–	11,998	1685[b]
$Mn_{c,\alpha} \to Mn_{c,\beta}$	532.02	–	980[b]
$Mn_{c,\beta} \to Mn_{c,\gamma}$	507.17	–	1361[b]
$Mn_{c,\gamma} \to Mn_{c,\delta}$	449.09	–	1412[b]
$Mn_{c,\delta} \to Mn_l$	–	2881.93	1519[b]

[a] G. H. Geiger; Handbook on Material and Energy Balance Calculations in Metallurgical Process, Revised Edition, 1993
[b] FactSage 5.2 Pelton

$$\Delta C_p = C_{p,Fe} + C_{p,CO} - C_{p,FeO} - C_{p,C}$$

Replacing the heat capacities for each temperature range:

Table 2.13 Heat of reaction at 298 K

Reaction	$\Delta H^o_{R, 298}\ \frac{cal}{mol}$
$FeO_{DRI} + \underline{C} \rightarrow Fe + CO$	37,118.84
$Fe_3C_{DRI} \rightarrow 3Fe + \underline{C}$	− 5400.00
$2FeO + \underline{Si} \rightarrow 2Fe + SiO_2$	− 89,558.49
$\underline{Si} + O_2 \rightarrow SiO_2$	− 216,619.35
$\underline{C} + \frac{1}{2}O_2 \rightarrow CO$	− 26,416.58
$Fe + \frac{1}{2}O_2 \rightarrow FeO$	− 63.535.43
$C_{(gr)} \rightarrow \underline{C}$	5100.00
$Si_l \rightarrow \underline{Si}$	31,100.00

$$\Delta H^o_R = \Delta H^o_{R, 298}$$

$$+ \int_{298}^{1000} \left(14.74 + 0.21 \times 10^{-3}T + 82,788T^{-2} - 732T^{-0.5} + 7663T^{-1} + 1,578,578T^{-3}\right) dT$$

$$+ \int_{1000}^{1033} \left(23.44 - 1.93 \times 10^{-3}T + 969,255T^{-2} - 968T^{-0.5} + 7663T^{-1}\right) dT$$

$$+ \int_{1033}^{1179} \left(30.47 - 9.03 \times 10^{-3}T + 926,255T^{-2} - 968T^{-0.5} + 7663T^{-1}\right) dT$$

$$+ \int_{1179}^{1644} \left(24.92 - 6.03 \times 10^{-3}T + 926,255T^{-2} - 968T^{-0.5} + 7663T^{-1}\right) dT$$

$$+ \int_{1644}^{1674} \left(4.31 + 1.28 \times 10^{-3}T + 320,782T^{-2} - 609T^{-0.5} + 7663T^{-1}\right) dT$$

$$+ \int_{1674}^{1700} \left(9.76 - 1.71 \times 10^{-3}T + 320,782T^{-2} - 609T^{-0.5} + 7663T^{-1}\right) dT$$

$$+ \int_{1700}^{1803} \left(-1.38 - 0.228 \times 10^{-3}T - 2,091,223T^{-2} - 154T^{-0.5} + 4993T^{-1}\right) dT$$

$$+ \int_{1803}^{T} \left(-1.68 - 0.228 \times 10^{-3}T - 2091,223T^{-2} - 154T^{-0.5} + 4993T^{-1}\right) dT$$

$$+ \Delta H_{Fe,\alpha - \beta} + \Delta H_{Fe,\beta - \gamma} + \Delta H_{Fe,\gamma - \delta} + \Delta H^m_{Fe,\delta - l} + \Delta H^m_{FeO} + \Delta H^d_C$$

After integration:

$$\Delta H_R^o = \Delta H_{R,298}^o$$

$$+ \left(14.74T + 0.105 \times 10^{-3}T^2 - 82,788T^{-1}\right.$$

$$\left. -1465T^{0.5} + 7663 \ln T - 789,289T^{-2}\right)\Big|_{298}^{1000}$$

$$+ \left(23.44T - 9.67 \times 10^{-3}T^2 - 969,255T^{-1} - 1936T^{0.5} + 7663 \ln T\right)\Big|_{1000}^{1033}$$

$$+ \left(30.47T - 4.51 \times 10^{-3}T^2 - 926,255T^{-1} - 1936T^{-0.5} + 7663 \ln T\right)\Big|_{1033}^{1179}$$

$$+ \left(24.92T - 3.01 \times 10^{-3}T^2 - 926,255T^{-1} - 1936T^{0.5} + 7663 \ln T\right)\Big|_{1179}^{1644}$$

$$+ \left(4.31T + 0.64 \times 10^{-3}T^2 - 320,782T^{-1} - 1218T^{0.5} + 7663 \ln T\right)\Big|_{1644}^{1674}$$

$$+ \left(9.76T - 0.85 \times 10^{-3}T^2 - 320,782T^{-1} - 1218T^{-0.5} + 7663 \ln T\right)\Big|_{1674}^{1700}$$

$$+ \left(-1.38T - 0.114 \times 10^{-3}T^2 + 2,091,223T^{-1} - 309T^{0.5} + 4993 \ln T\right)\Big|_{1700}^{1803}$$

$$+ \left(-1.68T - 0.114 \times 10^{-3}T^2 + 2,091,223T^{-2} - 309T^{-0.5} + 4993 \ln T\right)\Big|_{1803}^{T}$$

$$+ \Delta H_{Fe,\alpha-\beta} + \Delta H_{Fe,\beta-\gamma} + \Delta H_{Fe,\gamma-\delta} + \Delta H_{Fe,\delta-l}^m + \Delta H_{FeO}^m + \Delta H_C^d$$

In the previous analysis C is in the form of C_{gr}. After considering the heat of dissolution: $\underline{C} \to C_{gr}$, and defining T as 1600 °C:

$$\Delta H_R^o = 37,118.843 - 1118.17 - 13.4463 - 137.538$$
$$- 1623.183 - 113.454 - 86.132$$
$$- 341.501 - 253.31 + 410 + 210$$
$$+ 110 + 3700 + 7490 - 5100$$

$$\Delta H_R^o = 40,252.107 \tfrac{cal}{mol}$$

$$\Delta H_{R,C} = \left(\tfrac{1000}{71.8464}\right)\Delta H_{R,ToK}^o = 560,252.246 \tfrac{cal}{kg\,FeO}$$

$$= 559\,kcal/kg\,FeO$$

(2) $Fe_3C_{DRI} \to 3Fe + \underline{C}$

$$\Delta H_R^o = \Delta H_{R,298}^o + \int_{298}^{463} \Delta C_p dT + \int_{463}^{1000} \Delta C_p dT + \int_{1000}^{1033} \Delta C_p dT$$

$$+ \int_{1033}^{1179} \Delta C_p dT + \int_{1179}^{1500} \Delta C_p dT + \int_{1500}^{1674} \Delta C_p dT + \int_{1674}^{1803} \Delta C_p dT$$

$$+ \int_{1803}^{T} \Delta C_p dT + \Delta H_{Fe,\alpha-\beta} + \Delta H_{Fe,\beta-\gamma}$$

$$+ \Delta H_{Fe,\gamma-\delta}^s + \Delta H_{Fe,\delta-l}^m + \Delta H_{Fe_3C,\alpha-\beta} + \Delta H_{Fe_3C,\beta-l}^m + \Delta H_C^d$$

where:

$$\Delta C_p = 3C_{p,Fe} + C_{p,C} - C_{p,Fe_3C}$$

Replacing the heat capacities for each temperature range:

$$\Delta H_R^o = \Delta H_{R,298}^o$$

$$+ \int_{298}^{463} \left(5.09 - 0.61 \times 10^{-3}T + 326,315T^{-2} - 244T^{-0.5} - 1,578,578T^{-3}\right) dT$$

$$+ \int_{463}^{1000} \left(-0.88 + 1.638 \times 10^{-3}T + 326,315T^{-2} - 244T^{-0.5} - 1,578,578T^{-3}\right) dT$$

$$+ \int_{1000}^{1033} \left(-9.58 + 18.52 \times 10^{-3}T - 560,151T^{-2} - 37.94T^{-0.5}\right) dT$$

$$+ \int_{1033}^{1179} \left(11.5 - 2.77 \times 10^{-3}T - 689,151T^{-2} - 37.94T^{-0.5}\right) dT$$

$$+ \int_{1179}^{1500} \left(-5.14 + 6.22 \times 10^{-3}T - 689,151T^{-2} - 37.94T^{-0.5}\right) dT$$

$$+ \int_{1500}^{1674} \left(-8.52 + 9.22 \times 10^{-3}T - 689,151T^{-2} - 37.94T^{-0.5}\right) dT$$

$$+ \int_{1674}^{1803} \left(7.82 + 0.228 \times 10^{-3}T - 689,151T^{-2} - 37.94T^{-0.5}\right) dT$$

$$+ \int_{1803}^{T} \left(6.92 + 0.228 \times 10^{-3}T - 689{,}151T^{-2} - 37.94T^{-0.5}\right) dT$$

$$+ \Delta H_{Fe,\alpha-\beta} + \Delta H_{Fe,\beta-\gamma} + \Delta H^{s}_{Fe,\gamma-\delta} + \Delta H^{m}_{Fe,\delta-l}$$

$$+ \Delta H_{Fe_3C,\alpha-\beta} + \Delta H^{m}_{Fe_3C,\beta-l} + \Delta H^{d}_{C}$$

After integration:

$$\Delta H^{o}_{R} = \Delta H^{o}_{R,298}$$

$$+ \left(5.09T - 0.308 \times 10^{-3}T^2 - 326315T^{-1}\right.$$

$$\left. -489T^{0.5} + 789{,}289T^{-2}\right)\Big|_{298}^{463}$$

$$+ \left(-0.88T + 0.819 \times 10^{-3}T^2 - 326{,}315T^{-1} - 489T^{0.5} + 789{,}289T^{-2}\right)\Big|_{463}^{1000}$$

$$+ \left(-9.58T + 9.26 \times 10^{-3}T^2 + 560{,}151T^{-1} - 18.97T^{0.5}\right)\Big|_{1000}^{1033}$$

$$+ \left(11.5T - 1.38 \times 10^{-3}T^2 + 689{,}151T^{-1} - 18.97T^{0.5}\right)\Big|_{1033}^{1179}$$

$$+ \left(-5.14T + 3.11 \times 10^{-3}T^2 + 689{,}151T^{-1} - 18.97T^{0.5}\right)\Big|_{1179}^{1500}$$

$$+ \left(-8.52T + 4.61 \times 10^{-3}T^2 + 689{,}151T^{-1} - 18.97T^{0.5}\right)\Big|_{1500}^{1674}$$

$$+ \left(7.82T + 0.114 \times 10^{-3}T^2 + 689{,}151T^{-1} - 18.97T^{0.5}\right)\Big|_{1674}^{1803}$$

$$+ \left(6.92T + 0.114 \times 10^{-3}T^2 + 689{,}151T^{-1} - 18.97T^{0.5}\right)\Big|_{1803}^{T}$$

$$+ \Delta H_{Fe,\alpha-\beta} + \Delta H_{Fe,\beta-\gamma} + \Delta H^{s}_{Fe,\gamma-\delta} + \Delta H^{m}_{Fe,\delta-l}$$

$$+ \Delta H_{Fe_3C,\alpha-\beta} + \Delta H^{m}_{Fe_3C,\beta-l} + \Delta H^{d}_{C}$$

In the previous analysis C is in the form of C_{gr}. After considering the heat of dissolution: $C_{gr} \rightarrow \underline{C}$, and defining T as 1600 °C:

$$\Delta H^{o}_{R} = -5400 - 896.121 + 1388.118$$

$$+ 6934.418 + 1107.595 + 817.484$$

$$+ 975.455 + 1001.629 + 484.253 + 410$$

$$+ 210 + 110 + 3700 + 180 + 12300 + 5100$$

$$\Delta H^{o}_{R} = 28{,}422.83 \ \tfrac{cal}{mol}$$

$$\Delta H_{R,C,T°K} = \left(\tfrac{1000}{179.552}\right) \Delta H^{o}_{R,T°K} = 158{,}298.598 \ \tfrac{cal}{kg\,Fe_3C}$$

(3) $2FeO + \underline{Si} \rightarrow 2Fe + SiO_2$

$$\Delta H^{o}_{R} = \Delta H^{o}_{R,298} + \int_{298}^{1033} \Delta C_{p}dT + \int_{1033}^{1179} \Delta C_{p}dT$$

$$+ \int_{1179}^{1644} \Delta C_p dT + \int_{1644}^{1674} \Delta C_p dT + \int_{1674}^{1685} \Delta C_p dT$$

$$+ \int_{1685}^{1803} \Delta C_p dT + \int_{1803}^{T} \Delta C_p dT$$

$$+ \Delta H_{Fe,\alpha-\beta} + \Delta H_{Fe,\beta-\gamma} + \Delta H_{Fe,\gamma-\delta}$$

$$+ \Delta H_{Fe,\delta-l}^m + \Delta H_{FeO}^m + \Delta H_{Si}^m + \Delta H_{Si}^d$$

where:

$$\Delta C_p = 2C_{p,Fe} + C_{p,SiO_2} - 2C_{p,FeO} - C_{p,Si}$$

Replacing the heat capacities for each temperature range:

$$\Delta H_R^o = \Delta H_{R,298}^o$$

$$+ \int_{298}^{1033} \left(29.86 - 1.35 \times 10^{-3}T + 794,468T^{-2} - 807T^{-0.5} + 66,938,861T^{-3} \right) dT$$

$$+ \int_{1033}^{1179} \left(43.92 - 15.54 \times 10^{-3}T + 708,468T^{-2} - 807T^{-0.5} + 66,938,861T^{-3} \right) dT$$

$$+ \int_{1179}^{1644} \left(32.82 - 9.556 \times 10^{-3}T + 708,468T^{-2} - 807T^{-0.5} + 66,938,861T^{-3} \right) dT$$

$$+ \int_{1644}^{1674} \left(-8.39 + 5.074 \times 10^{-3}T - 502,478T^{-2} - 89.55T^{-0.5} + 66,938,861T^{-3} \right) dT$$

$$+ \int_{1674}^{1685} \left(2.508 + 0.925 \times 10^{-3}T - 502,478T^{-2} - 89.55T^{-0.5} + 66,938,861T^{-3} \right) dT$$

$$+ \int_{1685}^{1803} \left(2.667 + 1.007 \times 10^{-3}T - 586,845T^{-2} - 89.55T^{-0.5} + 66,938,861T^{-3} \right) dT$$

$$+ \int_{1803}^{T} \left(2.077 + 1.007 \times 10^{-3}T - 586,845T^{-2} - 89.55T^{-0.5} + 66,938,861T^{-3} \right) dT$$

$$+ \Delta H_{Fe,\alpha-\beta} + \Delta H_{Fe,\beta-\gamma} + \Delta H_{Fe,\gamma-\delta} + \Delta H_{Fe,\delta-l}^m + \Delta H_{FeO}^m + \Delta H_{Si}^m + \Delta H_{Si}^d$$

After integration:

$$\Delta H_R^o = \Delta H_{R,\,298}^o$$

$$+ \left(29.8T - 0.678 \times 10^{-3}T^2 - 794{,}468T^{-1} - 1614T^{0.5} - 33{,}469{,}430T^{-2}\right)\Big|_{298}^{1033}$$

$$+ \left(-43.92T - 7.778 \times 10^{-3}T^2 - 708{,}468T^{-1} - 1614T^{0.5} - 33{,}469{,}430T^{-2}\right)\Big|_{1033}^{1179}$$

$$+ \left(32.82T - 4.778 \times 10^{-3}T^2 - 708{,}468T^{-1} - 1614T^{0.5} - 33{,}469{,}430T^{-2}\right)\Big|_{1179}^{1644}$$

$$+ \left(-8.39T + 2.537 \times 10^{-3}T^2 + 502{,}477T^{-1} - 179.107T^{0.5} - 33{,}469{,}430T^{-2}\right)\Big|_{1644}^{1674}$$

$$+ \left(2.508T - 0.462 \times 10^{-3}T^2 + 502{,}477T^{-1} - 179.107T^{0.5} - 33{,}469{,}430T^{-2}\right)\Big|_{1674}^{1685}$$

$$+ \left(2.667T - 0.503 \times 10^{-3}T^2 - 586{,}845T^{-1} - 179.107T^{0.5} - 33{,}469{,}430T^{-2}\right)\Big|_{1685}^{1803}$$

$$+ \left(2.077T - 0.503 \times 10^{-3}T^2 - 586{,}845T^{-1} - 179.107T^{0.5} - 33{,}469{,}430T^{-2}\right)\Big|_{1803}^{T}$$

$$+ \Delta H_{Fe,\alpha-\beta} + \Delta H_{Fe,\beta-\gamma} + \Delta H_{Fe,\gamma-\delta} + \Delta H_{Fe,\delta-l}^m + \Delta H_{FeO}^m + \Delta H_{Si}^m + \Delta H_{Si}^d$$

After considering the heat of dissolution of Si in liquid steel: $Si_l \rightarrow \underline{Si}$. and defining T as 1600 °C:

$$\Delta H_R^o = -89558.497 - 1277.104 + 410 + 210$$
$$+ 110 + 3700 + 7490 + 11{,}998 + 31{,}100$$
$$\Delta H_R^o = -35{,}817.513 \; \tfrac{cal}{mol}$$
$$\Delta H_{R,SiO_2} = \left(\tfrac{1000}{28.0855}\right)\Delta H_{R,T°K}^o = -1{,}275{,}302.69 \; \tfrac{cal}{kg\,Si}$$

(4) $\underline{Si} + O_2 \rightarrow SiO_2$

$$\Delta H_R^o = \Delta H_{R,\,298}^o + \int_{298}^{1000} \Delta C_p dT + \int_{1000}^{1685} \Delta C_p dT$$

$$+ \int_{1685}^{T} \Delta C_p dT + \Delta H_{Si}^m + \Delta H_{Si}^d$$

where:

$$\Delta C_p = C_{p,SiO_2} - C_{p,Si} - C_{p,O_2}$$

Replacing the heat capacities for each temperature range:

$\Delta H_R^o = \Delta H_{R,\,298}^o$

$$+ \int\limits_{298}^{1000} \left(8.07 - 4.98 \times 10^{-3}T - 557,288T^{-2} - 70.63T^{-0.5} + 66,938,861T^{-3} + 16.17 \times 10^{-7}T^2\right) dT$$

$$+ \int\limits_{1000}^{1685} \left(-6.92 - 0.578 \times 10^{-3}T + 3,962,791T^{-2} + 896.71T^{-0.5} + 66,938,861T^{-3} - 22,897T^{-1}\right) dT$$

$$+ \int\limits_{1685}^{T} \left(-7.97 + 0.345 \times 10^{-3}T + 3,878,423T^{-2} + 896.71T^{-0.5} + 66,938,861T^{-3} - 22,897T^{-1}\right) dT$$

$$+ \Delta H_{Si}^m + \Delta H_{Si}^d$$

After integration:

$\Delta H_R^o = \Delta H_{R,\,298}^o$

$$+ \left(8.07T - 2.49 \times 10^{-3}T^2 + 557,288T^{-1} - 141.26T^{0.5} - 33,469,430T^{-2} + 5.39 \times 10^{-7}T^3\right)\Big|_{298}^{1000}$$

$$+ \left(-6.92T - 0.289 \times 10^{-3}T^2 - 3,962,791T^{-1} + 1793.17T^{0.5} - 33,469,430T^{-2} - 22,897\ln T\right)\Big|_{1000}^{1685}$$

$$+ \left(-7.97T + 0.172 \times 10^{-3}T^2 - 3,878,423T^{-1} + 1793.17T^{0.5} - 33,469,430T^{-2} - 22,897\ln T\right)\Big|_{1685}^{T}$$

$$+ \Delta H_{Si}^m + \Delta H_{Si}^d$$

After considering the heat of dissolution of Si in liquid steel: $Si_l \rightarrow \underline{Si}$ and defining T as 1600 °C:

$$\Delta H_R^o = -216,619.357 + 924.222 + 1313.533 + 426.119 + 11,998 + 31,100$$

$$\Delta H_R^o = -170,857.483 \tfrac{cal}{mol}$$

$$\Delta H_{R,\,SiO_2} = \left(\tfrac{1000}{60.0843}\right)\Delta H_{R,\,T°K}^o = -2,843,629.417 \tfrac{cal}{kg\ SiO_2}$$

(5) $\underline{C} + \tfrac{1}{2}O_{2(g)} \rightarrow CO_{(g)}$

$$\Delta H_R^o = \Delta H_{R,\,298}^o + \int\limits_{298}^{1000} \Delta C_p dT$$

$$+ \int\limits_{1000}^{1700} \Delta C_p dT + \int\limits_{1700}^{T} \Delta C_p dT + \Delta H_C^d$$

where:

$$\Delta C_p = C_{p,CO} - C_{p,C} - \tfrac{1}{2}C_{p,O_2}$$

Replacing the heat capacities for each temperature range:

$$\Delta H_R^0 = \Delta H_{R,298}^0$$
$$+ \int_{298}^{1000} \left(3.85 - 1.6 \times 10^{-3}T - 593,090T^{-2} - 364.6T^{-0.5} + 1,578,579T^{-3} + 8 \times 10^{-7}T^2 + 7663T^{-1}\right) dT$$
$$+ \int_{1000}^{1700} \left(5.05 - 1.54 \times 10^{-3}T + 2,553,416T^{-2} - 116.22T^{-0.5} - 3785T^{-1}\right) dT$$
$$+ \int_{1700}^{T} \left(-6.102 - 0.056 \times 10^{-3}T + 141,411T^{-2} + 338.35T^{-0.5} - 6455T^{-1}\right) dT + \Delta H_C^d$$

After integration:

$$\Delta H_R^0 = \Delta H_{R,298}^0$$
$$+ \left(3.85T - 0.8 \times 10^{-3}T^2 + 593090T^{-1} - 729T^{0.5}\right.$$
$$\left. -789289T^{-2} + 2.7 \times 10^{-7}T^3 + 7663 \ln T\right)\Big|_{298}^{1000}$$
$$+ \left(5.05T - 0.773 \times 10^{-3}T^2 - 2553416T^{-1} - 232.45T^{0.5} - 3785 \ln T\right)\Big|_{1000}^{1700}$$

In the previous analysis C is in the form of C_{gr}: $C_{gr} + \frac{1}{2}O_{2(g)} \rightarrow CO_{(g)}$. After considering this reaction: $\underline{C} \rightarrow C_{gr}$ and defining final T as 1600 °C:

$$\Delta H_{R,1873} = -26,416.587 - \cdots - 5100$$
$$\Delta H_R^0 = -26,416.587 - 348.472 - 1116.191 - 305.663 - 5100$$
$$\Delta H_R^0 = -33,286.91 \tfrac{cal}{mol}$$
$$\Delta H_{R,1873} = \left(\frac{1000}{28.01}\right) \Delta H_{R,T}^0 = -1,188,393.79 \frac{cal}{kg\,CO}$$

(6) $Fe + \frac{1}{2}O_2 \rightarrow FeO$

$$\Delta H_R^0 = \Delta H_{R,298}^0 + \int_{298}^{1000} \Delta C_p\, dT + \int_{1000}^{1033} \Delta C_p\, dT$$
$$+ \int_{1033}^{1179} \Delta C_p\, dT + \int_{1179}^{1644} \Delta C_p\, dT + \int_{1644}^{1674} \Delta C_p\, dT$$
$$+ \int_{1674}^{1700} \Delta C_p\, dT + \int_{1700}^{1803} \Delta C_p\, dT$$

$$+ \int_{1803}^{T} \Delta C_p \, dT + \Delta H_{Fe,\alpha-\beta} + \Delta H_{Fe,\beta-\gamma}$$

$$+ \Delta H_{Fe,\gamma-\delta} + \Delta H_{Fe,\delta-l}^{m} + \Delta H_{FeO}^{m}$$

where:

$$\Delta C_p = C_{p,FeO} - C_{p,Fe} - \tfrac{1}{2} C_{p,O_2}$$

Replacing the heat capacities for each temperature range:

$$\Delta H_R^o = \Delta H_{R,298}^o$$

$$+ \int_{298}^{1000} \left(10.89 - 1.8 \times 10^{-3} T - 589,878.7 T^{-2} + 368.18 T^{-0.5} + 8.085 \times 10^{-7} T^2\right) dT$$

$$+ \int_{1000}^{1033} \left(-18.39 + 0.388 \times 10^{-3} T + 1,670,161 T^{-2} + 851.855 T^{-0.5} - 11,448 T^{-1}\right) dT$$

$$+ \int_{1033}^{1179} \left(-25.42 + 7.48 \times 10^{-3} T + 1,670,161 T^{-2} + 851.855 T^{-0.5} - 11,448 T^{-1}\right) dT$$

$$+ \int_{1179}^{1644} \left(-19.87 + 4.48 \times 10^{-3} T + 1,670,161 T^{-2} + 851.855 T^{-0.5} - 11,448 T^{-1}\right) dT$$

$$+ \int_{1644}^{1674} \left(0.732 - 2.82 \times 10^{-3} T + 2,232,634 T^{-2} + 493.133 T^{-0.5} - 11,448 T^{-1}\right) dT$$

$$+ \int_{1674}^{1803} \left(-4.717 + 0.173 \times 10^{-3} T + 2,232,634 T^{-2} + 493.133 T^{-0.5} - 11,448 T^{-1}\right) dT$$

$$+ \int_{1803}^{T} \left(-4.417 + 0.173 \times 10^{-3} T + 2,232,634 T^{-2} + 493.133 T^{-0.5} - 11,448 T^{-1}\right) dT$$

$$+ \Delta H_{Fe,\alpha-\beta} + \Delta H_{Fe,\beta-\gamma} + \Delta H_{Fe,\gamma-\delta} + \Delta H_{Fe,\delta-l}^{m} + \Delta H_{FeO}^{m}$$

After integration:

$$\Delta H_R^o = \Delta H_{R,298}^o$$

$$+ \left(10.89 T - 0.906 \times 10^{-3} T^2 + 589,878.7 T^{-1} + 736.36 T^{0.5} + 2.695 \times 10^{-7} T^3\right)\Big|_{298}^{1000}$$

$$+ \left(-18.39 T + 0.194 \times 10^{-3} T^2 - 1,670,161 T^{-1} + 1703.71 T^{-0.5} - 11,448 \ln T\right)\Big|_{1000}^{1033}$$

$$+ \left(-25.42 T - 4.51 \times 10^{-3} T^2 - 1,670,161 T^{-1} + 1703.71 T^{0.5} - 11,448 \ln T\right)\Big|_{1033}^{1179}$$

$$+ \left(-19.87T + 2.24 \times 10^{-3}T^2 - 1{,}670{,}161T^{-1} + 1703.71T^{0.5} - 11{,}448\ln T\right)\Big|_{1179}^{1644}$$

$$+ \left(0.732T - 1.41 \times 10^{-3}T^2 - 2{,}232{,}634T^{-1} - 986.26T^{0.5} - 11{,}448\ln T\right)\Big|_{1644}^{1674}$$

$$+ \left(-4.717T + 0.086 \times 10^{-3}T^2 - 2{,}232{,}634T^{-1} - 986.26T^{0.5} - 11{,}448\ln T\right)\Big|_{1674}^{1803}$$

$$+ \left(-4.417T + 0.086 \times 10^{-3}T^2 - 2{,}232{,}634T^{-1} - 986.26T^{0.5} - 11{,}448\ln T\right)\Big|_{1803}^{T}$$

$$+ \Delta H_{Fe,\alpha-\beta} + \Delta H_{Fe,\beta-\gamma} + \Delta H_{Fe,\gamma-\delta} + \Delta H_{Fe,\delta-l}^m + \Delta H_{FeO}^m$$

for $T = 1600\ °C$:

$$\Delta H_R^o = -63{,}535.43 + 3972.287 - 30.624 - 79.546$$
$$+ 867.628 + 61.801 + 31.561$$
$$+ 128.46 + 410 + 210 + 110$$
$$+ 3700 + 7490$$
$$\Delta H_R^o = -49{,}663.96\ \tfrac{cal}{mol}$$
$$\Delta H_{R,FeO,T°K} = \left(\tfrac{1000}{71.8464}\right)\Delta H_{R,T°K}^o = -691{,}250.501\ \tfrac{cal}{kg\,FeO}$$

Table 2.14 summarizes the heat of reaction of the few selected chemical reactions.

Comparing values for the heat of reaction requires a clear definition of the physical state of reactant species and temperature, for example, for the C–O reaction in the previous table it is clearly indicated that carbon is dissolved in liquid steel and oxygen is in the form of a gas. The result is different if carbon is solid or oxygen is also dissolved in liquid steel. Table 2.15 indicates values reported by Pfeifer and Kirschen [5] and Jones [7] for the heat of oxidation reactions.

Table 2.14 Heats of reaction of selected chemical reactions

	Reaction	$\Delta H_{R,1873}^0$, cal/mol	$\Delta H_{R,1873}^0$, cal/kg oxide	$\Delta H_{R,1873}^0$, kWh/kg oxide	$\Delta H_{R,1873}^0$, kWh/kg element
ΔH_{R1}	$FeO_{DRI} + \underline{C} \rightarrow$ $Fe + CO$	40,252.1	560,252.2	0.65	
ΔH_{R2}	$Fe_3C_{DRI} \rightarrow 3Fe + \underline{C}$	28,422	158,298.5	0.18	
ΔH_{R3}	$2FeO + \underline{Si} \rightarrow$ $2Fe + SiO_2$	− 35,817.5	− 1,275,302.6	− 1.48	
ΔH_{R4}	$\underline{Si} + O_2 \rightarrow SiO_2$	− 170,857.4	− 2,843,629.4	− 3.30	− 7.07
ΔH_{R5}	$\underline{C} + \tfrac{1}{2}O_{2(g)} \rightarrow CO_{(g)}$	− 33,286.91	− 1,188,393.7	− 1.38	− 3.22
ΔH_{R6}	$Fe + \tfrac{1}{2}O_2 \rightarrow FeO$	− 49,663.86	− 691,250.5	− 0.80	− 1.03

Table 2.15 Heats of oxidation reactions

Reaction	ΔH_R^0, kWh/kg element [5]	ΔH_R^0, kWh/m³ O₂ [5]	ΔH_R^0, kWh/m³ O₂ @1593 °C, [7]
$Si + O_2 = SiO_2$	− 8.70	− 10.92	− 9.182
$C + 0.5O_2 = CO$	− 2.55	− 2.73	− 3.178
$C + O_2 = CO_2$	− 9.10	− 4.88	
$Fe + 0.5O_2 = FeO$	− 1.32	− 6.58	− 6.003
$2Fe + 1.5O_2 = Fe_2O_3$	− 2.03	− 4.74	
$2Al + 1.5O_2 = Al_2O_3$	− 8.61	− 13.86	− 11.936
$Mn + 0.5O_2 = MnO$	− 1.95	− 9.56	− 8.404
$S + O_2 = SO_2$	− 2.75	− 3.94	

Chemical energy from burners can be represented by the following chemical reaction:

$$CH_{4(g)} + 2O_{2(g)} = CO_{2(g)} + 2H_2O_{(g)}$$
$$\Delta H_{1823\,K} = -404\,kJ/mol = 10\,kWh/m^3$$

Chemical energy from post-combustion can be represented by the following chemical reaction:

$$CO_{(g)} + \tfrac{1}{2}O_{2(g)} = CO_{2(g)}$$
$$\Delta H_{1823\,K} = -556\,kJ/mol$$

Sims [31] provided the following values for the heat of formation of oxides. The large differences cannot be attributed, in principle, to the thermodynamic databases but most probably to the physical states of the reactants which is not always clearly stated in most references.

Oxide	CaO	MgO	SiO₂	FeO	Fe₂O₃	MnO	Al₂O₃	P₂O₅	CO
ΔH_f^0, kWh/kg	− 3.148	− 4.149	− 4.206	− 1.028	− 1.429	− 1.509	− 4.567	− 3.007	− 1.097

Table 2.16 Heat of formation of dicalcium silicate

Phase	H, cal/mol	Range, K	R^2
$2CaO \cdot SiO_2$	$0.11907\,T - 30, 225.50197$	298–970	1.000
	$3.433 \times 10^{-3}T^2 - 5.632322T - 27, 406.500666$	970–1710	0.9983
	$4.049168T - 30, 493.638427$	1710–2000	0.9995

E. G. King, *J. Am. Chem. Soc.*, 73, 656, 1951.
J. F. Elliot, M. Gleiser, V. Ramakrishna, *Thermochemistry for Steelmaking*, Vol. 2, 1963, pp. 301.

Table 2.17 Heat of formation of slag species, after Sims [31]

Phase	ΔH_f^0, kWh/kg	Source
$2CaO \cdot SiO_2$	-5.846×10^{-1}	SiO_2
$2FeO \cdot SiO_2$	-6.637×10^{-2}	FeO
$Fe_2O_3 \cdot 2CaO$	-5.389×10^{-2}	Fe_2O_3
$4CaO \cdot P_2O_5$	-1.315×10^0	P_2O_5
$3CaO \cdot Al_2O_3$	-4.220×10^{-2}	Al_2O_3

The slag is formed from several sources; oxides present in the gangue content of DRI, scrap dust, oxides formed due to oxidation of elements in scrap, etc. The slag is not composed by pure oxides but defined by a complex multi-component equilibrium. Table 2.16 summarizes some correlations to define the heat of formation of dicalcium silicate and Table 2.17 provides additional information from other oxide species.

The melting point of slag components is in general above the normal steelmaking temperature (1600 °C). Since the slag is in fact liquid below that temperature is an indication of the presence of a complex multicomponent system, dominated by the presence of FeO, CaO and SiO_2 which form lower melting point compounds. To simplify the calculations is it assumed an average chemical composition. Daiconu et al. estimated a melting enthalpy, ΔH_m^0, of 5.8×10^{-2} kWh/kg.

2.3.5 Case Studies with Static Energy Balances

Prakash et al. [2] reported mass and energy balances to study the effect of the fraction of DRI and hot metal in the metallic charge on energy consumption. In a heat balance the energy inlets are electric and chemical energy and the energy outlets are liquid steel, liquid slag, the off-gas, water cooling, heat losses through the walls, electric losses and other heat losses. Their results suggest an increment in electric energy of 10 kWh/ton for every 10% increment in DRI. The highest consumption of electric energy and lowest input of chemical energy, 83% and 17% respectively, was found for the case using 100% scrap. They suggested to increase the fraction of hot metal or an increase in the carbon content in DRI to decrease the consumption of electric energy.

Arzpeyama et al. [30] applied energy balances to study the effect of HBI on oxygen injection, energy consumption and slag volume. They estimated the metal and slag composition using distribution factors which depend on the equilibrium partition ratios. Their results indicate that increasing 1% HBI decreases the consumption of oxygen and increases both energy consumption and the amount of slag by 0.16 Nm^3/ton, 1.29 kWh/ton and 34 kg/ton, respectively.

Diaconu et al. [19] described a detailed analysis of heat losses for a small EAF of 7 ton using scrap. They reported a low thermal efficiency of about 44% due to large heat losses, in particular the heat losses due to water cooling, however their numbers are inconsistent.

Kirschen et al. [40] evaluated heat balances for 16 EAF's charged with different ratios of scrap and DRI; 11 EAF's with 100% scrap and 5 EAF's with 70–90% DRI. They defined efficiency factors for heat transfer coming from the electric and chemical energies. Heat losses for each term was computed by the difference (1-η_i)(ΔH_i):

$$Q_m + Q_s = \eta_{elec}\Delta H_{elec}\tau + \eta_{chem}\Delta H_{chem}\tau + \eta_{burner}\Delta H_{burner}\tau + \eta_{pc}\Delta H_{pc}\tau$$

where: Q_m is the sensible heat in liquid steel, Q_s is the sensible heat in liquid slag, τ is time.

The electric efficiency term included not only Joule losses but also radiation at the electric arcs ($\eta_{el}\eta_{arc}$).

By assigning the efficiencies indicated in Table 2.18 they were able to reproduce the experimental values. These results are very useful, however, it should also be noticed that each heat evaluated was assigned a different set of efficiencies. The efficiency factors change for every heat. For example, similar heats with 100% scrap reported an oxygen efficiency from 76 to 90%. In other words, this information is only a reference which cannot be employed to reproduce any heat. Also notice that the values on efficiencies for burners and post-combustion for heats employing high amounts of DRI is wrong because those methods only apply for scrap-based operations.

Considering the values reported in Table 2.18 for the combined electric efficiency, $\eta_{el}\eta_{arc}$, it can be estimated the arc efficiency. Pfeifer et al. [6] reported electric efficiency at the secondary side of the transformer in the range from 0.88 to 0.92. Heat transfer efficiency from the arc to the molten metal depends on the slag foaming conditions and can fluctuate from 0.36 to 0.93. According with the data in the table,

Table 2.18 Heat transfer efficiencies using scrap and DRI in the EAF [40]

	$\eta_{el}\eta_{arc}$	η_{chem}	η_{oxygen}	η_{burner}	$\eta_{post-comb}$
70–90% DRI	65–68	73–75	82–90	55	10
100% DRI	67–80	70–75	76–90	55	10
Range	65–80	70–75	76–90	55	10

Table 2.19 Total energy consumption and distribution of energy (%)

Refs.	Metallic charge	TE	EE	CE	Steel	Slag	OffG	ee	wc	rad
[2]	60DRI/40S	672	83	17	43.2	7.0	13.6	11.7	5.8	17.5
[2]	70HM/ 20DRI/10S	674	62	37	43.2	5.3	14.8	11.6	5.8	17.5
[2]	40HM/ 20DRI/20PI/ 20S^{real}	767	45	54	38	6	16	12	7	19
[2]	40HM/ 20DRI/20PI/ 20S^{mod}	674	47	57	43	5.3	14.8	11.6	5.8	17.5
[2]	Scrap	597	83	17	48.8	4.9	10.6	11.6	5.8	17.5
[5]	Scrap	758	62	38	53	5.4	16.9	3.8	18.9	
[5]	Scrap	810	48	52	53	3.8	20.7		13	
[19]	Scrap				44	8	6.7	3	22	
	Scrap	762	61.2	38.8	51.4	5.5	15.9	4.6	15	7.6
[40]	Scrap/DRI	650–850	40–65	22–60	45–60	5–10	15–35		10–20	

TE = total energy, EE = electric energy, CE = chemical energy, Steel = sensible heat in liquid Steel, Slag = sensible heat ion liquid slag, OffG = sensible heat in the off-gas, ee = heat losses electric circuit, wc = heat losses due to water cooling, rad = heat losses due to radiation

the combined efficiency is about 65–80%. Taking an average value for $\eta_{el} = 0.9$, the average value of $\eta_{arc} = 0.7$–0.9 for the heats evaluated.

Table 2.19 summarizes reported energy distributions using different raw materials and proportions of chemical energy from different sources. It can be observed that heat losses in the off-gas range from 10 to 20%, water cooling from 6 to 20% and radiation from 17 to 19%, this last term strongly influenced by the number of charges of scrap baskets. Chemical energy for normal scrap and DRI operation without burners and post-combustion is in the order of 17%, increasing up to 70% as hot metal, burners and post-combustion is included in the heat balance, for the group of data in this table.

2.4 Dynamic Energy Balance

A dynamic energy balance involves rate phenomena, defines the instantaneous changes in chemical composition of all phases (liquid steel, liquid slag, off-gas, dust), their temperature and their mass. It involves a much more complex analysis in comparison with a static model. In addition to thermodynamics the rates of heat, momentum and mass should be considered. A dynamic model describes the rates of change in all the variables and therefore the problem involves differential equations which can be solved using standard methods or computational fluid-dynamics (CFD) software, in all cases involving the development of mathematical models. Realistic modelling of each single operation, for example; scrap melting, slag foaming, decarburization, etc. is usually impossible because of limitations on computing time or lack of information on kinetic parameters, and the only way to make a model is to simplify the number of variables and make assumptions. Even under these conditions, complex dynamic models are not adequate for control on-line due to the long time to run a single simulation. Model validation is also a challenge due to the difficulty to obtain direct information from the actual process, therefore validations are based on physical models or partial data taken from the real process.

Hay et al. [46] reviewed mathematical process models up to 2021. The majority of the models developed have focused on single operations. Several dynamic models based on proprietary software that include multiple operations have been reported [47–50], however the lack of details makes difficult to evaluate its real capabilities. In one of those cases, the full energy losses were not computed but using lump values which clearly is far from a phenomenological approach. Much more useful, in terms of free knowledge, are the reports from academic groups [51–55]. A summary of these works is described below which provides a better idea of the development of a dynamic model.

The work carried out at ESIQIE-IPN [51–53] in Mexico was one of the first research works reporting integral models that described in detail the reaction kinetics involving multiple operations using DRI in the EAF, in particular slag foaming. For full details the reader should check those references.

Figure 2.4 shows a typical EAF with continuous DRI feeding, oxygen and carbon injection. Their model describes the change in chemical composition of both slag and steel, the temperature of liquid steel and a novel dynamic foaming index. The DRI melting rate was described using an empirical relationship based on plant data.

Figure 2.5 shows schematically the general structure of the mathematical model using 100% DRI. Energy consumption is initially estimated from an energy balance but heat losses are largely under-estimated, because the off-gas composition is assumed to be formed only by CO and other heat-losses were also not considered.

The mass balance defines the mass of DRI, fluxes, coke and oxygen injection. Using the following empirical relationships for the DRI melting rate and lime dissolution rate it is possible to calculate the melting rate of DRI and rate of dissolution of fluxes:

$$\frac{dW_{DRI}}{dt} = -226{,}403.57 + 5016.6(\%Met) - 28.8(\%Met)^2$$

$$\frac{dW_{lime}}{dt} = f(T) \cdot X_{CaO}^{lime} \frac{dW_s}{dt}$$

where: W_{lime} is the rate of dissolution of lime, W_{DRI} is the mass of DRI, %Met is the degree of metallization of DRI, W_s is the mass of slag, $f(T)$ is a function of temperature.

The rate of formation of CaO, SiO_2, MgO and Al_2O_3 depends on the rate of melting of DRI and the rate of dissolution of lime. Oxidation of elements is neglected because

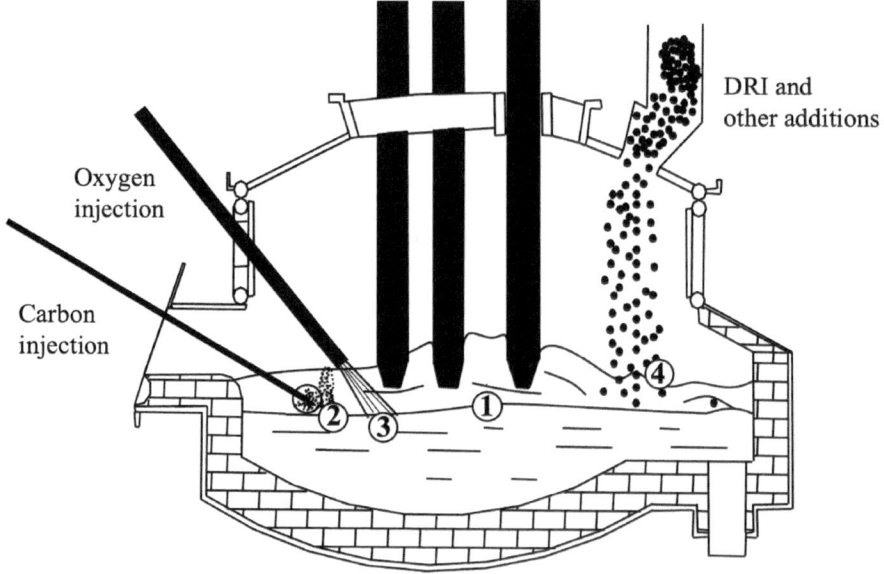

Fig. 2.4 EAF showing the reaction interfaces; (1) slag-metal interface, (2) carbon particles-slag interface, (3) oxygen injection-slag interface, (4) melting of additions

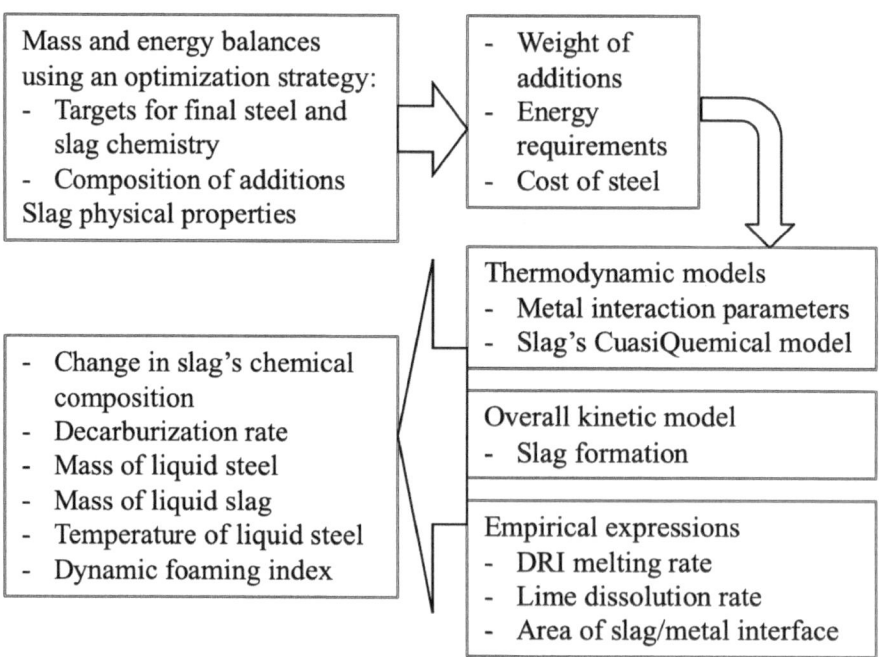

Fig. 2.5 Overview of dynamic model. After ref. [51–53]

there is no scrap in the metallic charge, only DRI.

$$\frac{d(X_{MO}^s \cdot W_s)}{dt} = X_{MO}^{lime}\frac{dW_{lime}}{dt} + X_{MO}^{DRI}\frac{dW_{DRI}}{dt}$$

where: X_{MO}^s is the fraction of oxide in the slag, X_{MO}^{DRI} is the fraction of the same oxide in DRI.

The final mass of iron oxide in the slag results from the FeO added by DRI and FeO due to oxidation minus the FeO reduced with carbon particles. The reduction rate of FeO by carbon particles is considered to take place by two simultaneous reactions; reduction at the CO/slag interface and consumption of carbon particles by the Boudouard reaction.

$$\frac{d(X_{FeO}^s \cdot W_s)}{dt} = X_{MO}^{DRI}\frac{dW_{DRI}}{dt} + \frac{M_{FeO}}{M_O}K_{O_2}W_{O_2}$$

$$- M_{FeO}A_{sm}r_{CO}'' - \frac{M_{FeO}}{M_O}Q_s\theta_p^g\left[1 - \left(\frac{r_p^f}{r_p^0}\right)^3\right]$$

where: M is the molecular mass, K_{O_2} is the rate of oxidation of liquid iron, r_{CO}'' is the rate of formation of CO, Q_s is the air flow rate to inject carbon particles, θ_p^g is the fraction of particles that are able to penetrate into the slag, r is the radius of the carbon particles.

The rate of formation of slag is then given by the net formation of all oxides:

$$\frac{dW_s}{dt} = \frac{d(X_{FeO}^s \cdot W_s)}{dt} + \frac{d(X_{CaO}^s \cdot W_s)}{dt}$$

$$+ \frac{d(X_{SiO_2}^s \cdot W_s)}{dt} + \frac{d(X_{MgO}^s \cdot W_s)}{dt} + \frac{d(X_{Al_2O_3}^s \cdot W_s)}{dt}$$

The expression that defines the fraction of carbon particles that penetrate into the slag, θ_p^g, is critical to the calculation of both iron oxide reduction and the formation of CO. The expression developed was the following [51–53]:

$$\theta_p^g = 1 - \frac{12\sigma_{lg}(f - \cos\beta)}{\left(U_p^g\right)^2 d_p\rho_p}$$

where: f is the difference between the gas–liquid interface area while the particle is passing through the interface and the area of the undisturbed interface, σ_{lg} is the gas phase interfacial energy, β is the solid–liquid contact angle, d_p is the carbon particle diameter and ρ_p is the carbon particle density. The expression indicates that in order to maximize the fraction of particles crossing into the slag, the variables in the denominator of the second term on the right side of the equation should be increased as much as possible and/or those in the numerator decreased. This result has significant consequences because with one single equation is possible to identify how to manipulate the process variables to improve the slag foaming conditions.

Another dynamic model reported by Bekker et al. [56] in 1999 was one of the first dynamic models that described the scrap melting rate and the composition of the off-gas. This work is based on thermodynamics and a large number of heat transfer approximations, giving a very rough representation of the real process. For example, the melting rate of solid scrap is estimated based on the temperature difference between solid scrap and liquid steel multiplied by a heat transfer factor proportional to the ratio of the same temperatures, in turn the temperature of liquid steel is estimated by dividing the total power in the furnace by the specific heat capacity of all materials heated to the melting temperature:

$$\frac{dW_{scrap}^{solid}}{dt} = -\frac{M_{Fe}k_tk_AW_{scrap}(T_1 - T_s)\left(\sqrt{\frac{T_s}{T_1}}\right)}{\left(\lambda_{Fe} + C_{p,Fe}(T_1 - T_s)\right)} = -\frac{dW_{steel}^{liquid}}{dt}$$

$$\frac{dT_1}{dt} = \frac{\sum net\ power}{\sum \frac{C_{pi}}{M_i}}$$

where: dW/dt is the rate of scrap melting, k_t is a heat transfer coefficient (0.4 kW/ Km^2), k_A is the specific surface area of steel scrap (0.005 m^2/kg), T_l is the temperature of liquid steel and T_s is the temperature of scrap, λ_{Fe} is the latent heat of melting, C_p represents the heat capacity.

In their model, they assume the slag/metal system is far from equilibrium and the exact value of the equilibrium constants is not critical [56, 57]. This assumption eliminates the use of thermodynamic models to compute activity coefficients. They suggested an original approach to estimate the equilibrium composition. For example, for the following chemical reaction:

$$(FeO) + [C] = [Fe] + CO_g$$

The solubility product at equilibrium is:

$$k_{C\%} = (\%FeO)(\%C)_{eq}$$

$$k_{C\%} = \left(X_{FeO}\frac{M_{FeO}}{M_{slag}}100\right)\left(X_C^{eq}\frac{M_C}{M_{Fe}}100\right)$$

$$k_{C\%} = X_{FeO}X_C^{eq}\left(\frac{M_{FeO}M_C}{M_{slag}M_{Fe}}\right)100^2$$

From the previous expression, an expression for the equilibrium carbon concentration can be derived:

$$X_C^{eq} = \frac{k_{C\%}}{X_{FeO}}\left(\frac{M_{slag}M_{Fe}}{M_{FeO}M_C100^2}\right)$$

The parameters which are constant are defined as a dimensionless equilibrium concentration:

$$k_{XC} = k_{C\%}\left(\frac{M_{slag}M_{Fe}}{M_{FeO}M_C 100^2}\right)$$

$$k_{XC} = X_{FeO} \times X_C^{eq}$$

In this way, these authors calculated the equilibrium carbon concentration as follows:

$$X_C^{eq} = \frac{k_{XC}}{X_{FeO}}$$

The value of $k_{C\%}$ was taken from Turkdogan [58]. The equilibrium constant given by Turkdogan can be calculated by the following expression:

$$\log K_{XC} = \log\left(\frac{p_{CO}}{[\%C]a_{FeO}}\right) = \frac{5730}{T} + 5.096$$

At 1600 °C, $K_{XC} = 108.8$. Turkdogan defined a_{FeO} in terms of (%FeO), assuming a slag with a binary basicity of 3.2, $\gamma_{FeO} = 1.3$. He reported the following calculations.

$$a_{FeO} = \gamma_{FeO}X_{FeO} = 1.3X_{FeO} = 1.3\frac{(\%FeO)}{72 \times 1.65} = 0.11 \times (\%FeO)$$

$$K_{XC} = \frac{p_{CO}}{[\%C]a_{FeO}} = \frac{1.5}{[\%C] \times 0.11 \times (\%FeO)}$$

$$[\%C] \times (\%FeO) = \frac{1.5}{0.11 \times K_{XC}} = \frac{1.5}{0.11 \times 108.8} = 1.25$$

This calculation has probably a typographical mistake because in both cases the correct math with those values is not 0.11 and 1.25. There is no attempt to correct it because these values have been employed subsequently in other models. With the previous result, is now possible to define the value k_{XC} in Bekker's model:

$$k_{C\%} = [\%C] \times (\%FeO) = 1.25$$

$$M_{FeO} = 0.0718\,kg/mol$$

$$M_C = 0.012\,kg/mol$$

$$M_{Fe} = 0.0558\,kg/mol$$

$$M_{slag} = 0.0606\,kg/mol$$

$$k_{XC} = k_{C\%}\left(\frac{M_{slag}M_{Fe}}{M_{FeO}M_C 100^2}\right)$$

$$= 1.25 \times \left(\frac{0.00338}{8.62}\right) = 490 \times 10^{-6}$$

$$k_{XC} = X_{FeO} \times X_C^{eq} = 490 \times 10^{-6}$$

This value is about one order of magnitude lower than the value reported by Turkdogan ($1/108 = 925 \times 10^{-5}$).

Bekker [57] and Oosthuizen et al. [59] extended the previous EAF model to study the off-gas composition and exit temperature. Bekker [57] originally assumed the off-gas is formed of $CO + CO_2 + N_2$. The net production rate of CO is the result of CO formed due to oxygen injection and reduction of iron oxide minus CO transformed to CO_2 due to post-combustion inside the EAF due to air ingress both at the slag door and the slip gap. It is assumed all the carbon particles are consumed.

$$\frac{dCO}{dt} = f\left[k_{DeC}\left(X_C - X_C^{eq}\right)\right] + f(\dot{m}_c) - f(k_{PR}) - f(D_{4th}h)$$

where: k_{DeC} is a decarburization rate constant (12 kg/s), X_C is the concentration of carbon in liquid steel, $\dot{m}_c$ is the rate of injection of carbon particles, k_{PR} is a constant which depends on the area of the slag door and the relative pressure inside the furnace, D_{4th} is the diameter of the fourth hole and the air-gap distance. The net production rate of CO_2 results from CO post-combustion inside the furnace and at the slip gap. Nitrogen depends on air ingress through the slag door and at the slip gap, and also air employed as a carrier gas for carbon injection. They estimated that for every 150 kg of injected graphite, there is 1 kg of nitrogen in the carrier gas.

In order to improve the accuracy of the model predictions on the exit temperature of the off-gas, Oosthuizen et al. [59] reviewed in more detail its heat balance considering the heat released due to combustion and the heat extracted due to convection through the duct sides. The amount of heat extracted was calculated using the Petukhov equation for turbulent convection heat transfer.

Matson and Ramirez [60] developed a dynamic model considering equilibrium at all the interfaces with the rates of reaction depending on transport limitations among the phases. The scrap melting rate was estimated assuming it was formed by spheres.

MacRosty and Swartz [54] developed a dynamic model aiming to achieve a phenomenological approach, capable to predict the nonlinear behavior of many variables, numerically robust and using the least amount of plant data. It consisted of 4 modules to study the melting rate of scrap, production of liquid steel, slag/metal reactions and the gas phase. All chemical reactions were assumed to be in equilibrium at the slag/metal interface, estimating the mass transfer coefficients with industrial data. The gas phase was composed by a complex mixture containing CO, CO_2, N_2, CH_4, H_2, H_2O and C_9H_{20}. The slag phase consists of FeO, MnO, MgO, Al_2O_3, SiO_2 and CaO. Due to the large number of equations in this model, the authors reported heat and mass balances at the interfaces in a general way. The first mass balance corresponds to the scrap zone where the rate of melting is computed:

$$\frac{dm_{ss}}{dt} = \dot{M}_{scrap} - \dot{M}_{melt}$$

where: m_{ss} is the mass of solid scrap, $\dot{M}_{scrap}$ is the mass of added scrap and $\dot{M}_{melt}$ is the melting rate of scrap. The melting rate was computed using a similar approach to that proposed by Bekker [57].

The mass balance at the molten metal zone, slag/metal interface, and gas zone are defined by the following general expression:

$$\frac{db_{k,z}}{dt} = F_{k,z}^{in} - F_{k,z}^{out}$$

where: b_k is the molar amount of element k in zone z, $F_{k,z}^{in}$ and $F_{k,z}^{out}$ are the flow rates of element k into and out of zone z.

An important suggestion from this work was the transformation into logarithmic space the molar quantities because the large differences in values would create numerical problems.

Heat balances were also reported with general expressions. The temperature of liquid steel was calculated similar to Bekker [57]. Due to the limited amount of experimental data (electrical and material inputs, off-gas chemistry, endpoint carbon and slag composition) the model predictions were compared with the off-gas, indicating a reasonably good accuracy.

The work from Logar et al. [25, 61] is one of the most updated dynamic models so far, extensively based on the previous work by Bekker [57] and MacRosty and Swartz (McR-S) [54] but described in much more detail and with additional improvements. It also describes the evolution in the chemical energy supplied over an entire heat. The slag is assumed as an ideal solution, formed by FeO, SiO_2, CaO, MnO, MgO, P_2O_5, Cr_2O_3 and Al_2O_3. Similar to McR-S, the geometric domain includes; the solid scrap zone, the liquid steel zone, the liquid slag zone, the gas zone, additionally the wall zone and roof zones were added. Every zone is assumed to be thermal and chemically homogeneous (perfect mixing conditions). This model does not compute the initial steel and slag chemistry for a given metallic charge but are provided as initial conditions.

The change in chemical composition for the metal, slag and gas was computed assuming that the driving force for a chemical reaction is the difference between the actual concentration and the equilibrium values, similar to Bekker:

$$\dot{m}_i = \frac{dm_i}{dt} = \frac{\Delta m_i}{\Delta t} = k_i \left(X_i - X_i^{eq} \right)$$

The net change in concentration varies for each chemical species depending on its sources and consumption rates. Carbon is added as carbon particles and other materials in the metallic charge, later on it can be consumed by FeO, MnO and oxygen injection. Silicon is a simpler case, it is consumed by oxidation with FeO, O_2 and MnO, as follows:

$$2(FeO) + [Si] = 2[Fe] + (SiO_2)$$
$$[Si] + O_{2(g)} = (SiO_2)$$

$$2(MnO) + [Si] = 2[Mn] + (SiO_2)$$
$$\dot{m}_{Si} = \left[-k_{Si1}\left(X_{Si} - X_{Si}^{eq}\right)\right]$$
$$+ \left[-k_{Si2}\left(X_{Si} - X_{Si}^{eq}\right)\right]\left[k_{O_2-SiO_2}O_2\right]$$
$$+ \left[-k_{Si3}\left(X_{MnO} - X_{MnO}^{eq}\right)\frac{M_{Si}}{M_{MnO}}\right]$$

Once the mass of the different species for liquid metal, liquid slag and gas phase is defined as a function of time, then $\dot{m}_I$, the rate of chemical energy supplied and rate of energy consumed can be computed. Figure 2.6 shows the results reported by Logar et al. [54] for scrap melting. It shows the contribution to chemical energy by exothermic reactions (chemical), organic (combustible), oxidation of electrodes (electrodes) and burners (CH_4). According with these results, the main contribution is due to exothermic reactions of dissolved elements in the molten bath, about 26%. Oxidation of electrodes and burning of organic matter represent 2% and 1%, respectively. The chemical energy from oxy-fuel burner in this example was 9.2%.

The sensible heat in liquid steel, off-gas and slag was 51.4%, 15.9% and 5.5%, respectively. Heat losses by radiation was computed from the values of radiosity of N radiating surfaces, as follows:

$$Q_{i\text{-rad}} = A_i \sum_{j=1}^{N} VF_{ij}\left(J_i - J_j\right)$$

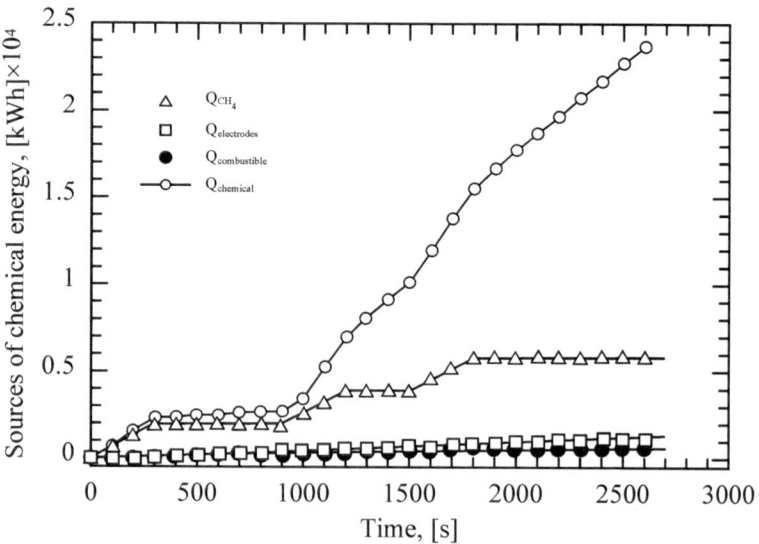

Fig. 2.6 Contribution of chemical energy throughout a heat. After Logar et al. [54]

$$J_i = \varepsilon_i \sigma_{SB} T_i^4 + (1 - \varepsilon_i) \sum_{j=1}^{N} \left(VF_{ij} - J_j \right)$$

where: J_i is the radiosity of body j, ε_i is the body's emissivity (0–1), σ_{SB} is the Stefan-Boltzmann constant, T_i is the body's temperature, VF_{ij} is the view factor from surface I to surface j (0–1). Heat transfer due to radiation to the water-cooled panels was estimated as 15%.

Other heat losses reported were; electric losses 4.6% and miscellaneous losses 7.6%.

According with Logar et al. [25, 61], their model accurately describes the evolution in chemical composition of steel, slag and the off-gas, however it should be noticed that agreement was largely influenced by the empirical parameters included in the model, valid for the specific conditions where the model was applied. Most of the empirical parameters are related to the mass transfer coefficients.

In addition to the extensive use of empirical parameters, it is quite clear that previous dynamic models have employed unrealistic assumptions like an ideal slag solution and perfect mixing conditions in the EAF which gives a tremendous room for further improvements. A research group at Aachen university [62] has incorporated the use of two slag models to predict thermodynamic activities of the slag components but their predictions show considerable deviations from the actual experimental values.

The Centre de recherches métallurgiques (CRM) in Belgium has reported the development of dynamic models for scrap melting [47, 63] and more recently have also included DRI in the metallic charge [64, 65]. Nyssen et al. [47, 63] provided general details about the dynamic model for scrap-based operations.

1. Scrap melting model: EAF is divided into 15 sectors, each sector with different layers
2. Bath model: Includes oxygen injection, oxidation reactions and follows the chemical, thermal and mass evolution of liquid steel. The model assumes perfect mixing conditions.
3. Slag model: Follows the chemical, thermal and mass evolution of liquid slag. The real slag foaming height is adjusted based on the slag leaving the furnace.
4. Refractory model: Calculates heat losses and erosion rate. The erosion rate is employed to compute the increment in volume available for slag and liquid steel.
5. Arc model: Defines heat transfer to different part of the EAF during scrap melting.
6. Scrap pre-heating: Computes heat exchange between off-gas and solid scrap.
7. Post-combustion: Calculates post-combustion including air ingress
8. Burners' module: Calculates the combustion of natural gas
9. Water panels module: Computes heat transfer to water cooled panels.
10. Skulls model: Evaluates solidification of hot heel due to long delays.

The exhaust flow rate is estimated with data at the quenching tower, using the amount of sprayed water and gas temperature before and after quenching, according with the following heat balance:

$$Q_{\text{off-gas}} C_p^{\text{off-gas}} \left(T_{\text{off-gas}}^f - T_{\text{off-gas}}^0 \right) = Q_{\text{water}} \Delta H_{\text{water}}^{\text{vaporization}}$$

where: $Q_{\text{off-gas}}$ is the off-gas flow rate, Q_{water} is the flow rate of sprayed water.

The slag model predicts the foaming conditions. The slag volume is controlled with data from the refractory model that predicts the increment in volume due to erosion and the slag flow rate leaving the furnace. The furnace volume increases from heat to heat, for example considering an initial furnace capacity of 100 ton, after 280 heats they reported a furnace volume higher than 200 ton. It was reported that the slag volume remains almost constant during the heat. The apparent slag density of the foaming slag was calculated with the following equation:

$$\rho_s^{\text{app}} = \rho_s e^{-\frac{V_{CO}}{V_{CO}^{\text{ref}}}}$$

where: ρ_s^{app} is the apparent slag density, ρ_s is the slag density as a function of slag chemical composition, V_{CO} is the volume generated by the reduction of iron oxide V_{CO}^{ref} is a reference volume of CO.

The dynamic model developed by CRM using DRI assumes that DRI is heated instantaneously to 950 °C and FeO in DRI is self-reduced about 92% [65]. The models developed have been applied online for process control which might suggest the use of a large number of empirical factors obtained by trial and error at each plant.

2.5 Optimization Model Involving an Energy Balance

Another group of models is the use of optimization tools. The optimization schemes are usually focused on minimization or maximization of variables. There are two groups of optimization, single-objective optimization (SOO) and multiple-objective optimization (MOO). The simplest case is SOO, for example minimization of the price of steel, minimization of emissions of CO_2, minimization of energy consumption, etc. These types of problems have a single solution and easier to understand. More frequently the problem to be solved involves multiple-objective optimization (MOO) of several variables, for example minimization of the price of steel, minimization of energy consumption and minimization of emissions of CO_2. Deb [66] illustrates the following example which describes the differences between SOO and MOO: How to buy a car based on price or based on comfort, or both. If the decision is based on the minimum price, then the selection will be on the car with the lowest comfort, and vice versa. The two-objective optimization problem involves buying a car in which there is a trade-off between price and comfort. The trade-off involves sacrificing objectives. Trade-off solutions are optimal solutions to a MOO problem, forming a front on an objective space, also called Pareto-optimal front.

Chapter 3 includes a summary of several optimization models. The following section describe in more detail the optimization models, in particular the one reported by Cardenas et al. [15].

2.5.1 Single-Objective Optimization (SOO)

The typical SOO in the steel industry has focused on the minimization of the production cost of steel. Once the objective function is defined the rest of the variables should align to reach this objective. Producing steel with the lowest production cost usually would affect other critical variables such as minimization of emissions of CO_2 and steel quality.

One of the first optimization models using linear programming was reported by Geiger in 1980 [67] which employed mixtures of scrap and DRI in the metallic charge. Cárdenas et al. [15] updated Geiger's model, and applied the model to study energy consumption with 100% DRI, applying the SIMPLEX method. Some of the improvements include the formation of CO_2, the addition of metallurgical and dolomitic lime, removing the addition of ferroalloys in the EAF and including DRI temperature. The system of equations is formed by the objective function, mass balances for the components in steel, slag and the off-gas, restrictions which consider special conditions such as availability of raw materials, maximum concentration of residual elements, etc., and a heat balance equation. Mass and heat balances are equalities and the restrictions are inequalities. The objective function is defined by the sum of costs for all materials and electric energy, as follows:

$$\text{Min } z = c_1 x_1 + c_2 x_2 + \cdots + c_n x_n$$

Mass balances and restrictions:

$$A_{11} x_1 + A_{12} x_2 + \cdots + A_{1n} x_n ? b_1$$
$$A_{21} x_1 + A_{22} x_2 + \cdots + A_{2n} x_n ? b_2$$
$$\vdots \quad \ldots \quad \vdots$$
$$A_{m1} x_1 + A_{m2} x_2 + \cdots + A_{mn} x_n ? b_m$$

where: z is the cost of steel, c_i represents the cost of all materials and electric energy supplied, A_{ii} and A_{ij} are coefficients related to the fraction of a given chemical specie in the raw materials for the mass balances or specific energy consumption to raise the temperature of different raw materials as well as chemical energy, ? represents an operator $\leq$, $=$ or $\geq$.

The following example consists of 6 mass balances, 6 constraints and one energy balance. Mass balances include Fe, O, Si, Al, CaO, MgO, CO_2 and a general mass balance. Si–Al and CaO–MgO are combined into one equation in each case. The mass balances for carbon and sulphur were handled as restrictions. Carbon is allowed to range between a minimum value to reach the carbon specifications in liquid steel and allow for the reduction of iron oxide in DRI to a maximum value which in addition to the previous two terms also include an amount of carbon for decarburization. The energy balance states input of electric energy (considering an electric efficiency of 83%) and the energy released due to exothermic reactions is equal to the heat required to raise the temperature of all materials to the steelmaking temperature plus sensible heat in the off gas, sensible heat in the slag, endothermic reactions and heat losses. Heat losses are considered in a lump parameter which depends on an empirical thermal efficiency. All equations are summarized in Table 2.20.

The commercial optimization software LINDO API® was employed. This software package solves linear programming, integer programming, nonlinear programming, stochastic programming and global optimization. The mass and energy balances were developed in a computer program using Visual Basic version 6.0 (currently obsolete) which has an interface to connect with the optimization software. The thermochemical data base was taken from the commercial software package FactSage®. The steelmaking temperature is defined as an inlet condition. The total number of unknowns is 10, including the weight of DRI and types of scrap, fluxes, oxygen and carbon requirements, weight of slag, weight of off-gas and consumption of electric energy. The solution defines the metallic charge that ensures minimum production cost for one ton of liquid steel. From a perspective of energy consumption, the model can evaluate the effect of chemical composition of raw materials, use of chemical energy, tapping temperature, scrap temperature and DRI temperature. Sensitivity analysis was carried out. A detailed explanation of the model is indicated in the thesis by Cárdenas et al. [44].

The model developed consists in three Graphical User Interfaces (GUI). The first one inputs the inlet conditions; chemical composition of all materials added to the EAF (DRI, scrap, fluxes and coke), the target chemistry of liquid steel, liquid slag and dust, tapping temperature, initial DRI temperature, maximum percentage of scrap and DRI (restrictions), cost of both, materials and electric energy in USD/kg and the thermal efficiency. The first interface computes all the coefficients in the system of equations, displayed on a second interface. This interface relates to the calculations using the SIMPLEX method and the results are shown on a third interface. The results provide the minimum cost per ton of steel, the amount of materials needed to produce one ton of steel and its electric energy consumption, it shows the mass balance, heat due to exothermic reactions, sensible heat in both liquid steel, liquid slag and the off-gas, and heat losses. The model can easily be adjusted to reach a higher accuracy in the thermal balance by changing the value of the thermal efficiency.

Figure 2.7 shows the results of a simulation describing the influence of the DRI ratio in the metallic charge on energy consumption, as a function of metallization of DRI and thermal efficiency. Increasing DRI in the metallic charge has a strong effect on energy consumption, however the intensity in this effect can be drastically reduced by increasing either the metallization of DRI or the thermal efficiency. The values shown in this figure indicate an increment of about 150 kWh/ton when the

Table 2.20 Set of equations for the optimization model

Impurity restriction: all impurities, including residual elements. This example; 0.4%
$X_{DRI}^{Imp} W_{DRI} + X_{Scrap}^{Imp} W_{Scrap} \leq 0.4\%$
Carbon restriction: Maximum C needed for DeC, as alloying element and to reduce FeO
$X_{DRI}^{C} W_{DRI} + X_{Scrap}^{C} W_{Scrap} + X_{Coke}^{C} W_{Coke} \leq 30 W_{Slag} + X_{Steel}^{C} W_{Steel} + X_{DRI}^{FeO}\left(\frac{12}{71.85}\right) W_{DRI}$
Scrap restriction: In the case that a given scrap needs to be below certain limit
$W_{Scrap} \leq Max$
Sulphur restriction: Maximum limit
$X_{DRI}^{S} W_{DRI} + X_{Scrap}^{S} W_{Scrap} + X_{Coke}^{S} W_{Coke} \leq X_{Steel}^{S} W_{Steel} + X_{Slag}^{S} W_{Slag}$
DRI restriction: Controls the fraction of DRI, from 0 to 100%
$W_{DRI} \leq Max$
Fe balance
$\left[X_{DRI}^{Fe} + \left(\frac{55.85}{71.85}\right) X_{DRI}^{FeO}\right] W_{DRI} + X_{Scrap}^{Fe} W_{Scrap} = \left[X_{Steel}^{Fe} + \left(\frac{55.85}{71.85}\right) X_{Steel}^{FeO}\right] W_{Steel} + \left(\frac{55.85}{71.85}\right) X_{Slag}^{FeO} W_{Slag} + \left(\frac{55.85}{71.85}\right) X_{Fume}^{FeO} W_{Fume}$
Si + Al balance: Si and Al in the charge are oxidized
$\left[\left(\frac{28.08}{60.08}\right) X_{DRI}^{SiO_2} + \left(\frac{53.96}{101.96}\right) X_{DRI}^{Al_2O_3}\right] W_{DRI} + X_{Scrap}^{Si+Al} W_{Scrap} = X_{Slag}^{Si+Al} W_{Slag}$
CaO + MgO Balance: CaO and MgO charged increase the volume of slag
$X_{DRI}^{CaO+MgO} W_{DRI} + X_{Cal}^{CaO+MgO} W_{Cal} = X_{Slag}^{CaO+MgO} W_{Slag}$
Oxygen balance: Oxygen in DRI and injected leaves in the gas phase and slag phase
$\left[\left(\frac{16}{71.85}\right) X_{DRI}^{FeO} + \left(\frac{32}{60}\right) X_{DRI}^{SiO_2} + \left(\frac{48}{102}\right) X_{DRI}^{Al_2O_3}\right] W_{DRI} + W_{O_2} = X_{CO}^{O} W_{CO} + \left[\left(\frac{16}{71.85}\right) X_{DRI}^{FeO} + \left(\frac{32}{60}\right) X_{DRI}^{SiO_2} + \left(\frac{48}{102}\right) X_{DRI}^{Al_2O_3}\right] W_{Slag}$
CO_2 balance: Ash content in lime and volatile matter in the coke
$X_{Cal}^{PxC} W_{Cal} + X_{Coke}^{M.V.} W_{Coke} = W_{CO_2}$

(continued)

Table 2.20 (continued)

Impurity restriction: all impurities, including residual elements. This example: 0.4%

Carbon requirements: Minimum carbon requirements

$$X_{DRI}^C W_{DRI} + X_{Scrap}^C W_{Scrap} + X_{Coke}^C W_{Coke} \geq X_{Steel}^C W_{Steel} + X_{DRI}^{FeO}\left(\frac{12}{71.85}\right) W_{DRI}$$

Total mass balance

$$W_{DRI} + W_{Scrap} + W_{Cal} + W_{O_2} + W_{Coke} = W_{Slag} + W_{CO} + W_{CO_2} + W_{Steel} + W_{Fume}$$

Total energy balance

$$\Delta H_{DRI} W_{DRI} + \Delta H_{Scrap} W_{Scrap} + \Delta H_{Coke} W_{Coke} + \Delta H_{Cal} W_{Cal} + \Delta H_{O_2} W_{O_2}$$

$$- \Delta H_{Fume} W_{Fume} + \Delta H_{R_{Endo}} - \Delta H_{R_{Exo}} + (1 - Efic.)\left(\sum \Delta H_{R_{Exo}}\right) = [1 - (1 - Efic.)]0.83 \, E.E.$$

Objective function: minimization

$$C = C_{DRI} W_{DRI} + C_{Scrap} W_{Scrap} + C_{Coke} W_{Coke} + C_{Cal} W_{Cal} + C_{O_2} W_{O_2} + C_{E.E.} E.E.$$

Where: X_{ij}, mass fraction specie i in material j; W_{slag}, slag weight; W_A, alloys weight; W_{Scrap}, scrap weight; cal, fluxes; E.E., electric energy; C_i, materials cost or energy per ton of steel; ΔH_i, sensible heat, reaction heat and energy losses

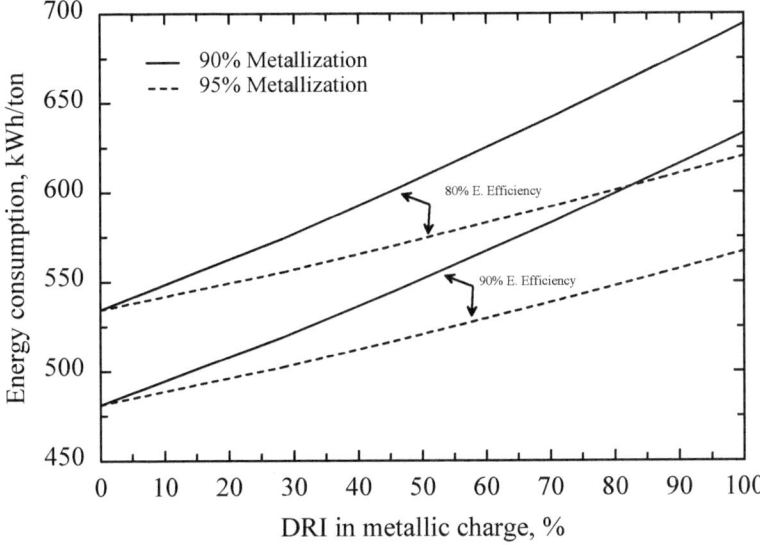

Fig. 2.7 Effect of % DRI on electric energy consumption, as a function of metallization of DRI and thermal efficiency [15, 44]

metallic charge changes from 100% scrap to 100% DRI with 90% metallization and 90% thermal efficiency. If the metallization is improved to 95% the increment on energy consumption is reduced by half to 80 kWh/ton, in other words the impact factor ranges from about 0.8 to 1.5 kWh/ton for every 1% of DRI. In practice a value of 1 kWh/ton can be applied.

The influence of metallization and carbon content in DRI on energy consumption is shown in Fig. 2.8. The impact factors of metallization and carbon content on energy consumption are 11 and 32 kWh/ton, respectively, for every increment of 1% in each variable. This result indicates a stronger effect of the carbon content in DRI.

Figure 2.9 shows the effect of carbon in DRI for different degrees of metallization on the consumption of electric energy consumption. Carbon has a strong effect to decrease energy consumption because is a source of chemical energy. Before the 1990s, the typical carbon concentration in DRI was about 1%. Suppliers of DRI technology had the idea that DRI should be produced with as much metallic iron units and possible, in this way increasing carbon would dilute the concentration of iron and was against the production of value. Progressively from the 1990s with the interest to decrease the consumption of electric energy and replace it by chemical energy there was a full change in philosophy, it became accepted the need to increase the concentration of carbon to higher values, in the range from 2 to 4%C. The optimization model suggests that an increment of 1%C in DRI decreases energy consumption by about 32 kWh/ton.

Figure 2.10 shows the effect of oxygen injection on energy consumption. The model predictions indicate that for each 1 Nm^3 of oxygen per ton of steel there is

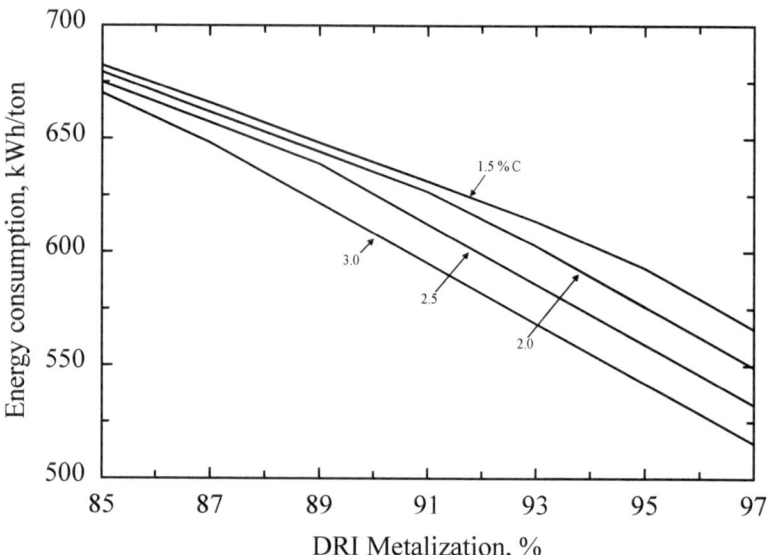

Fig. 2.8 Effect of metallization and carbon content of DRI on electric energy consumption [15, 44]

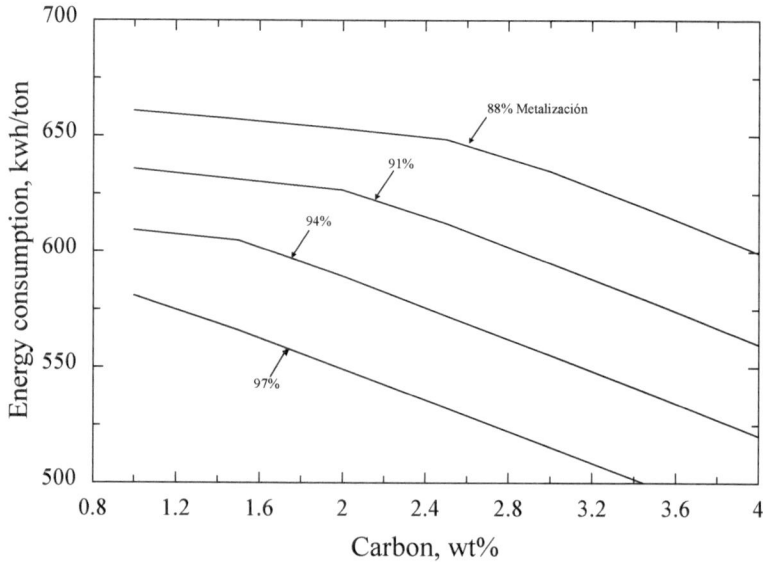

Fig. 2.9 Effect of carbon in DRI on energy consumption at different degrees of metallization [15, 44]

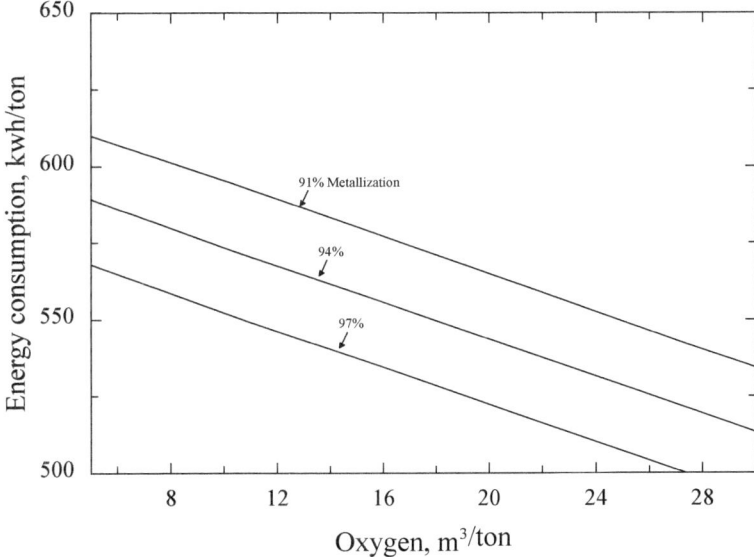

Fig. 2.10 Effect of oxygen injection on energy consumption at different degrees of metallization [15, 44]

a decrease in energy consumption of 3 kWh/ton. Oxygen injection is described in detail in Chap. 4.

Figure 2.11 shows the negative effect of the gangue content in DRI on energy consumption. The gangue content in DRI is one of the most critical aspects on DRI quality, in particular the amount of acid gangue, because its effects are multiplied. Its first negative effect is the energy for melting high melting point oxides and second, the additional energy to melt the additional fluxes required to control the slag chemistry. The model predictions indicate that increasing 1% gangue in DRI increases energy consumption by 15 kWh/ton.

Figure 2.12 reports Sankey diagrams describing total energy input and outputs using 100% DRI of 90 and 95% metallization and a thermal efficiency of 55%. The input of chemical energy is in the order of 20–25%. The heat losses in the off-gas are underestimated. CO_2 is produced only from volatile matter in the coke and calcination losses in lime, which explains the predicted low value of the off-gases and its sensible heat underestimated, from 10 to 12%.

It should be noticed that the impact factors obtained with this model result from the variation in the variables involved, in other words specific for the plant where the data was taken, however the model offers the potential to make a good sensitivity analysis of each variable, allowing to define its impact on both the production costs and energy consumption.

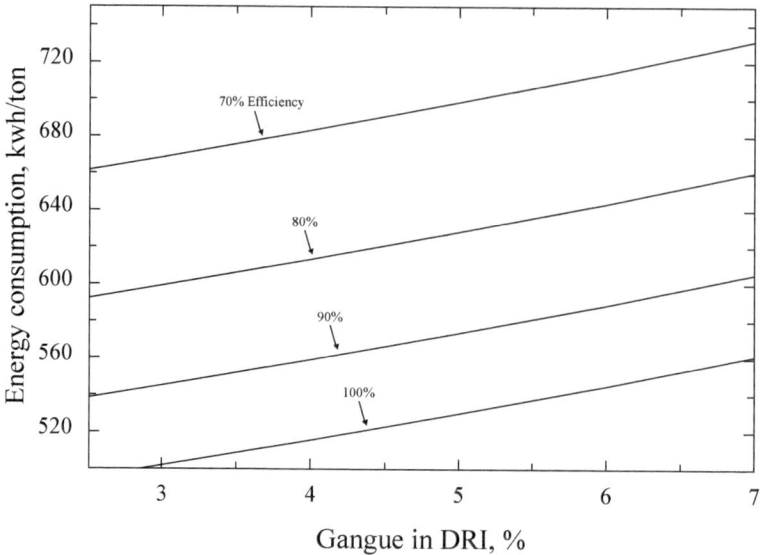

Fig. 2.11 Effect of gangue content in DRI on energy consumption at different thermal efficiencies [15, 44]

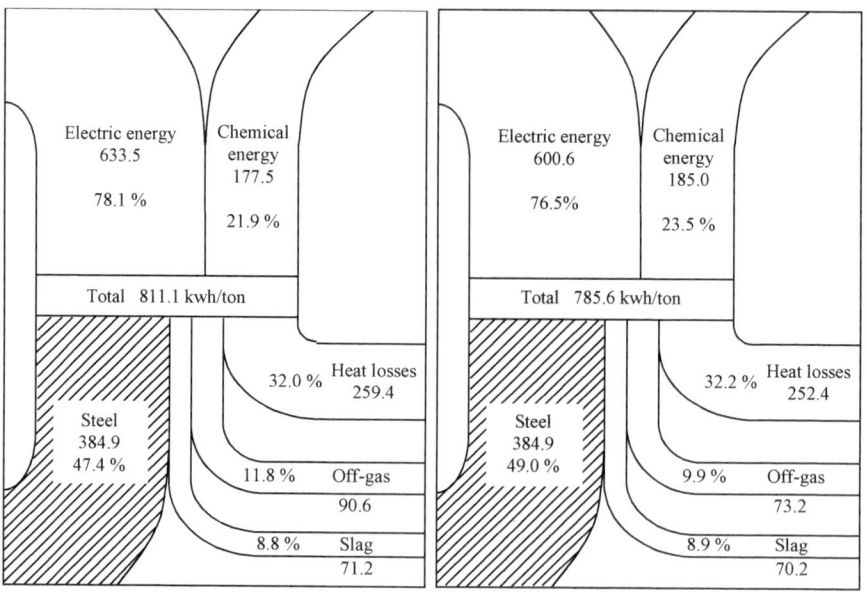

Fig. 2.12 Sankey diagrams indicating total inputs and outputs of energy for DRI of different metallization; 90% (left) and 95% (right) [15, 44]

References

1. Morris A, Geiger G, Fine HA (2011) Handbook on material and energy balance calculations in materials processing, 3rd edn
2. Prakash S, Mukherjee K, Singh S, Mehrotra SP (2007) Simulation of energy dynamics of electric furnace steelmaking using DRI. Ironmak Steelmak 34:61–70. https://doi.org/10.1179/174328107X155169
3. Madias J, Bilancieri A, Hornby S (2017) The influence of metallics and EAF design. Steel Times Int 2017:35–40
4. BP (2020) Statistical review of world energy 2020. https://www.bp.com/content/dam/bp/business-sites/en/global/corporate/pdfs/energy-economics/statistical-review/bp-stats-review-2020-full-report.pdf
5. Pfeifer H, Kirschen M (2002) Thermodynamic analysis of EAF energy efficiency and comparison with statistical model of electric energy demand. In: 7th European electrical conference, 26–29 May 2002, Venice, Italy
6. Pfeifer H, Kirschen M, Simoes JP (2005) Thermodynamic analysis of EAF electrical energy demand. In: 8th European electrical conference, 9–11 May 2005. Birmingham UK
7. Jones JAT (2002) Understanding energy use in the EAF. In: 60th electrical furnace conference, San Antonio, TX, USA, 10–13 Nov 2002, pp 141–54
8. Bowman B (1982) Optimaler Einsatz von Graphitelektroden im Lichtbogenofen. Stahl Eisen 102:1153–1158
9. Bowman B (1995) An updated model of electrode consumption
10. Prum C, Werner C, Wirling J (2005) Reducing dioxin emissions in electric steel mills. MPT Int: 36–41
11. Birat JP, Arion A, Faral M, Baronnet F, Marquaire PM, Rambaud P (2001) Abatement of organic emissions in EAF exhaust flue gas. Rev Metall 98:839–854. https://doi.org/10.1051/metal:2001132
12. Bonte L, Buttiens K, Fournelle R, Merchiers G, Pieters M (2001) A new coal injection installation for dioxin reduction at the Sidmar sinterplants. Rev Metall 98:321–326
13. Dopico M, Gómez A, Dopico M, Gómez A (2015) Review of the current state and main sources of dioxins around the world. J Air Waste Manage Assoc 65:1033–1049. https://doi.org/10.1080/10962247.2015.1058869
14. Conejo A (2014) Steel slags: characcterization and alternatives of recycling. CONAC, Monterrey NL Mexico, 23–26 Mar 2014, AIST Mexico, p 12
15. Cárdenas JGG, Conejo AN, Gnechi GG (2007) Optimization of energy consumption in electric arc furnaces operated with 100 % DRI. METAL, Hradec Nad Moravici, Czechia, 22–24 May 2007, pp 1–7
16. Simon MJ (1989) The thermal performance of water cooled panels in electric arc steelmaking furnaces. Sheffield Hallam University, UK
17. Opfermann A, Riedinger D (2008) Energy efficiency of electric arc furnace. In: 9th European electrical conference, Krakow Poland, 19–21 May 2008
18. d'Entremont JC, Englebrecht ML (1979) Computer simulated usage of direct reduced iron in electric arc furnace operations. In: ISS-AIME (ed) Ironmaking conference, vol 38, pp 279–284
19. Diaconu B, Anghelescu L, Cruceru M (2020) Analysis of energy balance for a steel electric arc Furnace. WSEAS Trans Environ Dev 16:48–56. https://doi.org/10.37394/232015.2020.16.6
20. Pearce J (1986) Development trends in EAF steelmaking. JOM 38:38–45. https://doi.org/10.1007/BF03257895
21. Hornby-Anderson S, Kempe M, Clayton J (1998) Comparison of shaft and conventional furnace combustion efficiency. In: 56th electric furnace conference on proceedings, New Orleans, LA, USA, 15–18 Nov 1998. ISS
22. Pujadas A, McCauley J, Tada Y, Mathis G, Iacuzzi M (2002) Electric arc furnace energy optimization at Nucor Yamato steel. In: 7th European electric steelmaking conference, Venice, Italy, 26–29 May 2002, pp 1.399–1.411

23. Kirschen M, Pfeifer H, Wahlers F-J (2002) Mass and energy balances of stainless steel EAF. In: 7th European electric conference, Venice, Italy, 13–15 June 2002, p 10
24. Kirschen M, Risonarta V, Pfeifer H (2009) Energy efficiency and the influence of gas burners to the energy related carbon dioxide emissions of electric arc furnaces in steel industry. Energy 34:1065–1072. https://doi.org/10.1016/j.energy.2009.04.015
25. Logar V, Dovžan D, Škrjanc I (2012) Modeling and validation of an electric arc furnace: part 2, thermo-chemistry. ISIJ Int 52:413–423. https://doi.org/10.2355/isijinternational.52.413
26. Pesamosca A, Kuran O, Olivieri L, Sellan R (2014) Heat size upgrade in Kroman Celik EAF. In: 45th ABM steelmaking seminar, Porto Alegre, Brazil. May 25–28 2014, ABM, pp 764–771
27. Thekdi A, Nimbalkar S, Keiser J, Storey J (2015) Preliminary results from electric arc furnace off-gas enthalpy modeling. In: Iron steel Technology conference exposition. Oak Ridge National Lab. (ORNL), Oak Ridge, USA, p 15
28. Yang LZ, Zhu R, Ma GH (2016) EAF gas waste heat utilization and discussion of the energy conservation and CO_2 emissions reduction. High Temp Mater Process 35:195–200. https://doi.org/10.1515/htmp-2014-0183
29. Liu Y, Wei G, Tian B (2023) Analysis and optimisation on the energy consumption of electric arc furnace steelmaking. Ironmak Steelmak: 1–15. https://doi.org/10.1080/03019233.2023.2172826
30. Arzpeyma N, Gyllenram R, Jönsson PG (2020) Development of a mass and energy balance model and its application for HBI charged EAFS. Metals 10. https://doi.org/10.3390/met10030311
31. Sims CE (1963) Electric furnace steelmaking. Volume II. Theory and fundamentals. Interscience, New York, USA
32. Madhavan N, Brooks GA, Rhamdhani MA, Rout BK, Overbosch A (2021) General heat balance for oxygen steelmaking. J Iron Steel Res Int 28:538–551. https://doi.org/10.1007/s42243-020-00491-0
33. Adams W, Alameddine S, Bowman B, Lugo N, Paege S, Stafford P (2002) Total energy consumption in arc furnaces. MPT Int: 44–50
34. Abraham S, Chen S (2007) EAF energy and material balance modeling. In: AISTech 2007 conference, pp 733–737
35. Jones J, Safe P, Wiggins B (1999) Optimization of EAF operations through offgas system analysis. In: Electric furnace conference, Pittsburgh, PA USA, 14–16 Nov 1999, pp 459–480
36. Abuluwefa HT (2017) Mass and heat balance of steelmaking and refining process using electric arc furnace. Tech J 1:53–66
37. Ameling D, Petry J, Sittard M, Ulbrich W, Wolf J (1986) Untersuchungen zur Schaumschlackenbildung im Elektrolicht bogenofen. Stahl Eisen: 625–630
38. Qi G, Shan F, Li Q, Yu J (2013) Process analysis of the electric arc furnace melting for alumina spinel production. Adv Mater Res 602–604:2104–2107. https://doi.org/10.4028/www.scientific.net/AMR.602-604.2104
39. Electric KK, Furnaces A (2005). In: Starck A, Muhlbauer A (eds) Handbook thermoprocessing technology. Vulkan-Verlag, Germany, pp 207–222
40. Kirschen M, Badr K, Pfeifer H (2011) Influence of direct reduced iron on the energy balance of the electric arc furnace in steel industry. Energy 36:6146–6155. https://doi.org/10.1016/j.energy.2011.07.050
41. Sankey HR (1898) The thermal efficiency of steam engines. Proc Inst Civ Eng 134:278–312
42. Madias J (2014) Electric furnace steelmaking. Treatise Process Metall 3:271–300. https://doi.org/10.1016/B978-0-08-096988-6.00013-4
43. Fruehan R, Fortini O, Paxton H (2000) Theoretical minimum energies to produce steel for selected conditions. US Department of Energy Office of Industrial Technology
44. Cardenas JG (2008) Analysis on the consumption of electric energy in the EAF. M.Sc. thesis, Morelia Technological Institute, Mexico (in Spanish)
45. Conejo AN, Cardenas JGG (2006) Energy consumption in the EAF with 100% DRI. AISTech Conf 1:529–535

46. Hay T, Visuri VV, Aula M, Echterhof T (2021) A review of mathematical process models for the electric arc furnace process. Steel Res Int 92:1–22. https://doi.org/10.1002/srin.202000395
47. Nyssen P, Colin R, Junqué JL, Knoops S (2004) Application of a dynamic metallurgical model to the electric arc furnace. Rev Metall Cah D'Informations Tech 101:317–326. https://doi.org/10.1051/metal:2004203
48. Kleimt B, Köhle S, Kühn R, Zisser S (2005) Application of models for electrical energy consumption to improve EAF operation and dynamic control. In: 8th European electric steelmaking congress, Birmingham, UK, pp 183–197
49. Natschlager S, Dimitrov S, Stohl K (2008) EAF process optimization: theory and real results. Arch Metall Mater 53:373–378
50. Ojeda C, Ansseau O, Nyssen P, Baumert JC, Thibaut JC, Lowry M (2015) EAF process optimization tool using CRM dynamic model. ESTAD, Düsseldorf, Germany, p 2
51. Conejo A, Morales R, Rodriguez H (2001) Mathematical modeling of the EAF process using direct reduced iron. In: 59th electric furnace conference, Phoenix, AZ, USA, 11–14 Nov 2001, pp 797–810
52. Morales RD, Rodríguez-Hernández H, Conejo AN (2001) A mathematical simulator for the EAF steelmaking process using direct reduced iron. ISIJ Int 41
53. Morales RD, Conejo AN, Rodríguez HH (2002) Process dynamics of electric arc furnace during direct reduced iron melting. Metall Mater Trans B Process Metall Mater Process Sci 33
54. MacRosty RDM, Swartz CLE (2005) Dynamic modeling of an industrial electric arc furnace. Ind Eng Chem Res 44:8067–8083. https://doi.org/10.1021/ie050101b
55. Opitz F, Treffinger P (2016) Physics-based modeling of electric operation, heat transfer, and scrap melting in an AC electric arc furnace. Metall Mater Trans B Process Metall Mater Process Sci 47:1489–1503. https://doi.org/10.1007/s11663-015-0573-x
56. Bekker JG, Craig IK, Pistorius PC (1999) Modeling and simulation of an electric arc furnace process. ISIJ Int 39:23–32. https://doi.org/10.2355/isijinternational.39.23
57. Bekker JG (1998) Modelling and control of an electric arc furnace off-gas process. University of Pretoria
58. Turkdogan ET (1996) Fundamentals of steelmaking. The Institute of Materials, London UK. https://doi.org/10.1179/095066096790151213
59. Oosthuizen DJ, Viljoen JH, Craig IK, Pistorius PC (2001) Modelling of the off-gas exit temperature and slag foam depth of an electric arc furnace. ISIJ Int 41:399–401. https://doi.org/10.2355/isijinternational.41.399
60. Matson S, Ramirez WF (1999) Optimal operation of an electric arc furnace. In: 57th electric furnace conference, Pittsburgh, PA, 14–16 Nov 1999, pp 719–728
61. Logar V, Škrjanc I (2012) Modeling and validation of the radiative heat transfer in an electric arc furnace. ISIJ Int 52:1225–1232. https://doi.org/10.2355/isijinternational.52.1225
62. Hay T, Reimann A, Echterhof T (2019) Improving the modeling of slag and steel bath chemistry in an electric arc furnace process model. Metall Mater Trans B Process Metall Mater Process Sci 50:2377–2388. https://doi.org/10.1007/s11663-019-01632-x
63. Nyssen P, Colin S, Knoops S, Junque J-L (2002) On-line EAF control with a dynamic metalurgical model. In: AIM (ed) 7th European electric steelmaking conference, Venice, Italy, 26–29 May 2002, pp 293–304
64. Ojeda C, Ansseau O, Nyssen P, Baumert JC, Thibaut JC, Lowry M (2015) EAF process optimisation tool using CRM dynamic model. In: 2nd ESTAD, Düsseldorf, 15–19 June 2015
65. Pierret J-C, Ojeda C, Nyssen P, Baumert J-C, Thibaut J-C, Lowry M et al (2020) Predictive EAF model for optimisation of melting at ArcelorMittal Lazaro Cardenas. In: Future steel forum, 8–9 Dec 2020 (Online)
66. Deb K (2005) Multi-objective optimization. In: Burke EK, Graham K (eds) Search methodology introduction tutorials optimization decision support technology, UK, pp 273–316
67. Geiger G (1980) Process engineering involved in the use of direct reduced iron. In: Stephenson RL, Smailer RM (eds) Direct reduction iron technology economics production use. Iron and Steel Society, USA, pp 149–159

Chapter 3
Predictive Models on Energy Consumption

Predicting energy consumption is a valuable tool to design metallurgical practices. Energy consumption models have several classifications, one of those classifications makes the difference between static and dynamic models. Static models predict outputs depending on the information provided into the model and in general apply an empirical approach. Dynamic models can describe the process in real time and can involve a mechanistic formulation of the transport phenomena. Static models can be more simple and easier to use but are valid for specific conditions. Dynamic models are more complex to formulate but are more general in its application.

In an even more general classification, energy consumption models can fall into the following groups; statistical models, machine learning models, optimization models and mathematical models. Statistical and machine learning models can be classified in a larger group as Multivariate Statistical Process Control (MSPC or SPC) models. All these models can also be classified into two major groups, mechanistic and empirical models. A mechanistic model describe its behavior by differential equations [1].

Statistical regression models are usually applied to obtain linear relationships. For non-linear relationships there is a relatively new approach focused on machine learning (ML) methods.

3.1 Multi Linear Models (MLR, PCA, PCR, PLS)

General description of multilinear models: MLR, PCA, PCR, PLS

- Multiple Linear Regression (MLR) is an extension of Ordinary Least Square (OLS) regression analysis. The least-squares method was developed by Legendre (1805) and Gauss (1809) [2]. MLR is the most common statistical method. It is best suited when the number of observations is higher than the number of variables.

A. N. Conejo, *Electric Arc Furnace: Methods to Decrease Energy Consumption*,
https://doi.org/10.1007/978-981-97-4053-6_3

105

In the following discussion, the dependent variable (y) is energy consumption as a function of a number of independent variables (x). Each sample is defined by the following function:

$$y = \sum a_0 + a_i x_i + e$$

where e is the absolute error.

The equation matrix $(X)(b) = (Y)$ is defined by minimization of the absolute error. MLR seeks to find a single factor for each variable that best correlates with the predicted variable.

- Principal Component Analysis (PCA): Finds new variables that are linear functions of those in the original dataset that successively maximize variance. Once the new variables are found the solution reduces to solve an eigenvalue/eigenvector problem [3].
- Principal Component Regression (PCR): Finds factors that capture the greatest amount of variance in the predictor variables
- Partial Least Square regression (PLS): Finds the fundamental relations between two matrices (X and Y) that maximizes variance. It is best suited when there are more variables than observations.

Performance indicators of statistical models: Good performance indicates that the differences between the experimental values and the model's predicted values are small. The performance of statistical models can be measured in terms of several statistical parameters; (i) The coefficient of multiple determination (R^2) gives the percentage variation in "y" explained by x-variables. In general, as R^2 increases the predicted value is close to the actual value, therefore $R^2 \cong 1$ is usually desired, however its value can be artificially increased adding excess independent variables or due to collinearity. Collinearity is a condition in which some of the independent variables are highly correlated. Therefore, R^2 itself cannot explain the accuracy of a regression model [4]. The adjusted R^2 compensates for the addition of variables and only increases if the new predictor enhances the model above what would be obtained by probability. (ii) The Root Mean Square Error (RMSE) measures the deviations from the true value. It is desired that RMSE $\cong 0$. (iii) The Mean Absolute Error (MAE) is an arithmetic average of the absolute errors. Chai and Draxler [5] compared RMSE and MAE arguing that a combinations of metrics is better to evaluate statistical models. (iv) The Pearson correlation coefficient (r_{xy}) measures the extent of linear correlation between two variables with values between $+ 1$ to $- 1$, where $+ 1$ represents total positive linear correlation. It works well only if the data are linearly associated.

Several new approaches to measure dependencies among variables have been proposed [6, 7]. Szekely et al. [6] in 2005 reported an improved model, called distance correlation (dCor). When dCor $= 0$ indicates that the variables x and y are truly independent. Reshef et al. [7] proposed a model based on information theory called Maximal Information Coefficient (MIC). De Siqueira Santos et al. [8] compared six methods and concluded that its selection depends on the type of data and the number of observations. (v) Standard Error of the Mean (SEM) measures the deviations from the mean, defined as the ratio $\sigma/\sqrt{n}$, where σ is the standard deviation.

Importance of input variables in MLR: The objective of a multiple linear regression statistical model is to define a functional relationship between the dependent variable (electric energy consumption) and a number of independent variables. The selection of the input variables (also called predictors) is critical to improve the performance of the model. Including additional variables can never decrease the value of R^2 [9], on the contrary, sometimes adding irrelevant independent variables can even increase R^2 because of random effects but in this case the model's predictability will be distorted if applied to a different set of data. Its selection should be made based on experience and strong knowledge on the subject [10].

Table 3.1 shows 24 variables that affect energy consumption in the EAF. The total number of variables can be further increased considering the following additional variables: Additional materials: HBI, coke, lime, dolomite; number of scrap buckets; steel grades; Delays can be due to a large number of factors, as shown in [11], reporting the following 11 sources: (i) reports on chemical analysis, (ii) temperature measurements, (iii) peak electricity tariff, (iv) changes of steel grade, (v) tapping problems, (vi) adjust chemistry before tapping for special grades of steel, (vii) adjust tapping temperature, (viii) scrap charging, (ix) electrode replacements, (x) lining maintenance, (xi) other delays; Energy losses (slag, water cooling, de-dusting system, off-gas), etc. The current MLR models commonly employ from 6 to 16 input variables [12].

Table 3.1 Variables that affect energy consumption in the EAF

	Variable	Units		Variable	Units
1	Energy consumption	kWh/ton	13	DRI metallization	%
2	Weight of scrap	Tons/tonls	14	DRI carbon content	%
3	Weight of DRI	Tons/tonls	15	DRI gangue content	%
4	Weight of hot metal	Tons/tonls	16	Weight of liquid steel	Ton
5	Fluxes	Tons/tonls	17	Weight of slag	Ton
6	Power-on time	min	18	Scrap density	Ton/m^3
7	Power-off time	min	19	Hot heel	Ton
8	Tap-to-tap time	min	20	Slag foaming operation	kg C/ton
9	Delays	min	21	Continuous/disc operation	–
10	Tapping temperature	°C	22	Energy losses	kWh/ton
11	Oxygen injection	Nm^3/tonls	23	EAF stirring	Nm^3/tonls
12	Burner fuels	Nm^3/tonls	24	Preheating metallics	°C

Tonls means tons of liquid steel

Two important limitations of MLR models are the range of application and its empirical nature. In principle, the functional relationships reported should be applied to similar operational conditions, therefore, this information should be reported for each investigation. Both limitations indicate that they are expected to be valid for individual steel plants.

Importance of sample size and its quality in MLR: The minimum number of samples for MLR has been defined as a function of the number of input variables, by a large number of conventional rules in the form: sample size $\geq$ C $\times$ number of input variables, where C ranges from 2 to 30 [13, 14]. The sample size should be large enough to be representative of the population, therefore the larger, the better. However, it is not only the number of samples but data quality which is also important. Since the primary objective is on the energy consumption per ton of liquid steel, the number of heats/year can be used as the total population of samples. This number can fluctuate in a broad range, from 1×10^3 to 20×10^3, depending on the number of EAF's in the steel shop and their nominal capacity. Due to the large scale of this population and the large number of factors that affect the quality of the data, it is important to eliminate heats with wrong data. Data filtering could be even more than 50%. At industrial scale the following factors can be used as criteria to eliminate wrong data: (1) unusual long delays affect other variables such as energy consumption, tap-to-tap time, etc., (2) Bad calibration in the measurement systems such as temperature, weighing systems, chemical composition, etc., (3) abnormal weight of liquid steel, for example heats close to the peak hour (4) abnormal energy consumption, etc.

Discussion of relevant statistical models: Carlsson et al. [12] reported a general review about statistical models used to predict energy consumption in the EAF. They reported 23 models collected from the last 30 years, involving 10 groups of researchers, which means that one group compared more than one model. Most of these groups have focused on MLR models. The number of heats and input variables included in the models is diverse but in general, the number of samples and input variables in machine learning (ML) models is higher than for MLR models. The number of heats included ranged from 60 to 4,000 in MLR models and 400–20,000 for ML models. The input variables ranged from 6 to 16 for MLR models and 5–127 for ML models. One important parameter is the correlation coefficient (R^2) that was reported in some of those models and varied in a large range from 0.3 to 0.96. It was also observed that MLR models are transparent in contrast to ML models that are a black box. In those models, the main iron unit in the raw materials was scrap, few works have been reported for the cases involving DRI and/or hot metal.

Köhle et al. [15–21] published a series of regression equations from 1992 to 2005. The first regression equation from 1992 was based on data from Alternating Current (AC) scrap-based EAF's. It was a small data base including 14 EAF's and the mean values of 6 input variables (ton scrap/tonls, ton fluxes/tonls, tapping temperature, tap-to-tap time, methane and oxygen consumption). Scrap was not preheated. Half of the heats were charged by two scrap baskets. Using the mean values of the 14 EAF's they analyzed the effect of energy consumption on electrode consumption, confirming a decrease in electrode consumption when energy consumption is also decreased [16]. Bowman [22] applied the first regression equation reported by Köhle et al. to compare 11 AC and 11 DC furnaces and reported a new expression whose only change was a small difference in one constant but the standard deviation was higher. One conclusion from this comparison was the important finding that energy consumption in AC and DC EAF's, was similar. The second regression equation from Köhle et al. [17] in 1997 included post-combustion and data involving mean values taken from 7 EAF's, with and without post-combustion. The coefficient for post-combustion in the regression equation was 2.8 kWh/Nm^3. Comparing with the theoretical heat released from combustion between CO at 1600 °C and O$_2$ at 25 °C, equal to 6.23 kWh/Nm^3, would indicate a heat transfer efficiency of 42%. In all the cases reported, electric energy consumption was always lower applying post-combustion. The third regression equation from 1999 added four additional variables to account for Direct Reduced Iron (DRI), hot metal, power-on time and power-off time, additionally a constant term was introduced to account for continuous of discontinuous operation. This is the first regression equation to account for all the main sources of metallics in the EAF, however, from the global database that included annual averages of 35 EAF's, only in four of them DRI was partially employed, from 10 to 40% [18]. The fourth and last regression equation from Köhle et al. resulted from application of the previous ones to single heats in 4 steel plants [19, 20] including the addition of measurements on heat losses and the weight of all ferrous materials. Its application resulted in major changes of some coefficients. The accuracy of the regression equation was low in general and varied from plant to plant.

Because of the empirical nature of statistical models, one model cannot accurately predict heats for multiple furnaces of varying designs, metallic charge and operational conditions [12]. Table 3.2 summarizes a large number of regression equations together with R^2 and σ values. Notice the magnitude and sign of the coefficients. A positive value indicates that the variable increases electric energy consumption and vice versa. Its magnitude is the impact factor of each variable, for example, the coefficient 80 in Kohle's equation indicates that 1% of pre-reduced materials increases energy consumption 0.8 kWh/ton. From all the regression equations reported by Köhle et al., the first one showed the highest prediction capabilities. This is due to the small dataset and use of average annual values. They eliminated one data point to report the standard deviation as 5.1 kWh/ton. The standard deviation in the regression equation using post combustion increased two-fold and five-fold when DRI was added as variable. In the last case, one reason was probably the lack of information on the chemical composition of DRI. The performance of the last regression equation was poor when it was applied to single heats in different steel plants, with R^2 values less than 0.8 and an average σ value of 25 kWh/ton. If the formula is applied to the same plant but comparing monthly average values and values from single heats, the accuracy is higher using mean values. Table 3.3 summarizes the range of the variables in the models already described.

UCAR reported several regression equations for the total energy consumption [23–25], also shown in Table 3.2. The signs in the terms that contribute with chemical energy are positive.

Table 3.2 Statistical regression equations that predict electric energy consumption in the EAF

	Year	Authors	R^2	σ	Regression equation
1	1992	Köhle	0.98	8.1	$\dfrac{kWh}{ton} = 300 + 900\left(\dfrac{W_{Fe}}{W_{ls}} - 1\right) + 1600\left(\dfrac{W_{flux}}{W_{ls}}\right)$ $+0.7(T_{ls} - 1600) + 0.85t_{tt} - 8Q_{ng} - 4.3Q_{O2}$
2	1995	Bowman	–	11.2	$\dfrac{kWh}{ton} = 300 + 900\left(\dfrac{W_{Fe}}{W_{ls}} - 1\right) + 1600\left(\dfrac{W_{flux}}{W_{ls}}\right)$ $+0.7(T_{ls} - 1600) + 0.85t_{tt} - 10Q_{ng} - 4.9Q_{O2}$
3	1997	Köhle	0.89	17.9	$\dfrac{kWh}{ton} = 300 + 900\left(\dfrac{W_{Fe}}{W_{ls}} - 1\right) + 1600\left(\dfrac{W_{flux}}{W_{ls}}\right)$ $+0.7(T_{TAP} - 1600) + 0.85t_{tt} - 8Q_{ng} - 4.3Q_{O2} - 2.8Q_{PC}$
4	1999	Köhle	0.76	40	$\dfrac{kWh}{ton} = 300 + 900\left(\dfrac{W_{Fe}}{W_{ls}} - 1\right) + 80\left(\dfrac{W_{DRI}+W_{HBI}}{W_{ls}}\right) + 0.7(T_{TAP} - 1600)$ $-300\left(\dfrac{W_{HM}}{W_{ls}}\right) + 1600\left(\dfrac{W_{flux}}{W_{ls}}\right)$ $+0.85(t_{tt} + t_{off}) - 8Q_{ng} - 4.3Q_{O2}$ $-2.8Q_{PC} - 15CON$
5	2002	Köhle	0.1–0.6	15–29	$\dfrac{kWh}{ton} = 375 + 400\left(\dfrac{W_{Fe}}{W_{ls}} - 1\right) + 80\left(\dfrac{W_{DRI}+W_{HBI}}{W_{ls}}\right)$ $-50\left(\dfrac{W_{shs}}{W_{ls}}\right) - 350\left(\dfrac{W_{HM}}{W_{ls}}\right) + 1000\left(\dfrac{W_{flux}}{W_{ls}}\right)$ $+0.3(T_{TAP} - 1600) + 1(t_{tt} + t_{off}) - 8Q_{ng}$ $-4.3Q_{O2} - 2.8Q_{PC} - K(H_{loss} - \overline{H}_{loss})$

(continued)

Table 3.2 (continued)

	Year	Authors	R^2	σ	Regression equation
6	2000	Bowman			$E_{total} = \frac{kWh}{ton} + 127\left(\frac{W_{PI}}{W_{Fe}}\right) - 100\left(\frac{W_{DRI/HBI}}{W_{Fe}}\right) + 4Q_{O2} + 10.5Q_{ng}$
7	2001	Adams			$E_{total} = \frac{kWh}{ton} + 110\left(\frac{W_{PI}}{W_{Fe}}\right) - 100\left(\frac{W_{DRI/HBI}}{W_{Fe}}\right)$ $+ 450\left(\frac{W_{HM}}{W_{Fe}}\right) + 10.5Q_{ng} + 5.2(Q_{O2} - 2Q_{ng})$
8	2002	Adams		50	$E_{total} =$ $\frac{kWh}{ton} + 110\left(\frac{W_{PI}}{W_{Fe}}\right) - 100\left(\frac{W_{DRI/HBI}}{W_{Fe}}\right) + 450\left(\frac{W_{HM}}{W_{Fe}}\right) + 10.5Q_{ng} + 11Q_{oil} + 8Q_{LPG} + 5.2\left(Q_{O2} - 2Q_{ng} - 2Q_{oil} - 1.5Q_{LPG}\right)$
9	2003	Baumert			$E_{total} = \frac{kWh}{ton} + (9.14 \times 0.88)(C_{ch} + C_{inj})$ $+ (9.14 \times 0.99)C_{el} + (9.14 \times 0.04)W_{PI}$ $+ \left(0.007 \times 9.43\frac{kWh}{kgSi}\right)W_{PI} + (10.25)(Q_{ng})$ $+ 6.69\{Q_{O2} + Q_{PC} + Q_{ng} - 4 - (22.4/12)[(0.88)(C_{ch} + C_{inj}) + (0.99)C_{el} + (0.04)W_{PI}] - 2Q_{ng}\}$
10	2006	Conejo	0.92	18	$\frac{kWh}{ton} = 42.27 + 384.3\left(\frac{W_{DRI}}{W_{ls}}\right) + 1303\left(\frac{W_{flux}}{W_{ls}}\right)$ $+ 0.218(T_{TAP} - 1600) + 2.05t_{tt} + 0.135t_{off} + 0.001t_d$ $- 2.69Q_{O2} - 0.391M_{DRI} - 17.8C_{DRI} + 4.296G_{DRI}$

(continued)

Table 3.2 (continued)

	Year	Authors	R^2	σ	Regression equation
11	2017	Kirschen			$\frac{kWh}{ton} = -152.5 + 4.21(W_{Fe}) - 5.08(W_{ls})$ $- 1.44(W_{DRI}) - 1.3(W_{CDRI}) - 1.97(W_{scr})$ $+ 3.09\left(\frac{W_{ls}}{W_{Fe}}\right) + 0.48t_{tt} + 4.86t_{on}$ $- 0.468t_{off} - 0.319t_p - 0.004V_{O2} + 6.87C_{ch}$ $+ 4.79C_{inj} + 3.95W_{CaO} + 1.36W_{Dol} + 3.87P_{avg}$
12	2019	Elkoumy	0.82		$\left(\frac{kWh}{ton}\right)_{refining} = -1.77 + 8.8t_{on} + 1.07\,Q_{O2}$ $- 0.33C_{inj} + 0.404W_{CaO}$
13	2022	Moskal			$\frac{kWh}{ton} = 169.9 + 48.79\left(\frac{kg\,HM}{kg_{scr}}\right) + 3.27\left(\frac{kg\,O_2}{kg_{scr}}\right)$ $- 1.98\left(\frac{kg\,CH_4}{kg_{scr}}\right) + 0.11\left(\frac{kg_{flux}}{kg_{scr}}\right) + 0.26t_{tt}$

where: E_{total} is the total energy consumption per ton of liquid steel in kWh/ton; kWh/ton is the electric energy consumption, W_{Fe}, W_{scrap}, W_{shs} W_{DRI}, W_{HBI}, W_{HM}, W_{PI}, represent weight of all metallics, scrap, shredded scrap, DRI, HBI, hot metal and pig iron, respectively, in ton; W_{ls} is the weight of liquid steel before tapping in ton, W_{flux} is the weight of fluxes in ton, T_{TAP} is the tapping temperature, in °C, Q_{ng} is the specific consumption of natural gas by the burners, in Nm³/tonls, Q_{PC} is the specific consumption of natural by the burners for post-combustion, in Nm³/tonls, Q_{oil} is the specific consumption of fuel oil by the burners, Q_{LPG} is the specific consumption of liquid propane by the burners, Q_{O2} is the specific consumption of oxygen injection, in Nm³/tonls, Q_{O2T} is the volume of oxygen injection in m³, is the specific consumption of oxygen for post-combustion, in Nm³/tonls, t_{tt} is the tap-to-tap time in min, t_{on} is the power-on time in min, t_{off} is the power-off time, t_p is the preparation time in min, t_d is the delays time in mins, P_{avg} is the average power in MW, C_{ch}, C_{inj} and C_{el} is carbon in the charge, injected and C from the electrodes, respectively, in kg/ton; M_{DRI}, C_{DRI} and G_{DRI} represent metallization, carbon and gangue content in DRI in %. CON is a dummy variable with values $+1$ or -1 for continuous and discontinuous operation, respectively. K is a constant (0.2–0.4), H_{loss} and $\overline{H}_{loss}$ represent measured and average heat losses, in kWh/ton

Table 3.3 Range of the variables in the models proposed by Köhle et al. [15–21]

		1992	1997	2002
1	Energy consumption, kWh/ton	375–600	329–478	580–680
2	Weight of liquid steel, ton	64–147	60–95	210
3	Scrap, ton/tonls	1.05–1.23	1.07–1.17	–
4	Direct reduced iron, ton/tonls	–	–	1.16
5	Hot metal, ton/tonls	–	–	–
6	Fluxes, ton/tonls	0.02–0.05	0.03–0.04	0.05
7	Tapping temperature, °C	1610–1725	1625–1660	1671
8	Tap-to-tap time, min	52–140	41–65	81
9	Power on time, min			57
10	Power off-time, min			24
11	Delays, time			46
12	Natural gas (burners), Nm^3/tonls	0.6–6	0–12	–
13	Oxygen injection, Nm^3/tonls	2.4–28	9.2–25	22
14	Post-combustion O_2, Nm^3/tonls		0–23.6	–
15	DRI metallization, %			94
16	DRI carbon, %			2.3
17	DRI gangue, %			5.5

The original terms for the metallic charge involved units in %. They were changed to weight units, for example; % DRI $= \frac{W_{DRI}}{W_{Fe}} 100$. The coefficients on these terms were estimated from thermodynamic calculations [24, 25]. The coefficient associated with oxygen injection was based on a standard case, considering that for stoichiometric combustion, about 2.0 m^3 O_2 is required for each Nm^3 of natural gas. For hot metal the largest energy contribution is due to sensible heat, followed by oxidation of Si, Mn and C. The coefficient in the term for DRI, -1.0 kWh/t per 1% replacement of scrap, was calculated based on the assumption of high quality DRI; 92–93% metallization and 4–5% gangue. For lower quality DRI that value can be up to -2.0 kWh/t.

Moskal et al. [26] compared Köhle's models to an industrial database and found a large discrepancy, from 10 to 50%. They reported a new relationship using MLR, however such relationship is wrong because in principle, both oxygen and hot metal increased energy consumption, which is not true.

Baumert et al. [27] reported a regression equation developed for conditions where scrap was the dominant raw material, however statistical indicators to define its performance were not provided.

In order to provide scientific basis to the previous empirical regression equations, in particular those from Köhle et al., a research group from Aachen University explained the value of the coefficients on the basis of the first law of thermodynamics [28, 29]. Their results are summarized as follows; (i) they applied empirical efficiency values on heat transfer; for example, the electric efficiency (η_{el}) in the power system was taken from 90 to 95%, the efficiency of heat transfer from the arc (η_{arc}) largely depends on the foaming conditions and can vary from 36 to 93%. The authors choose to define the total efficiency factor, $\eta_{el}\eta_{arc}$, to range from 60 to 80%, assuming foaming conditions from partial coverage to full coverage of the arc, representing the common practice. This efficiency range has a high impact on the final results of their thermodynamic calculations. (ii) The constant term (+ 375) in Kohle (2002) was related to the minimum energy requirements for melting regular scrap. The coefficient for total metallics was related to the enthalpy in the dust (10–25 kg/ton), assuming that 80–90% of this dust comes from the non-ferrous metals in the metallic charge. Shredded scrap has a negative coefficient because charging this material with respect to regular scrap helps to decrease energy consumption. Hot metal contributes with sensible heat which results from the oxidation of Si, Mn and C. (iii) the coefficients for the tapping temperature and tap-to-tap time were derived from individual relationships with energy consumption. These authors have made an important observation indicating that Köhle's equations cannot be applied to single heats because of the large differences in the values of the variables in different plants, however, if mean values are used, for example monthly values, their prediction capability can be satisfactory.

In the previous regression equations the effect DRI on energy consumption has been accounted, however the amount of data on DRI used in the regression analysis was extremely limited; out of the annual averages of 35 EAF's, only in four of them DRI was partially employed, from 10 to 40% [18], furthermore, its chemical and physical composition was not included. For practical purposes this is good enough considering that worldwide steel scrap is the main source of metallics, however several steel plants operate with 100% DRI, especially in countries with large availability of natural gas. In order to define a regression equation valid for 100% DRI, Conejo et al. [30] reported an equation based on 1122 heats, after a filtering process that removed almost 50% of the initial heats due to extremely large delays, short heats, peak hour, etc. The large number of parameters that affect the quality of the data at the industrial scale has been described before, therefore it should not be a surprise to find this level of data filtering. Figure 3.1 compares the experimental data and the calculated values. The regression equation, from this work is shown in Table 3.2, indicating a high prediction capability. Its limitation however is that is valid for steel shops that operate under similar conditions and use similar DRI quality. Cold DRI was employed with a high metallization and on average 2.3%C. Electric energy consumption in most of the heats ranged from 580 to 680 kWh/ton. In the analysis a few heats with energy consumption below 500 kWh/ton were included because it helped to improve the R^2 value. The EAF was AC type, with a nominal capacity of 220 ton. Since the main source of metallics is DRI, burners were not needed and post-combustion trials didn't provide important benefits. Metallic yield was about 86%.

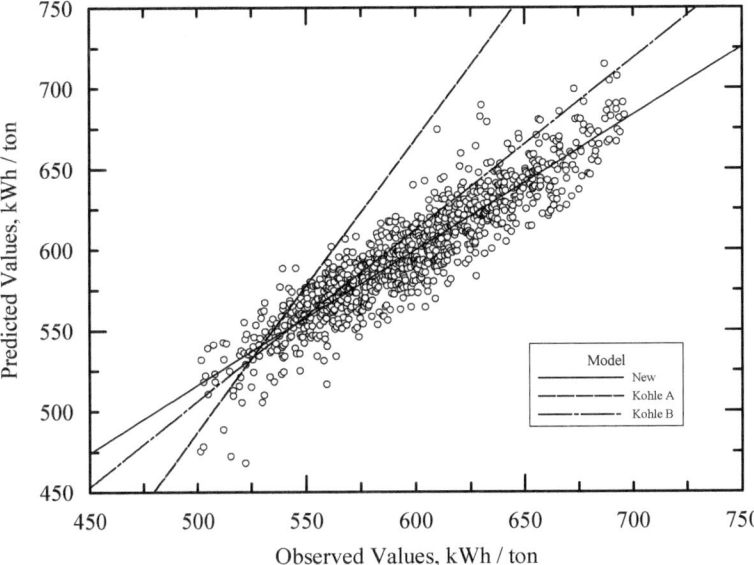

Fig. 3.1 Prediction of energy consumption using 100% DRI [30]

Elkoumy et al. [31, 32] reported an expression for the electric energy consumed during the refining period, once the metallic charge is fully melted, also included in Table 3.2. This expression was obtained from 100 heats in an EAF with a nominal capacity of 225 tons, employing DRI in the metallic charge. They estimated the electric energy in the refining period is about 4–12% of the total energy consumed.

One of the most valuable findings from the regression equations is the meaning of each coefficient. Its magnitude defines the impact factor of each one of the variables involved and therefore can be used as a criterion to decide the quality of raw materials and design of metallurgical practices. Table 3.4 summarizes impact factors from different regression equations. There is good agreement on the impact factors for fluxes, oxygen injection, natural gas injection and hot metal. Comparing with the latest equation from Köhle et al. [19], there is also good agreement with the coefficients for metallic yield and superheat, however it should be clear that not all the coefficients from Köhle and Conejo can be compared because one case is fundamentally for scrap-based operations and the other one for DRI-based operations, therefore the magnitude of the constant term and the coefficient for DRI have different meaning on their corresponding equations. Furthermore, since the expression developed by Conejo only included heats with 100% DRI, the first term doesn't represent the effect of increasing 1% DRI.

Hot Briquetted Iron (HBI) in the electric arc furnace is a common replacement of DRI. In order to safely transport DRI it is compressed to decrease its large porosity. Arzpeyma et al. [33] reported a mass and energy balance using HBI in the EAF. For each 1% increase of HBI, they found the following effects: (i) A decrease on oxygen

Table 3.4 Impact factors of different variables on energy consumption, increment or decrement in kWh/ton

	Kohle	Bowman	Adams	Conejo	Pfeifer
Yield, 1%	+ 4, + 9			+ 3.84	
DRI, 1%	+ 0.8	+ 1	+ 1	(+ 3.84)*	− 0.7 to + 0.8
Fluxes, 1%	+ 10 to + 16			+ 13	− 5.5 to + 7.3
1 Nm^3 O$_2$/ton	− 4.3	−.0	− 5.2	− 2.688	− 4.3 to − 6.9
1 Nm^3 natural gas	− 8.0	− 10.0	− 10.5		− 6.9 to − 11.5
1 min tap-to-tap	+ 0.85			+ 2.18	+ 0.4 to + 1.7
Hot metal, 1%	− 3 to − 3.5	− 4.5	− 4.5		− 3.2 to − 5.1
Shreded Scrap, 1%	- 0.5				− 0.32 to − 0.82
Pig iron, 1%		− 1.27	− 1.1		
1 °C superheat	+ 0.3 to + 0.7			+ 0.218	+ 0.29 to + 0.38
0.1%C$_{DRI}$				− 1.785	
1% M$_{DRI}$				− 0.391	
1% G$_{DRI}$				+ 4.296	

* Expression obtained for heats with 100% DRI.

consumption of 0.16 Nm^3/ton, and (ii) An increase in energy consumption of 1.29 kWh/ton.

The latest developments on MLR models have been reported by Kirschen et al. [34–36]. They have adapted Kohle's regression model to different EAF's including bottom argon stirring. They have reported that Kohle's regression model can be used if the coefficients are adjusted. In fact, they report drastic changes in some of the coefficients depending on the operational conditions. The equation reported in Table 3.2 includes the effect of hot DRI, carbon charged and injected as well as the average active power. Although a term to evaluate argon stirring was not included, they reported a minimum decrease in energy of consumption from 3 to 10 kWh/ton.

Partial Least Square regression (PLS): Sandberg et al. [37] applied Principal Component Analysis (PCA) and Partial Least Square regression (PLS) to study different variables that reduce energy consumption. The description of the models was very general and it appears that only in the case of post-combustion they obtained satisfactory results, suggesting optimum effects injecting the equivalent of 7 kWh/Nm^3 during the first scrap basket and 1.5 kWh/Nm^3 during the second scrap basket. In subsequent studies they applied PLS to study EAF energy consumption from two steel plants [38–40]. The data base consisted of 7553 heats. Data filtering was less than 20%. Some of the variables included the grade of steel, weight of hot heel and the number of scrap buckets. The values of R^2 and σ for the two plants varied from 0.46 to 0.50 and 8–12 kWh/ton, respectively. One of the causes of the low value of R^2 was attributed to variations in the weight of hot heel [37].

3.2 Machine Learning (ML) Models

General description of machine learning models: ML is a statistical tool but is more generally recognized as a branch of Artificial Intelligence (AI). Machine learning algorithms are programs that can learn from data and improve from experience, without human intervention. Banzhaf et al. [41] cited the definition from R. Friedberg as follows: In ML, a machine should "perform a task event though we could not describe a precise method for performing it…in short, although it might learn to perform a task without being told precisely how to perform it, it would still have to be told precisely how to learn".

The number of models or algorithms of ML is huge but only about 5 of them have been applied to study EAF energy consumption. The main advantage of ML models is its capability to study non-linear relationships among the variables. The fact that MLR models show high R^2 is an indication that most of the variables that affect energy consumption follow a linear relationship, however it is important to mention that MLR models include a relatively small number of variables (< 20) and only a few of them exhibit non-linear behavior, however if the number of variables is increased (< 100) the possibility to include non-linear relationships increases. The amount of data in ML models should be high to obtain a good training algorithm. Since this is a specialized field for computer programmers, the detailed explanation of the models is out of the scope of this review, given only in a general way.

The following is a brief description of some of the ML algorithms:

- Artificial Neural Networks (ANN): Is a self-learning computing system. The processing units are called neurons which are interconnected by nodes. The neural network attempts to learn about the information presented, based on an internal weighting system to produce one output report. Backwards propagation of error successively improves their output results. The first layer is the input layer and the last layer is the output layer. The intermediate layers are the hidden layers. The method is also called multi-layer perceptrons.
- Deep Neural Network (DNN): A Deep Neural Network is an artificial neural network with multiple hidden layers.
- Support Vector Machine (SVM): Is a machine learning algorithm that analyzes data for classification and regression analysis. It gives one global solution.
- Decision tree (DT): Is a supervised learning algorithm to create a training model that can be used to predict the value of the target variable. The tree defines a hierarchical set of rules that form the internal nodes.
- Random Forest (RF): Consists of a multitude of decision trees.

Big data: Sagiroglu and Sinanc [42] provided the dramatic increment in information being generated every day by human activities that would require about 20 billion PC's to store all of the world's data for the information available in 2012. That information would double every two years. Big data is concerned with the analysis of huge databases, both big data and AI complement each other. With the existence of huge databases, it becomes necessary to apply algorithms to identify wrong data (outliers).

Case studies using ML: Baumert et al. [43–45] in 2002 were pioneers in the application of ANN to study energy consumption in the EAF. The motivation to introduce ANN was based on non-linear interactions between some input variables and energy consumption, such as burner gas consumption, post-combustion energy, energy losses, etc. They included 10 variables including burner pre-heating energy for the first and second scrap baskets, post-combustion energy and energy losses (water cooling and fumes). Initially they applied a static ANN model and compared the results with Kohle's MLR model. The standard deviation with MLR was 6.4 MWh/heat and 1.9 MWh/heat with the static ANN model, approximately 40 and 12 kWh/ton, respectively, clearly indicating the advantages of ANN. Further improvements were added using dynamic ANN. In this case the number of variables was extended by a factor of 10, including measurements of those variables over time, four times during the first scrap basket and six times during the second basket. Applying the time-series ANN model the standard deviation was 1.3 MWh/heat.

Gajic et al. [46] investigated the influence of scrap chemical composition on energy consumption, keeping relatively constant the other variables from a set of 46 heats. They used the software statistica vs. 8.0 to develop the ANN model, applying a Multilayer perceptron (MLP) algorithm. The optimal MLP architecture consisted of 5 inputs, 5 neurons in the hidden layer and 1 output (5-5-1) with $R^2 = 0.91$. The main disadvantage of this model was their dependence on the accuracy of the measured scrap chemical composition, usually low due to their large heterogeneity. Haupt et al. [47] evaluated the effect of scrap quality on energy consumption. Scrap was preheated in a finger-shaft EAF. Heats with a weight of liquid steel out of the range 78–90 tons were not included. Their results indicated a strong effect of scrap quality on energy consumption; variations of about 40% comparing low-quality scrap (HMS) with shredded scrap.

Kovacic et al. [11] compared MLR and Genetic Programming (GP) models but a comparison between predicted and actual values on energy consumption was not reported. The MLR model had 26 variables (including a large number of different types of delays). The equation reported had a low R^2 value of 0.63. The sample size consisted of 3248 heats. The accuracy of the large number of different delays employed as variables is hard to evaluate in a steel plant. Chen et al. [48] compared MLR with several data mining algorithms (DNN, SDT, SVM). The algorithm Keras in the commercial code Python was used to develop the model. Based on the performance indicators, their results indicate that DNN has superior performance.

Carlsson et al. [49] applied ANN, including additional statistical modeling tools to assess model performance, in particular to assess the relevance of the input variables. The modeling tools were Feature Importance (FI), Kolmogorov–Smirnov (KS) test and Distance Correlation (dCor). A higher FI value indicates a stronger effect of the input variable on the output variable. These authors made an important observation; The value of the output variable in ML models is a value based on maximum probability, the physicochemical relation between variables is not relevant, therefore, a variable that remains almost constant can be excluded. Their models were based on input data from 35 variables, from an AC-EAF charged with scrap, including scrap chemistry for 8 types of scrap. One strategy to remove data was based on values $\pm 3\sigma$ for variables normally distributed. A total of 20,160 model types were trained with different number of variables. Model performance was based on the values of the adjusted R^2. They found that increasing the number of input variables or the number or hidden layers does not automatically increase the performance of the model. They found the following variables have a strong effect on energy consumption (high FI values); total weight of materials, metal weight, delays and process time.

The reliable prediction of the temperature of the molten bath is critical to decrease energy consumption in the EAF and decrease the superheat as much as possible. ML models have also been used to predict the temperature of liquid steel [50–52].

Comparison of MLR and ML models: Several investigations have compared MLR and ML models [53–55]. Reimann et al. [53] compared MLR and two ML models (ANN and GPR). The highest R^2 value corresponded to Gaussian Process Regression (GPR), indicating that the quality of the measured data has a big influence on this value especially if the number of variables increases. WEKA is an open-source software developed by Waiko University in New Zealand that contains several ML algorithms. León et al. [54] applied this software to predict energy consumption in the EAF comparing four types of ML algorithms and MLR. The initial data set consisted of 3493 heats and 21 variables and after pre-processing and debugging was reduced to 2328 heats and 17 variables. Missing data were filled out with average values. The highest correlation coefficient was obtained with Decision Tree (DT) models (M5P and J48); 0.87 and 0.84, respectively, followed by ANN, MLR and Decision Table (DT) with 0.68, 0.63 and 0.60, respectively.

Tomazic et al. [55] compared three ML methods (k-NN, Evolving and conventional fuzzy modelling) and MLR, employing a small data base with 577 heats and only 13 variables. Their results indicate better predictability with the fuzzy model, however the range in all four cases for the coefficient of determination was low, in the range from 0.44 to 0.53.

3.3 Optimization Models

Larsson and Dahl [56] studied in a macro-scale the reduction in energy consumption involving all major processes of an integrated steel plant with a commercial optimization solver (ILOG CPLEX 7.1). They compared several optimization cases; individual processes and the whole system, and found that the optimum solution was achieved when the whole system is considered. This study reported several ways to decrease energy consumption; increase PCI and DRI in the blast furnace, increase pre-heated scrap in the BOF, etc. The integral approach suggested by the authors is probably the best way, but it depends on the objective function. In this case they choose to decrease energy and obviously this can lead to situations that are unfeasible, for example the use of high-quality raw materials that would increase the production costs. In this direction, a similar case was reported by Riesbeck et al. [57]. They optimized the scrap mix to reduce energy consumption, resulting in a higher ratio of shredded scrap. They even suggested to avoid DRI because of higher energy requirements. Some of their conclusions are unrealistic because it would be un-economical to operate only on the basis of optimum high quality raw materials. Larsson et al. [58] updated their model to include CO_2 and recycling of by-products. Such flexibility is possible because the core of the model is a mass an energy balance for the entire system. Sutherland and Haapala [59] defined as objective functions the minimization of electric energy, slag amount and off-gas volume in order to consider the environment, applying a generalized reduced gradient nonlinear optimization method, developed by Lasdon et al. [60]. One important conclusion is that the decrease of electrical energy consumption can be made by increasing the input of chemical energy, however the amount of slag and the volume of off-gases increase, which obviously imply negative effects on the environment. An updated model by Haapala et al. [61] analyzed electric energy consumption and CO_2 in 5 scenarios. The EAF model was extended to include ferroalloy additions in the ladle and two of the constrains were related to a fixed tapping temperature or allowed to fluctuate in a broad range. The model computes the values of 23 input variables (including various types of scrap and ferroalloys). They reported a drastic change in the input variables depending on the objective function. Using their model, they claimed important reductions on electric energy and CO_2. Wang et al. [62] also reported an optimization model for an integrated steel plant. They considered three cases involving multi-objective optimization, their solutions were obtained applying fuzzy linear programming as well as Hybrid Petri Nets (HPN) to define the constraints.

Gosiewski and Wierzbicki [63] applied the Maximum Principle (MP), an optimal control theory. The objective function was to minimize the production cost, considering the cost of the electric energy and time. Dutta et al. [64] applied linear programming in a case of multi-objective optimization in an integrated steel plant, which included maximization of profits, cost minimization and product maximization. Camdali [65] reported an academic exercise using linear programming, optimizing scrap type and steel chemical composition and then comparing their exergy values.

The previous optimization models have not included DRI. Although scrap is the main source of metallics in the EAF, substitution with DRI requires responses to the following questions: is there an optimal replacement ratio from a technical and economical viewpoint? what are the differences in the metallurgical practices?. In an attempt to answer those questions, d'Entremont and Englebrecht [66] were pioneers in the late 1970s to develop a computer code to perform mass and energy balances involving scrap, DRI and their costs. The costs for refractory and electrode consumption were not included. In that work they suggest the use of linear programming. Geiger [67] in 1980 reported a detailed linear programming model using DRI. This model was updated by Cárdenas et al. [68, 69], adding the formation of CO_2, including the addition of metallurgical and dolomitic lime, removing the addition of ferroalloys in the EAF and including DRI temperature. Details of all calculations can be found in ref. [69]. References [68, 69] include a large sensitivity analysis describing the influence of process variables on energy consumption. In spite of the model's limitation to estimate the thermal efficiency, the results are within the values observed in current practice and more important it is a valuable model to assess the impact of the quality of raw materials and their cost on energy consumption. Table 3.5 summarizes the impact factor of several process variables and also includes the predicted values from Geiger [67] and a previous statistical model by the same authors.

The agreement between Cardenas et al. [68, 69] and Geiger [67] was expected because both models have the same basis. From Table 3.5, it is noteworthy to identify which process variables have the stronger effect. On one side, carbon in DRI appears to have the highest potential to decrease energy consumption and on the other hand, the gangue content in DRI shows a dramatic increase on energy consumption. The gangue content, in particular acid gangue has multiple effects because increasing the acid gangue automatically increases the requirements of fluxes to control slag basicity. Every 1%C in DRI can potentially decrease energy consumption from 18 to 35 kWh/ton, on the other hand, every 1% in gangue content increases energy consumption by about 21 kWh/ton. The large decrease in energy consumption due to C in DRI increases its value in use (VIU).

Qiu and Zhang [70] superficially described a performance function in terms of three parameters; maximizing chemical energy based on the amount of CO produced, maximizing yield based on the concentration of FeO in the slag and minimizing the process time. They suggested the use of dynamic programming but didn't report results.

Saboohi et al. [71] reported a multi-objective optimization model that includes both maximization of the useful energy (energy in the main product, liquid steel) and profits per heat, being the first optimization model to account for both energy consumption and production costs. The optimization problem was divided into three groups; optimal slag foaming height, optimal slag basicity and optimal energy transfer. The optimal energy transfer is equivalent to maximize the useful energy.

Table 3.5 Impact factor of process variables on electric energy consumption

Factor	Geiger [67] (kWh/ton)	Statistical model [30] (kWh/ton)	Optimization model [68] (kWh/ton)
1%C	− 32	− 17.85	− 35.24
1% Metallization	− 11	− 0.39	− 13.37
1% gangue	+ 15	+ 4.29	+ 21.11
1 m^3 O$_2$/ton	− 3	− 2.68	− 2.95
10 °C T$_{tapping}$	+ 4	+ 4.36	+ 4.17
10 °C DRI	− 3.86	–	− 4.03
10 °C scrap	–	–	− 3.28
1% LOI	–	–	+ 1.82
1 ton fluxes	–	+ 6.15	+ 5.28

The model is quite complex because involves a multiphase system of physicochemical reactions, that evolved from previous studies by Logar et al. [72–80]. Since this is a multi-objective optimization problem, the use of only one optimization method can result in an unfeasible solution, therefore multiple solvers were employed. One example of the optimized model gave a higher useful energy (40.5 vs. 38.2 MWh) and lower total input cost per heat. The cost of oxygen and carbon increased in the optimized model, however the chemical energy and foaming conditions improved which led to lower energy consumption.

Bin and Lee [81] reported an optimization model using 100% scrap, with and without a carbon tax of 1.5 USD/ton CO_2 per ton of liquid steel. Not including the carbon tax might decrease the cost of steel but CO_2 emissions are increased.

Bai [82] reported an optimization model to minimize the production cost of liquid steel, which was described by a quadratic function and solved using a Linear Quadratic Regulator (LQR) model. They included 7 variables; electric energy, electrode consumption, scrap and chemical energy sources. The reported cost savings ranged from 7 to 22% but no details were disclosed.

3.4   Dynamic Models

Chapter 2, Sect. 2.4 previously discussed dynamic models that include energy balances. This section provides few additional details. Static models are based on thermodynamics and dynamic models in addition to thermodynamics also include kinetics. Static models predict the output, dynamic models predict the instantaneous change of the variables. The instantaneous energy consumption is one variable in a dynamic model.

Nyssen et al. [83, 84] reported a general description of a dynamic model for an EAF charged mainly with scrap. The model includes the following modules; (i) Scrap melting. The EAF was divided into 15 sections. Calculates the melting rate, scrap motion and temperature. (ii) liquid metal module. Calculates the chemical and thermal evolution of liquid steel, assuming perfect mixing. (iii) Slag module. Calculates the chemical and thermal evolution of slag. The wasted slag is deduced to account for the actual volume of foaming slag. (iv) Electric arc module. Defines the distribution of electric power. (v) Post-combustion module. Thermal exchange inside the EAF including air ingress. (vi) Burner's module. (vii) Scrap preheating with off-gas module. (viii) Refractory module. The available volume is based on refractory consumption. Thermal losses are calculated. (ix) Water cooled panels (WCP) module. Computes thermal losses by water cooling. (x) The hot metal module calculates the temperature changes due to additions of hot metal. Figure 3.2 describes the structure of the model and the 15 sectors. Each sector has different layers or scrap, as shown in Fig. 3.3. Part of the solid scrap is immersed in either liquid steel and liquid slag where is subject to a high rate of heat transfer. Simulation of the scrap charging operation requires knowledge on its density, this value was obtained by MLSR assuming the scrap basket is full. The scrap density ranged from 0.6 to 1.6 ton/m^3. The discharge of slag was measured using a video camera, the flow rate ranged from 100 to 650 kg/min. The amount slag in the furnace was kept in the range from 10 to 15 ton. The results reported by the authors show a good agreement between predicted and experimental values for the evolution of temperature of the off gas, slag and steel chemistries, heat losses and energy consumption. The heat losses and free volume were used to estimate the moment to charge the next basket of scrap. One on-line application in a steel plant reported an std. error or 5 kWh/ton in the prediction of electric energy consumption.

Kleimt et al. [21] from BFI a German Research institute, reported a dynamic model based on the continuous monitoring of energy losses. The model involved measuring the temperature, chemical composition and flow rate of the off-gas, energy

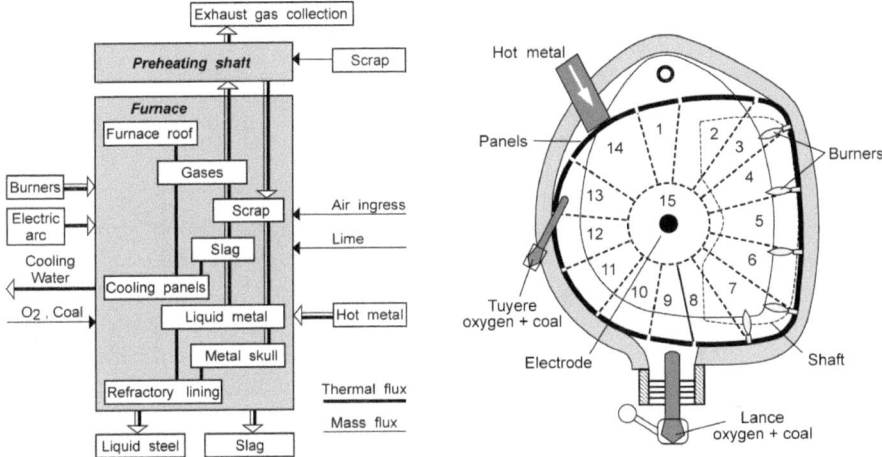

Fig. 3.2 Dynamic model developed by CRM. (Left) basic structure, (right) division of zones. After [83, 84]

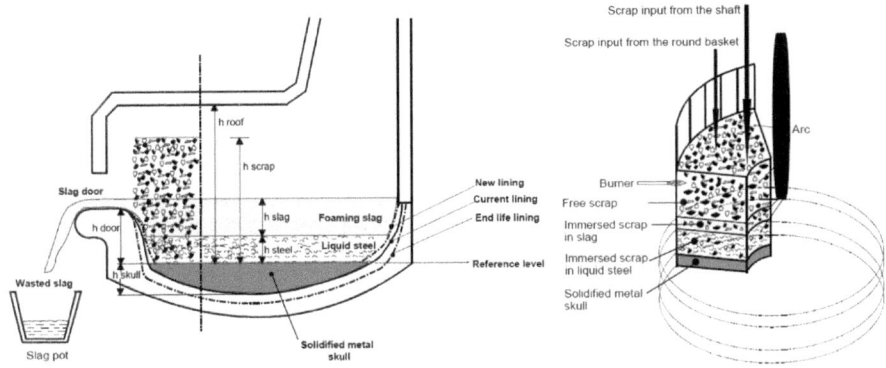

Fig. 3.3 Structure of each sector containing steel scrap. After [83, 84]

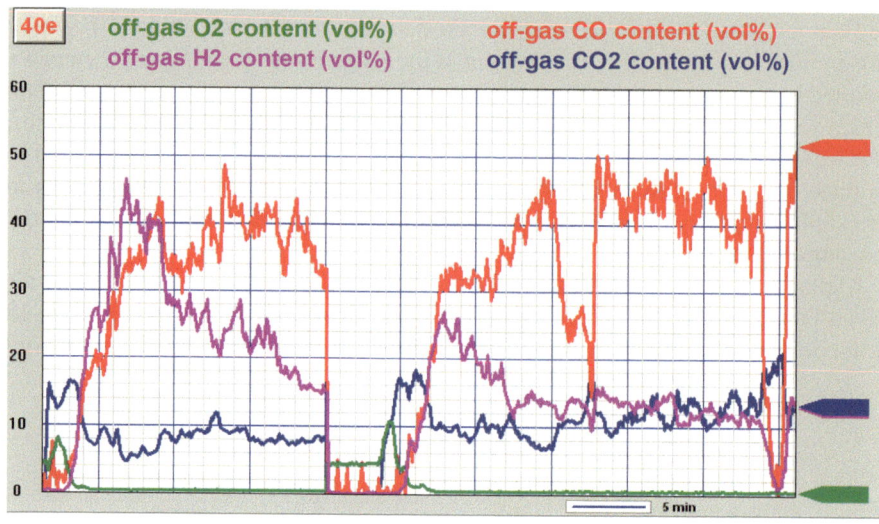

Fig. 3.4 Example of evolution of the off-gas composition during a heat. After [21]

for water cooling and energy radiated from the furnace hearth. The energy input was calculated from the sum of electric energy, chemical energy and sensible heat from the hot heel. To account for the complex number and extent of chemical reactions, they included empirical "gain-factors" to compute the effective chemical energy from burners, post-combustion and oxygen injection. The energy in the melt results from the difference between the energy input and energy losses. The temperature of liquid steel is computed from the energy in the melt. Figure 3.4 shows one example of the measurements of the off-gas composition during a heat, during 50 min. Scrap from the first scrap bucket is melted in the first 20 min. Oxygen shows a small peak at the beginning but once CO is formed its concentration is negligible. The highest concentration is for CO, which fluctuates in the range from 30 to 50%. In the example the second major component is hydrogen. During melting of the first scrap bucket, it reached a peak of 50%, then decreases to less than 20%. CO_2 concentration is in the order of 10%. The model predicts the consumption of electric and chemical energy.

3.5 Exergy Models

An energy balance using the first law of thermodynamics defines the total distribution of energy, however it cannot define which fraction of that energy can be used and exploit its commercial value. This difference is especially important to define the usable energy from the outlet streams. Exergy is defined as the maximum theoretical work that a system can achieve when it comes into equilibrium with the environment [85]. In general, the energy input into the EAF is also the same in terms of exergy.

Due to endothermic chemical reactions and energy losses, energy and exergy can show big differences for the outlet streams. Exergy is lost in the EAF. Exergy analysis involves application of the second law of thermodynamics. Exergy provides more detailed information about the real value on energy recovery. Bisio [86] provides several examples comparing energy and exergy analysis in the steel industry. Gandt et al. [87] reported exergy analysis for the production of steam using the exergy contained in the off-gas, with different amounts of false air, indicating the potential to produce 25–28 ton steam/heat for an EAF of 145 ton.

3.6 Mathematical Models

3.6.1 Mathematical Models with Analytical Solutions

The number of variables that affect energy consumption is quite large as has been already indicated in previous sections. Any one or some of those variables can be investigated in detail defining the physics for each rate phenomenon; oxygen injection and its related chemical input of chemical energy, scrap-preheating temperature, post-combustion, slag foaming, evolution of liquid steel temperature, etc. The evolution of the thermal and chemical changes as a function of time can be described by differential equations and its analytical solution is obtained by common numerical algorithms.

Conejo et al. [88–92] reported an integral model that describes changes in steel and slag composition during a heat charging 100% DRI and additionally predicted a dynamic slag foaming index to evaluate the instantaneous foaming conditions. Their modeling approach starts with the optimization of mass and energy which defines the amount of input materials. This work defines the optimum injection conditions to improve slag foaming, considering carbon particle size, injection flow rate, particle's residence time, carbon reactivity, slag physical and chemical properties as well as the generation rate of CO. This was one of the earliest integral EAF models. Logar et al. in addition to taking into consideration the thermochemical processes also included electrical aspects and reported with good accuracy the temperature of liquid steel [72–75]. Prediction of the off-gas composition and temperature is extremely useful because the chemical and thermal state of liquid steel can also be predicted. Meier et al. [93] reported a numerical model which accurately predicts the off-gas composition.

3.6.2 Mathematical Models Involving CFD

Computational Flow Dynamics (CFD) is a necessary tool in any current research work nowadays. It provides a deeper insight of the physics associated with how the variables affect energy consumption. Due to the complex nature of a holistic approach, available studies focus on specific targets. Odenthal et al. [94] in 2018 reviewed ten aspects on mathematical modeling work involving CFD on the EAF, each one provides information on process optimization, for example; design of geometry and flow rates in water cooled panels to obtain a more homogeneous temperature distribution and thicker slag coating which would decrease heat losses, exact position and number of burners or injection lances to enhance the scrap melting rate, post-combustion and slag foaming, temperature and composition of the off-gas, scrap melting rate, etc. One of the challenges to perform a holistic approach is in principle the current limitation of supercomputers to solve problems involving millions of cells in a relatively short time. Later on, Hay et al. [95] in 2021 focused on mathematical models involving process modelling and according to them there are still some gaps in this field, such as; (i) validation of the foaming index for industrial conditions, (ii) more accurate modeling of the electric arc, (iii) implementation of new measurement techniques to assist the validation of the models.

References

1. Wold S, Lasse K (1999) Method for monitoring multivariate processes. US patent 5949678
2. Stigler S (1981) Gauss and the invention of the least squares. Ann Stat 9:465–474
3. Jollife IT, Cadima J (2016) Principal component analysis: a review and recent developments. Philos Trans R Soc A 374. https://doi.org/10.1098/rsta.2015.0202
4. Hahn GJ (1973) The coefficient of determination revisited. Chem Technol 3:609–612. https://doi.org/10.1007/978-94-009-3591-4_13
5. Chai T, Draxler RR (2014) Root mean square error (RMSE) or mean absolute error (MAE)? Arguments against avoiding RMSE in the literature. Geosci Model Dev 7:1247–1250. https://doi.org/10.5194/gmd-7-1247-2014
6. Székely GJ, Rizzo ML, Bakirov NK (2007) Measuring and testing dependence by correlation of distances. Ann Stat 35:2769–2794. https://doi.org/10.1214/009053607000000505
7. Reshef D, Reshef Y, Finucane H, Grossman S, Mcvean G, Turnbaugh P et al (2011) In large data sets. Sci Transl Med 334:1518–1524. https://doi.org/10.1126/science.1205438
8. Santos SS, Takahashi DY, Nakata A, Fujita A (2013) A comparative study of statistical methods used to identify dependencies between gene expression signals. Brief Bioinform 15:906–918. https://doi.org/10.1093/bib/bbt051
9. Schroeder L, Sjoquist D, Stephan P (1986) Understanding regression analysis. Sage
10. Schneider A, Hommel G, Blettner M (2010) Linear regressions analysis. Dtsch Arzteblatt Int 107:776–782. https://doi.org/10.3238/arztebl.2010.0776
11. Kovačič M, Stopar K, Vertnik R, Šarler B (2019) Comprehensive electric arc furnace electric energy consumption modeling: a pilot study. Energies 12:1–13. https://doi.org/10.3390/en12112142
12. Carlsson LS, Samuelsson PB, Jönsson PG (2019) Predicting the electrical energy consumption of electric arc furnaces using statistical modeling. Metals 9. https://doi.org/10.3390/met9090959
13. Brooks GP, Barcikowski RS (2012) The PEAR method for sample sizes in multiple linear regression. Mult Linear Regres Viewpoints 38:1–16. https://doi.org/10.1007/978-94-009-3591-4_13
14. Austin PC, Steyerberg EW (2015) The number of subjects per variable required in linear regression analyses. J Clin Epidemiol 68:627–636. https://doi.org/10.1016/j.jclinepi.2014.12.014

15. Köhle S (1992) Effects on the electric energy consumption of arc furnace steelmaking. In: 4th European electric steel congress, Madrid, Spain
16. Köhle S (1992) Variables influencing electric energy and electrode consumption in electric arc furnaces. Metall Plant Technol Int 6:48–53
17. Kleimt B, Köhle S (1997) Power consumption of electric arc furnaces with post-combustion. Metall Plant Technol Int 3:56–57
18. Köhle S (1999) Improvements in EAF operation practices over the last decade. In: Electric furnace conference, Pittsburgh, PA USA, 14–16 Nov 1999, pp 3–14
19. Köhle S (2002) Recent improvements in modelling energy consumption of electric arc furnaces. In: 7th European electric steelmaking conference, Venice Italy, 26–29 May 2002
20. Köhle S, Hoffmann J, Baumert JC, Picco M, Nyssen P, Filippini E (2003) Improving the productivity of electric arc furnaces. Luxembourg. ECSC report 20803
21. Kleimt B, Köhle S, Kühn R, Zisser S (2005) Application of models for electrical energy consumption to improve EAF operation and dynamic control. In: 8th European electric steelmaking congress, Birmingham UK, pp 183–197
22. Bowman B (1995) Performance comparison update-AC vs DC furnaces. Iron Steel Eng: 26–29
23. Bowman B, Lugo N, Wells T (2000) Influence of tap carbon and arc voltage on electrode and energy consumption. In: 58th electric furnace conference, pp 649–657
24. Adams W, Alameddine S, Bowman B, Lugo N, Paege S, Stafford P (2001) Factors influencing the total energy consumption in arc furnaces. In: 59th electric arc furnace conference, 11–14 Nov 2001, Phoenix, AZ, USA, pp 691–702
25. Adams W, Alameddine S, Bowman, B, Lugo N, Paege S, Stafford P (2002) Total energy consumption in arc furnaces. MPT Int: 44–50
26. Moskal M, Migas P, Karbowniczek M (2022) Multi-parameter characteristics of electric arc furnace melting. Materials 15:1–14. https://doi.org/10.3390/ma15041601
27. Mathy C, Terho K, Chouvet M, Le Coq X, Baumert J-C, Engel R, Hoffmann J (2003) Production of steel at lower operating costs in EAF. Luxembourg. ECSC report 20895
28. Pfeifer H, Kirschen M (2002) Thermodynamic analysis of EAF energy efficiency and comparison with statistical model of electric energy demand. In: 7th European electric conference, 26–29 May 2002, Venice, Italy
29. Pfeifer H, Kirschen M, Simoes JP (2005) Thermodynamic analysis of EAF electrical energy demand. In: 8th European electric conference, 9–11 May 2005, Birmingham UK
30. Conejo AN, Cardenas JGG (2006) Energy consumption in the EAF with 100% DRI. AISTech Conf 1:529–535
31. Elkoumy M, El-Anwar M, Fathy AM, Megahed GM, El-Mahallawi I, Ahmed H (2018) Simulation of EAF refining stage. Ain Shams Eng J 9:2781–2793. https://doi.org/10.1016/j.asej.2017.10.002
32. Elkoumy MM, Fathy AM, Megahed GM, El-Mahallawi I, Ahmed H, El-Anwar M (2019) Empirical model for predicting process parameters during electric arc furnace refining stage based on real measurements. Steel Res Int 90:1–10. https://doi.org/10.1002/srin.201900208
33. Arzpeyma N, Gyllenram R, Jönsson PG (2020) Development of a mass and energy balance model and its application for HBI charged EAFS. Metals 10. https://doi.org/10.3390/met10030311
34. Kirschen M, Hanna A, Zettl KM (2016) The application of benchmark models for EAF energy efficiency with a focus on process improvements by EAF gas purging. In: AISTech—iron steel technology conference proceedings, vol 1, Cleveland, OH, USA, pp 843–856
35. Kirschen M, Zettl K-M, Echterhof T, Pfeifer H (2016) Analysis of benchmark models for EAF energy efficiency with application to process improvements by EAF gas purging. In: 11th European electric conference, 25–27 May 2016, Venice, Italy
36. Kirschen M, Zettl K-M, Echterhof T, Pfeifer H (2017) Models for EAF energy efficiency. Steel Times Int: 2–4
37. Sandberg E, Bentell L, Undvall P (2002) Energy optimisation of electric arc furnaces by statistical process evaluation. In: 7th European electric steelmaking conference, Venice, Italy, 26–29 May 2002, pp 435–443

38. Sandberg E (2005) Energy and scrap optimisation of electric arc furnaces by statistical analysis of process data. Thesis. Lulea University. Lulea, Sweden
39. Sandberg E, Lennox B, Marjanovic O, Smith K (2005) Multivariate process monitoring of EAFs. Ironmak Steelmak 32:221–225. https://doi.org/10.1179/174328105X45884
40. Sandberg E (2007) Scrap management by statistical evaluation of EAF process data. Control Eng Pract 15:1063–1075
41. Banzhaf W, Nordin P, Keller R, Francone F (1998) Genetic programming—an introduction. On the automatic evolution of computer programs and its applications. San Francisco-Heidelberg
42. Sagiroglu S, Sinanc D (2013) Big data—a review. Int Conf Collab Technol Syst: 42–47. https://doi.org/10.26634/jit.6.1.13507
43. Baumert J-C (2002) Artificial neural networks model the electric arc furnace process. In: 7th European electric steelmaking conference, Venice, Italy, 26–29 May 2002, pp 1.255–1.264
44. Baumert JC, Engel R, Weiler C (2002) Dynamic modelling of the electric arc furnace process using artificial neural networks. Rev Metall 99:839–849. https://doi.org/10.1051/metal:2002144
45. Baumert J-C, Rendueles J-L, Nyssen P, Schaefers J, Schutz G, Gille S (2005) Improved control of electric arc furnace operations by process modelling. Luxembourg. ECSC report 21411
46. Gajic D, Savic-Gajic I, Savic I, Georgieva O, Di Gennaro S (2016) Modelling of electrical energy consumption in an electric arc furnace using artificial neural networks. Energy 108:132–139. https://doi.org/10.1016/j.energy.2015.07.068
47. Haupt M, Vadenbo C, Zeltner C, Hellweg S (2017) Influence of input-scrap quality on the environmental impact of secondary steel production. J Ind Ecol 21:391–401. https://doi.org/10.1111/jiec.12439
48. Chen C, Liu Y, Kumar M, Qin J (2018) Energy consumption modelling using deep learning technique—a case study of EAF. Procedia CIRP 72:1063–1068. https://doi.org/10.1016/j.procir.2018.03.095
49. Carlsson LS, Samuelsson PB, Jönsson PG (2020) Using statistical modeling to predict the electrical energy consumption of an electric arc furnace producing stainless steel. Metals 10. https://doi.org/10.3390/met10010036
50. Mesa Fernández JM, Cabal VÁ, Montequin VR, Balsera JV (2008) Online estimation of electric arc furnace tap temperature by using fuzzy neural networks. Eng Appl Artif Intell 21:1001–1012. https://doi.org/10.1016/j.engappai.2007.11.008
51. Blachnik M, Mączka K, Wieczorek T (2010) A model for temperature prediction of melted steel in the electric arc furnace (EAF). Lecture notes computer science (including subseries lecture notes artificial intelligence lecture notes bioinformatics), 6114 LNAI, 371–378. https://doi.org/10.1007/978-3-642-13232-2_45
52. Kordos M, Blachnik M, Wieczorek T (2011) Temperature prediction in electric arc furnace with neural network tree. Lecture notes computer science (including subseries lecture notes artificial intelligence lecture notes bioinformatics), 6792 LNCS, 71–78. https://doi.org/10.1007/978-3-642-21738-8_10
53. Reimann A, Hay T, Echterhof T, Kirschen M, Pfeifer H (2021) Application and evaluation of mathematical models for prediction of the electric energy demand using plant data of five industrial-size EAFS. Metals 11. https://doi.org/10.3390/met11091348
54. León-Munizaga N, Aguirre-Munizaga M, Lagos-Ortiz K, Del Cioppo-Morstadt J (2020) Prediction of energy consumption in an electric arc furnace using Weka. In: 6th international conference on technology innovation, Guayaquil, Ecuador, 30 Nov–3 Dec 2020, pp 58–70
55. Tomažič S, Andonovski G, Škrjanc I, Logar V (2022) Data-driven modelling and optimization of energy consumption in EAF. Metals 12. https://doi.org/10.3390/met12050816
56. Larsson M, Dahl J (2003) Reduction of the specific energy use in an integrated steel plant—the effect of an optimisation model. ISIJ Int 43:1664–1673. https://doi.org/10.2355/isijinternational.43.1664
57. Riesbeck J, Lingebrant P, Sandberg E, Wang C (n.d.) Energy system optimization for a scrap based steel plant using mixed integer linear programming: 1676–1683

58. Larsson M, Wang C, Dahl J (2006) Development of a method for analysing energy, environmental and economic efficiency for an integrated steel plant. Appl Therm Eng 26:1353–1361. https://doi.org/10.1016/j.applthermaleng.2005.05.025

59. Sutherland JW, Haapala KR (2007) Optimization of steel production to improve lifecycle environmental performance. CIRP Ann Manuf Technol 56:5–8. https://doi.org/10.1016/j.cirp.2007.05.003

60. Lasdon LS, Waren AD, Jain A, Ratner M (1978) Design and testing of a generalized reduced gradient code for nonlinear programming. ACM Trans Math Softw 4:34–50. https://doi.org/10.1145/355769.355773

61. Haapala KR, Catalina AV, Johnson ML, Sutherland JW (2012) Development and application of models for steelmaking and casting environmental performance. J Manuf Sci Eng Trans ASME 134:1–13. https://doi.org/10.1115/1.4007463

62. Wang P, Jiang Z, Liu Z, Fu S (2013) Modeling and optimizing energy utilization of steel production process: a hybrid petri net approach. Adv Mech Eng 2013. https://doi.org/10.1155/2013/191963

63. Gosiewski A, Wierzbicki A (1970) Dynamic optimization of a steel-making process in electric arc furnace. Automatica 6:767–778. https://doi.org/10.1016/0005-1098(70)90024-5

64. Dutta G, Sinha GP, Roy PN, Mitter N (1994) A linear programming model for distribution of electrical energy in a steel plant. Int Trans Oper Res 1:17–29. https://doi.org/10.1016/0969-6016(94)90042-6

65. Çamdali Ü (2005) Determination of the optimum production parameters by using linear programming in the AC electric arc furnace. Can Metall Q 44:103–110. https://doi.org/10.1179/cmq.2005.44.1.103

66. d´Entremont JC, Englebrecht ML (1979) Computer simulated usage of direct reduced iron in electric arc furnace operations. In: ISS-AIME (ed) Ironmaking conference, vol 38, pp 279–284

67. Geiger G (1980) Process engineering involved in the use of direct reduced iron. In: Stephenson RL, Smailer RM (eds) Direct reduced iron technology economic production use. Iron and Steel Society, , USA, pp 149–159

68. Cárdenas JGG, Conejo AN, Gnechi GG (2007) Optimization of energy consumption in electric arc furnaces operated with 100 % DRI. METAL, Hradec Nad Moravici, Czechia, 22–24 May 2007, pp 1–7

69. Cardenas JG (2008) Analysis on the consumption of electric energy in the EAF. M.Sc. Thesis, Morelia Technological Institute, Mexico (in Spanish)

70. Qiu D, Zhang DJ (2010) The research of energy balance dynamic model on electric arc furnace. In: International conference on information, networking automation, pp 507–511

71. Saboohi Y, Fathi A, Skrjanc I, Logar V (2019) Optimization of the electric arc furnace process. IEEE Trans Ind Electron 66:8030–8039. https://doi.org/10.1109/TIE.2018.2883247

72. Logar V, Dovžan D, Škrjanc I (2011) Mathematical modeling and experimental validation of an electric arc furnace. ISIJ Int 51:382–391. https://doi.org/10.2355/isijinternational.51.382

73. Logar V, Dovžan D, Škrjanc I (2012) Modeling and validation of an electric arc furnace: part 1. Heat and mass transfer. ISIJ Int 52:402–412. https://doi.org/10.2355/isijinternational.52.413

74. Logar V, Dovžan D, Škrjanc I (2012) Modeling and validation of an electric arc furnace: part 2, thermo-chemistry. ISIJ Int 52:413–423. https://doi.org/10.2355/isijinternational.52.413

75. Logar V, Škrjanc I (2012) Modeling and validation of the radiative heat transfer in an electric arc furnace. ISIJ Int 52:1225–1232. https://doi.org/10.2355/isijinternational.52.1225

76. Logar V, Škrjanc I (2012) Development of an electric arc furnace simulator considering thermal. Chem Electr Aspects 52:1924–1926

77. Fathi A, Saboohi Y, Škrjanc I, Logar V (2015) Low computational-complexity model of EAF arc-heat distribution. ISIJ Int 55:1353–1360. https://doi.org/10.2355/isijinternational.55.1353

78. Logar V (2016) Modelling and simulation of the melting process in electric arc furnace: an overview. Simul Notes Eur 26:91–98. https://doi.org/10.1002/srin.201500141

79. Logar V, Fathi A, Škrjanc I (2016) A computational model for heat transfer coefficient estimation in electric arc furnace. Steel Res Int 87:330–338. https://doi.org/10.1002/srin.201500060

80. Fathi A, Saboohi Y, Škrjanc I, Logar V (2017) Comprehensive electric arc furnace model for simulation purposes and model-based control. Steel Res Int 88:1–22. https://doi.org/10.1002/srin.201600083

81. Bin YW, Lee I-B (2020) Systematic optimization using mathematical model of electrical arc furnace producing liquid steel. J Chem Eng Japan 53:533–539. https://doi.org/10.1252/jcej.17we361

82. Bai E (2014) Minimizing energy cost in electric arc furnace steel making by optimal control designs. J Energy 2014:1–9. https://doi.org/10.1155/2014/620695

83. Nyssen P, Colin S, Knoops S, Junque J-L (2002) On-line EAF control with a dynamic metalurgical model. In: AIM (ed) 7th European electric steelmaking conference, Venice, Italy, 26–29 May 2002, pp 293–304

84. Nyssen P, Colin R, Junqué JL, Knoops S (2004) Application of a dynamic metallurgical model to the electric arc furnace. Rev Metall Cah D'Informations Tech 101:317–326. https://doi.org/10.1051/metal:2004203

85. Dincer I, Cengel YA (2001) Energy, entropy and exergy concepts and their roles in thermal engineering, vol 3. https://doi.org/10.3390/e3030116

86. Bisio G (1993) Exergy method for efficient energy resource use in the steel industry. Energy 18:971–985. https://doi.org/10.1016/0360-5442(93)90007-Z

87. Gandt K, Meier T, Echterhof T, Pfeifer H (2016) Heat recovery from EAF off-gas for steam generation: analytical exergy study of a sample EAF batch. Ironmak Steelmak 43:581–587. https://doi.org/10.1080/03019233.2016.1155812

88. Conejo A, Morales R, Rodriguez H (2001) Mathematical modeling of the EAF process using direct reduced iron. In: 59th electric furnace conference, Phoenix, AZ, USA, 11–14 Nov 2001, pp 797–810

89. Morales RD, Rodríguez-Hernández H, Conejo AN (2001) A mathematical simulator for the EAF steelmaking process using direct reduced iron. ISIJ Int 41

90. Rodriguez HH, Conejo AN, Morales RD (2001) Theoretical analysis of the interfacial phenomena during the injection of carbon particles into EAF slags. Steel Res 72

91. Morales RD, Conejo AN, Rodríguez HH (2002) Process dynamics of electric arc furnace during direct reduced iron melting. Metall Mater Trans B Process Metall Mater Process Sci 33

92. Morales RD, Rodríguez-Hernández H, Vargas-Zamora A, Conejo AN (2002) Concept of dynamic foaming index and its application to control of slag foaming in electric arc furnace steelmaking. Ironmak Steelmak 29:445–453

93. Meier T, Gandt K, Echterhof T, Pfeifer H (2017) Modeling and simulation of the off-gas in an electric arc furnace. Metall Mater Trans B Process Metall Mater Process Sci 48:3329–3344. https://doi.org/10.1007/s11663-017-1093-7

94. Odenthal HJ, Kemminger A, Krause F, Sankowski L, Uebber N, Vogl N (2018) Review on modeling and simulation of the electric arc furnace (EAF). Steel Res Int 89:1–36. https://doi.org/10.1002/srin.201700098

95. Hay T, Visuri VV, Aula M, Echterhof T (2021) A review of mathematical process models for the electric arc furnace process. Steel Res Int 92:1–22. https://doi.org/10.1002/srin.202000395

Chapter 4
Oxygen Injection

The oxygen injection practice is essential to steelmaking. Steelmaking is an oxidation process, in particular the removal of carbon and other elements. Previously, it has been clearly stated that the total energy requirement is fixed once the amount and chemical composition of the metallic charge is fixed and this energy requirement cannot be changed. Decreasing energy consumption mainly refers to the decrease of expensive electric energy by cheaper sources of energy, like chemical energy.

The definition of the oxygen injection practice involves the following aspects:

- Compute the optimum volume and gas flow rate of oxygen injected
- Compute the decarburization rate
- Compute chemical energy due to exothermic chemical reactions as a function of steel chemical composition
- Design of the off-gas system to evacuate a higher volume of CO and CO_2 when the volume of oxygen injection is increased
- Consider air infiltration in the calculations.

This chapter provides a summarized but comprehensive analysis of the oxygen injection practice in the EAF. An optimized injection practice is challenging because it involves the decrease of not only the consumption of electric energy but also higher metallic yield, improved slag foaming conditions, etc.

4.1 Introduction

4.1.1 Early Methods on Decarburization of Liquid Steel

Before the 1940s the main source of oxygen was the addition of iron ore or mill scale. This practice is now obsolete because it increases energy consumption and the melting time. Saberifar et al. [1] compared a base case without mill scale and two

© The Author(s), under exclusive license to Springer Nature Singapore Pte Ltd. 2024   131
A. N. Conejo, *Electric Arc Furnace: Methods to Decrease Energy Consumption*,
https://doi.org/10.1007/978-981-97-4053-6_4

other cases using mill-scale. They reported disadvantages using mill scale; a higher power-on time in the range from 3 to 6 min, an increment on energy consumption in the range from 4.5 to 8%, higher FeO, from about 21% to about 26% an increment in carbon in the range from 7 to 21%, and its benefit was a decrease in oxygen consumption in the range from 18 to 36%. In spite that oxygen was available since late 1920s it was widely applied in the EAF until the late 1960s when became cheaper.

4.1.2 Commercial Availability of Oxygen

Oxygen was available in large quantities after 1928 when the Linde-Frankl process was developed. The Linde process is a multi-cycle step that transforms air from gas to liquid. Carl von Linde patented the idea in 1895 in Germany while simultaneously W. Hampson did that in the UK. Almqvist [2], in its history of industrial gases, also includes the American Charles Tripler among the pioneers.

A more recent development is the separation of oxygen from air using Pressure Swing Adsoprtion (PSA) processes which today makes commercial oxygen cheaper than that obtained by the Linde-Frankl process. Oxygen purity ranges from 85 to 95%. The original idea about PSA comes from the early 1930s when Finlayson and Sharp [3] applied adsorption theory to separate the components of a gas mixture like air, however Skarstrom is credited with its commercial application in 1960 [4]. Santos et al. describe in more detail the sequence of developments that placed PSA as a competitive technology of both high purity oxygen and nitrogen since the 1980s.

4.2 Oxygen Injection Jets

4.2.1 Consumable Lances

The first technology for oxygen injection employed consumable lances, introduced through the slag door. These lances had the advantage that could penetrate into the molten bath, however the working conditions were very hard and dangerous. They were operated manually. Consumable lances have the advantage of immersing deep into the molten bath, change direction and position but also have many disadvantages like long delays due to the need to replace them even if they had a special coating to resist higher temperatures.

Shver proposed a consumable lance tip made of refractory material [5] that extends its service life. Consumable lances were subsequently replaced by water-cooled lances that use supersonic jets. This type has a longer life and can inject simultaneously gas and powders but its position cannot be inside of the molten bath. Despite advantages in automation with water-cooled lances, consumable lances are still employed due to investment costs. Jamnik et al. claim some advantages when consumable lances are used in the production of stainless steels [6]. A further development by BSE in 1985 introduced multiple consumable lances using a machine but still involved some manual labor, in addition to this there was the need to keep the slag door open which increases air ingress.

4.2.2 Supersonic Lances

The velocity of the sound is 343 m/s (1 Mach). Supersonic lances inject oxygen at velocities higher than 487 m/s (Mach 1.5), such velocity produces a jet that crosses the slag and penetrates into liquid steel.

The vertical and horizontal angle are very important in the operation of an injection lance. Vertical lances are inconvenient because when the height of molten bath is shallow the jet creates a high erosion of the refractory hearth. Figure 4.1 shows an example of the position of a supersonic lance. The vertical angle is 37° and the horizontal angle 30°.

The supersonic lances are water cooled. There is some risk of explosions under the following conditions: the lance reaches or penetrates the molten bath, when water cooling is suspended or using lances in bad conditions. Since the tip is water cooled it cannot be submerged into liquid steel. Some designs included a replaceable tip. If the lance is not immersed into the liquid, part of the oxygen increases the oxidizing conditions inside the furnace.

In 2001 Shver [5] invented a supersonic lance whose tip was made of refractory material and allowed to be immersed into liquid steel.

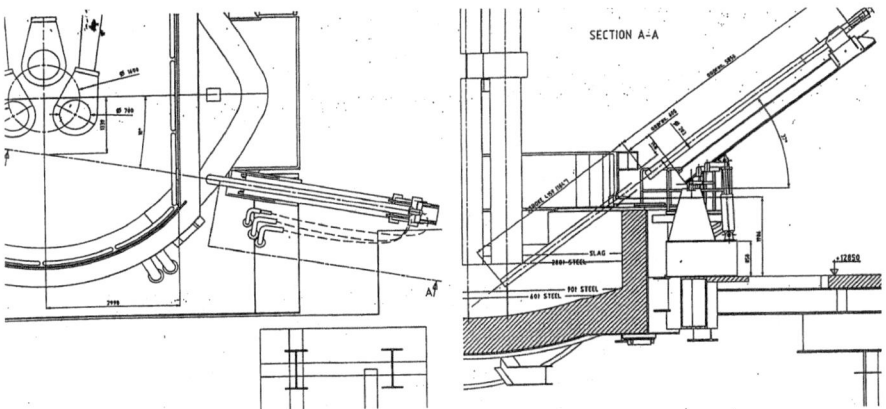

Fig. 4.1 Position of supersonic lance

4.2.3 Coherent Jets

The most recent development is the coherent jet [7, 8] characterized by sustaining high flow velocities at least twice in comparison with supersonic jets. Coherent jets are also supersonic. The concept of a coherent jet evolved from the previous lance design developed by Savard and Lee in 1966 and applied on the BOF for bottom oxygen injection (OBM) in 1972. Anderson and Farrenkopf [9] from Praxair (currently Linde) reported in 1998 the length of the coherent jet is about 50 times the nozzle diameter, about 4 times the length of the oxygen supersonic jet. The nozzle diameter varies from 12 to 50 mm. The velocity of the oxygen jet at the discharge point is about 300–450 m/s. Figure 4.2 compares the coherent distance for supersonic and coherent jets based on the experimental data from Anderson and Farrenkopf [9]. In this case, the coherent distance is about 50 times the diameter of the nozzle, in comparison with a supersonic jet with a value of 10 times the diameter of the oxygen nozzle.

The basic idea of a coherent jet is the formation of a shrouding flame around the oxygen jet. The temperature around the oxygen jet increases with respect to that in a conventional supersonic lance allowing to maintain higher velocities over a longer distance and thus promoting higher penetration depths of the oxygen jet into liquid steel, achieving both higher decarburization rates and stirring energy.

Currently, depending on the stage of the heat, injectors can play multi-functional roles, as follows:

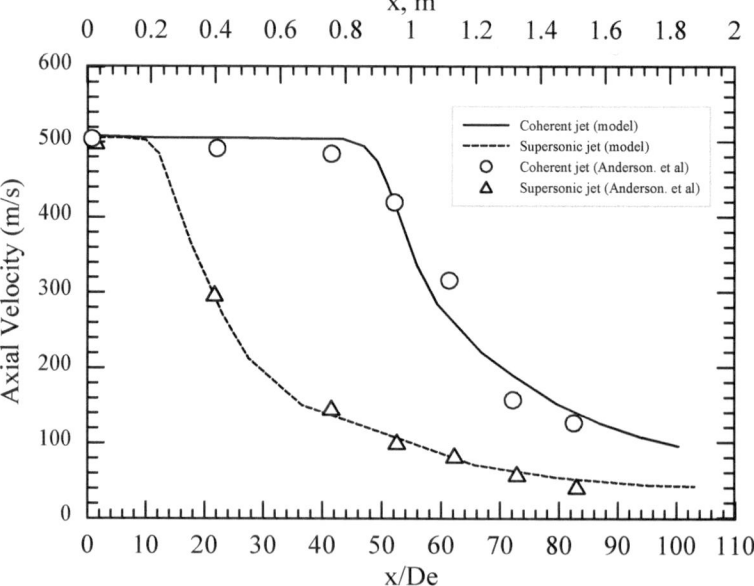

Fig. 4.2 Comparison of supersonic and coherent jets. After [9]

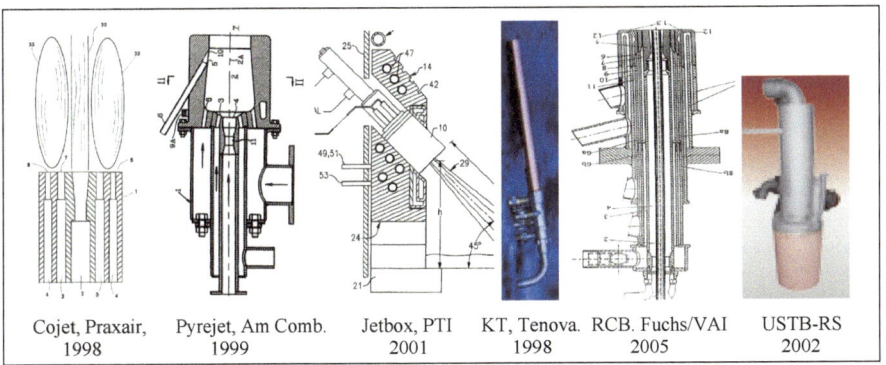

| Cojet, Praxair, 1998 | Pyrejet, Am Comb. 1999 | Jetbox, PTI 2001 | KT, Tenova. 1998 | RCB. Fuchs/VAI 2005 | USTB-RS 2002 |

Fig. 4.3 Some designs of coherent jets

- Decarburization, operating as oxygen injectors
- Scrap pre-heating, operating as burners
- Slag foaming as carbon injectors
- Slag basicity control as flux injectors

A multi-functional device is not as efficient as that designed for a given specific purpose and for that reason even if there are installed multi-functional coherent jets, specific devices to fulfill specific functions can also be installed in the EAF.

The variety of available designs is huge and each supplier claims its own advantages. Figure 4.3 summarizes six designs developed by different companies. Praxair who created the design for the EAF in 1998 it merged in 2018 with Linde. In 2001 American Combustion merged with Air Liquide. In 2005 Fuchs/VAI merged with Siemens and in 2014 Mitsubishi merged with Siemens forming Primetals. In addition to these designs there are more suppliers such as MORE, Danieli, etc.

Mathur and Messina [10] reported results obtained at several steel plants which include a higher volume of oxygen and a decrease in electric energy consumption as well as tap-to-tap time.

Cates et al. [11] compared the benefits of replacing supersonic lances with coherent jets in an EAF of 100 ton, using 30% hot metal. The use of hot metal is more challenging in terms of oxygen injection because it requires a higher total volume and higher flow rates, using supersonic jets it can lead to large skulling formation and violent eruptions because the jet has a smaller penetration depth which then oxidizes the slag and ejects more droplets. The jet penetration depth using coherent jets was estimated in the range from 20 to 50 cm and the stirring energy from 2 to 15 W/ton. The results at a Japanese plant indicated that replacing supersonic lances with coherent jets allowed an increment in oxygen injection from 24 to 34 Nm^3/ton without increasing the slag volume (93 kg/ton) or affecting the metallic yield (92.9%) and resulted in a decrease in energy consumption from 349 to 304 kWh/ton and productivity from 88 to 91 ton/h.

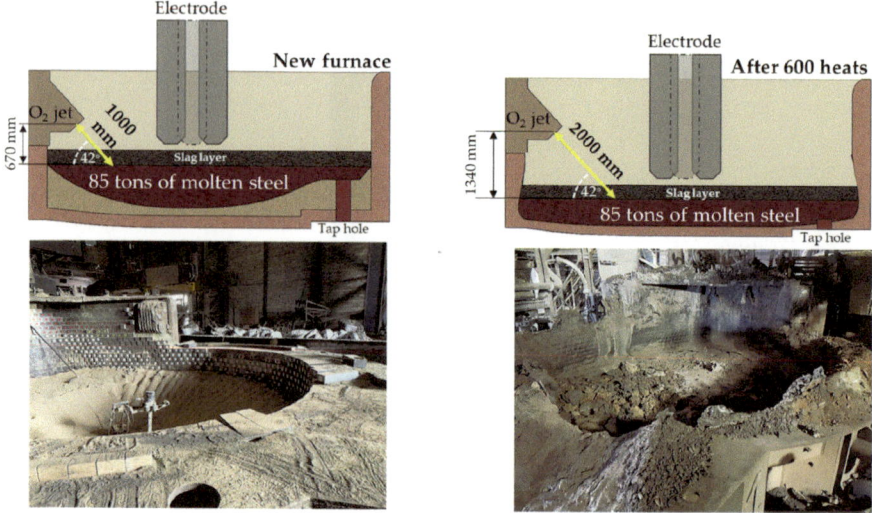

Fig. 4.4 Distance from tip of oxygen lance to batch surface due to refractory wear. (Left) new refractory lining, (right) old refractory lining. After [12]

Thongjitr et al. [12] compared the coherent length for conventional and coherent jets and the importance of a higher coherent length due to refractory wear. As shown in Fig. 4.4 the tip from the lance to the batch surface can drastically increase due to refractory wear, in this case from 1 to 2 m.

These authors estimated the jet penetration depth for three cases; conventional jet, coherent jet with natural gas as the shrouding gas and another coherent jet with a shrouding gas consisting of natural gas and oxygen. The jet penetration depth was higher with a shorter distance from the lance tip to the bath surface, for example, for 1 m, the values for the three previous cases were; 389, 720 and 728 mm, respectively, the jet penetration depth for a distance of 2 m decreased to the following values; 286, 378 and 427 mm, respectively. The results indicate the importance to have a higher coherent length especially as the refractory life increases.

Numerical modelling of oxygen injection should take into consideration the differences between supersonic and coherent jets. The k-ε turbulence model is inaccurate to predict highly turbulent jets. To achieve a better representation of a shrouding jet, Alam et al. [13] modified an strategy developed previously by Abdol-Hamil et al. to define the constant involved in the calculation of the turbulent viscosity.

4.2.4 Submerged Lances

Wei et al. [14–17] have suggested submerged lances for oxygen injection while co-injecting CO_2. The decarburization reaction with carbon dioxide as well as the decomposition of the shrouding gas are endothermic and helps in the formation of a mushroom on the tip front of the injection element which enhances its life in service and reduces refractory erosion. These authors reported improvements in refractory life using mixtures CO_2–O_2 in comparison with pure oxygen, 410 and 220 heats, respectively. The process was called COMI which stands for Coherent CO_2–O_2 Mixed gas Injection. CO_2 decreases the fire spot temperature from about 2500 °C with pure oxygen to 1700 °C employing 20% CO_2. In addition to gas injection the system can inject carbon and lime powders. Contrary to argon stirring, stirring with CO_2 is more intense because the decarburization reaction produces two moles of CO for every mole of CO_2. The fraction of CO_2 injection should be limited because if CO_2 increases in the gas mixture there as an increase in the carbon concentration and a decrease in the temperature of liquid steel, as shown by You et al. [18]. Water model experiments indicated that increasing the lance vertical injection angle, from 0 to 15°, increased the vertical penetration distance but also decreased the horizontal penetration distance, at a given gas flow rate [15]. Further analysis of mixing time investigated the arrangement of two submerged nozzles with three modes [16]. In mode A the central axis of the two submerged injectors point to the EAF center, in mode B both of their central axis rotates clockwise 20 degrees and in mode C their central axis rotates 20° in the opposite directions. It was found that mode B produced shorter mixing times and higher liquid velocities. Mode B also produced less refractory erosion because fluid flow is more homogeneous with less volume of dead zones. It was also reported that increasing the immersion depth from 30 to 60 mm, mixing time was decreased from 135 to 112 s. Mode B was numerically and experimentally investigated for an EAF of 75 ton [17]. The system was composed by two coherent jets with a vertical angle of 44° and two submerged jets in mode B with a vertical angle of 10° immersed 240 mm, using a gas mixture of 10% CO_2-90% O_2, as shown in Fig. 4.5. The gas flow rate for the coherent jets was 33 Nm^3/min and for the submerged jets ranged from 3 to 10 Nm^3/min. The refractory thickness around the submerged jets was higher than the conventional refractory thickness, 750 and 500 mm, respectively, as shown in Fig. 4.6. The numerical model predictions indicate that the main contribution to the average velocity 100 mm below the surface is due to the submerged jets. At a plane 400 mm below the surface, the average velocity obtained with submerged jets was 2.7 cm/min and a total value of 2.9 cm/min using both types of jets. Comparison using coherent jets and a combination involving submerged jets reported higher approach to equilibrium conditions in the combined system which provides lower values of FeO and a lower [C][O] solubility product. FeO was decreased from 25.6 to 23.5% and the solubility product from 0.00318 to 0.00252.

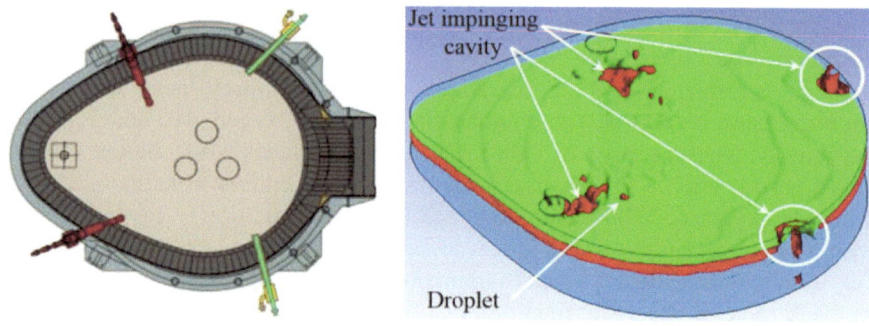

Fig. 4.5 Combined system using coherent and submerged jets [17]

Fig. 4.6 Refractory thickness in the location of submerged jet [16]

4.3 Fluid Flow Due to Oxygen Injection

4.3.1 Cavity Formation

When the oxygen jet impinges on the liquid metal a cavity is produced. Molloy [19] defined three types of cavities; dimpling, splashing and penetrating. Hwang and Irons [20] defined the dimpling/splashing and splashing/penetrating transitions in terms of the Blowing number, approximately 1 and 4, respectively, equivalent to impact velocities of 13 and 28 m/s, respectively. The liquid displaced moves along the cavity surface producing a wave that moves along the free surface [21]. This effect is the same independently of the injection of reactive or non-reactive gases [22].

Lee et al. [23] investigated the effect of two types of injection angles on splashing phenomena, employing a 1/3 scale water model; the vertical (inclination) angle and the horizontal (offset) angle. The offset angle is zero when the lance is perpendicular to the center of the pitch circle. Low and high inclination angles promote splashing. Low angles promote front splashing and high angles, back splashing. Both are detrimental to the furnace equipment. These authors suggest an optimum vertical angle of 50° from the horizontal. Additionally, they also indicated that the vertical angle has a small effect in bath homogenization in comparison with the horizontal angle, suggesting an operation by offsetting the lance at least 10° from the furnace central line. Alam et al. [13] concluded that by increasing the vertical angle from the horizontal and by decreasing the height from the tip of the oxygen lance to the surface of molten steel, splashing is decreased. Therefore, to inhibit splashing the oxygen lance should be placed as close as possible to the surface of molten steel.

It has been reported that the penetration depth increases by increasing the injection angle from the horizontal [24–27]. Collins [24] reported variations in the inclination angle from 15 to 90°. Yang et al. [26] reported variations to the inclination angle from 25 to 35° using a water model with a geometric scale 1:10. Li et al. [27] defined the inclination angle from the vertical. In this case, they made small changes to the inclination angle from 14 to 17.5°. Their results also show that increasing the inclination angle from the horizontal, the penetration depth increases. Few of these studies have included the effect of the slag layer during oxygen injection [27–30]. Ek and Sichen [29] reported a decrease in the penetration depth because a fraction of the momentum is dissipated into the slag, Cao et al. [30] additionally indicated that the slag layer can restrain splashing.

4.3.2 Velocity Fields Due to Oxygen Injection

Chapter 12 discusses in detail EAF stirring, which includes fluid flow behavior due to all different stirring phenomena, including stirring due to oxygen injection. Table 4.1 summarizes the average velocity of liquid steel due to oxygen injection. It has been reported velocities of liquid steel ranging from 0.6 to 2 cm/s, which depend on the gas flow rate, number of lances, type of lance, inclination angles, etc.

Table 4.1 Average velocity of liquid steel due to oxygen injection

Year	Authors	EAF, ton	Nm^3O_2 per min	Average velocity, cm/s	Velocity at cavity region, cm/s	Jet type	Vertical angle, degrees
2000	Guo et al. [31]	190	51	2	–	S (1)	90
2010	Wang et al. [32]	150	50	0.6–0.7	10–50	S (4)	42
2010	Han et al. [33]	60	33	1.5	21	–	–
2011	He et al. [34]	150	33	–	21	–	–
2020	Chen et al. [35]	–	–	1.4	–	CJ (1)	40

S = Conventional supersonic jet, CJ = Coherent jet

4.4 Fe–O Equilibrium

Boom [36] discussed some of the multiple and opposing roles of oxygen; First, in ironmaking oxygen should be removed from iron ores using carbon, then during steelmaking oxygen is injected to remove carbon, finally oxygen should be removed as much as possible to produce clean steels.

Taylor and Chipman (1943) measured the Fe–O equilibrium, reporting the solubility of oxygen in liquid iron as a function of temperature as follows:

$$\log[\%O] = -\frac{6329}{T} + 2.734$$

Contrary to the high solubility of carbon in liquid iron, oxygen has a smaller solubility. The maximum solubility is 0.23%O at 1600 °C and increases to 0.48%O at 1800 °C.

The Fe–O equilibrium is defined by the following expression:

$$Fe_{(l)} + O = FeO_{(l)}$$
$$\Delta G_f^o = -109,710 + 45.89 \, T(J)(1800 - 2500 \, K)$$
$$K = \frac{a_0}{a_{FeO}}$$
$$a_0 = K \times a_{FeO}$$

The activity coefficient for oxygen in liquid steel and the equilibrium constant can be calculated:

$$\log f_O = -0.1 \times \%\underline{O}$$
$$\log K = -\frac{5730}{T} + 2.397$$

However, to compute the concentration of oxygen in equilibrium at the slag/metal interface, it is necessary to know the activity of iron oxide

$$(a_{FeO} = \gamma_{FeO} \cdot X_{FeO})$$

Basu et al. [37] reported the following expression to compute the activity coefficient of iron oxide:

$$\log \gamma_{FeO} = -0.7335 \log X_{FeO} - 0.2889$$

4.5 Decarburization Thermodynamics

Oxygen injected is dissolved in liquid steel and then reacts with dissolved carbon to form CO and CO_2, however under equilibrium conditions the concentration of CO_2 is negligible, as shown below. The following table summarizes the free energy of formation of CO (cal/mol), using data reported by Chipman (1954).

Reaction	ΔG_f^0, cal/mol (T)
$C_{(gr)} + \frac{1}{2}O_{2(g)} = CO_{(g)}$	$-26,700 - 20.95$
$C_{(1wt\%)} = C_{(gr)}$	$-5400 + 10.1$
$O_{(1wt\%)} = \frac{1}{2}O_{2(g)}$	$28,000 + 0.69$
$\underline{C} + \frac{1}{2}O_{2(g)} = CO_{(g)}$	$-32,100 - 10.85$

There are three possible decarburization reactions. The first one is between carbon particles in the slag and oxygen injected, the second one is between dissolved carbon and oxygen injected and the third one between CO_2 and carbon dissolved in liquid steel.

The free energy of formation for CO and CO_2 using data reported by Gokcen (1956) and Richardson (1953) is given below. The equilibrium constant is obtained using the Gibb's isotherm:

$$\Delta G^0 = -2.302RT \log K = -19.1T \log K; \ (R = 8.314 \text{ J/gmole } ^\circ K)$$

Reaction	ΔG_f^0, J/mol (T)	log K
$\underline{C} + \underline{O} = CO_{(g)}$	$-20,112 - 40.79$	$\log K_{CO} = \frac{1053}{T} + 2.135$
$CO_{(g)} + \underline{O} = CO_{2(g)}$	$-154,480 + 84.76$	$\log K_{CO_2} = \frac{8088}{T} - 4.438$
$CO_{2(g)} + \underline{C} = 2CO_{(g)}$	$134,368 - 124.796$	$\log K = -\frac{7035}{T} + 6.574$

Alternate ways to form CO_2 are:

$$\underline{C} + 2\underline{O} = CO_{2(g)}$$
$$CO_{(g)} + \frac{1}{2}O_{2(g)} = CO_{2(g)}$$

The table below summarizes all the decarburization reactions previously described. As will be shown below, in order to define the final decarburization rate is important to include the effect of iron oxide

Reaction	ΔG_f°, J/mol [38] (T)
$C_{(gr)} + \frac{1}{2}O_{2(g)} = CO_{(g)}$	$-114{,}400 - 85.8$
$\underline{C} + \frac{1}{2}O_{2(g)} = CO_{(g)}$	$-117{,}934 - 84.01$
$\underline{C} + \underline{O} = CO_{(g)}$	$-20{,}112 - 40.79$ $-22{,}200 - 38.34$
$CO_{2(g)} + \underline{C} = 2CO_{(g)}$	$144{,}700 - 129.5$

4.5.1 Carbon–Oxygen Equilibrium

The equilibrium concentration of CO and CO_2, assuming an ambient pressure equal to one atmosphere and carbon follows Henry's law ($h_C = \%C$), is calculated as follows:

$$K = \frac{p_{CO}^2}{p_{CO_2} \cdot h_C}$$

$$K_{(1600\,^\circ C)} = 657$$

$$p_{CO} + p_{CO_2} = 1$$

$$K = \frac{p_{CO}^2}{(1 - p_{CO}) \cdot h_C}$$

The previous expression can be re-arranged as follows:

$$p_{CO}^2 = 657(h_C - p_{CO} \cdot h_C)$$

$$p_{CO}^2 + (p_{CO})(657\% C) - 657\% C = 0$$

This expression gives the equilibrium concentration for CO and CO_2 as a function of the concentration of carbon dissolved in liquid steel, as shown below:

%C	p_{CO}	p_{CO_2}
0.01	0.88	0.110
0.05	0.97	0.028
0.10	0.98	0.014
0.50	0.99	0.003

It is observed that for concentrations above 0.05%C, the concentration of CO_2 is negligible, therefore it is valid to assume that the main product from the decarburization of liquid steel is only CO.

Under steelmaking conditions, the oxygen potential is controlled by carbon dissolved in liquid steel. This is very important because once the oxygen potential is defined it is possible to predict the concentration of other chemical species in the slag/metal system. The equilibrium C–O is defined from the following reaction:

$$\underline{C} + \underline{O} = CO_{(g)}$$

$$h_C \cdot h_O = [f_C(wt\% \, C)] \times [f_O(wt\% \, O)] = \frac{p_{CO}}{K}$$

where: h_C and h_O are the Henrian activities for carbon and oxygen, respectively, f_C and f_O are the activity coefficients for carbon and oxygen, respectively. The activity coefficients for carbon and oxygen activities in liquid iron are computed using interaction parameters, as follows:

$$\log f_j = \sum e_j^i(wt\% \, i)$$

Interaction parameters at 1600 °C:

e_C^i	e_O^i
$e_C^C = 0.23$	$e_O^O = 0$
$e_C^O = -0.24$	$e_O^C = -0.32$

$$\log f_C = e_C^C(wt\%C) + e_C^O(wt\%O)$$
$$= 0.23(wt\%C) - 0.24(wt\%O)$$
$$\log f_O = e_O^O(wt\%O) + e_O^C(wt\%C)$$
$$= -0.32(wt\%C)$$

Replacing these values, it is obtained an equation with two unknowns, carbon and oxygen:

$$\log \%C + \log \%O - 0.09\%C - 0.24\%O = \log p_{CO}K^{-1}$$

A solution is obtained defining the value for the atmospheric pressure, temperature and concentration of one of the chemical species, either carbon or oxygen. The following expression is obtained for 1%C, 1 atm and a temperature of 1600 °C.

$$\log\%O - 0.24\%O + 2.61 = 0$$

this is a transcendental equation, solved by a numerical method, gives:

$$O = 0.0024\%O = 24.5 \, ppm \, O$$

Alternatively, in a more simplified route, assuming $f_i = 1$, then $h_i = \%i$, it is obtained a relatively close approximation:

$$\%O = \frac{1}{K_{CO} \cdot \%C}$$
$$O = 11.8 \, ppm \, O$$

4.5.2   Carbon–FeO Equilibrium

Equilibrium under steelmaking conditions is defined by the carbon concentration. Once the carbon concentration is defined it would be possible to predict the composition of both liquid steel and liquid slag. The FeO–C equilibrium is defined by the following chemical reaction.

$$(FeO) + \underline{C} = CO_{(g)} + \underline{Fe}$$
$$K_{FeO-C} = \frac{a_{Fe}p_{CO}}{h_C a_{FeO}} = \frac{p_{CO}}{h_C a_{FeO}}$$

The value of the equilibrium constant can be defined from the following elemental reactions:

$$\underline{Fe} + \underline{O} = (FeO)$$
$$\log K_{FeO} = \frac{6150}{T} - 2.604$$
$$\underline{C} + \underline{O} = CO_{(g)}$$
$$\log K_{CO} = \frac{1160}{T} + 2.003$$

Combining the previous reactions:

$$\log K_{FeO-C} = -\frac{4990}{T} + 4.607$$

The solubility product [C]–(FeO) describes the equilibrium line and can be defined, knowing the values in the following relationship:

$$X_C X_{FeO} = \frac{p_{CO}}{f_C \gamma_{FeO} K_{FeO-C}}$$

After conversion of mole fraction to wt%:

$$\%C\%FeO = \frac{p_{CO}}{Cf_C\gamma_{FeO}K_{FeO-C}}$$

where: C is a constant that results from converting mole fractions into wt%'s.

The previous expression indicates that the concentration of FeO in equilibrium depends on the activity coefficient of FeO, dissolved carbon in liquid steel, ambient pressure and temperature.

At 1600 °C, K_{FeO-C} has a value of 87.66. The minimum pressure at the slag/ steel interface if the free surface is exposed to ambient pressure would be about 1.2 atmospheres considering the thickness of the slag layer. Turkdogan [39] summarized the activity coefficient of FeO as a function of basicity, indicating a maximum value of about 3 for a basicity of 1.8.

4.6 Decarburization Rate

4.6.1 Decarburization Rate

Figure 4.7 compares the decarburization rate in the BOF and a conventional EAF [40]. It can be observed that the rate of decarburization in the BOF is at least 4–10 times higher compared with the EAF. Cicutti et al. [41] reported for the BOF a maximum decarburization rate of 0.35%C/min decreasing to 0.05%C/min at the end of the blow, similar values were also reported by Shukla et al. [42], from 0.3 to 0.35%C/min by increasing the oxygen flow rate from 500 to 600 Nm^3/min, using 76% hot metal with 4.5%C. In contrast to the high DeC rate in the BOF, Lee and Sohn [43] reported the decarburization rate in the EAF, in the order of 0.01%C/min using scrap with 0.1%C. The DeC rate increases to 0.04%C/min by increasing carbon in the metallic charge to about 1.2%C. The DeC rate in the EAF can be increased using hot metal. These authors reported an increment in the DeC rate from 0.04%C/min to 0.08%C/min increasing carbon in the metallic charge from 1.2 to 2%C. Gottardi et al. [44] reported an average DeC rate of 0.14%C/min using 35% hot metal, a molten bath with an average of 2%C and an oxygen flow rate of 133 Nm^3/min. These results clearly indicate that the DeC rate strongly depends on the initial carbon content in the molten bath and the oxygen flow rate. Chapter 10 discusses in more detail the effects of using hot metal in the EAF.

Using hot metal in the EAF has increased the typical levels of decarburization using scrap or DRI. Figure 4.8 shows that increasing the size of the furnace increases the decarburization rate as a result of higher injection intensity of oxygen. In this plot plant A with a metallic charge of 100% scrap injects 7200 Nm^3/h reaching a decarburization rate about 130 kg C/min, plant B which charges 40% hot metal injects 12,000 Nm^3/h then its decarburization increases to 212 kg C/min. Plant C has the highest decarburization rate at 322 kg C/min injecting 18,000 Nm^3 O_2/h.

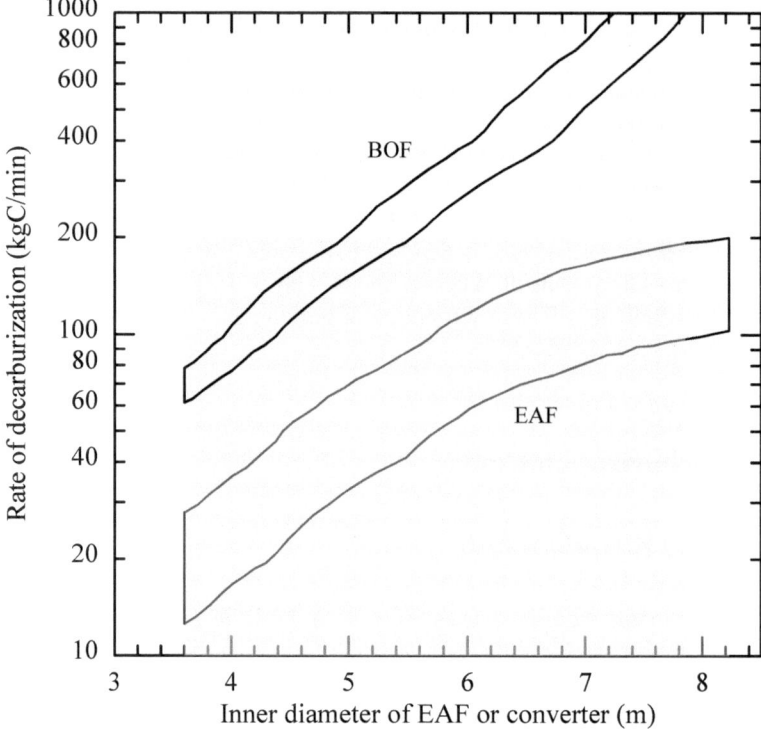

Fig. 4.7 Comparison on decarburization rates among BOF and EAF [40]

From a thermodynamic view point oxygen should react first with carbon and then with iron, however the oxidation rate of iron is faster than with carbon due to the larger availability of iron atoms.

Decarburization can be explained through a sequence of different chemical reactions, as shown below with two different reaction mechanisms.

Mechanism 1: Formation of iron oxide and the reduction of iron oxide by carbon dissolved in liquid steel. The formation of iron oxide with injected oxygen required that molecular oxygen dissolves in liquid steel to form FeO

$$\frac{1}{2}O_{2(g)} = \underline{O}$$
$$\underline{Fe} + \underline{O} = (FeO)$$

Then FeO in the slag reacts with dissolved carbon:

$$(FeO) + \underline{C} = CO_{(g)} + \underline{Fe}$$

Adding the three elementary reactions gives the global reaction:

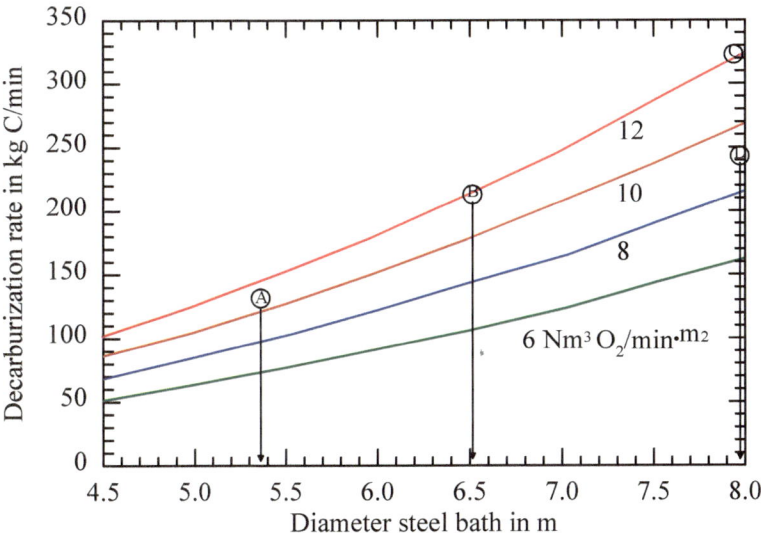

Fig. 4.8 Decarburization rate and generation of CO as a function of furnace diameter and oxygen injection. Plant A 100-ton, Plant B 180-ton, Plant C 300-ton. After [45]

$$\underline{C} + {}^1/_2 O_2 = CO_{(g)}$$

In this reaction mechanism the elementary reactions involve iron oxide but that is not shown on the global reaction.

Mechanism 2: Oxygen dissolves in liquid steel and then reacts with carbon dissolved in liquid steel. This mechanism assumes there is no formation of iron oxide.

$$^1/_2 O_{2(g)} = \underline{O}$$
$$\underline{C} + \underline{O} = CO_{(g)}$$

In both cases the final product is a gas phase. The formation and growth of a bubble inside of a liquid need to overcome huge forces. Kozakevitch (1953) defined the magnitude of those forces; If the bubble grows, the pressure inside the bubble should be larger than the combined ferrostatic pressure and pressure due to surface tension:

$$P_{CO} = \text{ferrostatic pressure} + \text{pressure due to Surface tension}$$

$$P_{CO} = p_o + \frac{2\pi r \sigma}{\pi r^2} = p_o + \frac{2\sigma}{r}$$

The surface tension at 1600 °C was defined as follows (Kozakevitch 1953):

$$\sigma = 1600 - (100 \times \%C); \; [\text{mN/m}]$$

The magnitude of the pressure required for a bubble to grow can be compared for three cases 1 mm, 1 μ and 6 Å. The ferrostatic pressure 36 cm below the surface of liquid steel is about 1.25 atmospheres and it will be assumed liquid steel contains 1%C. For the largest bubble the pressure due to surface tension is only 0.06 atm, for the small bubble is 58 atmospheres but for the initial nucleation of a bubble the pressure is about 10^4 atmospheres. This calculation indicates that homogeneous nucleation would not be possible. The only possibility is heterogeneous nucleation and the formation of CO bubbles at an interface, for example the refractory walls or through the walls of the bubbles being formed:

$$CO_{(g)} + (FeO) = CO_{2(g)} + \underline{Fe}$$
$$CO_{2(g)} + \underline{C} = 2CO_{(g)}$$

The global reaction, then becomes:

$$(FeO) + \underline{C} = CO_{(g)} + \underline{Fe}$$

From the previous reactions, $CO_{2(g)} + \underline{C} = 2CO_{(g)}$ is the slowest reaction. This gas/liquid reaction indicates adsorption of CO_2 at the slag/metal interphase. It has been studied extensively in the temperature range from 1200 to 1600 °C. It has been found that the reaction is first order with respect to the partial pressure of CO_2 in the pressure range from 0.1 to 0.5 atm. and independent of the carbon concentration, below 1%C, because carbon has a small effect on the interfacial tension of liquid iron. On the basis of interfacial control, the following reactions are relevant in the description of the overall decarburization process; CO_2 is adsorbed and dissociates at the interphase releasing oxygen atoms and CO, the oxygen atoms react subsequently with adsorbed carbon atoms adsorbed at the interface, forming CO.

$$CO_{2(g)} = CO_{2(ad)}$$
$$CO^{\circledR}_{2(ad)} O_{(ad)} + CO_{(ad)}$$
$$\underline{C} = C_{(ad)}$$
$$\underline{C} + O_{ad} = CO_{(ad)}$$
$$2CO_{(ad)} = 2CO_{(g)}$$

Alternatively, dissociative chemisorption of CO_2 can occur as follows:

$$CO_{2(g)} \rightarrow CO_{2(ad)}$$
$$CO_{2(ad)} = O_{(ad)} + CO_{(ad)}$$

If the decarburization process is controlled by interfacial reactions, tensoactive species block the reaction. Sulphur is a tensoactive specie. It has been reported that small concentrations, such as 0.05%S decrease the mass transfer coefficient. Sain and Belton [46] proposed the following relationship to describe the effect of sulphur on the reaction rate constant:

$$k_a = k_o[(1 - \theta_s) + 0.014\theta_s]$$

where: θ_s is the fraction of sites occupied by sulphur atoms.

Paul et al. [47] developed a kinetic model to describe the reduction rate of iron oxide by carbon in liquid steel. They estimated the activity coefficient of FeO using the regular solution model. They found that the controlling mechanism for the rate of reduction of iron oxide depends on the concentration of FeO. At low values, below 5%, the reaction is controlled by mass transfer of FeO in the slag and chemical reaction at the slag/metal interface. At intermediate values, from 5 to 40% is a mixed control mechanism, involving; mass transfer of FeO in the slag, chemical reaction at the gas/metal and chemical reaction at the gas/slag interface. For concentrations above 40%FeO, the reduction rate is controlled by chemical reactions both at the slag/metal and gas/slag interfaces.

As will be shown below, the stoichiometric requirements of oxygen as a function of carbon in liquid steel can defined as follows:

$$\text{kg-mol } O_2 = \left(\frac{W \times \%C}{\eta_{O_2} \times 100 \times 2 \times 12} \right)$$

re-arranging:

$$\%C = \left(\frac{\text{kg mol } O_2 \times 100 \times 2 \times 12}{W} \right) \times \eta_{O_2}$$

A simple expression for the rate of decarburization can now be defined:

$$\frac{d\%C}{dt} = -\left(\frac{\dot{n}_{O_2} \times 100 \times 2 \times 12}{W} \right) \times \eta_{O_2}$$

where: $\dot{n}_{O_2}$ is the rate of oxygen injection in kg mol/min, W is the mass of liquid steel in kg and η_{O_2} is the fraction of oxygen reacting with carbon to form CO.

The rate of decarburization changes during the heat. In the beginning there is a large concentration of carbon which makes oxygen diffusion the rate controlling mechanism but at the end carbon is low, then the rate of decarburization becomes limited by the diffusion of carbon, both cases indicate a mass transfer control mechanism. When the rate is controlled by mass transfer of carbon from the bulk to the reaction interface, it can be defined as follows:

$$V_m \frac{d\%C}{dt} = -k_c A(\%C_t - \%C_e)$$

where: k_c is the mass transfer coefficient in liquid steel in min^{-1}, A is the reaction interface, C_t is the concentration of carbon at time t and C_e is the carbon concentration at the slag/metal interface and assumed to be the carbon equilibrium concentration. After integration, from t = 0, % C = % C_0 and t = t, % C = % C_t:

$$\ln \frac{(\%C_t - \%C_e)}{(\%C_0 - \%C_e)} = -\frac{k_c A}{V_m} t$$

This equation can be re-arranged as follows:

$$[\%C_t] = [\%C_e] + \{[\%C_0] - [\%C_e]\} \times \exp\left(-\frac{k_c A}{V_m} t\right)$$

The reaction interface plays a big role on the decarburization rate. Increasing the stirring conditions will increase the decarburization rate, on the other hand surface active species like sulphur in steel or P_2O_5 and SiO_2 in the slag will decrease the decarburization rate.

Also, depending on the concentration of FeO, the reaction can become limited due to diffusion of FeO from the bulk of the slag to the reaction interface, then a similar expression can be derived. In this case the mass transfer coefficient of FeO in liquid slag, to simplify the calculations is taken as 1/10 of the mass transfer coefficient of carbon in liquid steel.

$$V_s \frac{d\%FeO}{dt} = -k_{FeO} A \left(a_{FeO}^t - a_{FeO}^{eq}\right)$$

Wei et al. [48] investigated the decarburization rate in high carbon melts, 4.4%C, at 1300 °C. They found a linear relationship between the activity of iron oxide and the decarburization rate, as shown in Fig. 4.9 The activity of FeO was calculated using the following relationship, valid from 0.5 to 15% FeO.

$$a_{FeO} = 0.0047(\%FeO) + 0.0005(\%FeO)^2 + 0.0067$$

The efficiency of oxygen injection on the DeC rate under isothermal conditions depends on the following variables; lance inclination angle, distance from the tip of the nozzle to the liquid metal, gas flow rate, number of lances, density and surface tension of both slag and liquid steel, as well as slag thickness. In the BOF the high rates of oxygen injection promote slag emulsification and this phenomenon was first defined by Mayer et al. as the main zone for DeC [49]. Schoop et al. reported the importance of the hot spot region as the primary DeC zone [50]. This zone has a much higher temperature, about 500–800 °C, in comparison with the bulk temperature. Some reports identify either the impact of slag-metals zones as more relevant [51].

DeC kinetics follows two principal mechanisms depending on the carbon content dissolved in liquid steel [52–54]; at high carbon contents, above 0.2–0.4%C, the large availability of carbon makes the oxygen supply as the rate limiting step, on the other hand, at low carbon concentrations the diffusion of carbon is the rate limiting step,

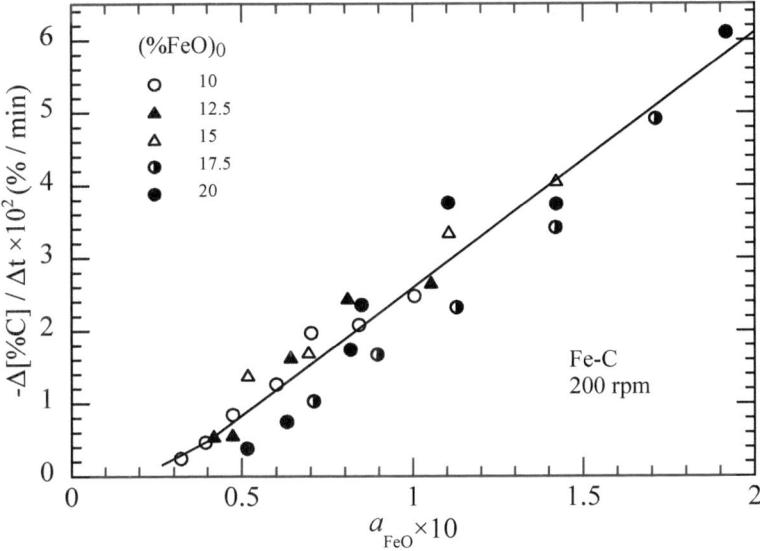

Fig. 4.9 Effect of the activity of iron oxide on the decarburization rate [48]

therefore, to compute the decarburization rate, two equations are required to satisfy any carbon content. The refractory can also be a source of oxygen, MgO shows a higher DeC rate compared with Al_2O_3 or CaO [55].

Memoli et al. [56] described a DeC model that included the oxygen injection conditions, however, the description of relevant information is missing. He et al. [57] developed a mathematical model to describe the effect of the height of the oxygen lance on the rate of decarburization in a 150 ton EAF with a fixed injection angle of 42°. They reported an optimum lance height of 0.45 m as a result of the opposing effect between lance height and exposed area for DeC; by increasing lance height, the penetration depth increases but the area exposed for DeC decreases. Li et al. [58] incorporated a term in the rate of decarburization to define the mixing degree and found that by decreasing the height from the tip of the oxygen lance to the surface of molten steel, there is an increase in the extent of mixing in the bath and the rate of decarburization increases. Chen et al. [35] reported a mathematical model for DeC using four coherent oxygen jets with a fixed inclination angle of 45°. In previous works the motion of the fluid is primarily associated with the momentum of the jet. In this work the authors reported the large difference when bubble stirring is included in the analysis; the average velocity of the liquid was 0.014 m/s due to momentum stirring and increased 10 times to 0.14 m/s when the effect of CO stirring was included. Due to lower stirring conditions in the EBT region, the carbon concentration was higher with respect to the rest of the liquid in the EAF, in addition to this, they also observed the lowest carbon concentration around the cavities formed by the impingement jets. Consistent with the improved stirring conditions due to CO stirring, mixing time decreased from 1665 s to 174 s. without and with bubble stirring,

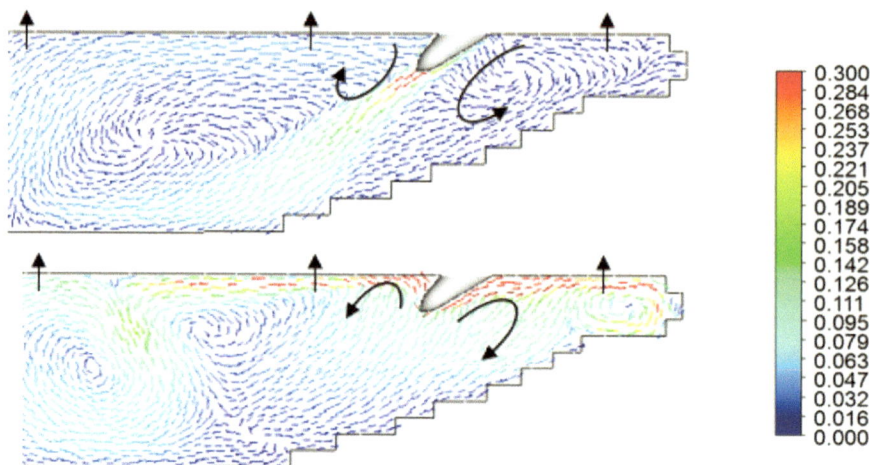

Fig. 4.10 (Top) only O$_2$ jet, (bottom) O$_2$ jet + CO bubbles. After [35]

respectively [59]. The flow pattern is also different. A flow pattern with a defined high velocity flow along the jet direction from the tip of the cavity that reaches the bottom of the EAF is produced when the jet momentum is the only driving force. This pattern disappears when the bubbles are taken into account, leading to more homogeneous and enhanced mixing conditions, as shown in Fig. 4.10.

The injection angle of the oxygen lance (s) plays a big role in fluid flow and consequently on oxygen efficiency and the rate of DeC. There are many investigations on DeC of liquid steel in the EAF [60–65], however previous research has assumed a fixed lance inclination angle. Few research works have been carried out to investigate the role of the injection angle on the rate of DeC. A high inclination angle from the vertical promotes splashing and electrode wear. A low inclination angle also promotes back splashing to the nozzle contributing to its early failure. An inclination angle of 40° from vertical has been reported to minimize splashing [13, 23]. Ramirez et al. [66] found that increasing the inclination angle from the vertical, from 49 to 80°, the decarburization rate increases due to higher velocity fields and stronger stirring conditions in the bulk of liquid steel, however the splashing conditions are more severe. In this work in addition to the injection angle also included the effect of gas flow rate and number of lances on the rate of DeC of liquid steel. Variations in lance efficiency can result in a variation of electric energy consumption of ± 30 kWh/ton [67].

4.6.2 Effect of DeC on N₂ and H₂

An additional benefit of decarburization is the removal of both hydrogen and especially nitrogen. The maximum solubility of nitrogen in pure iron is 450 ppm but can be reduced to 80–100 ppm. Higher decarburization rates improve the removal of both N_2 and H_2. Jones [68] indicated "it has been demonstrated that decarburizing at a rate of 1% per hour can bring the hydrogen content from 8 to 2 ppm in 10 min". The nitrogen concentration in scrap-EAF operations is on average higher than in the BOF, 40–110 pm and 10–40 ppm, respectively [69]. The nitrogen content in scrap can range from 30 to 120 ppm and in DRI from 20 to 30 ppm.

Using DRI decreases nitrogen due to the stirring effect created by the reduction of FeO. Anderson et al. [69] reported that increasing DRI from 0 to 100% can decrease nitrogen from 75 to 20 ppm, as shown in Fig. 4.11.

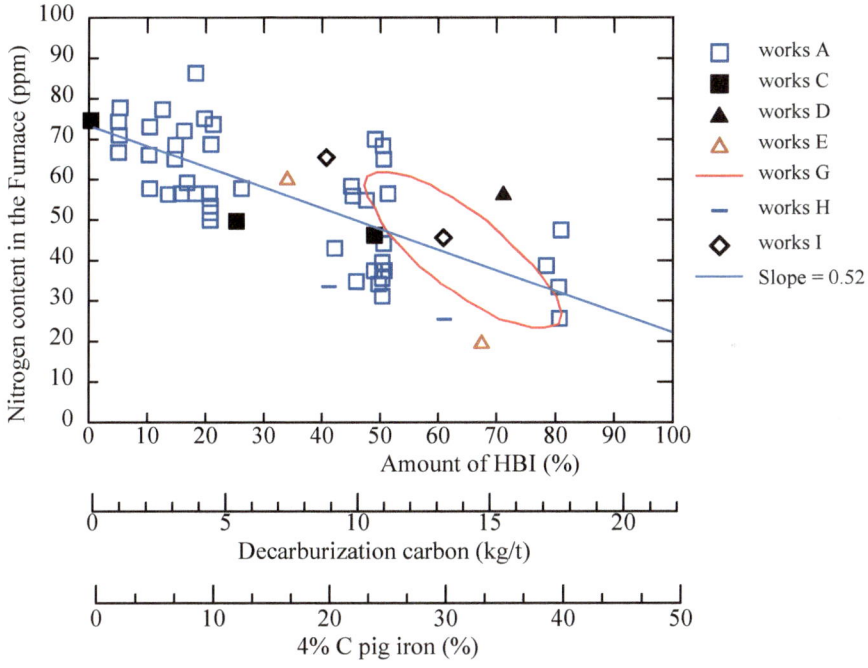

Fig. 4.11 Nitrogen content as a function of DRI in the metallic charge. After [69]

	Nm³/ton	Nm³/min·ton	Chemical energy, % P_{total}
1980s	3–8		
1990s	> 22		20–30
2000s	18–27	0.4–0.9	10–30

Table 4.2 Typical oxygen consumption and equivalent chemical energy, according with Jones et al. [14–16]

4.7 Oxygen Consumption; Trends, Theoretical Calculation and C–O Balance

4.7.1 Trends in Oxygen Consumption

Jones et al. [70–72] compared the rapid changes on oxygen injection from the late 1980s to the late 1990s. In the late 1980s oxygen consumption was less than 10 Nm³/ton, usually from 3 to 8 Nm³/ton. In the late 1990s reached levels of 35 Nm³/ton and up to 70 Nm³/ton considering oxygen for post-combustion. This amount of oxygen increased the input of chemical energy up to 30–40%. The values reported for oxygen injection in the early 2000s ranged from 18 to 27 Nm³/ton [72]. The chemical energy due to exothermic reactions was equivalent to 10–30% of the total energy input. Optimum values have been suggested in the range from 28 to 35 Nm³/ton. A flow rate of 30 Nm³/ton is equivalent to a power input of 11 MW, assuming combustion of C to CO_2. The specific gas flow rate ranges from 0.4 to 0.9 Nm³/min·ton but can be as high as 1.4 Nm³/min·ton in order to increase the input of chemical energy. Table 4.2 summarizes the previous data.

4.7.2 Volume of Oxygen Required for DeC

The total value of oxygen includes oxygen for decarburization (DeC), burners and post-combustion:

$$O_{total} = O_{DeC} + O_{burner} + O_{PC}$$

The minimum volume requirements is obtained from the following chemical reaction:

$$\underline{C} + {}^1\!/_2 O_{2(g)} = CO_{(g)}$$

According with the stoichiometry of this reaction, 12 kg-mol of carbon react with ($\frac{1}{2} \times 32$) kg-mol of oxygen.

$$Nm^3O_2 = (kg\,C)\left(\frac{\frac{1}{2} \times 32}{12}\right)\left(\frac{kgmol\,O_2}{32\,kg}\right)\left(\frac{22.4\,Nm^3}{1\,kgmol\,O_2}\right)$$

$$Nm^3O_2 = \left(\frac{kgsteel \times \%C}{100}\right)\left(\frac{1.33}{1.428}\right)$$

$$Nm^3O_2 = \left(\frac{W \times \%C}{100}\right)\left(\frac{1.33}{1.428}\right)\left(\frac{1}{\eta_{O_2}}\right)$$

where: 12 is the atomic mass of carbon and 32 the atomic mass of molecular oxygen, W is the total amount of liquid steel in kg, %C is the concentration of carbon dissolved in liquid steel in wt% and η_{O_2} is the efficiency of oxygen.

This analysis assumes that all oxygen injected forms CO. This is true for carbon concentrations above 0.05%C but for ultralow carbon the formation of CO_2 can be as high as 20%. The fraction of oxygen that reacts with carbon (η_{O_2}) is estimated to be about 0.75.

In practice is common to define the volume of oxygen using the concept of points. One point of an element, for example carbon, is equal to 0.01%C, which is equivalent with 100 g/ton:

$$1\,point\,C = 0.01\%C = 0.1\,kg/ton = 100\,g/ton$$

Using this unit, according with the previous Eq. 1 point of C requires 93.3 L of oxygen per ton of steel. To decarburize 20 points of carbon in a furnace with 220 ton of liquid steel, it would require 410 Nm^3 O_2 with an efficiency of 100%:

$$93.3 \times 20\,points\,C \times 220\,ton = 410.5\,Nm^3O_2.$$

An alternate expression for the oxygen requirements is:

$$kg\text{-}mol\,O_2 = (kgC)\left(\frac{\frac{1}{2} \times 32}{12}\right)\left(\frac{kgmol\,O_2}{32\,kg}\right) = \left(\frac{kgsteel \times \%C}{100}\right)\left(\frac{kgmol\,O_2}{2 \times 12}\right)$$

$$kg\text{-}mol\,O_2 = \left(\frac{W \times \%C}{\eta_{O_2} \times 100 \times 2 \times 12}\right)$$

where: W is the weight of liquid steel in kg.

Jones et al. [71] indicates that the typical oxygen gas flow rates range from 30 to 100 Nm^3/min depending on the furnace size, as shown in Fig. 4.12, corresponding to 0.85 and 0.78 Nm^3/min ton for an EAF of 40 and 106 ton capacity, respectively. Using hot metal in the EAF requires a much higher volumes of oxygen, about 280 Nm^3/min, closer to the values employed in the BOF. In a conventional BOF, the specific gas flow rate ranges from 2.5 to 3.8 Nm^3 O_2/ton min and using combined blowing from 3.5 to 5.0 Nm^3 O_2/ton min [40].

It is important to notice that the volume of off-gases increases due to a higher generation of chemical energy based on fossil fuels. The capacity of the off-gas

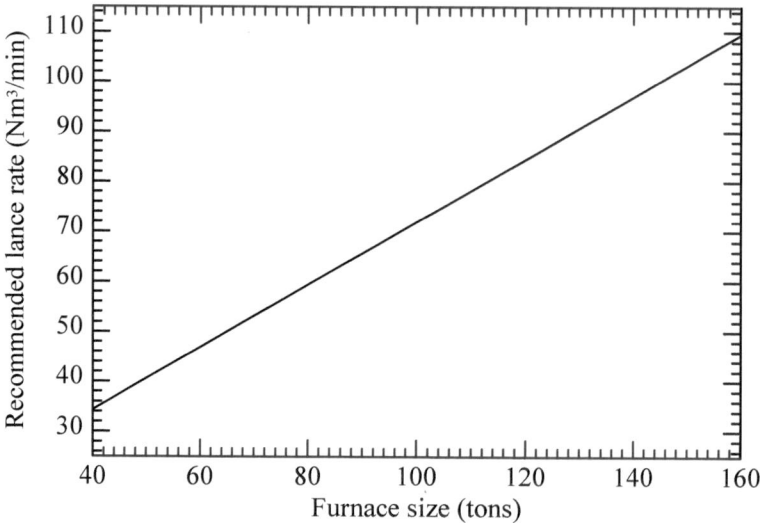

Fig. 4.12 Gas flow rate of oxygen injection as a function of EAF capacity [71]

system is an important factor that should be taken into account when chemical energy is increased in existing facilities. Considering that a fraction of 0.75 of the oxygen injected reacts with carbon, every Nm^3 O_2 will produce 1.5 Nm^3 CO, as shown below:

$$\underline{C} + {}^1\!/_2 O_{2(g)} = CO_{(g)}$$

$$Nm^3 CO = Nm^3 O_2 \times \frac{C}{\frac{1}{2}O_2} = 0.75 \times \frac{1}{0.5} = 1.5$$

In addition to CO produced due to DeC, the iron oxide formed can be reduced by either injected carbon particles into the slag or carbon at the slag/liquid interface, increasing the total volume to about 2.5 Nm^3 CO/Nm^3 O_2 [71].

4.7.3   Balance C–O

The most important aspect when defining the maximum volume of oxygen injection for decarburization (O_{DeC}) is to keep a C–O balance. Using an excessive volume of oxygen will lead to higher iron oxidation and lower metallic yield. The maximum volume of oxygen depends on the type of metallic charge and in addition to carbon, the concentration of other oxidable elements like Si and Mn. The amount of carbon increases from scrap to DRI to hot metal. Defining limits is relative to the amount of those oxidable elements in the metallic charge. The contribution of chemical energy due to oxygen injection in the range from 18 to 36 Nm^3/ton, can be in the order of 50–100 kWh/ton and a decrease in tap-to-tap time from 3 to 6 min.

In addition to decarburization (DeC) of liquid steel, oxygen injection serves two other purposes; replace electric by chemical energy and cutting steel scrap. Oxygen injection is the most important variable to replace electric energy by chemical energy in the EAF.

The chemical energy released due to oxidation of the elements dissolved in hot metal in the BOF sustains high temperatures of the molten liquid without the need of electric energy. The gas flow rate of oxygen injection in the BOF can fluctuate in a broad range, from 560 to 1000 Nm^3/min [73].

Current technologies can operate without hot metal in the BOF. The Energy Optimizing Furnace (EOF) from Brazil and the Z-BOP from Russia have the potential to operate with 100% pre-heated scrap using the chemical energy supplied by oxygen and carbon injection. Using 18% scrap, which is a typical value from a conventional BOF, the EOF process has an specific oxygen consumption of 62 Nm^3/ton [74], however in the case of a 100% scrap operation, oxygen consumption can reach from 150 to 170 Nm^3/ton and that of carbon or coke from 95 to 110 kg/ton [75]. Such high levels of carbon will increase sulphur in liquid steel.

Oxygen injection is the second most important source of energy in the EAF. In order to increase oxygen injection carbon should also be added accordingly. Torres et al. [76] discussed in detail the limits of oxygen injection in order to avoid negative effects on the metallic yield using 100% DRI. If DRI of 94.5% metallization and 1%C is charged there is no need to inject oxygen because the 6.4% FeO in the DRI is enough to reduce carbon in the DRI. About 15 Nm^3/ton is needed if carbon in DRI increases to 2%. Their results show that oxygen injection in the order of 30 Nm^3/ton produce about 44% FeO, which is a clear indication of a deficit in carbon input. Adding excessive oxygen injection without a proper C–O balance can be extremely detrimental to metallic yield because the formation of iron oxide severely increases, furthermore, decarburization below 0.04%C under atmospheric conditions exponentially increases the oxidation conditions of liquid steel; liquid steel with 0.04%C is in equilibrium with 500 ppm O and for 0.02%C is 1000 ppm O. It is very simple to identify an operation where there is no C–O balance, expressed in very high values of FeO (more than 30%) or dissolved oxygen (more than 800 ppm).

A high concentration of iron oxide has many detrimental effects, not only decreases metallic yield, high FeO reduces slag viscosity which affects slag foaming. A fluid slag is more aggressive to the refractory, also decreases slag basicity affecting the removal of phosphorous which in turn would demand more fluxes and also, the decrease in basicity increases the saturation concentration of MgO [77]. Therefore, the concentration of FeO should be kept as close as possible to equilibrium values and strictly avoid values above 25–30%. At ambient pressure, the FeO–C equilibrium depends on the activity of FeO, which in turn depends on the slag chemical composition. The product %C × %FeO = constant. The value of this constant has been reported by different authors, as follows:

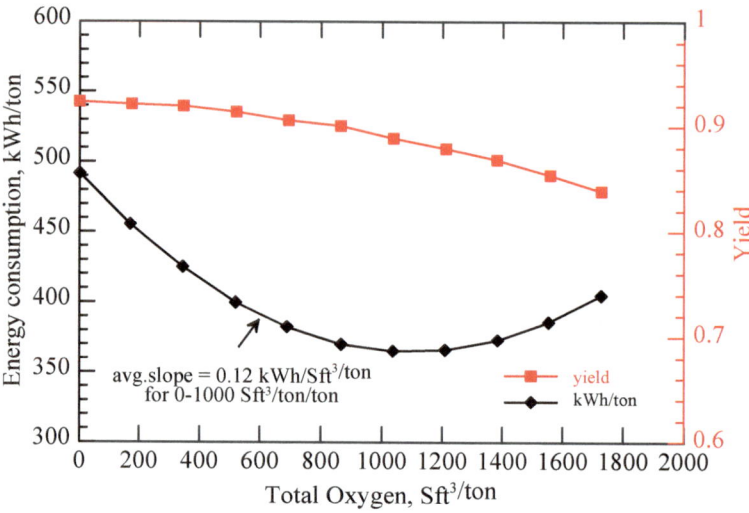

Fig. 4.13 Effect of oxygen injection on energy consumption and metallic yield. After [71]

References	[78]	[79]	[80]	[39]
[%C] × (%FeO)	1	0.55	0.33	
$\sqrt{\%C}(\%FeO)$				4.2 ± 0.3

Real values are usually above equilibrium. Turcotte et al. [81] proposed the following relationship to predict FeO in a real practice.

$$(\%FeO) = \frac{100}{points\ C} + 10$$

where: points C = %C × 100

Inagaki et al. [71] have suggested an optimum range on oxygen injection from 20 to 30 Nm^3/ton to avoid excessive yield losses, as shown in Fig. 4.13.

4.8 Chemical Energy from Oxygen Injection

4.8.1 Chemical Energy from Oxygen Injection

Oxygen injection is employed for decarburization and dephosphorization of liquid steel. Carbon and phosphorous and not iron, is the intended purpose of oxygen injection, furthermore, from an equilibrium view point the reaction between carbon and oxygen is more spontaneous, comparing free energies of formation of − 23 kJ/mol for FeO and − 96 kJ/mol for CO, however, since steel is primarily made of iron atoms, it is unavoidable that oxygen is primarily in contact with iron atoms rather than carbon atoms and therefore FeO will always form.

The chemical energy due to oxygen injection derives from the exothermic reactions, as shown below. Oxidation of carbon releases a maximum value of 2.8 kWh/ Nm^3 O_2 assuming an efficiency of 100%. Since actual energy savings due to oxygen injection are in the order of 2–4 kWh/Nm^3 O_2 with an average value of 3.5 kWh/ Nm^3 O_2, it is an indication that oxygen reacts with both carbon and iron [71]. The fraction of oxygen that reacts with carbon has been estimated in the order of 70–80% [71].

Chemical reaction	ΔH (kWh/Nm^3 O_2)
$Fe + \frac{1}{2} O_{2(g)} = (FeO)$	− 6.0
$Si + \frac{1}{2} O_{2(g)} = (SiO_2)$	− 11.0
$C + \frac{1}{2} O_{2(g)} = CO_{(g)}$	− 2.8
$CO_{(g)} + \frac{1}{2} O_{2(g)} = CO_{2(g)}$	− 5.8
$Mn + \frac{1}{2} O_{2(g)} = (MnO)$	− 13.0

The theoretical requirement of oxygen to produce 10 kWh is about 2.78 Nm^3, however due to efficiency the actual value is in the range from 3.17 to 8 Nm^3 with an average value of 4.3 Nm^3, which represents an average input of chemical energy of 2.32 kWh/Nm^3 O_2. Considering the heat of combustion from C to CO_2, an oxygen flow rate of 30 Nm^3/min would release 11 MW of heat.

The company BSW [70] reported several benefits increasing the consumption of oxygen from 9 to 27 Nm^3/ton; a decrease in electric energy consumption of 137 kWh/ ton (consumption decreased from 494 to 357 kWh/ton) and higher productivity, from 32 to 85 ton/h.

The impact factor of oxygen injection on the specific input of chemical energy is shown in Table 4.3, based on references [67, 82–88]. The range covers from 2.69 to 6.9 kWh/ton. This value depends on the chemical composition of the metallic charge and efficiency of oxygen injection. The impact factor using DRI appears to be lower in comparison with scrap. This can be attributed to the fact that charging DRI introduces oxygen in the form of FeO and therefore in DRI operations the values of oxygen injection can be lower. Adams et al. [67] reported value is based on 25 m^3/ton oxygen injection, 14 kg total carbon/ton and assumes that 50% SiO_2 in the slag comes from the refractories. Inagaki et al. [82] reported values correspond to specific oxygen injection values of 15 and 30 Nm^3/ton, respectively.

Pesamosca and Patrizio [89] in 2016 reported a different perspective about the use of oxygen as a source of chemical energy, considering data from 8 different steel plants that employed 100% scrap. Their analysis considered the following aspects:

- Is a fact that increasing chemical energy also increases the total energy input; the total energy increased 38% adding 250 kWh/ton of chemical energy and 48% adding 350 kWh/ton, however, the actual energy transferred to liquid steel decreased, from about 59–53%. In other words, the efficiency of electric energy is higher in comparison with chemical energy.

Table 4.3 Impact factors of oxygen injection on savings on energy consumption, kWh/Nm^3 O$_2$

Scrap-based						DRI-based		
Inagaki, 1988	Kohle, 1992	Bowman, 1995	Adams, 2001	Memoli, 2004	Pfeifer, 2005	Geiger, 1980	Conejo, 2006	Cardenas, 2007
− 4.7 to − 6.8	− 4.3	− 4.0	− 5.2	3.2	− 4.3 to − 6.9	− 3	− 2.69	− 2.95

- 1 Nm^3 oxygen injection/ton without compensating with carbon results in 4.2 kg of iron losses and high metallic yield losses (0.38%) but even if there is compensation, 1.35–1.47 kg C/Nm^3, there is still some metallic yield losses (about 0.03%) because the reduction of FeO by C is lower compared to the rate of formation of FeO.
- Increasing use of oxygen results in higher production of CO$_2$, in the range from 7 to 18 kg CO$_2$/ton steel

Their suggestion was to decrease the input of chemical energy, in particular if the productivity demand cannot be achieved with additional electric energy.

4.8.2 The Ultra High Chemical Power EAF (UHCP-EAF) or Fuel Arc Furnace (FAF)

The Ultra High Chemical Power (UHCP)-EAF, according to Concast (Currently SMS) is characterized by three features: higher capacity oxy-fuel burners, higher capacity transformers and scrap pre-heating. Gottardi et al. [90–93] reported performance results from three UHCP-EAF's in the USA and Turkey, as shown in Table 4.4. The EAF with a higher productivity has higher consumption of oxygen. The UHCP injects 20–30% more oxygen than a conventional EAF, obtaining savings in electric energy in the order of 35 KWh/ton.

These authors reported operational details of two steel shops; Icdas I and Icdas II. At Icdas I, the first charge of scrap is preheated and a second scrap bucket is cold scrap. The first scrap bucket charges about 60% of the total amount of scrap. The EAF has a total burners capacity of 39 MW, with a total oxygen flow rate of 21,000 Nm^3 O$_2$/h. The oxygen delivered by the oxy-fuel burners with respect to the upper steel surface ranges from 100 to 300 Nm^3/h m^2. The value for an UHCP with hot metal increases to 47 Nm^3/ton. The design values for Icdas II include 8 burners of 8 MW each separated every 3 m. Its power density is (8 × 8)/27 = 2.4 MW/m of shell circumference. UHCP operate with energy densities up to 3.2 MW/m.

Fuel Arc Furnace (FAF): Toulouevski and Zinurov [94, 95] have reported a two stage heating process. In the first stage heating is carried out only with chemical energy until about 70% of the total energy input is introduced, followed by a second stage where electric energy accounts for only about 14% of the total energy and

Table 4.4 Performance comparison between three different EAF plants. After [90–93]

	A	B	C		A	B	C
Nominal capacity, tons	82	175	220	Power-on time, min	24		32
Shell diameter, m	6.7		8.7	Power-off time, min	8		
Shell volume, m^3	140		280	Tap-to-tap time, min	32	47	41
Height of WCP, m	3.1			Heats/day	45		
Transformer, MVA	110	168	230	Productivity, ton/h	150	230	320
Transformer, MVA/ton	1.3	0.96	1.04	Productivity, ton/h MVA	1.36	1.37	1.39
Reactance, Ω/phase	2.5		1	Energy (cold scrap), kWh/ton	380	325	330
Active power, MW			145	Energy (hot scrap), kWh/ton		290	295
Oxy-fuel burner, MW	5		8	Oxygen cons, Nm^3/ton	30	47	44
Number of burners	4		8	Natural gas, Nm^3/ton	4.8	6	6.1
Oxygen flow rate, Nm^3/h	2500		3500	Carbon injection, kg/ton	14		
O$_2$ PC mode, Nm^3/h	1000			Electrode cons, kg/ton	1.5	0.9	1.0
Oxygen in Nm^3/h ton	121		127				
Chemical energy, %			41	Fluxes, kg/ton			35
Hot heel, %	23/30			Metallic yield, %			90
Number scrap buckets	1		2	Area WCP per ton, m^2/ton			0.4
Scrap density, t/m^3	0.73			Radiation index, kWV/cm^2			282
Scrap charging time, s	50–70			Geometry, H/D			0.41
Max C injec rate, kg/min	80		80	Total power, MW			240
Prod. steel, Mton/year	1.1	1.6	2.3				

Notes %hot hill with respect to tapping weight, oxygen flow rate per burner

the remainder is also chemical energy. Their proposal includes charging preheated scrap and novel high-power rotary (HPR) burners. Carbon materials (coal or coke) are charged with the scrap to decrease iron oxidation during post-combustion. When the scrap reaches a critical temperature that could increase iron oxidation, about 800 °C, corresponding to 75–80% of the melting process, burners are turned-off. These authors reported a reduction in energy costs of about 5.4 USD/ton [94]. This cost is based on a price of 0.035 USD/kWh for electric energy and 0.09 USD/m^3 for natural gas. The specific consumption of natural gas increases from 8 m^3/ton using conventional burners to 20 Nm^3/ton in the FAF process, similarly, oxygen injection increases from 40 to 82 Nm^3/ton. Increasing chemical energy due to oxygen injection increases the production of CO_2.

Table 4.5 Average chemical composition of tires [96, 97]

	EU	Lightweight trucks	Lightweight automobiles
Rubber (elastomeric compounds)	47	40–50	35–55
Carbon	21.5	22–30	20–30
Steel	16.5	5–17	5–16
Fiber (nylon, polyester)	5.5	2–10	2–8
Zinc oxide	1	1–2	1–4
Sulphur	1	0.5–1.5	1–3
Softener (petroleum oil, etc.)		2–11	2–12

4.8.3   Additional Sources of Chemical Energy

Tires: Tires are a carbon substitute. Gorez et al. [96] reported the feasibility to replace shredded tires (10 cm) by about 1.7 kg tires/kg carbon and a maximum level of addition of 8–12 kg tires/ton steel which is equivalent to the levels of addition of carbon in the EAF, in the order of 5–12 kg C/ton. The chemical energy derived from combustion of tires comes from the oxidation of both carbon and iron. Nakao and Yamamoto [97] reported the heating value of tires in the range from 7500 to 8100 kcal/kg. Table 4.5 shows an average tire composition, noticing that sulphur is also similar to other carbon sources.

The upper limits on tire consumption are defined by the excessive generation of fumes and the capacity of the fume extraction system. It has been recommended to pay special attention to the charging method and avoid direct contact of the tires with either liquid steel or burners, because the volume of fumes and bad smells severely increases.

References

1. Saberifar S, Jafari F, Kardi. H, Jafarzadeh MA, Mousavi SA (2014) Recycling evaluation of mill scale in EAF. J Adv Mater Process 2:73–78
2. Almqvist E (2003) History of industrial gases. Springer Science, New York, USA
3. Finlayson D, Sharp AJ (1932) Concentration of "vapours" by a pressure swing adsorption over a charcoal adsorbent. UK Patent 365092
4. Skarstrom CW (1960) A method and appartus for fractioning gaseous mixtures by adsorption. US 2944627
5. Shver VG (2001) Reusable lance with consumable refractory tip. US Patent 6212218
6. Jamnik M, Gemo L, Partyka A, Miani S, Rahm C (2018) Application of EAF wall injectors for high-alloy steel production. Iron Steel Technol 15:42–49
7. Anderson JE, Mathur PC, Selines RJ (1998) Method for introducing gas into a liquid. US Patent 5814125
8. Malfa E, Maddalena F, Giavani C, Memoli F (2005) Numerical simulation of a supersonic oxygen lance for industrial application in EAFs. MPT Metall Plant Technol Int 28:44–50
9. Anderson JE, Farrenkopf DR (1998) Coherent gas jet. US Patent 5823762

10. Mathur P, Messina C (2001) Praxair CoJet technology—principles and actual results. AISE Steel Technol 78:21–25
11. Cates L, Fujimoto K, Okada Y, Okano H (2008) Installation of Praxair's CoJet® gas injection system at Sumikin steel and other EAFs with hot metal charges. In: AISTech 2008 proceedings, PIttsburgh, PA, USA, pp 1–9
12. Thongjitr P, Kowitwarangkul P, Pratumwal Y, Otarawanna S (2023) Optimization of oxygen injection conditions with different molten steel levels in the EAF refining process by CFD simulation. Metals 13. https://doi.org/10.3390/met13091507
13. Alam M, Irons G, Brooks G, Fontana A, Naser J (2011) Inclined jetting and splashing in electric arc furnace steelmaking. ISIJ Int 51:1439–1447. https://doi.org/10.2355/isijinternational.51.1439
14. Wei G, Zhu R, Wu X, Dong K, Yang L, Liu R (2018) Technological innovations of carbon dioxide injection in EAF-LF steelmaking. J Met 70:969–976. https://doi.org/10.1007/s11837-018-2814-3
15. Wei G, Zhu R, Tang T, Dong K, Wu X (2019) Study on the impact characteristics of submerged CO_2 and O_2 mixed injection (S-COMI) in EAF steelmaking. Metall Mater Trans B Process Metall Mater Process Sci 50:1077–1090. https://doi.org/10.1007/s11663-018-1482-6
16. Wei G, Zhu R, Han B, Yang S, Dong K, Wu X (2020) Simulation and application of submerged CO_2–O_2 injection in electric arc furnace steelmaking: modeling and arrangement of submerged nozzles. Metall Mater Trans B Process Metall Mater Process Sci 51:1101–1112. https://doi.org/10.1007/s11663-020-01816-w
17. Wei G, Zhu R, Yang S, Wu X, Dong K (2021) Simulation and application of submerged CO_2–O_2 injection in EAF steelmaking: combined blowing equipment arrangement and industrial application. Ironmak Steelmak 48:703–711. https://doi.org/10.1080/03019233.2021.1896068
18. You X, He S, Zhang M, Zeng J, Li L, Wang Q et al (2020) Thermodynamic discussion of CO_2 injection in molten steel. Steel Res Int 91:1–5. https://doi.org/10.1002/srin.201900450
19. Molloy NA (1970) Impinging jet flow in a two phase system: the basic flow pattern. J Iron Steel Inst 208:943–950
20. Hwang HY, Irons GA (2012) A water model study of impinging gas jets on liquid surfaces. Metall Mater Trans B 43:302–315. https://doi.org/10.1007/s11663-011-9613-3
21. Whitney V (2003) Physical modelling of fluid flow in electric arc furnace caused by impinging gas jets. Ironmak Steelmak 30:209–213. https://doi.org/10.1179/030192303225001793
22. Van der Lingen TW (1966) Penetration of inclined gas jets into liquid baths. J Iron Steel Inst 204:320–325
23. Lee M, Caffery G, Molloy N (2001) Metallurgical performance of the water-cooled lance at the BHP Sydney steel mill: a physical modelling study. Scand J Metall 30:220–224. https://doi.org/10.1034/j.1600-0692.2001.300404.x
24. Collins RD, Lubanska H (1954) The depression of liquid surfaces by gas jets. J Appl Phys 5:22–26
25. Holden C, Hogg A (1960) The physics of oxygen steelmaking. J Iron Steel Inst 196:318–332
26. Yang W, Zhou J, Wang M, Ding Y, Yu P, Xu D (2002) Simulative experiment of oxygen utilization for large EAF steelmaking and its industrial application. J Iron Steel Res 14:1–5 (in Chinese)
27. Li M, Li Q, Kuang S, Zou Z (2016) Determination of cavity dimensions induced by impingement of gas jets onto a liquid bath. Metall Mater Trans B 47:116–126. https://doi.org/10.1007/s11663-015-0490-z
28. Qian F, Mutharasan R, Farouk B (1996) Studies of interface deformations in single- and multi-layered liquid baths due to an impinging gas jet. Metall Mater Trans B 27:911–920. https://doi.org/10.1007/s11663-996-0004-0
29. Ek M, Sichen D (2012) Study of penetration depth and droplet behavior in the case of a gas jet impinging on the surface of molten metal using liquid Ga–In–Sn. Steel Res Int 83:678–685. https://doi.org/10.1002/srin.201100336
30. Cao L, Liu Q, Sun J, Lin W, Feng X (2019) Effect of slag layer on the multiphase interaction in a converter. JOM 71:754–763. https://doi.org/10.1007/s11837-018-3243-z

31. Guo D, Gu L, Irons GA (2000) Evaluation of stirring in electric arc furnaces. In: 58th electric furnace conference, Orlando, Fl. USA, 12–15 Nov 2000, pp 223–233
32. Wang L, Zhu R, He C (2010) Two-phase mathematical model of oxygen jet impinging on molten steel bath surface in EAF. Chin J Process Eng 10:625–631 (in Chinese)
33. Han J, Chen Y, Li S (2010) Study of molten stirred by oxygen jet in EAF. Ind Heat 39:30–33 (in Chinese)
34. He C, Zhu R, Liu C, Li J (2011) Analysis on decarburization rate based on numerical simulation in EAF Steelmaking. Steelmaking 27:41–45 (in Chinese)
35. Chen Y, Silaen AK, Zhou CQ (2020) Penetration and decarburization in EAF refining process. Processes 8
36. Boom R (2011) Oxygen and iron-and steelmaking: best friends and arch-enemies. J Iron Steel Res Int 18:16–27
37. Basu S, Lahiri AK, Seetharaman S (2008) Activity of iron oxide in steelmaking slag. Metall Mater Trans B Process Metall Mater Process Sci 39:447–456. https://doi.org/10.1007/s11663-008-9148-4
38. JSPS (1988) Steelmaking data sourcebook. Gordon and Breach, Montreux Switzerland
39. Turkdogan ET (1996) Fundamentals of steelmaking. The Institute of Materials, London UK. https://doi.org/10.1179/095066096790151213
40. Fritz E, Gebert W (2005) Milestones and challenges in oxygen steelmaking. Can Metall Q 44:249–260. https://doi.org/10.1179/cmq.2005.44.2.249
41. Cicutti C, Valdez M, Pérez T, Donayo R, Petroni J (2002) Analysis of slag foaming during the operation of an industrial converter. Lat Am Appl Res 32:237–240
42. Shukla AK, Deo B, Millman S, Snoeijer B, Overbosch A, Kapilashrami A (2010) An insight into the mechanism and kinetics of reactions in BOF steelmaking: theory vs practice. Steel Res Int 81:940–948. https://doi.org/10.1002/srin.201000123
43. Lee B, Sohn IL (2015) Effect of hot metal on decarburization in the EAF and dissolved sulfur, phosphorous, and nitrogen content in the steel. ISIJ Int 55:491–499. https://doi.org/10.2355/isijinternational.55.491
44. Gottardi R, Miani S, Partyka A (2008) The hot metal meets the electric ARC furnace steelmaking route. Arch Metall Mater 53:517–522
45. Abel M, Hein M (2008) The use of scrap subtitutes like cold/hot DRI and hot metal in electric arc furnaces. Arch Metall Mater 53:353–357
46. Sain DR, Belton GR (1978) The influence of sulfur on interfacial reaction kinetics in the decarburization of liquid iron by carbon dioxide. Metall Trans B 9:403–407. https://doi.org/10.1007/BF02654414
47. Paul A, Deo B, Sathyamurthy N (1994) Kinetic model for reduction of iron oxide in molten slags by iron-carbon melt. Steel Res 65:414–420. https://doi.org/10.1002/srin.199401186
48. Wei P, Sano M, Hirasawa M (1991) Kinetics of carbon oxidation reaction between molten. ISIJ Int 31:358–365
49. Meyer HW, Porter WF, Smith GC, Szekely J (1968) Slag-metal emulsions and their importance in BOF steelmaking. J Met 20:35–42. https://doi.org/10.1007/bf03378731
50. Schoop J, Resch W, Mahn G (1978) Reactions occuring during the oxygen top-blown process and the calculation of metallurgical control parameters. Ironmak Steelmak 2:72–79
51. Kadrolkar A, Dogan N (2019) Model development for refining rates in oxygen steelmaking: impact and slag-metal bulk zones. Metals 9:309. https://doi.org/10.3390/met9030309
52. Chou K-C, Pal UB, Reddy RG (1993) A general model for BOP decarburization. ISIJ Int 33:862–868
53. Jiang ZH, Gong W, Wang WZ (2006) A mathematical model of the decarburization in UHP electric arc furnace charged with hot metal. Dev Chem Eng Miner Process 14:429–438. https://doi.org/10.1002/apj.5500140310
54. Matsuura H, Manning CP, Fortes RAFO, Fruehan RJ (2008) Development of a decarburization and slag formation model for the electric arc furnace. ISIJ Int 48:1197–1205. https://doi.org/10.2355/isijinternational.48.1197

55. Sano M, Yetao H, Sawada T, Kato M (1994) Decarburization reaction of molten iron of low carbon concentration with solid oxides. Trans ISIJ 34:649–656

56. Memoli F, Mapelli C, Ravanelli P, Corbella M (2004) Simulation of oxygen penetration and decarburisation in EAF using supersonic injection system. ISIJ Int 44:1342–1349. https://doi.org/10.2355/isijinternational.44.1342

57. He CL, Zhu R, Dong K, Qiu YQ, Sun KM, Jiang GL (2011) Three-phase numerical simulation of oxygen penetration and decarburisation in EAF using injection system. Ironmak Steelmak 38:291–296. https://doi.org/10.1179/1743281210Y.0000000011

58. Li GH, Wang B, Liu Q, Tian XZ, Zhu R, Hu LN et al (2010) A process model for BOF process based on bath mixing degree. Int J Miner Metall Mater 17:715–722. https://doi.org/10.1007/s12613-010-0379-4

59. Chen Y, Wang Y, Tang G, Silaen AK, Vanover K, Zhou CQ (2019) Numerical investigation of decarburization reaction characteristics in electric arc furnace steelmaking process. Association for Iron & Steel Technology. AISTech 2019 Proceedings, 789–796

60. Morales RD, Conejo AN, Rodríguez HH (2002) Process dynamics of electric arc furnace during direct reduced iron melting. Metall Mater Trans B Process Metall Mater Process Sci 33

61. Risonarta VY, Voj L, Pfeifer H, Jung HP, Lenz S (2008) Optimization of electric arc furnace process at Deutsche Edelstahlwerke. Arch Metall Mater 53:1–7. https://doi.org/10.1017/CBO9781107415324.004

62. Logar V, Dovžan D, Škrjanc I (2012) Modeling and validation of an electric arc furnace: part 2, thermo-chemistry. ISIJ Int 52:413–423. https://doi.org/10.2355/isijinternational.52.413

63. Dong K, Zhu R, Liu W (2015) Simplified calculation kinetic model for solid metal melting and decarburization process. High Temp Mater Process 34:447–456. https://doi.org/10.1515/htmp-2014-0045

64. Ma G, Zhu R, Dong K, Li Z, Liu R, Yang L et al (2016) Development and application of electric arc furnace combined blowing technology. Ironmak Steelmak 43:594–599. https://doi.org/10.1080/03019233.2016.1144547

65. Odenthal HJ, Kemminger A, Krause F, Sankowski L, Uebber N, Vogl N (2018) Review on modeling and simulation of the electric arc furnace (EAF). Steel Res Int 89:1–36. https://doi.org/10.1002/srin.201700098

66. Ramirez-Argaez MA, Conejo AN, Nava E (2012) Mathematical modeling of the decarburization kinetics in the electric arc furnace: C–O reaction. In: 5th international congress science technology steelmaking, Dresden Germany, 1–3 Oct 2012, pp 1–8

67. Adams W, Alameddine S, Bowman B, Lugo N, Paege S, Stafford P (2001) Factors influencing the total energy consumption in arc furnaces. In: 59th electric arc furnace conference, 11–14 Nov 2001, Phoenix, AZ, USA, pp 691–702

68. Jones JAT (2008) Electric arc furnace steelmaking. AISI, 1–9

69. Hornby SA, Trotter D, Varcoe D, Reeves R (2002) Use of DRI and HBI for nitrogen control of steel products. In: Electric furnace conference, pp 687–702

70. Jones J, Safe P, Wiggins B (1999) Optimization of EAF operations through off gas system analysis. In: Electric furnace conference, Pittsburgh, PA, USA, 14–16 Nov 1999, pp 459–480

71. Jones JAT, Bowman B, Lefrank PA (1998) Electric arc furnace steelmaking, 11th edn. In: Fruehan RJ (ed) Making, shaping treating steel, steelmaking, refining, Pittsburgh, PA USA, pp 525–660

72. Jones JAT (2002) Understanding energy use in the EAF. In: 60th electric furnace conference, San Antonio TX USA, 10–13 Nov 2002, pp 141–154

73. Miller TW, Jimenez J, Sharan A, Goldstein DA (1998) Chapter 9: oxygen steelmaking processes, 11th edn. In: Fruehan R (ed) Making, shaping treating steeling, steelmaking refining, pp 475–524. https://doi.org/10.1017/CBO9781107415324.004

74. Vidhyasagar M, Murali G, Balachandran G (2020) Thermo-kinetics, mass and heat balance in an energy optimizing furnace for primary steel making. Ironmak Steelmak 9233. https://doi.org/10.1080/03019233.2020.1737790

75. Weber R, Nosé D, Morsoletto L, Pfeifer HC (1994) Last achievements with the EOF process. Rev Metall: 439–444

76. Torres R, Lule R, López F, Conejo AN (2003) Analysis of metallurgical operations affecting metallic yield using 100% DRI in electric arc furnaces. In: 14th IAS steelmaking conference, Buenos Aires, Argentina, 13 Nov 2003, pp 547–558
77. Pujadas, A. McCauley, J. Tada, Y. Mathis, G. Iacuzzi M (2002) Electric arc furnace energy optimization at Nucor Yamato steel. In: 7th European electric steelmaking conference, Venice, Italy, 26–29 May 2002, pp 1.399–1.411
78. Ghosh A, Chatterjee A (2008) Ironmaking and steelmaking theory and practice, vol 20
79. Barati M (2014) Chapter 3.3. Application of slag engineering fundamentals to continuous steelmaking. Treatise Process Metall Process Phenom 2:305–357
80. Turkdogan ET (1984) Physicochemical aspects of reactions in ironmaking and steelmaking processes. Trans Iron Steel Inst Japan 24:591–611. https://doi.org/10.2355/isijinternational1 966.24.591
81. Turcotte S, Marquis H, Dancy T (1985) The use of direct reduced iron in the electric arc furnace. In: Taylor C (ed) Electric furnace steelmaking. The Iron and Steel Society, Inc., USA, pp 115–126
82. Inagaki, E. Izumi, K. Ichikawa M (1988) Integrated oxygen enrichment control to attain maximum overall economy in steelmaking furnaces. In: Electroheat congress, Madrid Spain, Oct 1988, p A6.2
83. Köhle S (1992) Effects on the electric energy consumption of arc furnace steelmaking. In: 4th European electric steel congress, Madrid, Spain
84. Bowman B (1995) Performance comparison update-AC vs DC furnaces. Iron Steel Eng: 26–29
85. Pfeifer H, Kirschen M, Simoes JP (2005) Thermodynamic analysis of EAF electrical energy demand. In: 8th European electric conference, 9–11 May 2005, Birmingham UK
86. Geiger G (1980) Process engineering involved in the use of direct reduced iron. In: Stephenson RL, Smailer RM (eds) Direct reduction iron technology economics production use. Iron and Steel Society, USA, pp 149–159
87. Conejo AN, Cardenas JGG (2006) Energy consumption in the EAF with 100% DRI. AISTech Conf 1:529–535
88. Cárdenas JGG, Conejo AN, Gnechi GG (2007) Optimization of energy consumption in electric arc furnaces operated with 100 % DRI. METAL, Hradec Nad Moravici, Czechia, 22–24 May 2007, pp 1–7
89. Pesamosca A, Patrizio D (2016) Latest trends in EAF optimization of scrap-based melting process: Balancing chemical and electrical energy input for competitive and sustainable steelmaking. 47° Semin. aciaria, vol 47, Rio de Janeiro Brazil, 26–30 Sept 2016. ABM, pp 169–182. https://doi.org/10.5151/1982-9345-27626
90. Gottardi R, Miani S, Partyka A (2006) Ultra high chemical power EAF. AISTech 2006 conference, Cleveland, OH, USA, 1–4 May 2006, pp 1–10
91. Gottardi R, Miani S, Partyka A (2007) New giant generation EAF. AISTech 2007 conference, Indianapolis IN, USA, 7–10 May 2007, pp 1–10
92. Gottardi R, Miani S, Partyka A, Engin B (2008) Ultra high chemical power EAF for 320 t/h. Rev Metall Cah D'Informations Tech 105:596–600. https://doi.org/10.1051/metal:2009003
93. Gottardi R, Miani S, Partyka A (2008) A faster, more efficient EAF. Arch Metall Mater 53:475–482
94. Toulouevski YN, Zinurov IY (2002) Outlook for reduction in energy consumption of electric arc furnaces. In: 7th European electric steelmaking conference, Venice, Italy, 26–29 May 2002, pp 1.57–1.64
95. Toulouevski YN, Zinurov IY (2017) Fuel arc furnace (FAF) for effective scrap melting: from EAF to FAF. Springer Nature, Singapore. https://doi.org/10.1007/978-981-10-5885-1
96. Gorez JP, Gros B, Birat JP, Grisvard C, Huber JC, Le Coq X (2003) Charging tires in the EAF as a substitute to carbon. Rev Metall Cah D'Informations Tech 100:17–23. https://doi.org/10.1051/metal:2003143
97. Nakao Y, Yamamoto KK (2002) Waste tire recycle and its collection system. Nippon steel technical report no. 86, 21–24

Chapter 5
Burners

5.1 Introduction

5.1.1 Overall Benefits Using Burners

There are several ways to increase the scrap melting rate. Using chemical energy to replace electrical energy includes three options: oxygen injection, oxy-fuel burners and post-combustion. Previously different devices carried out each one of the three previous functions, currently the same device can perform the three functions, if needed, however due to their differences sometimes can be separated according with their optimum position in the furnace.

Chemical energy with oxy-fuel burners can be increased as follows:

1. Increasing the number of burners
2. Increasing burner power
3. Increasing its thermal efficiency.

Increasing the number of burners promotes a more uniform scrap pre-heating but also the maintenance costs increase. A typical installation includes three burners located at the cold spots, as shown in Fig. 5.1.

Burners are very useful to preheat scrap located in cold spots. Adoloph et al. [2] suggested the areas of cold spots illustrated in Fig. 5.2. The main cold spots are located in the EBT area, the fourth hole, slag door and adjacent regions to the hot spots.

The overall benefits using oxy-fuel burners are:

- Enhances the supply of chemical energy, thus reducing electric energy consumption
- Eliminates cold spots
- Enhances scrap melting
- Increments productivity.

A. N. Conejo, *Electric Arc Furnace: Methods to Decrease Energy Consumption*, https://doi.org/10.1007/978-981-97-4053-6_5

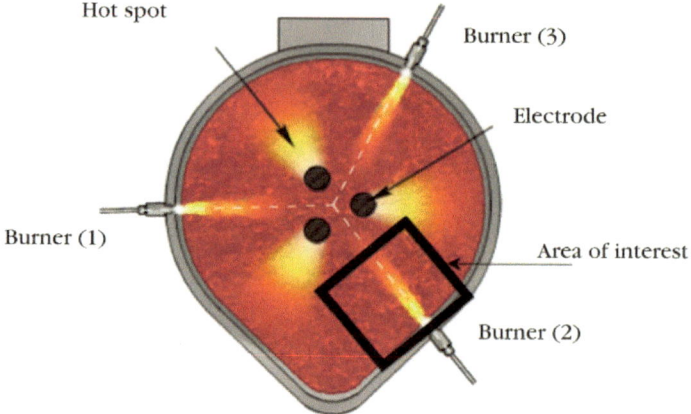

Fig. 5.1   Typical burner's location, after Ref. [1]

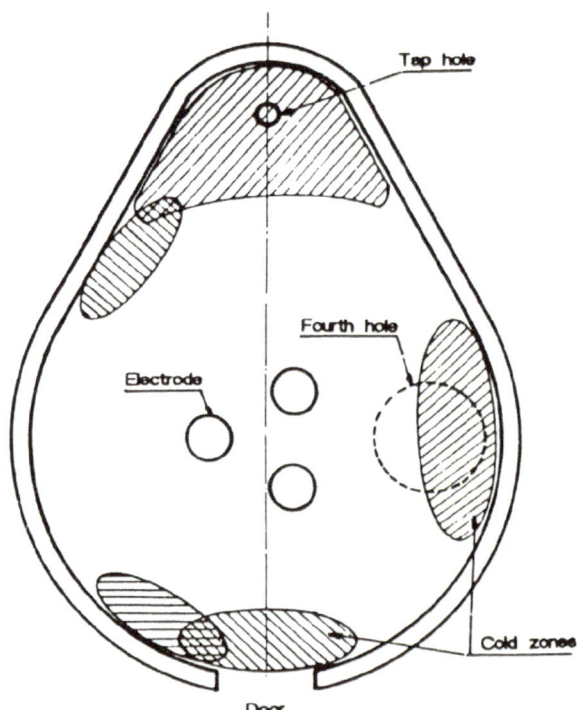

Fig. 5.2   Cold spots in an EAF with EBT according to Ref. [2]

The contribution of chemical energy to the total input energy depends on the burner power, time of operation and especially on its thermal efficiency. Values reported from energy balances from a large number of EAF's range from 2 to 10% [3]. Other reports have indicated values ranging from 6.2 to 9.2% [4, 5]. The benefits on productivity have been reported from 5 to 20% [6].

The chemical energy from a burner depends on the fuel employed. The typical fuel employed is natural gas.

$$CH_{4(g)} + 2O_{2(g)} = 2H_2O_{(g)} + CO_{2(g)}$$

Stoichiometrically, the chemical reaction indicates that for every m^3 of oxygen it is required 0.5 m^3 of methane.

Pittam and Pilcher [7] reported the following heats of combustion for the following fuels (@25 °C and 1 atm):

$$\Delta H_c^0(\text{Methane}) = -890.71 \, \text{kJ/mol} = -11.03 \, \text{kWh/m}^3$$
$$\Delta H_c^0(\text{Propane}) = -2219.17 \, \text{kJ/mol} = -27.5 \, \text{kWh/m}^3$$

In a typical burner, the fuel is burned by air. Oxygen in the air reacts with hydrogen and carbon in the fuel to form water vapor and carbon dioxide, releasing heat in this process. Nitrogen in the air dilutes the oxygen and carries away energy. Oxy-fuel burners replace air by oxygen, improving heat transfer and decreasing energy losses. Combustion efficiency largely depends on the design of the burner, in particular the length of the premix chamber.

5.1.2 Burner Efficiency

The chemical energy from a burner depends on its power (P_b), time of operation (t_b) and thermal efficiency (η_b) as shown by the following expression:

$$Q_b = \eta_b P_b t_b$$

The heating value of natural gas ranges from 9.77 to 11.48 kWh/m^3 [8] depending on its chemical composition with an average value of 11 kWh/m^3. A burner operating under stochiometric conditions consumes 2.0 m^3 O$_2$ for each 1 m^3 of natural gas. The burner power (P_b) or heating value corresponds to the heat of combustion of natural gas.

The average efficiency of the burner at any time t is defined by the following general expression [9]:

$$\eta_b = \frac{\int_0^t \dot{Q}_b dt}{\dot{P}_b t}$$

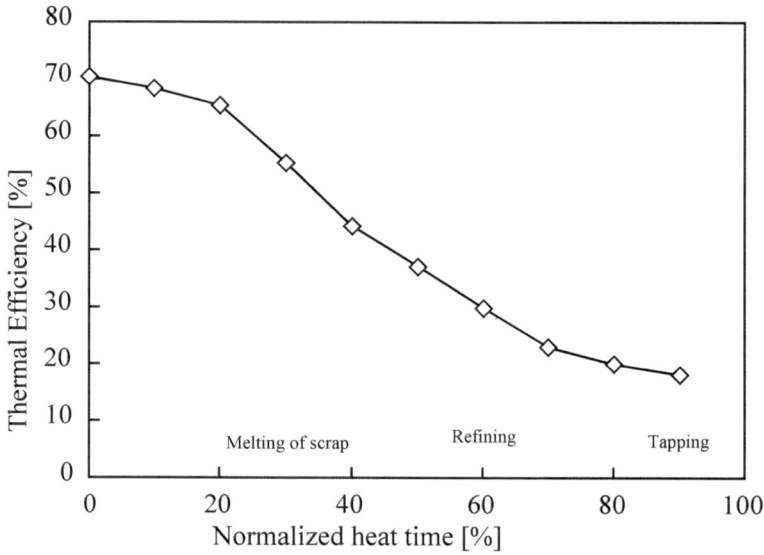

Fig. 5.3 Efficiency of oxy-fuel burners according to Bergman and Gottardi [10]

The thermal efficiency of an oxy-fuel burner decreases as the scrap gets pre-heated. Bergman and Gottardi [10] reported that the maximum heat transfer rate occurs at the beginning when the scrap is cold, about 70%, then decreases below 35% at about 50% into the heat and close to the end is about 20%, as can be seen in Fig. 5.3. This figure also shows a sudden drop in energy efficiency after 20% of the tap-to-tap time, about 50% of the melting period. Burners should be turned off when its efficiency is below 50%. Reported values vary from 50 to 80% depending on time of operation and energy savings but considering the decrease in tap to tap time that value falls in the range from 45 to 65%, assuming a reference of energy consumption of 2 kWh/ton for each minute during meltdown [6].

The decrease in thermal efficiency implies that the heat of combustion is not transferred to the scrap and therefore the off-gas temperature increases, consequently the off-gas temperature can be a good parameter to estimate the drop in thermal efficiency.

Baumert et al. [11] studied the thermal efficiency of burners during scrap-preheating. The thermal efficiency was calculated from Köhle's regression equation. They found that the thermal efficiency increases during preheating and melting of the first scrap basket but then drastically decreases with the second scrap basket. Furthermore, they also found that increasing the specific volume of natural gas decreases the thermal efficiency, for example with a specific consumption of 6.6 Nm³/ton the energy contribution was 11.7 kwh/Nm³ and decreased to 4–5 kwh/Nm³ when the specific consumption exceeded 12 Nm³/ton. These results indicate that burners should be optimized using an optimum specific consumption of natural gas from

10 to 12 Nm³/ton for the first bucket and decrease the specific consumption for the second basket, around 2 Nm³/ton.

Mandal and Irons [1, 12, 13] reported experimental measurements and numerical model predictions on heat transfer. The numerical model computed the burners' thermal efficiency. For the experimental work they built a heating furnace provided with a burner using three power levels (8.3, 12.8 and 17.3 kW), charged with three types of scrap; small shredded, busheling and some smaller heavy metal pieces. Its size and density were measured, as well as the void fraction. The scrap temperature was measured with thermocouples. Combustion was studied with the off-gas composition. The study included measurements on scrap oxidation. They found that using 10% excess oxygen over the required stoichiometry it was possible to obtain full combustion. The degree of combustion was measured based on the ratio of carbon in the off gas (CO_2) and carbon in the fuel (propane in this case). The extra energy provided due to scrap oxidation was estimated from 5 to 7%. It was neglected due to its relatively low value. Based on the scrap temperature distribution they found improved heat transfer for scrap beds with a smaller void fraction, obtained with smaller size scrap, as shown in Fig. 5.4. Arranging scrap of larger diameter at the bottom and an upper layer of smaller particle size decreased the overall void fraction of the bed.

These authors indicated that the void fraction in the scrap bed is the most critical parameter that defines the extent of heat transfer from the burner to the solid scrap. The porosity of a particulate mixture strongly depends on the scrap size, shape and distribution. Muller [14] proposed the following expression to estimate the radial porosity distribution in randomly packed fixed beds of uniformly sized spheres in cylindrical containers, an exponentially decay function, valid for $D/d \geq 2.61$

$$\varepsilon = \varepsilon_b + (1 - \varepsilon_b)ar^*e^{-b/r^*}$$
$$r^* = r/d$$
$$a, b, \varepsilon_b \Rightarrow f(D/d)$$

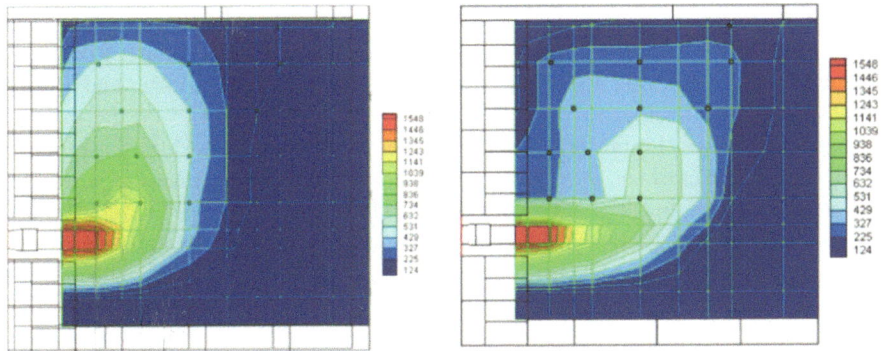

Fig. 5.4 Temperature profiles of scrap after 60 min using small shredded scrap (left) and large shredded scrap (right). Burner power of 12.8 kW. After Ref. [12]

where: ε is the bed porosity, ε_b is the porosity in the bulk region of the fixed packed bed, d is the diameter of the spherical particles. According with the measurements by Mandal and Irons [12], they also obtained an exponential decay function but only for scrap with a smaller particle size (shredded scrap), from 1 at the walls to about 0.84 at the center. Scrap of higher diameter showed a more uniform porosity distribution. The fact that porosity is highest at the walls offers gas channeling that decreases the residence time of the off gas, decreasing heat transfer efficiency.

The numerical model reported by Mandal and Irons [12] to describe burner efficiency was the first one reported in the literature. The model is about heat transfer through porous media, however, it has distinctive and challenging features, as follows:

- Scrap particles are non-uniform and much larger than in other applications
- Gas velocities are high giving a porous Re number higher than unity (for incompressible fluids), for which Darcy's law applies, therefore the fluid is non-Darcian and turbulent. Darcy's velocity is the superficial gas velocity defined by the ratio Q/A (m/s).
- Porosity is much higher than normal porous media, non-constant, non-uniform and changes with time due to localized scrap melting.
- Different temperatures for gas and solids. The gas temperature is very high

Due the extreme complexity of the problem, the authors indicated that no commercial software was available to solve it in 2013. These authors reported that the Darcy-Lapwood-Forchheimer-Brinkman flow model was capable to consider both inertial and/or viscous effects. On the other hand, the heat transfer coefficient between the gas and solid scrap was estimated using a relationship defined by Kitaev et al. based on previous work by Furnas at 1100 °C:

$$h_V = h_{fs}A_S = \frac{Af(\omega)V_g^{0.9}T^{0.9}}{d_p^{0.75}}$$

where: h_V is the volumetric heat transfer coefficient, h_{fs} is the gas-solid heat transfer coefficient, A is a coefficient which depends on bed properties (about 80 for lump materials), $f(\omega)$ is a function of void fraction (about 0.5), V_g is the superficial gas velocity (m/s), T is the temperature in °C, and d_p is the particle diameter in mm.

The model predictions indicate a negligible effect of the scrap size and a large effect of porosity on the thermal efficiency. Decreasing the porosity, the thermal efficiency increases. Figure 5.5 shows the thermal efficiency for different types of scrap. Shredded scrap has the highest thermal efficiency, not because of a smaller particle size but due to a higher bed density with lower porosity. The model indicates a maximum thermal efficiency of about 90% and decreases to 55% in 60 min with an exponential decay in temperature. This temperature profile is similar to the temperature profile reported by Bergman and Gottardi [10], in particular at the beginning. In this model the thermal efficiency decreases immediately in contrast to the experimental measurements which indicate an initial thermal efficiency of about 70% and remains constant for about 20% of the heat. The authors suggested to control the

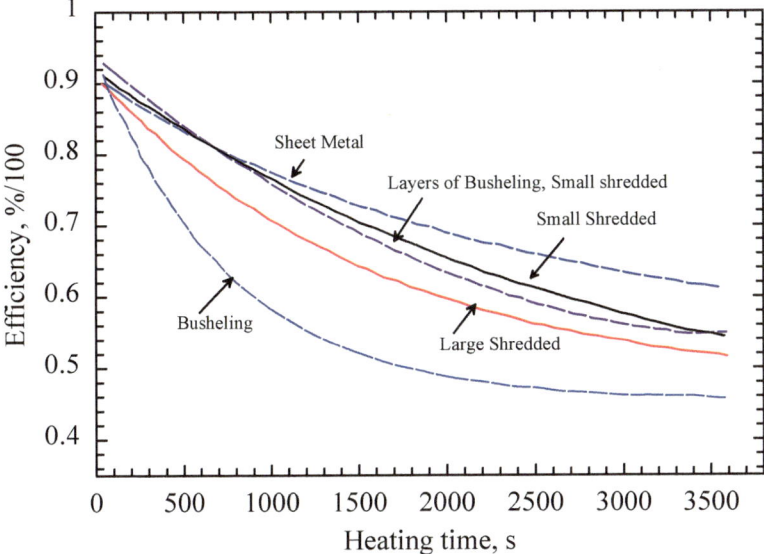

Fig. 5.5 Thermal efficiency of burners with different type of scrap, after [13]

scrap density, placing higher density scrap around the burners area to enhance the thermal efficiency, also, to avoid placing large pieces in this area because the gases can be deflected back on to the burner and refractory, creating a blow back effect.

The same model by Mandal and Irons [1, 12, 13] in 2010 was extended by another group of researchers in 2019 to include scrap oxidation, however in this case the authors employed the commercial software Ansys-Fluent. The details are discussed further below.

Logar et al. [15] employed the experimental values from Bergman and Gottardi [10], suggesting a hyperbolic fitting function:

$$Q_b^{tot} = \eta_b^0 P_b \times f(T_{sc}) = \eta_b^0 P_b \times \left[0.35 + 0.65 \tanh\left(\frac{1300}{T_{sc}} - 1\right)\right]$$

where: $f(T_{sc})$ is a hyperbolic-tangent approximation which depends on the temperature of solid scrap, in °C and η_b^0 is the initial burner efficiency (0.7). Figure 5.6 compares predictions of the thermal efficiency using the previous equation. In this plot the second x-axis represents the temperature of solid scrap.

Hernandez et al. [9] has reported the lack of fundamental studies on oxy-fuel burners and the prevalence of the empirical results from Bergman and Gottardi [10]. They developed a mathematical model assuming solid scrap in front of the oxy-fuel burner is a hollow cylinder instead of a cone frustum, the hottest part of the flame is always surrounded by solid scrap (view factor equal to one), the length of the flame was estimated as 25 times its diameter (others have employed 60 times) and

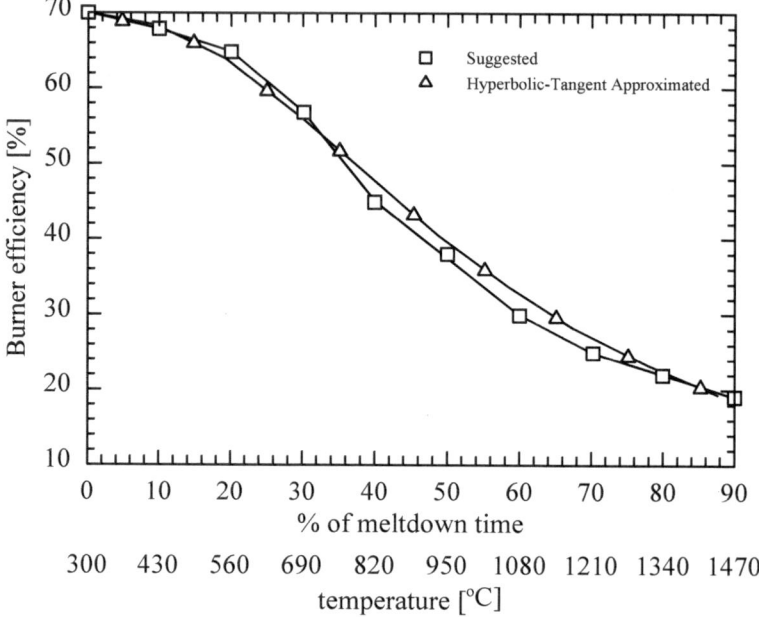

Fig. 5.6 Comparison of burners thermal efficiency using the approximation reported by Logar et al. [15].

in order to involve convection the cylinder was divided into 15 sub-sections which exchange heat with the surrounding atmosphere. Calculation of the Adiabatic Flame Temperature (AFT) assumes perfect combustion. This temperature is in the order of 3200 K. The model estimates both the temperature and mass of both solid and liquid scrap. Radiation is computed as follows:

$$\dot{Q}_{rad_i} = \frac{V_{T_{fg}} - V_{T_{fg}}}{R_{rad_{sc}} + R_{VF_{fg-sc}} + R_{rad_{fg}}}$$

$$V_{T_i} = \sigma_b T_i^4$$

$$R_{rad_i} = \frac{1 - \varepsilon_i}{\varepsilon_i A_i}$$

$$R_{VF_{fg-sc}} = \frac{1}{VF_{fg-sc} A_i}$$

where: $\dot{Q}_{rad_i}$ is heat transfer by radiation, σ_b is the Stefan-Boltzmann constant (5.67×10^{-8} W/m^{2}K^4), ε is the emissivity, for the flue gas is equal to 0.5 and for the surface of solid scrap is equal to 0.7, VF is the view factor (the view factor describes the fraction of energy that leaves from the emitting surface and is absorbed by the receiver surface), subindex "fg" indicates flue gas and "sc" scrap. $VF_{fg-sc} = 1$.

Convection was computed using the following relationships:

$$\dot{Q}_{conv_i} = 2A_{conv}h_{conv_i}\left(T_{out_i} - T_{sm}\right)$$

$$h_{conv_i} = \frac{Nu_i k_{fg}}{\left(\frac{L_{tub}}{20}\right)}$$

where: Nu is the Nusselt number, k_{fg} is the thermal conductivity of the flue gas, h_{conv_i} is the convective heat transfer coefficient, T_{out_i} is outlet temperature of the gas at each subsection, T_{sm} is the temperature of solid scrap, L_{tub} is the length of the hypothetical plate created from the cylinder.

The model results considering 3 burners in an EAF of 100-ton capacity and a natural gas flow rate of 200 Nm^3/h, are shown in Fig. 5.7. This model underpredicts/overpredicts the thermal efficiency below/above 35% in the melting process, respectively, with respect to the experimental data from ref [10]. The results shown for Logar's model [15] are wrongly reproduced as can be seen in their original results in Fig. 5.6.

Hernandez et al. [9] reported that radiation represents about 33% of the total heat transferred from the burners to the scrap while convection heat transfer decreases from 35% at the beginning to 0% at about 75% of the melting process.

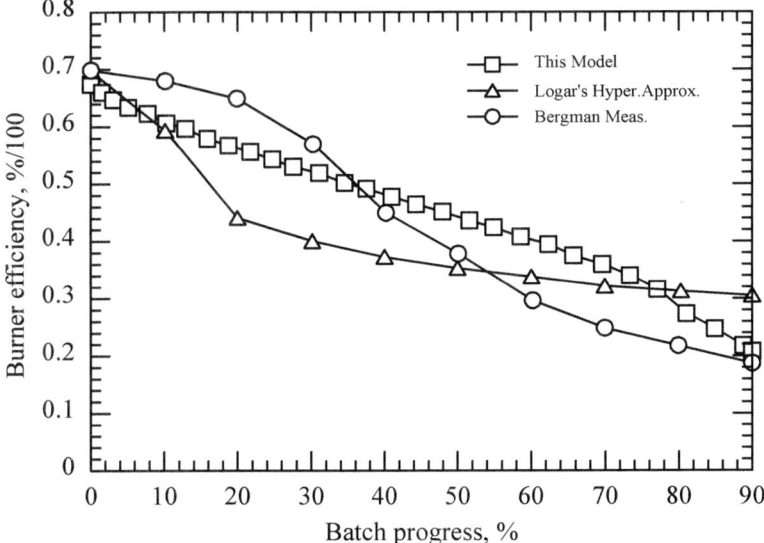

Fig. 5.7 Comparison of model results on burner efficiency. After [9]

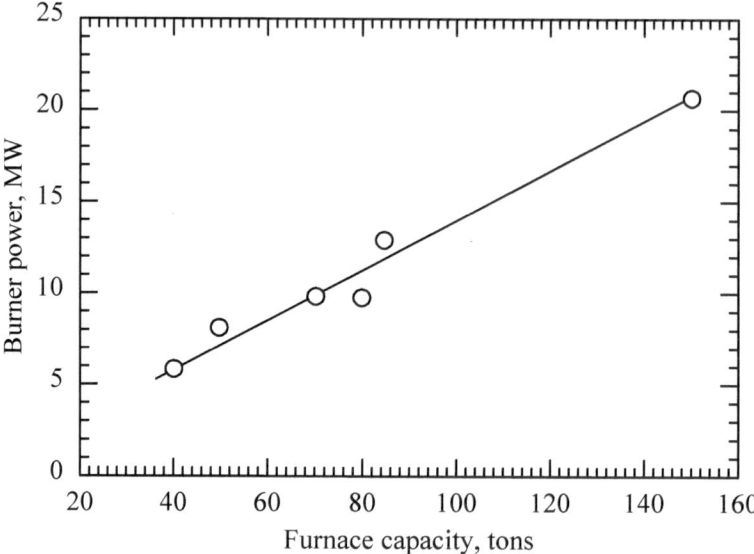

Fig. 5.8 Relationship between burner power and furnace capacity, after [10]

5.1.3 Burner Capacity

Bergman and Gottardi [10] reported a relationship between furnace capacity and burner capacity, shown in Fig. 5.8. Their relationship corresponds to 0.133 MW per ton. This plot would suggest three burners of 4–5 MW each for an EAF of 100 ton.

Adolph et al. [2] in an older report commented that burner capacity is not proportional to furnace size or transformer power but is a function of the heat distribution in the furnace.

5.1.4 Developments on Oxy-Fuel Burners

The heating process is the most intensive in energy consumption. The rate of energy supply defines the tap-to-tap time. Oxy-fuel burners have the purpose to enhance the supply of energy, in particular in those areas where the temperature is lower, allowing a more uniform melting rate. The heating time period consumes the largest amount of time and energy in a heat.

Hinds [16] in 1965 proposed the installation of oxy-fuel burners on the roof of the EAF. This location was suggested to compensate the large temperature differences between the region around the electrodes and the scrap temperature closer to the walls. The number of burners and gas flow rates depended on the nominal capacity of the furnace as shown in Table 5.1. In the table it can be observed that the total demand for oxygen significantly increases with the nominal capacity, reaching values from 4500 to 5500 Nm³/h for larger furnaces with a capacity in the order of 200 ton, creating supply problems due to space limitations to place the burners on the roof. The burners proposed by Hinds are not vertical but with an inclination angle with the vertical, as shown in Fig. 5.9, showing a flame overlapping and forming an annular zone of flames. The inventor reported a decrease in melting time of 21% and also a decrease of 17.6% in energy consumption in an EAF of 140 ton using 12 burners. The melting time was decreased from 2 h 57 min to 2 h 19 min and energy consumption was decreased from 519 to 428 kWh/ton. Wunsche et al. [17] indicated three limitations for this design: lack of free space above the roof to mount several burners, insufficient support structure and space for the oxygen and fuel lines.

Modern burners are commonly placed on the furnace walls with the burner axis inclined downwardly at an angle up to 30° with respect to the horizontal axis to avoid contact with the furnace electrodes and are also displaced into the furnace a very short distance. The tip of the lance has a limited life. To add a new tip, the previous one is cut away and the new one welded.

Wunsche et al. [17, 18] in 1987 designed an oxy-fuel burner that integrated three functions; burner, oxygen injection and particle injector (carbon particles or lime particles). If needed, gas and solids can be introduced through the same pipe. In addition to this, the lance can follow the direction of the scrap in the vertical direction as it melts, because solid scrap occupies a much larger volume in comparison to when it is melted, as shown in Fig. 5.10. The tip has a curved (arcuate) shape which allows oxygen to be injected vertically.

CRM [19] developed a method to measure the distance between the tip of the burner and the front of solid scrap using fiber optic. Information on this distance allows to follow the evolution in scrap melting in front of the burner, optimizing the burner operation and switching to oxygen injection mode. The signal can also be used to prevent abnormal conditions that might lead to blow-back conditions causing panel overheating.

The conventional burner design consists of burners attached to the wall with a fixed vertical inclination angle. The company Badische developed a tiltable burner in 2012 [20, 21]. This design includes an angle in burner mode of about 20° and about 40° for oxygen injection mode. The challenges reported to make this design were

Table 5.1 Number of burners and flow rates for oxygen and fuel as a function EAF nominal capacity

Furnace size, tons	Number of burners	Flow rate/burner, Nm³/h		Total gas flow rate, Nm³/h	
		Oxygen	Fuel	Oxygen	Fuel
50	6	169.9	94.3	1019.4	566.3
80	6	212.4	118.1	1274.3	707.9
100	9	226.5	126.0	2038.8	1132.7
150	12	236.2	118.1	2831.7	1415.8

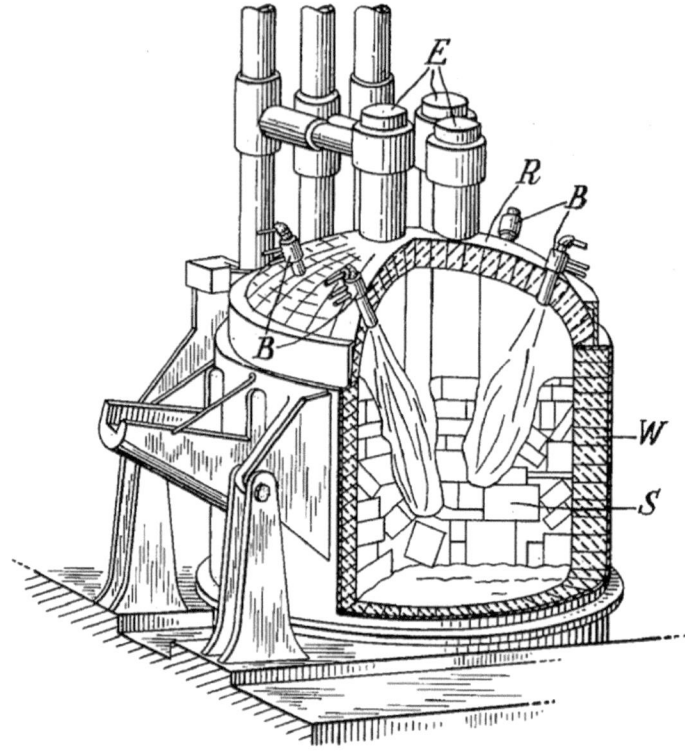

Fig. 5.9 Old burners location design proposed by Hinds in [16]

related to developing a tight sealing between the movable part and the water-cooled box, capable to withstand high heat, flames and slag splashing.

Sung et al. [22] reported a design consisting of a protruding water-cooled copper jacket which allows to reduce the distance between the jet nozzle and the molten steel and using separate lances for the burner, oxygen injection and carbon injection, the vertical angle can also be separated, using 40° for oxygen injection and 25° for the burners. Figure 5.11 shows a conventional oxygen lance and the modifications with the new lance.

The outlet stream pressure is also an important design parameter in oxy-fuel burners. Megahed et al. [23] reported a higher jet length by increasing the outlet stream pressure from 8 to 12 bar, changing the nozzle diameter from 22 to 18.7 mm at a gas flow rate of 1800 Nm³/h. These variables are related by the following expression:

$$Q = \frac{822 \times e \times P_1 \times d_0^2}{T_1^{1/2}}$$

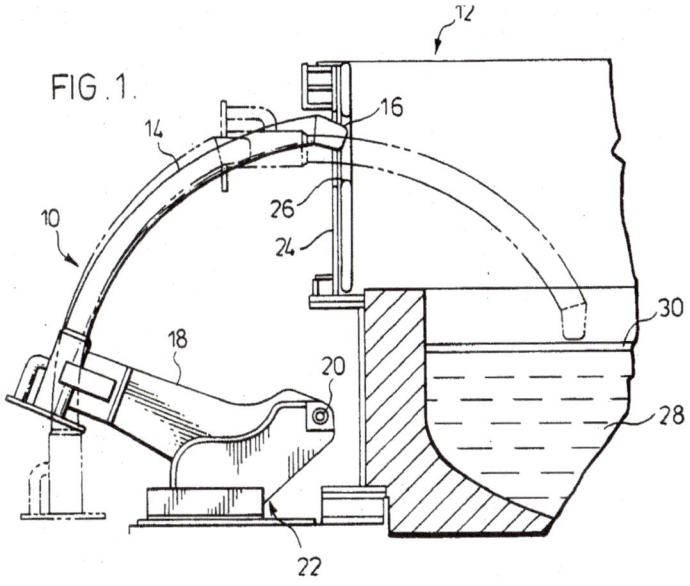

Fig. 5.10 Movable burner proposed in 1987. After [17, 18]

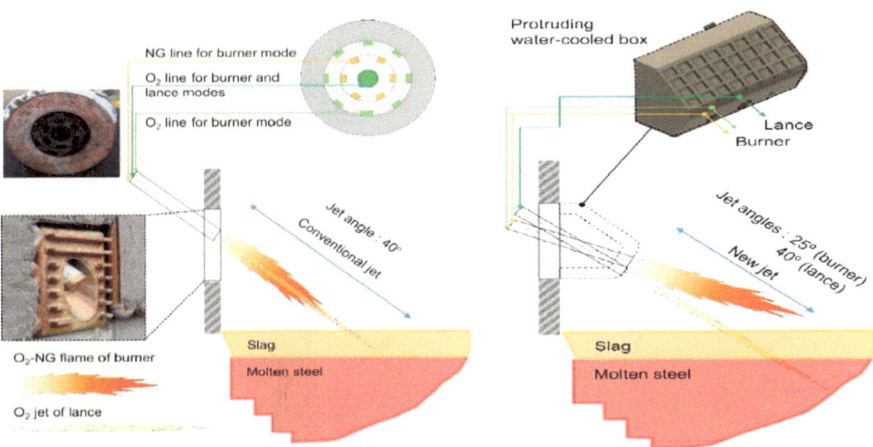

Fig. 5.11 (Left) conventional lance, (right) modified lance protected with a water-cooled box. After [22]

where: Q is the oxygen flow rate in Nm^3/h, e is the restricting coefficient, P_1 is the absolute pressure upstream the nozzle, d_0 is the nozzle diameter in cm and T_1 is the absolute temperature upstream the nozzle in K.

5.2   Heat Transfer During Scrap Pre-heating with Oxy-Fuel Burners

5.2.1 Fundamentals

Newton's law of cooling: The main heat transfer mechanism using oxy-fuel burners is convection heat transfer. The rate of convective heat transfer is defined by Newton's law of cooling. This law was reported anonymously in Latin by Newton in 1701 with the title; "Scala Graduum Caloris" ("A scale of the degrees of heat"). The original text has been discussed in detail because no equation was provided, only the experimental results, which can be defined in the form of an equation [24, 25]. It is considered the first theoretical rate equation in the history of physics for convection heat transfer. Newton carried out experimental observations on the rate of cooling of hot iron placed in a uniformly blowing wind (forced convection) and found that for equal intervals of time there was a proportional change in temperature where the time scale was in an arithmetic progression and the temperature scale in a geometric progression, defined by the following linear first order differential equation:

$$\frac{dT}{(T - T_g)} = - \, kdt; \, T(0) = T_0$$

Which is equivalent to:

$$\frac{dT}{dt} = - \, k \left(T_s - T_g \right)$$

This expression indicates that the rate of cooling is proportional to the temperature difference between the solid and the fluid. Newton in his original analysis didn't define the convective heat transfer coefficient and mixed heat with temperature since its difference was still not defined at that time. In modern notation Newton's equation or Newton's law of cooling is defined as follows:

$$\dot{Q} = hA \left(T_s - T_g \right)$$

where; T_s is the temperature of the solid as a function of time, $\dot{Q}$ is the rate of heat transfer in Watts, h is the convective heat transfer coefficient in $W/m^2 \cdot K$ and A is the surface area in m^2.

The rate of heat transfer or heat supplied from the oxy-fuel burners to the scrap can be calculated from the following relationship:

$$\dot{Q} = \dot{q}A = \dot{m}C_p \left(T_g^e - T_g^i \right)$$

where; $\dot{q}$ is the heat flux in W/m^2, $\dot{m}$ is the fluid's mass flow rate in kg/s, C_p is the fluid's heat capacity in kJ/kg K, T_g^e is the exit fluid's temperature in °C and T_g^i is the inlet fluid's temperature in °C.

The two previous equations can be employed to study scrap pre-heating due to convection heat transfer, for specific conditions (boundary conditions). Usually, in order to simplify the analysis, it is required to simplify the problem, for example a complex scrap geometry can be assumed to have the form of spheres or cylinders, it can be assumed the heat flux is constant, it can be assumed that the solid's temperature remains constant, etc. Cengel and Ghajar [26] provided detailed examples. One of those examples is to define an expression to compute the exit fluid's temperature, T_g^e, in a circular pipe, assuming the solid's temperature remains constant. A heat balance in a control volume, for temperatures below the melting point of the steel scrap, can be expressed as follows:

$$\dot{m}C_pd\left(T_g^e - T_g^i\right) = h(T_s - T_g)\, dA$$

Separating variables and after integration:

$$T_g^e = T_s - \left(T_s - T_g^i\right) \exp\left(-hA/\dot{m}C_p\right)$$

In an actual scrap pre-heating process in a shaft the scrap's temperature can be assumed to be constant once it reaches a certain height but it's temperature changes as a function of height. This analysis involves more equations, as shown by Ibrahim and Al-Qassimi [27].

Convective heat transfer coefficient: The previous analysis requires the value of the convective heat transfer coefficient. This value can be obtained from numerous experimental correlations available. The scrap geometry can be simplified as a packed bed of spheres, plates or cylinders. It is worth to mention the work conducted by Zukauskas [28, 29] because he reported a large number of correlations for a bank of cylinders arranged in-line or staggered. Applying dimensional analysis, it can be found that heat transfer can be expressed in terms of three dimensionless numbers; Nusselt (Nu), Reynolds (Re) and Prandtl (Pr) numbers, where Nu = f(Re, Pr). Once the Nusselt number is defined, the value of the heat transfer coefficient can be calculated. Zukauskas has reported his results with expressions that have the following general form:

$$Nu_f = \frac{hL}{k} = CRe_f^m Pr_f^{0.36} \left(\frac{Pr_f}{Pr_w}\right)^{0.25}$$

where: Nu, Re and Pr are Nusselt, Reynolds and Prandtl numbers, respectively, f represents the main fluid and w the wall. The values of C and m depend on the Re number. Cengel and Ghajar [26] summarized Zukauskas' correlations. The Prandtl number depends only on the properties of the fluid, its value can vary from 1 to 1000 for most fluids. Liquid metals have extremely low Pr numbers, in the order of 0.003 to 0.030. The Pr number can be calculated using the following expression:

$$Pr_f = \frac{\nu}{\alpha} = \frac{\mu/\rho}{k/\rho C_p} = \frac{\mu C_p}{k}$$

where: ν is momentum diffusivity (kinematic viscosity) in m^2/s, α is the thermal diffusivity in m^2/s, μ is the dynamic viscosity in Pa s ($N\ s/m^2$), k is the thermal conductivity in W/m K, C_p is the heat capacity in J/kg K and ρ the fluid's density in kg/m^3.

In the case of a bank of pieces of scrap, the Reynolds number uses the maximum velocity (U_{max}) instead of the approaching velocity (U). The maximum velocity depends on the distances between pieces of scrap; S_T is the vertical distance, S_L is the horizontal distance and S_D is the diagonal distance.

$$Re_f = \frac{\rho U_{max} L}{\mu} = \frac{U_{max} L}{\nu}$$

Kitaev et al. [30] summarized a large amount of experimental work on heat transfer in shaft furnaces carried from the 1930s until the 1960s, starting with the work reported by C. Furnas for the blast furnace. The vast majority of research work reviewed is from Russian authors and predominate the correlations involving Nu with the Re number. Furnas [31] reported in 1930 the following expression, valid for temperatures higher than 1000 °C, which has been subsequently modified:

$$h_v = C \frac{U_g^{0.7} T_g^{0.3} f(\omega)}{d_p^{0.9}}$$

where: h_v is the volumetric heat transfer coefficient in $W/m^3 K$, C is a constant characteristic of the material (about 160), U_g is the gas velocity referred to 0 °C in m/s, T_g is the average gas temperature in K, d_p is the particle diameter in m and $f(\omega)$ is the void fraction (about 0.5).

Wakao et al. [32] re-evaluated a large number of experimental data, usually for temperatures below 1000 °C and were able to fit the data to the following equation, which is valid for Re numbers from 15 to 8500:

$$Nu_f = 2 + 1.1 Re_f^{\frac{1}{3}} Pr_f^{0.6}$$

5.2.2 Scrap Pre-heating Time with Off-Gas in the Shaft

Toulouesvki and Zinurov [33] estimated the pre-heating time in the shaft of a Quantum EAF. This section is based on their original analysis. They employed the following data: EAF with 100 ton of nominal capacity, shaft's cross sectional area of 4.6 × 2.6 m and height of 3.5 m, volume 42 m³, shaft scrap capacity of 27.5 ton (scrap with a bulk density of 0.65 ton/m³), scrap diameter 0.025 m, gas thermal conductivity 0.0896 W/m °C, scrap density 7900 kg/m³, gas kinematic viscosity 170 × 10⁻⁶ m²/s, scrap heat capacity 0.54 kJ/kg °C. It was estimated a gas flow rate of 5 Nm³/s and a gas velocity of about 3 m/s. The flame temperature was estimated at 1650 °C and an exit temperature of 500 °C. With this data they obtained an average scrap pre-heating temperature of 400 °C. The Pr number can be defined as 0.739 for an estimated average gas temperature above 1000 °C. These authors employed the following correlation to estimate the Nu number:

$$\mathrm{Nu_f} = 0.40\,\mathrm{Re_f^{0.6}Pr_f^{0.36}}$$

$$\left(\frac{hL}{k}\right) = 0.40\left(\frac{3 \times 0.025}{170 \times 10^{-6}}\right)^{0.6}(0.739)^{0.6} = 0.40(441)^{0.6}(0.739)^{0.6} = 13.85$$

From this equation, the heat transfer coefficient is:

$$h = 13.85\left(\frac{0.0896}{0.025}\right) = 49.64\,\mathrm{W/m^2\,°C}$$

The preheating time is obtained from a similar heat balance, as shown before. The authors indicate that the inlet gas temperature is 1650 °C and leave at 500 °C, taking 1075 °C as an average gas temperature. The initial scrap temperature is 25 °C and is preheated to 600 °C:

$$t = 0.64\frac{d \cdot \rho \cdot C_p}{2h}\log\left(\frac{\overline{T}_g - T_s^i}{\overline{T}_g - T_s^e}\right)$$

$$= \frac{0.64 \times 0.025 \times 7900 \times 0.54}{2 \times 49.64}\log\left(\frac{1075 - 25}{1075 - 600}\right)$$

$$= 14.21\ \mathrm{min}$$

In the original calculation the authors reported a pre-heating time of 8 min employing as initial scrap temperature 0 °C and an average scrap pre-heating temperature of 400 °C. Those results were artificially manipulated to decrease the preheating time. The same authors repeated the calculations indicating the inlet gas temperature is 1400 °C and leave at 1150 °C, then the maximum scrap preheating temperature would be 700 °C. For these conditions they report a heat transfer coefficient 3 times higher without an explanation on how the velocity was increased to yield a Re number as high as 987,000. Evidently, the much higher heat transfer coefficient allows to shorten the preheating time to 5 min for each batch of scrap in the shaft. It is also mentioned that the final scrap pre-heating temperature increases from 700 to 800 °C due to 1.5% iron oxidation into Fe_3O_4. The corresponding replacement in electric energy due to scrap pre-heating up to 800 °C is about 145 kWh/ton:

$$Q = mC_p(\Delta T)$$
$$= (1000 \, kg)(0.67 \, kJ/kg\,^{\circ}C)(800 - 25)\,^{\circ}C$$
$$= 520{,}000 \, kJ = 145 \, kWh$$

Burner power calculation: Scrap pre-heating in the shaft from room temperature to 400 °C is carried out by the off-gas, then the burners should raise the pre-heating temperature to about 700 °C. The specific energy required is about 59 kWh/ton, for a batch of scrap of 27.5 ton the total energy is about 1622 kWh, assuming 100% efficiency. If the burner efficiency is in the range from 50 to 70%, the final value is about 2704 kWh. Since this energy should be supplied in 5 min (0.083 h), the burner power corresponds to 32.6 MW. The volume of natural gas, assuming a calorific value of 10.3 kWh/m^3 correspond to $2704/10.3 = 262$ m^3 and the specific consumption for 27.5 ton is 9.54 Nm^3/ton. The Quantum EAF taken as a reference for the calculations is charged 4 times, so in total it takes about 20 min for scrap melting using their special burners. These authors mention a preheating time per batch of 7 min with the conventional design without burners in the shaft, indicating that the F-EAF can shorten the tap-to-tap time by 8 min.

Benefits of proposed F-EAF: Some performance parameters for the conventional Quantum EAF are the following; scrap preheating temperature 400 °C, tap-to-tap time 36 min, productivity 167 t/h, electric energy consumption 280 kWh/ton, consumption of natural gas 4.4 m^3/ton, transformer capacity 80 MVA. Assuming that it is possible to increase the preheating temperature to 800 °C in a shorter period of time using 10.5 m^3/ton of natural gas, their proposed F-EAF Quantum would have the following improvements: tap-to-tap time 28 min, productivity 214 t/h, electric energy consumption 200 kWh/ton and a transformer capacity of 70 MVA.

5.2.3 Scrap Pre-heating with Burners

Tang et al. [34] extended the model developed by Mandal and Irons described previously in Sect. 5.1.2, in their case, they employed the commercial software Ansys-Fluent. Scrap was treated as a porous medium using energy source terms for both gas and solid phases with UDF's. The Discrete Ordinates (DO) radiation model was employed to predict heat transfer with a new Weighted Sum of Gray Phases Model (WSGPM) to predict radiation heat transfer from the combustion gases. Combustion was studied using the Eddy Dissipation Concept (EDC) model. The scrap oxidation rate was predicted considering the following reaction:

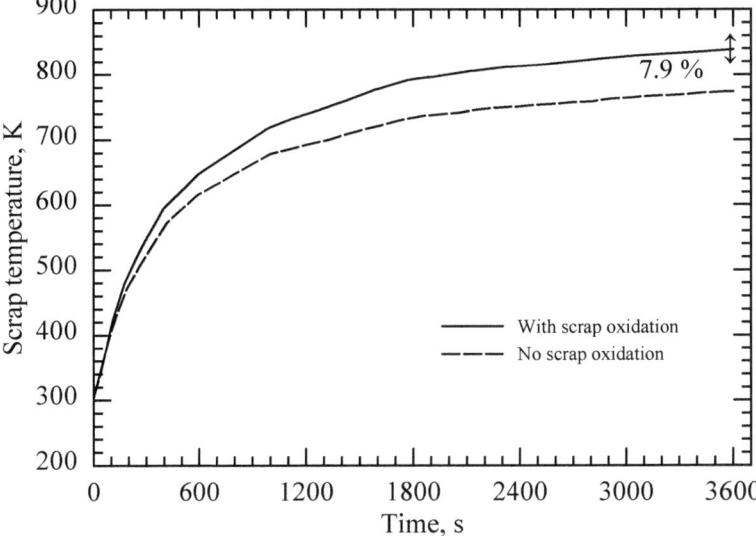

Fig. 5.12 Scrap temperature predictions based on the model developed by Tang et al. [34]

$$Fe_{(s)} + {}^1/_2 O_{2(g)} = FeO_{(s)}$$

The model validation was carried out using the experimental data from Mandal and Irons. Figure 5.12 shows their numerical model predictions. It can be observed that burners promote scrap oxidation, since this oxidation reaction is exothermic, the scrap temperature increases. The figure shows an increment in the scrap temperature of about 8% in 60 min, however scrap is melted in a much shorter time. The increment in temperature for 10 min is lower, about 3%.

Zhang and Oeters [35] reported a small degree of scrap oxidation, of about 0.2% by scrap pre-heating using post-combustion gases. This low value was attributed to the short residence time of the gas with solid scrap.

Chen et al. [36] in 2022 further expanded the previous model by Tang et al. [34] to study the effect of burner power, scrap porosity, scrap-preheating and scrap blockage on burner thermal efficiency, the scrap melting rate and cavity size. This work is very comprehensive and is worth to describe in detail. The research group is led by Prof Chenn Zhou from Purdue University.

The new model involves two sub-models; jet model and a scrap pre-heating model. The conservation equation for gas species in the gas phase for the jet model included the net rate of production of species using the eddy dissipation model. Radiative heat transfer due to the shrouding combustion flame in the gas phase was estimated using the Discrete Ordinates (DO) radiation model. A temperature correction method suggested previously by Alam et al. to the turbulence model which modifies the gas turbulence viscosity was employed because the standard κ-ε model does not account for the influence of temperature gradients on the turbulent mixing zone within the gas phase. The scrap melting model considers the scrap as a porous medium using the dual-cell approach which allows to estimate the fraction of solid and liquid in each cell. The thermal conductivity is computed according with the cell temperature, for three ranges that cover solid, liquid and intermediate temperatures. The heat transfer coefficient between gas and solid scrap was estimated for two temperature regions, below and above 1100 °C. The model validation, using one burner of 3.2 MW with an initial porosity of 0.86 and 0.3 million cells, included both the jet characteristics and the scrap heating rate. Figure 5.13 shows validation of the jet model using experimental data reported by Anderson (U.S. Patent 5823762). Additionally, Fig. 5.14 shows validation of the scrap heating model using experimental data from Mandal and Irons [1, 12, 13]. In both cases it is shown a good agreement of model predictions with experimental data.

The model captures a more realistic representation of the scrap melting rate by considering re-solidification, which indicates that liquid iron does not immediately sink to the molten pool of steel but exchanges heat with the scrap below the cavity and then re-solidifies forming a low porosity region below the cavity. When the cavity grows to a critical size the scrap above will collapse into the cavity due to insufficient support and colder scrap refills the empty cavity, repeating a new cycle of scrap pre-heating and melting. The simulation time was 5 min which was estimated is the duration of those cycles. The penetration distance was defined for an isotherm of 600 K.

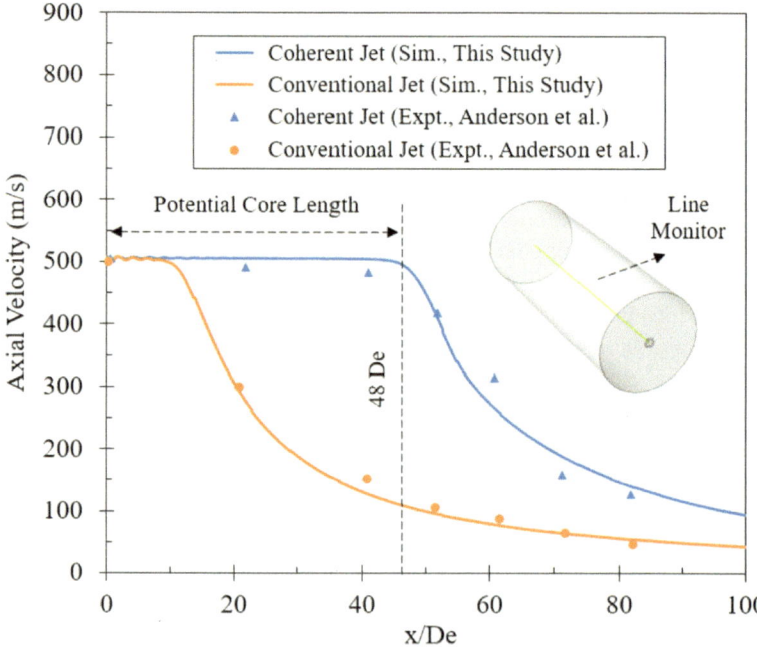

Fig. 5.13 Jet model validation using experimental data from Anderson, after Ref. [36]

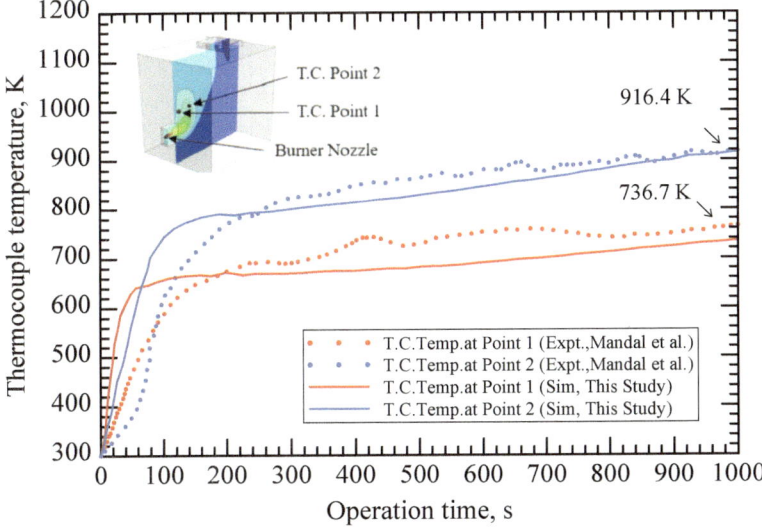

Fig. 5.14 Scrap model validation using experimental data from Mandal and Irons after ref [36]

The effect of burner power on the burner thermal efficiency, scrap heating rate and cavity size is shown in Fig. 5.15. Three cases were investigated, 2.4, 3.2 and 4.0 MW, an increment in burner power of almost 70%. Figure 5.15a shows that increasing burner power has a small effect, below 5%, on increasing the burner efficiency. The authors attributed this behavior to the similar ratio heat transfer to scrap/total heat input when the burner power increases. When the burner power increases, it increases the heat transfer coefficient but the flame temperature remains almost the same. On the other hand, increasing the burner power has a strong effect on the scrap melting rate, as shown in Fig. 5.15b, c. Both the volume average scrap temperature and volume of the cavity increases with burner power. In other words, the same cavity volume is formed faster using a higher power burner. For example, a cavity of 0.2 m^3 is formed in 90 s using a burner power of 4 MW and 150 s with a burner power of 2.4 MW.

Figures 5.16 and 5.17 shows the effect of scrap porosity and scrap-preheating temperature on burner thermal efficiency, scrap heating rate and cavity size. It is shown that scrap porosity increases both the burner thermal efficiency and the scrap melting rate. Increasing the scrap porosity from 0.81 to 0.91 increases the burner thermal efficiency by about 20%, also the average scrap temperature increases about 50 °C and furthermore the volume cavity formed in 240 s increases from about 0.34 to 0.52 m^3. As scrap porosity increases the gas is free to move and exchange heat with other pieces of scrap, also the burner flame penetrates a higher distance. On the other hand, the scrap pre-heating temperature has negative effects on both the burner thermal efficiency and scrap heating rate but not on the cavity size.

Scrap pre-heating affects both the burner thermal efficiency and the average scrap temperature but increases the scrap melting rate. The burner thermal efficiency

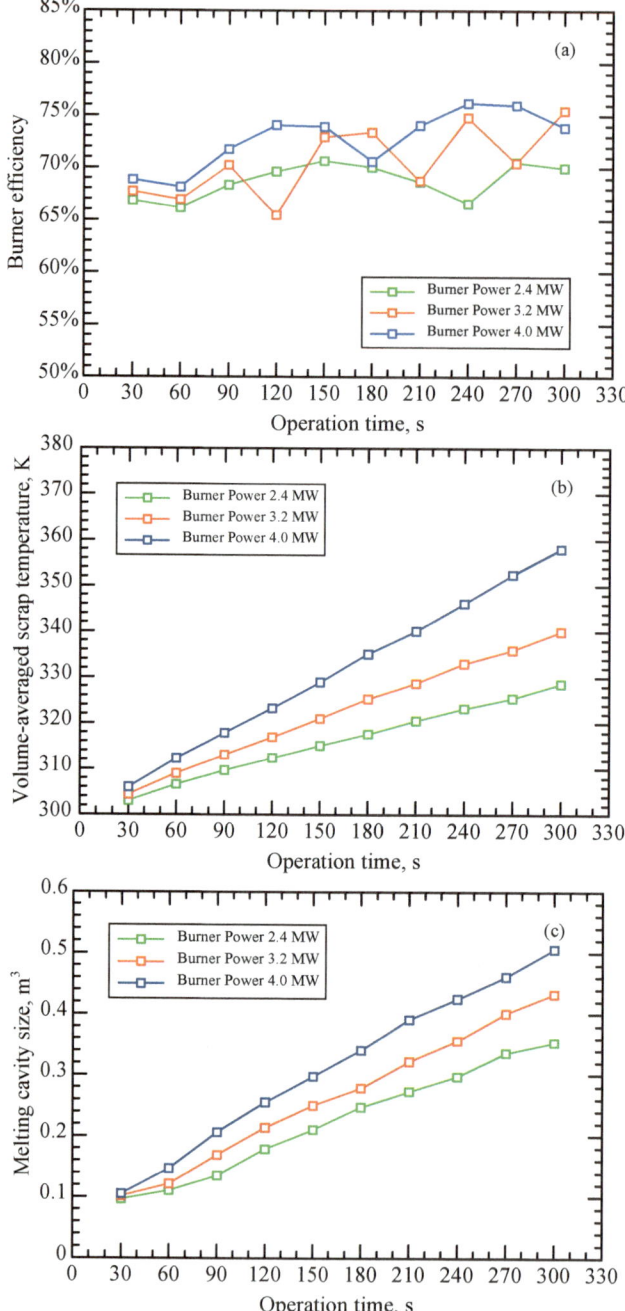

Fig. 5.15 Effect of burner power on **a** burner thermal efficiency, **b** scrap heating rate and **c** cavity size, after Chen et al. [36]

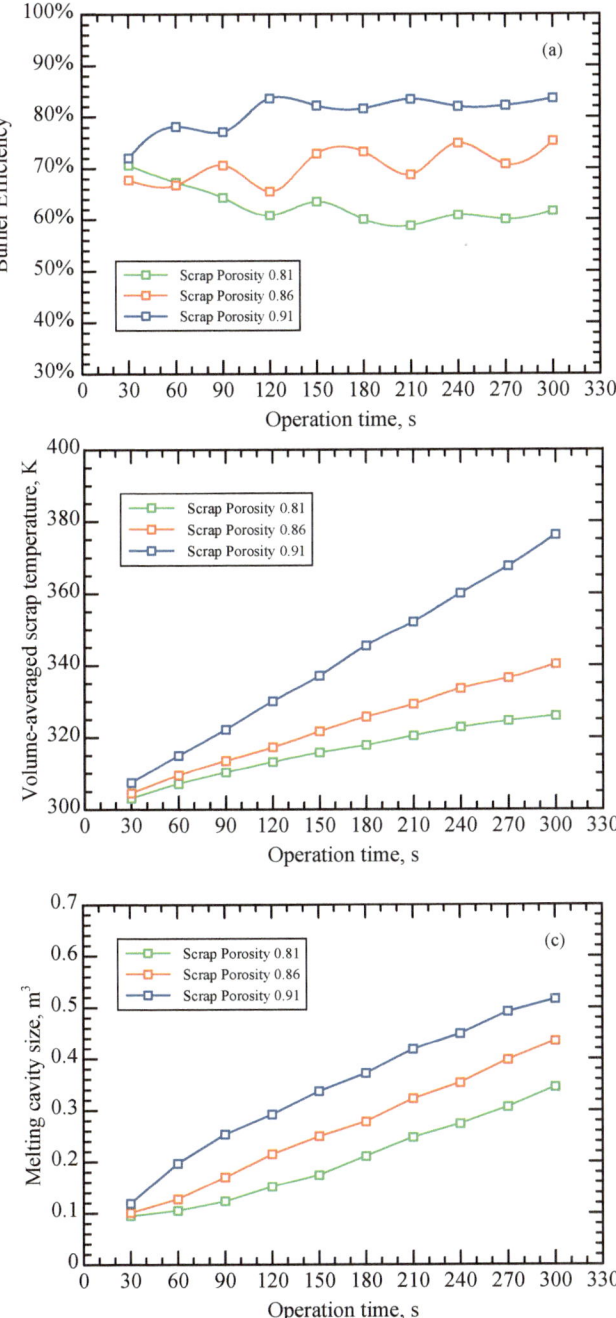

Fig. 5.16 Effect of scrap porosity **a** burner thermal efficiency, **b** scrap heating rate and **c** cavity size, after Chen et al. [36]

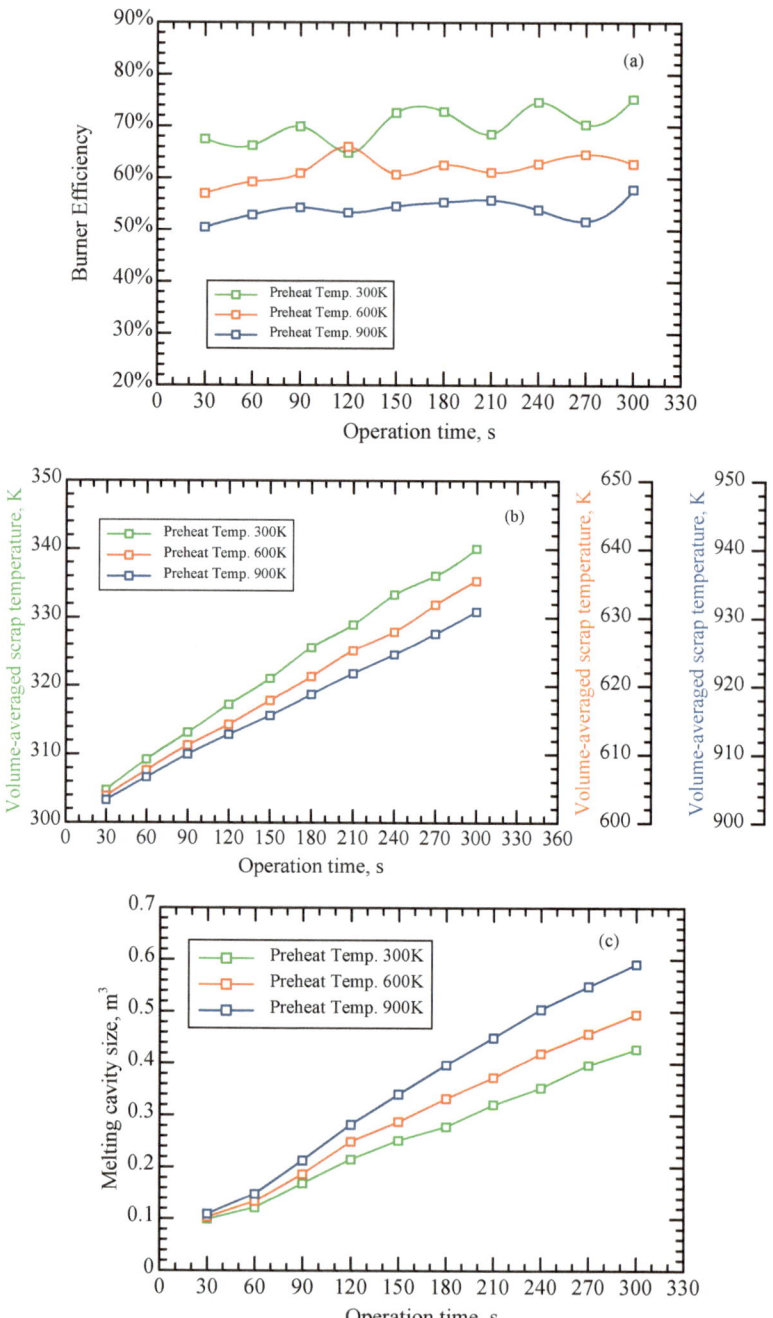

Fig. 5.17 Effect of scrap pre-heating on **a** burner thermal efficiency, **b** scrap heating rate and **c** cavity size, after Chen et al. [36]

decreases from 75 to 59% when the scrap pre-heating temperature increases from 300 to 900 K, as shown in Fig. 5.17a. Figure 7.17c shows that scrap pre-heating accelerates the volume of the cavity which then increases the scrap melting rate. These authors didn't discuss the convenience to use burners if the EAF has scrap pre-heating facilities but it can be speculated that due to a much lower burner efficiency this technology would be highly inefficient under those conditions. Finally, conditions that lead to back flow were also investigated in their numerical predictions. When the scrap is not properly prepared and large pieces of scrap are located in front of the burners, blocking the flame, the heat accumulates and overheats not only the burner but also the refractory and water-cooled panels. The burner thermal efficiency abruptly decreases from about 70–55% and the wall temperature abruptly increases by about 765 °C.

5.2.4 From Burner to Oxygen Lance

Oxyfuel burners can be operated in different modes depending on the stage of the heat. At the beginning they can be used to preheat scrap and cut large pieces of scrap but as the scrap is preheated, they should be changed from burner mode to oxygen injection mode. Opfermann and Riedinger [37] from BSW described a profile of the changes in the oxygen/fuel ratio for different modes of operation of a burner of 3 MW, as shown in Table 5.2. Figure 5.18 shows the large differences in the shape of the flame and resulting velocity and temperature fields. In mode 1 the velocity decreases at a short distance but the volume heated is large. In mode 2, oxygen is increased, the flame has a higher velocity and gets deeper into the scrap, in mode 3 the flow rates are similar but the main oxygen flow rate is now increased, in this state heat is transferred down to the scrap and the bath surface. With a higher superheat in the liquid bath, scrap melting is more uniform. They also found that decreasing the burners vertical angle from 42° to 38°, the distance for scrap preheating increased.

Table 5.2 Changes in O_2/gas ratio for different burner modes of operation. After [37]

	Main O_2, Nm^3/min	Second O_2, Nm^3/min	Natural gas, Nm^3/min
Mode 1 (burner)	5.8	5.0	5.0
Mode 2 (burner)	1.7	9.2	5.0
Mode 3 (burner)	9.2	1.7	5.0
Burner + O_2 lancing	16.7	3.3	3.3

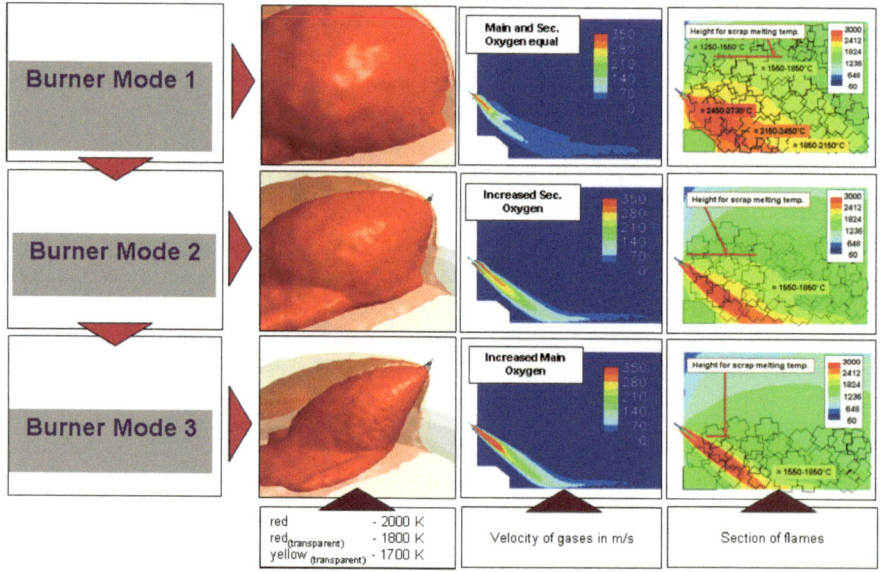

Fig. 5.18 flame shape, gas velocity and temperature fields depending on the flow rates of oxygen and gas in an oxyfuel burner of 3 MW. After [37]

5.3 Chemical Energy From Burners

5.3.1 The Ultra-High Chemical Power EAF (UHCP-EAF)

This is the name given by Concast (currently SMS) to the EAF with higher chemical energy using oxy-fuel burners and post-combustion in addition to scrap pre-heating and high-capacity transformers. Gottardi et al. [38–41] reported the following characterization of UHCP-EAF's:

1. Intensified melting rate due to:

   1.1. Increasing the transformer capacity. This solution has a direct impact only on the power-on time however if the power off-time is high the final effect on the tap-to-tap time can be drastically affected.

   An increment in the transformer capacity will increase the arc voltage and its electric power capacity, beneficial to increase the melting rate but at the same time the radiation index increases affecting the thermal load on the water-cooled panels. This change will demand a much higher control of the foaming conditions.

   1.2. Increasing the supply rate of chemical energy. There are two options in terms of oxy-fuel burners; increase its number or increase its power. The first option improves a more homogeneous scrap pre-heating but also involves higher maintenance costs.

2. A higher supply rate of chemical energy requires EAF modifications:

    2.1. To improve heat transfer from the gas to solid scrap, the gas residence time and its temperature should be increased. This requires a change in the geometry of the EAF. Increasing the shell height and decreasing the cross-sectional area, is equivalent with a decrease in the A_{shell}/V ratio. This ratio decreases by increasing the EAF capacity, from about 0.5 for an EAF of 25 ton to 0.10 for an EAF of 180 ton. Decreasing the A/V ratio decreases the slag/metal interface, affecting the removal of impurities, therefore this change should be compensated with additional stirring.

    2.2. High power burners require less number compared to low power burners; however, the water-cooled panels are exposed to much higher thermal loads.

3. Scrap-preheating: Scrap pre-heating can be carried out using a separate ladle, employing the off-gas from the EAF.

5.3.2 The Fuel-EAF (F-EAF) Concept

Toulouesvki and Zinurov invented a new type of oxy-fuel burners that produce a lower flame temperature in comparison with conventional burners [33] by mixing natural gas with coke oven gas. This mixture is cheaper and has an intermediate calorific value of 5.5 kWh/m^3 which prevents very high flame temperatures, a cause of two major problems; (1) localized scrap melting prevents a smooth descend from the shaft due to sticking of both the scrap pieces and the fingers and (2) higher iron oxidation. They have proposed the concept of Fuel EAF which basically consists in increasing the scrap pre-heating temperature using their burners in a shaft EAF like Quantum EAF. Part of the off gas on top of the shaft is forced to recirculate entering the burner and mixing with both oxygen and fuel, as shown in Fig. 5.19. The recirculated off-gas helps to decrease the flame temperature to values in the order of 1400 °C.

5.3.3 Analysis on the Full/Partial Replacement of Electric Energy by Chemical Energy

The replacement of electric energy by chemical energy in the EAF should take into account not only the amount of fossil fuels consumed in the EAF but also the amount of fossil fuels employed to produce electricity. The image of the EAF as a reactor that produces less CO_2 is based on the use of scrap and electricity. The amount of CO_2 is estimated only on the basis of oxygen and fossil fuels injected during the steelmaking process (carbon and natural gas). This analysis hides the amount of CO_2 involved in the production of electric energy. Currently the production of electricity is still dominated by fossil fuels with a global share in 2020 of 61.3%,

Fig. 5.19 a Low
temperature oxy-fuel
burners, **b** installed on EAF
quantum (Taken from
Toulouesvki and Zinurov,
pp. 83–84 [33])

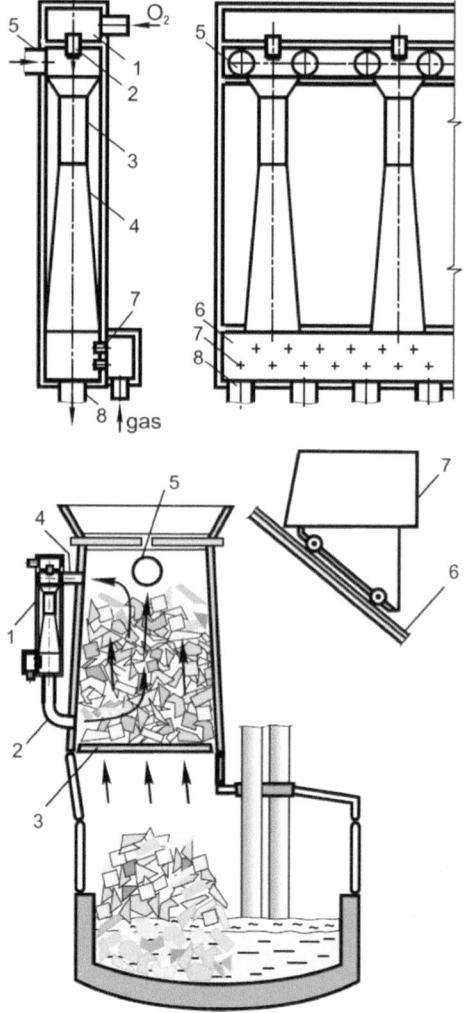

influenced by China, the largest producer of electricity using about 66% of fossil fuels
[42]. A parameter that defines both the intensity of fossil fuels and its efficiency is
the Aggregate Carbon Intensity (ACI) defined as the ratio of total CO_2 emissions
from fossil fuels to the total amount of electricity produced, reported in kg CO_2/
kWh. The ACI value for China in 1990 was 0.9 but decreased to 0.69 ion 2013.
In this year the USA reported an ACI value of 0.48 [43]. A more recent report for
2015 indicates an ACI value for China of 0.53 kWh/ton [44]. These achievements
have been made not because of a much larger share of electricity production from
renewable energies but due to improvements in thermal efficiency. It is estimated
that the share of electricity produced from renewable energy will reach 86% by the
year 2050 [45]. This background information indicates that at least for the next two

decades the carbon foot print due to the production of electricity by fossil fuels should be considered in the analysis to assess the replacement of electric energy by chemical energy in the EAF.

A full analysis on the replacement of electric energy by chemical energy requires some information unavailable and therefore in order to provide a quick analysis, major simplifications will be employed.

On the other hand, it is important to emphasize that the analysis involves prices of fossil fuels which can considerably fluctuate from country to country and through time. What follows is a summary of the analysis carried out by Wunsche in 2003 [18] and more recently in 2013 by Toulouesvki and Zinurov [46]. They invented new types of burners and promote the use of chemical energy in the EAF. Wunsche suggests to fully dismantling the transformer. Toulouesvki and Zinurov suggest a partial replacement of electric energy by chemical energy.

Wunsche's analysis [18]: Wunsche compares the cost of fuels and emissions of CO_2 for two cases, with 100% electric energy and 100% chemical energy using oxy-fuel burners.

Case 1: Using 100% electric energy. Cost of primary fuel and emissions of CO_2 associated to the production of electric energy using a power plant running on natural gas.

Electric energy consumption, kWh/ton	400
Primary energy at the power plant to produce electricity (400 kWh/ton) with an efficiency of 32%, kWh/ton	1250
Calorific value of natural gas, kWh/m^3	10.3
Price of natural gas, USD/1000 ft^3 (USD/m^3)	5 (0.176)
Price to operate an air fan, USD/1000 ft^3 (USD/m^3)	0.2 (0.007)

Cost of primary fuel to produce 1 ton of steel from cold scrap:

$$= \frac{1250\,\text{kWh}}{\text{ton}} \times \frac{\text{m}^3}{10.3\,\text{kWh}} = \frac{121.35\,\text{m}^3}{\text{ton}} \times \frac{0.176\,\text{USD}}{\text{m}^3} \equiv 21.3\frac{\text{USD}}{\text{ton}}$$

Cost of operating air fan for the combustion of fuel. Considering pure methane:

$$CH_{4(g)} + 2O_{2(g)} + 7.53N_{2(g)} = 2H_2O_{(g)} + CO_{2(g)} + 7.53N_{2(g)}$$

$$= \frac{121.35\,\text{m}^3\,\text{fuel}}{\text{ton}} \times \frac{9.5\,\text{m}^3\,\text{air}}{\text{m}^3\,\text{fuel}} \times \frac{0.007\,\text{USD}}{\text{m}^3} \equiv 8.09\frac{\text{USD}}{\text{ton}}$$

Amount of CO_2 emitted by power plant $= 121.35 \times 1.74 = 211$ kg/ton (The author employed a density of CO_2 of 1.74 kg/m^3)

Case 2: Full replacement of electric energy by chemical energy. Cost of oxy-fuel and emissions of CO_2.

Electric energy consumption, kWh/ton	400
Efficiency of chemical energy, %	100
Price of oxygen, USD/1000 ft^3 (USD/m^3)	2 (0.070)

Cost of fuel to produce 1 ton of steel from cold scrap:

$$= \frac{400\,\text{kWh}}{\text{ton}} \times \frac{\text{m}^3}{10.3\,\text{kWh}} = \frac{38.83\,\text{m}^3}{\text{ton}} \times \frac{0.176\,\text{USD}}{\text{m}^3} \equiv 6.83\frac{\text{USD}}{\text{ton}}$$

Cost of oxygen.

$$CH_{4(g)} + 2O_{2(g)} = 2H_2O_{(g)} + CO_{2(g)}$$
$$= \frac{38.83\,\text{m}^3\,\text{fuel}}{\text{ton}} \times \left(\frac{2}{1}\right) \times \frac{0.07\,\text{USD}}{\text{m}^3} \equiv 5.46\frac{\text{USD}}{\text{ton}}$$

Amount of CO_2 by oxy-fuel burners $= 38.83 \times 1.74 = 70.3\,\text{kg/ton}$
(The author employed a density of CO_2 of $1.80\,\text{kg/m}^3$)

According with the previous calculations, both, the cost of to produce one ton of liquid steel in the EAF using electric energy and the amount of emission of CO_2 is higher in comparison with chemical energy, which would suggest a full replacement of electric energy by chemical energy, as summarized below:

Type of energy	USD/ton	kg CO_2/ton
100% electric energy	29.42	211
100% chemical energy	12.29	70.3

Wunsche's calculation is biased because he assumes a low thermal efficiency of the natural gas at the power station (32%) which increases its cost and an unrealistic thermal efficiency of 100% using oxy-fuel burners which decreases its cost.

Toulouesvki and Zinurov's analysis [46]: These authors estimate the overall efficiency using natural gas to produce electricity and its further use in the EAF. The overall efficiency includes the thermal efficiency of natural gas to produce electricity at the power station (η_1), efficiency to transport that electricity (η_2) and the efficiency of the electric energy in the EAF (η_3).

$$\eta_T = \eta_1 \cdot \eta_2 \cdot \eta_3$$

These authors suggest the following values $\eta_1 = 0.7$, $\eta_2 = 0.92$ and $\eta_3 = 0.41$, therefore the total efficiency is about 0.26. They concluded that using electric energy is equivalent to use natural gas but with an efficiency of about 25%. With this reference, they argue that oxy-fuel burners have a larger efficiency and therefore replacing electric energy by chemical energy would consume less amount of fossil fuels.

The two previous analyses have been oversimplified, however are a good reference to pay attention to the primary emissions of CO_2 and the need to add them in the analysis of CO_2 emissions by the EAF. On the other hand, the replacement of oxy-fuel burners has been reported to have not only a low thermal efficiency but also the rate of heat transfer is lower in comparison with electric energy. Therefore, even if the electric energy can be fully replaced by chemical energy it decreases the melting rate. A better way is to use oxy-fuel burners under conditions that yield a high thermal efficiency and turned-off after that.

CO_2 per kWh at a power plant: Kirschen et al. [3] analyzed the replacement ratio of chemical energy from oxy-fuel burners comparing the amount of total CO_2 produced from burning one m^3 of fuel with the amount of CO_2 generated to produce 1 kWh of electric energy in different countries. The amount of CO_2 in a power plant depends on the fuels employed, ranging from 0.22 to 0.77 kg CO_2/kWh. Producing electric energy with a lower amount of CO_2 would require a higher thermal burner efficiency to achieve similar or lower emissions of CO_2 than the power plant.

The total of CO_2 produced from burning one m^3 of fuel come from burning fuel and from the CO_2 involved to produce oxygen:

$$CH_{4(g)} + 2O_{2(g)} = 2H_2O_{(g)} + CO_{2(g)}$$

From this reaction, 1 m^3 CH_4 produces 1 m^3 of CO_2. The density of CO_2 at one atmosphere decreases from 1.951 kg/m^3 at 0 °C to 1.784 kg/m^3 at 25 °C [47]. For STP conditions, the amount of CO_2 produced is 1.951 kg/m^3 which can be rounded-off to 2.0 kg CO_2/m^3 fuel.

The chemical reaction also consumes 2 m^3 O_2/m^3 fuel. The authors estimated an additional value of 0.2 kg CO_2/m^3 O_2.

Total CO_2 using chemical energy from burners $= 2 + (2 \times 0.2) = 2.4$ kg CO_2/m^3 fuel. kWh equivalent considering a calorific value for the natural gas of 10 kWh/m^3:

$$\text{kWh equivalent} = \frac{2.4 \text{ kg } CO_2}{m^3 \text{ fuel}} \times \frac{m^3 \text{ fuel}}{10 \text{ kWh}} = 0.24 \frac{\text{kg } CO_2}{\text{kWh}}$$

This analysis indicates that burning 1 m^3 of fuel is equivalent to produce electric energy that produces only 0.24 kg CO_2 per kWh. Table 5.3 shows the amount of CO_2 produced at power plants of different countries in kg CO_2/kWh and the minimum replacement ratio in kWh/m^3 fuel. The ratio of the CO_2 produced with 1 m^3 of fuel using oxy-fuel burners to the CO_2 produced at a power plant to produce one kWh of electric energy using different fuels gives the minimum replacement of electric energy. These results indicate that the cleaner the electric energy is produced, the thermal efficiency using burners should be larger, with replacement values that range, for most cases, from 3.1 to 6.6.

Figure 5.20 illustrates the example for power plants that produce electric energy generating 0.365 kg CO_2/kWh, therefore the minimum chemical energy from burners should be 6.6 kWh/m^3 fuel. Values below this level would be considered an inefficient replacement.

Table 5.3 CO_2 emissions to produce electric energy. Ref. [3]

Country	kg CO_2/kWh	kWh/m³ fuel
Germany, gas combustion	0.365	6.6
Germany, total mix	0.596	4.0
Great Britain	0.473	5.1
China	0.771	3.1
Canada	0.224	10.7
U.S.A.	0.575	4.2

Analysis based on a natural gas that releases 10 kWh/m³

$$\frac{kWh}{m^3 \text{ fuel}} = \frac{kWh}{0.36 \text{ kg } CO_2} \times \frac{2.4 \text{ kg } CO_2}{m^3 \text{ fuel}} = 6.6 \frac{kWh}{m^3 \text{ fuel}}$$

The added energy contribution from burners should not be limited to the operation of the burner itself, there is an additional energy contribution due to the shortening of the melting time. Jones et al. [48] cites a report from Krupp which suggest to add 2 kWh/ton per min due to faster melting rate.

High power rotary (HPR) burners: Toulouevski and Zinurov [49] described the development of HPR burners in Russia. They indicate that conventional EAF burners have two limitations; low power (3.5–5.5 MW) and the flame has both only one fixed direction and position. HPR burners have a power of 20–25 MW and can be moved up and down and rotate. They claim that using HPR burners scrap is preheated more

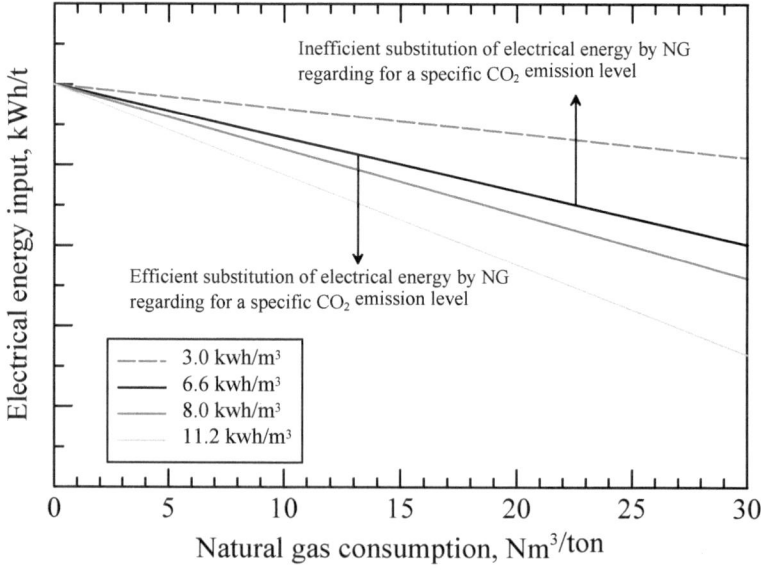

Fig. 5.20 Minimum replacement of chemical energy from burners. After Ref. [3]

homogeneously and a lower number of burners is required, for example in a 120-ton EAF is common to place eight conventional burners but using HPR burners only four would be needed.

Coherent jet technology: Conventional supersonic jets maintain its original velocity and diameter for distances equivalent to less than 35 times their nozzle diameters, compared with 70 times their nozzle diameters for coherent jets [50]. Burners can be switched from burner mode to oxygen injection mode to help cut the scrap. Oxygen ignites if the scrap is properly preheated. The cutting process is stopped at 50–60% of meltdown [6].

5.3.4 Limitations to Increased Chemical Energy in the EAF

In principle is possible to use pure chemical energy to melt 100% scrap. In Chap. 4 the EOF process which uses pure chemical energy was briefly described. Jepson [51] also described another example in the foundry industry, the oxy-fuel rotary furnace which is charged with iron scrap, castings returns, pig iron, steel scrap and carbon. This furnace prevents any air infiltration and burners operate with some degree of post-combustion. The oxygen/fuel ratio was increased from about 1.8 to 3.43 using burners of 5 MW in a furnace of 20 tons. Oxygen increased from 130 to 133 Nm^3/ton and natural gas was decreased from 70 to 38 Nm^3/ton. Carbon additions increased from 16 to 32 kg/ton. The huge amount of oxygen for the oxy-fuel rotary furnace is in contrast to the levels employed in a conventional EAF, as shown in Fig. 5.21. This figure also shows results using the K-ES system. The K-ES system consist of oxygen and carbon injection at high rates to provide for an enhanced supply of chemical energy. It was developed by Klockner Technology and Tokyo steel, later owned by VAI (currently Primetals).

Jepson [51], from Air Liquide, argues that although full replacement of electric by chemical energy in the EAF is possible there are the following practical limitations:

- Lower production rates using only chemical energy: A typical production rate for an EAF of 100 ton is at least 100 ton/h using electric energy, in comparison with an oxy-fuel rotary furnace the productivity is less than 50 ton/h. Electric energy, due to its high energy density is a more efficient energy for melting by contrast chemical energy involves gases which need time and space to transfer that energy.
- The EAF has limitations to inject high flow rates of oxygen. Increasing the oxygen flow rate at one point of injection can create splashing problems.
- Enhanced use of chemical energy requires higher capacity of the off-gas exhaust system, capable to handle higher gas volumes and heat loads.

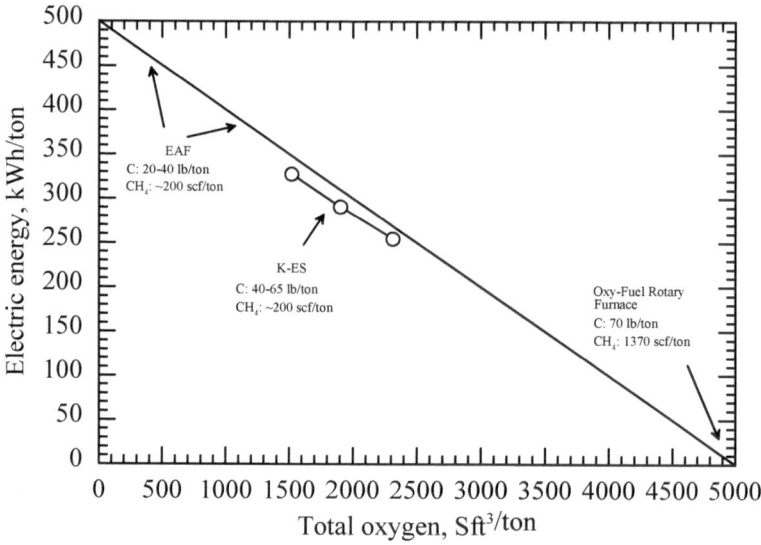

Fig. 5.21 Electrical versus chemical energy for ferrous scrap melting. After Ref. [51]

5.3.5 Savings on Energy Consumption Using Burners

Kirschen et al. [3] evaluated 70 EAF's. The effect of burners on the total and electrical energy consumption is shown in Fig. 5.22. These results show a very small contribution of burners on the total energy consumption, but on the other hand, they have an important effect to decrease electric energy consumption. The chemical energy added from burners replaces electric energy. The effective chemical energy from burners which replaces electric energy can be estimated from the theoretical calorific value of the fuel (9.28–10.70 kWh/m^3), burner thermal efficiency (0.60–0.70) and also the electric efficiency (0.80–0.95), which would suggest a range from 6.9 to 11.5 kWh/m^3 fuel.

$$Q_{b,el} = \eta_b\,\eta_b\,\Delta H_b$$

$$Q_{b,el} = (0.6 - 0.7) \times (0.90 - 0.95) \times (9.28 - 10.70) = 6.9 - 11.5\frac{kWh}{m^3}$$

Additionally, they reported detailed results for two furnaces, as shown in Fig. 5.23. These results indicate a decrease in energy consumption of 9.6 kWh/m^3 when the consumption of fuel is below 10 m^3/ton. Other similar values have been reported; Kohle 8 kWh/m^3, Klein and Schlinder 8.5 kWh/m^3 and Kuhn 11.2 kWh/m^3.

It is important to be careful on the reports on the decrease in energy consumption due to burners in kWh/ton because of its multi-functional work both as oxygen injectors, burners and post-combustion and the decrease in energy is not always defined for one function. Sung et al. reported the decrease due to one burner indicating

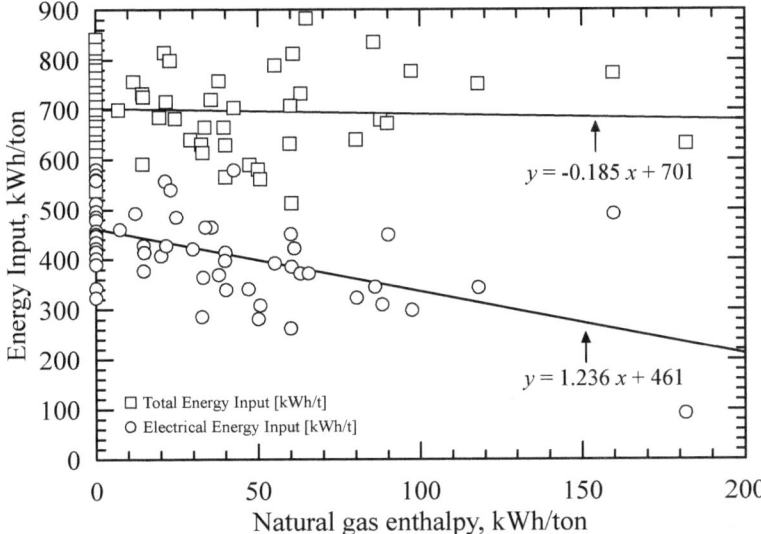

Fig. 5.22 Effect of burners chemical energy on energy input, after Kirschen et al. [3]

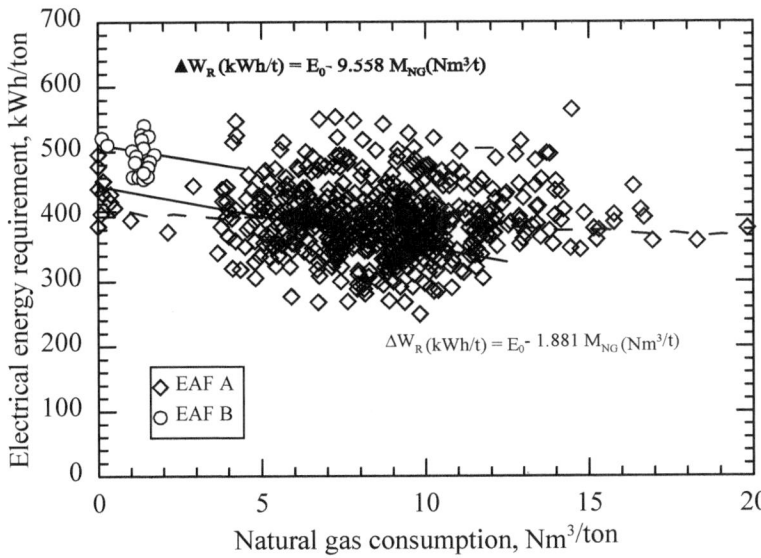

Fig. 5.23 Effect of natural gas on energy consumption, after Kirschen et al. [3]

a decrease from 396 to 391 kWh/ton, total savings in consumption of electric energy of 5 kWh/ton and an increment in productivity from 120 to 123.1 ton/h. Several sources cite much higher savings in electric energy consumption but it can be due not only because of a larger number of burners but also the effects of oxygen injection (oxygen flow rate and injection pressure). Megahed et al. [23] made improvements to optimize burner's operation in an EAF of 185 ton with a transformer of 133 MVA. The number of injectors were eight; 4 as oxy-fuel burners, 1 burner and 3 carbon injectors. All of them start operating as burners. The decrease in energy consumption varied from 38 to 64 kWh/ton by increasing DRI in the metallic charge from 20 to 60%. The highest energy savings (12%) were obtained due to improved foaming conditions. Cantacuzene et al. [52] reported a decrease in electric energy consumption from 427 to 392, 35 kWh/ton, corresponding to 6.3% energy savings by increasing the number of oxy-fuel burners from 4 to 10. Thomson et al. [53] reported energy savings due to changes in the oxygen/natural gas ratio of the oxy-fuel burners, indicating a decrease of 4% when the ratio O_2:fuel increased from 3:1 to 8:1. Increasing the oxygen ratio increased post-combustion. J. Jones, cited by Worrell et al. [54] reported energy savings from 20 to 40 kWh/ton with oxy-fuel burners, which suggest contribution from both oxygen injection and burners.

Economics also has influence on the extent of replacement of electric energy by chemical energy using oxy-fuel burners. Locations with a high price ratio of electric energy/natural gas are more suitable to enhance the use of burners. Sung et al. [22] summarized the prices of natural gas and electricity, indicating ranges from 1.05 to 7.27 cent/kWh for natural gas and 6.81–17.12 cent/kWh for electricity. These numbers indicate large differences in prices in different countries.

References

1. Mandal K (2010) Modeling of scrap heating by burners. McMaster University, Ontario CA
2. Adolph H, Paul G, Klein KM, Lepoutre E, Vuillermoz JC, Devaux M (1990) A new concept for using oxy-fuel burners and oxygen lances to optimize electric arc furnace operation. La Rev Metall 87:47–54
3. Kirschen M, Risonarta V, Pfeifer H (2009) Energy efficiency and the influence of gas burners to the energy related carbon dioxide emissions of electric arc furnaces in steel industry. Energy 34:1065–1072. https://doi.org/10.1016/j.energy.2009.04.015
4. Pfeifer H, Kirschen M, Simoes JP (2005) Thermodynamic analysis of EAF electrical energy demand. In: 8th European electric conference, 9–11 May 2005, Birmingham, UK
5. Logar V, Dovžan D, Škrjanc I (2012) Modeling and validation of an electric arc furnace: part 2, thermo-chemistry. ISIJ Int 52:413–423. https://doi.org/10.2355/isijinternational.52.413
6. Jones JAT (2002) Understanding energy use in the EAF. In: 60th electric furnace conference, San Antonio TX USA, 10–13 Nov 2002, pp 141–154
7. Pittam D, Pilcher G (1972) Measurements of heats of combustion by flame calorimetry. Part 8. J Chem Soc Faraday Trans 68:2224–2229
8. Vogel, W. Kalb H (2010) Large scale solar thermal power. Wiley-VCH Weinhem Germany
9. Hernandez J, Onofi L, Engell S (2019) Model of an electric arc furnace oxy-fuel burner for dynamic simulations and optimisation purposes. Int Fed Autom Control: 30–35
10. Bergman K, Gottardi R (1990) Design criteria for the modern UHP electric arc furnace with auxiliaries I. Ironmak Steelmak 17:282–287

11. Mathy C, Terho K, Chouvet M, Le Coq X, Baumert J-C, Engel R, Hoffmann J (2003) Production of steel at lower operating costs in EAF. Luxembourg. ECSC report 20895
12. Mandal K, Irons GA (2013) A study of scrap heating by burners. Part I: experiments. Metall Mater Trans B 44:184–195
13. Mandal K, Irons GA (2013) A study of scrap heating by burners: part II—numerical modeling. Metall Mater Trans B Process Metall Mater Process Sci 44:196–209. https://doi.org/10.1007/s11663-012-9752-1
14. Muller GE (1991) Prediction of radial porosity distributions in randomly packed fixed beds of uniformly sized spheres in cylindrical containers. Chem Eng Sci 46:706–708
15. Logar V, Dovžan D, Škrjanc I (2012) Modeling and validation of an electric arc furnace: part 1. Heat and mass transfer. ISIJ Int 52:402–412. https://doi.org/10.2355/isijinternational.52.413
16. Hinds GW (1965) Method of operating electric arc furnace. US 3197539
17. Wunsche ER, Wunsche AA, Kosanovich M (1987) Multi-purpose pyrometallurgical process enhancing device, US Patent 4653730, Apr 24, US Patent 4653730
18. Wunsche E (2003) Multi-purpose, multi-oxy-fuel, power burner/injector/oxygen lance device, US Patent 2003/007584, 24 Apr 2003
19. Mathy C, Nyssen P, Brimmeyer M, Gualtieri D, Rigoni D, Baumert JC (2008) Innovative technique for reliable operations and blow-back prevention of EAF annular burners, combined burners and injectors. Arch Metall Mater 53:469–473
20. Grosse A, Opfermann A, Libera K, Gökce F, Schweikle R, Volkert A (2014) The next generation of chemical energy application of Badische. AISTech Proc 1:957–965
21. Volkert A, Wohlfarrt S, Libera K, Grosse A, Opfermann A, Dieter GS et al (2016) The first tiltable burner of Badische for higher energy efficiency and less refractory wear. 47° Semin. aciaria, Rio de Janeiro Brazil, 26–30 Sept 2016, pp 401–412. https://doi.org/10.5151/1982-9345-27765
22. Sung Y, Lee S, Han K, Koo J, Lee S, Jang D et al (2020) Improvement of energy efficiency and productivity in an electric arc furnace through the modification of side-wall injector systems. Processes 8. https://doi.org/10.3390/PR8101202
23. Megahed GM, Fathy AM, Morsy MA, Abdelaziz EA (2010) Improving EAF performance by chemical energy optimisation at Ezz flat steel. Ironmak Steelmak 37:445–451. https://doi.org/10.1179/030192310X12690127076596
24. Cheng KC, Fujii T (1998) Isaac newton and heat transfer. Heat Transf Eng 19:9–21. https://doi.org/10.1080/01457639808939932
25. Besson U (2012) The history of the cooling law: when the search for simplicity can be an obstacle. Sci Educ 21:1085–1110. https://doi.org/10.1007/s11191-010-9324-1
26. Cengel YA, Ghajar A (2011) Chapter 7. External forced convection. Heat mass transfer fundamental application, 5th edn., McGraw-Hill, New York, USA, pp 446–452
27. Al-Haj Ibrahim H, Al-Qassimi M (2010) Simulation of heat transfer in the convection section of fired process heaters. Period Polytech Chem Eng 54:33–40. https://doi.org/10.3311/pp.ch.2010-1.05
28. Zukauskas A (1972) Heat transfer from tubes in cross flow. Adv Heat Transf 8:93–160
29. Žukauskas A, Ulinskas R (1985) Efficiency parameters for heat transfer in tube banks. Heat Transf Eng 6:19–25. https://doi.org/10.1080/01457638508939614
30. Kitaev BI, Yaroshenko YG, Suchov VD (1967) Heat exchange in shaft furnaces. Pergamon Press, UK
31. Furnas CC (1930) Heat transfer from a gas stream to a bed of broken solids-II. Ind Eng Chem 22:721–731
32. Wakao N, Kaguei S, Funazkri T (1979) Effect of fluid dispersion coefficients on particle-to-fluid heat transfer coefficients in packed beds. Correlation of Nusselt numbers. Chem Eng Sci 34:325–336. https://doi.org/10.1016/0009-2509(79)85064-2
33. Toulouevski YN, Zinurov IY (2017) Fuel arc furnace (FAF) for effective scrap melting: from EAF to FAF. Springer Nature, Singapore. https://doi.org/10.1007/978-981-10-5885-1
34. Tang G, Chen Y, Silaen AK, Krotov Y, Zhou CQ (2019) Effects of steel scrap oxidation on scrap preheating process in an electric arc furnace. In: 10th international symposium high-temperature metallurgical process, 453–465. https://doi.org/10.1007/978-3-030-05955-2_43

35. Zhang L, Oeters F (1999) Possibilities of counter-current scrap pre-heating with melting by use of 100% fossil energy. Steel Res 70:296–308. https://doi.org/10.1016/b978-0-12-248291-5/50008-0

36. Chen Y, Luo Q, Ryan S, Busa N, Silaen AK, Zhou CQ (2022) Effect of coherent jet burner on scrap melting in electric arc furnace. Appl Therm Eng 212. https://doi.org/10.1016/j.applthermaleng.2022.118596

37. Opfermann A, Riedinger D (2008) Energy efficiency of electric arc furnaces. In: AISTech 2008 proceedings, pp 795–808

38. Gottardi R, Miani S, Partyka A (2006) Ultra high chemical power EAF. In: AISTech 2006 conference, Cleveland, OH, USA. 1–4 May 2006, pp 1–10

39. Gottardi R, Miani S, Partyka A (2007) New giant generation EAF. In: AISTech 2007 conference, Indianapolis IN, USA, 7–10 May 2007, pp 1–10

40. Gottardi R, Miani S, Partyka A, Engin B (2008) Ultra high chemical power EAF for 320 t/h. Rev Metall Cah D'Informations Tech 105:596–600. https://doi.org/10.1051/metal:2009003

41. Gottardi R, Miani S, Partyka A (2008) A faster, more efficient EAF. Arch Metall Mater 53:475–482

42. British Petroleum (2021) Statistical review of world energy, 1–72. https://www.bp.com/en/global/corporate/energy-economics/energy-outlook/energy-outlook-downloads.html.

43. Ang BW, Su B (2016) Carbon emission intensity in electricity production: a global analysis. Energy Policy 94:56–63. https://doi.org/10.1016/j.enpol.2016.03.038

44. Zhao Y, Cao Y, Shi X, Zhang Z, Zhang W (2021) Structural and technological determinants of carbon intensity reduction of China's electricity generation. Environ Sci Pollut Res 28:13469–13486. https://doi.org/10.1007/s11356-020-11429-0

45. Mostafaeipour A, Bidokhti A, Fakhrzad MB, Sadegheih A, Zare Mehrjerdi Y (2022) A new model for the use of renewable electricity to reduce carbon dioxide emissions. Energy 238. https://doi.org/10.1016/j.energy.2021.121602

46. Toulouevski YN, Zinurov IY (2013) Innovation in electric arc furnaces. Scientific basis for selection, 2nd edn. Springer-Verlag, Berlin, Germany

47. Anwar S, Carroll J (2016) Density (kg/m^3) of carbon dioxide as a function of temperature and pressure. Carbon Dioxide Thermodyn Prop Handb: 9–148

48. Jones JAT, Bowman B, Lefrank PA (1998) Electric arc furnace steelmaking, 11th edn. In: Fruehan RJ (ed) Making, shaping treating steeling, steelmaking refining, Pittsburgh, PA, USA, pp 525–660

49. Toulouevski YN, Zinurov IY (2002) Outlook for reduction in energy consumption of electric arc furnaces. In: 7th European electric steelmaking conference, Venice, Italy, 26–29 May 2002, pp 1.57–1.64

50. Mathur P (2004) Coherent jets in steelmaking: principles and learnings. Praxair Tech. Inc., USA

51. Jepson S (2000) Chemical energy in the EAF: benefits and limitations. Electr Furn Conf Proc: 3–14

52. Cantacuzene S, Grant M, Boussard P, Devaux M, Carreno R, Laurence O et al (2005) Advanced EAF oxygen usage at Saint-Saulve steelworks. Ironmak Steelmak 32:203–207. https://doi.org/10.1179/174328105X45811

53. Thomson MJ, Kournetas NG, Evenson E, Sommerville ID, McLean A, Guerard J (2001) Effect of oxyfuel burner ratio changes on energy efficiency in electric arc furnace at Co-steel lasco. Ironmak Steelmak 28:266–272. https://doi.org/10.1179/030192301678136

54. Worrell E, Martin N, Price L (1999) Energy efficiency and carbon dioxide emissions reduction opportunities in the U. S. iron and steel sector. NBNL report

Chapter 6
Post-combustion

6.1 Introduction

6.1.1 Origins of Post-combustion

Prior to the 1990s, the CO in the off gas of EAF's was fully combusted in the gas collection system with air coming from the combustion gap. The energy released was totally wasted. Post-Combustion (PC) was a general practice in the BOF since first introduced in 1981 by Klockner in the development of the K-OMB and KMS processes which allow to melt a higher ratio of scrap from 30 to 50% [1]. Hirai et al. [2] reported a model for PC which includes a parameter to describe soft and hard blowing.

PC is employed in EAF scrap-based operations and represents the complete combustion of a partially combusted compound like CO or H_2 formed due to decarburization, evaporation or incomplete combustion of hydrocarbons, before these gases leave the EAF:

$$CO_{(g)} + {}^1/_2O_{2(g)} = CO_{2(g)}$$
$$H_{2(g)} + {}^1/_2O_{2(g)} = H_2O_{(g)}$$

Post combustion employs a fuel naturally produced through decarburization. The heat released can be transferred to both solid scrap and liquid steel. This practice is especially recommended for EAF operations that include a higher amount of carbon in the metallic charge, for example due to metallic charges including scrap and hot metal. It has also been recommended to use a low velocity for the oxygen jet to promote mixing with the gases, reduce scrap oxidation and oxygen rebound from the scrap to the water cooled panels [3].

For a BOF, the specific volume and temperature of the off-gas is in the order of 100 m^3/ton and 1220–1650 °C [4], with CO and CO_2 as the main gas components.

© The Author(s), under exclusive license to Springer Nature Singapore Pte Ltd. 2024
A. N. Conejo, *Electric Arc Furnace: Methods to Decrease Energy Consumption*,
https://doi.org/10.1007/978-981-97-4053-6_6

Yang et al. [5] for the EAF reported an off-gas generation rate of 500–800 Nm^3/ ton for UHP-EAF and from 1000 to 1200 Nm^3/ton for UHP-EAF using oxy-fuel burners, with an off-gas chemical composition consisting of 1–34% CO, 12–20% CO_2, 5–14% O_2, 45–74% N_2.

The production of CO in the EAF depends on the raw materials employed, usually is in the range from 0.3–2.6 kg CO/ton [6]. Jiang et al. [7] reported a value of 2.1 kg CO/ton using 100% scrap and 8.23 kg CO/ton using 100% hot metal. Li and Fruehan [8] reported a different amount, about 18 kg per ton of steel. The initial concentration of CO is important because it influences the efficiency of PC. PC decreases the concentration of CO below 10%.

Post-combustion increases the consumption of oxygen, reaching values as high as 75 Nm^3/ton [9]. It is another way to carry out scrap pre-heating taking advantage, primarily of the CO released due to decarburization of liquid steel. The heat released due to oxidation from C to CO is less than half of that released when CO is finally oxidized to CO_2. The heat of combustion from CO to CO_2 is potentially very high; a work cited by Jones et al. [9] indicates a potential of 75 kWh/ton, however only part of this chemical energy is transferred to pre-heat the scrap and the rest is transferred to the water-cooled panels. Without post-combustion the oxidation from CO to CO_2 occurs at the ducts of the gas collection system, enhancing its thermal load. The presence of oxygen and CO_2 enhance the oxidation rate of electrodes and steel scrap. Scrap oxidation provides chemical energy equivalent to 12 kWh/ton for every 1% of steel oxidized. The supply of chemical energy from PC contributes to decrease electric energy consumption, about 1 kWh/ton [9]. It is estimated that one minute decrease in tap-to-tap time can decrease electric energy consumption by about 2–3 kWh/ton [9].

Some advantages and disadvantages for post-combustion are summarized in Table 6.1

The heat capacity of CO_2 is higher than CO which means that CO_2 will transfer more heat per unit volume, for the same temperature of the off gas [9]. This heat is in the gas phase and should be transferred to the solid scrap or liquid metal. Since the gas residence time is low, the thermal efficiency is low, from 30 to 50%. An average value for the heat input is about 2.5 kWh/m^3. The final value depends on the

Table 6.1 Some advantages and disadvantages of post-combustion [9]

	Advantages	Disadvantages
Energy savings	20–40 kWh/ton	
Energy input	3–4 kWh/m^3 O_2	
Shorter tap to tap time	Less than minute	
Higher pressure operation	Decreases air infiltration	
Scrap oxidation		2–3%
Higher electrode consumption		Increases
Overheating of WCP		Increases

applied post-combustion profile with respect to the chemical and thermal conditions prevailing in the furnace.

6.1.2 Position of Oxygen Lances for PC

There are two ways to promote post-combustion in the EAF:

- Direct method using oxygen lances: these lances can be from oxy-fuel burners operated with a higher oxygen/fuel ratio. If the oxygen is injected above the free surface more heat can be transferred to liquid steel.
- Indirect method through the slag layer: soft blowing of oxygen into the slag layer. This practice can be applied during and after the scrap meltdown process. It requires a thick slag layer to avoid a larger generation of dust. With this method a higher amount of heat is transferred to liquid steel.

PC studies carried out in the BOF by Takashiba et al. [10] recommended oxygen injection just above the molten bath and more than 1 m from the refractory walls.

In one of the first designs for PC in the EAF, Brotzmann and Fritz [11] proposed multiple lances on top of the EAF chamber, intended to cover a larger volume. Mathur et al. [12] disclosed a patent for PC, shown in Fig. 6.1. In this figure, (5) indicates the primary oxygen injection for decarburization, lance (6) represents a secondary lance for PC which should be located at a point where CO is generated. In one example of its application in a 60-ton EAF, secondary oxygen is in contact with scrap and later, as melting progresses, within the foamy slag. Energy savings of 40 kWh/ton were reported equivalent to a heat transfer of 4.75 kWh/Nm^3 O$_2$ and HTE of 81%.

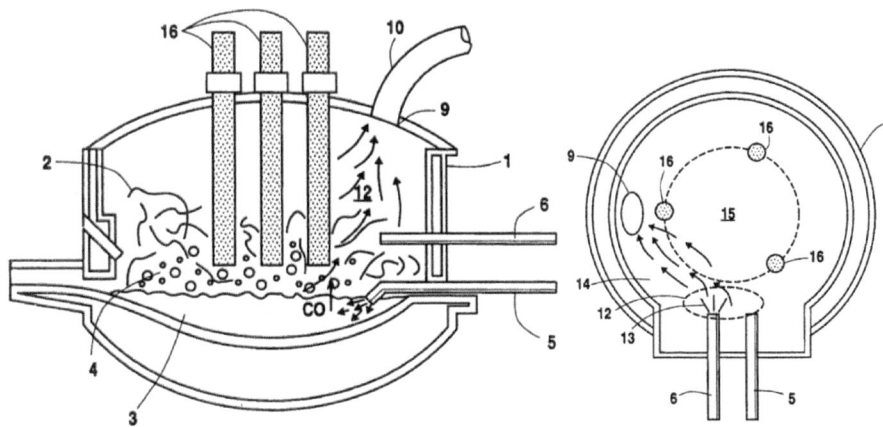

Fig. 6.1 Postcombustion lance according to Praxair [12]

6.2 Post-combustion Efficiency

6.2.1 Definitions

An important definition is the post-combustion ratio which defines the fraction of CO_2 in the off-gas, as follows:

$$PCR = \text{post combustion ratio} = \frac{\%CO_2}{\%CO + \%CO_2}$$

Including the secondary post-combustion of H_2 to H_2O, a more general definition is:

$$PCR = \frac{\%CO_2 + \%H_2O}{\%CO + \%CO_2 + \%H_2O} \times 100\%$$

Higher levels of PC indicate higher production of CO_2 and H_2O, therefore a higher production of chemical energy. Several reports from steel companies have reported PCR values in the range from 40 to 80% [9]. There are several reasons why post-combustion should not be 100% [9]:

1. First, in order to reach the minimum flammability limits at the combustion gap, CO in the off gas should be at least 5–10%. If CO is lower it will not react to produce CO_2, consequently 100% PCR is not recommended.
2. Second, high PCR involve very high temperatures in the off-gas affecting both the refractory and ducts in the off-gas system.
3. Additionally, the presence of CO helps to eliminate NO:

$$2CO_{(g)} + NO_{(g)} = CO_{2(g)} + N_{2(g)}$$

The need to decrease the emission of nitric oxide (NO) is because it combines with water vapor in the atmosphere to form nitric acid, which is one of the components of acid rain.

The PC ratio only defines the conversion of CO to CO_2 but not the efficiency to transfer heat to solid scrap. HTE defines the heat transfer efficiency, as follows:

$$HTE = \frac{(kWk/ton)_{No\text{-}PC} - (kWk/ton)_{PC}}{\Delta H_{PC}^{theory}}$$

where: $(kWk/ton)_{No\text{-}PC}$ is the energy consumption without PC, $(kWk/ton)_{PC}$ is the energy consumption using PC and ΔH_{PC}^{theory} is the maximum chemical energy produced considering the amount of oxygen injected for PC. A hypothetical HTE of 100% is reached when all the chemical energy produced by PC is transferred before leaving the furnace.

Gou et al. [13] reported the effect of lance height (hard or soft blowing) on PC in a K-OBM converter, indicating that raising the lance height decreased the decarburization rate and more primary oxygen is available to react with CO, increasing PCR, but increasing PCR decreases both the primary oxygen for DeC and the heat transfer efficiency, as shown in Table 6.2 and Fig. 6.2.

Farrand et al. [14] estimated the HTE for the KOBM process based on a correlation between tapping temperature and PCR. Assuming 100% HTE, the temperature increment in liquid steel would be 10.4 °C for 1% PCR. Their measurements reported a temperature increment in the tapping temperature of 4.9 °C for an increment of 1% PCR which gives a HTE value of 43.8%. They also observed a peak in the PC ratio at the beginning of the decarburization process and then continuously decreased, in a similar way to the thermal burner efficiency. The increment in the furnace volume due to refractory erosion increased the off-gas residence time and improved the PCR.

Post-combustion can be carried out with oxygen injection in the slag or in the freeboard. When post-combustion is carried out in the free board the gas transfers heat by radiation. When post-combustion is carried out in the slag, oxygen is introduced

Table 6.2 PC results in K-OBM at 8 min in the blow. After [13]

Lance height, m	DeC rate, kg/min	PCR, %	HTE, %	T_bath, °C
4, early campaign	740	15	63.4	1390
4, early campaign	722	16.3	63.2	1390
5, late campaign	678	22.6	55.5	1390

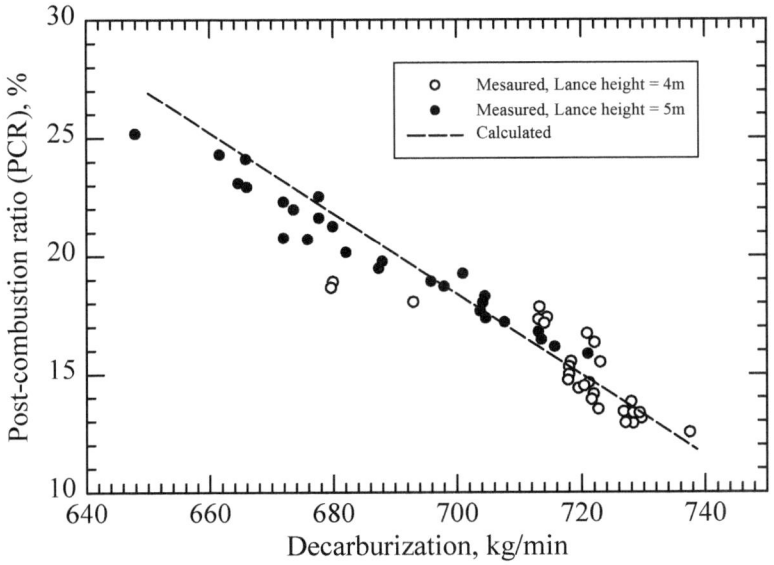

Fig. 6.2 Relationship between the decarburization rate and PCR in the BOF, after [13]

at low velocities. Heat transfer is by convection. To avoid contact of O_2 with the molten metal, the slag should be thick [9].

The adiabatic flame temperature of CO_2 due to post-combustion is similar to the burning of natural gas, about 2800 °C [9]. If this high thermal load is not transferred to the scrap it will transfer to the water-cooled panels.

The reaction product from post-combustion is an oxidizing gas. This gas can react with the scrap, forming FeO, decreasing the metallic yield. The oxidizing potential of iron by CO_2 increases with temperature; the equilibrium value of CO_2 decreases from 24% at 1377 °C to 8.6% at 1800 °C [9]. Once the layer of FeO is formed it acts as a barrier, decreasing the oxidation rate but this layer is melted if the temperature of the scrap reaches 1300 °C [9] Opferman et al. [15] reported a decrease in metallic yield using 2 kWh/Nm^3 of oxygen for post-combustion in a 90 ton EAF, but also an increase in productivity, decreasing the tap-to-tap time about 1.4 min. Guan et al. [16] suggest a maximum value of 15 Nm^3/ton for post-combustion to avoid a severe impact on metallic yield.

Air infiltration is a natural source of oxygen for post-combustion, however, in addition to oxygen air also brings nitrogen (79%) which is inert. 50% of the heat released by post-combustion is used to heat the nitrogen [17]

The best conditions for post-combustion are in the early stages of scrap melting when the scrap is cold and there is a high thermal gradient. In order to initiate oxygen injection for post-combustion CO should be generated due to the decarburization reaction. Oxygen injection for decarburization cannot be initiated until there is a pool of liquid steel. The formation of CO has been reported in the range from 0.3 to 2.6 kg/ton [9]. Its rate of formation depends on the rate of melting.

An oxy-fuel burner can be used as a burner and also as an oxygen source for PC if it is operated using super-stochiometric ratios of oxygen to fuel however this practice is inefficient compared with a separate operation of burners and oxygen injectors for post-combustion [17].

6.2.2 Off-Gas Composition and Temperature

Grant [17] indicates that designing a post-combustion system requires mass and energy balances of a number of heats to study the off-gas composition as a function of the metallic charge, temperature, etc. Depending on the production rate of CO and its changes during the heat, the PC system can be designed.

An important component of the post-combustion system is an on-line system to measure the chemical composition and temperature of the off-gas. The off-gas composition can be obtained using two systems: Extractive and in situ laser systems, described below.

1. Extractive system [18]: Extractive systems use a vacuum pump and a filter to continuously extract a sample of process off-gas through a probe positioned in the fume duct. The probe is water cooled. The probe position is located before

the air gap to avoid air dilution. Some disadvantages include a response time from 20 to 40 s, higher installation costs, needs regular analyzer calibration and separate analyzers for each off-gas component. Further details can be found in a patent by Evenson in [19].

The analytical methods include mass spectrometry, non-dispersive infra-red (NDIR) method (CO and CO_2), a solid-state electrochemical cell (O_2 analysis) and thermal conductivity (H_2). Measurement of water vapor (H_2O) helps to identify water leaks. N_2 provides information about the extent of air infiltration, it can be estimated subtracting from 100% all the previous components.

One commercial system is EFSOP (Expert Furnace System Optimisation Process) developed by Canadian H. Goodfellow in the late 1990s. In this system the extracted gases are transported, via a heated sample line. Initially the off-gases are filtered, dried and then analyzed in real-time. Figure 6.3 show the location of the sensor probe [20].

It has been reported the off-gas composition of several plants [20, 21]. The major components are CO and H_2, in the order of 30 and 20%, respectively as shown in Fig. 6.4. The concentration of CO_2 is in general below 15%. These results indicate insufficient post-combustion.

2. In situ laser system [18]: In situ laser systems transmit a single beam or a combined beam or multiple individual beams within the visible, near and mid IR range through the off-gas as it flows in the fume duct for subsequent pick-up by an optical detector(s). The typical laser is a tunable diode (TD) laser. The laser's wavelength is modulated around a particular spectroscopic line of the gaseous species to be measured. The amount of beam absorption is related to the concentration. Because its location in situ, the response time is in the order of 2 s, are self-calibrated and do not need and environmentally protective analyzer

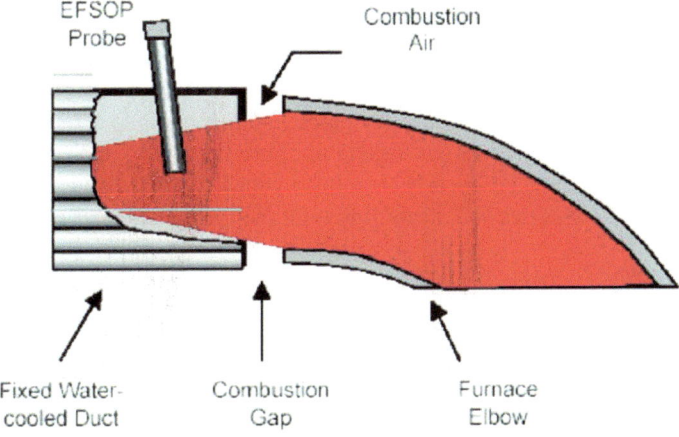

Fig. 6.3 Location of sensor probe to measure off-gas composition [20]

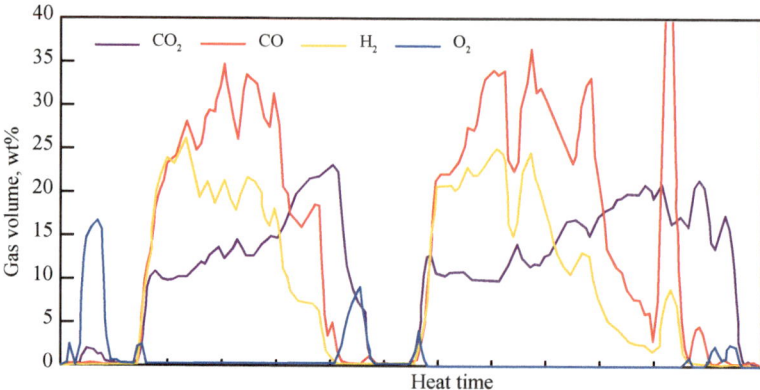

Fig. 6.4 Reducing conditions in the off-gas due to insufficient post combustion [20]

room. However, some of its disadvantages include the need of separate lasers; one for both CO_2 and H_2O and another one for CO. Because the CO and CO_2 absorption peaks begin to overlap as the off-gas temperature increases above about 300 °C, in situ laser systems need to employ one near-IR range laser with a suitable wavelength for CO_2 and a separate second near-IR range laser for CO. Lasers cannot be used to analyze mononuclear diatomic gases (H_2 and N_2). The laser to measure O_2 is costly and difficult to couple to a fiber optic and is usually omitted. Due to the previous limitations in situ laser systems only involve CO, CO_2 and H_2O. In addition to this, beam transmission is affected by the dust and can interrupt the signal. Industrial reports indicate about 50% of the EAF heats experience some degree of lost in situ signals from beam attenuation. The dust load in the EAF can range from 9 to 18 kg/ton.

Considering the previous advantages and disadvantages of the two systems it has been suggested a hybrid solution that guarantees a fast response and includes all gas species. Scipolo et al. describe further details in a patent from 2016 [22].

Krassnig et al. [23] reported a laser-based system which measures CO, O_2 and temperature of the off-gas. The laser is placed behind the gap after the elbow. In addition to this an IR/VIS system is used to measure H_2, CO_2 and CH_4. The concentrations of CH_4 and H_2 were linearly correlated with CO, then a criterion based only on the CO/O_2 ratio was defined, according with the following reaction:

$$aCO_{(g)} + bH_{2(g)} + cCH_{4(g)} + dO_{2(g)} = xCO_{2(g)} + yH_2O_{(g)}$$

$$\frac{CO}{O_2} = \frac{0.458}{\%CO - 0.54} + 0.848$$

If the values are below the criterion for the CO/O_2 ratio, the injectors will switch off. Another correlation was also defined based on the same ratio to control the gas flow rate of oxygen. With this method one of the results was an important decrease in oxygen consumption, about 40%, from 6.8 to 3.3 Nm^3/ton with the on-line system. A decrease on energy consumption of 8 kWh/ton was also reported.

6.3 Thermodynamics and Kinetics of Co Post-combustion

6.3.1 Thermodynamics of Post-combustion

The adiabatic flame temperature of PC is similar to that for combustion of natural gas with oxygen, in the order of 2800 °C.

Post combustion represents the oxidation of carbon monoxide into CO_2:

$$2CO_{(g)} + O_{2(g)} = 2CO_{2(g)}$$

The free energy of reaction reported by Gaskell [24] is as follows:

$$\Delta G^o_T(J) = -564,800 + 173.62\,T$$

This reaction is exothermic, indicating that CO_2 is promoted at lower temperatures. The equilibrium temperature is 705 °C. It means that above 705 °C, CO_2 is unstable and therefore its decomposition or de post-combustion can take place.

The equilibrium constant for PC is defined as follows:

$$K_{CO-CO_2} = \frac{p^2_{CO_2}}{p^2_{CO} \cdot p_{O_2}}$$

The equilibrium concentration of CO_2 in liquid steel is given by:

$$CO_{(g)} + O = CO_{2(g)}$$

$$K_{CO-CO_2} = \frac{p_{CO_2}}{p_{CO}a_O} = \frac{p_{CO_2}}{p_{CO}\%O}$$

The equilibrium constant reported by Turkdogan [25] is:

$$\log K_{CO-CO_2} = \frac{8620}{T} - 4.68$$

The oxygen concentration during most part of the blowing process for the KMS furnace is low due to the large amount of carbon, below 0.01%O. When carbon decreases to values in the order from 0.02 to 0.03%C, the oxygen concentration rises to about 0.04%O. Inserting these reference values, the obtained concentration of CO_2 is less than 0.8%, at the end raises to 1.9%. Vensel et al. [26] carried out thermodynamic calculations to compute %CO and %CO_2 in the BOF, as shown in

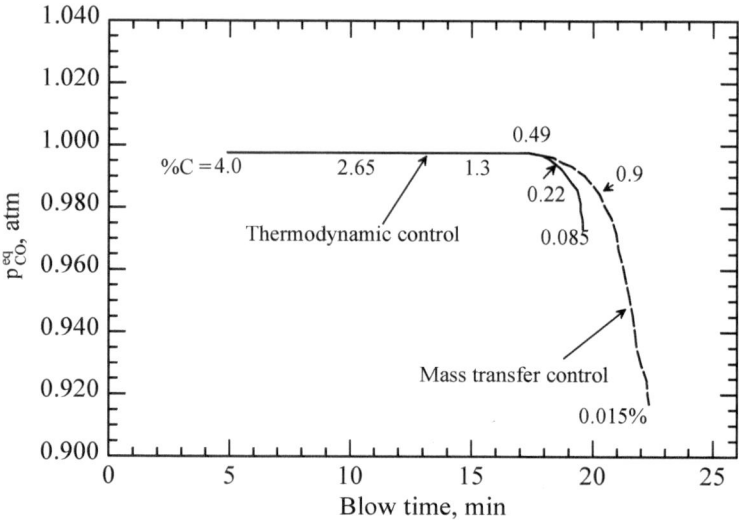

Fig. 6.5 Equilibrium partial pressure of CO as a function of carbon and blowing time [26]

Fig. 6.5. It can be observed that the main gas during most of the blowing time is CO, only when the carbon concentration decreases below about 0.3%, the %CO decreases from almost 100 to 93% CO. These results simplify the analysis on PC because it can be assumed that all CO_2 in the off gas is due to PC.

The CO_2 produced due to PC can take part in the decarburization of liquid steel. In this case, the contribution of chemical energy is off-set because this reaction is highly endothermic:

$$CO_{2(g)} + C = 2CO_{(g)}$$
$$\Delta G_T^o(J) = 144,682 - 129.495\,T$$

Table 6.3 summarizes reported values for the heat of post-combustion reactions in kWh/m³ O_2.

The heat of combustion can be directly computed at any temperature using enthalpy functions (kJ/kmol):

$$(H_T - H_{298})_{CO} = 28.42 + 2.0510^{-3}\,T^2 + 0.4610^5\,T^{-1} - 8802$$

Table 6.3 Heat of combustion in kWh/m³ O_2 at different furnace temperatures

Reaction/furnace temp	800 °C	1600 °C	1648 °C
$CO_{(g)} + \frac{1}{2}O_{2(g)} = CO_{2(g)}$	6.20 [27]	5.80 [27], 5.89 [9], 6.23 [25]	6.42 [17]
$H_{2(g)} + \frac{1}{2}O_{2(g)} = H_2O_{(g)}$	5.40 [27]	5.20 [27]	5.89 [17]
Ref. [27] injected oxygen at 25 °C			

$$(H_T - H_{298})_{CO_2} = 44.15 + 4.52/10^{-3} \, T^2 + 8.5410^5 \, T^{-1} - 16245$$
$$(H_T - H_{298})_{O_2} = 29.97 + 2.09/10^{-3} \, T^2 + 1.67/10^5 \, T^{-1} - 9676$$

Enthalpies of formation in kJ/kmol, at 298 K:

$$\Delta H^o_{CO_2} = -395,469$$
$$\Delta H^o_{CO} = -114,426$$

6.3.2 Kinetics of Post-combustion

The rate of oxidation of carbon monoxide with oxygen has been studied extensively for more than a century. In 1930 Topley [28] reported that the oxidation rate of CO with O_2 was drastically modified with the presence of water vapor, for example, at 580 °C the rate was 5.5%/min with a partial pressure of water of 1.8 mm, increasing to 49%/min when the water pressure increased to 18 mm. He suggested that the reaction involved a complex chain mechanism because of an increment in the temperature coefficient as the temperature raised. In 1932 Hadman et al. [29] found that the rate of oxidation of CO is almost proportional to the concentration of H_2O and CO and inversely proportional to the concentration of oxygen. They proposed that the first stage is the interaction of H_2O with O_2 to form OH ions and then those ions react with CO to form CO_2, as follows:

$$H_{2(g)} + O_{2(g)} = 2OH$$
$$OH + CO_{(g)} = CO_{2(g)} + H$$
$$H + O_{2(g)} + CO_{(g)} = CO_{2(g)} + OH$$

In 1967, Brokaw [30] reported that even very small concentrations of water vapor, about 20 ppm, have a large effect of the oxidation rate of CO. In 1977 Dixon-Lewis and Williams [31] summarized work done on the oxidation rate of CO. One possible set of chain reactions comprises the following steps:

$$CO_{(g)} + O_{2(g)} = CO_{2(g)} + O$$
$$O + H_2O_{(g)} = OH + OH$$
$$OH + H_{2(g)} = H_2O_{(g)} + H$$
$$H + O_{2(g)} = OH + O$$
$$O + H_{2(g)} = OH + H$$
$$OH + CO_{(g)} = CO_{2(g)} + H$$

These authors reported the rate of production of CO_2 obtained by Hottel et al. as follows:

$$\frac{d[CO_2]}{dt} = 1.2 \times 10^8 \exp\left(-\frac{8000}{T}\right) X_{CO} X_{H_2O}^{0.5} X_{O_2}^{0.3} \left(\frac{P}{RT}\right)^{1.8}$$

Where: $CO_{(g)} + O = CO_{2(g)}$ rate is in mole $\times l^{-1} \times s^{-1}$, P is in atmospheres, T in K, R in $l \times atm \times mole^{-1} K^{-1}$

In addition to CO, hydrogen is also present in the PC chamber due to combustion of grease and oil in the scrap. The hydrocarbons produce hydroxyl ions which enhance the rate of PC [9].

Zhang and Oeters [32–35] reported early fundamental studies from 1991 to 1993 on both thermodynamics and kinetics of PC in the converter. Their first paper [32] was focused on describing the fundamentals on heat transfer from the post-combusted gas to the melt and re-oxidation. The chemical reactions suggested in their model involve three possibilities for oxygen to react with:

- With C to produce CO: $O_{2(g)} + \underline{C} = 2CO_{(g)}$
- With CO to produce CO_2: $O_{2(g)} + 2CO_{(g)} = 2CO_{2(g)}$
- With other elements or compounds to generate slag components: $O_{2(g)} + 2\underline{M} = 2(MO)$

Oxygen in the second reaction is PC oxygen. De-PC or re-oxidation can happen when CO_2 oxidizes elements from the molten bath, for example C and Fe:

$$CO_{2(g)} + \underline{C} = 2CO_{(g)}$$
$$CO_{2(g)} + \underline{Fe} = (FeO) + CO_{(g)}$$

Heat transfer is assumed to occur through iron or slag droplets suspended in the gas phase, by radiation or convection heat transfer. Zhang and Oeters in their second paper [33] reported their model results indicating PC in the converter was below 20% and HTE below 50% due to reoxidation.

Using preheated air, these authors reported lower combustion temperatures, lower reoxidation and higher PCR which depended on the % of slag droplets involved.

6.4 Post-combustion Due to Air Infiltration

Natural PC occurs due to air infiltration, especially if the slag door is open. In these conditions the values of PCR can be in the range from 20 to 60% and a concentration of CO_2 in the order of 15–40%. Using PC the PCR values can increase to 60–80% [9].

Air infiltration decreases during meltdown because the furnace is filled with scrap but as the meltdown progresses it increases, if the slag door is open.

Trivellato and Labiscsak [36] reported a CFD model focused on the effect of the air gap, specifically to define the opening coefficient (ζ) which ensures complete combustion of H_2 and CO:

$$\zeta = \frac{A_{gap}}{A_{gap} + A_{inlet}}$$

where: A_{gap} is the area of the gap through which ambient air enters the PC chamber and A_{inlet} is the area through which exhaust fumes enter the PC chamber. They assumed an off-gas volume of 152 Nm^3/s entering the PC chamber, air ingress at a temperature of 200 °C, water temperature for the WCP was 80°C and the inlet pressure was $-10Pa$. The off-gas composition formed by 10% H_2 and 30% CO.

The PC reactions involved were:

$$2CO_{(g)} + O_{2(g)} = 2CO_{2(g)}$$
$$2H_{2(g)} + O_{2(g)} = 2H_2O_{(g)}$$

Prediction of the resulting gas composition is obtained from a kinetic analysis for the reaction rates. For combustion reactions it can be done applying the Eddy Dissipation Model (EDM). Trivellato and Labiscsak employed the kinetic model derived by Frassoldati et al. which consists of 37 reactions, the rate constant of each one defined by the following general expression:

$$k = AT^n e^{-\left(\frac{E_a}{RT}\right)}$$

Their results indicated that an opening coefficient of 0.52 guarantees the complete conversion of H_2 and CO. Depending on the extent of conversion, this value can range from 0.40 to 0.52.

Labiscsak et al. [37] also focused on the optimum air gap because if it is too large decreases the thermal efficiency but if it is too short oxygen can be low to promote the combustion of CO and H_2. Their model results indicate an air gap from 30 to 40 cm to reach lower concentrations of CO, below 0.15%.

Chan et al. [38] investigated the formation of NO_x due to air infiltration. They found that NO_x mainly originates from air infiltration through the slag door or roof ring gap flowing into the high temperature regions near the burners. They suggested that NO_x should not be controlled by temperature but by limiting air infiltration, closing the slag door and increasing the purity of the oxygen supply. NO_x is primarily composed by nitric oxide (NO) since N_2O can be an order of magnitude lower. They extended the mechanism proposed by Zeldovich for the formation of NO, as follows:

$$O + N_2 = N + NO$$
$$N + O_2 = O + NO$$
$$N + OH = H + NO$$

Table 6.4 Off-gas composition leaving the furnace, after [38]

	Heat 1	Heat 2	Heat 3	Heat 4
O_2	0	4	0	6
CO	18	1	31	1
CO_2	23	18	18	16
H_2O	15	24	15	24
H_2	11	0	8	0
N_2	32	54	27	54

Measured off-gas concentrations leaving the EAF are shown in Table 6.4.

These authors reported that most of the NO_x formed during a short time during the heat, its formation is proportional to both N_2 and O_2 and inversely proportional to CO and H_2. NO_x is primarily formed by air infiltration.

6.5 Post-combustion and Heat Transfer

6.5.1 Heat Transfer

Fruehan and Matway [39] reported that in order to achieve a higher heat transfer rate into liquid steel, the location of PC should be as close as possible to the steel surface, however this also worked on the opposite direction due to de-postcombustion:

$$CO_{2(g)} + C = 2CO_{(g)}$$
$$CO_{2(g)} + Fe = CO_{(g)} + (FeO)$$
$$CO_{2(g)} + Fe_{(s)} = CO_{(g)} + FeO_{(s)}$$

The reaction of CO_2 with iron can occur with both liquid steel or with solid scrap. In both cases the metallic yield is affected. Fruehan and Matway found that iron oxidation in the temperature range from 1100 to 1500 °C is proportional to the CO_2 partial pressure and temperature. The rate constant is shown in Fig. 6.6. Once solid FeO is formed the rate is controlled by diffusion. The rate increases above 1350 °C when FeO liquid is formed and the rate is controlled by chemical reaction.

Li and Fruehan [8] developed a 3D heat transfer model which included fluid flow, post-combustion, de-postcombustion and both radiative and convective heat transfer in order to identify how and where the PC takes place, how much and where the heat is transferred. The EAF investigated had a capacity of 165 tons with 5.0 m of internal diameter and 3.0 m of internal height, the pitch circle radius was 0.605 m, it consisted of 4 PC injectors at 45° with oxygen flow rates of 903 Nm^3/h per injector, located on the wall at a height of 2 m from the bottom. The inlet velocity and temperature of the oxygen jet were 100 m/s and 298 K, respectively. Carbon monoxide was

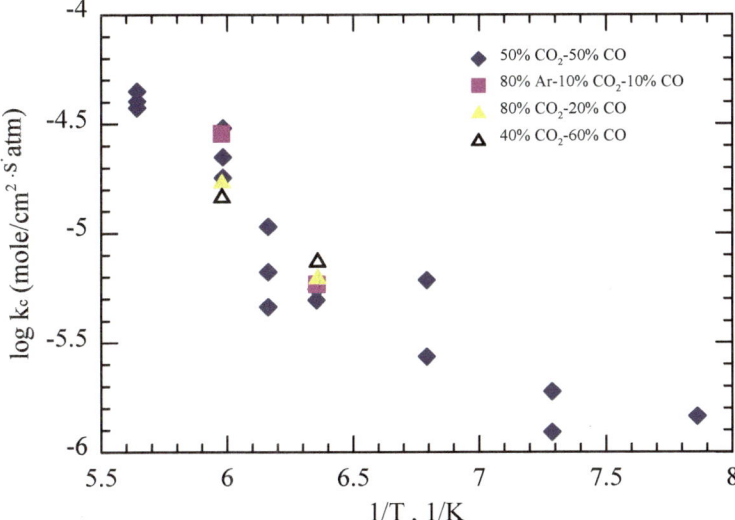

Fig. 6.6 Rate constant for iron oxidation with CO_2 mixtures [39]

estimated considering a decarburization rate of 0.039%/min. This decarburization rate represents 2.5 kg/s of CO.

$$0.039\frac{\%C}{min} \times \frac{\frac{10\,kg}{ton}}{\%C} = 0.39\frac{kg\,C}{min\,ton} \times 165 \text{ ton}$$

$$= 64\frac{kg\,C}{min} \times \frac{min}{60\,s} = 1.07\frac{kg\,C}{s} \times \frac{28}{12} = 2.5\frac{kg\,CO}{s}$$

In the energy equation, the source term includes the heat generated by PC:

$$\Delta H_R^T = \Delta H_R^{298} + \int_{298}^T \left[C_p^{CO_2} - C_p^{CO} - \frac{1}{2}C_p^{O_2} \right]$$

The adiabatic flame temperature (AFT) is a reference value because the highest temperature in the system cannot go above this temperature, also is a way to limit PC according with equilibrium. The authors noticed a problem to define this value because the theoretical value is extremely high. Combustion of CO and O_2 at 298 K produces an AFT of 5058 K. Including the endothermic dissociation reactions for CO_2 and O_2, the AFT is 2979 K. Finally, the authors decided to define its own equations for the rate of combustion considering the forward and backward reactions, as follows:

$$r = k_f C_{CO} X_{O_2}^{\frac{1}{2}} - k_b X_{CO_2}$$

$$k_f = 2.2 \times 10^8 e^{\left(\frac{-20.135}{T}\right)}$$

$$k_b = 5.78 \times 10^{12} e^{\left(\frac{-53.668}{T}\right)}$$

The de-postcombustion reactions involve reaction between CO_2 and carbon or iron, which can be dissolved carbon in liquid steel, solid carbon from the electrodes or carbon in steel scrap. The reaction with carbon is highly endothermic:

$$CO_{2(g)} + C = 2CO_{(g)}$$
$$\Delta H_R^o = 172 \, kJ/mol$$

The previous rate of de-postcombustion is proportional to the CO_2 partial pressure:

$$r = k\left(p_{CO_2}^s - p_{CO_2}^e\right)$$

where: $p_{CO_2}^s$ is the partial pressure of CO_2 at the interface, $p_{CO_2}^e$ is the equilibrium partial pressure of CO_2 (almost zero), and k a rate constant. The rate constant for the reaction between CO_2 and dissolved carbon (k_1), carbon from electrodes (k_2) and steel scrap (k_3) is given by the following expressions:

$$\log k_1 \left(mol/m^2 atms\right) = 2.4 - \frac{5080}{T}$$
$$\log k_2 \left(mol/m^2 atms\right) = 3.23 - \frac{4820}{T}$$
$$\log k_3 \left(mol/m^2 atms\right) = 5.704 - \frac{10,110}{T}$$

The jet velocity at the center axis decreased from 100 m/s to about 15 m/s above the metal surface. It was observed a slight increment in velocity at the nozzle outlet, attributed to gas heating. In parallel, the temperature of the oxygen jet increased from room temperature to about 2100 °C when both radiation and de-PC are included. The thermal behavior drastically changes if radiation or de-PC are not included, however the model is more realistic considering radiation and de-PC. Under these conditions the volumetric average temperature of the whole furnace was estimated as 1655 °C and the average off-gas temperature was 1574 °C.

The heat transfer for flat bath conditions is shown in Fig. 6.7, indicating that the largest amount (53.6%) is lost to the walls and roof. Its value is equivalent to 227 and 155 kW/m², respectively. The heat transferred to liquid steel was only 13.4% (3267 kW), indicating low HTE. However, at the beginning when steel scrap is colder, HTE is much higher, about 37%, as shown in Fig. 6.8.

Air ingress is detrimental for PC. It provides some extent of natural PC but at the expense of bringing a larger amount of inert gas. In the model, the authors estimated air ingress at a rate of 68 Nm³/h ton (11,220 Nm³/h) with a slag door measuring 1 m high and 1.22 m wide. They indicated that this value corresponds to 1.13 Nm³/ton, a moderate value considering that a high air ingress could be as much as 2.83 Nm³/ton. The net effect of air ingress is to decrease PC.

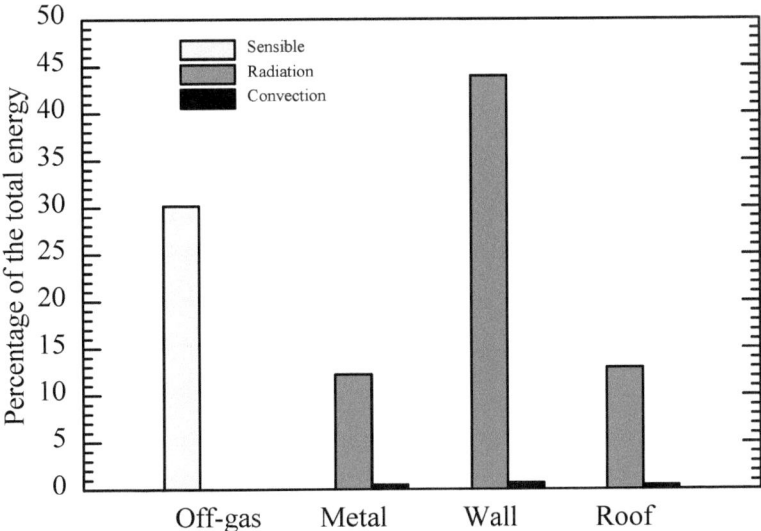

Fig. 6.7 Heat transfer from PC during flat bath conditions, after [8]

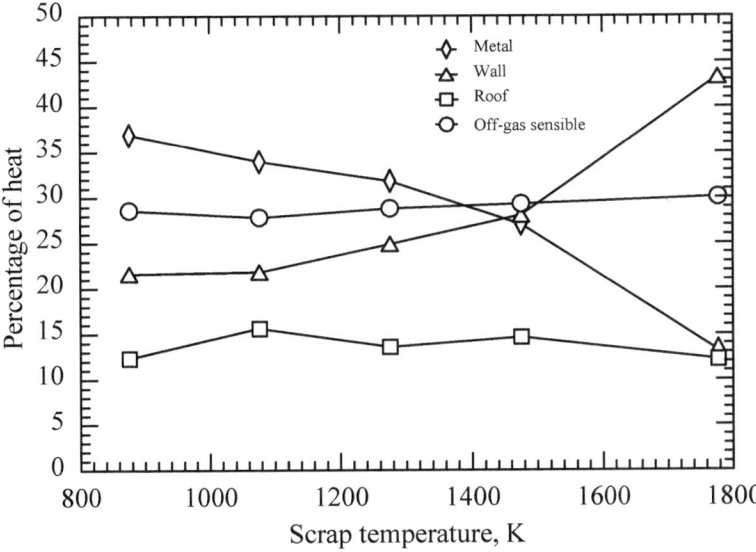

Fig. 6.8 Heat transfer from PC during scrap melting, after [8]

6.5.2 Post-combustion by CFD Modelling

Several CFD models on post-combustion have been reported. Since PC originated in the BOF and then transferred to the EAF, there are studies from the BOF which provide a fundamental understanding of this phenomena. One of those studies is from Gou et al. [13] who reported a PC model for the KOBM converter in 1993. They reported that increasing the lance height resulted in soft blowing by which less oxygen was available for decarburization but instead more primary oxygen for PC, however the distance to the surface of molten metal increases, therefore the HT decreases.

Zhonghua et al. [40] reproduced the same model developed by Li and Fruehan and conducted a sensitivity analysis to describe the effect of the number and location of the oxygen injectors and oxygen flow rate. Four layouts were evaluated with 4, 5 and 6 injectors, as shown in Fig. 6.9. All injectors placed on the wall had a height of 2 m from the bottom and a vertical angle of 45°, except case B. In this case the injectors were perpendicular to the wall.

To evaluate the performance of different layouts, they defined the following relationships, CO consumption and O_2 usage:

$$\varepsilon_1 = \frac{Q_{CO}^{tot} - Q_{CO}^{outlet}}{Q_{CO}^{tot}}$$

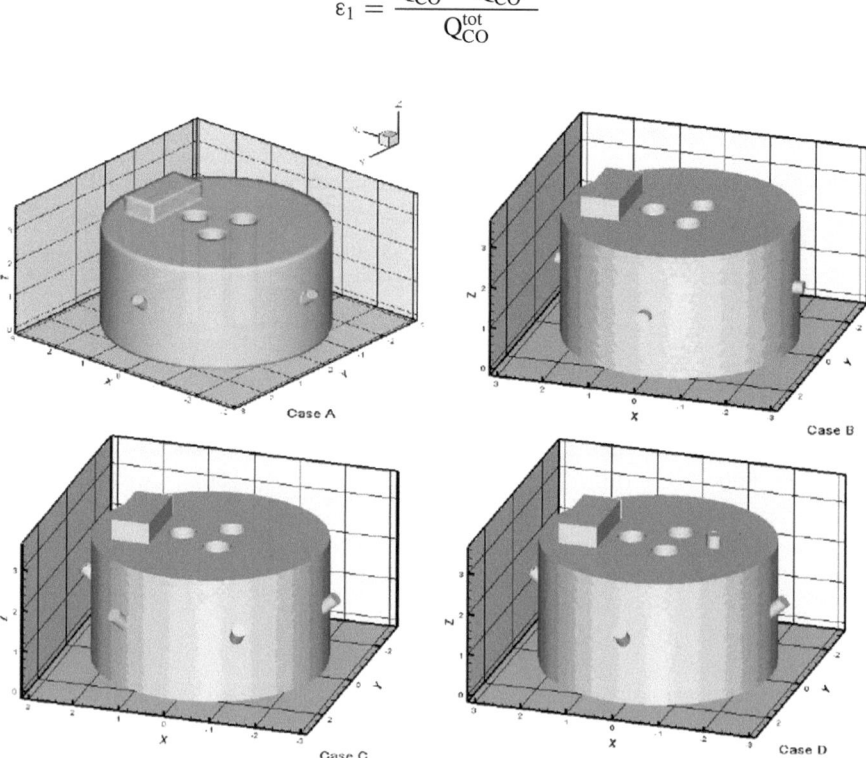

Fig. 6.9 Layouts of oxygen injection investigated in ref [40]

$$\varepsilon_2 = \frac{Q_{O_2}^{tot} - Q_{O_2}^{outlet}}{Q_{O_2}^{tot}}$$

where: the difference $Q_{CO}^{tot} - Q_{CO}^{outlet}$, represents CO consumed and the difference $Q_{O_2}^{tot} - Q_{O_2}^{outlet}$ is also the oxygen consumed. Higher values indicate a higher PCR. Table 6.5 summarizes their results. Case C with 6 lances reported the best case with higher values for the consumption of CO and O_2, suggesting that increasing the number of lances increases PCR. Comparing cases, A and B, suggests that inclined lances are better in comparison with horizontal lances.

Tang et al. [41] developed a mathematical model, which is also replica of the previous model by Li and Fruehan [8]. Li and Fruehan employed two commercial codes, FIDAP and CFX, indicating that FIDAP provided incomplete results [39]. Tang et al. [41] employed the commercial code Fluent. In the energy equation the source term included the heat of chemical reactions and radiation. Radiation heat transfer was estimated using the Discrete Ordinates (DO) radiation model. The combustion reactions were predicted using the Eddy Dissipation turbulence combustion model.

$$\nabla(\rho \overline{U} Y_i) = -\nabla \vec{J}_i + R_i$$

where: R_i is the net rate of production of species i by chemical reactions using the Eddy Dissipation model. $\vec{J}_i$ ◁ is the diffusion flux term of species i due to gradients of concentration and temperature, as follows;

$$\vec{J}_i = -\left(\rho D_{i,m} + \frac{\mu_t}{Sc_t}\right)\nabla Y_i - D_{T,i}\frac{\nabla T}{T}$$

where: $D_{i,m}$ is the diffusion coefficient for specie i in the mixture, $D_{T,i}$ is the thermal diffusion coefficient. Sc_t is the turbulent Sc number, assigned a value of 0.7. The pre-exponential factor (A) and activation energy (E) in the rate equations are given below:

Reaction	A (1/s)	E (J/mol)
$CO_{(g)} + \frac{1}{2} O_{2(g)} = CO_{2(g)}$	2.239×10^{12}	1.7×10^8
$CO_{2(g)} = CO_{(g)} + \frac{1}{2} O_{2(g)}$	5.0×10^8	1.7×10^8

Table 6.5 Summary of results. Ref. [40]

Case	CO, kg/s	O_2, kg/s	O_2 excess ratio	ε_1, %	ε_2, %
A	2.51	0.325×4	1.43	81.91	90.53
B	2.51	0.325×4	1.43	71.9	78.54
C	2.51	0.2167×6	1.43	87.3	96.78
D	2.51	0.26×5	1.43	82	90.51

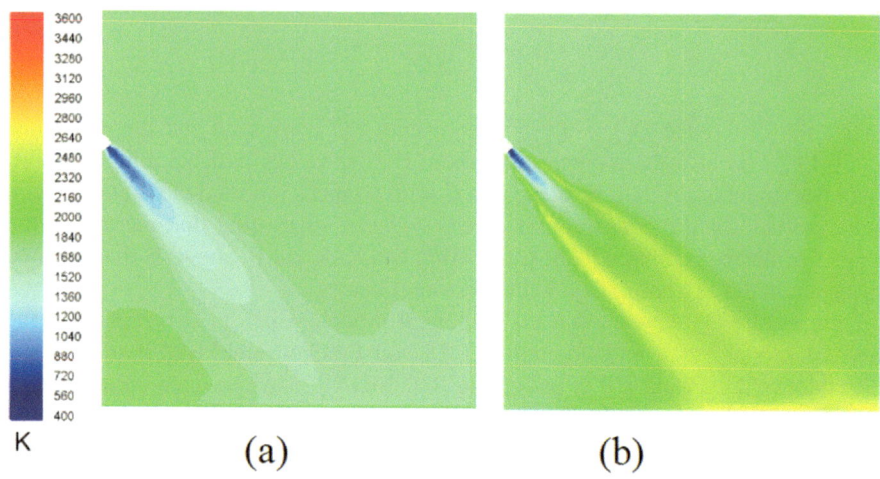

Fig. 6.10 Oxygen jet without PC (left) and with PC (right). After [41]

The EAF dimensions and boundary conditions are similar to the work by Li and Fruehan. The temperature of the slag free surface was taken as 1500 °C. Post-combustion oxygen was injected at an angle of 45°.

Figure 6.10 shows the oxygen jet with and without post-combustion. It can be observed the higher temperatures when post-combustion takes place, especially at the free surface with temperatures in the order of 2367 °C (2640 K).

They reported an increment in the furnace temperature of 19% compared to the case without PC, from 1300 to 1600 °C with an injection flow rate of 1.38 kg O_2/s, similarly the off-gas temperature also increased from 1050 to 1260 °C. Figure 6.11 shows the resulting CO concentration in the PC chamber by increasing the oxygen flow rate, from 1.1 to 1.6 kg/s. CO decreases in the PC chamber except in the dead zone below the oxygen injector.

	Case 1	Case 2	Case 3	Case 4	Case 5
O_2 (kg/s)	1.10	1.24	1.38	1.52	1.66
Avg. T, K	1826	1852	1882	1885	1852

Arzpeyma et al. [42] studied post-combustion using three burners, designed with outer secondary oxygen outlets, separated 120°, one of them placed on the wall below the 4th hole. The burners were operated in burner as well as in burner + oxygen lancing mode. Oxygen from the primary jet is excluded from post-combustion. It was assumed a constant generation rate of CO, about 0.04%/min, equivalent to 0.5 kg CO/s during burning mode and twice when operated in burner + lancing mode. The vertical injection angle was 40°. They found that increasing secondary oxygen 70%, from 0.063 to 0.10 kg/s, PCR increased 17%, from 0.58 to 0.68.

Yigit et al. [43] investigated heat transfer in the PC chamber of the EAF using 3 lances with an injection angle of 45° co-injecting simultaneously oxygen and carbon

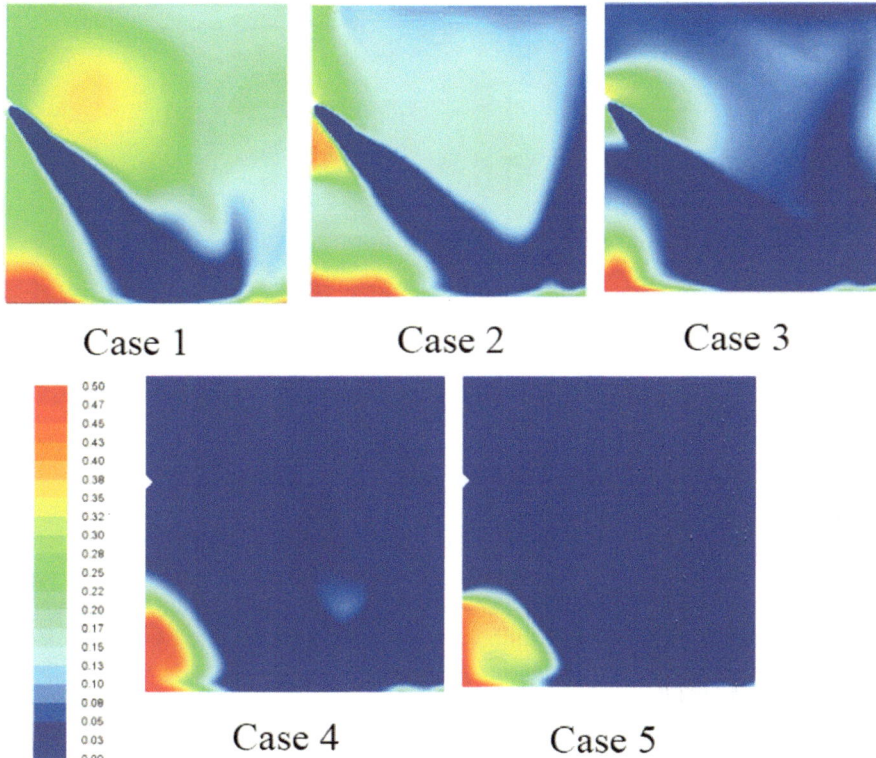

Fig. 6.11 CO concentration in the PC chamber, after [41]

particles. The oxygen and carbon flow rates were 1.29 and 0.33 kg/s, simultaneously. The velocity of the air oxygen-enriched jet was 480 m/s with a mass fraction of oxygen of 0.38 at 1227 °C and coal particles with a velocity and temperature of 0.76 m/s and 27 °C, respectively. The electric power supplied by the three electrodes was 100 MW. The oxygen jet reached the surface with a velocity of 75 m/s. An important reported result is the dominant role of radiation compared to convection heat transfer, as shown in Fig. 6.12. Their model calculations indicated the total energy generated by radiation and combustion was 100.75 MW and 5 MW, respectively. Most of this heat is radiated to the roof and walls, 91.26 MW, equivalent to 86%. Only a small fraction is transferred to the metal-slag surface, 8.9 MW and even a small amount of energy is in the off-gas, 2.34 MW. Heat transfer by convection is shown on the right side of the figure, in total about 1.2 MW. Radiation has been confirmed as the dominant heat transfer mechanism.

Coskun et al. [44] studied the optimum oxygen injection conditions (vertical angle, separation angle between injectors and injector length) to increase the surface temperature of the molten metal. The range of the variables investigated was − 35° to − 45° for the vertical angle, separation angle from 60° to 80° and injector

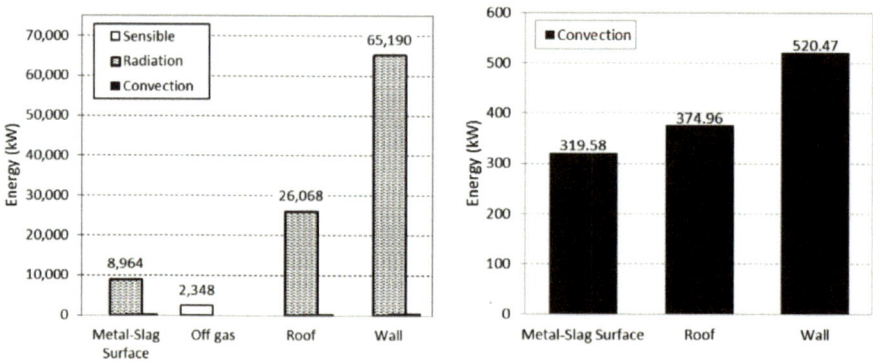

Fig. 6.12 Heat transfer in the PC chamber. **a** Radiation, **b** convection. After Ref. [43]

length from 414 to 614 mm. The standard case consisted of − 45°, 70° and 514 mm, respectively. Three injectors were located 1.07 m above the melt's surface. Oxygen was injected at a velocity of 137 m/s. The total electric power was 100 MW, which corresponds to 113 MW/m² at the bottom of each electrode with a radius of 0.305 m. Carbon particles are added to the slag to form a foamy slag, causing the formation of 0.021 kg/m² s of CO. Post-combustion kinetics was analyzed using the Eddy dissipation model. Initially the meshing structure was also investigated, comparing tetrahedral and polyhedral cells. They reported similar results comparing 13.7 million tetrahedral cells and 2.6 million polyhedral cells. After comparison of their numerical predictions, the authors concluded the optimum oxygen injection conditions were the following; vertical angle of 45°, separation angle of 60° and injector length of 614 mm. the temperature distribution comparing the standard case and optimized case is shown in Fig. 6.13.

Post-combustion is an important process in smelting reduction. In smelting reduction pre-reduced particulate iron ores and coal particles react at the gasifier. Air or oxygen is injected to react with the bath-generated gas, thus promoting PC. Becker

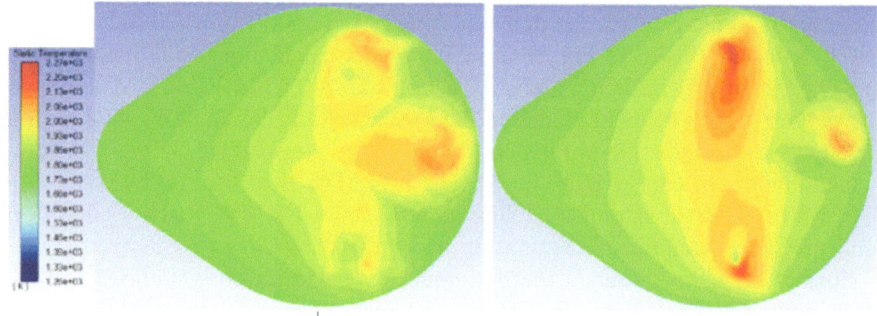

Fig. 6.13 Temperature distribution above the melt's surface for **a** standard case, **b** optimum case, after [44]

and Tacke [45] reported a PC model for the HIsmelt process. They reported that the position of the oxygen lance can be controlled to avoid hot spots inside the melter-gasifier. In addition to the conventional PCR, they also proposed the following relationship for smelting reduction conditions:

$$PCR = PCR_s - \frac{n_{FeO}^o}{n_C^A + n_{H_2}^i}$$

$$PCR_s = \frac{n_O - n_C^A}{n_C^A + n_{H_2}^i}$$

where: $n_C^A = n_C^i - n_C^o$, A is active, i means initial, o means outlet, PCR_s represents a stoichiometric PC degree in the idealized case of no iron oxide output.

Koria and Barui [46] reported a PC model for the COREX process, identifying a number of important findings from their model, as follows:

1. PCR can be increased by increasing the fraction of oxygen in air or using pure oxygen. The gas temperature increases from 2230 °C for pure air to 3420 °C with pure oxygen and 50% PC when no heat is transferred to the bath. The high temperature affects the refractory and therefore a higher HTE is necessary. The gas temperature decreases from 3419 °C for 0% HTE to 2014 °C for a HTE of 100%.
2. Increasing PCR and HTE decreased the coal ratio from 1064 to 540 kg/thm.
3. PC decreases the amount of reducing gas and therefore decreases metallization. The achievable metallization decreases from 87% at 10% PCR to 32% at 50% PCR.

Tang et al. [47] also studied post-combustion, however its work focused on PC in the exhaust system rather than inside the EAF.

6.6 K-ES Process

The Klöckner electric steelmaking process (K-ES) was a promising technology developed in the 1980s to replace electric energy by chemical energy injecting oxygen and coal from the bottom and intense post-combustion of the released CO. This technology was developed by Klöckner in collaboration with Tokyo steel, later acquired by VAI, then Siemens-VAI. de Beer et al. [48] summarized the following performance indicators; electric energy consumption of 300 kWh/ton using 22–30 kg coal/ton and theoretically 0 kWh/ton when coal and oxygen were increased to 255 kg coal/ton and 300 Nm^3 O$_2$/ton. The fist plant was built at Tokyo steel in 1986, retrofitting a 30-ton EAF, proving its commercial operation. Later on, two plants were built in Italy (Ferriere-Nord, 88 ton; Acciairia Venete, 82 ton). Heat transfer from the gases to solid scrap is inefficient and the large volume of gas is also an environmental problem. Figure 6.13 shows the K-ES process Fig. 6.14.

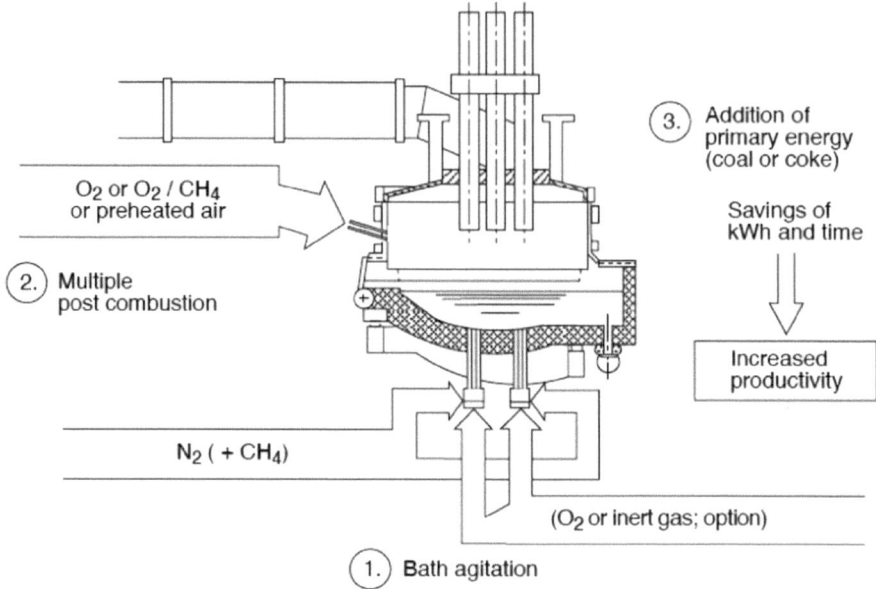

Fig. 6.14 K-ES process. After [9]

The bottom injection of reactive species is a technology developed in the BOF. Using a shrouding gas (methane, N_2 or CO_2) promotes the formation of a mushroom which protects the injection devices, as shown in Fig. 6.15.

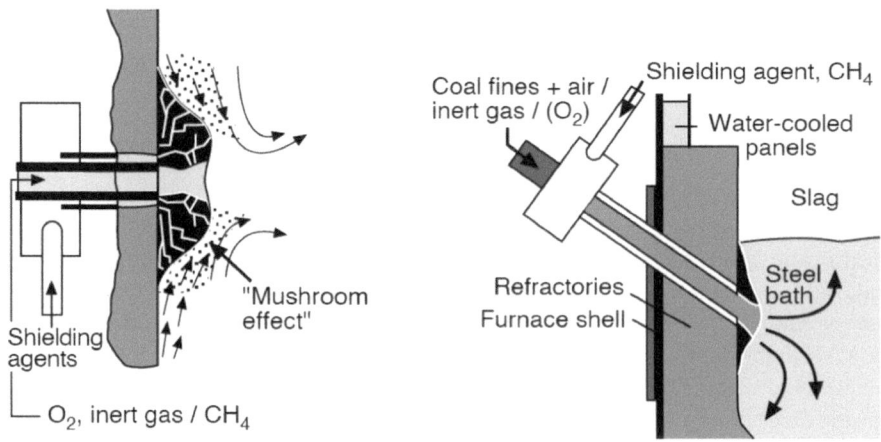

Fig. 6.15 Injection of oxygen (left) and carbon (right) in the K-ES process. After [9]

Table 6.6 Effect of O_2 flow rate for PC on energy consumption [51]

O_2 (PC), Nm^3/h	C, kg/ton	O_2, Nm^3/h	kWh/ton	Tap-to-tap, min
100	34	26	313	54
200	33	27	306	53
300	36	26	298	52

Table 6.7 Effect of O_2 flow rate for PC on FeO in liquid steel [51]

O_2 (PC), Nm^3/h	C, wt%	FeO, %	kWh/ton	Tap-to-tap, min
300	0.10	26	305	52
500	0.08	29	300	51
700	0.04	37	296	50

6.7 Post-combustion and Energy Consumption

Kleimt and Koehle [49] reported a decrease in energy consumption of 2.8 kWh/m^3 O_2 injected for post-combustion. This value is in agreement with savings of 64 kWh/ton reported by Gregory et al. [50]. Cantacuzene et al. [3] reported benefits on energy consumption using both burners and PC. Energy consumption was decreased from 427 to 392 kWh/ton, a decrease of 35 kWh/ton using 3 burners and 3 PC injectors. Oxygen and natural gas consumption was increased from 27 to 27 Nm^3/ton and 3 to 5 Nm^3/ton, respectively. Power-on time was decreased 4 min. Kim et al. [51] reported an example of PC in an EAF-ECOARC of 120 ton designed for scrap preheating. This example, EAF with scrap preheating, shows the optimum conditions to maximize PC and increase the HTE. Their results are based on a CFD model assuming scrap is spherical with a size, density and porosity of 0.078 m, 7750 kg/m^3 and 0.93, respectively. The temperature of scrap in the shaft increased by about 165 °C due to PC. Without PC the scrap temperature in the mid-shaft and bottom shaft regions were 100 and 815 °C, respectively, increasing to 200 and 980 °C, respectively, with PC and with an oxygen flow rate of 300 Nm^3/h. The decrease in electric energy consumption was about 19 kWh/ton. Table 6.6 shows the results obtained when the oxygen flow rate was increased from 100 to 300 Nm^3/h. It was also found that excessive oxygen in the PC lances could lead to severe oxidation of the metal, as shown in Table 6.7. The slag thickness was assumed to be 400 mm, for this value a jet with a flow rate of 700 Nm^3/h penetrated the slag, indicating an excessive value.

References

1. Hofer F, Patel P, Selenz HJ (1992) Fundamentals of post-combustion in steelmaking vessels. Steel Res 63:172–178. https://doi.org/10.1002/srin.199200494

2. Hirai M, Tsujino R, Mukai T, Harada T, Omori M (1987) Mechanism of post combustion in the converter. Trans Iron Steel Inst Japan 27:805–813. https://doi.org/10.2355/isijinternation al1966.27.805

3. Cantacuzene S, Grant M, Boussard P, Devaux M, Carreno R, Laurence O et al (2005) Advanced EAF oxygen usage at Saint-Saulve steelworks. Ironmak Steelmak 32:203–207. https://doi.org/10.1179/174328105X45811

4. Madhavan N, Brooks G, Rhamdhani M, Rout B, Scharma F, Overbosch A (2022) Analysis of heat loss in oxygen steelmaking. Iron Steel Technol: 42–47

5. Yang LZ, Jiang T, Li GH, Guo YF, Chen F (2018) Present situation and prospect of EAF gas waste heat utilization technology. High Temp Mater Process 37:357–363. https://doi.org/10.1515/htmp-2016-0218

6. Bender M (1994) Control of carbon monoxide emissions from electric arc furnaces. Iron Steelmak 21:67–73

7. Yang LZ, Jiang T, Li GH, Guo YF (2017) Discussion of carbon emissions for charging hot metal in EAF steelmaking process. High Temp Mater Process 36:615–621. https://doi.org/10.1515/htmp-2015-0292

8. Li Y, Fruehan RJ (2003) Computational fluid-dynamics simulation of postcombustion in the electric-arc furnace. Metall Mater Trans B Process Metall Mater Process Sci 34:333–343. https://doi.org/10.1007/s11663-003-0079-9

9. Jones JAT, Bowman B, Lefrank PA (1998) Electric arc furnace steelmaking, 11th edn. In: Fruehan RJ (ed) Making, shaping treating steeling, steelmaking refining. Pittsburgh, PA, USA, pp 525–660

10. Takashiba N, Kojima S, Take H, Okuda H (1989) Post combustion of converter

11. Brotzmann K, Fritz E (1986) Process for increased energy input in electric arc furnaces. EP 0257 450 A2

12. Mathur PC, Du Z, Selines RJ (1996) Electric arc furnace post combustion method. US Patent 5,572,544

13. Gou H, Irons GA, Lu WK (1993) Mathematical modeling of postcombustion in a KOBM converter. Metall Trans B 24:179–188. https://doi.org/10.1007/BF02657884

14. Farrand BL, Wood JE, Goetz FJ (1992) Post combustion trials at Dofasco's KOBM furnace. In: 75th steelmaking conference on proceedings, Toronto CA, 5–8 Apr 1992, pp 173–179

15. Opfermann A, Riedinger D, Baumgartner S, Grosse A (2009) Improvement of energy efficiency in EAF steelmaking. Milen Steel: 65–72

16. Guan L, Kuan P, Meng C, Memoli F, Negru O (2007) The practice of carbon injection and postcombustion in order to achieve a metallic yield recovery in the electric arc furnace: the experience of AMSTEEL mills (Malaysia). In: AISTech conference, Indianapolis, IN, USA

17. Grant MG (2000) Principles and strategy of EAF post-combustion. In: 58th electric furnace conference, 1–14

18. Zuliani SJ (2017) Next generation off-gas analysis. Steel Times Int 41:33–40

19. Evenson EJ (1998) Method and apparatus for sampling and analysis of furnace off-gases. US Patent 5777241

20. Scipolo V, Khan M, Patil S, Holmes G (2008) Optimization of the EAF process at Cape Gate (Pty) Ltd. (Davsteel division) using Goodfellow EFSOP® technology. Arch Metall Mater 53:657–663

21. Ferro L, Palma M, Maiolo J, Memoli F (2006) EAF process control. Millenium Steel: 93–96

22. Scipolo V, Zuliani DJ, Avishekh P, Ovidiu N (2016) System and method for analyzing dusty industrial off-gas chemistry. WIPO 2016/023104

23. Krassnig H-J, Kleimt B, Voj L, Antrekowitsch H (2008) EAF post-combustion control by online laser-based off-gas measurements. Arch Metall Mater 53:455–462

24. Gaskell DR (2003) Introduction to the thermodynamics of materials, 4th edn. Taylor & Francis

25. Turkdogan ET (1996) Fundamentals of steelmaking. The Institute of Materials, London UK. https://doi.org/10.1179/095066096790151213

26. Vensel D, Henein H, Dauby PH (1988) A thermodynamic analysis of decarburization and post combustion in the BOP. In: Neumatic steelmaking conference, pp 51–58

27. Thomson MJ, Kournetas NG, Evenson E, Sommerville ID, McLean A, Guerard J (2001) Effect of oxyfuel burner ratio changes on energy efficiency in electric arc furnace at Co-steel lasco. Ironmak Steelmak 28:266–272. https://doi.org/10.1179/030192301678136

28. Topley B (1930) The homogeneous isothermal reaction $2CO + O_2 = 2CO_2$ in the presence of water vapour. Nature 125:560–561

29. Hadman G, Thompson H, Hinshelwood C (1932) The oxidation of carbon monoxide. Proc R Soc Lond A 137:87–101

30. Brokaw RS (1967) Ignition kinetics of the carbon monoxide-oxygen reaction. In: 11th symposium combustion, pp 1063–1073

31. Dixon-Lewis G, Williams DJ (1977) The oxidation of hydrogen and carbon monoxide. Compr Chem Kinet: 1–248

32. Zhang L, Oeters F (1991) Model of post-combustion in iron-bath reactors, part 1: theoretical basis. Steel Res 62:95–106. https://doi.org/10.1002/srin.199101258

33. Zhang L, Oeters F (1991) Model of post-combustion in iron-bath reactors, part 2: results for combustion with oxygen. Steel Res 62:107–116. https://doi.org/10.1002/srin.199101259

34. Zhang L, Oeters F (1993) Model of post-combustion in iron-bath reactors, part 3: theoretical basis for post-combustion with pre-heated air. Steel Res 64:542–548. https://doi.org/10.1002/srin.199301569

35. Zhang L, Oeters F (1993) Model of post-combustion in iron-bath reactors, part 4: results for post-combustion with preheated air. Steel Res 64:588–596. https://doi.org/10.1002/srin.199301576

36. Trivellato F, Labiscsak L (2015) The post-combustion chamber of steelmaking plants: role of ambient air in reactant exhaust fumes. Appl Math Model 39:19–35. https://doi.org/10.1016/j.apm.2014.09.016

37. Labiscsak L, Straffelini G, Corbetta C, Bodino M (2011) Fluid dynamics of a post-combustion chamber in electric arc steelmaking plants. WIT Trans Modelling Simul 51:205–214. https://doi.org/10.2495/CMEM110191

38. Chan E, Riley M, Thomson MJ, Evenson EJ (2004) Nitrogen oxides (NO_x) formation and control in an electric arc furnace (EAF): analysis with measurements and computational fluid dynamics (CFD) modeling. ISIJ Int 44:429–438

39. Fruehan RJ, Matway RJ (2004) Optimization of post combustion technology. AISI/DOE Technology Roadmap Program

40. Zhonghua W, Mazumdar D, Mujumdar AS (2009) Optimization of post combustion in an electric arc furnace for advanced steelmaking. In: 3rd international conference on processing material property, PMP III, vol 1, Bangkok, Thailand, 1–10 Dec 2009, TMS, pp 261–266

41. Tang G, Liu W, Silaen AK, Zhou CQ (2017) Modeling of post-combustion in an electric arc furnace. In: International mechanical engineering congress exposition, Tampa FL, USA, pp 1–6

42. Arzpeyma N, Ersson M, Jönsson PG (2019) Mathematical modeling of postcombustion in an electric arc furnace (EAF). Metals 9. https://doi.org/10.3390/met9050547

43. Yigit C, Coskun G, Buyukkaya E, Durmaz U, Güven HR (2015) CFD modeling of carbon combustion and electrode radiation in an electric arc furnace. Appl Therm Eng 90:831–837. https://doi.org/10.1016/j.applthermaleng.2015.07.066

44. Coskun G, Sarikaya C, Buyukkaya E, Kucuk H (2023) Optimization of the injectors position for an electric arc furnace by using CFD simulation. J Appl Fluid Mech 16:233–243. https://doi.org/10.47176/jafm.16.02.1352

45. Becker-Lemgau U, Tacke KH (1996) Mathematical model for post combustion in smelting reduction. Steel Res 67:127–137. https://doi.org/10.1002/srin.199605469

46. Koria SC, Barui MK (2000) Influence of post-combustion heat transfer efficiency on fuel reduction in COREX process. Ironmak Steelmak 27:348–354. https://doi.org/10.1179/030192300677642

47. Tang X, Kirschen M, Abel M, Pfeifer H (2003) Modelling of EAF off-gas post combustion in dedusting systems using CFD methods. Steel Res Int 74:201–210. https://doi.org/10.1002/srin.200300182

48. De Beer J, Worrell E, Blok K (1998) Future technologies for energy-efficient iron and steel making. Annu Rev Energy Environ 23:123–205. https://doi.org/10.1146/annurev.energy.23.1.123

49. Kleimt B, Köhle S (1997) Power consumption of electric arc furnaces with post-combustion. Metall Plant Technol Int 3:56–57

50. Gregory DS, Ferguson DK, Slootman E, Viraize F, Luckhoff J (1996) Results of ALARC-PC post-combustion at cascade steel rolling mills. Iron Steelmak 23:49–54

51. Kim DS, Jung HJ, Kim YH, Yang SH, You BD (2014) Optimisation of oxygen injection in shaft EAF through fluid flow simulation and practical evaluation. Ironmak Steelmak 41:321–328. https://doi.org/10.1179/1743281213Y.0000000143

Chapter 7
Slag Foaming

7.1 Introduction

Slag foaming is the most important variable to control and reduce energy consumption in the EAF, not only because is focused on achieving a higher thermal efficiency which decreases the total energy consumption but due to additional multiple benefits. In spite of the ancient knowledge on foaming, this technique was not applied extensively in the EAF before the 1990s. Perhaps it contributed to a negative impression the vast experience from steelmaking in converters, the BOF in particular. In the BOF slag foaming is a common phenomenon but mostly known for its worst part due to slopping. Slopping is the sudden formation of foamy slag which violently erupts both slag and steel from the converters, causing extreme damages to the equipment and creating highly unsafe working conditions. Therefore, rather than promoting foaming, in the 1970s and 1980s researchers focused on ways to suppress/control its formation.

The earliest research reports on slag foaming, according with Kozakevitch [1] were carried out by K. Fellcht and by K. Ludemann, both in 1955. Fellcht classified slags as foaming and non-foaming depending on its chemical composition and rates of decarburization. The review paper by Kozakevitch [1] in 1969 shows that since those years there was a clear understanding of the main variables controlling its formation and stability. He defined the slag viscosity, slag surface tension, presence of suspended particles and its wetting properties, slag chemical composition (basicity, SiO_2 and P_2O_5) and decarburization rate as relevant variables controlling slag foaming. He also noticed that the viscosity of open-hearth slags was too low and was the most important variable because allowed faster drainage and then lower foam stability. Hara and Ogino [2] in their review paper from 1992 cited several reports that discussed slag basicity values to reach a maximum foaming height, one work citing a CaO/SiO_2 equal to 1.6, another one a value lower than 1.2 and another one which suggested partial precipitation of CaO. Based on these discrepancies they pointed out that slag foaming was the result of two phenomena; slag properties and the rate of generation of CO.

The foamy slag practice in the EAF originated from DRI-based operations together with other developments. The reduction of iron oxide and carbon contained in DRI produces CO that helps the formation of a foamy slag. In addition to this, the early work carried out by Professor Fruehan, which started in 1989, and the large experience in the field by Japanese researchers, contributed to visualize foamy slag as a benefit rather than a threat in the EAF. Ogawa et al. [3] in 1992 carried out one of the first investigations on slag foaming using carbon particles, although its objective was to define the mechanism by which large carbon particles can suppress foaming, it was the basis of the foamy slag practice from today.

7.1.1 Slag Foaming Benefits

One of the most important consequences of operating the EAF under slag foaming conditions is the full switch to long-arc operation. For almost 80 years the EAF operated under short-arc operation conditions, limiting the active electric power. In those years if the EAF was operated with a long-arc operation and a conventional thin slag, under flat-bath conditions, the heat radiated to the refractory walls would create large and intense hot spots that would destroy the refractory even in one single heat. In the 1980s the development of water-cooled panels made possible to apply a long-arc operation to some extent, however without a foamy slag that practice was inefficient due to large heat losses. In 1986 Ameling et al. [4] reported a classical graphical description of the effect of slag foaming on energy recovery in the EAF, shown in Fig. 7.1. This figure indicates that under no foamy slag the heat losses are more than 90%. When the electric arc is fully covered the heat losses are less than 7%. Wunsche, an inventor of oxy-fuel burners, together with Simcoe in 1984 had reported a similar plot [5], shown in Fig. 7.2.

The main objective of slag foaming is to minimize heat losses from the electric arc to the water-cooled panels. The heat losses are time dependent but also can be much higher during flat-bath conditions due to inefficient slag foaming. Adams et al. [6] have reported the heat losses shown in Table 7.1. Heat losses during melting are lower because the arcs are covered by scrap.

Slag foaming has multiple benefits and this is why is so important to understand how to optimize this practice. The main benefits are the following:

- Increase in metallic yield
- Improves electric power quality:

    - Increase in arc stability and lower harmonics. Arc voltage wave forms become almost sinusoidal, yielding a reduction in harmonics
    - Decrease in reactance which increases the active power: Improved arc stability decrease the operational reactance, which increases the active power
    - Decrease in noise: The larger slag volume absorbs the acoustic power

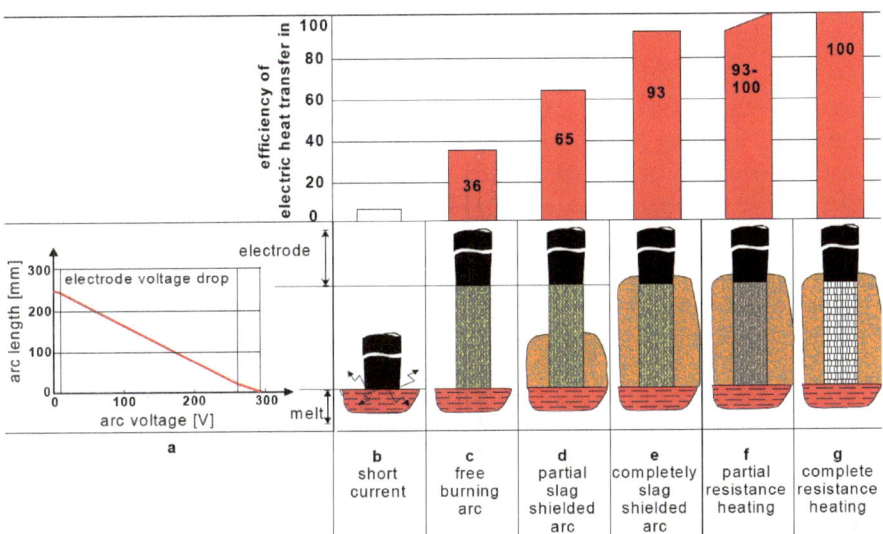

Fig. 7.1 Graphical representation on the effect of foamy slag height on the efficiency of electric heat transfer. After [4]

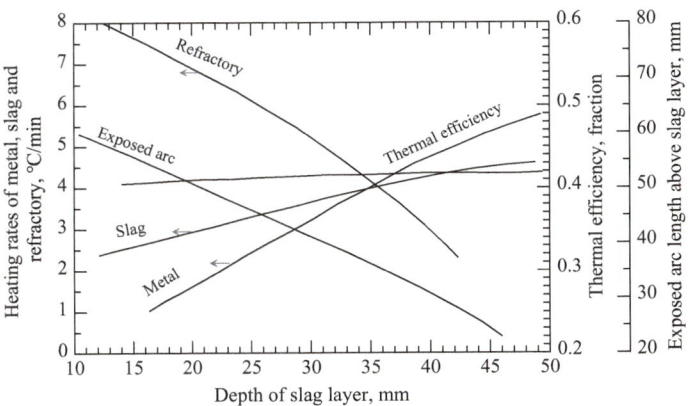

Fig. 7.2 Effect of foamy slag on the heating rates of metal, slag and lining. After [5]

Table 7.1 Heat losses in the EAF. After [6]

	kWh/ton min
During melting	0.4
During refining (flat bath)	1.7
Between heats (< 30 min)	0.5
Between heats (> 30 min)	0.2

– Decrease in flicker
– These benefits increase electric power efficiency

- Increase in productivity
- Decrease in energy losses due to heat radiated to water cooled panels and roof; less maintenance of EAF panels, longer life of panels and delta roof
- Decrease in melting time; lower specific electrode consumption
- Decrease in nitrogen
- Decrease NO_x
- Lower electric energy consumption.

Due to the large number of benefits, slag foaming is now commonly employed in all steel shops. The bases of this practice are simple:

- Promote the slag physical properties that improve slag foaming
- Produce enough CO gas to sustain the foaming conditions, which involves a controlled reduction of FeO with injected carbon particles and decarburization with oxygen injection.

Although the bases are quite simple, its achievement in practice involves multiple variables that should be controlled simultaneously.

7.1.2   Volume Changes Due to Slag Volume

Bowman and Krüger [7] estimated the volume increment due to generation of CO bubbles for two cases; decarburization of liquid steel and reduction of FeO by carbon in the DRI. Their calculations are based on the rate of formation of CO and residence time of the bubbles in the slag. It takes as a reference an EAF of 80 ton with a diameter of 4.73 m. The residence time is based on the velocity of CO bubbles rising in liquid steel. Zhang and Taniguchi [8] summarized a number of equations that define the terminal velocity of gas bubbles in liquid steel as a function of the bubble diameter, suggesting for industrial conditions a range from 33 to 43 cm/s for bubbles in the range from 9 to 19 mm. Bowman and Krüger employed a value reported by Szekely of 30 cm/s. Assuming a slag depth of 40 cm, the residence time is 40/30 = 1.33 s.

Assuming a decarburization rate of 0.05% C/5 min, equivalent to 0.133 kg C/s, the production rate of CO under standard conditions is 0.31 kg CO/s or 0.25 Nm³ CO/s. Therefore, at any instant in time the volume of CO in the slag is:

$$V_{CO}(STP) = 1.33\,s \times 0.25\,Nm^3/s = 0.33\,Nm^3$$

To define the volume of CO under steelmaking conditions, it is assumed the formation of CO occurs at the metal/slag interface, with a ferrostatic pressure of 1.3 bar and the temperature of liquid steel is 1600 °C:

$$V_{CO} = 0.33\,m^3 \times \frac{1873\,K}{300\,K} \times \frac{1.0\,bar}{1.3\,bar} = 1.58\,m^3$$

This volume is equivalent to a height of 9 cm in the 80-ton EAF, in other words the original slag level will be raised by 9 cm.

The reduction rate of FeO is now estimated for the case employing DRI, with a feeding rate of 1820 kg/min and a carbon content of 1.5%. This represents $1820 \times (1.5/100) = 27\,kg\,C/min = 0.45\,kg\,C/s$. Assuming all the carbon is reduced by FeO, the generation rate of CO is:

$$V_{CO} = 0.45 \tfrac{kg\,C}{s} \times \tfrac{28.01}{12} = 1.06\,Nm^3$$
$$V_{CO} = 1.06\,m^3 \times \tfrac{1873\,K}{300\,K} \times \tfrac{1.0\,bar}{1.3\,bar} = 5.1\,m^3$$

This volume is equivalent to a height of 29 cm in the 80-ton EAF, in other words the original slag level will be raised by 29 cm. On the basis of these values, Bowman and Krüger suggested that the typical slag volume increment is by a factor of 2–5. This increment in slag volume would indicate that the total volume in the foaming slag ranges from 50 to 80%. Their calculation is within the expected values of foam height for short to long arc operation. Karbowniczek [9] suggested a volume increase by a factor of 10–20. This value would be possible only if it is assumed there is no continuous slag extraction, which is not what happens in practice.

7.1.3 Differences Between Foams, Froths and Emulsions

Foams, froths and emulsions all involve a dispersed phase. Foams and emulsions contain two phases. The difference between a foam and an emulsion is that in an emulsion the dispersed phase is a liquid, for example droplets of liquid steel in the slag and in a foam the dispersed phase are gas bubbles. A froth is considered a three-phase system; gas, water and dispersed particles, for example in froth flotation the particles are the particles of iron ore concentrate. Figure 7.3 is a schematic representation of a foam, an emulsion and a froth.

Foams involve highly dynamic processes. Kitchener [10] suggested that in order to prevent excessive thinning of a bubble, the elasticity forces should react in a very short time, 1–00 ms in aqueous systems.

Millman et al. [11] suggested that the formation of emulsions in the BOF follows several steps; impinging oxygen jet, bubble rise from metal and drag a metal film into the slag, turbulence at the slag/metal interface. Steel droplets are very small, in the range from 0.15 to 1.2 mm. The fraction of liquid steel in the emulsion can be more than 30% of the total metallic charge. The surface area is about 600–800 m^3/ton hot metal, about two thirds of the total carbon is removed in the emulsion and only one third in the bulk metal. The decarburization rate of the metal droplets is low at the beginning because the slag is viscous and silicon is oxidized before than carbon, this oxidation increases the temperature to about 1565 °C, higher than the bulk, which is

Fig. 7.3 Foam (left), emulsion (center) and froth (right)

about 1300 °C. The average retention time of the droplets in the emulsion is about 2–3 min, due to higher viscosity in the beginning the droplets have higher residence time. The emulsion collapses when there is not enough CO to sustain it.

In the BOF process both foam and emulsion are important but the emulsion plays a larger role because it directly affects the rate of decarburization. The IRSID continuous steelmaking process developed in the late 1960s has an even more intense formation of slag foaming and emulsification. In this process the whole system liquid slag/liquid steel and gas form a colloidal suspension and due to higher interfacial area, the rates are expected to be higher.

7.1.4   Foaming in Bath-Smelting

This chapter is mainly focused on slag foaming in the EAF, however, foaming in other processes is controlled by the same variables and their results can be valuable information to understand slag foaming in the EAF. Iron bath-smelting basically consists of two steps; pre-reduction of iron ore particles or pellets and final reduction and melting in a melter-gasifier. COREX was the first successful process, developed by Korf engineering in the late 1970s and started commercial operation in 1988 in South Africa. Iron bath smelting involves large amounts of slag, injection of carbon particles and oxygen into the slag for final reduction of FeO in the liquid state, which create the conditions for the formation of large volumes of foamy slag. During coal gasification, the volume of CO is twice the volume of oxygen injected, as can be seen in Fig. 7.4. The CO produced is post-combusted at the free board of the melter-gasifier. The intensity of slag foaming in iron bath smelting is much higher in comparison with the EAF. The foamy slag in iron bath smelting is essential for a rapid heat transfer from the post-combustion heat to the metal bath below.

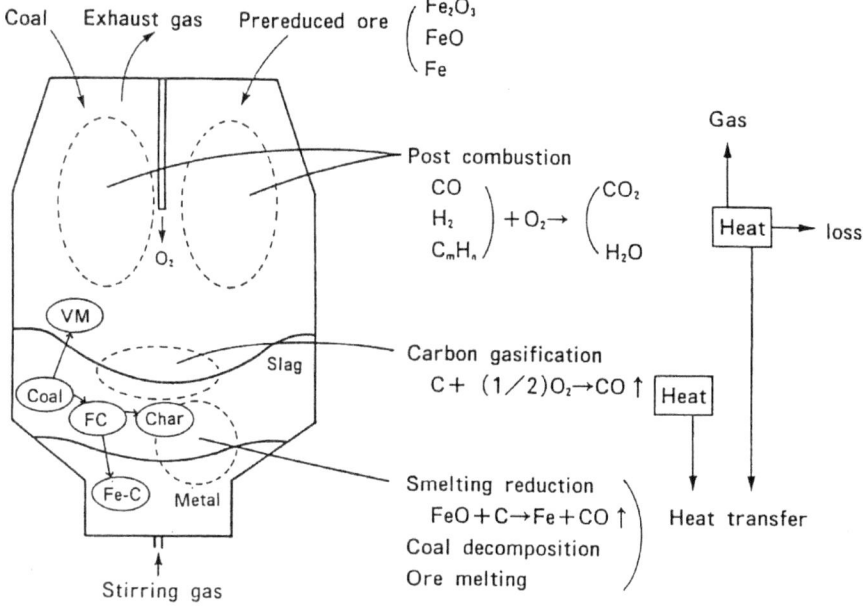

Fig. 7.4 Iron bath-smelting reactions. After [12]

7.1.5 General Scope of Variables that Influence Slag Foaming

Many variables influence the formation of a foam. Slag foaming represents the slag volume expansion due to retention of a gas phase. Its formation requires two principal conditions, first, adequate slag physicochemical properties and second, the presence of a gas phase. If either one of these conditions is not fulfilled it is not possible to form a stable foam. In addition to this, conditions which lead to a stable and uniform reduction of iron oxide in the slag, forming CO, provides conditions for a stable formation of foamy slag. Figure 7.5 shows schematically the three groups of variables and its final effect is the electric arc coverage.

7.2 General Foaming Fundamentals

This section describes foaming fundamentals derived from aqueous systems; however, the concepts apply for any foaming phenomena including slag foaming which is described further below. There are excellent books and reviews on foaming. The following authors have reported good books; Bikerman, 1973 [13], Stevenson, 2012 [14], Pugh, 2016 [15]. In addition to books there are review papers such as; papers series by DeVries, 1958, on foam stability [16–20], Kitchener, 1959 and 1964

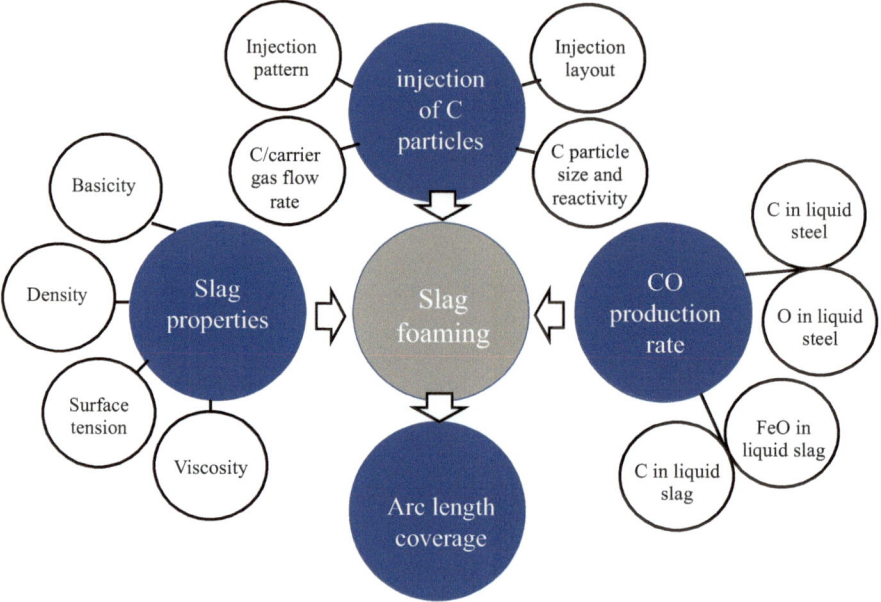

Fig. 7.5 Overall picture of variables that affect slag foaming in the EAF

[10, 21], Walstra, 1989 [22], Schramm and Wassmuth, 1994 [23], Nexhip et al. 2004 [24] and Wang et al. 2016 [25]. The review by Nexhip covers all the fundamentals on foams in general and also the physicochemical properties for slag foaming. In spite that foaming of aqueous and non-aqueous systems follow the same fundamental principles for its formation, slag foaming has its own particularities. Calhoun et al. [26] without including slags pointed out the large differences between aqueous and non-aqueous systems; Aqueous and Non-aqueous systems have important differences, non-aqueous systems generally exhibit much lower surface tensions (15–30 mN/m) than aqueous systems (72 mN/m), have much less electrostatic repulsion and also lower dielectric constants. These differences in the system properties influence differences in foam stability.

It is very important to have a basic understanding of the concepts and equations that control foaming phenomena, summarized in this section. It helps to better visualize foam formation, foam decay, foam stability, etc. However, if the reader is only interested in the more practical aspects, Sect. 7.2 can be initially skipped. Without this basic knowledge the foundations to understand foaming will be ignored.

7.2.1 Pioneers on Foam Films

Gochev et al. [27] reviewed from a historical perspective the work carried by all pioneers on foam films until the 1960s. In this review, according to Plateau the first scientists studying foam films were Boyle and Hooke in the second half of the seventeenth century. Robert Boyle from 1660 to 1672 studied the interference colors in soap films and described the appearance of holes ("black films"). Newton in the late seventeenth century studied soap films and attributed foam drainage to be caused by gravity, also he was the first one to attempt to measure the thickness of the thin films, reporting a range from 107 to 143 nm. This review also described the first attempts to measure surface tension since Leidenfrost in 1756 first suggested the existence of a contractile force of soap bubbles.

In the nineteenth century, Young and Laplace formulated the principles governing contact between fluids. Thomas Young (1773–1829) was a British physician and polymath, described as the "Last man who knew everything". He devised a rule of thumb for determining a child's drug dosage. Young's Rule states that the child dosage is equal to the adult dosage multiplied by the child's age in years, divided by the sum of 12 plus the child's age. In 1814 he completely translated the "enchorial" text of the Rosetta Stone. His work on elasticity defined the Young's modulus. In 1804 developed the theory of capillary phenomena on the principle of surface tension, basis of the Young–Laplace equation. Pierre Simon Laplace (1749–1827) gave a formal mathematical description to the relationship described earlier by Young. Laplace was a French polymath. He was Napoleon's examiner when Napoleon attended the École Militaire in Paris in 1784. From 1780 to 1784, Laplace and French chemist Antoine Lavoisier collaborated on several experimental investigations, designing their own equipment. In 1783 they published their joint paper, Memoir on Heat, in which they discussed the kinetic theory of molecular motion. In their experiments they measured the specific heat of various bodies, and the expansion of metals with increasing temperature. Laplace's early published work in 1771 started with differential equations and finite differences, one of them was the Laplace's method for approximating integrals, also at that time had developed his ideas on the mathematical and philosophical concepts of probability and statistics.

Joseph Plateau (1801–1883), was a Belgian physicist who conducted extensive studies of soap films, describing the structures formed by such films in foams. He suffered from uveitis which made him blind at the age of 42, the second half of his life. Plateau pointed out a different viscosity between the surface and the bulk, indicating that; "in some liquids, the surface viscosity is stronger than the interior viscosity".

Carlo Marangoni (1840–1925) was an Italian physicist. He graduated in 1865 with a thesis that discussed the Marangoni effect. This effect was formally reported in two publications in 1871 and 1878. After graduation, he worked as a high school physics teacher at the Liceo classico Dante in Florence Italy during 45 years until his retirement in 1916. Scriven and Sternling [28] attribute to James Thompson the first report on the Marangoni effect in 1855 on a paper titled "On certain curious motions observable at the surfaces of wine and other alcoholic liquors". Wilhelmy [29] in 1863 proposed a method to measure surface tension using a rectangular plate. He obtained a PhD degree from Heidelberg University in 1846 and worked there only for a short time. One year before his death, he published a paper describing his method.

Bikerman developed fundamentals studies on foaming in the early 1930s. Joseph Jacob Bikerman (November 8, 1898 in Odessa to June 11, 1978 in Cleveland, Ohio) was a Russian-American chemist [30]. He studied chemistry from 1916 to 1921 at the University of Saint Petersburg. In 1921 was teaching assistant at the chemistry department of St. Petersburg University. In 1921 emigrated to Germany with his family. From 1924 to 1934 worked as an assistant in the Kaiser Wilhelm Institute for Physical Chemistry and Electrochemistry in the Herbert F. Freundlich department. After the National Socialists came to power, Bikerman was ousted from the Kaiser Wilhelm Institute because of his Jewish descent. He emigrated to Great Britain and was classified as an enemy of the German state and put in a list of people to be killed. In the UK he received a fellowship at Manchester University (1935–1937) and then at Cambridge University (1937–1939). From 1939 to 1945 he was researcher in three different private companies. In March 1945, he went to the United States, where he worked for Merck & Co. until 1951. From 1951 to 1956 he worked for Yardney Electrical Company Inc. in New York. From 1956 to 1964 was researcher at the Adhesives Laboratory at the Massachusetts Institute of Technology. From 1964 to 1970 he worked as a senior research associate for Horizons Inc. in Cleveland. He was Adjunct Professor at Case Western Reserve University in Cleveland from 1974 until his death in 1978.

7.2.2 Physicochemical Properties of Interfaces

A foam is a disperse system, consisting of highly concentrated dispersions of gas bubbles (dispersed phase) in a liquid (continuous phase). Bubbles are separated from each other by thin liquid films. When a foam forms, the surface energy increases depending on the size and number of bubbles, each bubble with its own surface area and surface tension. Since the gas/liquid surface energy increases the formation of a foam is not spontaneous requiring the supply of energy to sustain the foam. Pure liquids cannot foam unless a surface-active material is present. The capacity of a surfactant solution to form a foam is called foamability and its lifetime as foam stability. There are plenty of applications of foams in daily life, for example the foam to make a cappuccino, the foam of a beer, soap, shaving cream, etc.

Foams are a particular case of colloidal dispersions one in which a gas is dispersed in a continuous liquid phase. The period of stability strongly depends on the properties of the interface separating the dispersed and continuous phases [31]. Kitchener [10] indicated that five properties are required to completely characterize the behavior of an interface:

- Surface shear viscosity
- Surface dilational viscosity
- Surface shear elasticity
- Surface dilational elasticity
- The equilibrium surface tension

The study of these properties is the field of surface rheology. Rheology was defined by Bingham in 1928 to describe the deformation of a fluid when a force is applied. Viscosity in a Newtonian fluid is constant, independent on the shear rate. For a non-Newtonian fluid, the viscosity changes depending on the shear rate, this is due to the interaction between the components in a fluid.

An interface can be considered as a new phase because it has different properties in comparison with the bulk phases. The interface thickness is arbitrary, it is usually defined with the thickness of a monolayer, because of this, its structure, chemical composition and states of energy, as well as interfacial tension are different from the bulk phases.

Surface tension: Surface tension is the property of the liquid in contact with a gas phase (usually air) and interfacial tension is the property between liquid–liquid, liquid–solid or solid–air. The surface and interfacial tensions are defined as the work necessary for the formation of 1 cm^2 of a new surface or interface, the higher the surface tension the higher the energy requirement for foaming, therefore a low surface tension would be better to promote foaming. The interface contains surface-active molecules which control its surface tension. Surface tension is the main variable that control the bubble size.

In 1867 der Mensbrugghe [32] designed an experiment to measure surface tension placing a small ring in contact with a foam film, two concentric films are formed by moving up the ring. This system reaches mechanical equilibrium when the strain film resists rupturing. In this condition the force balance is given by the following equation:

$$2\pi R \cos \theta \cdot \sigma = w$$

where w is the weight of the suspended object, σ is the surface tension, R is the radius of the suspended ring and θ is an angle. In 1919 DuNouy proposed an improved device. Drelich et al. [33] provide a summary of the methods to measure surface and interfacial tension. Figure 7.6 describes the Wilhelmy and Du Noüy methods.

Surface viscosity: Plateau in 1869 was the first to recognize the difference between bulk and surface viscosity. Boussinesq in 1913 was also the first to define two types of surface viscosities; surface shear viscosity and surface dilational viscosity [34]. Joly [35] has reviewed the experimental techniques to measure surface shear viscosities. El Omari et al. [36] reviewed both dilational and shear viscosities. Miller et al. [37] have reported a review pointing out the dynamic nature of surface rheology.

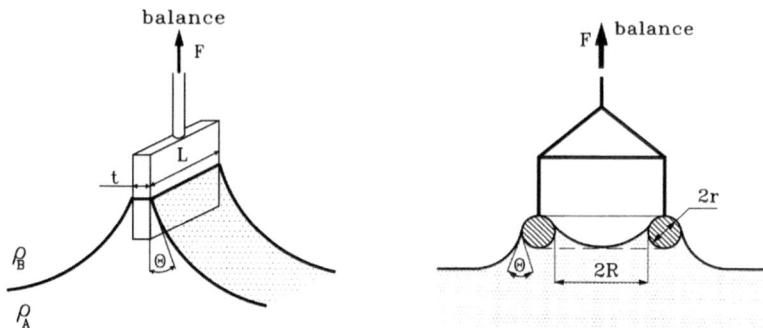

Fig. 7.6 Wilhelmy plate method (left) and Du Noüy ring method (right). After [33]

The motion of molecules can be quite different depending on the duration and force applied. The result involves two different phenomena; surface elasticity and molecular diffusion. The interface can be enlarged by a displacement of molecules from the bulk into the interface. The adsorption process takes time depending on the surfactant concentration. Surface dynamics involve two types of motion; dilation and shear. The surface dilational viscosity can be produced by any relaxation mechanism driving the dynamic surface tension back to its equilibrium value, in other words it measures the dissipation into heat of mechanical work done against surface tension forces, through any relaxation mechanism that affects the surface tension. Boussinesq gave the following definition for the surface dilational viscosity:

$$\mu_d^s = \frac{\Delta\sigma}{(dA/dt)A} = \frac{\Delta\sigma}{d\ln A/dt}$$

where μ_d^s is the surface dilational viscosity in Pa s m.

The surface dilational viscosity is measured by changes on the surface tension when the area of the interface changes with time. The higher the magnitude of μ_d^s the higher is the foam stability. In many systems the surface dilational viscosity is much larger than the surface shear viscosity, for example, Wasan et al. [38] found the surface dilational viscosity was 6 orders of magnitude larger than the surface shear viscosity for an organic solution, their results shown in Fig. 7.7 illustrate higher foam stability increasing the surface dilational elasticity of these organic solutions.

Joly [35] defined the surface shear viscosity as a small element (dx dy) of a monolayer flowing an adjacent monolayer, the surface viscosity is the shear viscosity of the monolayers spread or adsorbed either on the surface of a liquid or at the interface between two liquids:

$$\mu_s^s = \frac{\sigma_{xy}}{\left(dU_y/dU_x\right)}$$

Fig. 7.7 Effect of surface dilational viscosity on foam stability. After [38]

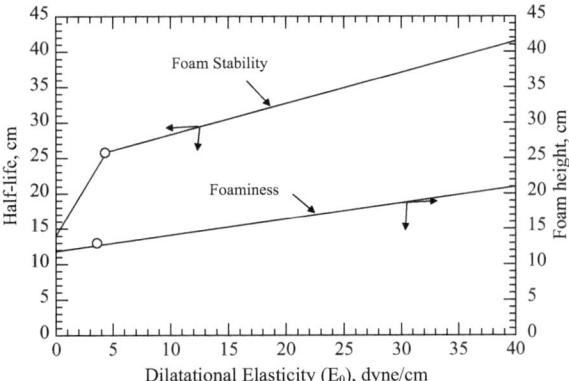

where A monolayer is in its plane (xy) moving at a velocity U(y) in the direction x. Since a monolayer has by definition one molecule thickness, it is a two-dimensional system. The dimensions of surface viscosity are MT^{-1} instead of $ML^{-1}T^{-1}$ for bulk viscosity.

The surface shear viscosity is measured by deforming the interface (without changing its area), moving an object with variable geometry (needle, bi-cone, ring, etc.). The resistance of the interface is recorded by the sensor of a rheometer.

Surface or film elasticity: Bubbles in a foam behave like a weak plastic solid. At small stresses they undergo reversible elastic deformation and at higher stresses they undergo plastic flow, with slip of bubbles. The surface dilational elasticity has frequently been either directly or indirectly implicated as the most important dynamic surface property for creating stable foam film surfaces [22]. Gibbs defined surface elasticity as a ratio between the stress divided by the strain for a unit area. When a bubble film is stretched by an area dA, its surface tension rises and its thickness decreases as follows:

$$E = \frac{2}{A}\frac{d\sigma}{dA} = 2\frac{d\sigma}{d\ln A}$$

where E is surface elasticity, film elasticity or surface dilational modulus, $d\sigma$ is the increment in surface tension and dA the increment in surface area. The factor 2 indicates two interfaces. Copper and Kitchener suggested that any liquid with a positive value for the surface elasticity should be capable of foaming.

Later on, Scheludko proposed the following expression:

$$E = \frac{4\Gamma^2 RT}{\delta C}$$

where δ is the thickness of the lamellae and C is the surfactant concentration. If the thickness of the lamellae or the surfactant concentration E decreases, an indication that film elasticity is inversely proportional to E.

7.2.3   Foam Structure

Foams are colloidal systems formed by a gas in a dispersed medium. The surface
of a bubble in a foam is formed by a liquid film. Since a foam is composed by a
network of bubbles, their surfaces form a thin film containing a liquid. All foams are
thermodynamically unstable (due to the high interfacial free energy). The volume
fraction of a gas in a foam ranges from 0.5 to 0.97 [22]. The bubble size is in the
range from 0.1 to 0.3 mm, with a total of about 10^3 bubbles/ml [22]. To estimate
the foam density is based on the mass of the liquid divided by the volume of the
foam because the mass of the gas and volume of the liquid can be neglected. Foam
densities typically vary from about 0.02 to about 0.5 g/mL [23].

Walstra defines the dynamics of foaming into several stages [22]:

  (i)   Bubble formation
 (ii)   Smaller bubbles dissolve, while bigger ones may grow in size, by diffusion of
        gas through the continuous phase, a process called Ostwald ripening
(iii)   If the void fraction is above 0.75 the bubbles deform one another leading to a
        polyhedral foam
 (iv)   Liquid drains from the foam to the bulk
  (v)   Thin films rupture, leading to bubble coalescence. When a lamellae becomes
        too thin it can reflect light appearing black, at this stage is called black film,
        first reported by Newton.

The most stable shape of a single gas bubble surrounded by liquid is the sphere
because it has the smallest surface area for a given volume of gas, however when they
are polydispersed the pressure differences between small and larger bubbles creates
instability. When the gas void fraction is above 74% the spherical shape becomes a
polyhedral shape [39].

The corners or intersection of the thin films are called Plateau borders, shown in
Fig. 7.8. Due to the large density difference, the bubbles rise to the top, its motion
promotes deformation of the polyhedral structures. Due to deformation of the thin
film its thickness becomes thinner. The surface tension increases, causing film instability. Mechanical equilibrium is achieved when a maximum of three bubbles can be
connected at the Plateau borders, forming an angle of 120°.

As indicated previously, above a gas fraction of 0.75 and due to the Ostwald
ripening effect, which is explained below, the foam has a polyhedral form. Mechanical equilibrium is achieved with a maximum of three bubbles connected at the
Plateau borders, forming an angle of 120° at the intersection of the Plateau border,
as shown in Fig. 7.9. With this geometry the surface tension forces acting on each
Plateau border junction are equal.

The surface tension is a contracting force at the surface which has a tendency
to minimize the surface area, therefore bubbles and drops tend to adopt a spherical
shape, which reduces the total surface free energy. One of the equations that define
the force balance between a bubble and the surrounding liquid is the Young–Laplace
equation. If σ represents the interfacial tension, the total surface free energy of the

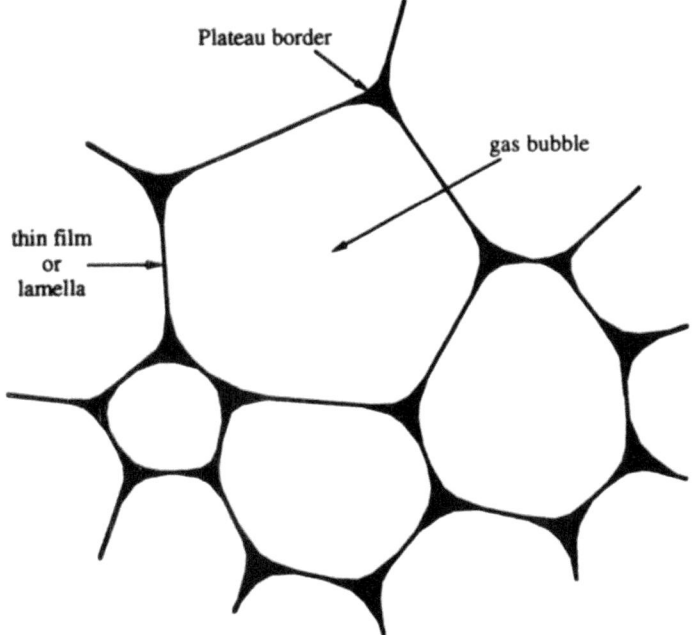

Fig. 7.8 Schematic representation of a foam. After [40]

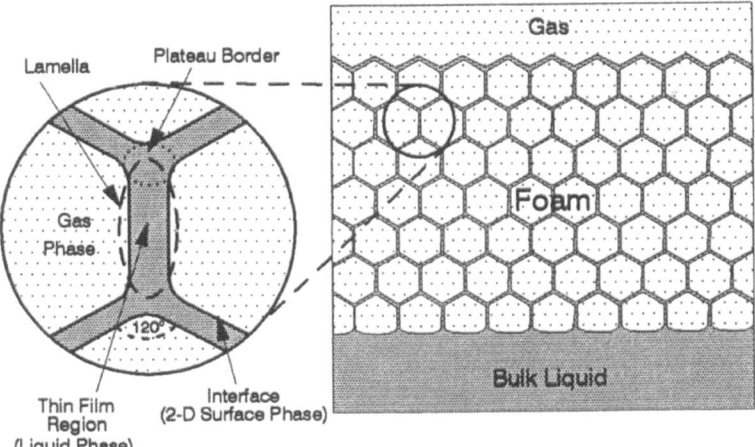

Fig. 7.9 2-D structure of polyhedral foam. After [23]

bubble is $4\pi r^2\sigma$. The symbol γ is also commonly used to define the surface tension. When the bubble size changes there is a change in its radius. If the radius increases by dr, the change in surface free energy is from $4\pi r^2\sigma$ to $4\pi(r+dr)^2\sigma$, this difference can be simplified as $8\pi r dr\,\sigma$. Increasing its size increases the surface energy. This change is balanced by a pressure difference across the film, ΔP (between the inside relative to the outside). The work done due to this pressure difference is $dW = \Delta P dV$, the volume, $V = \frac{4}{3}\pi r^3$, differentiating, it is obtained $dV = 4\pi r^2 dr$, then $dW = \Delta P 4\pi r^2 dr$, which is equal to the decrease in surface free energy. The relationship between pressure and surface tension is finally defined as follows:

$$\Delta P \cdot 4\pi r^2 dr = 8\pi r dr\,\sigma$$
$$\Delta P = \frac{8\pi r dr g}{4\pi r^2 dr} = 2\frac{\sigma}{r}$$
$$\Delta P = P_g - P_l = 2\frac{\sigma}{r}$$

This equation shows that the pressure inside a spherical surface is always greater than the pressure outside, as shown in Fig. 7.10, but the difference decreases to zero as the radius becomes infinite (when the surface is flat). On the contrary, the pressure difference increases if the radius becomes smaller and tends to infinite when r tends to zero. In other words, the internal pressure in bubbles of smaller radius is higher than that in larger bubbles and is a driving force for the diffusion of gas from smaller to larger bubbles, reducing the number of bubbles and increases drainage. The shrinking in size causes a decrease in the surface energy.

The simplicity of the Young–Laplace equation is only apparent and somewhat deceiving as only two of the four properties can be measured directly, the contact angle and the liquid surface tension. There are only a few cases where all four properties of the Young–Laplace equation have been independently measured.

Since the pressure difference tends to infinite when r tends to zero, homogeneous nucleation of bubbles is very difficult to achieve. The smallest nuclei that can be formed spontaneously by homogeneous nucleation has a radius of about 2 nm, requiring a pressure gradient of about 1000 atmospheres [22].

Another important contribution from Thomas Young was a fundamental relationship that describes contact angle and surface tension. This work was reported in

Fig. 7.10 Pressure gradient in a foam. After [22]

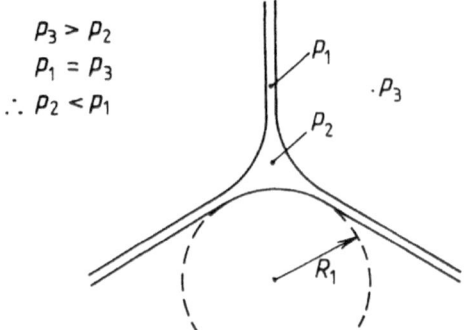

$$P_3 > P_2$$
$$P_1 = P_3$$
$$\therefore P_2 < P_1$$

Fig. 7.11 Geometric relationship between surface tension and contact angle

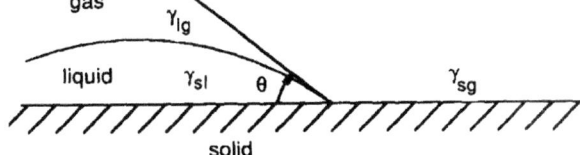

1805 as an essay on the cohesion of fluids [41]. The equation defines the bubble size depending on the surface tension of the phases involved and the contact angle. In the simple case where a liquid droplet is on a solid surface surrounded by a gas phase, Fig. 7.11 shows a case where the contact angle is less than 90°, the conventional upper limit for wetting conditions.

Assuming it is a slag droplet on the surface of liquid metal, the force balance is defined by the Young's equation:

$$\sigma_{metal\text{-}slag} = \sigma_{metal} - \sigma_{slag} \cos\theta$$

where $\sigma_{metal\text{-}slag}$ is the interfacial tension slag/metal, σ_{metal} is the surface tension of the metal, σ_{slag} is the surface tension of the slag and θ is the contact angle. From the previous expression the contact angle can be defined in terms of the surface tensions:

$$\cos\theta = \frac{\sigma_{metal} - \sigma_{metal\text{-}slag}}{\sigma_{slag}}$$

It can be observed that increasing the slag/metal interfacial tension and increasing the metal or slag surface tension have different effect on the contact angle. Since cos $0° = 1$ and cos $90° = 0$, a decrease in the contact angle corresponds with an increase in the value of cos θ. The following trends can thus be summarized as follows;

		$\cos\theta$	θ	wetting
σ_{metal}	↑	↑	↓	↑
$\sigma_{metal\text{-}slag}$	↑	↓	↑	↓
σ_{slag}	↑	↓	↑	↓

For bubbles at the slag/metal interface, the Young equation has to be modified because the geometry is more complex, however the basic relationship among the variables is similar, as shown in Fig. 7.12.

Fig. 7.12 Contact angles at
the joining point of a
three-phase system

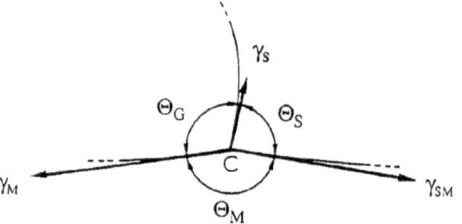

Cramb and Jimbo [42] reviewed the contact angle and interfacial tensions for slags containing $CaO\text{-}SiO_2\text{-}CaF_2\text{-}Al_2O_3(+FeO)$ in contact with liquid iron. Figure 7.13 shows that oxygen decreases the contact angle and sulphur has the opposite effect.

Figure 7.14 shows that both oxygen and sulphur decrease the interfacial tension.

SiO_2 and FeO have also a large effect to decrease the interfacial tension, as shown in Fig. 7.15.

Fig. 7.13 Change in contact
angle as a function of
a oxygen and **b** sulphur.
After [42]

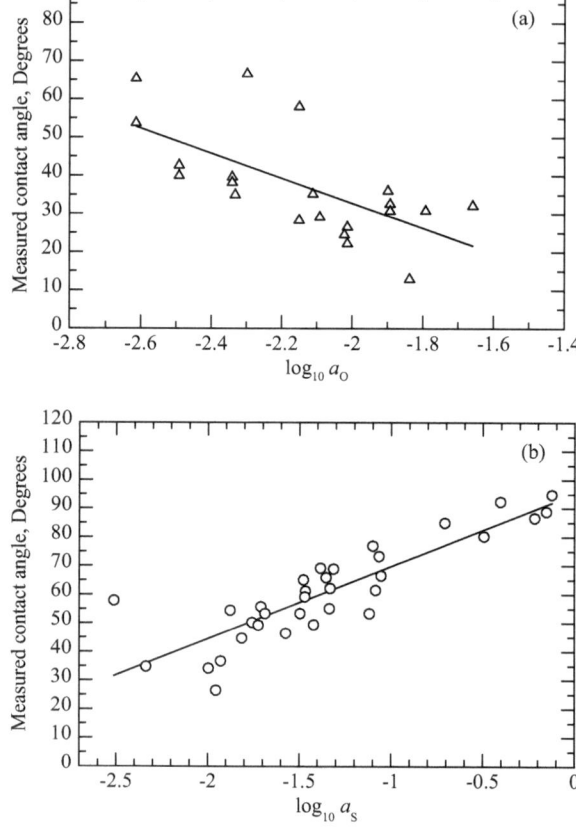

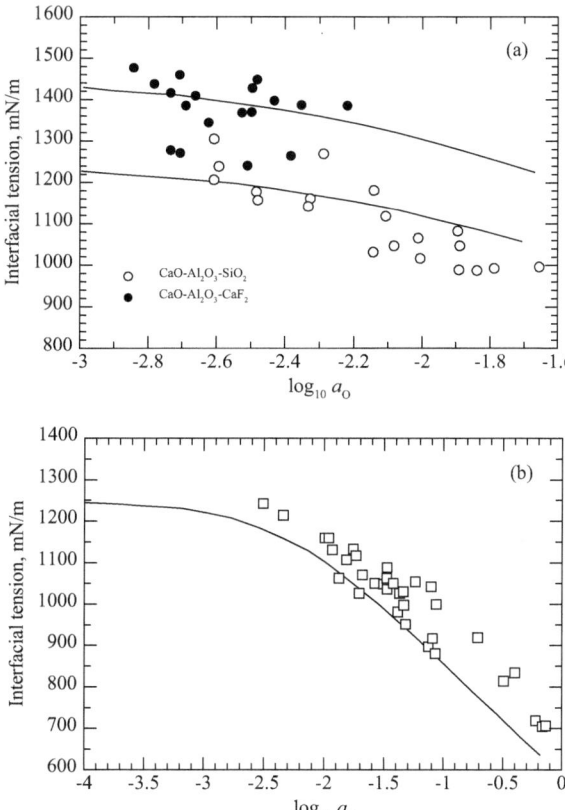

Fig. 7.14 Effect of **a** oxygen and **b** sulphur on the interfacial tension. After [42]

The bubble shape can be predicted by the Weber number, which is a ratio of inertial to surface tension forces;

$$We = \frac{\text{Inertial forces}}{\text{Surface tension forces}} = \frac{\rho U^2 L}{\sigma}$$

The inertial forces tend to distort the bubble and the surface tension forces to restore the bubble shape. Loth [43] reported a diagram showing the shape of the bubbles depending on the We number. For We < 1 the bubbles are spherical, indicating that surface tension forces dominate over the inertial forces.

Fig. 7.15 Effect of **a** SiO$_2$ and **b** FeO on the interfacial tension. After [42]

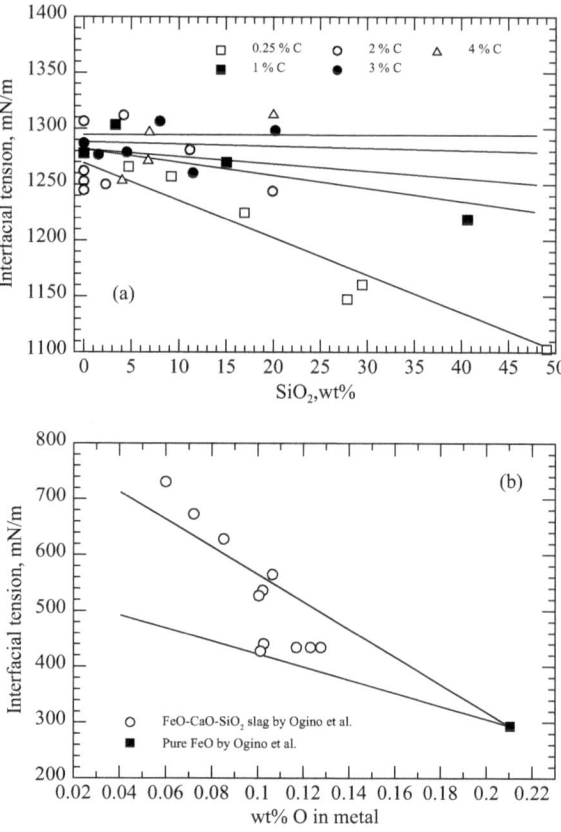

7.2.4   Foam Stability

Foam stability is controlled by three processes [44]; bubble drainage, Ostwald ripening (bubble coarsening) and bubble coalescence. These processes are inter-related.

(i) *Liquid drainage*

Liquid drainage mainly occurs along the Plateau borders driven by two forces; gravity and Plateau borders suction. Mysels et al. concluded that gravity has a small contri-bution because of the nature of capillary flow and it is important only in rigid films [45]. The Plateau borders are curved sections, the pressure inside this region is lower than the pressure in the flat region of the liquid film, the pressure difference creates a suction force inside the Plateau borders, its value given by the following expression:

$$\Delta P = \frac{2\sigma}{r} - \Pi$$

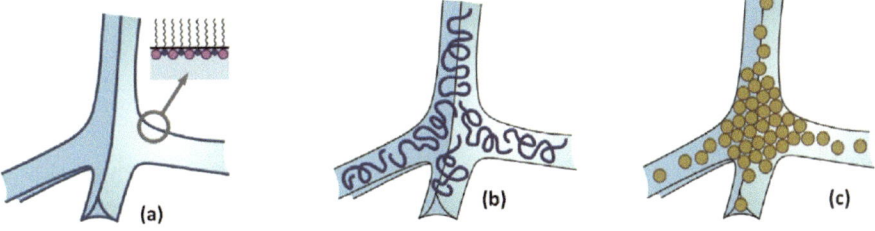

Fig. 7.16 Liquid drainage control; **a** Marangoni effect, **b** solute which increases liquid viscosity and **c** trapping of particles in the Plateau borders. After [44]

where Π is the disjoining pressure.

Liquid drainage can be slowed down by three mechanisms;

- Marangoni effect,
- Solute additions which increase the liquid's viscosity,
- Trapping of particles or oil drops in the Plateau borders, as shown in Fig. 7.16.

The Marangoni effect is also called the Gibbs–Marangoni effect. The Marangoni effect is the transport of surfactant molecules from the adjacent bulk phase to the interface due to a gradient of the surface tension. Gibbs elasticity refers to the increase in the film surface tension derived from a decrease in the surfactant concentration. The gradient in surface tension creates tangential stresses at the walls which decreases the flow of liquid, as shown in more detail in Fig. 7.17. Without a gradient in surface tension the liquid behaves like a free fall droplet. This is why a pure liquid cannot be foamed.

The surface tension gradient induces flow of surfactant molecules from the bulk to the interface, and these molecules carry liquid with them. The gradients in surface tension can be caused by inhomogeneities in temperature or chemical composition. The Gibbs-Marangoni effect prevents thinning and disruption of the liquid film between the air bubbles, stabilizing the foam. In the case of temperature dependence, this phenomenon may be called Bénard–Marangoni convection (or thermo-capillary convection).

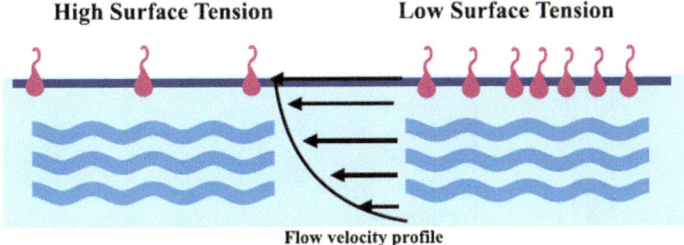

Fig. 7.17 Gibbs-Marangoni effect. After [46]

The relationship between the surfactant concentration and surface tension is defined by the Gibbs adsorption isotherm:

$$d\sigma = -RT\Gamma \, d(\ln a)$$

where σ is the change in surface tension, Γ is the surface excess or adsorbed concentration and a is the activity of the surfactant. The surface excess concentration is the amount of surfactant per unit surface area in excess of the amount that would be present if the surfactant would not have any preference to adsorb onto the surface. The activity is lower than its concentration because only a fraction is adsorbed.

Due to liquid drainage, it is expected the upper layers of bubbles to be much thinner in comparison with bubbles at the bottom, as shown in Fig. 7.18. The upper foam is called dry foam and the bottom foam is called wet foam.

Nexhip et al. [48, 49] using a thermogravimetric technique, withdrawing a rectangular foam with a Pt wire frame, were able to study the draining rate from a single foam at 1300 °C for $CaO\text{-}SiO_2\text{-}Al_2O_3$ slags, as shown in Fig. 7.19. It can be observed that the film thickness (defined as fringe frequency) becomes constant after a drain time of 30 s. In this moment most of the liquid has been drained, no further liquid is supplied to the Plateau borders from the film lamellae and the bubble film approaches a plane-parallel stage, before breaks up. This result indicates that foam drainage and collapse are controlled by the rate of fluid flow from the bubble lamellae. Further research work [50] applying a Michelson-type laser interferometer allowed to map the draining rate and using a laser absorption/transmission method based on the

Fig. 7.18 Dry and wet foams due to drainage. After [47]

Fig. 7.19 Film thickening as a function of time. After [49]

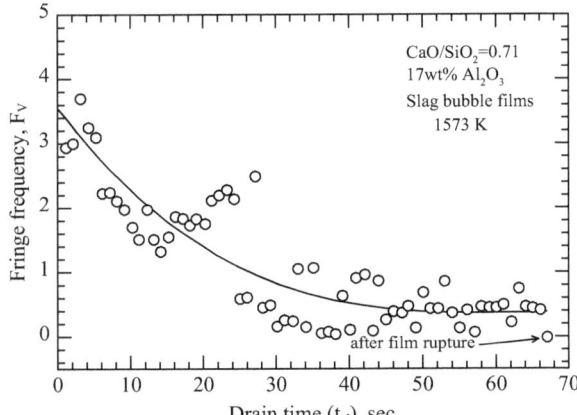

Beer-Lambert relationship they also measured the film thickness. They found the thickness of a well-drained film had a thickness from 0.1 to 0.4 μm (1000–4000 Å).

(b) *Ostwald ripening*

The process of bubble coalescence due to gas diffusion from a small to a bigger bubble is called Ostwald ripening. The rate of ripening increases also with the gas solubility and gas diffusivity in the continuous phase. Ostwald ripening can be fully stopped when the liquid becomes a solid. Due to Ostwald ripening the bubble size becomes homogeneous forming a polyhedral structure. Walstra [22] suggest Ostwald ripening is usually the most serious threat to foam stability.

Denkov et al. [44] summarizes 5 ways to slow down Ostwald ripening, as shown in Fig. 7.20:

(a) Increase film thickness, trapping dispersed additives that reduce gas solubility and diffusivity
(b) Addition of surfactants that promote the formation of highly viscoelastic layers that reduce gas permeability
(c) Formation of an irreversible layer of adsorbed particles around the bubbles
(d) Creating yield stress/elasticity in the continuous phase
(e) Using a gas mixture which contains a gas with poor solubility in the liquid.

de Vries [17] developed an expression which shows the shrinkage of smaller bubbles due to gas diffusion from the smaller to the larger bubbles. The driving force is the pressure difference given by Laplace equation:

$$\Delta P = 2\sigma \left(\frac{1}{r} - \frac{1}{R} \right)$$

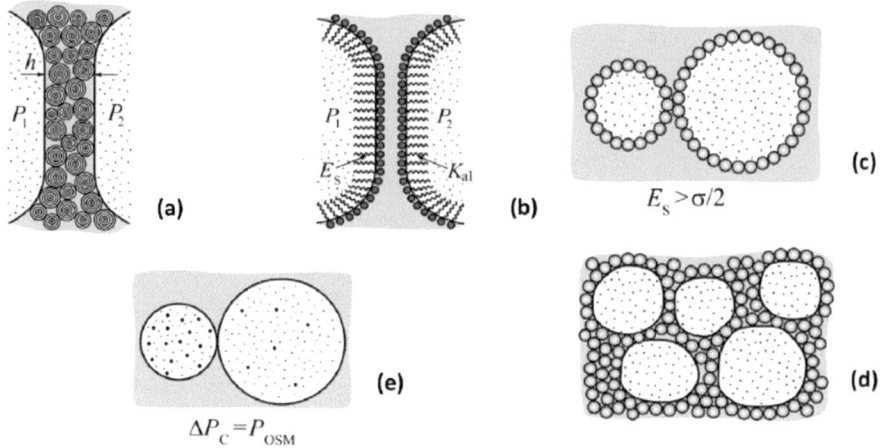

Fig. 7.20 Different ways to slow down Ostwald ripening. After [44]

where r is the radius of the smaller bubble and R is the radius of the larger bubble. When R is at least 20 times larger than r, the term 1/R can be neglected. In this case the pressure difference causing diffusion is equal to the excess pressure in the small bubble.

The diffusion of gas is given by the expression:

$$\frac{dn}{dt} = D \cdot A \frac{\partial C}{\partial x}$$

where dn/dt represents the number of moles of gas per unit time through the film separating the bubbles, D the diffusion constant, A is the area of the film through which the gas diffuses, approximately the surface area of the small bubble $(4\pi r^2)$, $\partial C/\partial x$ is the concentration gradient of the gas in the liquid film (approximately $\Delta C/\delta$) where δ is the average thickness of the liquid film and ΔC is the difference in concentration across the film from $x = 0$ to $x = \delta$. William Henry in 1803 proposed that, at constant temperature, the amount of a given gas that dissolves in a liquid is directly proportional to the partial pressure of that gas in equilibrium with that liquid; $\Delta C = S \Delta p = S(2\sigma/r)$, where S represents the average solubility. The diffusion equation can be recast as:

$$\frac{dn}{dt} = D \cdot (4\pi r^2) \frac{2\sigma S}{r\delta}$$

Additionally, the number of moles of gas in the small bubble can be derived assuming an ideal gas ($pV = nRT$), from which:

$$n = \frac{\frac{4}{3}\pi r^3}{RT} \left(p_0 + \frac{2\sigma}{r} \right)$$

where p_0 is the atmospheric pressure. Since $p_0 \gg (2\sigma/r)$, the term involving surface tension can be neglected.

Inserting the value of n in the diffusion equation and after integration, de Vries reported the following expression:

$$r_0^2 - r^2 = \left(\frac{4RT}{p_0} \cdot \frac{D\sigma S}{\delta}\right)t$$

where: r_0 is the radius at $t = 0$. This equation indicates that a plot r^2 versus time should give a straight line proportional to the diffusion coefficient and the solubility of the gas in the liquid, also proportional to the surface tension and inversely proportional to the average thickness of the liquid films separating the small vanishing bubbles from the larger ones. Experimental determination of the rate constant (term in parenthesis) allows to estimate the thickness of the liquid film. The linear relationship has been confirmed experimentally. Additionally, De Vries also estimated the energy required to start a hole of a critical size and found it to be $0.73d^2\sigma$, where d is the film thickness. Kitchener [10] indicated a value of d/3 from the work carried out by Gleim and Shelomov, indicating its inaccuracy due to the complexity of the interactions leading to bubble rupture.

(iii) *Bubble coalescence*

If thin films rupture, the consequence is bubble coalescence. Different factors can lead to bubble break up.

(1) In the case of thin film thinning, this effect is counterbalanced by diffusion of surfactant to restore the gradient in surface tension, however Gibbs elasticity depends on the surfactant concentration. Gibbs elasticity shows a maximum peak with concentration indicating that too low or too high surfactant concentrations can decrease the film elasticity
(2) Hydrophobic particles promote defoaming due to bubble break up, especially if the film has been drained to a certain extent
(3) Film thinning due to drainage ends up in bubble rupture
(4) Disjoining pressure: At close distances between interfaces, forces of intermolecular origin, both attractive (Van der Waal, hydrophobic) and repulsive (electrostatic and steric) define the disjoining pressure, a concept introduced by Derjaguin in 1937 to describe the force per unit area between the two interfaces of a liquid film.

   a. If the attractive forces prevail, the film is unstable and breaks up. The van der Waal forces between molecules are balanced in the bulk liquid except those in the interfacial region. The imbalance pulls the molecules of the interfacial region toward the interior of the liquid.
   b. On the other hand, repulsive forces, such as the electrostatic forces are produced because each one of the two interfaces has similar electric charge, inducing repulsive forces that oppose to the thinning process.

When films are thinned below 1000 Å (< 100 nm) the intermolecular forces (disjoining pressure) begin to dominate [23]. At the equilibrium thickness, the repulsive forces are balanced by the attractive forces and the disjoining pressure defined as:

$$P_c = \Pi_{vdw} + \Pi_{el}$$

where P_c is the disjoining pressure, Π_{vdw} represents van der Waal forces and Π_{el} electric forces.

Ghosh [51] reviewed in detail the film drainage theory and the stochastic theory of coalescence. His review summarized different expressions to calculate the coalescence time. One example is the equation reported by Chen et al.

$$t_c = 1.07 \frac{\mu a^{3.4} (\Delta\rho \cdot g)^{0.6}}{\sigma^{1.2} B^{0.4}}$$

where; t_c is the coalescence time, μ is the liquid viscosity, a is the bubble average radius, $\Delta\rho$ is the density difference between gas and liquid, σ is the surface tension, B is the Hamaker constant.

Kozakevitch in 1949 and Ogino in 1988, from Ref. [24], suggested the formation of a repulsion electric double layer at the gas/liquid interface. This layer is formed by ionic surfactants adsorbed in the bubble film surfaces; SiO_4^{4-}, $Si_2O_7^{6-}$ and PO_4^{3-}.

7.2.5 Surface-Active Molecules

There are two types of surface-active molecules; surfactants and polymers. Surfactants usually form a compact adsorbed layer with a low interfacial tension and polymers typically form a visco-elastic, irreversibly adsorbed layer.

Surfactants are molecules that form oriented monolayers at interfaces, decrease the surface tension and increase the interfacial viscosity, promoting foam stability. Increased interfacial viscosity provides a mechanical resistance to film thinning and rupturing. According with the Gibbs adsorption isotherm, the excess surface concentration is obtained from the slope of a plot of surface tension versus concentration (activity). Increasing the surfactant concentration almost linearly decreases surface tension up to a point where micelles are formed, called Critical Micelle Concentration (CMC). Above this concentration, an increment in surfactant concentration no longer decreases the surface tension or does it very slowly. The area per adsorbed molecule is obtained with the expression:

$$A_s = \frac{1}{N_A \Gamma}$$

where A_s is the area per adsorbed molecule, N_A is Avogadro's number and Γ is the surface excess concentration.

Swisher and McCabe [52] calculated the surface excess concentration in CaO-SiO$_2$-(Cr$_2$O$_3$) slags at 1600 °C to be equal to 10^{-11} mol/cm^2. This value was two orders of magnitude lower in comparison with aqueous systems (10^{-9} mol/cm^2), suggesting that the effect of surfactants to decrease the surface tension was not enough to guarantee foaming conditions.

7.2.6 Foaming Measurement of Aqueous Systems

Bikerman [53] in 1938 introduced a measure of foamability to describe the bubble's residence time in a foam. It was the first model that proposed an equation to predict the foam height. By that time, foaminess was measured in different ways and the results expressed in different dimensions; (1) time required for the complete collapse of a foam (T), (2) dilution until the solution ceases to froth when shaken (V/M), (3) height of the foam column (L) and (4) the inverse rate of drainage (T/M). T, V, M, L represent time, volume, mass and length, respectively.

The foaming column employed by Bikerman had been employed previously by Foulk and Miller [54], who indicated the original design had been made by Hansley in 1928. The column is shown in Fig. 7.21 (left), consists in a column of liquid, air is passed through soda lime to remove CO$_2$ and filtered through cotton wool. Bikerman noticed a liquid film creeping up the walls when using urine in his experiments and to avoid this he suggested to dry the tube with a current of warm air filtered through cotton wool. Most of his experiments were carried out with an aqueous solution containing 1% n-butyl alcohol because of its high foaming capacity. Its equipment is shown in Fig. 7.21 (right). In this figure, M indicates a porous glass membrane and C a cotton wool filter.

Bikermann conducted experiments on foaming using columns of three different diameters (20, 30 and 40 mm). He found that it was necessary to define the minimum volume of liquid above which foaming can be formed.

The average foam volume was defined as the difference between the average volume occupied by (liquid + foam) and the volume of the liquid at rest. Samples of liquid (usually 50 ml) were foamed in a thermally insulated glass column (70 cm high and 3 cm in diameter). Since some details are missing about how the experiments were conducted, the description given by Pugh [15] indicates that a typical experiment begins with a low flow rate and once dynamic equilibrium is reached, i.e., when the rate of bubble generation is equal to the rate of bubble collapse, which takes about 3–4 min, the average foam volume is measured, then the flow rate is slightly increased and again a time interval is required to reach the new equilibrium. This procedure is repeated several times until a fairly high flow rate is reached. The results correspond to the volume of the foam as a function of the gas flow rate. Bikerman found that the foam height was not a good indicator of the foaming capacity because that value

Fig. 7.21 Glass columns
employed by Foulk and
Miller (left) and Bikerman
(right)

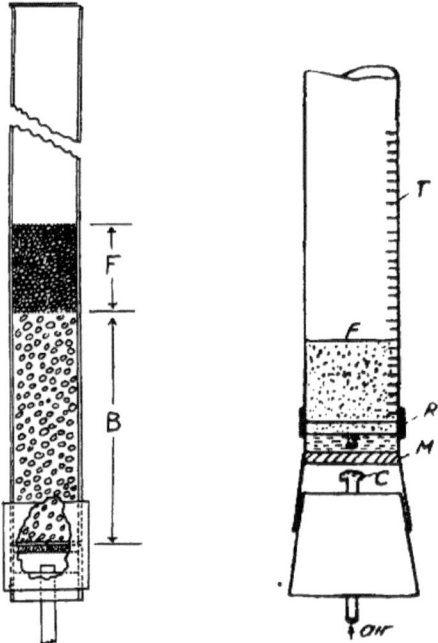

changed over time, as shown in some of his results shown in Table 7.2, however the
ratio $\left(\overline{V}_{\text{foam}} \cdot t / V_{\text{air}}\right)$ remained almost constant.

Based on the previous experimental work, Bikerman defined the following ratio
as the foaming index, Σ:

$$\Sigma = \frac{\overline{V}_{\text{foam}} \cdot t}{V_{\text{air}}} = \frac{\overline{V}_{\text{foam}}}{Q}$$

where; Σ is the foaming index in seconds, $\overline{V}_{\text{foam}}$ is the average volume of the foam,
V_{air} is the volume of air employed to form the foam and t is the time of the experiment.
Q is the air flow rate. Since the volume of foam produced is proportional to the height
of the foam and the flow rate is also proportional to the velocity of the gas injected,

Table 7.2 Foam height as a function of time. After [53]

Series I: 7 cm³	t, s	515	490	378	323	243	242	213	212	202
	H_{max}, cm	1.7	1.6	1.9	3.5	4.6	3.6	4.3	5.7	5.9
	$\overline{V}_{\text{foam}} \cdot t / V_{\text{air}}$, s	3.6	3.4	3.1	4.2	4.0	4.1	3.9	4.1	4.0
Series I: 10 cm³	t, s	660	360	330	330	245	225	197	192	182
	H_{max}, cm	1.4	3.5	2.9	3.9	5.3	4.8	6.6	5.8	6.9
	$\overline{V}_{\text{foam}} \cdot t / V_{\text{air}}$, s	3.0	4.3	4.0	4.4	4.4	4.7	4.7	4.7	4.7

an equivalent expression is:

$$\Sigma = \frac{H_{foam}}{U_g}$$

where; H_{foam} is the equilibrium height of the foam and U_g is the gas superficial velocity (Q/A), A is the cross-sectional area. Bikerman indicated that the previous equation was independent of the gas flow rate, shape and dimensions of the measuring tube and the amount of liquid present [13]. In another group of experiments, he studied the effect of the initial volume of the liquid in a column with a diameter of 30 mm, as shown in Fig. 7.22. It shows that it is required a minimum volume of liquid of 10 cm³ to obtain constant values of the foaming index. These results indicate that it is important to exceed a critical volume of the liquid to eliminate its influence on the foaming index. In this case, the height was about 15 mm, indicating that the height should be at least half the diameter of the column:

$$h_{liquid} \geq \frac{1}{2} D_{vessel}$$

Pugh [15] reported a review on the methods to measure foaming. Since foams are unstable, the method depends on the type of foam. The methods can be classified in two groups; dynamic and static. Dynamic tests measure the height or volume of the slag in a state of dynamic equilibrium between formation and decay. In the static methods the gas flow into the liquid is eliminated and the rate of collapse is measured. A dynamic method is convenient for transient foams and a static method is convenient for very stable foams. Most methods have become standardized but cannot be compared because of different conditions employed in the measurements.

Rudin [55] developed a method to measure beer foaminess. In this method the foam is formed by bubbling CO_2. When the foam is produced the gas is stopped

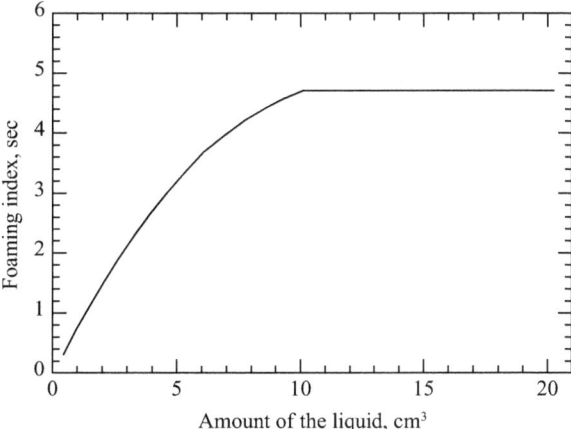

Fig. 7.22 Effect of liquid's volume on the foaming index. After [53]

and the decay of the foam is recorded by measuring the volume of liquid separated from the collapsed bubbles. In a typical experiment that records the collapsing of bubbles, the initial time corresponds to initial volume of liquid forming a foam (V_0), when the volume of drained liquid is half of V_0, the time is taken as the half-life time. The longer it takes for a foam to collapse, the longer its stability. By a proper design, the data are collected when the draining rate is logarithmic which allows a linear relationship between the volume of liquid and time. Figure 7.23 shows Rudin's method. The initial time is taken when the foam/liquid interface reaches the mark of 5 cm then the foam is allowed to collapse. When the interface moves to the mark of 7.5 cm, it corresponds to the half-life time. Rudin reported a residence time for beers, ranging from 90 to 150 s.

Rusanov et al. [56] suggested a foaminess method using a monolayer instead of a multilayer, using bubbles of the same size. In addition to the constant bubble size, in this method, the foam height is constant. The number of bubbles per unit time (n) to sustain the monolayer is measured for different experimental conditions. The total number of bubbles (N) is constant. The foaming index (τ) is measured as follows:

$$\tau = \frac{N}{n}$$

Using this method, they reported that when an aqueous solution reaches a critical solute concentration, which is the Critical Micelle concentration (CMC), foaminess is maximum, as shown in Table 7.3.

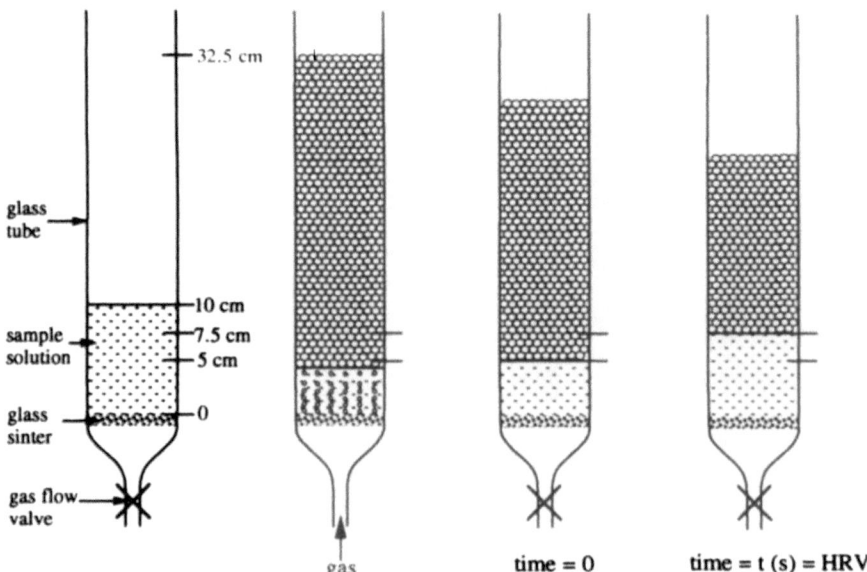

Fig. 7.23 Rudin's method to measure beer foaminess. After [40]

Table 7.3 Foaminess of aqueous sodium dodecylsulfate. After [56]

C, mol/m^3	0.95	1.33	3.0	3.4	6.1	6.8	7.3	7.6	13.6	34.7
τ, s	34	60	130	240	1600	4200	510	200	86	300

Jeelani et al. [57] in 1990 and Hartland et al. [58] in 1993 reported foaming studies using a water solution containing 10% glycerine. The physical properties of this solution are; density 1.014×10^3 kg/m^3 and viscosity 1.222×10^{-3} kg/m s. The surfactant added was Marlophene 89, a non-ionic alkyl phenol polyglycolether with a fixed concentration of 120 mg/l. The container was a glass column (D = 100 mm, H = 2 m), as shown in Fig. 7.24. Sparging was carried out with a porous glass frit of 40 mm diameter with an average pore size from 90 to 50 μm. A scale was attached at the outside of the column, at the top surface of the liquid. Different gases were injected (nitrogen, xenon, nitrous oxide and carbon dioxide). The system was first bubbled for about 10 h to ensure the complete saturation of the gas into the liquid, then the level of the liquid was brought down to 450 mm. The experiments started sparging gas and measuring the change in foam height. The maximum foam height is not constant, has some minor fluctuations. To account for this effect several measurements are made over time and the average height is reported.

7.2.7 Models of Equilibrium Foam Height

The prediction of foaming capacity has been investigated in the past and has resulted in several proposed models. Wang et al. [25] reviewed four of those models; Hartland and Barber from 1974, Hrma from 1990, Pilon from 2002 and Hutzler et al. from 2011. In this section only three of these models will be described. All foam models of interest to slag foaming are described in a subsequent section, including Pilon's model.

Hartland and Barber [59] from the Swiss Federal Institute of Technology in Zürich, developed a model which related the foam height to the film thickness, bubble diameter, gas velocity, and physical properties of the liquid (density, viscosity and surface tension). They assumed that foams may be represented by a tessellated structure of pentagonal dodecahedra in which liquid is carried upwards by the films forming the faces of the dodecahedra and returns to the bulk by gravitational flow in the Plateau borders. The foam height was defined by the following expression:

$$\frac{H_{max}}{d_b} = \frac{0.55\left(\mu U_g d_b\right)^{\frac{5}{4}}}{(\rho g)^{\frac{1}{4}} \sigma \delta^{\frac{7}{4}}}$$

where: H_{max} is the foam height, d_b is the bubble diameter, δ is the critical liquid film thickness, ρ, μ and σ represent the density, viscosity and surface tension of the liquid, respectively, U_g is the gas velocity. Hartland and Barber also compared their model

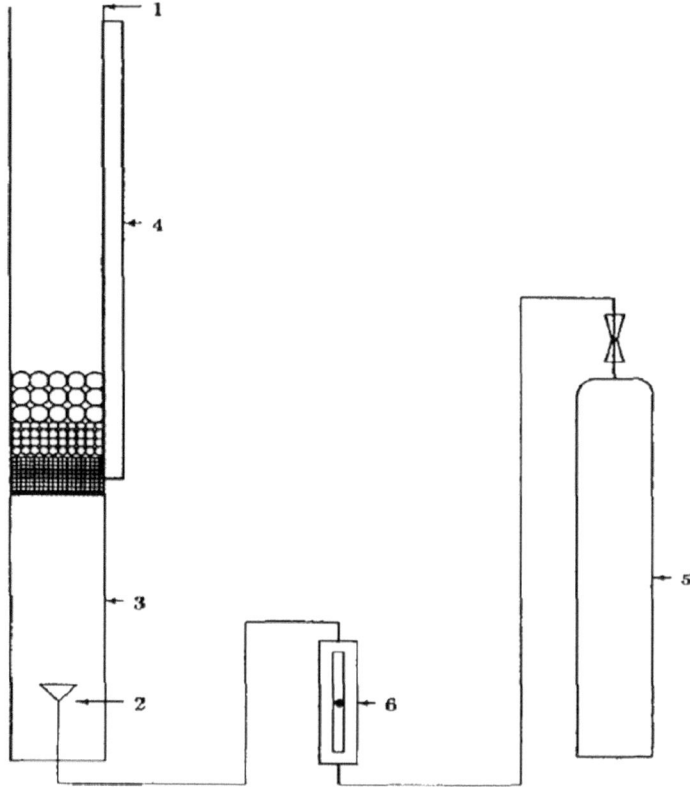

Fig. 7.24 Experimental set up; (1) glass column, (2) sparger, (3) liquid pool, (4) glass column. After [57]

with experimental data. The model satisfactorily predicted the experimental data. The liquid hold-up was measured using a radioactive tracer. Their model suggests that the foam height increases by increasing the liquid viscosity, gas velocity and bubble size and decreases as the liquid's density, surface tension and film thickness increase. According to [25] this model assumes that drainage of the films is represented by the axisymmetric drainage of liquid film between two flat disks, for which the Reynolds equation is applied, such assumption can be limited because the gas–liquid interfaces are mobile or partially mobile.

Hrma [60] from Case Western Reserve University developed a model that defined the foam height as a function of bubble size and superficial gas velocities assuming the height is controlled by gravitational drainage and the survival time of a critically thin film separating a top bubble from the atmosphere. The gas velocities are the critical superficial gas velocity $\left(U_g^c\right)$ and the minimum superficial gas velocity $\left(U_g^m\right)$. The minimum superficial gas velocity is the velocity required to generate a foam and the critical superficial gas velocity corresponds to the breakdown of steady state conditions, a velocity beyond which the foam height will grow without limit. The expression suggested is the following:

$$h_{max} = 2r_b \frac{\left(U_g^c - U_g^m\right)U_g}{\left(U_g^c - U_g\right)U_g^m}$$

where; r_b is the bubble radius, and U_g is the superficial gas velocity. This equation is subjected to three conditions: (1) the superficial gas velocity ranges from the minimum to the critical value, (2) the thin film breaks up as soon as the critical film thickness is reached, (3) the foam height is proportional to the superficial gas velocity. The expression can be simplified for three conditions; (1) when $U_g < U_g^m$ the gas is not sufficient to create a foam layer, (2) when $U_g = U_g^m$, the gas forms a monolayer of bubbles with a thickness $2r_0$, where r_0 is the average radius of the bubbles, (3) if $U_g^m < U_g < U_g^c$ the foam is steady and its height is defined by the previous equation.

Hrma reported that the expression suggested by Bikerman is only valid if the superficial gas velocity is much lower than the critical superficial gas velocity $\left(U_g \ll U_g^c\right)$ and the minimum superficial velocity is much lower than the critical superficial gas velocity $\left(U_g^m \ll U_g^c\right)$, under these conditions his equation reduces to:

$$\Sigma = \frac{h_{max}}{U_g} = \frac{2r_b}{U_g^m}$$

Hrma's model lacks a way to estimate the critical superficial gas velocity and the minimum superficial gas velocity.

Hutzler et al. [61] developed a model for steady state conditions with a constant superficial gas velocity, using an organic solution with a surfactant concentration of two times the micelle concentration. Their model employed a differential equation suggested by Verbist et al. to estimate the drainage rate of the foam. It involves parameters which cannot be easily estimated:

$$\Sigma = \frac{1}{0.16} \frac{\sigma\mu}{(\rho g)^2} \frac{f}{V_g} \left(\Phi_H^{-\frac{1}{2}} - \Phi_0^{-\frac{1}{2}}\right)^3$$

where: ρ, μ and σ represent the density, viscosity and surface tension of the liquid, respectively, V_g is the average bubble volume, g is the gravitational acceleration constant, f is a factor that defines boundary conditions for the flow of liquid through the Plateau border, Φ_H and Φ_0 are the liquid fraction at the top and bottom of the foam column.

Wang et al. [25] indicted that the previous models estimate the equilibrium foam height assuming there is no bubble coalescence on top of the foam, which is not realistic.

7.2.8 Models of Foam Growth or Decay

Barbian et al. [62–64] investigated the foam growth during froth flotation. They measured the bubble life time based on Bikerman's foaming index. Their results are summarized in Fig. 7.25. It is shown that as the flow rate increases, it also increases the maximum froth height.

Barbian et al. summarized the foam height growth with the following expression:

$$H = H_{max}\left(1 - e^{\frac{t}{\Sigma}}\right)$$

Iglesias et al. [65] reported a relationship between the foam height and time, using a linear scale for the height and a logarithmic scale for time. This relationship is then employed to define the differential equation that describes the decrease of height in a foam as a function of time.

Fig. 7.25 a Foaming index as a function of air flow rate, **b** equilibrium froth growth as a function of time for different flow rates. After [62]

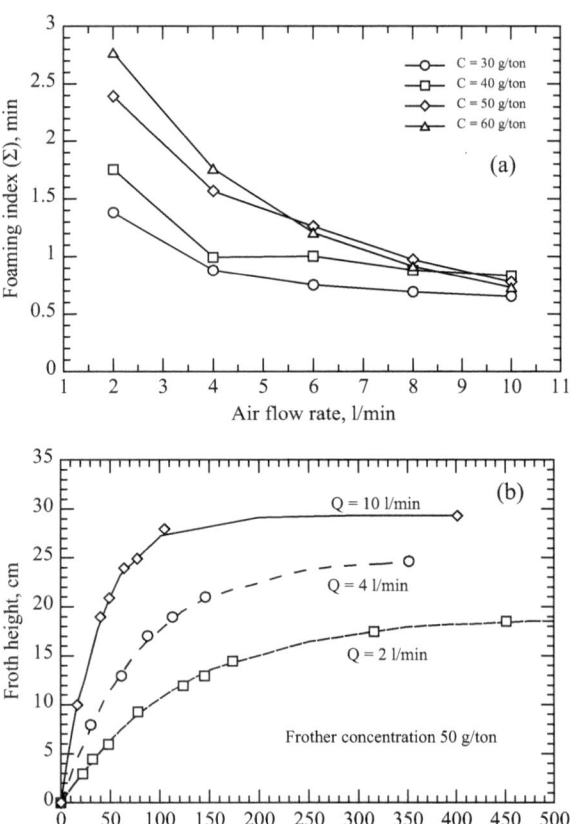

Fig. 7.26 Foam decay model results according to Iglesias et al. [65]

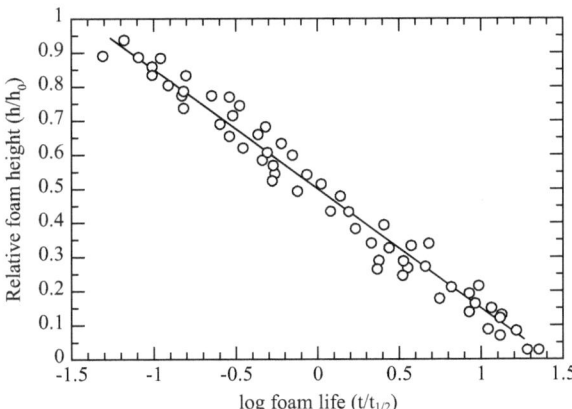

$$H = -a \log t + b$$
$$\frac{dH}{dt} = -\frac{k}{t}$$

Since $t = 0$ cannot be employed in this function, a reference time is employed, in this case the half-life time. Upon integration, from $t = t$ and $H = H_{(t)}$ to $t = t_{1/2}$ and $H = 0.5 \, H_{max}$;

$$\frac{H_{(t)}}{H_{max}} = -\alpha \log\left(\frac{t}{t_{1/2}}\right) + 0.5$$

The value of α ranges from 0.3 to 0.4. According with this expression, it is to be expected that all the experimental data give a value for the ratio $H_{(t)}/H_{max} = 0.5$ when $\log(t/t_{1/2}) = 0$. Figure 7.26 confirms this conclusion. This equation includes two important parameters, the maximum foam height and the half-time.

7.2.9 Anti-foaming

In many industrial processes foaming is a problem rather than a desired phenomenon. For example, in the oil industry the separation of gas from oil, in the BOF converter excessive slag foaming promotes slopping, which is a dangerous phenomenon in the steel shop.

Kitchener [10] suggests that to prevent foaming, its surface elasticity must be destroyed and this can be done by introducing surface layers containing an excess of a compound of low surface tension and virtual insolubility, like silicones.

Garret [66–68] indicated that the simplest defoaming or antifoaming material is the presence of hydrophobic particles. These particles create an unbalanced capillary force in the vicinity of the particle leading to an enhanced drainage rate and film collapse. The best air/water contact angle depends on the geometry of the particle.

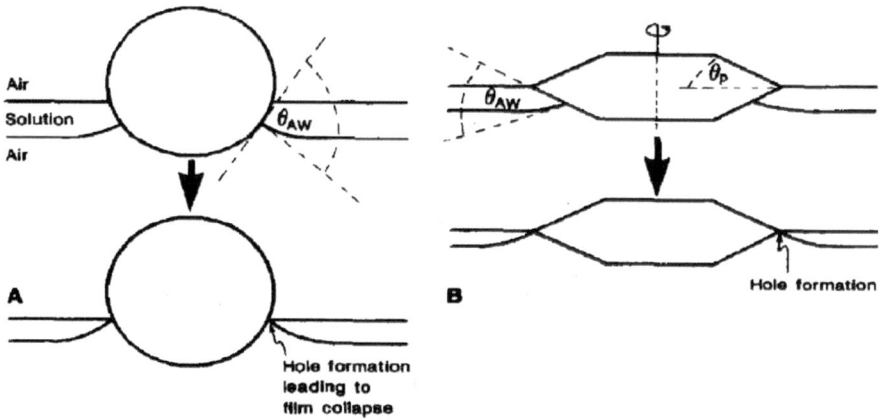

Fig. 7.27 Antifoaming due to the presence of hydrophobic particles. **a** Spherical particles, **b** Sharpe edge particles. After [66]

For spherical particles is better a contact angle higher than 90° but for particles with sharp edges a contact angle below 90° is still adequate, as shown in Fig. 7.27. The addition of particles for defoaming purposes always contaminates the system to be defoamed. In cases where this is a problem an option is to employ mechanical means, for example ultrasound. It has been found that there is an optimum ultrasound frequency which is directly related to the bubble size. The book by Garret is fully devoted to defoaming [67].

Slag foaming can be suppressed by changing the slag physical properties that makes a slag foam unstable either by promoting a higher drainage rate or promote bubble coalescence. This phenomenon is relevant in the BOF where violent foaming (slopping) can occur but it is usually not of interest in the case of EAF operations.

Yi and Rhee [69] studied the addition of C and Al particles to suppress foaming in blast furnace slags. They reported that Al particles of less than 1 mm were more effective than carbon particles because FeO was reduced faster and causing a sudden decrease in slag viscosity. The exothermic reaction and increment in temperature contribute to the decrease in viscosity. They also found that the particle shape was important because round pellets containing Al didn't suppress foaming.

Carbon particles of large size have been employed to suppress slag foaming. Hara and Ogino [2] reported that addition of coke can destroy the foam because the small bubbles can coalesce at the surface, as shown in Fig. 7.28.

Komarov et al. [70] investigated the use of sound waves to suppress slag foaming. They reported that using a frequency of 400 Hz could suppress foaming in CaO-SiO_2-FeO slags at 1300 °C. The sound waves produced by a loudspeaker were propagated to the foamed slag surface through the gas atmosphere. The sound waves propagate through the foam and produce bubble oscillations which enhance liquid drainage.

Fig. 7.28 Foam suppression
due to coke additions. After

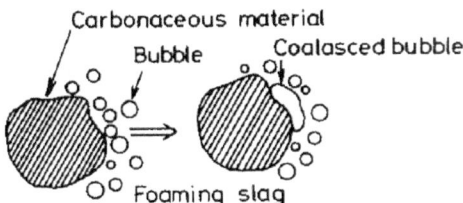

7.3 Slag Foaming Fundamentals

7.3.1 Pioneers on Slag Foaming

One of the earliest reports on slag foaming was carried out by Joseph A. Kitchener et al. [10, 21, 71–73] in 1959. Kitchener was a British scientist born in London in 1916, who worked on mineral processing, in particular on froth flotation. Upon completion of his secondary school in 1934 he completed his PhD in only three years and nine months. He was awarded a position on physical chemistry at Imperial College, working on colloid chemistry. In 1961 he shifted to do research work on mineral processing in the department of mining and mineral technology. He worked 40 years at Imperial College, until 1978 but kept working as a Senior Research fellow for seven more years until 1985. He died at the age of 93 in 2009. Photo taken from [73].

Kitchener et al. reported two reviews on foaming in 1959 [21] and 1964 [10] where he discussed the complexity to describe foaminess in a simple way because it depends on many variables and the methods employed to measure foaming operate under different conditions leading to different results. Even for the same method, variations in bubble size and impurities yield different results.

Paul Kozakevitch carried an extensive amount of experimental work measuring the thermophysical properties of steelmaking slags even before he became researcher at the former Institut de recherche de la sidérurgie (IRSID, currently ArcelorMittal Maizières Research) in 1951. A large amount of research work on slag foaming has been carried out by Japanese researchers. Shigeta Hara and Kazumi Ogino from Osaka University provided a large number of fundamental reports on interfacial properties of slags since the early 1970s. The review papers from both Kozakevitch [1] and Hara and Ogino [2] are classic papers on slag foaming.

Richard J. Fruehan carried our extensive research on slag foaming for about 20 years, from 1989 to 2009.

Richard Fruehan was born in Scranton PA in USA on 1942 and died on 2022 at the age of 80 years old in Saint Augustine Beach, FL. He completed his BSc and PhD degrees at the University of Pennsylvania in 1963 and 1966, respectively. His PhD supervisor was Prof. G. Belton. He became postdoctoral researcher at Imperial College working with Prof. F. Richardson for one year. After that, in 1967 he started working as researcher at U.S. Steel working with Dr. E. Turkdogan and Dr. L. Darken, until 1980, when he became Professor at Carnegie Mellon University. He worked at CMU for 36 years. Since 1985 he became the director for the Center for Iron and Steelmaking Research, the most important research center on steelmaking in the USA. He retired in 2016.

Slag foaming has been extensively investigated in steelmaking since 1959. Slag foaming is a well know phenomena to steelmakers but it was known for its negative effects on converters. In the early 1980's a large number of Japanese researchers investigated slag foaming, in particular to understand its suppression mechanisms. When it was applied to the EAF in the late 1980's it got a massive attention becoming a fundamental slag property with multiple benefits. Prof Fruehan started doing research on slag foaming in 1989. Today there are plenty of reviews on the subject, however all of them lack of a comprehensive integration of the whole available literature. There are about 10 reviews and the maximum number of papers reviewed varies from 20 to 63 papers in most of them with the exception of the review paper by Nexhip et al. with 86 references published up to 2004. Some reviews which include a larger number of references cover metallurgical slags in general and foaming is only a fraction of those reviews. The first review on slag foaming in steelmaking was made by Kozakevitch [1] in 1969. The review by Hara and Ogino [2] is a good review that includes most of the work carried out in Japan until 1992. In the last 20 years several reviews have been reported [24, 74–81].

7.3.2 The Electric Arc Under Foaming Conditions

In 1990 Bowman [82] built a model to describe an arc submerged in foaming slag, obtaining the following results:

(1) The arc still remains under foaming conditions
(2) The experience of underwater welding shows that an arc can operate inside a bubble of vaporized liquid
(3) Current swings are reduced in amplitude.

Bowman and Krüger [7] indicted that a basic slag improves arc stability. Calcium has a lower ionization potential (6.11 eV) than iron (7.90 eV). CaO decreases the slag viscosity and this improves the electric conductivity. A similar effect is obtained by the carbon particles.

Schroeder [83] summarized some of the advantages of foamy slag indicating that decreases harmonics and the immersed electrodes increase its electric power by 6–9%. It can further be improved to 15% if the electrodes are immersed 30–60 cm deep into the slag.

Wunsche and Simcoe [5] reported a decrease in the voltage drop as the slag basicity increases, as shown in Fig. 7.29. As the voltage gradient drops, the arc length increases. They suggested a minimum basicity (CaO + MgO/SiO$_2$) equal to 1.8.

Increasing arc voltage increases arc length. If the arc voltage is increased without a foamy slag, energy losses will increase. It has been reported energy losses of 30 kWh/ton per 100 V and has also been suggested to avoid arc voltages above 450 V for AC and 600 V for DC without a foamy slag [84]

7.3.3 Governing Variables of Slag Foaming

Ogawa et al. [85, 86] in the early 1990s developed a physical model that described the governing variables of slag foaming. Their physical model is described in Sect. 7.4.1. This model considers the formation of CO due to decarburization, excluding the

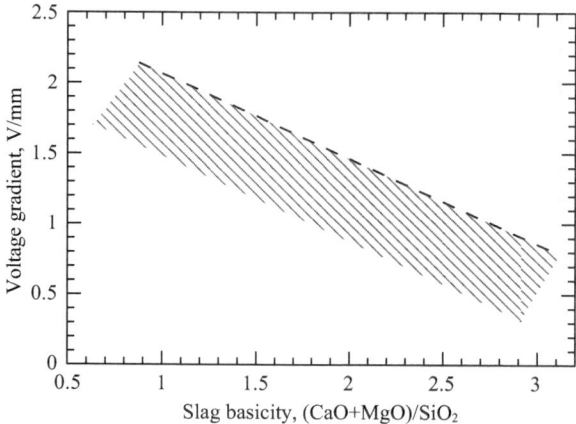

Fig. 7.29 Effect of slag basicity on the voltage drop at the electric arc. After [5]

formation of CO due to injection of carbon particles. Their concept of slag foaming is described in terms of the following three principal stages, also shown schematically in Fig. 7.30.

1. Generation rate of CO bubbles at the slag metal/interface and their detachment
2. Foam formation due to bubble rise in the slag and their accumulation under the free surface
3. Coalescence and gas escape rate at the top slag surface. The gas escape rate depends on the bubble size, number of bubbles and bubble break-up rate. The rate of bubble break-up depends on the bubble size, film thickness and slag physical properties.

Conceptually, slag foaming is a mass balance between the rate of bubbles created at the slag/metal interface and the rate of bubbles destroyed at the slag/air interface.

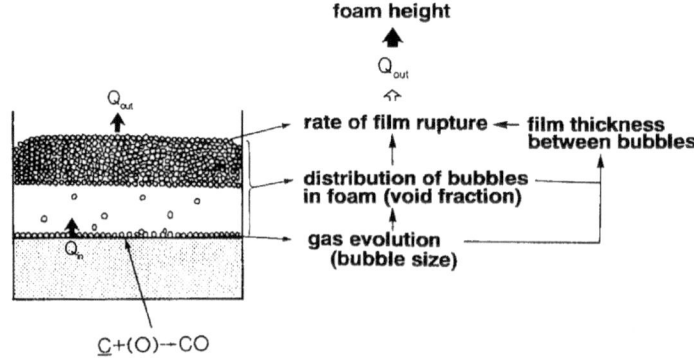

Fig. 7.30 Slag foaming process. After [85, 86]

Consequently, a foaming model should be capable to calculate the size of bubbles at the slag/metal interface, distribution of bubbles in the foam (void fraction) and rate of bubble break-up. The balance of bubbles generated/destroyed defines the foam height.

7.3.3.1 Size of Bubbles Generated at the Slag/Metal Interface

The bubble size results from a force balance of bubbles at the slag/metal interface, as shown in Fig. 7.31.

The following force balance and final calculation of bubble size is based on the work carried out by Terashima et al. [87]. At equilibrium, the pressure difference between the bubble and the surrounding phases is in balance with the static pressure which depends on the following three variables; position on the interface, the difference between the densities of the two phases and gravity.

$$\sigma_{12}\left(\frac{1}{R_1} + \frac{1}{R_2}\right) = \left[(\rho_1 - \rho_2)g(z - z_0)\right] + \sigma_{12}\left(\frac{1}{R_{10}} + \frac{1}{R_{20}}\right)$$

where: σ_{12} is the interfacial tension between the two phases, R_1 and R_2 are the two principal radii of curvature of the interface, ρ_1 and ρ_2 are the densities of the two phases, g is the gravity acceleration force and O is a point of reference where the principal radii of curvature are known. The bubble size is defined for the maximum bubble size detached from the slag/metal interface. This size depends on the wettability between the liquids. The bubble size increases with a decrease of wettability (the contact angle increases). Figure 7.32 shows the predicted bubble size as a function of the contact angle. In this figure the contact angle was increased (wettability decreases) by increasing the slag/metal interfacial tension. The contact angle is increased by decreasing both oxygen and FeO or by increasing sulphur.

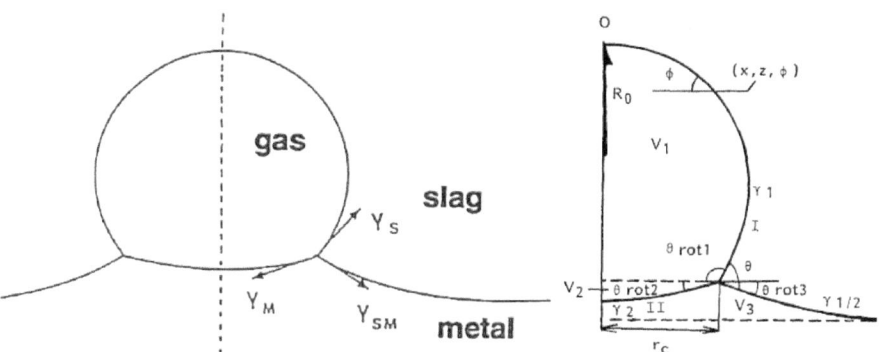

Fig. 7.31 Force balance at the slag/metal interface. After [85–87]

Fig. 7.32 Predicted bubble size based on the model developed by Terashima et al. [87]

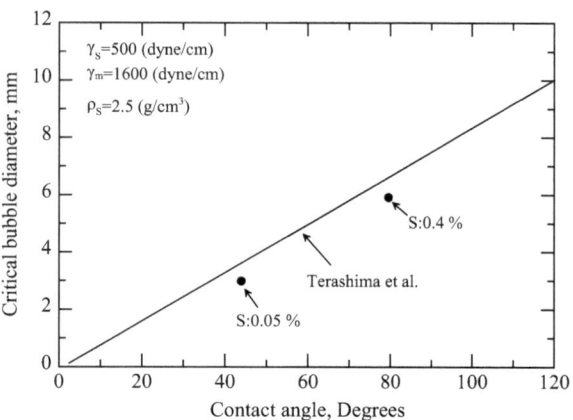

7.3.3.2 Bubble Distribution

The bubble distribution was calculated from the void fraction for a two-phase flow using the model proposed by Wallis in 1963. Increasing the gas flow rate increases the void fraction. Wallis reported the following expression to estimate the void fraction in a system with two layers; foam layer (top) and dispersed gas layer (bottom). The dispersed gas slayer is the layer where bubbles ascend freely.

$$\alpha_1 = \frac{1}{2}\left(1 + \sqrt{1 - 4J/U_\infty}\right)$$

where J is the superficial gas velocity and U_∞ is the rise velocity, defined by the modified Stokes law:

$$U_\infty = \frac{(\rho_L - \rho_G)gR^2}{3\mu_L}$$
$$\alpha_1 = 1 - \alpha_1$$

where ρ_L and ρ_G are densities of liquid slag and gas phases, μ_L is the slag's viscosity, α_1 and α_2 are the void fractions.

7.3.3.3 Gas Escape Rate

The gas escape rate was estimated from measurements on the foam decay after the gas supply was stopped and also from a numerical model. Ogawa et al. found that the gas escape rate increases by increasing bubble size, which suggests that the foam life increases by decreasing the bubble size.

Ogawa et al., based on the previous discussion, summarized the governing variables on slag foaming;

Table 7.4 Summary on the influence of key variables on slag foaming. After [85, 86]

		Foam height	Bubble break- up	Bubble size	Foam void fraction
Slag viscosity	↑	Increase	↓		↓
Slag surface tension	↑	Decrease	↑	↑	↑
Slag/metal interf. tension	↑	Decrease	↑	↑	↑
Metal surface tension	↑	Increase	↓	↓	↓

(i)   Slag viscosity: according to Stokes law, increasing the slag's viscosity the bubble rise velocity decreases, then, according with the Wallis model the foam void fraction decreases, consequently the film between two bubbles becomes thicker, decreasing the bubble break-up rate

(ii) Slag surface tension: increasing the slag surface tension increases the bubble size and the void fraction, then the film thickness decreases which lead to larger bubble break up and a decrease in foam life

(iii) Slag/metal interfacial tension; same effect as the slag surface tension

(iv) Metal surface tension: increasing the metal surface tension decreases the bubble size and consequently the foam height increases.

Table 7.4 schematically summarizes these results.

7.4   Slag Foaming Measurement

In a previous section some methods on foaming measurement have been described, these methods usually employ aqueous liquids, convenient to study foams for food, beer, etc. Most of these methods are based on visual observation of the foaming process, however, the steel/slag system involves high temperatures, the gas is created internally and the liquids are opaque, therefore the bubbles and the liquids cannot be observed.

For slag foaming measurements there are two groups of experimental measurements at low and high temperatures, as shown in Table 7.5. In this section the methods employed to study slag foaming are fully described, the experimental results using these methods are reported further below in this chapter.

An important difference between the experiments at low and high temperatures is that in the real process at high temperature, the gas is produced by chemical reactions at both the slag/metal and liquid slag/solid carbon interfaces. The bubble size at the metal/slag interface produces small bubbles which form stable foams, on the other hand, in aqueous low temperature systems the bubble size is larger and foams are less stable [24].

Table 7.5 Experimental methods to measure slag foaming

T range	Methods	External gas injection	Internal gas production by chemical reactions	Foam height	Foam life
Room temperature	Water models (physical models)	X	X	X	X
High temperature	Immersion of metallic rods	X	X	X	
	Electric sensor (probe)	X	X	X	X
	X-ray fluoroscopy (XRF)		X	X	
	Sessile drop method	X		X	
	Photoelectric method	X		X	
Low and high temperatures	Vibrations method	X	X	X	
	Radar method	X	X	X	
	Acoustic method	X		X	
	Statistical method	X	X	X	

Gas hold-up

The foam height and foam life are characteristic for specific systems and cannot be compared. Several researchers have suggested to employ the gas hold-up as a better criterion for slag foaming [88–90]. This parameter is dimensionless and varies from 0 to 1. The gas holdup is a ratio between the volume of the gas and the volume of the foaming slag:

$$\varepsilon = \frac{V_g}{V_s} = 1 - \frac{V_o}{V_s} = 1 - \frac{h_o}{h}$$

where V_g is the volume of gas in the slag, V_s is the apparent volume or foaming slag volume, V_o is the volume of dense slag before foaming, h_o is the height of slag before foaming, h is the height of foaming slag. With the previous equation, the experimental data of foam height is transformed to gas hold up.

$$\bar{\varepsilon} = 1 - \frac{h_o}{\bar{h}}$$

where: $\bar{\varepsilon}$ is the average hold-up (volume fraction of gas in the foaming slag), h_0 is the initial height of slag before foaming and $\bar{h}$ is the average foaming slag height from time 0 to t.

Yi and Kim [90] designed a box to collect slag samples, the solid sample was weighed and crushed, the slag was microscopically analyzed to estimate the gas hold up.

Gas holdup is not a new method to measure slag foaming but an alternative to report the experimental data on foam height in dimensionless form. The gas holdup ranges from 0 to 1 and can be used to compare the slags foaming power.

Yoshida and Akita [91] reported experimental data on the gas holdup as a function of the superficial gas velocity. Yi and Kim [90] reported those data in the following relationship, from which the superficial velocity can be estimated if the gas holdup is known:

$$\log(1 - \varepsilon)^{-1} = 0.146 \log(1 + U_g) - 0.06$$

7.4.1 Foaming Measurement

The measurements on slag foaming can be expressed in different ways:

(1) Foaming index
(2) Maximum foam height
(3) Average foam height
(4) Foam life or foam decay rate

Bikerman defined a foaming index, Σ, as a ratio between the foam height to the superficial gas velocity. This ratio is equivalent to the average time for the gas to cross through the slag layer. The foam height as a function of time can be divided into three stages; foaming growth, equilibrium or steady state and foam collapse, as shown in Fig. 7.33. In the steady state condition, the rate of bubble generation is equal to the rate of bubble break up.

Foaming stability can measured for any of the three sections in the previous figure; (1) foaming stage; maximum height at a given gas flow rate, (2) equilibrium stage: gas is injected to keep steady-state conditions, then the steady-state foam volume of foam height is measured, (3) decaying stage; the half-life time is measured when the initial foam volume is decreased by half.

Foaming index: The foaming index for a given experiment can be estimated from a ratio between the maximum foam height and the superficial gas velocity.

$$\Sigma = \frac{h_{foam}^{max}}{U_g}$$

Fig. 7.33 Dynamic behavior of foaming, after [92]

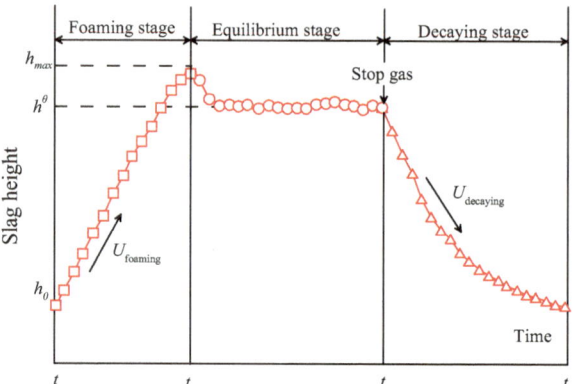

Ozturk and Fruehan [93] have suggested that a more accurate way to define the foaming index is to plot the foam height at different superficial velocities and the slope is equal to the foaming index, as shown in Fig. 7.34.

$$\Sigma = \frac{dh_{foam}}{dU_g} = slope$$

Both King [94] and Zhu [95] studied the foam height as a function of the superficial gas velocity. They reported that there is indeed a linear relationship, however at a critical velocity there is a large change in slope, as shown in Fig. 7.35. The change in slope occurs above 0.2 m/s, a velocity that exceeds the superficial gas velocity in the EAF. This change happens due to bubble coalescence because introducing more gas decreases the distance between bubbles.

Luz et al. [74] made a general remark indicating that one single linear relationship is only obtained at high temperatures (> 1500 °C). The equation proposed from Lin

Fig. 7.34 Relationship between the gas superficial velocity and foaming height. After [93]

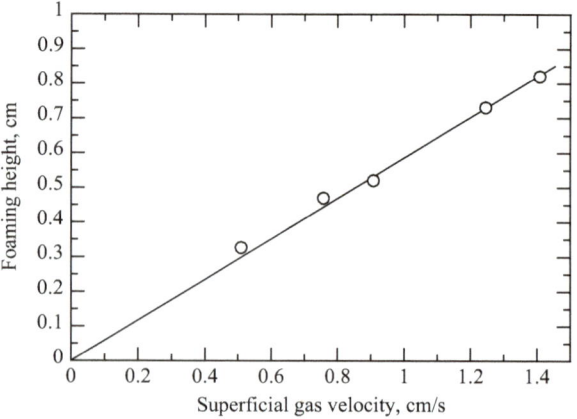

Fig. 7.35 Effect of the superficial gas velocity on the foam height. After [94, 95]

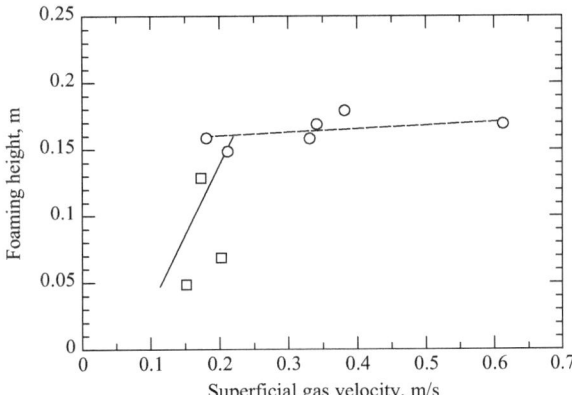

and Guthrie, taken from ref [96] suggests that the foaming index also depends on the initial slag height and bubble diameter, as follows:

$$\Sigma = \frac{h + \Delta h}{\left[U_g + (0.5gd_b)^{0.5}\right]}$$

where h is the initial height of slag, Δh is the foam height, d_b is the bubble diameter.

Previously, it has been pointed out that a linear relationship between the superficial gas velocity and foam height is strictly valid for high temperature systems but not for all aqueous systems. Zhu and Sichen [97] carried out physical modelling studies with silicon oil. The bubble column had a diameter of 80 mm and height of 350 mm. Nitrogen was injected through the bottom using a glass filter with a pore size between 20 and 40 µm. The maximum foam height and bubble size were measured using a video camera. They carried out experiments with three different oil heights; 2.5, 3.75 and 5.0 cm. The physical properties of the oils employed are indicated in Table 7.6.

They results indicate several findings; (a) formation of a two-layer foam at low superficial gas velocities, (b) the two-layer foam produces short foam heights, (c) the foam height and the superficial gas velocity do not show a linear relationship. Figure 7.36, taken from another investigation in the same laboratory, shows the foam transitions as the superficial velocity increases. At high superficial gas velocities, the bubble size increases. Figure 7.36b corresponds to the foaming regime.

Figure 7.37, for an initial oil height of 3.75 cm, shows a very different behavior for the three oils. Oils A and B have similar behavior but in oil C after the velocity

Table 7.6 Oils employed in the experiments. After [97]

	μ_l, mPa s	σ_l, 10^{-5} N/m	ρ_l, kg/m^3
A	19.2	47.8	960
B	48.0	55.6	960
C	97.0	47.7	970

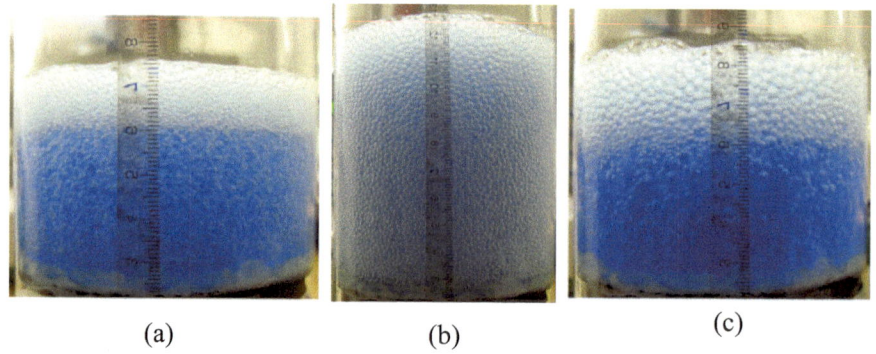

Fig. 7.36 Foam behavior for different superficial gas velocities (mm/s); **a** 0.18, **b** 0.72, **c** 1.44. After [98]

reaches 1.25 cm/s the foam height reaches a peak and then decreases, attributed to growth of the bubble size. Oil B has two layers and the lowest foam height. One of the relevant features to observe in this plot is that the linear relationship between superficial gas velocity and foam height is not linear.

Foam life: the foam life results from the difference between the foam formation rate and foam collapse rate. Ito and Fruehan [99] made the following analysis assuming the foam formation rate is proportional to the gas flow rate and the foam collapse rate is also proportional to its height, this balance is expressed as follows:

$$\frac{dh}{dt} = C_1 Q - C_2 h$$

Under steady-state conditions dh/dt = 0 and $C_1 Q = C_2 h$ but in addition to this expression, if the gas flow rate is stopped once it reaches steady state;

Fig. 7.37 Relationship between superficial gas velocity and foam height. After [97]

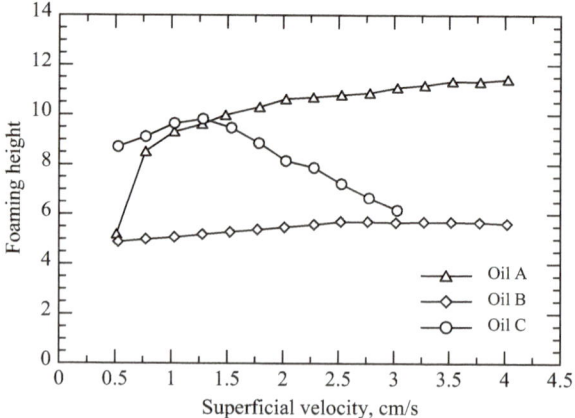

$$\frac{dh}{dt} = -C_2 h$$

Separating variables and integrating from $t = 0$ and $h_{max} = h_0$ to $t = t$ and $h = h$:

$$C_2 t = -\ln\left(\frac{h}{h_{max}}\right)$$

When the foam collapses to a certain height (for example $h_{max}/2$), t is defined as the average foam life and C_2, the slope in a plot $-\ln\left(\frac{h}{h_{max}}\right)$ versus time, is the foam decay rate.

7.4.2 Slag Foaming Measurement Using Physical Models at Room Temperature

In Sects. 7.4.2 to 7.4.4 details of the experimental measurements are provided. The results are discussed in detail in Sects. 7.5 to 7.12

7.4.2.1 Visual Measurement of Foam Height and Collapse Under Steady-State Conditions

Due to the opacity of the slag-steel system at steelmaking temperatures is not possible to observe the flow structure inside these liquids. An equivalent system at room temperatures that can replace liquid steel is a water model, also called physical model. Water at room temperature has the same kinematic viscosity of liquid steel at high temperatures. Using a water model allows to visualize fluid flow and in the case of slag foaming is possible to observe the formation and collapse of a foam.

Kleppe and Oeters [100] were probably the first group to design a physical model to study slag foaming in 1977. They employed two systems; water-glycerol and water–methanol. The water-glycerol solutions had different viscosities, from 5 to 43 cP. Glycerol or glycerine, $C_3H_8O_3$, is a colorless, odorless, viscous liquid that is sweet-tasting and non-toxic, it has a huge variety of applications; in the food industry can be used as a humectant, solvent, and sweetener, and may help preserve foods, is mildly antimicrobial used to treat wounds and also can be used as a fuel. Its density at 20 °C is 1.26 g/cm^3, melts at 17.87 °C, is miscible in water and has a viscosity of 1.412 Pa s (648 cP).

Ogawa [85, 86] in the early 1990's reported a physical model to study foaming. The physical model was designed to test a hypothesis about the two main driving forces that define the foaming height; rate of bubble formation and rate of bubble break-up. They designed two types of experiments using the experimental set up shown in Fig. 7.38. Experiment I was designed to verify the ability of the Wallis model to describe the foaming phenomena (specifically to calculate void fractions using Wallis model) and experiment II to compare the film life of a bubble at the top of a foam with respect to that at the top of the slag without foaming. In experiment I, an initial height of liquid (water + ethanol + saccharose and water + ethanol) was put in a vertical cylindrical tube (D = 37 mm, 1 m height) and nitrogen was injected from the bottom through a gas filter. In experiment II, argon was injected at different gas flow rates through the aqueous solution containing 1% gelatin. The pore size of the glass filter was changed. The foam height and bubble size were measured using a high-speed camera.

It is important to notice a major difference between a physical model and the real process; the bubble size in the high temperature argon/steel system is larger in comparison with the bubble size for the low temperature water/air system. This is due to higher surface tension and lower wettability. Mirsandi et al. [102] compared both cases, as shown in Fig. 7.39. The differences become more significant at lower gas flow rates.

Gudenau et al. [103] conducted water modelling to study slag foaming in bath-smelting and reported some details of their experimental work. Plexiglass vessels with variable diameters (180, 440 and 630 mm) and a height of 900 mm were employed. Water was mixed with a surfactant and on top of this liquid olive oil was employed to simulate the slag. Coal gasification was simulated using CO_2. Air gas was injected from the bottom. Using hydrogen peroxide (H_2O_2) and muriatic acid (HCl) the bubbles were spherical, small and stable as shown in Fig. 7.40a. Figure 7.40b shows a moment when the bubbles escape and disintegrate.

Kapoor and Irons [104] studied the effect of high superficial velocities on the flow of gas–liquid systems, in particular, they provided the first discussion about similarity between the liquids employed to simulate the slag and the real system. The experimental set up was formed by a Plexiglas column with an inner diameter of 150 mm and 2 m high. Gas was injected through the bottom through an orifice plate containing 52 holes, each 3.5 mm in diameter. The gas fraction along the column

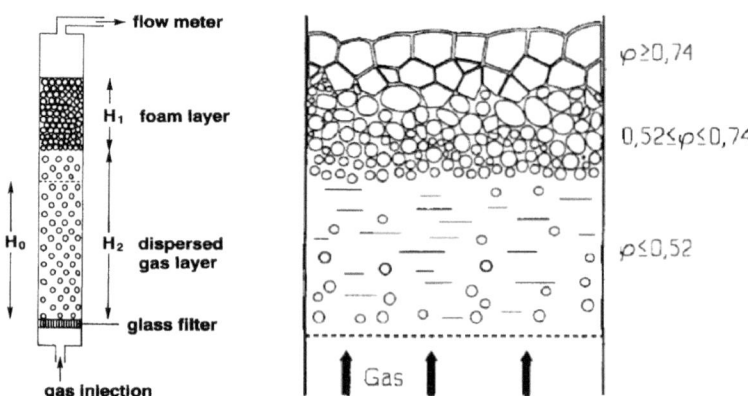

Fig. 7.38 (left) Physical model employed by Ogawa et al. [85, 86]. Right figure is an enlargement scheme of the two layers [101]

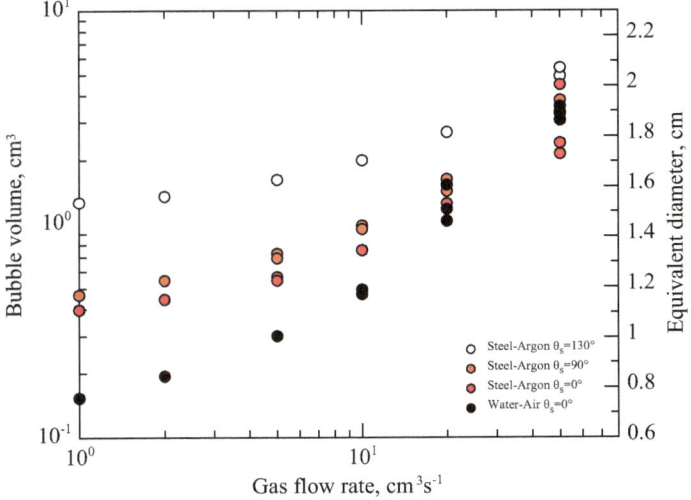

Fig. 7.39 Comparison in the detached bubble volumes in steel/argon and water/air systems. After [102]

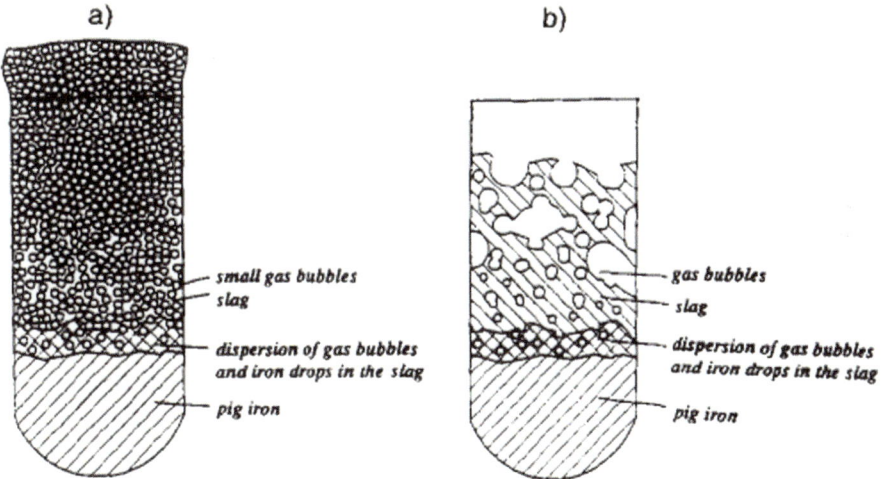

Fig. 7.40 Foaming observations using HCl and H_2O_2. After [103]

center line was measured with a narrow-beam gamma ray densitometer. In regard with similarity among the gas–liquid systems for the prototype and water model, they indicate that all the variables involved $\left(\rho_l, \rho_g, \mu_l, \mu_g, d_b, \sigma, u_s, u, H_o, g\right)$ can be described with the following six dimensionless numbers:

$$\alpha = f(Re, M, E_o, \psi, \varphi, d_b/H_o)$$

$$\text{where; } \alpha = \frac{u_s}{u}, Re = \frac{\rho_l u d_b}{\mu_l}, M = \frac{g\mu_l^4 \Delta\rho}{\rho_l^2 \sigma^3}, Eo = \frac{g\Delta\rho d_b^2}{\sigma}, \varphi = \frac{\rho_g}{\rho_l}, \psi = \frac{\mu_g}{\mu_l}.$$

In order to simply the analysis of the most relevant dimensionless numbers it was noticed that for single gas bubbles rising in liquids, the density and viscosity ratios are very small and can be neglected, in particular at low superficial velocities. The dimensionless numbers; Re, E_o and d_b/H_o all depend on the bubble diameter, which is not known in the real system, consequently is not possible to define full similarity. Upon this conclusion, Kapoor and Irons choose the Morton number to achieve partial similarity. In addition to this, in order to evaluate the effect of the density ratio, two gases were employed, nitrogen and helium. Table 7.7 shows the properties of the liquids employed and the Morton numbers compared with the prototype. The systems investigated consisted of N_2-tap water, N_2-ethanol, N_2-ethylene glycol, He-tap water and N_2-sodium oleate solution.

Urquhart and Davenport [105] briefly discussed the issue of similarity between an oil/water system and the real slag/system. Their general observations indicated that the oil/water system could be used to represent the slag/metal system because on one side, oils and slags both have an ionic structure which gives a low electrical conductivity, on the other hand, both water and liquid steel have a high electrical conductivity.

Ghag [106] investigated the foaming ability of aqueous solutions. He employed binary distilled water-AR grade glycerol solutions in a range of viscosities similar to steelmaking slags. The solution viscosities were measured using a viscometer (Brookfield model DV-III) and the surface tension was measured using the Wilhelmy plate method. The water glycerol solution didn't produce a foam and could be treated as a solvent. The viscosity and surface tension of this binary solution is shown in Fig. 7.41. As a surfactant, sodium dodecylbenzene (SDBS) was used to control the surface tension.

Foaming was carried out in a glass column with a diameter of 107 mm, in a temperature-controlled room (20 ± 2 °C) to avoid fluctuation on the physical properties of the aqueous solutions. Industrial grade compressed air was injected through the bottom using a sintered disk with a porosity labelled as 2, into 1.5 L of the aqueous solution. The foam height was measured with the aid of a scale attached to

Table 7.7 Similarity; comparison of Morton numbers. After [104]

Liquid	Gas	T, K	ρ_l, kg/m^3	σ, N/m	μ_l, Pa s	ρ_g, kg/m^3	M
Bath smelting slag[a]	CO	1773	2790	0.484	0.353	0.2	4.8×10^{-4}
Tap water	N_2	298	1000	0.034[b]	0.001	1.2	2.5×10^{-10}
Ethanol	N_2	298	789	0.023[b]	0.0012	1.2	1×10^{-9}
Ethylene–glycol	N_2	298	1109	0.047[b]	0.060	1.2	1.3×10^{-5}

[a] Based on estimation made by Jiang and Fruehan for a slag with CaO/SiO$_2$ = 1.25, 4% Al$_2$O$_3$ and 7.5% FeO
[b] Measured values

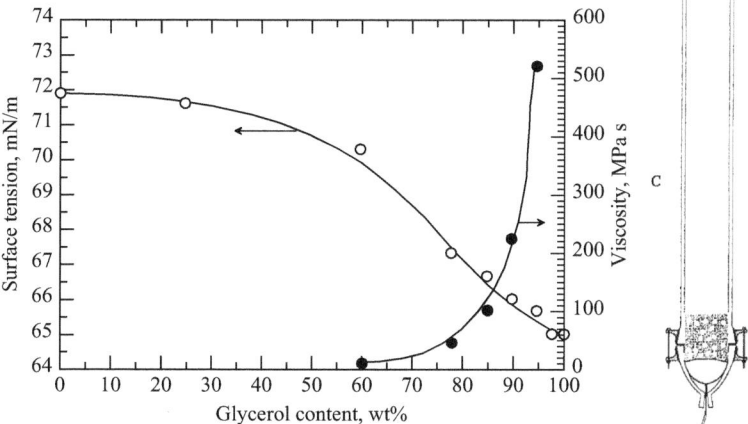

Fig. 7.41 Surface tension and viscosity of aqueous solutions. After [106]

the external surface of the glass column. The bubble size at the foam-liquid interface was measured with a video camera, assisted with a scale attached to the glass column. The mean bubble size was defined for the average of more than 200 bubbles. To avoid contamination of the glass column, before each experiment it was soaked with a solution of distilled water containing a commercial solvent (Decon 90) for 24 h. Once the gas was injected and the foam achieved steady state, the foam height was observed for a about one hour and then the procedure repeated.

Warczok and Utigard [107] carried out two groups of experiments with aqueous solutions. The first group with water-glycerol (0–75%) and water–ethanol (0–30%). These solutions by themselves do not foam. A detergent (Kodak photoflow 200) was added as a surfactant. The gas injected was air, through a glass tube, immersed vertically into the solution to a depth of 20 mm. The container was a plexiglass container (55 × 75 × 300 mm) holding 300 ml of the liquid mixture. The second group of experiments was carried out with water-alcohol mixtures (hexanol, heptanol, octanol, nonanol and decanol). The gas injected at specific depths. The container was filled with 500 ml of the liquid mixture. A video camera recorded the dynamic behavior of the foam. After the foam reached steady state (stable height) the foam height was measured using a scale attached to the container. Due to the dynamic behavior of the foam, the variation in foam height was in the range of ± 15%. The experimental values, foam height and superficial velocity, were employed to calculate the foaming index defined by Bikerman.

Zhu and Sichen [97] employed silicon oils of different viscosity, the injected gas was nitrogen. The container was a plexiglass cylinder (55 × 350 mm). Nitrogen was injected through the bottom using a glass filter with a pore size between 20 and 40 μm. The maximum foaming height and bubble size were measured by direct observation and with a digital camera. Once steady state was obtained for about 10 min, the height

Table 7.8 Oils employed in the experiments

Type	μ, mPa s	σ, 10^{-5} N m^{-1}	ρ
Oil A	19.2	47.8	960
Oil B	48.0	55.6	960
Oil C	97.0	47.7	970

was measured more than three times. The oils employed are presented in Table 7.8. All experiments were carried out at room temperature, between 21 and 21.5 °C.

Zhu and Sichen [108] in a second group of experiments employed two immiscible liquids; Sodium bicarbonate ($NaHCO_3$, 1 M) and silicon oil. Sodium bicarbonate is also called sodium hydrogen-carbonate and more popularly as baking soda or bicarbonate of soda. The experimental set up shown in Fig. 7.42 consists of a cylindrical vessel (80 × 400 mm). The initial height of silicon oil was in most experiments 25 mm. The generation of gas was made possible adding oxalic acid ($C_2H_2O_4$) powder (from 7.5 to 12.5 g), according with the following reactions:

$$C_2H_2O_{4(s)} + 2NaHCO_{3(l)} = Na_2C_2O_{4(l)} + 2CO_{2(g)} + 2H_2O_{(l)}$$

The gas generated caused a foam in the oil. The variation of the foaming height with time was visually measured every 3 s using a graduate scale on the outer surface of the vessel. The rate of CO_2 generated was also measured.

Kapilashrami et al. [109] employed a similar water model to study bubble formation but instead of silicon oil on the top, it was replaced with oleic acid. The sodium bicarbonate solution has a higher density than oleic acid. The camera was placed above the interface to cover a broader area of the oil/water interface. In this case

Fig. 7.42 Experimental set up using aqueous solution with baking soda and silicon oil. After [108]

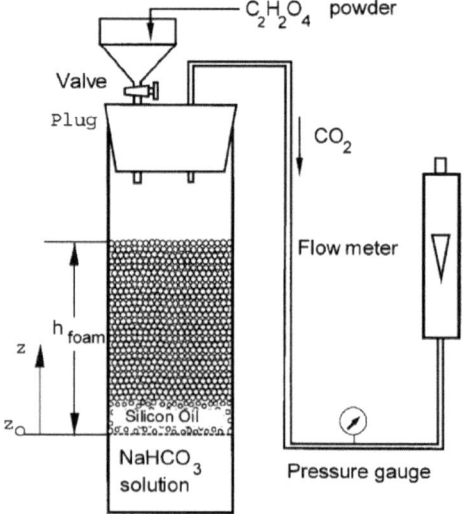

there is no need to add any extra powder, oleic acid reacts with baking soda to produce CO_2:

$$CH_3(CH_2)_7CH = CH(CH_2)_7COOH_{(l)} + NaHCO_{3(aq)} = CH_3(CH_2)_7CH$$
$$= CH(CH_2)_7COONa^+_{(aq)} + CO_{2(g)} + H_2O_{(l)}$$

Oleic acid is extracted from different oils, such as olive oil. The oleic acid employed had a purity of 90% (Sigma Aldrich). The aqueous solutions were prepared with distilled water containing baking soda in concentrations from 0.5 to 1.0 M at room temperature. Oleic acid was added slowly to prevent emulsification. The interface was monitored by CCD cameras at the side, slightly above the interface to cover the full depth of the interface. Once the oleic acid is added it starts to react forming a milky white layer in the interfacial zone. The bubble size was measured with a scale placed on the outer surface of the container. The images were evaluated using an image analysis program (Redlake, Motionscope Inc.). The container was a squared plexiglass, $80 \times 80 \times 80$ mm.

Zhang et al. [110] employed an aqueous solution containing approximately 55% glycerol. Sodium dodecylbenzene sulphonate (SDBS) was added as surfactant. Baking soda was added to this solution. CO_2 was produced by adding oxalic acid powder (9 g). In addition to this a small fraction of particles (0.25–1.25%) was added to study its effect on the viscosity. The viscosity was measured using a viscometer (Brookfield model DV-III). The foaming height was measured every 5 s using a graduated scale on the outer surface of the vessel. Subsequently, the same research group [111] included two types of particles; Polyphenylene sulphide (PPS) powder and sand grain wax (SGW. Their wettability was measured with a JF99A contact angle analyzer. The contact angle of PPS was 37.8° and for SGW, 82.8°. The experimental set up is shown in Fig. 7.43. The container had a diameter of 80 mm and height of 1000 mm. In a typical experiment, the basic solution was prepared using glycerol, water, SDBS, NaHCO$_3$ (8.4 g) and particles (PPS, SGW). The addition of oxalic acid was made slowly using a cushion valve at the bottom of the funnel. The height of the foam was measured visually, every 5 s during its formation, steady state and collapse conditions. The rate of gas produced was simultaneously measured.

Stadler [112, 113] carried out foaming experiments at low and high temperatures in 2002. Two groups of experiments were carried out, for two and three phase systems. The three-phase system includes addition of particles. The experiments al low temperature and a two-phase system consisted of water solutions containing surface tension modifiers as well as viscosity and density modifiers. Ethanol and Methyl Iso-Butyl Carbinol (MIBC) were employed as surfactants and both glycerol and sucrose to modify the viscosity and density. The experimental set up shown in Fig. 7.44 shows a glass column (two diameters 32, and 68 mm, height 280 mm) with porous sinter disks to distribute the gas, type 13 (13-micron pores) and type 2 (70-micron pores). For the three-phase system the glass column dimensions were enlarged (diameter 60 mm, height 2000 mm), water saturated with $MgSO_4$ or starch

Fig. 7.43 Experimental set up employed by [110, 111]

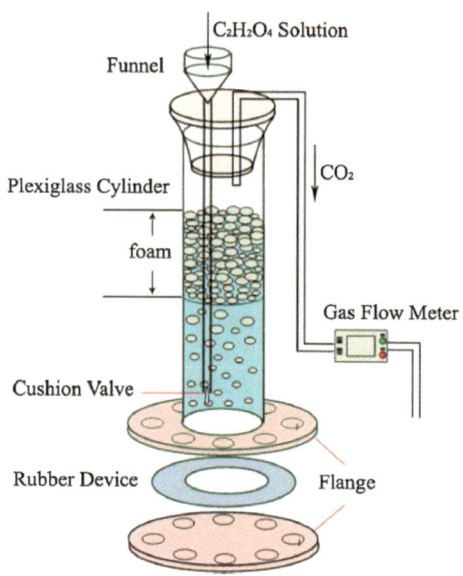

were employed as the liquid. $MgSO_4$ was employed to simulate an ionic melt. Supersaturation allowed to ensure a constant amount of solids in the foam. MIBC was added as surfactant to aqueous $MgSO_4$ solutions. The bubble size distribution was measured using an instrument that aspired the bubbles into a capillary tube. This method was developed by Tucker et al. [114]. The surface tension was measured using the Wilhelmy plate method, the viscosities using a Haake rotoviscometer. Air was injected from the bottom.

Wang et al. [92, 115–119] from NEU in China have reported several physical modelling studies using similar experimental set ups as those from KTH in Sweden. In one group of experiments they inverted the order of additions, instead of adding and acid to react with a carbonate dissolved in water, they dissolved an acid in water and added the carbonate, as shown in Fig. 7.45. The gas is formed according with the following reaction:

$$Na_2CO_{3(s).} + H_2SO_{4(aq)} = Na_sSO_{4(aq)} + CO_{2(g)} + H_2O_{(l)}$$

Another two groups of experiments were carried out with a glycerol aqueous solution, adding SDBS as surfactant. In one case, shown in Fig. 7.46, air was injected through the bottom. The length of the square bottom was 50 mm and a height of 1000 mm. The initial height of the water-glycerol solution had a height of 100 mm (250 ml). The flow rate was varied from 0 to 8 lt/min ad a pressure of 0.2 MPa. In a second case, shown in Fig. 7.47, air was injected from both top and bottom simulating the Q-BOP operation, in addition to this, additional gas was internally produced at the oil/water interface dissolving $NaHCO_3$ in water and adding oxalic acid from the top. The physical properties of the glycerol solution were measured; the dynamic

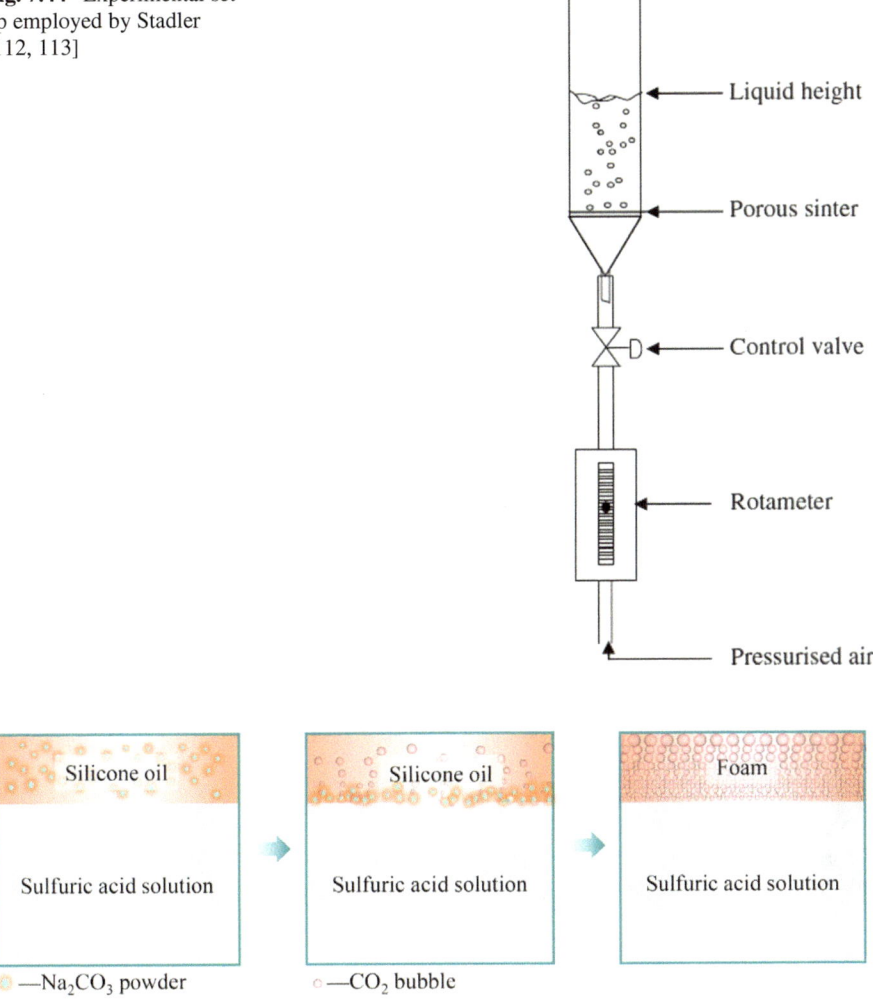

Fig. 7.44 Experimental set up employed by Stadler [112, 113]

Fig. 7.45 Physical modeling of slag foaming induced by solid/liquid reaction at the oil/water interface. After

viscosity was measured with a digital viscometer (manufactured by Brookfield USA, model DV-III) and the surface tension with a GBX surface tension tester (type 3S, France).

Previously it was discussed the similarity analysis presented by Kapoor and Irons, reporting one characteristic value for the Morton number for the fluid representing the slag. The research group from NEU provided additional information indicating a larger range for steelmaking slags and the values that can be achieved using glycerol solutions, as shown in Table 7.9. From this table is clear that a glycerol solution exhibits similar Morton numbers to that of steelmaking slags.

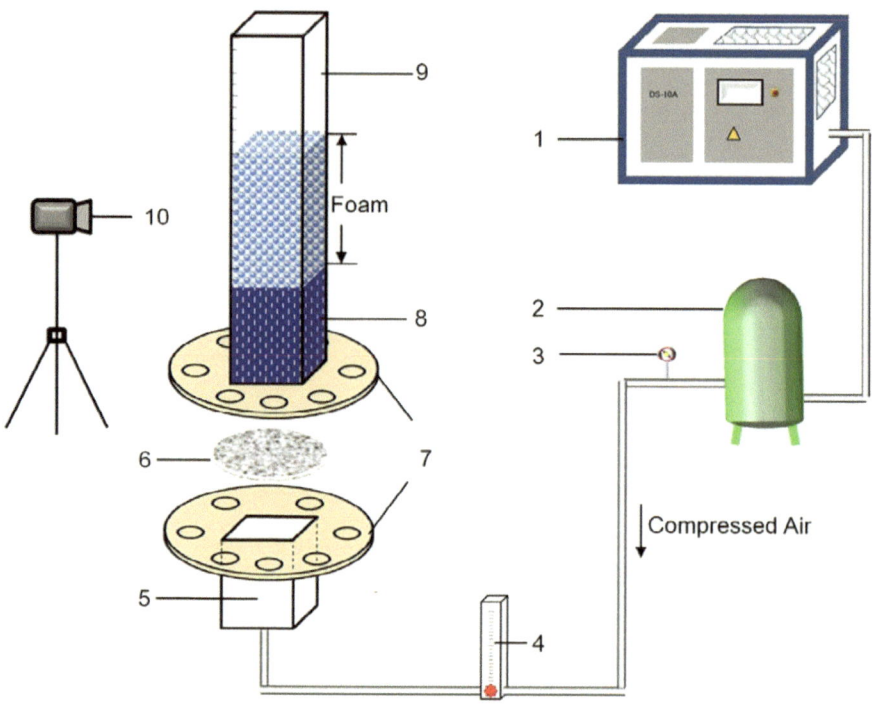

Fig. 7.46 Experimental set up to study slag foaming. After [92]

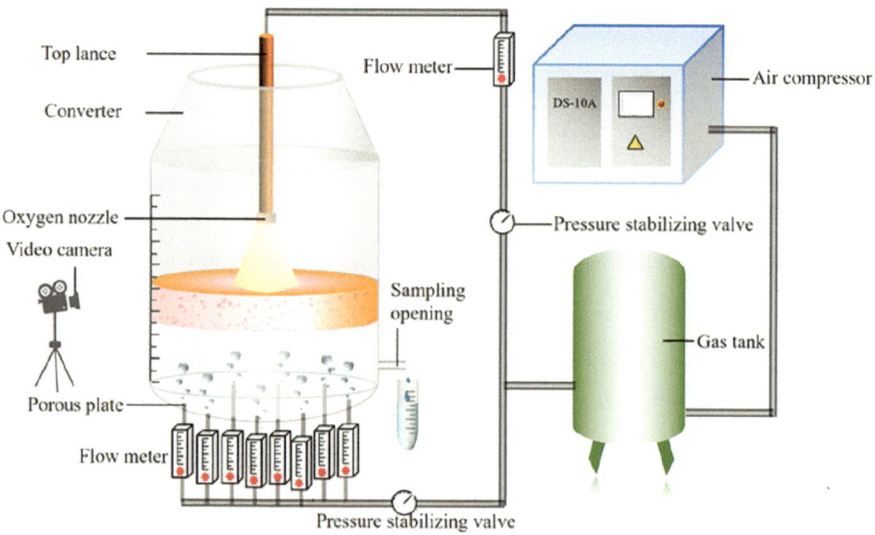

Fig. 7.47 Experimental set up to study slag foaming. After [119]

Table 7.9 Comparison of the Morton number values between BOF slags and glycerol solutions. After [92]

	T, °C	ρ_l, kg/m^3	μ_l, mPa s	σ, mN/m	M
BOF slag	1350	2800–3200	50–1200	200–600	9.45×10^{-8} to 9.07×10^{-1}
Glycerol solution	25	1209–1247	68–397	62	6.70×10^{-4} to 9.04×10^{-1}

7.4.2.2 Physical Model Using Acoustic Measurements

Birk et al. [120] in 2000 reported the first physical model using acoustic signals to measure foaming. The model consisted of a tube, 1 m long and 193 mm of inner diameter. The sonic meter is attached to an external inverted tube to avoid vibrations. Figure 7.48 shows the experimental set up. The aqueous solution was prepared with water and glycerin. The solutions prepared are shown below:

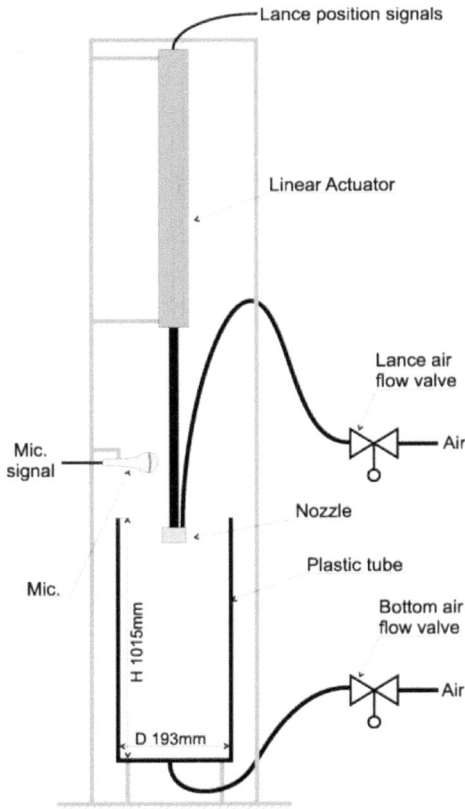

Fig. 7.48 Experimental set up to measure foam height using a sonic meter. After [120]

Mixture	Viscosity, cP	Foam life, s	Surface tension, N/m
A	1.0	1064	0.27
B	3.1	3448	0.27
C	1.0	189	0.49

They carried out two types of experiments; static and dynamic. Static means experiments conducted with a constant gas flow rate at different lance positions until the foam volume was constant. In the dynamic experiments the foam height was measured as the foam volume was increasing. They employed the following expression developed by Ingard which relates sound intensity with foam height:

$$I_t = I_o \exp(-\beta_w h_t)$$

where: I_t is the sound intensity at time t, I_o is the initial sound intensity, β_w is the frequency dependent attenuation coefficient and h_t is the foam height. The previous expression is recast to define the foam height as a function of the sound's intensity.

$$h_t = \frac{\ln I_o - \ln I}{\beta_w}$$

An algorithm was developed to process the data to obtain the value of β_w, using Discrete Fourier Transform (DFT). The frequency employed ranged from 2 to 11 Hz. Figure 7.49 compares the estimated foam height using the sonic meter and the measured values. It is clear that the acoustic method is reliable, according with this report.

Fig. 7.49 Comparison between the estimated (solid line) and measured (dashed line) foam height using acoustic measurements. After [120]

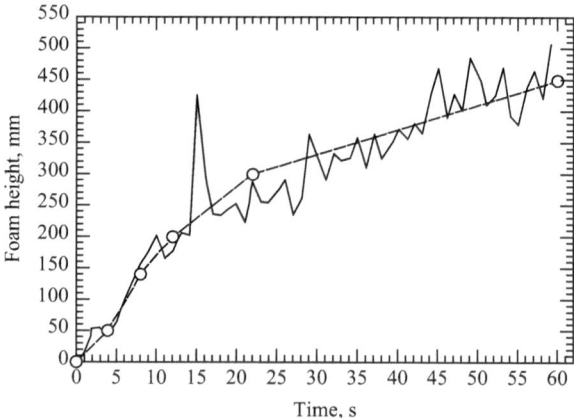

7.4.2.3 Measurement of Impedance

Harada et al. [121] found a relationship between the gas fraction in a foam and its impedance using Nyquist diagrams. The Nyquist diagram (or Strecker–Nyquist stability criterion) was a criterion independently developed by Felix Strecker in 1930 and Harry Nyquist in 1932. It is a graphical technique to define the electric stability of a dynamical system. Experiments were carried out with ultrapure water, 0.01 KCl aqueous solution, 2-propanol, soluble glass, glycerol and several glycerol aqueous solutions. Nitrogen was injected because has similar properties to CO. The container was a stainless-steel tube with 40 mm in diameter placed within an acrylic pipe and both were connected to the impedance analyzer (IM3570, Hioki EE Corp). The electrode was made of Pt–Rh with a diameter of 2 mm, immersed 10 mm into the solution. A manometer was placed on the sidewall to calculate the gas fraction. The authors found that the Nyquist plots show semicircular shapes with a diameter that increases as the gas fraction in the foam increases, however it seems that this method still needs to be further explored to guarantee a practical application to measure foam height.

Table 7.10 summarizes some details on the experimental conditions employed by different researchers on physical models to study slag foaming.

7.4.3 Slag Foaming Measurement at High Temperatures

7.4.3.1 Immersion of Metallic Rods

Iron or nickel rods have been employed to measure the foam height as a function of time. Kitamura and Okohira [122] immersed an iron rod every 60 s. and measured the slag adhesion length. Their experimental set up consisted of a Tamman furnace, electrolytic iron (400 g) and graphite powder (40 g) were added into a crucible of 50 mm in inner diameter and 180 mm in height, which simulated the formation of hot metal, as shown in Fig. 7.50. Once the hot metal was liquid, a mixture of oxides (180 g) without FeO was added. After 5 min FeO was added three times using paper bags in a period of time of 20 min that lasted each experiment. The first addition of FeO marked the beginning time for each experiment. The foaming height was calculated as the height difference between the measured slag height minus the initial slag height before the first addition of FeO. In general foaming decreased as a function of time, however the response was different for every experiment. Each FeO addition produced a different foam height. The results were analyzed in terms of the maximum slag height and the decreasing rate of the slag height. The maximum foaming height was defined as the average of the three values according to the order of foaming in each experiment.

Paramguru et al. [123] used a plasma reactor with a graphite crucible (diameter 100 mm and 150 mm height), its side walls were coated with zircon to allow reaction between FeO and carbon from the crucible's bottom. 500 g of dried slag were charged

Table 7.10 Experimental conditions by previous research on physical modeling

	Author	System	D, mm	H, mm	Extern gas	Internal Gas gene	h$_{liq}$, mm	H$_{foam}$
1977	Kleppe [100]	Water-glycerol/methanol						Visual
1992	Ogawa [85]	Water-ethanol	37	1000	N$_2$, Ar			Visual
1992	Gudenau [103]	Water-(H$_2$O$_2$-HCl)-surfactant and oil	180, 440 630	900	Air			Visual
1997	Kapoor [104]	Water-ethanol-ethylene glycol	150	2000	H$_2$, He		100	Visual
1996	Ghag [106]	Water-glycerol-SDBS	107		Air			Visual
1994	Warczok [107]	Water-glycerol-detergent Water-ethanol-detergent	55	300	Air		120	Visual
		Water-alcohol	55	300	Air		210	Visual
2000	Zhu [97]	Silicon oil	55	350	N$_2$			Visual
2001	Zhu [108]	Baking soda-silicon oil-oxalic acid power	80	400		Ox. acid	25	Visual
2005	Kapilashrami [109]	Water-Baking soda and oleic acid	80 × 80	80		Ol. acid		Visual
2015	Zhang [110]	Water-glycerol-surfactant-baking soda	80	1000		Ox. acid		Visual
2002	Stadler [112]	Water-glycerol-surfactant-baking soda	32,68	280	Air			Visual
		Water-MgSO$_4$-Butyl (MIBC)	60	2000	Air			Visual
2020	Wang [115]	Baking soda-silicon oil-oxalic acid power				B. Soda		Visual
		Water-glycerol-SDBS	50 × 50	1000	Air		100	Visual
2000	Birk [120]	Water-glycerol	193	1000				Acoustic
2017	Harada [121]	Water-glycerol	40		N$_2$			Impedance

Fig. 7.50 a Foaming height measurement using a steel wire and **b** typical experimental results. After [122]

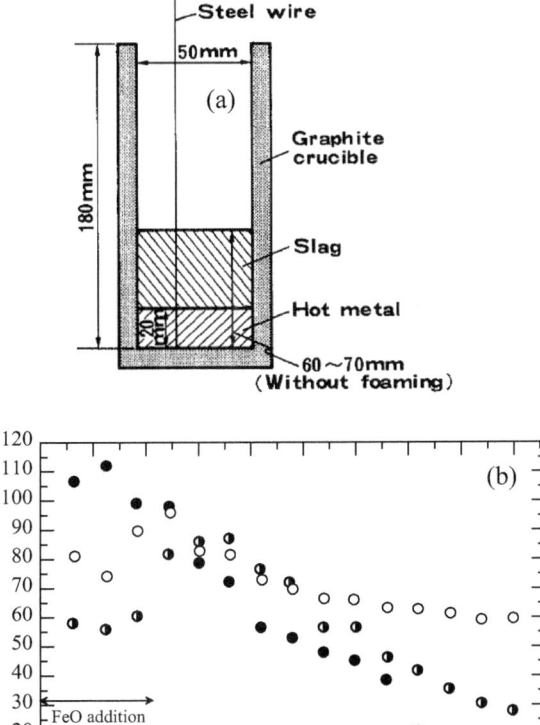

in the reactor, heated and melted, then iron ore was added. The dissolution time was 7 min. An iron rod was employed to measure the foam height, dipping the rod every 3–4 min to take samples. The length of the slag sticking to the rod was taken as the foam height. The slag composition was 40% CaO-34%SiO$_2$-21%Al$_2$O$_3$-5%MgO, analytical reagent grades, mixed, pelletized using molasses as a binder and dried before charging into the reactor. FeO was added in the form of pre-reduced iron ore containing 52% FeO and 25% Fe$_2$O$_3$.

Hong et al. [88] employed a nickel rod to measure the slag foaming height. Their experimental set up consisted of an alumina crucible (diameter 4 mm and height 150 mm) inside of a resistance SiC furnace, 75 or 90 g of primary slag were initially melted. They conducted two types of experiments, with external gas supply (comparing Ar and CO) using a J-shaped alumina tube and also with internal gas formation due to reaction between an immersed/rotating graphite rod (10 × 40 mm) and the slag. In addition to measure the foam height they also evaluated slag foaming using the gas hold up.

7.4.3.2   Electric Probes

The molten slag is placed on a crucible The initial height of liquid slag is detected
with an electric probe. The experiments are conducted at a constant flow rate. The
foaming height can be defined by setting the initial height of liquid slag as zero and
the final height when foaming achieves steady state for a given set of experimental
conditions. This technique has been employed by many researchers, for example;
Cooper and Kitchener [71], Hara et al. [124], Stadler et al. [113], Ito and Fruehan
[99, 125] and Kim et al. [126]. Cooper and Kitchener were the first ones to report
experimental results on slag foaming stability, measuring the collapse time of a foam
between two marked points, from 47.5 to 5.5 mm, after the injected gas is stopped.
Ito and Fruehan [99] employed a non-reactive system, i.e. argon gas was supplied
externally using a stainless steel pipe with a knife edged nozzle (inner diameter =
2.1 mm). An alumina crucible (D = 32–55 mm and H = 200 mm) was used to contain
the slag, its wall was coated with slag to reduce wall effects. Since the surface level of
the foamed slag showed periodic movement due to bubble break up, the foam height
was defined for the maximum surface level since it showed good reproducibility.
The surface position of the slag was measured with a stainless-steel electric probe.
Bubble frequency can be measured with a pressure transducer connected to the gas
injection nozzle. The mean bubble size was calculated from the gas flow rate and the
bubble formation frequency, assuming they are spherical, using the following two
equations:

$$V_b = \frac{Q_g}{f}$$
$$d_b = \left(\frac{6Q_g}{\pi \cdot f}\right)^{\frac{1}{3}}$$

where: V_b is the volume of the bubbles, f is the bubble frequency, Q_g is the gas flow
rate and d_b is the bubble size. The bubble frequency is measured using a pressure
transducer installed between the gas flow controller and the gas lance (also called
gas nozzle).

The foam life can also be defined as the time that it takes for the foam to collapse
to a reference height when the gas is turned off. In further experiments by Jiang and
Fruehan [127] they carried out two types experiments with different amount of slag
focused on bath-smelting. In the first group they employed a crucible made of alumina
of 45 mm diameter, 150 g of slag which was equivalent to an initial slag height of
40 mm, complying with the criterion defined by Bikerman, already explained. Argon
was injected using an iron pipe with a knife edged nozzle (2.1 mm ID) placed about
0.5–1.0 mm above the bottom of the crucible. The pick-up of alumina during the
experiments (30–60 min long) varied from 4 to 8%. A second group of experiments
at 1500 °C was carried out in an induction furnace using 1000 g of slag, using an
alumina crucible with 92 mm in diameter and 230 mm height. The nozzle diameter to
inject the gas had a diameter of 2.77 mm. Due to the larger dimensions in the second
group of experiments it was necessary to place a gas distributor at the bottom and
form a homogeneous foamy slag. The foam height was measured with an electric

probe and the bubble size observed using X-ray spectrometry, finding bubbles in the range from 5 to 20 mm. Since similar results were obtained with both groups experiments, they concluded the foaming index is independent of the container size.

Previously Jiang and Fruehan [127] investigated foaming of bath-smelting slags using large bubbles (20 mm), later on, Zhang and Fruehan [128] studied similar slags but with smaller bubble sizes from 1 to 15 mm. Since smaller bubbles could not be measured by X-ray due to poor images, its size was measured from the quenched slag foam after it was cooled down. The small bubbles were produced with a one-end-closed alumina pipe with four orifices on the sides, near the bottom, with less than 1 mm in diameter. The foam height was measured with an electric probe. Another group of experiments was carried out producing CO bubbles at the slag/metal interface with 40 g of hot metal containing Fe-C-S and 60 g of slag ($40\%CaO\text{-}40\%SiO_2\text{-}5\%FeO\text{-}15\%Al_2O_3$) at a temperature of 1450 °C. Sulfur in the range from 0.002 to 0.17% controlled the bubble size, decreasing the bubble size with a lower %S. Partial replacement of CaO with BaO improved image analysis using X-rays equipment. The foam height was measured with X-ray video photographic equipment, also estimated by the highest mark of the slag on the inner wall of the crucible after it was cooled down. The initial slag height was calculated by the liquid slag and hot metal densities and volumes. The estimated error in the measured foam height was about 10–15%. The production of CO was measured with a calibrated Matheson 8100 series digital flowmeter.

The experimental set up describing the electric probe method shown in Fig. 7.51 was employed by Hara et al. [124].

Corbari et al. [129] studied slag foaming at 1550 °C with a slag containing FeO and additions of different types of carbon. The foam height and production rate of CO were measured independently. The slags contained $CaO\text{-}SiO_2\text{-}FeO\text{-}MgO$ and spheres of different carbonaceous materials. The initial composition had a basicity of 2.0, 8%MgO and FeO was varied from 15 to 45%. FeO pellets were prepared reducing hematite with $CO\text{-}CO_2$ mixtures.

The foam height was measured using two molybdenum wire electrical probes (2-mm diameter). One probe was immersed in the slag and the second was a movable probe, which was first dipped until it touched the crucible bottom to define a reference position, then it was raised until it reached the top surface of the initial slag. At this point there was a flow of current when a constant voltage of 9 V was applied across the two probes. The initial slag height was the distance between the bottom of the crucible and the top slag surface before foaming. After foaming, the foam height was based on the difference between the measured height of the probe for the top of the slag and the initial slag height before foaming. The crucible was made of MgO. The furnace employed was a Lindberg blue box type resistance furnace capable to reach a maximum temperature of 1600 °C with a hot zone of about 40 mm. The furnace temperature was measured using a B-type thermocouple. Figure 7.52 (left) shows the electric probe method employed in this work.

The rate of production of CO was measured using a constant volume pressure increase (CVPI) technique. The furnace is sealed and pressure tested. A high accuracy pressure transducer PX811-030AV manufactured by Omega eng. USA was used

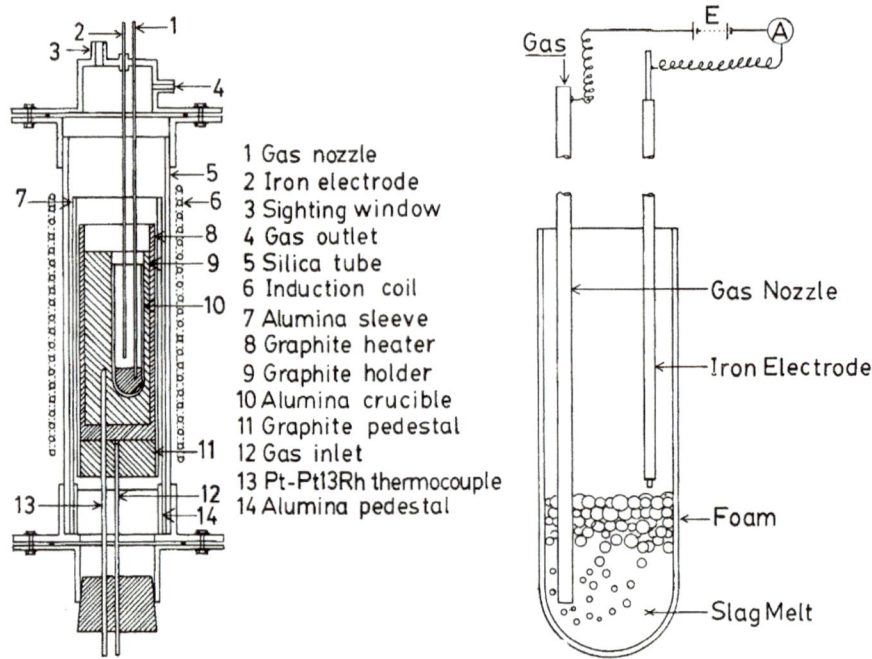

1 Gas nozzle
2 Iron electrode
3 Sighting window
4 Gas outlet
5 Silica tube
6 Induction coil
7 Alumina sleeve
8 Graphite heater
9 Graphite holder
10 Alumina crucible
11 Graphite pedestal
12 Gas inlet
13 Pt-Pt13Rh thermocouple
14 Alumina pedestal

Fig. 7.51 Foaming height measurement using the electric probe method. After [124]

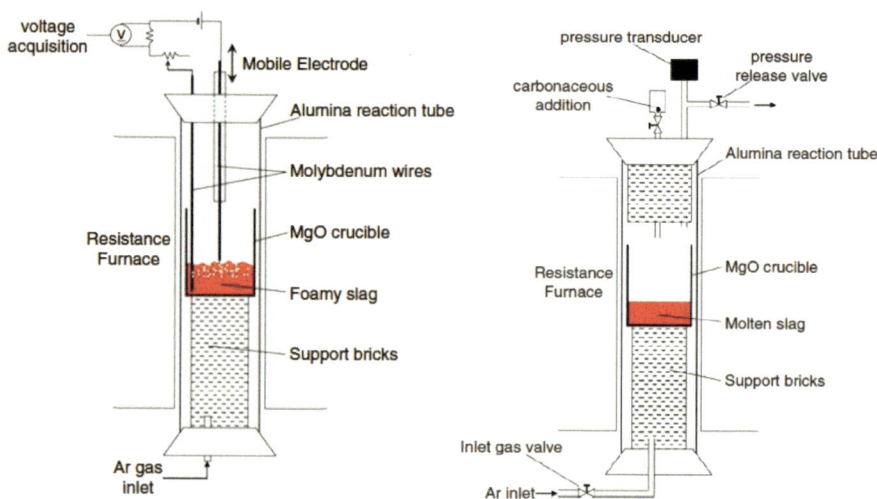

Fig. 7.52 Electric probe method (left) and CPVI technique (right). After [129]

to measure the pressure inside the reaction tube, the pressure increase was related the moles of CO generated. Slags were initially melted and then the carbonaceous material were added using a sealed valve. The moles of CO produced fairly agreed with the moles of carbon added. Figure 7.52 (right) shows the experimental arrangement.

Stadler et al. [113] measured the foaming index for acid CaO-SiO_2-FeO-Al_2O_3 slags with SiO_2 concentrations ranging from 41 to 53%, with a basicity range from 0.2 to 0.8 and total FeO from 20 to 28%. They employed a vertical tube furnace and alumina crucibles with a diameter of 45 mm and height of 280 mm. Argon gas introduced through an alumina pipe of 4.9 mm internal diameter. The slags were prepared using reagent grade oxides. Wüstite was prepared premixing Fe° and Fe_2O_3, pre-reacted as compacted pellets in a vertical tube furnace at 900 °C for 10 h in an argon atmosphere. The master slags were pulverized-remelted-pulverized and then mixed with wüstite. The final slag was remelted in a tube furnace under Ar atmosphere, allowing 8 h melting time before starting gas injection. The foam height was measured by the electric probe method.

Chang et al. [130] studied the effect of MgO on the foam life of CaO-SiO_2-MgO-$5Al_2O_3$-$30FeO$ slags at 1500 °C using a molybdenum crucible, 35 g of slags were added into the crucible and once the slag was melted the initial foam height was measured, then 1.2 g of carbon were added to the bottom of the slag through a Mo pipe. MgO was changed from about 4 to 13%. The slag was water quenched and analyzed. The slag viscosity was also experimentally measured in a separate group of experiments. The fraction of solid particles in the slags was estimated using Factsage.

Xiang et al. [131] studied foaming of blast furnace slags containing TiO_2. Iron ore from Panzhihua in China contains vanadium-titanomagnetite. These BF slags can have about 22% TiO_2 and 0.3% V_2O_5 with a basicity ratio about 1.1. These slags foam and create problems due to uneven gas distribution. Their experimental work included experiments at low temperatures to evaluate the foam height with the electric probe and the type of electrode as well as high temperature experiments. For the low temperature experiments, the authors used sodium bicarbonate, oxalic acid, glycerin and a surfactant. In this system CO_2 is internally generated. The authors reported that foam height has a linear relationship with the current, as shown by the following expression:

$$\frac{L}{L_0}\left[1 - \left(\frac{L - L_0}{L}\right)^{\frac{2}{3}}\right] = C\frac{I}{I_0} + D$$

where: L is the instantaneous foam height, L_0 is the initial slag height, C and D are constants, I is the current with foamy slag and I_0 is the initial current without foaming. The linear relationship was confirmed with the experimental results and also found a better correlation coefficient using a Mo sheet in comparison with a Mo wire, 0.92 and 0.68, respectively. The experiments at high temperature included two types of slags. Type A: (28–32%) CaO-(24–30%) SiO_2-13% Al_2O_3-7%MgO-(20–23%) TiO_2-(8–10%) C. Type B: include 15% FeO. The analytical grade reagents were compressed in the form of cylinders with a diameter of 40 mm, were dried for

8 h at 120 °C. 200 g of slag were placed in a graphite crucible (70 mm diameter, 500 mm height) and heated to 1500 °C, holding at this temperature for 3 h to ensure full melting. Argon was flushed during the experiments to prevent oxidation. 30 g of FeO were added when slag melting was completed. The initial height of slag was about 35 mm. FeO was prepared by heating hydrated ferrous oxalate ($FeC_2O_4 \cdot H_2O$) at 900 °C, according with the following reaction

$$FeC_2O_4 \cdot H_2O = FeO_{(s)} + H_2O_{(g)} + CO_{(g)} + CO_{2(g)}$$

7.4.3.3   X-Ray Imaging Apparatus

This technique is based on translucency differences of the materials, the denser material (metal) is more opaque in comparison with the foaming slag. X-Ray is a non-destructive technique that obtains images of opaque materials, therefore, allows the study in-situ at high temperatures. The X-Ray equipment includes an X-ray generator, a photon detector (receiver) a CCD camera and video equipment for data collection and processing. The furnace has aluminum windows that allows the passage of X-rays. The experimental set up is shown in Fig. 7.53. This technique was employed extensively by Japanese researchers who carried early studies on slag foaming [2, 132]. A disadvantage of this technique is a poor image quality [133]. With this method the foaming height can be measured as a function of time as well as the bubble size. Kapilashrami et al. [134] employed a graphite crucible (45 × 150 mm) with alumina-lined sides so that the reaction with the slag at 1550 °C occurred only at the bottom of the crucible. The slags investigated consisted of CaO-SiO_2-5%MgO-2%MnO-7%Al_2O_3-(5–15%) Cr_2O_3. Basicity kept constant at 1.5. The separate components were calcined and then pre-melted and quenched to obtain dense samples. The foam height was measured using images from a close-coupled device camera in the X-ray equipment, at time intervals of 30 s, using a scale inserted in the image, finally, the reported foam height was based on the average between the maximum and minimum values. Each experiment was terminated when the gas formation could no longer be observed. Slag samples were taken for chemical analysis using XRF. In their experimental work, the rate of reduction of FeO, the formation rate of CO and slag chemical composition were not constant, closer to what happens in the real system; hence, they identify their measurements with a dynamic foaming system, in contrast to steady-state conditions where those parameters are constant. As observed in Fig. 7.54, the slag foaming height is dynamic, it increases when the bubble formation is higher than the bubble burst rate and decreases when the bubble burst rate is higher than the bubble formation. It is clear that in a dynamic system there is a mis-match between the rate of formation and break up of gas bubbles. However, Kapilashrami et al. [134] results do not represent the real case because in their experiments the concentration of FeO decreases over time but in the real case due to continuous oxygen injection there is newly formed FeO, which sustain the foam life.

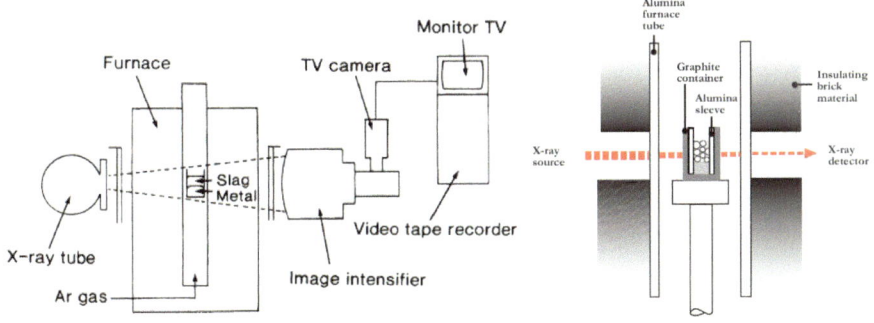

Fig. 7.53 Experimental set up for X-ray imaging. After [132] (left) and [133] (right)

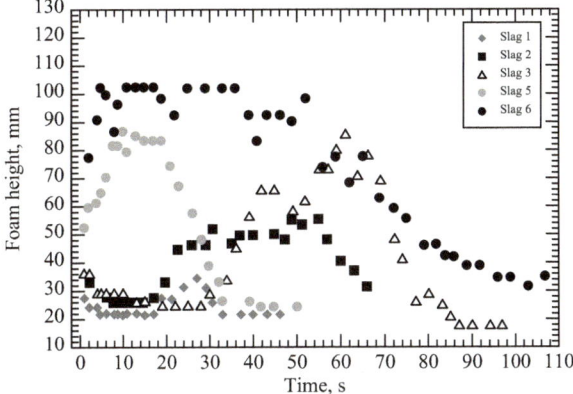

Fig. 7.54 Change in average foam height with time. Error bars indicate the fluctuations in foam height during the time interval of 30 s. Ref. [133]

7.4.3.4 Photoelectric Method

Koch and Ren [135] and Ren et al. [136] in 1994 reported a photoelectric method. This method continuously measures the volume of slag. The system consists of an alumina process tube (diameter 100 mm and length 1100 mm) with a quartz window in one of the water-cooled tube end caps. A video-camera records the foaming process. The video pictures are played back on a monitor after the end of the test. A change in slag volume is related with a change in brightness. The signals are converted by a photoelectric cell into voltage values. The foam volume is calculated from the voltage value. The procedure involves a calibration process before and after the test. Data processing assumed the projection area correspond to an ellipsoid or part of a sphere.

Two types of slags were prepared with a base composition that included CaO-SiO_2-Fe_2O_3, with additions of MgO in one group and alkalis in a second group. Fe_2O_3

ranged from 56 to 80%, basicity from 0.1 to 1.8, MgO from 0 to 10%, K_2O from 0 to 15% and Na_2O from 0 to 30%. The experiments were carried out in the range from 1250 to 1450 °C. The slag components were mixed, hydrostatically compressed into tablets of 3 g and pre-sintered for 3 h at 1100 °C under an argon atmosphere. Samples with alkalis were not pre-sintered due to risks of evaporation. The slags were placed on a solid substrate made of graphite. A reducing gas, $40\%CO$-$40\%CO_2$, was injected into the tube at a flow rate of 0.4 l/min. Figure 7.55 shows the experimental set up and Fig. 7.56 an example of their results.

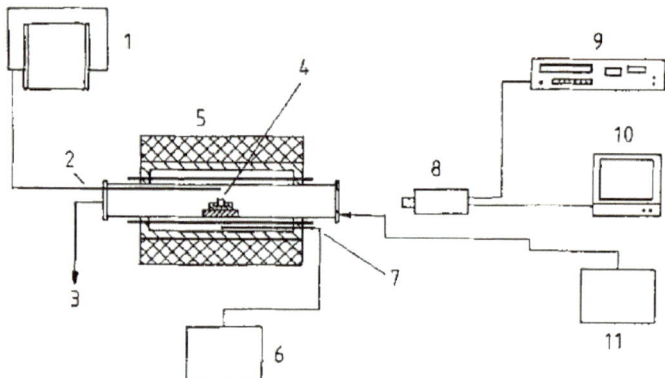

Fig. 7.55 Experimental set up for the photoelectric method; (1) plotter, (2) thermocouple, (3) exhaust gas line, (4) sample arrangement, (5) furnace, (11) gas supply. After [135]

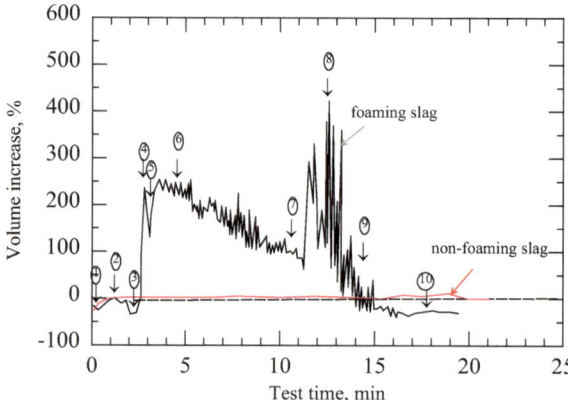

Fig. 7.56 Photoelectric method; (red line) no foaming slag, (black line) foaming slag. After [135]

7.4.3.5 Sessile-Drop Method

In 2007 Khanna et al. [137] reported the application of the sessile-drop method to study slag foaming. Figure 7.57 shows the experimental set up. The substrate was made of synthetic graphite with 99.3% fixed carbon. The weight of the slag was limited to 65 mg because with a higher mass of slag the evolution of gases showed a tendency to roll of the substrate. During one experiment the volume of the slag droplet decreases due to formation and release of CO. A numerical algorithm computes the actual volume of the slag droplet (V) and the ratio (V/V_0) is defined as a foaming ratio, shown in Fig. 7.58. Although this method provides information on slag foaming it use seems limited to fundamental aspects on slag foaming.

Tables 7.11 and 7.12 gives a summary of the experimental conditions in the previous research work on slag foaming. Table 7.11 provides details on the equipment

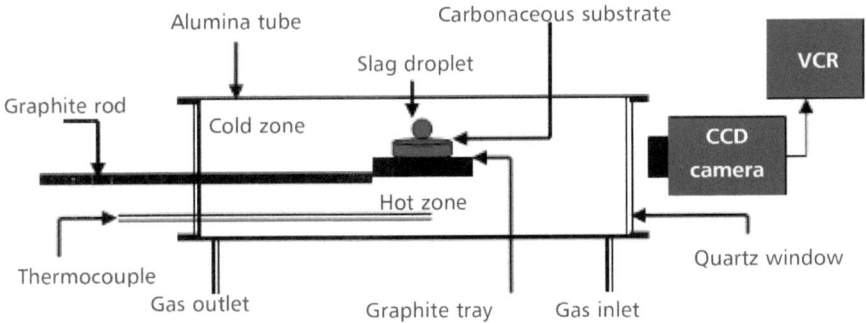

Fig. 7.57 Experimental set up of sessile drop method

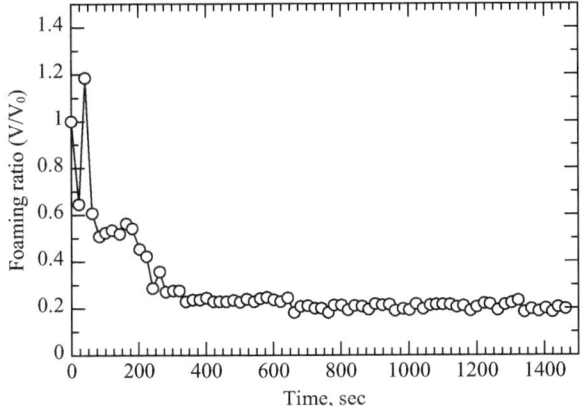

Fig. 7.58 Foaming ratio using the sessile drop method. After [137]

and Table 7.12 gives detailed information on the slag preparation to carry out the experimental work.

7.4.4   Slag Foaming Measurement at Both Low/High Temperatures

7.4.4.1   Microwaves Method

Malmberg et al. [145] reported a review on the use of microwaves in the steel industry. Microwaves are electromagnetic radiation which propagate through matter, changing phase and amplitude as well as polarization depending on the nature of the matter that it passes through. If the signal is transmitted and received by an interferometer in the aperture plane, a full three-dimensional structure of the volume can be reconstructed. This technology is generally insensitive to dust and fume and therefore it is well fitted for measurements in the harsh environment of steel shops. The principle of microwaves is the Frequency Modulated (FM) radar. The difference between radio waves and microwaves is that microwaves have a higher frequency, above 300 MHz and shorter wave length, approximately 10 m to 1 mm. Kobayashi et al. [146] and Maki et al. [147] in the early 1980s applied microwaves to predict foaming in the BOF. Ruuska et al. [148] reported accurate control of slag foaming in the BOF using microwaves. The position of the radar is illustrated in blue color in Fig. 7.59 and typical results on the height of steel and slag with this technique are shown in Fig. 7.60. In this figure it can be observed the rapid rise of slag height, about 1 m in 10 s.

Millman et al. [11] reported in detail the equipment and governing equations to design a control system using microwaves, which they called radio wave interferometry (RWI). The system requires one or more radio antennas. This equipment is sensitive to excessive heat and should be protected from heat radiation and metal splashing. It is placed over the hood and sealed with a ceramic window which should be transparent to the microwaves. Sakamoto et al. [149] in 1980 reported a patent to control the level of foaming slags in the BOF.

7.4.4.2   Acoustic Method

Sound has been employed in the BOF to study the slag's level since the 1970s [150]. The basic idea of a sonic meter is the attenuation of sound of the oxygen jet as the foaming slag increases its volume. The foaming height changes depending on the refractory wear, therefore is not an accurate indicator of the slag foaming height. Low noise is an indication of foaming conditions. To compare the acoustic measurements, heats should be grouped according to similar geometrical conditions. Heenatimulla et al. [151] recently reported a short review on the acoustic analysis of slag foaming

Table 7.11 Summary of experimental conditions to evaluate slag foaming

	Author	D_c, mm	H_c, mm	Crucible	T, °C	W_{slag}, g	h^o_{slag}, mm	Slag system	Method
1992	Kitamura [122]	50	180	MgO	1300–1400	180		C-S-F-A-M-M'-C'-T	Steel wire
1997	Paramguru [123]	100	150	Graphite	1500	500		C-S-F-F'-A-M	Iron rod
1998	Hong [88]	48	150	Alumina	1300	75, 90		C-S-F-A-M-L	Ni rod
2002	Ji [138–140]	191	295	MgO	1593–1745	3000–4000		C-S-F-A-M-C	Iron rod
1989	Ito [99, 125]	32–50	200	Alumina	1250, 1300, 1350, 1400		40–70	C-S-F-(P, C', S, M)	Electric probe
1991	Jiang [127]	45		Alumina	1500	150	40	C-S-F C-S-F-A-M	Electric probe
		92	230	Alumina	1500	1000		C-S-F C-S-F-A-M	Electric probe
1995	Zhang [128]	45		Alumina	1500			C-S-F-A	Electric probe
		45		Alumina	1450	60		C-S-F-A	X-Ray
1995	Zhang [141]	41–50	300	Alumina	1500	120		C-S-F-C'-C particles	Electric probe
1995	Ozturk [93]	41		Alumina	1450–1600	160	42	C-S-F-A	Electric probe
2009	Corbari [129]	44.5	127	MgO	1550			C-S-F-M	Electric probe
2021	Xiang [130]	70	500	Graphite	1500	230	35	C-S-F-A-M-T-C	Electric probe
2021	Chang [130]	45	110	Mo	1500	35		C-S-F-A-M-C	Electric probe
2023	Chang [142]	(45)	(110)	Mo	1550	35		C-S-F-M-A	Electric probe
2007	Stadler [112, 113]	45	280	Alumina	1136–1400	500		C-S-F-A	Electric probe
1994	Koch [135]			Graphite		3		C-S-F'-M C-S-F'-K-N	Photo-electric
2001	Kim [126]	45	150	MgO	1510			C-S-F-M-X	Electric probe
2006	Kapilashrami [143]	45	150	Graphite	1600	50		C-S-F-M'-A-Cr	X-ray

$C = CaO$, $S = SiO_2$, $F = FeO$, $F' = Fe_2O_3$, $A = Al_2O_3$, $M = MgO$, $M' = MnO$, $C' = CaF_2$, $T = TiO_2$, $L = Li_2O$, $S = S$, $Cr = Cr_2O_3$, $K = K_2O$, $N = Na_2O$

Table 7.12 Summary of slag preparation/experimental work on slag foaming

		Slag system	Slag preparation/experimental work
1	Kitamura [122]	CaO-SiO_2-Al_2O_3-P_2O_5-TiO_2-MnO-MgO-CaF_2	Once the hot metal was liquid, a mixture of oxides (180 g) without FeO was added. After 5 min FeO was added three times using paper bags in a period of time of 20 min that lasted each experiment. The first addition of FeO marked the beginning time for each experiment
2	Paramguru [123]	40% CaO-34%SiO_2-21%Al_2O_3-5%MgO	Slag components were mixed, pelletized (using molasses) and dried. Added to the crucible and melted in a plasma reactor. Then, FeO added as pre-reduced iron ore. Dissolved in 7 min. Only the crucible bottom reacted with FeO. An iron rod was employed to measure the foam height, dipping the rod every 3–4 min to take samples
3	Ji [138–140]	CaO (22–40%)-SiO_2 (7–15%)-FeO (31–43%)-MgO (7–20%)-Al_2O_3 (2–9%)	17 kg of ULC carbon steel melted, then 90 g of copper were added to identify iron droplets from steel and slag, then the slag was added (3–4 kg), in three batches and fully melted. Carbon particles were injected with nitrogen. Slags prepared mixing lime, mill scale, calcium aluminate and sand
4	Hong et al. [88]	(0.2–0.45) CaO-(0.3–0.5)SiO_2-(0.04–0.07) Al_2O_3-(0.1–0.6)Li_2O	Basic slag melted in a SiC resistance furnace, then FeO added and graphite electrode rotated. Initial FeO and S were varied in the ranges 3–15% and 0–0.68%, respectively. Nickel rod employed to measure foam height
5	Ito [99, 125]	CaO-SiO_2-FeO-(CaF_2, P_2O_5, S, MgO), particles	Slag melted in a resistance furnace. Argon introduced through a stainless-steel pipe (2.1 mm), foam height measured with electric probe. With a crucible diameter above 32 mm, its effect is eliminated. Foaming index using plot, height versus Ug

(continued)

Table 7.12 (continued)

		Slag system	Slag preparation/experimental work
6	Jiang [127]	CaO-SiO$_2$-FeO-Al$_2$O$_3$-MgO (45–30-FeO-15–10) FeO = 0 to 9%	FeO was prepared mixing Fe$_2$O$_3$ and pure Fe° stoichiometrically and sintering in an iron crucible under Ar gas protection at 1300 °C for 8 h
7	Zhang [128]	CaO-SiO$_2$-Al$_2$O$_3$-FeO	40 g hot metal and 60 g of slag were melted in an alumina crucible at 1450 °C. Foam height measured by X-rays. CO measured with a digital flowmeter
8	Zhang [141]	CaO-SiO$_2$-FeO-CaF$_2$-solid particles	The slag was melted and the foam height measured with an electric probe at 1500 °C with and without solid particles
9	Ozturk [93]	CaO-SiO$_2$-FeO-Al$_2$O$_3$	FeO was prepared reducing Fe$_3$O$_4$ in a CO-CO$_2$ gas mixture at 1000 °C. The slag was heated and stabilized at the desired temperature under argon atmosphere, then FeO was added
10	Zhang [144]	30%CaO-60%SiO$_2$-10%CaF$_2$	Slag melted in a resistance furnace. FeO avoided to eliminate reactions with the injected gases. Gases injected; Ar, H$_2$ and He
11	Corbari [129]	CaO-SiO$_2$-FeO-MgO Plus 5 types of C particles	Two groups of experiments. First one to measure CO generation rate using a pressure transducer. Second group measured dynamic foaming at constant temperature, slag basicity and 8% MgO A minimum of 1 h was allowed for the slag to melt, homogenize, and reach MgO saturation. FeO was varied from 15 to 45% Five different types of carbonaceous material were used
12	Kapilashrami [143]	CaO-SiO$_2$-(5–15%) Fe$_2$O$_3$-5%MgO-2%MnO-7%Al$_2$O$_3$-(5–15%) Cr$_2$O$_3$	The separate components were calcined and then pre-melted and quenched to obtain dense samples. The slag reacted with the crucible bottom

(continued)

Table 7.12 (continued)

		Slag system	Slag preparation/experimental work
13	Xiang [130]	(28–32%) CaO-(24–30%) SiO$_2$-13% Al$_2$O$_3$-7% MgO-(20–23%) TiO$_2$-(8–10%) C-15% FeO	Reagents mixed, compressed as cylinders, dried at 120 °C. 200 g added into crucible, melting completed in 3 h, then 30 g FeO added. FeO obtained by calcination at 900 °C of ferrous oxalate dihydrate. Foam height measured with electric probe using Mo sheet
14	Chang [130]	(31–45%) CaO-(16–23%) SiO$_2$-(4–13%) MgO-5Al$_2$O$_3$-30FeO	The slag components, 35 g, were added into a Mo crucible and once the slag was melted 1.2 g of carbon were added to the bottom of the slag through a Mo pipe. The foam life was measured
15	Chang [142]	40CaO-16SiO$_2$-FeO-5MgO-Al$_2$O$_3$	The oxides powders were melted at 1800 °C in a high-frequency furnace. FeO was produced by melting Fe$_2$O$_3$ powder at high temperature in a graphite crucible. 35 g of slag were melted and then 7 g of FeO added. When melted, 1.2 g of C were added for foaming
16	Stadler [112, 113]	SiO$_2$-Al$_2$O$_3$-CaO-FeO	Wüstite was prepared premixing Fe° and Fe$_2$O$_3$, pre-reacted as compacted pellets in a vertical tube furnace at 900 °C for 10 h in an argon atmosphere. The master slags were pulverized-remelted-pulverized and then mixed with wüstite. The final slag was remelted in a tube furnace under Ar atmosphere, allowing 8 h melting time before starting gas injection

(continued)

Table 7.12 (continued)

		Slag system	Slag preparation/experimental work
17	Koch [135]	SiO_2-Al_2O_3-CaO-Fe_2O_3-MgO SiO_2-Al_2O_3-CaO-Fe_2O_3-K_2O-Na_2O	The slag components were mixed, hydrostatically compressed into tablets of 3 g and pre-sintered for 3 h at 1100 °C under an argon atmosphere, then placed on a graphite crucible (with alkalis, not sintered). A reducing gas (CO-CO_2 injected into the tube furnace). Foam volume measured by photoelectric method
18	Kim [126]	CaO-SiO_2-FeO-MgO_{satd}-X (X = Al_2O_3, MnO, P_2O_5, CaF_2)	A master slag CaO–SiO_2–30FeO–X with V = 1.2 was MgO saturated in a MgO crucible for 8 h. Ar was injected through a pipe to produce the foam

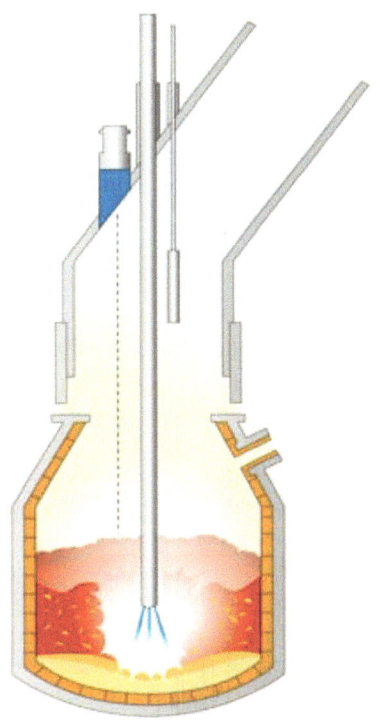

Fig. 7.59 Location of microwave radar in the converter. After [148]

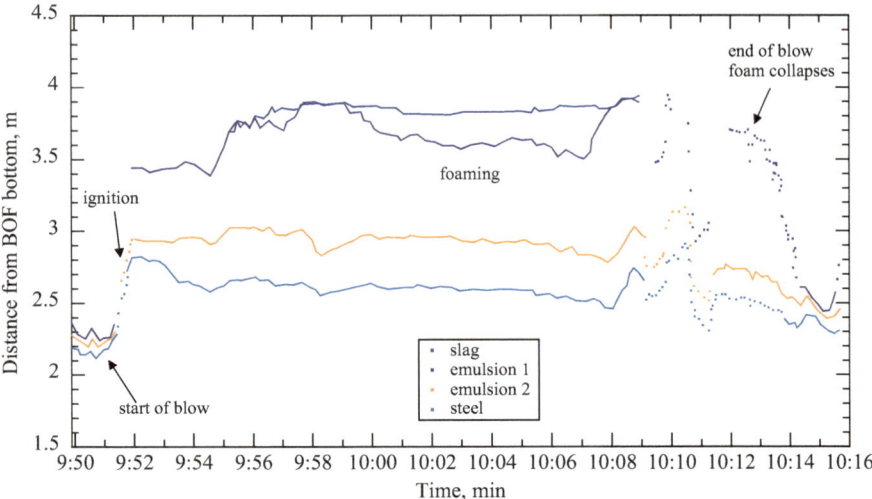

Fig. 7.60 Distance from BOF bottom (in m) as a function of time. After [148]

in the BOF. They concluded that this method successfully detects slopping but lacks accuracy to measure the slag height in real time.

de Vos et al. [152] carried out acoustic measurements of foamy slags in the BOF and attempted to find a correlation with previous foaming indices. Figure 7.61 shows two examples of two heats, showing that a foaming slag is formed at about 10% of the blow. In order to compare the acoustic measurements with foaming indices that are proportional to the slag's viscosity and inversely proportional to the product of the slag's surface tension times its density, they estimated these physical properties employing the final slag chemical composition. They found large changes in viscosity in comparison with the changes in density and surface tension; viscosities in the range from 10 to 30 mPa s, densities from 3.0 to 3.25 g/cm^3 and surface tension from 450 to 500 mN/m. The calculated foaming index was compared with the acoustic measurements during 10–30% of the blow. The reason to choose this range was because is in this range they found is critical for slopping. Figure 7.62 shows the relationship between the measured acoustic values and the calculated foaming index, observing a relatively poor relationship. Out of 5 groups, four groups gave a correlation coefficient below 0.44. The authors concluded that the foaming indices evaluated cannot be employed to predict foaming behavior of BOF slags arguing that those indices were obtained for homogeneous liquid slags using an inert gas, in contrast to the real case where the slag is not homogeneous and the gas is not inert. Most probably there are more reasons for the poor relationship they reported, for example, they didn't report the variation in slag chemical composition and temperature during the same period where the acoustic measurements were made. They assumed the temperature rise of liquid steel was linear and also the temperature of slag was the same for liquid steel.

Fig. 7.61 Acoustic pattern in two heats. After [152]

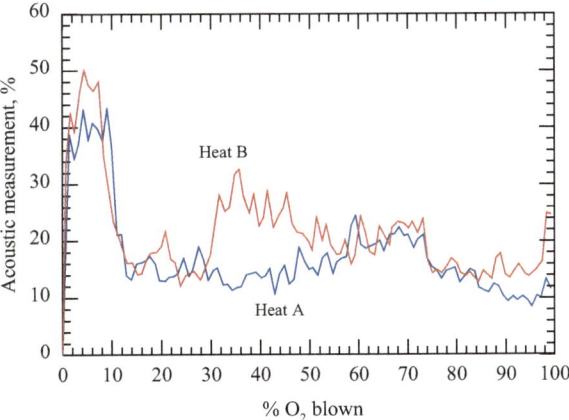

Fig. 7.62 Relationship between the measured acoustic values and the calculated foaming index. After [152]

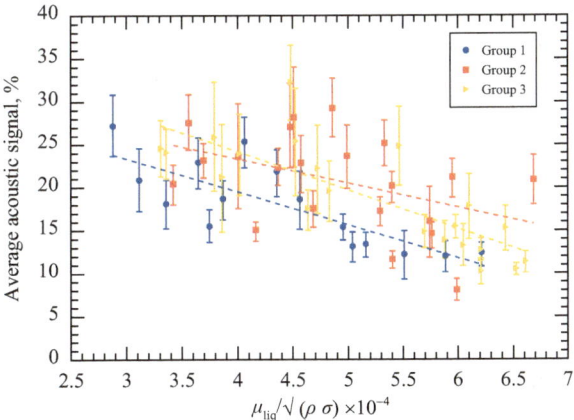

Deo et al. [153] indicted that the acoustic signals are extremely sensitive to changing conditions in lance height and rate of additions having a success rate of 70% or higher to control slopping, however leaves very little time to control it.

The acoustic method has also been applied in the EAF for slag foaming control. In 1978 Baumert [154] patented a method to measure the slag thickness using acoustic measurements. In 1998 Buydens et al. [155] from CRM, reported results of foamy slag control using acoustic measurements in a DC-EAF of 100 ton with EBT and bottom gas injection, charged with two scrap baskets through the roof. Oxygen and carbon injection were 23 Nm^3/ton and 6.15 kg/ton, respectively. Figure 7.63 shows the amplitude of the acoustic measurements as a function of frequency (Hz) for two conditions; (a) end of the first basket when the foamy slag is not yet formed and (b) 5 min after the beginning of foaming. Foaming sharply decreases the signals in the range from 350 to 400 Hz. This was the range selected to evaluate slag foaming.

Fig. 7.63 Preliminary trials to choose frequency range. After [155]

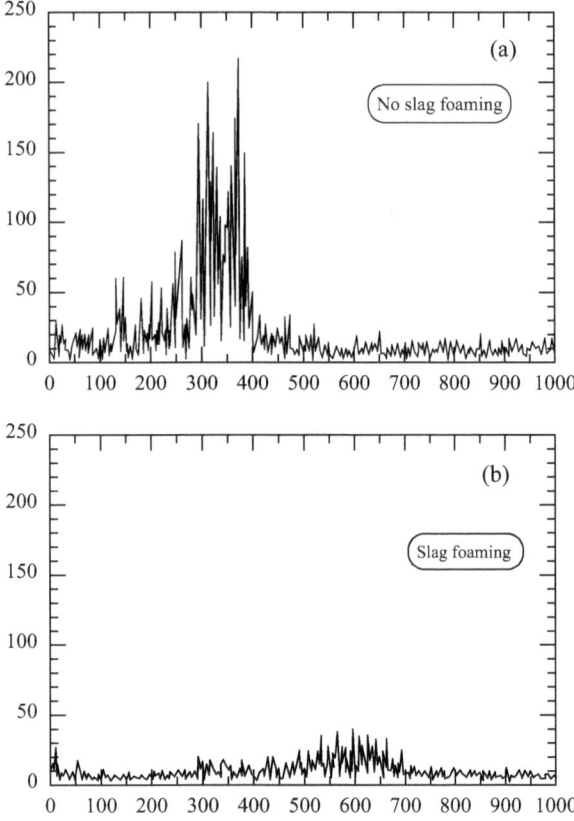

Figure 7.64 shows examples of two cases, for good and bad foaming conditions. It can be observed that a foamy slag is formed as soon as carbon particles are injected.

Dicker [156] reported the application of the acoustic method in the EAF. The system employed was a Rode® videomic directional video condenser microphone with a frequency range from 40 to 20,000 Hz and a maximum sound pressure level of 134 dB. The microphone was mounted unto a camera and positioned 10 m away from the furnace door. Sound data analysis was made using the software SignalScope Pro 1.8.4, which uses Fast Fourier Transformation. It was found that the primary audio frequencies emitted by the furnace were 100, 200 and 300 Hz.

7.4.4.3 Vibrations Method

An accelerometer is a device that converts the vibration energy into an electrical signal that is proportional to the momentary acceleration of the object. This method has been applied in the BOF since the 1970s [157, 158]. In the 1980s, Mucciardi and Barrera [159, 160] described in more detail its operation and application in the BOF,

Fig. 7.64 Examples of acoustic measurements for two cases; **a** good foamy slag, **b** poor slag foaming. After

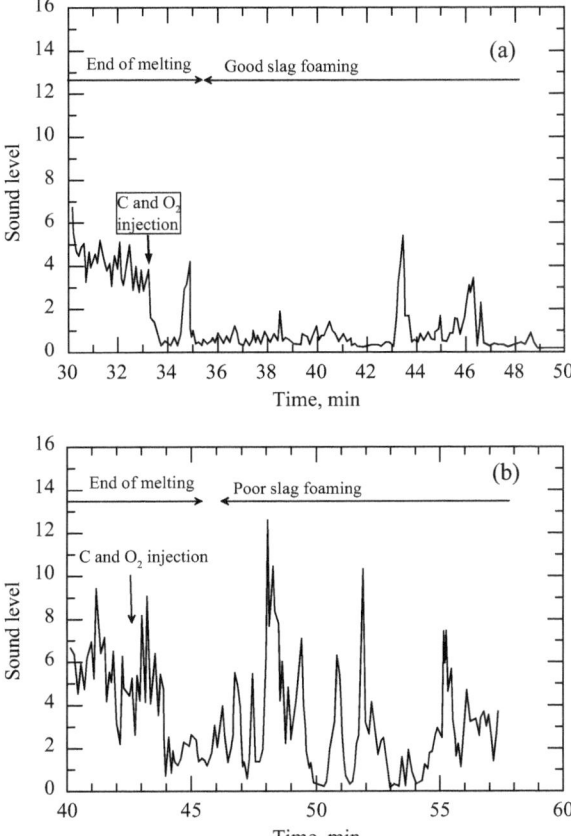

reporting the following relationship between stirring energy and the accelerometer signal, for molten lead:

$$\varepsilon = 148.5 Acc^{1.53}$$

where ε is the stirring power in W/ton and Acc is the slope of the integrated accelerometer signal. Further studies by the same research group, carried out measurements on mixing time. With this additional information the following relationship between mixing time and vibrations was proposed:

$$\sigma = 61.1 Acc^{-0.56}$$

Jeong et al. [161] reported in more detail the operation of an accelerometer and data processing using Fast Fourier Transform (FFT). They employed a piezoelectric vibration sensor with an operation frequency in the range from 5 to 10,000 Hz. Data processing includes data extraction with FFT and application of least mean squares to define the foam height.

Since 2005, Siemens (currently Primetals), carried out investigations to control slag foaming using vibrations coupled with an electrode control regulation system and an individual control of carbon injection through different lances [162–164]. Three accelerometers were placed on the furnace shell, opposite to the electrodes, as shown in Fig. 7.65. The signals are processed and the slag height estimated for each electrode. The electrode regulation system and carbon injection will be adjusted to entirely cover the electric arcs. Figure 7.66 shows that the total vibrations include air-borne sound and structure-borne sound. The supplier has reported benefits in terms of lower consumption of carbon particles and faster melting rate.

Heo and Park [165] in 2019 reported industrial measurements on the foam height from two Korean EAF's of 70 and 140 ton using the vibrations method. The slags were divided into two groups, A for fully liquid slags and B with MgO saturation and assumes solid precipitates, their average chemical composition is shown below:

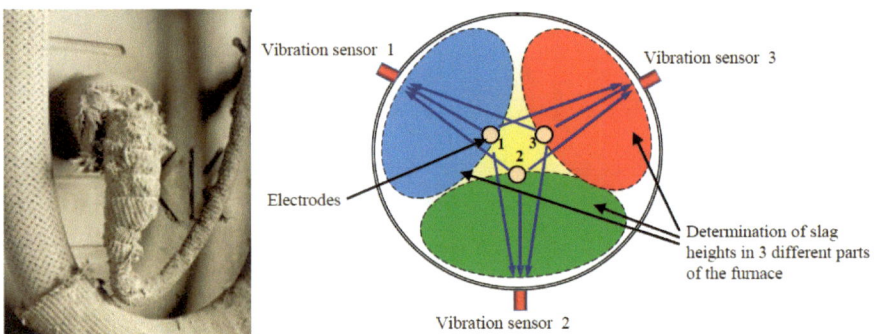

Fig. 7.65 (left) Accelerometer, (right) position of accelerometers. After [163]

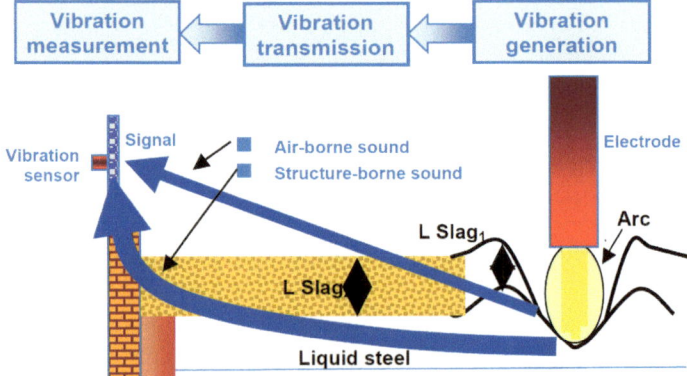

Fig. 7.66 Measurement principle with an accelerometer. After [164]

	SiO$_2$	Al$_2$O$_3$	FeO	CaO	MgO	MnO	P$_2$O$_5$	C/S
A	17.4($\pm$1.2)	15.7($\pm$1.5)	24.6($\pm$4.1)	24.7($\pm$3.4)	7.5($\pm$1.3)	9.9($\pm$0.8)	0.18($\pm$0.06)	1.42($\pm$0.2)
B	20.3($\pm$1.5)	8.4($\pm$0.7)	17.0($\pm$3.9)	32.1($\pm$2.9)	10.9($\pm$1.2)	11.0($\pm$0.5)	0.38($\pm$0.06)	1.59($\pm$0.2)

The weight of slag A was estimated as 7 ton and slag B as 14 ton. The initial slag height without foaming was computed with the initial mass, considering a furnace diameter of 5 m and a slag density of 3.63 ton/m^3. The height for the slag with 7 tons was 10.4 cm.

The foam height was measured with three vibration sensors installed in the outer wall of the furnace, the vibrational frequency was obtained from the voltage and current signals monitored in real-time. The source of errors came from a hetero-geneous slag composition, differences in temperature, differences in foaming in different regions of the furnace, etc.

The calculated foam height was obtained from the foaming index (Σ) reported by Kim [126] and the superficial velocity (U_g) suggested by Ito and Fruehan [166], inserting the reduction rate constant (k_r) reported by Sugata [167]:

$$\Sigma = 999 \frac{\mu}{\sqrt{\rho\sigma}}$$
$$U_g = \frac{Tk_r}{273} \cdot \frac{S}{A}$$
$$k_r = 2.88 \exp\left(-\frac{39700}{RT}\right) a_{FeO}$$

where: Σ is the foaming index in seconds, U_g is the superficial gas velocity in cm/s, S is the interfacial area between slag and solid carbon in cm^2, A is the cross-sectional area of the EAF, and k_r is the reduction rate constant for FeO with solid carbon in mol FeO/cm^2·s. The apparent viscosity was calculated with the Einstein-Roscoe equation. The ratio S/A was taken as 20 in the calculations.

$$H_{foam} = \Sigma \cdot U_g$$

Comparison between experimental and calculated values indicated an underpre-diction with the previous equations, for example a real value gave 50 ± 8.6 cm and the calculated value gave 10.4 cm. High discrepancies were expected due to different factors, such as the heterogeneous nature of the slags.

7.4.4.4 Statistical Method

Lee et al. [168] proposed a step-wise multiple regression method to predict the foam height, however the regression equation was not reported. They employed experimental data from 217 heats produced in a DC-EAF of 140 ton. The foam height was experimentally measured instantaneously through each heat using a vibration method. According with their authors the model predictions are quite satisfactory with an average $R^2 = 0.82$. Some of the reported data indicate no slag foaming during the first 25 min, increasing to 10 cm after that and remaining at that level for 25 more minutes and only from minute 66 to minute 83 the foam height reaches 40 cm. The arc length is not indicated but those values would indicate very bad foaming conditions for most part of the heat reported.

The following sections from 7.5 to 7.12, discuss in detail the effect of a large number of variables. It will be shown that the same variable can have opposite effects depending on the actual set of experimental conditions.

7.5   Effects of Gas Phase on Slag Foaming

7.5.1   *Void Fraction*

Gou et al. [169] developed a multiphase fluid dynamic model to calculate the void fraction as a function of the superficial gas velocity in order to understand the differences in slag foaming at low and high superficial gas velocities. In a previous review on bubble columns, Shah et al. reported the two best reliable correlations to estimate the void fraction, however those are valid for superficial velocities lower than 0.4 m/s. Gou et al. derived an expression to estimate the void fraction at high superficial gas velocities, above 1 m/s, valid for bath smelting operations. Figure 7.67 indicates that non-foaming aqueous systems have small void fractions and small superficial gas velocities, from 0.01 to 0.40 m/s. In bath smelting the superficial gas velocity is higher than 1 m/s, which gives a higher void fraction, from 0.7 to 0.9. One important difference between the foams formed at high superficial gas velocities and foams formed at low superficial gas velocities is that in the first case the foam collapses as soon as the gas is stopped and in the second case the foam remains some time depending on its stability. Zhu et al. [75] reported that most of the results on slag foaming at laboratory scale have been conducted at low superficial gas velocities, below 0.1 m/s, in comparison to the higher values in the industrial practice; 0.1–0.7 m/s for the EAF and 0.3–3 m/s for bath smelting.

Fig. 7.67 Void fraction as a function of the superficial gas velocity. After [169]

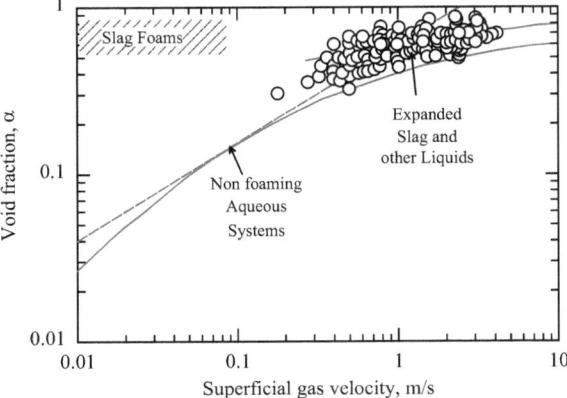

7.5.2 Minimum Superficial Gas Velocity for the Onset of Foaming

Hong et al. [88] reported that the slag didn't foam below a critical gas flow rate. They employed 75–90 g of a primary slag composed of Li_2CO_3, $CaCO_3$, SiO_2, Al_2O_3 and FeO in an alumina crucible (48 × 40 × 150 mm) placed in a resistance furnace. The argon flow rate was varied from 20 to 500 Nml/min, using an alumina pipe, immersed in the slag at a temperature of 1300 °C. FeO and S were changed from 3–15% and 0–0.68%, respectively. A graphite rod was rotated inside the slag at a speed of 1.67 rps to improve homogenization of the gas. The results shown in Fig. 7.68 indicate that foaming starts after a critical gas flow rate is reached, in this case about 3–7 Nml/s.

Pilon and Viskanta [170] developed a model to estimate the minimum superficial gas velocity, which marks the onset of foaming. Figure 7.69 shows three stages corresponding to non-foaming, onset of foaming and fully developed foam. This model is based in the one-dimensional drift-flux model for gravity driven flow with no wall shear developed by Wallis in 1969 and a drift-flux velocity derived by Ishii and Zuber in 1979.

The final expression to estimate the minimum superficial gas velocity, for viscous flow, is given by the following expression:

$$U_g^m = U_t \theta (1 - \theta)^2$$
$$U_t = \frac{2}{9} \frac{(\rho_f - \rho_g) g r^2}{\mu_f}$$

where: U_t is the terminal velocity calculated based on the Stokes's equation. In order to estimate the minimum superficial gas velocity, it is necessary to estimate the value of the void fraction. Figure 7.70 shows one example of the calculated gas velocity. The points correspond to experimental data.

Fig. 7.68 Effect of gas generation on the average void fraction. After

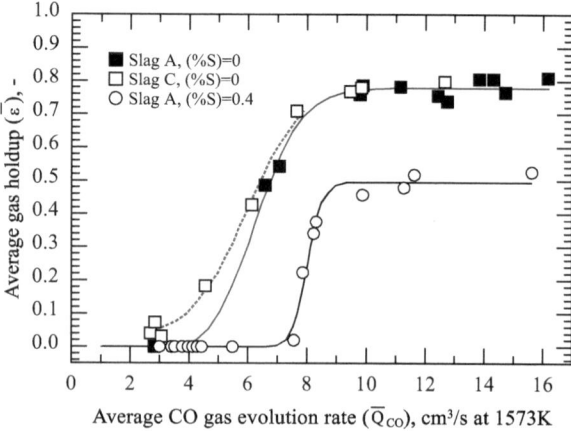

Average gas holdup ($\bar{\varepsilon}$), -

Slag A, (%S)=0
Slag C, (%S)=0
Slag A, (%S)=0.4

Average CO gas evolution rate ($\bar{Q}_{CO}$), cm³/s at 1573K

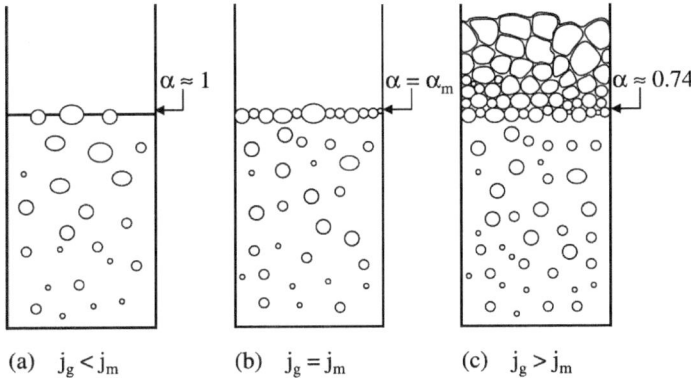

(a) $j_g < j_m$ (b) $j_g = j_m$ (c) $j_g > j_m$

Fig. 7.69 Foaming stages: **a** non-foaming, **b** onset of foaming, **c** developed foam. After [170]

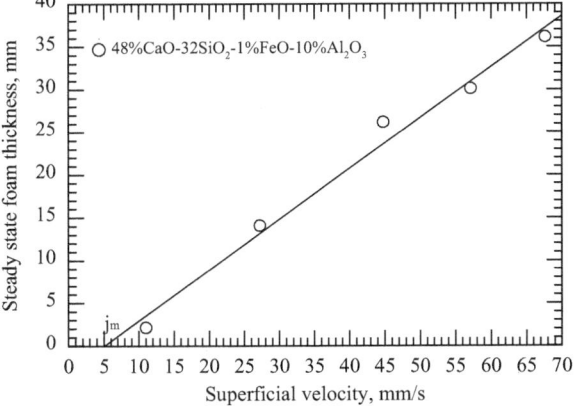

Fig. 7.70 Example of calculation of the minimum superficial gas velocity. After [170]

Hara et al. [124] reported the effect of the gas flow rate on the foam height for CaO-SiO$_2$-FeO slags at 1250 °C. The results are shown in Fig. 7.71. The numbers in those figures correspond to the slags shown in Table 7.13. It can be observed an increment in foam height as the gas flow rate increases.

7.5.3   Effect of Gas Phase Composition on Foaming

Hara et al. [124] reported that a small addition of hydrogen into argon increased the foaming height in slags containing 30%FeO, as shown in Fig. 7.72.

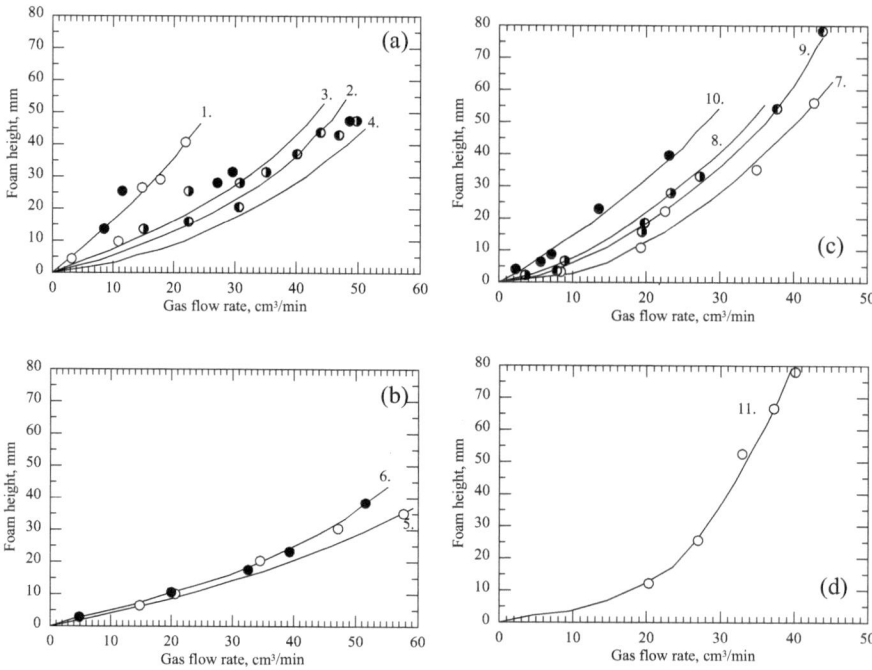

Fig. 7.71 Effect of gas flow rate on slag foaming. After [124]

Kitamura et al. [171] compared Argon, CO and CO$_2$ and found higher foaming conditions employing CO and CO$_2$, in comparison with argon, especially at higher gas flow rates, as can be seen in Fig. 7.73.

Zhang and Fruehan [144] compared argon, hydrogen and helium. They employed a slag containing 30%CaO-60%SiO$_2$-10%CaF$_2$. They excluded FeO to avoid any chemical reaction with the gas. Figure 7.74 shows the results. It is shown that the gas phase composition has an influence on the foaming height. Argon has better foaminess in comparison with helium and hydrogen, in this order. The effect of hydrogen is opposite to that reported by Hara et al. who indicated better foaming conditions adding small traces of H$_2$ to argon. Comparing the foaminess ratio, the ratio Ar/He is 2.19 and Ar/H$_2$ is 3.6. In comparison with argon, He decreases the foaming index by 20% and hydrogen by 70%. One difference among these gases is its density. The density ratio Ar/He and Ar/H$_2$ is 1.78/0.178 = 10 and 1.78/0.09 = 20, respectively. Another difference is the ratio of viscosities. At 1400 °C, the ratio is 1.2 and 2.7, respectively. It seems that is the gas phase viscosity the reason for the different results. Luz et al. [79] also indicated that the order of the effective diffusivity is H$_2$ < He < Ar, suggesting an additional reason for the observed behavior.

According to Zhang and Fruehan, Ogino et al. reported that the foam height of slags containing FeO was increased when the bubbling gas was used in the following

Table 7.13 Chemical composition of slags investigated by Hara et al. [124]

	FeO	SiO$_2$	CaO	Na$_2$O	MgO	CaF$_2$
1	65	35				
2	55	30	15			
3	70	30				
4	40	30	30			
5	80	20				
6	60	25	15			
7	30	45	25			
8	40	40	20			
9	45	40	15			
10	40	45	15			
11	26.5	41.9	32.5			
12	60	30		10		
13	50	30		20		
14	40	30	20	10		
15	40	30	20		10	
16	30	30	30		10	
17	30	30	30			10
18	20	30	30			20

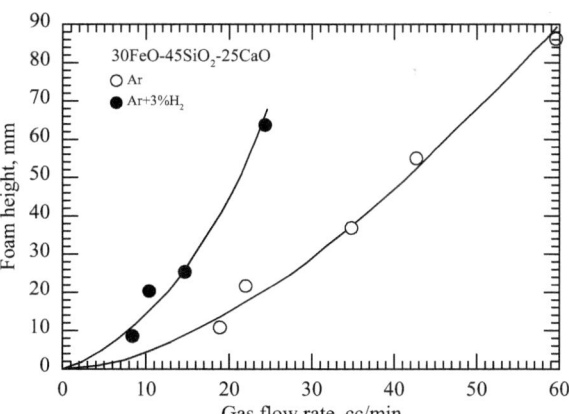

Fig. 7.72 Comparison of two gases on slag foaming. After [124]

order; Air, Ar, N$_2$-7%CO and this effect was attributed to a decrease in the bubble size.

It has been suggested that the previous linear relationship only holds for high temperature systems [79].

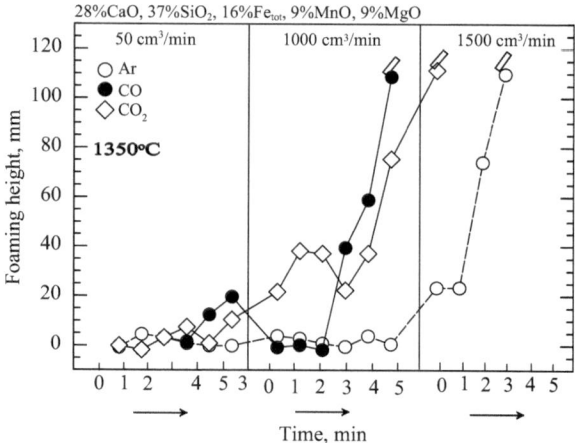

Fig. 7.73 Comparison of Ar, CO and CO₂ on slag foaming. After [171]

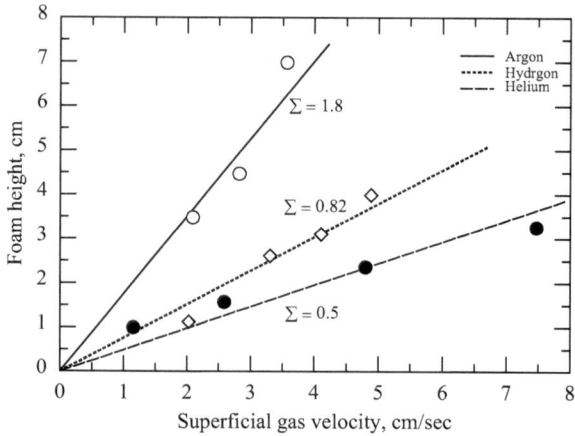

Fig. 7.74 Comparison of different gases on the foam height art 1500 °C. After [144]

7.5.4 Effect of Gas Pressure on Foaming

Zhang and Fruehan [144] compared slag foaming at 1.1 and 1.9 atmospheres. They employed X-rays to measure the bubble size and foam height at the higher pressure. The results are shown in Fig. 7.75. It can be observed that increasing the pressure from 1.1 to 1.9 atmospheres has no effect on slag foaming.

Ren et al. [136] reported different results. They compared foaming for 1 and 2 bars (almost 1–2 atm) using a photoelectric method and found that increasing pressure, also increased the foam volume, by about 150%. The slag employed had a basicity of 1 and 80% Fe_2O_3. This result is particular to the experimental conditions in this

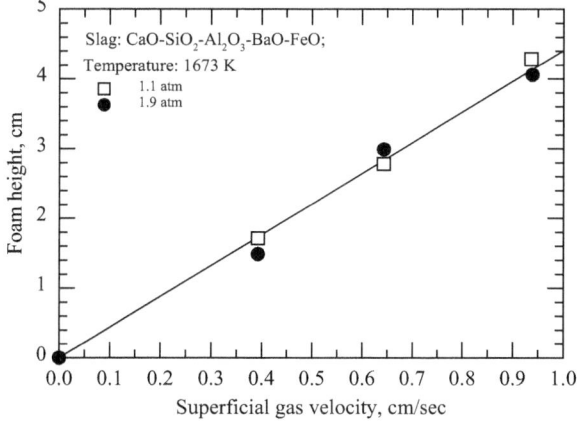

Fig. 7.75 Effect of gas pressure on slag foaming. After [144]

work because a reducing gas, CO, is passed through the reactor and increasing the pressure increases the kinetics for the reaction between CO and FeO in the liquid slag.

7.5.5 Critical Crucible Diameter

Ito and Fruehan [99] investigated the effect of the crucible diameter, from 25 to 50 mm, on slag foaming. Their results are shown in Fig. 7.76. They found that the foaming height was the same if the crucible diameter was above 32 mm, indicating that above this value the crucible diameter has no effect on the foaming index. Furthermore, it can also be observed that above a given flow rate or its equivalent velocity, the slope remains constant. This velocity was reported at about 1.0 cm/s. Ozturk and Fruehan [93] obtained a linear relationship between 0.5 and 1.5 cm/s.

7.6 Effect of Slag Chemical Composition on Slag Foaming

7.6.1 Effect of FeO

The following results will show the effect of FeO on the foaming index; however, it is important to mention that most of this work has been carried out without the injection of carbon particles, which is the common way to promote foaming. In a subsequent section it will be shown the importance of FeO to improve the foaming

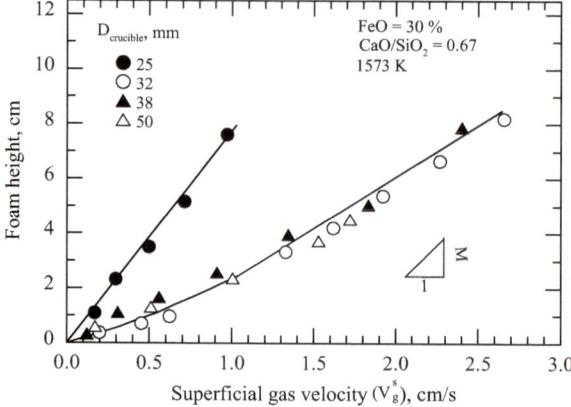

Fig. 7.76 Effect of gas flow rate and crucible diameter on foam height. After [99]

conditions, as long as there is enough carbon particles for its reduction and keep its concentration under control.

Buydens et al. [155] employed a foaming index $\left(\Sigma = 115 \frac{\mu}{\sqrt{\rho \sigma g}} \right)$ derived by Fruehan to evaluate the effect of iron oxide. Figure 7.77 shows that increasing FeO decreases the foaming index.

Jung and Fruehan [127] in 2000 reported the effect of iron oxide on the foaming index. Figure 7.78 includes their results as well as previous results by the same research group. It is clearly shown that increasing FeO decreases the foaming index. This effect is the result of a drastic decrease in the slag viscosity as FeO is increased, as shown in Fig. 7.79. For the slags investigated in their work the viscosity at 1440 °C (1713 K) sharply decreases when the iron oxide is increased from 0 to 20% but remains relatively constant when FeO is in the range from 20 to 30%.

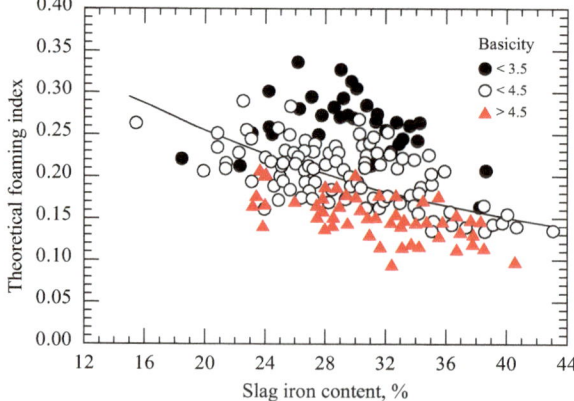

Fig. 7.77 Effect of iron oxide on the foaming index. After [155]

Fig. 7.78 Effect of iron on the foaming index. After [127]

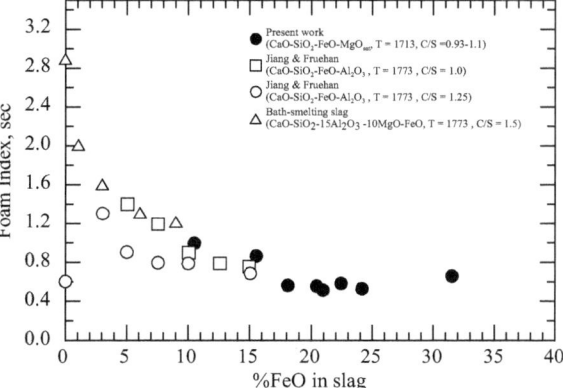

Fig. 7.79 Effect of iron on the slag's viscosity. After [127]

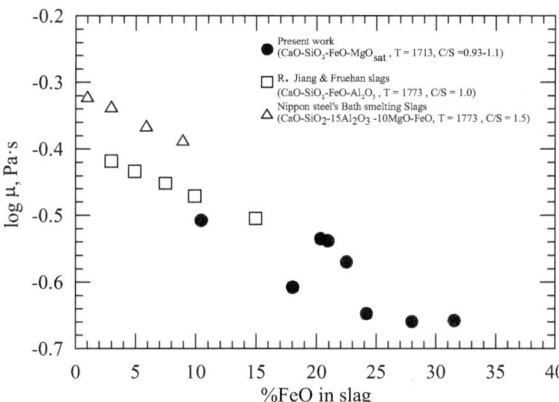

Luz et al. [79] made additional remarks to the effect of FeO on slag foaming. They noticed the equation for the rate of drainage is inversely proportional to the slag's viscosity and proportional to the density and bubble diameter;

$$t_D = \frac{\rho d_b^2 (1 - \theta)}{\mu}$$

where; t_D is the drainage rate in s^{-1}, ρ is the slag density, d_b is the bubble diameter, μ is the slag viscosity and θ is the void fraction. FeO decreases slag foaming not only due to a decrease in slag viscosity but also FeO increases the slag density. Figure 7.80 shows that FeO is an oxide with a higher density.

Bhoi et al. [172] found a linear decrease in the foaming index as FeO increased from 20 to 40%, as shown in Fig. 7.81. The foam height was measured dipping a mild steel rod of 12 mm in diameter at intervals of 2 min. The slags were prepared using mixtures of limestone, quartz and iron ore fines.

Fig. 7.80 Densities of various oxides. After [79]

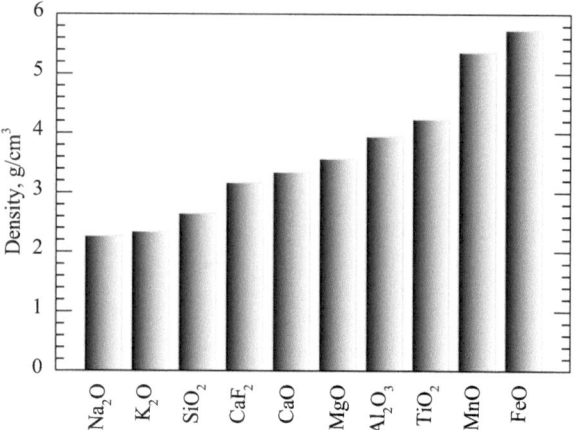

Fig. 7.81 Effect of FeO on the foaming index. After [172]

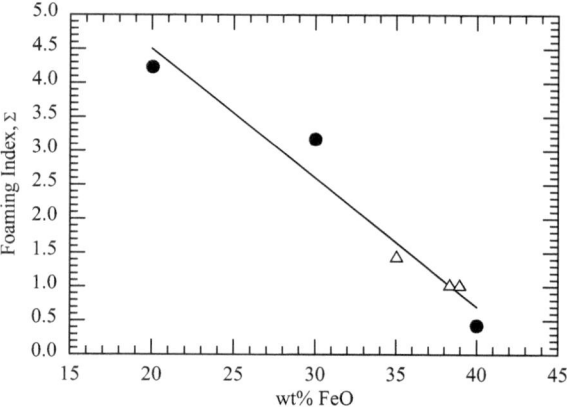

Jones et al. [173] reported a relationship between oxygen injection and iron oxide on the foam height from a work carried out by the center for materials production, shown in Fig. 7.82. According with this plot there is a linear relationship between oxygen injection, FeO and foam height. The foam height is increased by increasing the gas flow rate of oxygen and decreases if FeO is increased above 20%.

Hara et al. [124] in 1983 reported an extensive investigation on slag foaming of $CaO-SiO_2-FeO$ slags with high concentrations of FeO, from 20 to 80%. Some slags contained Na_2O, CaF_2 and P_2O_5. The chemical composition is shown in Table 7.13.

The addition of Na_2O, CaF_2 and P_2O_5 increased both foam life and foam height. These oxides all have in common that they decrease the slag surface tension. Figure 7.83 shows the direct relationship between foam height and foam life. Slags that form foams more stable also give higher foam heights. Figure 7.84 defines in a ternary diagram the foam height and foam life of the slags investigated, at 1250 °C,

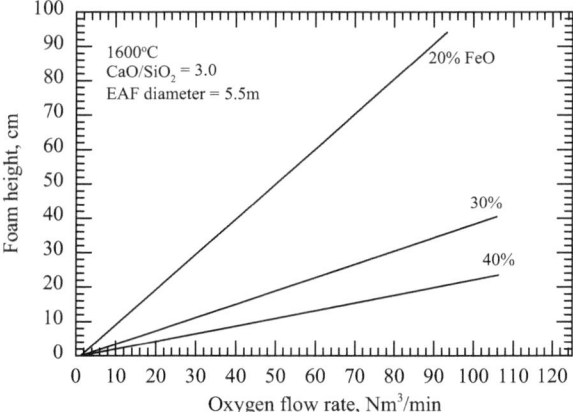

Fig. 7.82 Effect of oxygen flow rate and FeO on the foam height. After [173]

using a flow rate of argon of 30 cm^3/min. It can be observed that increasing SiO$_2$ increases both foam height and foam life. FeO has the opposite effect.

Kim et al. [126] investigated the effect of different components on slag foaming. They studied master slags containing CaO-SiO$_2$-30FeO-MgO$_{sat}$ (CSFM) with additions of MnO, Al$_2$O$_3$, CaF$_2$ and P$_2$O$_5$. They measured the foam height using the electric probe method. The crucible was made of MgO with 45 mm in diameter and 150 mm in height. Argon was introduced through a tube made of alumina with an internal diameter of 2 mm. The foam height was defined for the stable foam, reached after 10 min at 1510 °C. The foaming index was defined using the expression reported by Zhang and Fruehan.

Figure 7.85 shows the effect of FeO, Al$_2$O$_3$ and basicity on the foaming index with CSFA slags. The black symbols correspond to their experimental results and the hollow symbols correspond to Fruehan's group. The foaming index decreases when FeO increases from 0 to 20% and remains almost constant from 20 to 40%. Their

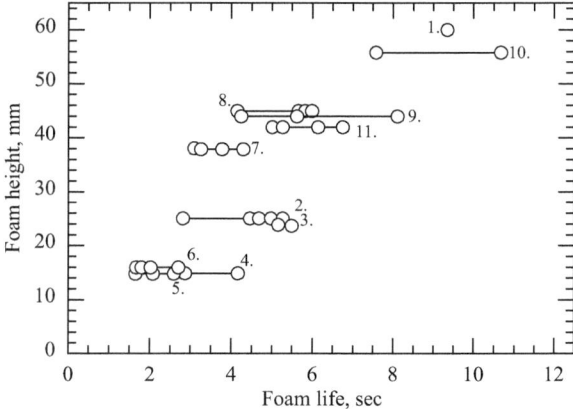

Fig. 7.83 Relationship between foam height and foam life. The numbers correspond to the slags shown in Table 7.13. After [124]

Fig. 7.84 Foam height (top) and foam life (bottom) for the slags indicated in Table 7.13. After [124]

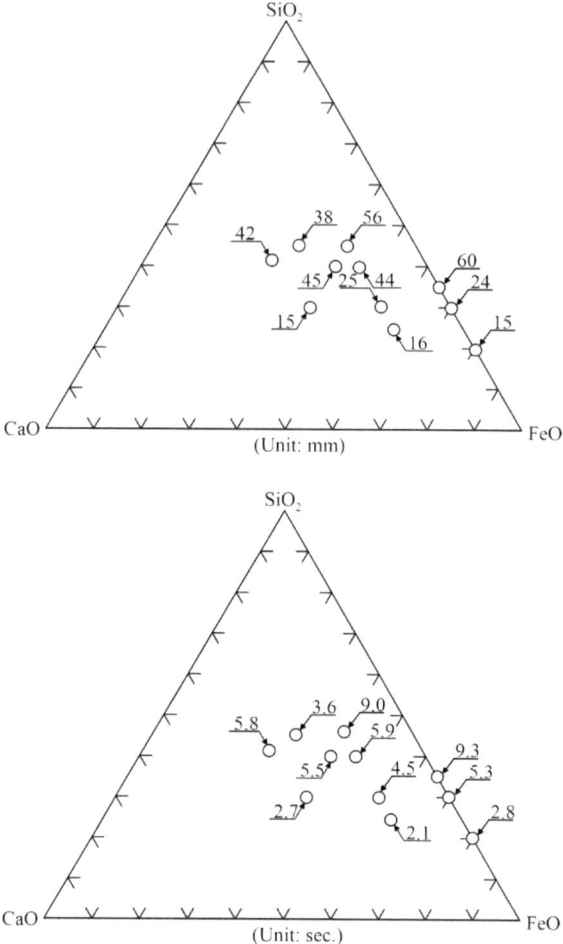

measurements on slag viscosity indicated that increasing FeO decreased the slag viscosity with a higher decrease rate from 0 to 10% and remained almost constant from 20 to 40%, therefore, a decrease in viscosity was responsible for the decrease in the foaming index. Their results also show that decreasing basicity from 1.2 to 0.93 increases the foaming index. It should be noticed that their slags were free of MgO.

Figure 7.86 shows completely different results when the previous slags involve MgO. The system involves CSFAM slags. Due to the presence of MgO, increasing FeO from 10 to 40% also increases the foaming index. This effect also increases by increasing Al_2O_3 from 10 to 30%. Since the addition of MgO increases the slag basicity the authors speculated this effect to FeO and Al_2O_3 acting as acid components that increase slag viscosity due to polymerization, furthermore, the MgO_{sat} value

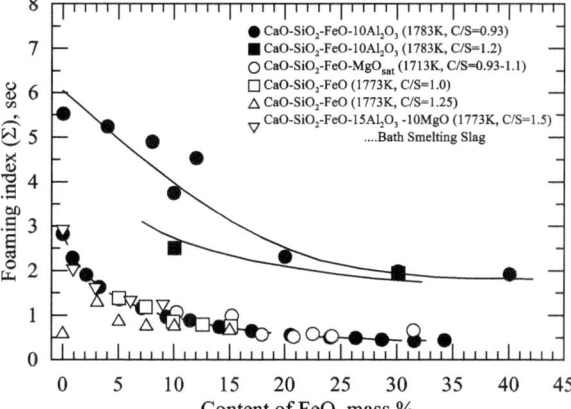

Fig. 7.85 Effect of FeO and slag basicity on the foaming index. After [126]

would be higher due to an enhancement in MgO solubility caused by the presence of Al_2O_3. Further experiments are needed to confirm this behavior.

It is important to take into account that iron oxide in iron and steelmaking slags is usually defined as wüstite, FeO, however in reality there are both Fe^{2+} and Fe^{3+} ions in liquid steel, therefore the presence of Fe_2O_3 should be considered for a more realistic analysis. Fe_2O_3 is an acid oxide. In some cases due to its low concentration, usually below 10% of the total iron oxides in the slag, it has a negligible effect, for example Tayeb et al. [174] reported it had an insignificant effect on MgO solubility.

Stadler et al. [113] investigated the effect of FeO in the range from 20 to 28% on slag foaming for acid $CaO-SiO_2-FeO-Al_2O_3$ slags with SiO_2 concentrations ranging from 41 to 53%, with a basicity range from 0.2 to 0.8. The foaming index was based on Bikerman's expression, obtained from the slope of a plot between superficial gas velocity and foam height, as shown in Fig. 7.87. It can be observed that the

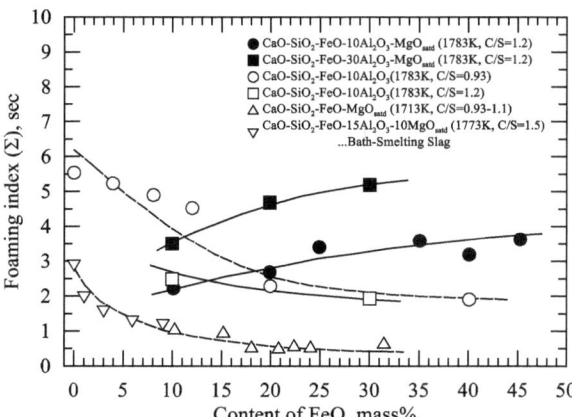

Fig. 7.86 Effect of FeO, MgO and Al_2O_3 on the foaming index. After [126]

Fig. 7.87 Calculation of foaming index. After [113]

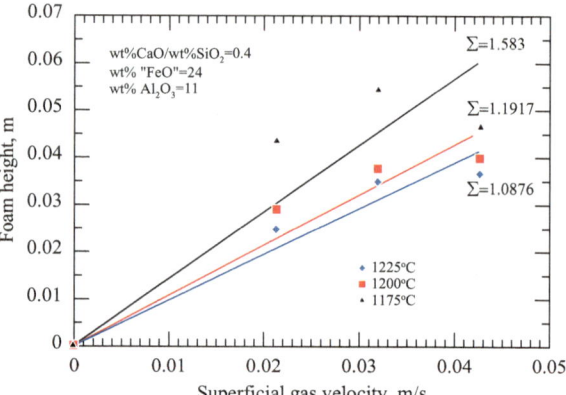

relationship is not perfectly linear which was attributed to the low gas superficial velocity.

Figure 7.88 shows that increasing FeO from 20 to 24% decreased the foaming index but slag foaming increases above 24%. They attributed this phenomena to the precipitation of solid particles as a cause for the increment in the foaming index. Their results show an abnormal behavior with respect to temperature because it shows a higher foaming index for higher temperatures. This behavior was not explained.

Heo and Park [165] reported industrial trials. The effect of FeO is shown in Fig. 7.89. It can be observed that in the range found in their industrial slags, from about 15 to 33% FeO, the foam height remains almost constant at about 18–23 cm but at the same time it also shows that viscosity decreases from about 600 to 400 cP. They explained these results due to two opposite effects; FeO, on one hand, it increases the gas generation rate which improves foaming but on the other hand it also decreases the slag viscosity, which decreases foaming. According with their results the foam height increased twice the original slag height.

Fig. 7.88 Effect of total FeO on the foaming index. After [113]

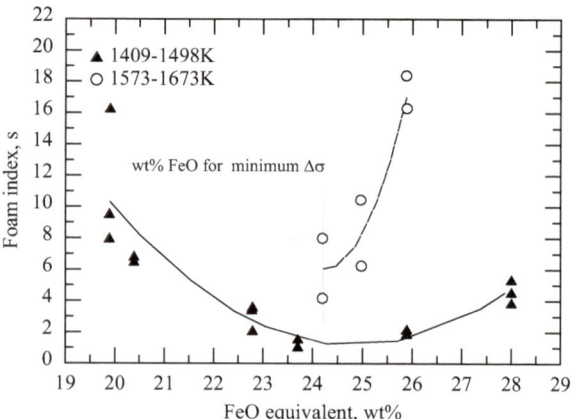

Fig. 7.89 Effect of FeO on slag foaming. After [165]

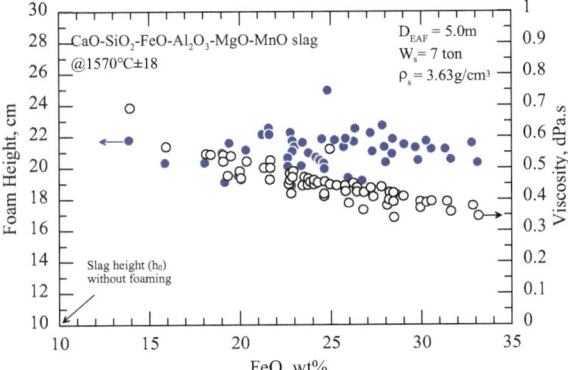

7.6.2 Effect of P_2O_5

Cooper and Kitchener [88] reported the first detailed experimental measurement on slag foaming in 1959. They reported that $CaO-SiO_2$ slags are non-foaming in spite that slags with 60% mol SiO_2 would have a surface composed mostly of SiO_2 and therefore a high surface viscosity. They attributed this effect to possible low film elasticity. However small additions of P_2O_5 (added as calcium phosphate) improved foam stability. They evaluated the effect of P_2O_5 at different temperatures and slag basicities. Foam stability was evaluated measuring the height of the foam decay after releasing the excess gas pressure used to produce the foam. Slags without P_2O_5 didn't form stable foams even with 66% SiO_2 (V = 0.53) indicating that a liquid film with only high viscosity is a poor foam stabilizer since it is not elastic enough. Figure 7.90 shows the effect of 0.2–2.1 mol% P_2O_5 at different temperatures and at a fixed binary basicity ratio of 0.77. It is clearly shown that P_2O_5 increases foam stability, also shows that increasing temperature has the opposite effect. They also reported that the surface tension had a small decrease, about 10 dyne/cm, per 1 mol% of P_2O_5.

Hara et al. [124] investigated slag foaming of $CaO-SiO_2-FeO$ slags. They reported that P_2O_5 from 0.92 to 1.84% increased slag foaming due to a decrease in the slag surface tension, as shown in Fig. 7.91.

Kim et al. [126] reported the effect of P_2O_5 on the foaming index of $CaO-SiO_2-30FeO-MgO_{sat}$ (CSFM) slags, as shown in Fig. 7.92. It can be observed that P_2O_5 increases the foaming index when its concentration increases from 0 to 3% but then it decreases and it remains constant after 6% P_2O_5. They attributed this effect to a decrease in the slag surface tension as P_2O_5 increases but upon reaching 3% P_2O_5, it reaches a maximum surface excess concentration.

Fig. 7.90 Foam stability of
CaO-SiO$_2$-P$_2$O$_5$ slags.
Effect of P$_2$O$_5$. After [88]

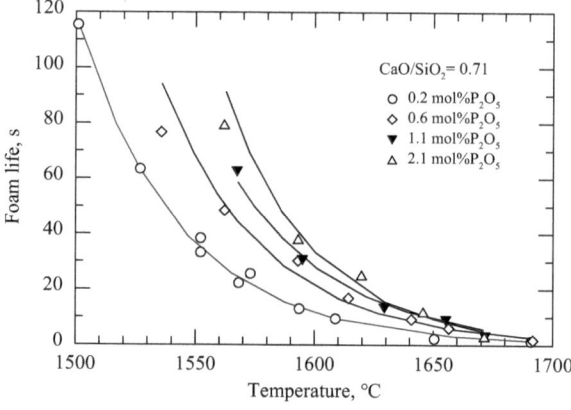

Fig. 7.91 Effect P$_2$O$_5$ on
slag foaming. After [124]

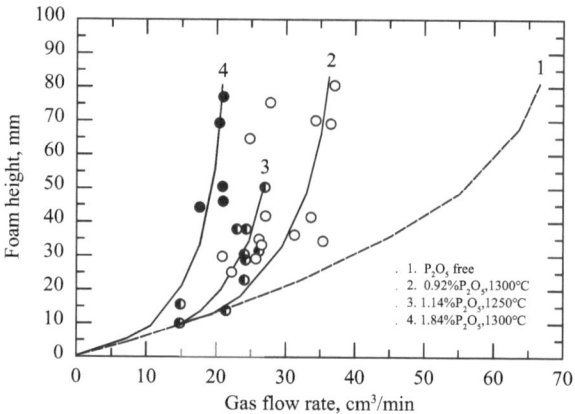

Fig. 7.92 Effect of P$_2$O$_5$ on
the foaming index. After

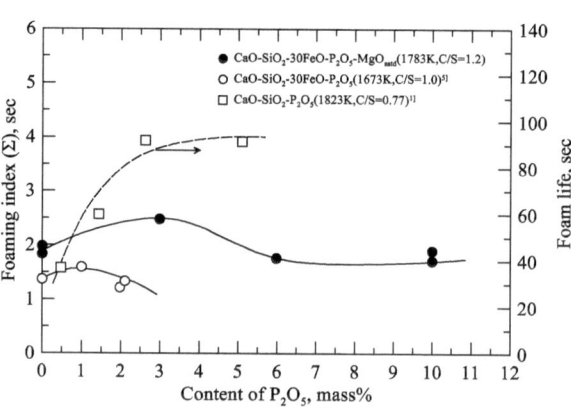

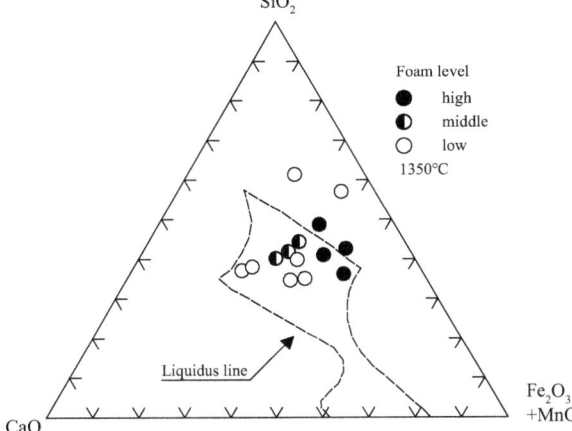

Fig. 7.93 Foam level in the CaO-SiO$_2$-MO system. After [171]

7.6.3   Effect of SiO$_2$

Hellbrugge and Endell [175] in 1941 were the first ones to report that SiO$_2$ increases the viscosity of slags. The viscosity of silicate melts increases as a function the degree of polymerization. They also provided arguments for the existence of the ionic theory. Bockris et al. [176] in 1955 provided further evidence.

Kozakevitch [1] indicates that one of the earliest measurements on slag foaming was a work carried out by Speith and Henrichs in 1953 using a FeO-SiO$_2$ system, indicating that foaming stability increased with silica, reaching a sharp maximum at 60% SiO$_2$, decreasing after that.

Kitamura et al. [171] reported the foam height for BOF slags at 1350 °C. They found that increasing silica increased the foam height, as shown in Fig. 7.93 but this effect decreased at high concentrations due to excessive viscosity.

7.6.4   Effect of Basicity

Fellcht, and Ludeman [1] both in 1955 studied open hearth slags, classifying into two groups; foaming and non-foaming slags. Table 7.14 is a section extracted by Kozakevitch from the work reported by Fellcht. It can be observed that foaming slags have four different characteristics; they are more acid slags (higher SiO$_2$), have a higher oxidation degree, a higher P$_2$O$_5$ and also resulted from a higher decarburization rate. Figure 7.94 plots the results from Fellcht and Ludeman. All of these results are outstanding considering it was the earliest report on the subject, and surprises to learn from Kozakevitch that their work was wrongly criticized.

Table 7.14 Foaming and non-foaming slags in the open-hearth furnace, according with Fellcht

	CaO	SiO$_2$	FeO	Fe$_2$O$_3$	MnO	P$_2$O$_5$	CaO/SiO$_2$	DeC, %/h
Foam	34.5	20.7	17.0	2.9	16.4	1.2	1.67	0.14
	32.1	27.0	9.5	1.4	15.5	0.9	1.19	0.17
	32.6	22.4	15.3	4.0	11.3	1.0	1.46	0.30
	33.8	27.0	9.1	1.5	16.8	1.1	1.25	0.24
	30.7	22.4	9.9	3.3	15.7	1.6	1.37	0.24
					13.1	1.3	1.31	0.18
	33.1	23.4	13.2	3.6	13.4	1.41	1.41	0.25
	37.8	22.8	11.9	0.8	10.4	1.65	1.65	0.15
	44.5	22.3	7.9	3.2	10.4	2.0	2.0	0.11
	39.2	19.4	14.6	3.2	10.7	2.0	2.0	0.07
	35.31	**23.41**	**11.67**	**2.75**	**13.39**	**1.51**	**1.51**	**0.18**
Non-foam	40.1	21.8	12.9	0.6	12.2	1.3	1.84	0.09
	40.8	22.9	11.0	1.6	12.4	1.4	1.78	0.04
	44.8	17.2	13.9	1.1	9.2	1.3	2.6	0.04
	42.2	26.3	9.2	2.0	10.6	1.2	1.65	0.16
	45.0	19.6	11.4	1.6	9.0	1.2	2.3	0.16
	43.8	15.7	9.7	1.0	19.4	1.0	1.32	0.18
	42.3	20.7	11.0	1.4	12.7	1.5	2.0	0.18
	50.9	16.0	12.0	1.8	7.3	0.8	3.19	0.24
	39.0	27.9	8.3	0.6	14.8	0.4	1.40	0.17
	45.9	17.0	12.7	1.4	12.2	1.3	2.70	0.24
	42.5	**21.51**	**11.32**	**1.31**	**11.96**	**1.14**	**2.0**	**0.15**

Taken from Ref. [1]

Fig. 7.94 Foaming and non-foaming slags depending on slag basicity. Taken from Ref. [1]

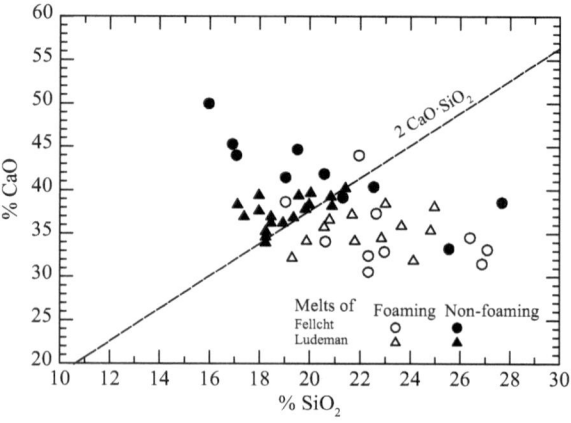

Fig. 7.95 Foam stability of CaO-SiO₂-P₂O₅ slags. Effect of basicity. After [88]

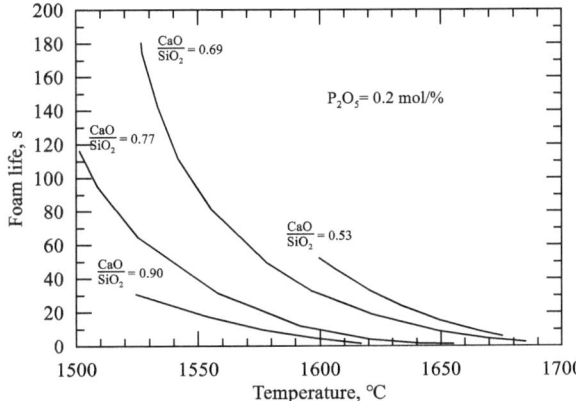

Cooper and Kitchener [88] reported that decreasing basicity (acid slags) show better foam stability, especially if the surface tension is also lowered, as shown in Fig. 7.95 with the addition of 0.2 mol % P₂O₅.

Ito and Fruehan [99] reported that increasing basicity in CaO-SiO₂-FeO-(Al₂O₃) slags at 1300 °C, the foaming index decreased until it reached a value of 1.2. In other words, to increase foaming the basicity of the slag had to be decreases below 1.2. At this point, due to the precipitation of particles, foaming is improved. When the temperature was increased to 1400 °C the foaming index was decreased. The basicity associated with this minimum slightly increased with respect to 1300 °C, to 1.22.

Jung and Fruehan [127] carried out additional experiments for CaO-SiO₂-FeO-MgO$_{sat}$ slags at 1440 °C, reporting a minimum foaming index for a basicity of 1.4. The increment in basicity values from about 1.2 to 1.4 was attributed to differences in Al₂O₃ and MgO in the two systems investigated, in the second case MgO is basic, increasing the basicity index employed. These results are shown in Fig. 7.96. If the slag is saturated in MgO, magnesio-wüstite [(Fe,Mg)O] could be the solid that first precipitates.

Yokoyama et al. [177] reported a similar behavior to that reported by Fruehan's group, as shown in Fig. 7.97.

Koch and Ren [135] reported the effect of basicity on slag foaming. The slag employed contained 60%Fe₂O₃-CaO-SiO₂, simulating a primary blast furnace slag. The crucible material was graphite and the gas supplied was 40%CO-60%N₂, with a gas flow rate of 400 Nml/min. For this type of slag, the foaming index increased with increasing basicity up to a maximum, for a basicity between 0.7 and 1.0, in the temperature range from 1250 to 1450 °C. The large concentration of Fe₂O₃ could form solid particles which would explain this behavior, however the authors didn't offer an explanation.

Ameling et al. [4] reported Fig. 7.98 which describes the effect of both FeO and basicity on slag foaming. This figure is taken from ref [77]. Region I (A to C) corresponds to low foamability due to large FeO even if it has low basicity. Region II (B) was estimated as high foamability due to low FeO and high viscosity. Region

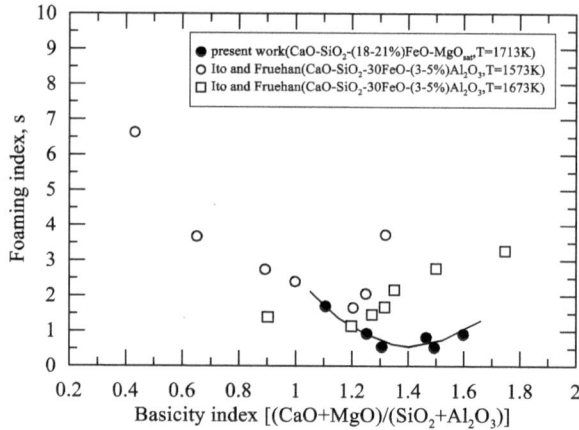

Fig. 7.96 Effect of basicity on the foaming index. After [99]

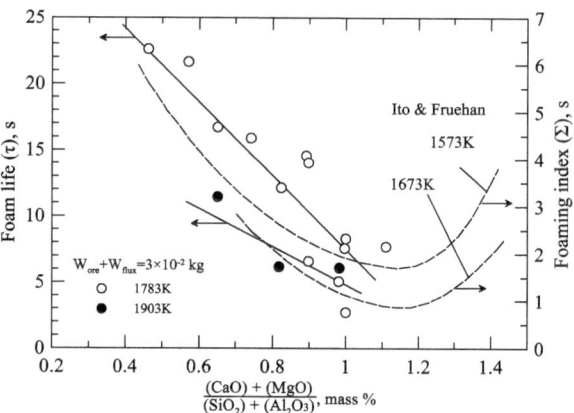

Fig. 7.97 Effect of basicity (B_4) on the foam life. After [177]

III (D,E) is also assigned as high foamability at high FeO but also high basicity which increases slag viscosity. This figure should be considered for characterization purposes because optimal foaming does not depend only on FeO and basicity.

Kim et al. [126] reported the effect of Basicity (B_7) on the foaming index of CaO-SiO_2-FeO-MgO_{sat}-X slags where X represents MnO, Al_2O_3, CaF_2 and P_2O_5. Basicity was defined as the ratio of ($CaO + FeO + MgO + MnO + CaF_2$)/($SiO_2 + Al_2O_3 + P_2O_5$). They also reported a decrease in the foaming index as basicity increases, reaching a minimum value and then increasing, as shown in Fig. 7.99. This behavior has been explained as a result of three different effects. In the first region, acid slags decrease the slag surface tension but predominates its effect to increase the slag viscosity. Acid slags are more viscous. Increasing basicity decreases the slag viscosity up to the saturation point where solid particles precipitate, then, due to an enhanced effective slag viscosity, the foaming index increases.

Min and Tsukihashi [78] summarized the effect of basicity from a large number of previous investigations on slag foaming in Fig. 7.100.

Fig. 7.98 Effect of basicity and FeO activity on foamability. After [4]

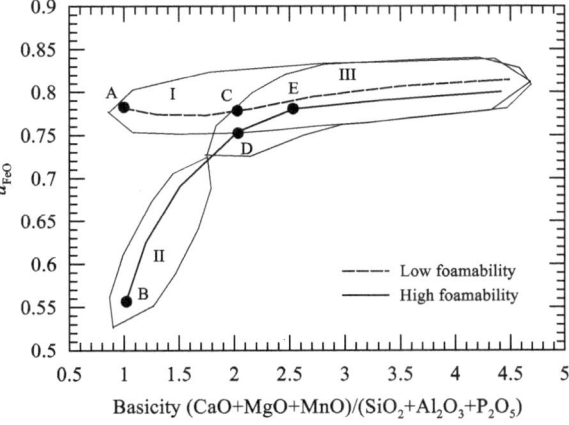

Fig. 7.99 Effect of basicity (B_7) on the foaming index. After [126]

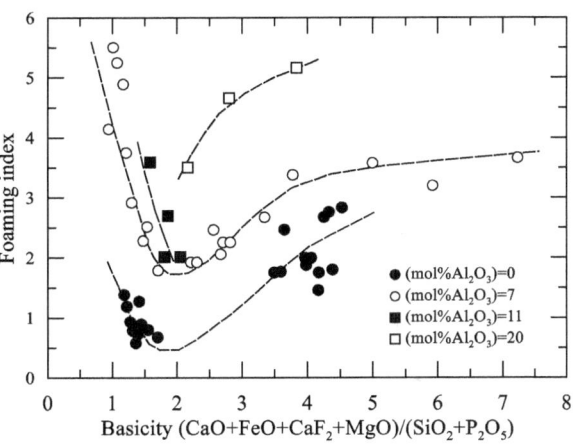

Fig. 7.100 Effect of basicity ratio on the foaming index. After [78]

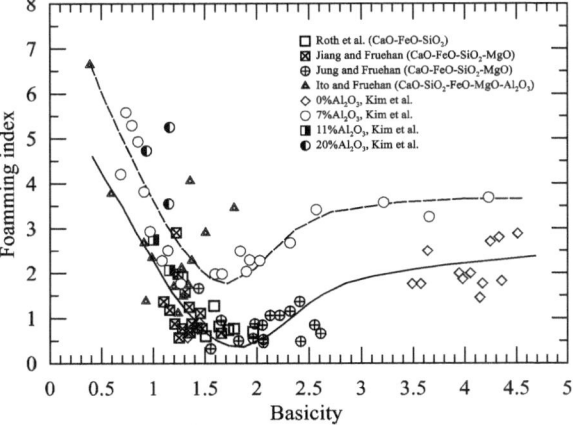

Heo and Park [165] reported industrial trials. The effect of basicity is shown in Fig. 7.101. The binary basicity ratio ranges from about 1.0 to 1.8. The industrial data show that as basicity increases from 1.0 to 1.8, there is a decrease in the foam height from 23 to 18 cm, respectively. Figure 7.102 shows in more detail the effect of both basicity and FeO. It shows that the foaming index is higher for slags with 22% FeO in comparison with slags with 27% FeO, for the same basicity.

Stadler et al. [113] reported the effect of basicity on the foaming index for acid $CaO\text{-}SiO_2\text{-}FeO\text{-}Al_2O_3$ slags, as shown in Fig. 7.103. Basicity was defined as a ratio $(CaO + Al_2O_3)/SiO_2$, alumina included as a basic oxide because of its amphoteric nature. It can be shown that in spite that slags are highly acid, they also report an initial decrease in the foaming index when the basicity increases to 0.7, after B_3 close to 0.7 it increases. They also report a higher foaming index at higher temperatures, which would be unexpected and was not explained. This work needs further confirmation.

Fig. 7.101 Effect of basicity on the foam height. After [165]

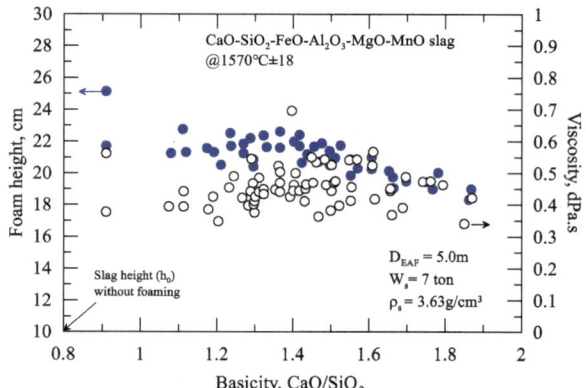

Fig. 7.102 Effect of basicity and FeO on the foam height. After [165]

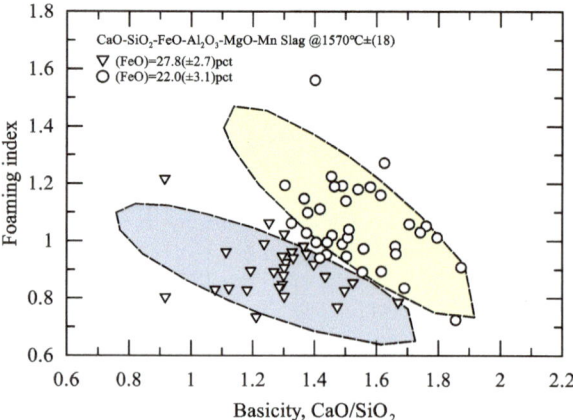

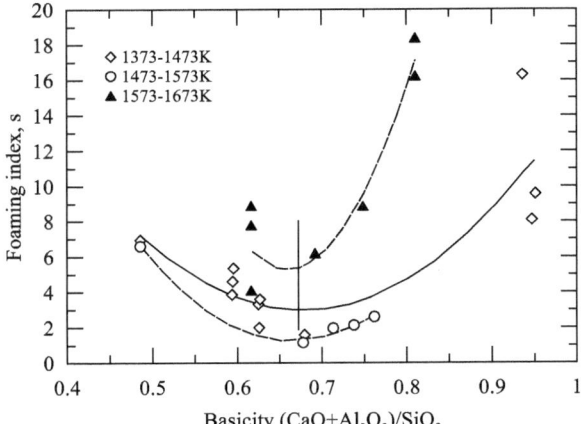

Fig. 7.103 Effect of basicity (B_3) on the foaming index. After [113]

In 1990 Capodilupo et al. [178]—taken from Ref. [179], reported the industrial data shown in Fig. 7.104. Their results indicate a peak in the foam height at a basicity of about 2.0–2.2, for 20% FeO.

Vieira et al. [180, 181] reported the effect of slag composition on energy consumption during the refining period, a stage where foaming conditions prevail and found the following optimum values: 27–30% FeO, basicity 3–3.2 and 7.5% MgO. Their binary basicity ratio is quite high for EAF operations. Based on their own results a more appropriate range would be from 2.6 to 2.8 and a slightly higher MgO saturation concentration.

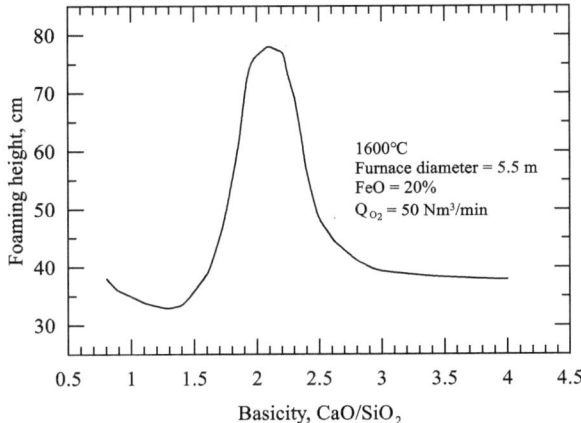

Fig. 7.104 Effect of basicity on the foam height. Industrial data. After [178]

7.6.5 Effect of MgO

Hara et al. [124] investigated slag foaming of $CaO-SiO_2-FeO$ slags. They compared slags without MgO and with 10% MgO obtaining the same results, as shown in Fig. 7.105, concluding that MgO had a neutral effect. They didn't mention the MgO saturation concentration.

Jung and Fruehan [127] studied the effect of MgO on the foaming index. They found that as MgO increased from 0 to 15%, the foaming index decreased from about 2 to 0.5 s, as shown in Fig. 7.106. Yokoyama et al. [177] reported the foam height for $CaO-SiO_2-Al_2O_3-MgO$ slags with similar results, as shown in Fig. 7.107.

Chang et al. [130] investigated the foam life of 6 slags with different MgO concentrations at 1500 °C and also measured their viscosity, as shown in Table 7.15. Slag 0

Fig. 7.105 Effect MgO on slag foaming. After [124]

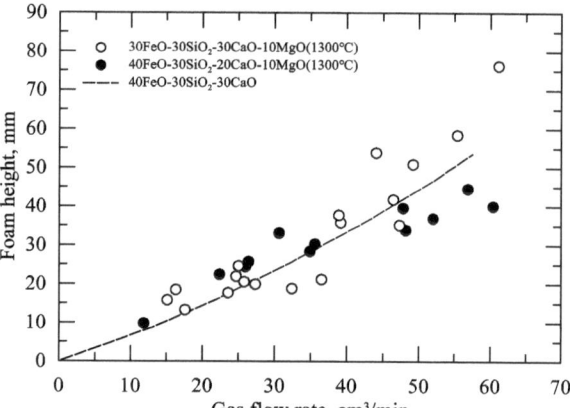

Fig. 7.106 Effect of MgO on the foaming index. After [127]

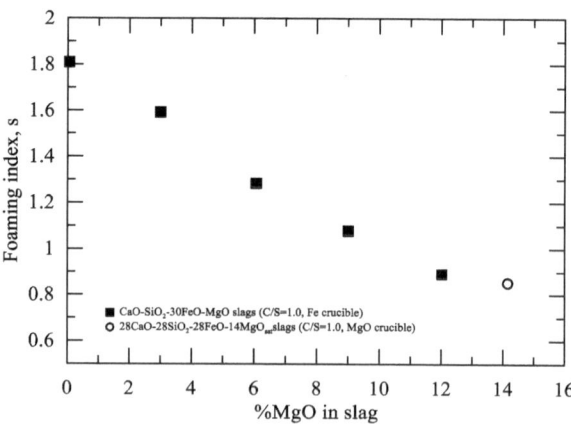

Fig. 7.107 Effect of MgO on the foam height. After [177]

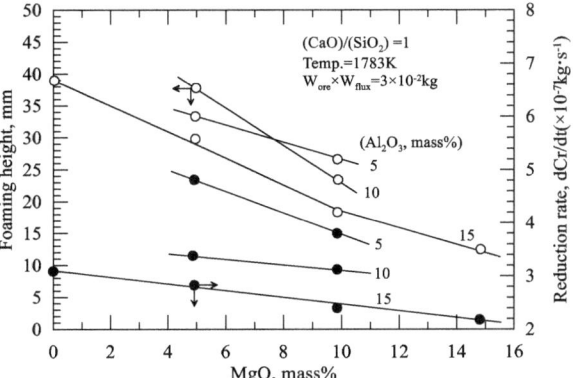

is the reference slag from a steel plant. Foam height and foam life were experimentally measured with an electric probe. Foam life was defined as the time taken for the foaming slag to go back to its original height.

From the previous table, it can be observed that as MgO increases the percentage of solids also increases, from 2.7% for 9.6% MgO to 11.7% for 13.6% MgO. The viscosity and melting temperature also increase. The results on foam life shown in Fig. 7.108 clearly indicate that once MgO increases and reaches the concentration to precipitate solid particles, slag foaming increases due an increase in the solid particles which also increase the effective slag viscosity, from 40 cP for 9.6% MgO to 1088 cP for 13.6% MgO.

The foam height alone is not a good reference to compare slags. As shown from the table, the slag with the longest foam life has the smallest foam height. This behavior was attributed to the slag's viscosity. The gas produced escapes faster in slags with lower viscosity but produces a higher foam height, on the contrary, slags with higher viscosity retain the bubbles a longer time but its maximum height is

Table 7.15 Chemical composition, viscosity and %solids of slags investigated by Chang et al. [130]

	CaO	SiO₂	MgO	Al₂O₃	FeO	1500 °C mPa s	1550 °C mPa s	T °C	% solids at 1500 °C	h^{max}_{foam}, cm	Foam life, min
0	45.6	16.4	4.7	4.4	29	22	21	1495	0	9	2.0
1	35.8	23.6	5.7	4.8	30.1	47	43	< 1300	0	8	5.5
2	35.1	23	7.4	5.1	29.4	40	38	1315	0	10.5	4.5
3	33.1	22	9.6	5.3	30	76	41	1495	2.7	10	9.5
4	32.5	21.4	11.5	5.4	29.2	363	126	1575	6.7	8	11.8
5	31.4	20.3	13.6	5.2	29.6	1088	567	> 1600	11.7	6	12.7

T = Critical viscosity temperature, °C

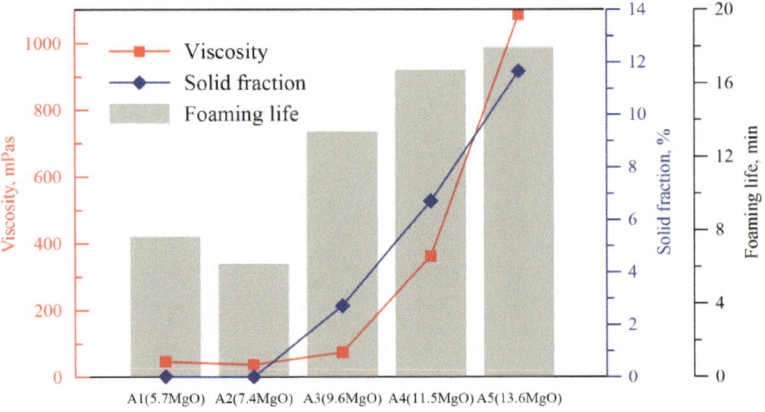

Fig. 7.108 Effect of MgO on the foam life, viscosity and solid fraction. After

lower. In this study the authors also reported results from Raman spectroscopy to evaluate the degree of polymerization of the slags.

Ren et al. [136] studied primary blast furnace slags containing 56–80% Fe_2O_3 and basicity ranging from 0.1 to 1.8. They analyzed the effect of MgO (0–10%) and alkalis (0–15% K_2O, 0–30% Na_2O). The effect of MgO on foaming changed depending on the slag basicity and temperature. Fe_2O_3 was constant at 60%. For highly acid slags (V = 0.1) additions of MgO from 0 to 10% increased foaming, this effect increased if the temperature increased from 1350 to 1450 °C. At lower temperatures, even in acid slags, foaming didn't happen. In basic slags (V = 1.4), increasing MgO from 0 to 10% decreased foaming. In general, small additions of alkali oxide intensified foaming of acid slags but as the additions increased foaming was suppressed. For basic slags, alkali additions decrease foaming. For blast furnace slags the extent of foaming is controlled by the degree of pre-reduction at the beginning of the cohesive zone.

7.6.6 Effect of MnO

Kim et al. [126] reported the foaming index of $CaO-SiO_2-30FeO-MnO-MgO_{sat}$ as shown in Fig. 7.109. It can be observed that MnO slightly decreases the foaming index up to 10%. This behavior is attributed to a decrease in the slag viscosity since MnO is a basic oxide. The magnitude in the foaming index, 3 times higher than that in previous reports, was attributed to using a much lower basicity slag (1.2 vs. 3 for the other cases reported in the plot).

Fig. 7.109 Effect of MnO on the foaming index. After [126]

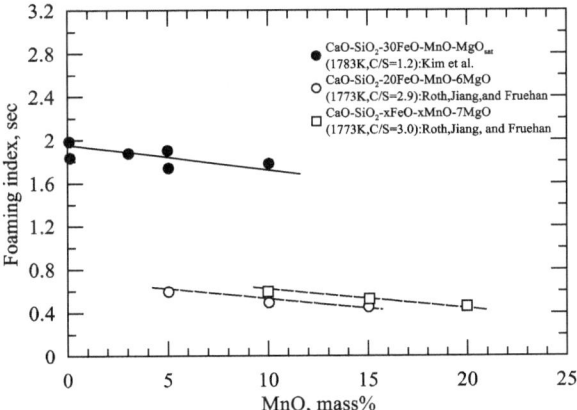

7.6.7   Effect of Al₂O₃

Chang et al. [142] investigated the effect of alumina in $40CaO$-$16SiO_2$-FeO-$5MgO$-Al_2O_3 slags with a constant basicity ratio of 2.5 on slag foaming at 1550 °C. The slags employed are shown in Table 7.16, alumina was varied in the range from 5 to 18%. It can be observed that the initial FeO also changed in each slag investigated in the range from 17 to 35%. Slags with the lower concentration of Al_2O_3 has the highest concentration of FeO. The amount of solids was estimated based on Factsage. It can be observed that most precipitated solid disappear above 1500 °C. The oxide powders were melted at 1800 °C in a high-frequency furnace. FeO was produced by melting Fe_2O_3 powder at high temperature in a graphite crucible protected with a nitrogen atmosphere, the melt was water quenched and the solid crushed to powder. FeO was confirmed by XRF. One salient aspect of this work is the experimental measurement of viscosity, surface tension and viscosity at high temperatures. Density was measured immersing a cone-shaped hammer of Pt into the slag using the immersion depth and its change in weight. Surface tension was measured using a Pt ring and measuring the force to pull it from the slag. Viscosity was measured using a Brookfield viscometer. The experimental viscosity measurements were compared with calculated values. The calculated values were obtained using Factsage, however since Factsage doesn't take into account the fraction of solids, the Einstein-Roscoe equation was included in the analysis. This equation also has simplifications because it assumes the particles are spherical and the solid is a single phase. The degree of polymerization was investigated using X-ray photoelectron spectroscopy (XPS). The authors reported a foaming efficiency (cm/min) based on the area under the curve of the plots of foam height versus time. This is a new proposal to define the foaming capacity of slags.

Figure 7.110 summarizes the effect of alumina on slag foaming. It can be observed that in the range from 5 to 15.7% Al_2O_3 the foaming efficiency increases. This is the result of an increase in the slag's viscosity and a decrease in both surface tension and density. Viscosity increases from about 17 to 60 cP and the decrease in surface tension

Table 7.16 Chemical composition, viscosity and %solids of slags investigated by Chang et al. [142]

	CaO	SiO$_2$	MgO	Al$_2$O$_3$	FeO	% solids @ 1400 °C	% solids @1450 °C	% solids @ 1500 °C	h$_{foam}^{max}$, cm	Foam life, min
0	39.1	15.4	4.9	5.1	35.5	15.4	3.4	0.7	12.4	3.0
1	41.7	16.4	4.5	8.6	28.2	5.9	0.2	0	10.8	4.1
2	41.5	16.8	5.7	11.9	24.1	2.1	0.9	0	10.5	6.3
3	40.9	16.6	6.0	15.7	20.8	1.4	0.2	0	10.7	10.0
4	42.4	17.0	5.3	18.3	17.0	0	0	0	8.7	10.3

and density is from 3.97 to 3.81 mN 10^{-2}/M and 3.02 to 2.97 g/cm^3, respectively. When the alumina concentration exceeds 15.7% the foaming efficiency decreases, which is now explained by a decrease in viscosity and an increase in surface tension. This work is very comprehensive; however, two aspects should be noticed, first; the results cannot be attributed only to alumina because FeO also has large changes in concentration and the second aspect is related to the large concentration of alumina in the slags that are supposed to replicate steelmaking slags. Aluminum is not added in the EAF and the concentration of alumina results from oxidation of aluminum on the steel scrap giving concentrations below 5%.

Previous results shown in Fig. 10.85 also report an increment in the foaming index when alumina is increased from 10 to 30%.

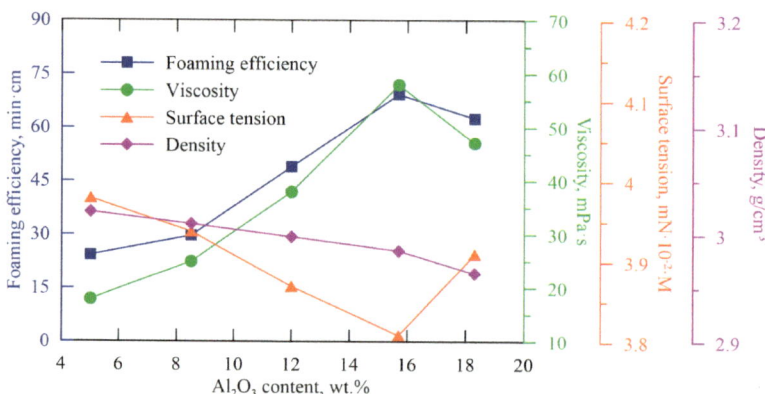

Fig. 7.110 Effect of Al$_2$O$_3$O on the foaming efficiency. After [142]

Fig. 7.111 Effect of TiO$_2$ on the foaming index. After [127]

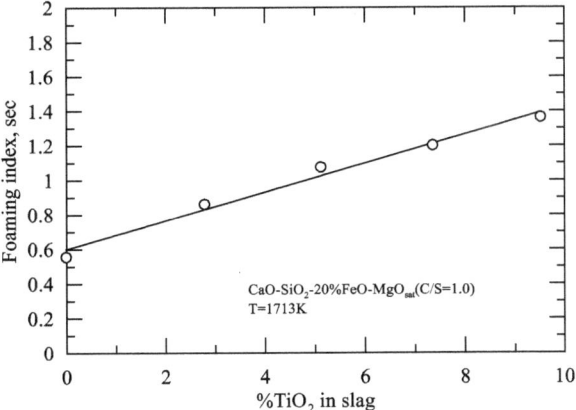

7.6.8   Effect of TiO$_2$

Jung and Fruehan [127] studied the effect of TiO$_2$ on the foaming index at 1440 °C. They found that as TiO$_2$ increased from 0 to 10% in CaO-SiO$_2$-FeO-MgO$_{sat}$ slags at 1440 °C, the foaming index increased from 0.5 to 1.5 s. TiO$_2$ like FeO and SiO$_2$ increases the viscosity and decreases the surface tension. Contrary to SiO$_2$, TiO$_2$, due to larger Ti^{4+} ions do not contribute to the formation of large networks due polymerization, instead it decreases viscosity in a similar way to CaO and FeO. The authors attributed its effect on viscosity to the precipitation of 2MgO·TiO$_2$ when TiO$_2$ combines with MgO dissolved from the crucible. Figure 7.111 shows the effect of TiO$_2$ on the foaming index.

The promotion of the slag foaming conditions with TiO$_2$ is a well-known problem in blast furnace operation [131]. China has large deposits of iron-ore containing vanadium and titanium in Panzhihua Sichuan, a south western region in China. This ore creates problems in the dripping zone with primary blast furnace slags, affecting the burden permeability.

Xiang et al. [182] investigated the effect of TiO$_2$ on slag foaming at 1500 °C, TiO$_2$ in the range from 20 to 32%. The slags contained (20–30%) CaO-(20–30%) SiO$_2$-13%Al$_2$O$_3$-7%MgO-(20–32% TiO$_2$)-(8.8–14.2%) C. They reported that TiO$_2$ increases foaming due to an increment in the slag's viscosity. TiO$_2$ is first transformed to TiC due to high carbon in the slag and then reduced by FeO which releases CO.

7.6.9   Effect of CaF$_2$ and Na$_2$O

Hara et al. [124] investigated slag foaming of CaO-SiO$_2$-FeO slags with additions of CaF$_2$ and Na$_2$O. They reported that 10–20% of both CaF$_2$ and Na$_2$O improve foaming, as shown in Fig. 7.112. They attributed this effect to their surfactant effects.

Fig. 7.112 Effect of **a** CaF$_2$
and **b** Na$_2$O on slag foaming.
After [124]

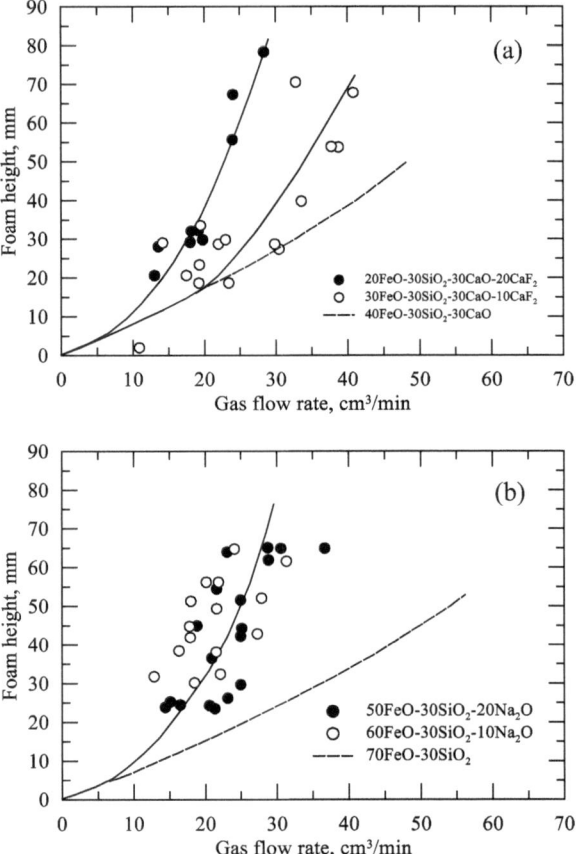

Kim et al. [126] reported the effect of 0–15% CaF$_2$ on foaming of CaO-SiO$_2$-30FeO-MgO$_{sat}$ as shown in Fig. 7.113. It can be observed that CaF$_2$ decreases the foaming index up top 5% but after this value, it increases the foaming index. Ito and Fruehan also reported a decrease in foaming index from 0 to 10% CaF$_2$. This effect was attributed to a decrease in the slag viscosity.

7.7 Effect of Slag's Physical Properties on Foaming

The review work by Nexhip [24] argues that individually each slag physical property cannot explain slag foaming by itself but it is their mutual interaction which could better explain this phenomena. He also emphasized the need to carry out experiments under similar experimental conditions in order to compare results because using different equipment with different geometry, different slag composition and temperature, different bubble size, different ways to generate the gas, different gas flow

Fig. 7.113 Effect of CaF$_2$ on the foaming index. After [126]

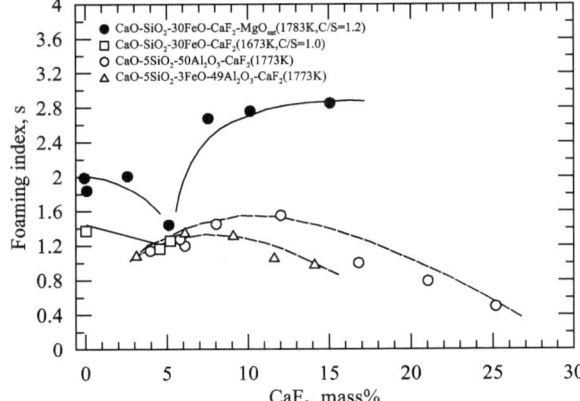

rates, etc., can lead to different results. They mention that the alternative proposed by Cooper and Kitchener to study individual bubbles was rejected because of sensitivity to contamination and difficulties to form controlled slag bubbles at high temperatures.

7.7.1 Effect of the Slag's Surface Tension

Kozakevitch [1] in 1949 reported the effect of silica on the surface tension of CaO-FeO-SiO$_2$ slags, as shown in Fig. 7.114. The results for P$_2$O$_5$ in Fig. 7.114 (right) were reported by Vishkarev et al. these results show a decrease in the slag's surface tension as both SiO$_2$ and P$_2$O$_5$ increase. The effect of P$_2$O$_5$ is more intense compared to SiO$_2$.

Hara et al. [124] reported that the addition of Na$_2$O, CaF$_2$ and P$_2$O$_5$ to primary CaO-SiO$_2$-FeO slags improved foam life and foam height. This was attributed to a decrease in the surface tension. Figure 7.115 shows that decreasing the surface tension from 500 to 360 dyn/cm, increases the foam life from 2 to 10 s. The numbers inside this figure correspond to the slags shown in Table 7.13. Figure 7.116 summarizes the effect of various oxides on the surface tension of slags.

Hara et al. [183] measured the surface tension depression and the surface viscosity of binary slags containing Na$_2$O-P$_2$O$_5$, BaO-B$_2$O$_3$, Na$_2$O-B$_2$O$_3$, and Na$_2$O-SiO$_2$ at 1000 °C. In two systems, BaO-B$_2$O$_3$ and Na$_2$O-B$_2$O$_3$ the surface viscosity was higher than in the bulk phase, due to the presence B$_2$O$_3$ which is a strong surfactant. Increasing the surface viscosity improves foaming but only if there is also a joint decrease in the surface tension. Figure 7.117 shows very poor foaming in spite of having high surface viscosities due to a low surface tension depression. The opposite is also true; high surface tension depression but low surface viscosities also give poor foam stability. The highest foam stability is reached when the slag has both, high surface viscosity and high surface tension depression.

Fig. 7.114 Effect of **a** silica and **b** SiO$_2$, P$_2$O$_5$ and Fe$_2$O$_3$ on slag's surface tension. After [1]

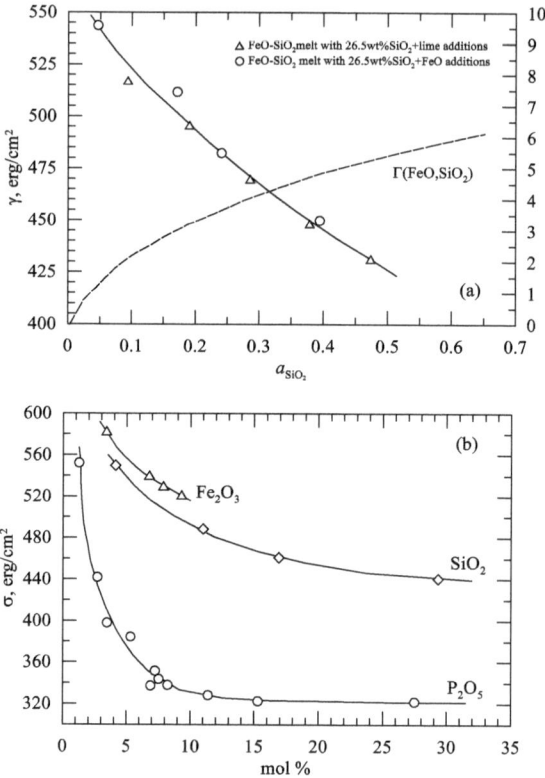

Fig. 7.115 Effect of surface tension on foam life. After [124]

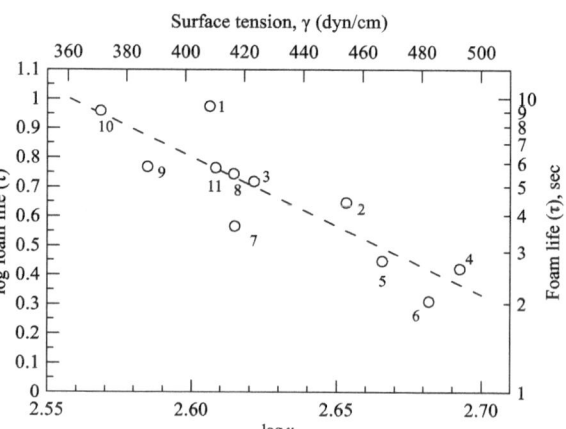

Fig. 7.116 Effect of different oxides on the surface tension of slags. After P.I. Popel[1]

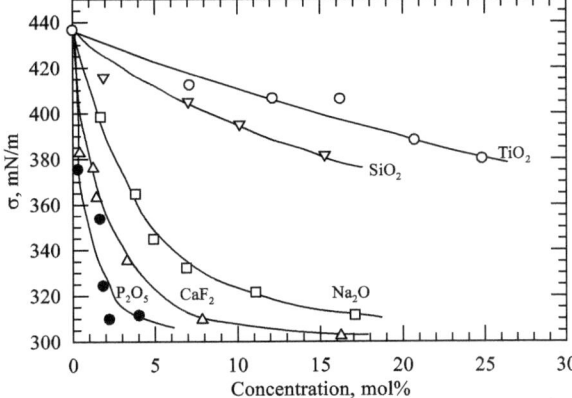

Fig. 7.117 Relationship between surface viscosity, surface tension depression and foam stability. Numbers indicate foam height in cm. After [183]

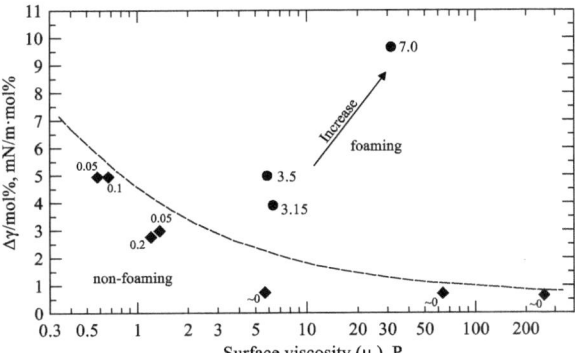

Skupien and Gaskell [184] carried out experimental measurements of surface tension of CaO-SiO_2-30%FeO slags from 1300 to 1435 °C using the dipping cylinder technique. This technique measures the maximum withdrawal force from the melt. They re-calculated the foaming index measured by Ito and Fruehan using their experimental values for surface tension and the results are shown in Fig. 7.118. As surface tension increases, the foaming index decreases, it should be noticed that a small increment from about 0.4 to 0.5 N/m, decreases the foaming index about 2.5 times.

Nekrasov et al. [77] made the observation that not all components that decrease the surface tension improve foaming automatically because, in general, a decrease in surface tension also causes a decrease in viscosity. Sulphur and oxygen, both decrease the metal/slag interfacial tension [185]. TiO_2 lowers the surface tension of FeO [1].

---

[1] S.I. Popel. Metallurgicheskie shlaki i primenenie ihk v stroit el'stve' (Metallurgical slags and their use in building), Akad - Stroit i Arkhitekt, SSSR, Ural'sk Filial, 1962, 97–127.

Fig. 7.118 Effect of surface
tension on the foaming
index. After [184] (1 N/m =
1000 dyne/cm)

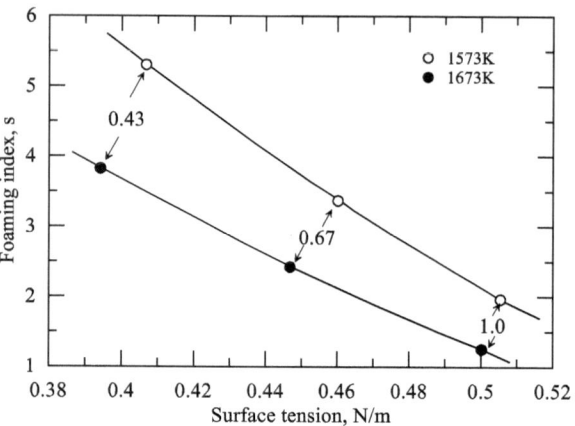

7.7.2   Effect of the Slag's Viscosity

Yamashita et al. [186] carried out foaming experiments using silicon oil and nitrogen,
however their primary objective was to measure the effect of the bubbles on the
relative viscosity. They reported an increment of the relative viscosity by increasing
the gas phase, this growth became exponential when the gas phase was above 60%.
The relative viscosity was defined as ratio of viscosity of the foam divided by the
viscosity of the liquid.

High bulk and surface slag viscosities improve the foaming conditions; however,
this physical property should be properly controlled. Hara and Ogino, as indicated
in Ref. [24], studied foaming of $30Na_2O$-$70SiO_2$ (mol%) at 1000 and 1300 °C, with
bulk viscosities of 177 and 180 Pa s, respectively, suggesting that viscosity didn't play
a significant role on slag foaming. This example illustrates a case with an excessive
surface concentration. The results from Swisher and McCabe also indicate similar
results. They found that above the solubility limit for Cr_2O_3, about 0.7 mol%, the
foam stability suddenly decreased.

Yokoyama et al. [177] reported foam life of CaO-SiO_2-Al_2O_3-MgO slags with
additions of chromite ore containing iron oxide and using a graphite crucible,
changing the slag composition and temperature. They reported the slag viscosity
had the largest effect to increase foam life, as shown in Fig. 7.119.

Chang et al. [187] studied the effect of viscosity on slag foaming. The slags
employed had the following chemical composition; 32–53% CaO, 12–24%SiO_2, 10–
43%FeO, 5–7%Al_2O_3, 4.7 to 7.1%MgO and a basicity range from 2.1 to 2.7. They
measured the slags melting point using both high temperature optical image analysis
and thermo-calc. The slag viscosity was measured with a viscometer (Brookfield DV
IIIRV manufactured by Ametek USA) from 1400 to 1575 °C. Figure 7.120 shows the
effect of FeO on slag's viscosity and compares with other studies which confirm that
increasing FeO decreases the slag viscosity. It is also important to mention the huge
effect; viscosity decreases by more than one order of magnitude, for example at 10%

Fig. 7.119 Effect of slag viscosity on foam life. After [177]

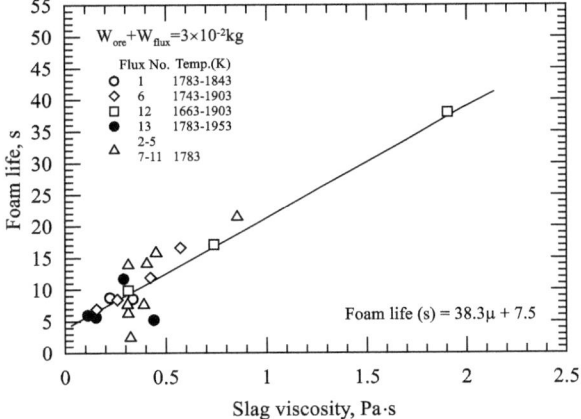

FeO the viscosity is 2300 cP and increasing to 20% FeO, the viscosity decreases to about 100 cP.

These authors focused only on the effect of slag viscosity on the foam height. They employed an electric probe to measure foam height and injected nitrogen to form the foam. Figure 7.121 shows their results. These results confirm that increasing the slag viscosity increases the foam height. The following linear relationship is valid for slags containing from 10 to 30%FeO, in the range from 25 to 1810 cP:

$$\frac{\Delta h}{h_0} = 2 \times 10^{-4}\mu + 0.6552$$

where $\Delta h/h_0$ represents the foam height ratio between foam height and original height and μ is the slag's viscosity in cP.

Min and Tsukihashi [78] summarized the effect of both viscosity and surface tension on slag foaming as shown in Fig. 7.122. Slags with a higher foaming index

Fig. 7.120 Effect of FeO on slag viscosity. After [187]

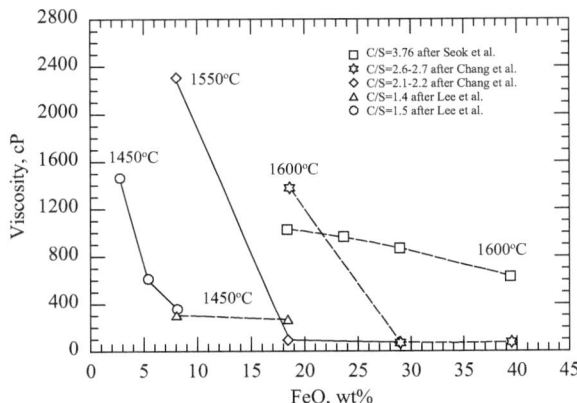

Fig. 7.121 Effect of slag viscosity on the foam height. After [187]

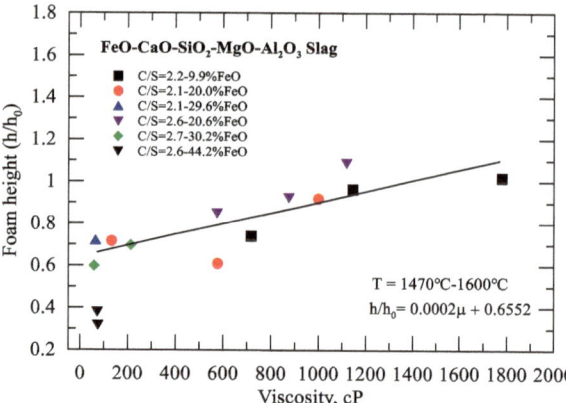

have higher viscosities from 3 to 5 Poise (300–500 cP) and lower surface tension, less than 450 mN/m.

Wu et al. [98] reported the effect of the kinematic viscosity on the foam height using different oils at room temperature, as shown in Fig. 7.123. The oils were selected, trying to achieve similarity of the Morton number with BOF slags, which have a Mo number around 0.006. The results indicate that the foam increases with the kinematic viscosity and reach a peak at almost the same kinematic viscosity value of 130×10^{-6} m²/s. After reaching this kinematic viscosity the foam height decreases, attributed to bubble coalescence. They also noticed the similarity of behavior between basicity and foam height and that of viscosity and foam height. In both cases there is a peak of maximum foaming corresponding to a low drainage rate but as the bubbles grow due to coalescence they are able to ascend faster to the top surface and break up.

Fig. 7.122 Effect of both viscosity and surface tension on the foaming index. After [78]

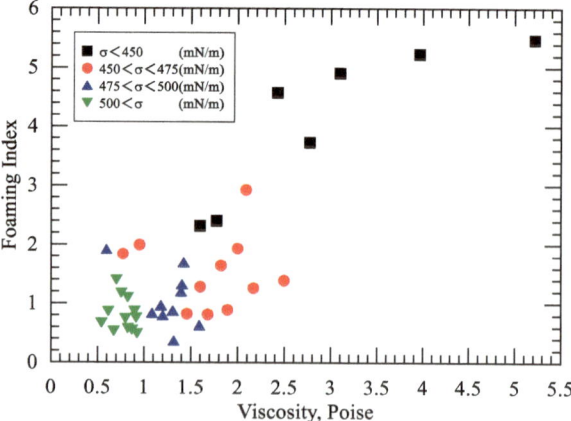

Fig. 7.123 Effect of the kinematic viscosity on the foam height. After

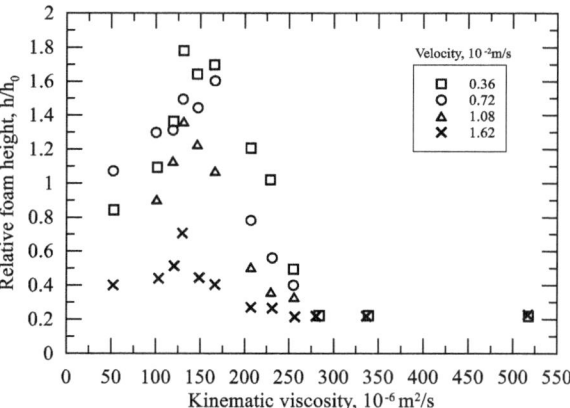

7.8 Effect of Temperature on Slag Foaming

Cooper and Kitchener [71] reported that foam stability decreases by increasing the temperature. This has been confirmed in numerous investigations, for example, Ito and Fruehan [99] studied foaming of CaO-SiO₂-FeO slags in the temperature range from 1250 to 1400 °C and later on Jung and Fruehan [127] from 1440 to 1550 °C. As temperature increases, the slag viscosity decreases and surface tension increases, both contributing to a decrease of the foaming index. Figure 7.124 reports their results. These results can be summarized by the following equation:

$$\log \Sigma = \frac{6610}{T} - 3.90$$

Fig. 7.124 Effect of temperature on the foaming index. After [99, 127]

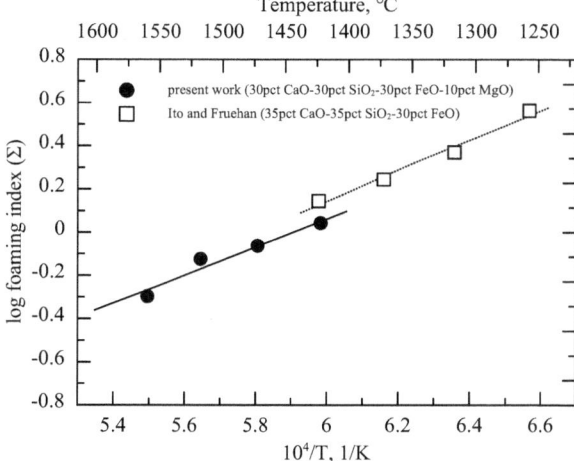

Ozturk and Fruehan [93] further expanded the temperature range to 1600 °C. They employed a slag with the following chemical composition; 48%CaO-32%SiO$_2$-10%Al$_2$O$_3$-10%FeO, in the temperature range from 1450 to 1600 °C. They employed 160 g of slag, equivalent to an initial depth of 4.2 cm. The reagent grade oxides were pre-melted in an alumina crucible. Their results are shown in Fig. 7.125 together with the slag viscosity values in the same temperature range. The foaming index as a function temperature is summarized by the following relationship:

$$\log \Sigma = \frac{16797}{T} - 4.75$$

The two previous plots represent an Arrhenius analysis, from their slopes the activation energy for the decay of the foam and the viscous flow can be estimated. As shown in Table 7.17, it can be observed that the activation energy for the foam decay and for viscous flow is similar.

Bhoi et al. [172] reported the effect of temperature on the foaming index, shown in Fig. 7.126.

Nexhip et al. [49] measured the rate of thinning of single bubbles as a function of temperature using a new technique, withdrawing individual rectangular shaped films with a platinum wire frame. Their results are shown in Fig. 7.127. From this figure, the activation energy for drainage was about 150 kJ/mol, very close to the activation energy for the same system, 160 kJ/mol.

Gaskell, cited in Ref. [24], reported the foaming index as function of temperature for a 35%CaO-35%SiO$_2$-30%FeO slag:

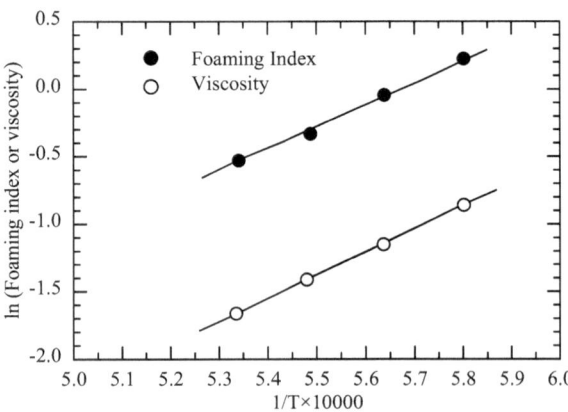

Fig. 7.125 Foaming index and slag viscosity as a function of temperature. After [93]

Table 7.17 Activation energies for foaming and viscous flow

First author, year	Gaskell, 1991	Ito, 1989	Ozturk, 1995	Jung, 2000	Nexhip, 2000
E$_a$, foaming kJ/mol	66	160.0	139.6	126.5	150
E$_a$, viscous flow kJ/mol			144.7		160

Fig. 7.126 Effect of temperature on the foaming index. After [172]

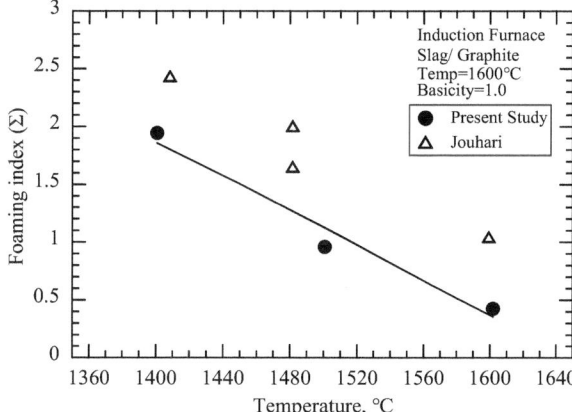

Fig. 7.127 Effect of temperature on the draining rate. After [49]

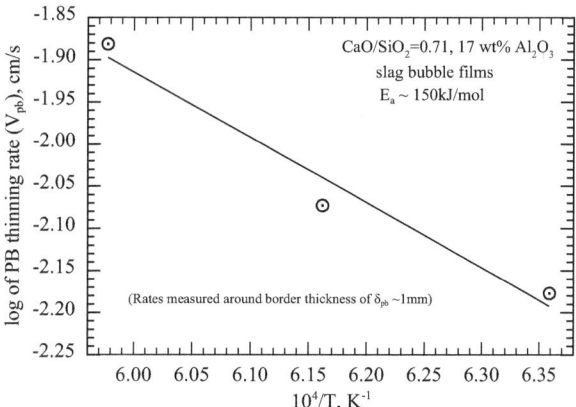

$$\log \Sigma = \frac{7750}{T} - 1.74$$

The activation energy reported by Gaskell was less than half of that reported by Fruehan's research group. Previous results by Cooper and Kitchener [71] and Swisher and McCabe [52] indicate a higher activation energy for foam decay in comparison with viscous flow, 4 times and 2.5 times higher, respectively. A much lower activation energy for viscous flow in comparison with the activation energy for foaming would suggest that viscous flow is not the most important variable in slag foaming. Gaskell suggest that viscosity is not the only variable controlling foaming and therefore it is possible to reach different scenarios; lower activation energies (his case), similar values (Fruehan's group and Nexhip) or higher activation energies. This subject still requires further examination using the same definition for the foaming index, slag chemistry and experimental measurement technique.

An important observation made by Nekrasov et al. [77] is that the reference temperature in the previous discussions is the temperature of liquid steel. The slag has a higher temperature because is in direct contact with the electric arcs. This difference should be considered when the slag physical properties are defined.

The following table summarizes with different expressions the effect of viscosity and temperature.

2021	Chang et al. [187]	$\frac{\Delta h}{h_0} = 2 \times 10^{-4}\mu + 0.6552$
2000	Jung and Fruehan [127]	$\log \Sigma = \frac{6610}{T} - 3.90$
1995	Ozturk and Fruehan [93]	$\log \Sigma = \frac{16797}{T} - 4.75$
1991	Gaskell [24]	$\log \Sigma = \frac{7750}{T} - 1.74$

7.9 Effect of Carbon Particles on Slag Foaming

7.9.1 Effective or Apparent Viscosity

The behavior of the solid particles on foam stability depends on the affinity or repulsion between the particle and the liquid. Hydrophobic particles attach to the bubbles and follow the bubbles (froth flotation), they adsorb to the gas–liquid interface and act as a barrier to promote bubble coalescence. Particles can stick to the bubble/liquid interface if the contact angle is in the range from 50 to 75° [1]. Hydrophilic nano particles can inhibit liquid drainage which also promotes foam stability; however, this effect depends on the size of the particles with respect to the maximum diameter of the circle inscribed in the Plateau border cross-section.

The presence of fine particles in the liquid increases the apparent viscosity. If the solid particles are smaller than the foam bubbles, they stabilize the foam, otherwise, large particles suppress the foam. Yoon and Shin, cited by Wu [188], reported in 1999 that increasing the carbon particle size from 0.8 to 4 mm, affected slag foaming by decreasing the foam height.

Lahiri et al. [189] discussed the relationship between bubble size and its stability when they reach a solid particle. Under equilibrium conditions the Young's equation defines the equilibrium contact angle and for this condition the net outward force is zero;

$$\sigma_{\text{solid-slag}} - \sigma_{\text{solid}} + \sigma_{\text{slag}} \cos \theta_e = 0$$

When there is no equilibrium, the contact angle $\theta = 180 - \theta_e$

$$\sigma_{\text{solid-slag}} = \sigma_{\text{solid}} + \sigma_{\text{slag}} \cos(180 - \theta_e)$$

Then, the outward force is:

$$F = \sigma_{slag}[\cos(180 - \theta_e) + \cos\theta]$$

The outward force increases by increasing θ_e and/or decreasing θ, when this happens the outward force will pull the bubble and make it flat. They found that as the bubble size increases the outward force also increases making the bubbles flat and breaking them up, on the contrary, for small bubbles the outward force is small and do not affect bubble break up.

There is a large number of equations that can be employed to compute the effective viscosity due to the presence of solid particles in the slag. Liu et al. [190] reviewed this subject. The first model was proposed by Einstein in 1906, valid for a small fraction of particles. In 1962 Rutgers reviewed 250 viscosity models. The model suggested by Roscoe in 1952 is the most accepted:

$$\mu_e = \mu(1 - 1.35\varphi)^{-2.5}$$

where: μ_e is the effective viscosity, μ is the bulk viscosity and φ is the fraction of solid particles. The factor 1.35 changes depending on the particle shape. The general expression with adjustable parameters is called Einstein-Roscoe equation:

$$\mu_e = \mu(1 - K\varphi)^{-B}$$

Kondratiev and Jak [191] fitted the previous expression using 800 slags containing CaO-SiO_2-"FeO"-Al_2O_3 and obtained $K = 2.04$ and $B = 1.29$. Wright et al. [192, 193] evaluated the effective viscosity of slags containing precipitated particles, reporting the following values:

CaO-SiO_2-MgO-Al_2O_3 slags; $K = 2.6 - 4.24$, $B = 1.28 - 2.5$
CaO-FeO slags; $K = 2.74 - 3.89$, $B = 2.5$

Increasing the effective viscosity is beneficial to foam stability because it decreases liquid's drainage. Figure 7.128 illustrates that increasing the effective viscosity improves the foam life until a certain limit is reached.

Martinsson and Sichen [195] measured the effective viscosity in a physical model using different particles and droplets: solid silica gel from 2 to 10 mm, solid pig iron particles of 2 mm and liquid metal droplets of 3 mm. They reported the importance to use the effective viscosity to analyze foaming. There are big differences between the effective and dynamic viscosity. The effective viscosity is about 4–5 times higher in comparison with the dynamic viscosity, as shown in Fig. 7.129.

Bindal et al. [196] also reported a saturation limit of colloidal particles when treating radioactive waste, which causes a peak in foaminess, as shown in Fig. 7.130.

Fig. 7.128 Effect of the effective viscosity on the foaming index. After [194]

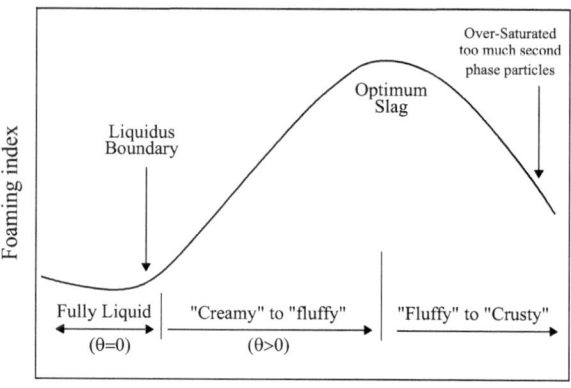

Fig. 7.129 Comparison of the dynamic viscosity for pure silicone oil and its effective viscosity containing particles. After [195]

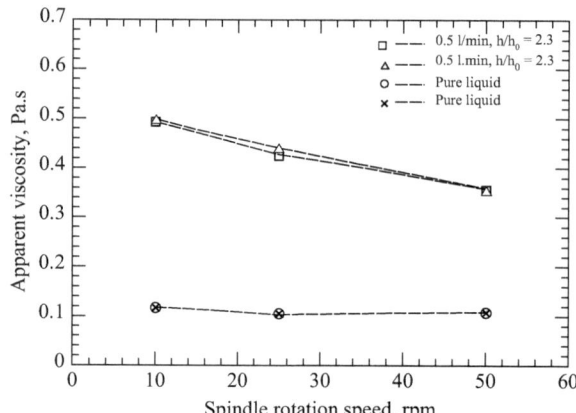

Fig. 7.130 Comparison of foaming with and without solid particles. After

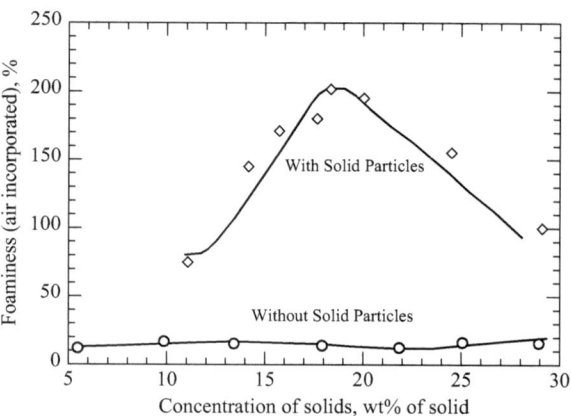

Fig. 7.131 Effect of slag viscosity on foam height. After [122]

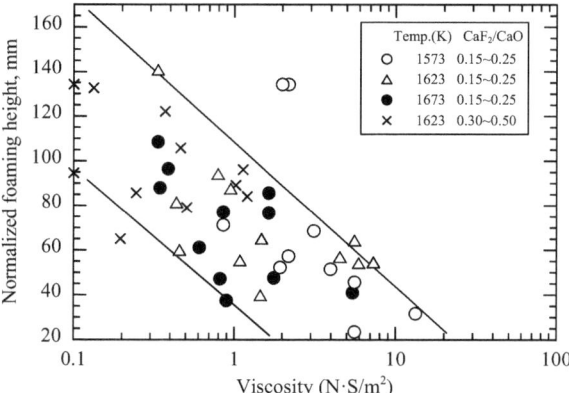

An example of extreme viscosities is the work reported by Kitamura and Okohira [122]. They employed CaO-SiO_2-FeO-$(1$–$2)MnO$-$(1.5$–$2.5)Al_2O_3$-$(2$–$3)MgO$-$(1.5$–$2.5)tiO_2$-$(5.5$–$7.5)P_2O_5$-CaF_2 slags with a basicity B_2 in the range from 0.8 to 5.0 and a CaF_2/CaO ratio from 0.15 to 1.0. Figure 7.131 shows how the foam height decreases by increasing viscosity but it should be noticed the scale of viscosity, from 100 to 10,000 cP. They also calculated with the cell model the fraction of solids as a function of basicity, temperature and CaF_2/CaO ratio. They found that increasing basicity above 1.2 increased the fraction of solid particles.

Coherent jets can operate as burners, oxygen lances and injectors of particles (carbon, fluxes, etc.). Mombelli et al. [197] compared two practices adding fluxes; as lump and injected particles in a 90 ton EAF. The flow rate of injected fluxes ranged from 40 to 100 kg/min. The dissolution efficiency is higher when fluxes are injected contributing to decrease the total amount of fluxes and energy consumption which decreased by about 20–30 kWh/ton. They evaluated improved foaming conditions using PSD diagrams, observing that the experimental values were closer to the saturation line.

7.9.2 Effect of Carbon Particles on Slag Foaming

Ito and Fruehan [99] in 1989 reported that solid particles increase the foaming index. They added $2CaO \cdot SiO_2$ or CaO particles with a diameter from 30 to 100 μm into the slag. The results are shown in Fig. 7.132, at two temperatures. In this figure Σ° represents the foaming index without solid particles.

In another investigation by Zhang and Fruehan [141] in 1995 they reported the suppressing effects of carbon particles on slag foaming. CaO-SiO_2-FeO-CaF_2 slags were foamed injecting argon through an alumina tube of 1.75 mm diameter and the foam height at 1500 °C was measured by an electric probe. Foaming was compared with and without solid particles. The solid particles consisted of three types; coke

Fig. 7.132 Effect of solid particles on the foaming index. After [99]

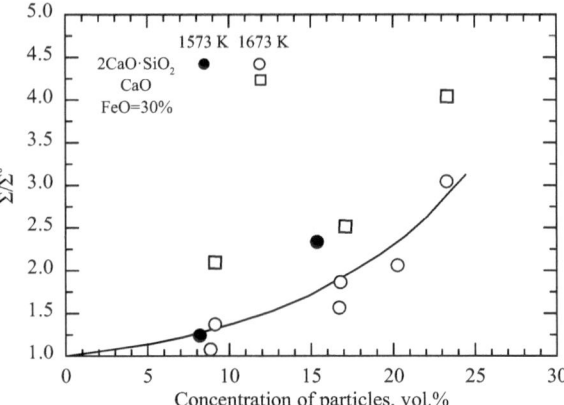

particles, FeO pellets with a diameter of 15 mm, alumina rods of 10 mm. The carbon particle size ranged from 3 to 10 mm for coke spheres, 10 mm for thin coke disks and 10 mm coal char lump particles. The reason why foaming was suppressed with the particles employed is due to the large size of the carbon particles; from 3 to 10 mm in diameter. This work is discussed in more detail in the section on foam suppression.

Corbari et al. [129] studied the reduction rate of FeO by solid carbon, its effect on slag foaming and the critical FeO concentration at which its effect on decreasing the slag viscosity is stronger than its effect on producing CO, representing the optimum FeO concentration. $CaO\text{-}SiO_2\text{-}FeO\text{-}MgO$ slags were prepared using reagent grades. Temperature, basicity and MgO were kept constant; 1550 °C, 2.0 and 8%, respectively. FeO was prepared using a 1:1 molar mixture Fe_2O_3 and Fe, the formed pellets were exposed to a $CO\text{-}CO_2$ gas at 1200 °C for 24 h and then cooled under argon. The pellets were pulverized and FeO confirmed by XRD. FeO was varied from 15 to 45%. Five different types of carbonaceous materials were tested; graphite, bituminous coal, bituminous char, anthracite coal and anthracite char, in amounts that varied from 0.25 to 0.285 g. Bituminous char and anthracite char were prepared by devolatilization of their respective coals under argon atmosphere at 1150 °C and 1050 °C, respectively. The carbonaceous materials were made into rough spheres. Due to a higher density, graphite, bituminous coal and anthracite coal were prepared in one piece but for bituminous char two pieces due to its lower density. Table 7.18 summarizes the physical properties of the carbonaceous materials.

The production of moles of CO as a function of time gave the rate of production of CO. The generation rate of CO increased as the concentration of FeO also increased. Figure 7.133 (left) shows the rate of production of CO for bituminous char is higher than graphite but due to higher surface area for bituminous coal (3.5 vs. 1.49 cm^2), the specific rates are similar, as shown in Fig. 7.133 (right). Anthracite coal gives the higher production rate of CO. The behavior was attributed to a high rate of devolatilization and subsequent carbon fragmentations when anthracite was in contact with the slag. The foam height as a function of time and different FeO concentrations, using anthracite char, is shown in Fig. 7.134. It is shown that as

Table 7.18 Properties of carbonaceous materials. After [129]

Type	C_{fix}, %	Volatiles, %	Ash, %	ρ, kg/m^3
Bituminous coal	60.0	34.0	6.0	–
Anthracite coal	85.6	3.4	11.0	
Bituminous char	90.0		9.0	560–720
Anthracite char	89.0		11.0	1260
Graphite	99.0		< 1	1450

FeO increases, from 15 to 45%, the foam height increases, remains stable and then decreases. The maximum height was reached for 25%FeO.

Figure 7.135 shows the foaming index as a function of FeO. The maximum value is obtained for 25% and then decreases. Graphite yields the highest foaming index with 25%FeO. This index was calculated excluding the bubble size because it was not measured in the experimental work.

Fig. 7.133 Rate of generation of CO, **a** mole/s and **b** mol/s cm^2. After [129]

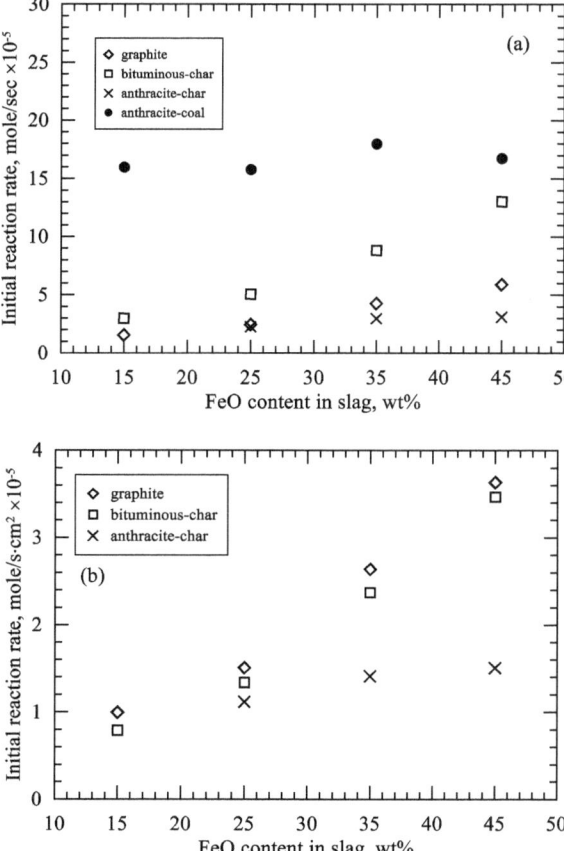

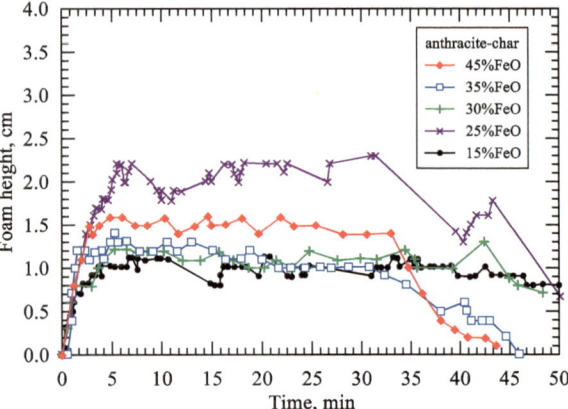

Fig. 7.134 Foam height as a function of time for different FeO concentrations. After [129]

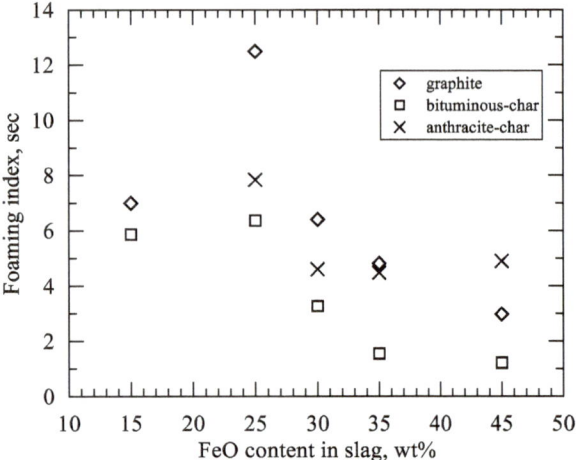

Fig. 7.135 Effect of the initial FeO on the foaming index for different carbonaceous materials. After [129]

The effect of coals was not clear in their experimental work. In principle, coals produce additional gases which increase the foam height, however the bubble size is large and the foam height decays in the case of anthracite coal. The behavior of bituminous coal and bituminous char was similar.

Replacement of coal by charcoal contributes to decrease the overall production of CO_2 [198].

Bianco et al. [6] investigated the replacement of coal by char obtained from biomass. Their results indicated that char can replace coal as a foaming agent, however from a more general report it was concluded mixed results from three plants [199]. DiGiovanni et al. [200] compared 4 types of carbon, shown in the table below, on slag foaming. The bio-briquette was made of black spruce softwood biochar and bio-oil. For the experiments, 20 kg of steel (0.2%C) and 1.13 kg of EAF slag (23%FeO, 39% CaO, 14% SiO_2, 10% MgO, 8% Al_2O_3) were melted in an induction furnace using a crucible made of alumina. The experiments were carried out in a temperature range from 1620 to 1650 °C. In each experiment carbon was added in 10 batches every 6 s. Carbon was arranged in packets of 6 g, in each experiment 60 g C were added. Once carbon was added the slag was stirred to mix completely.

	Nut coke	Industrial injection carbon (IIC)	Loose biochar pellets	Bio-briquette
Fixed carbon	88.9	91.3	87.8	67.7
Volatile matter	0.86	6.03	6.01	31.1

The maximum slag foaming height was higher with nut coke. Taking Industrial injection carbon (IIC) as a reference (100%), nut coke gave 5 times higher foam height and bio-briquette 2 times higher. Biochar was the inferior foaming agent. The authors attribute the good performance of the bio-briquette due to the high volatile matter. Considering a higher fixed carbon content, it would be expected a better performance of the carbon labelled IIC but its particle size was not indicated. In the case of bio briquette the authors found better foaming capacity with particles in the range from 0.15 to 0.6 mm in comparison with much larger particles of 6 mm.

Kieush et al. [201] compared 6 types of carbon sources on slag foaming; three conventional carbon sources and wood biochar. The slag was prepared mixing powders. The base composition was 31.6% CaO, 19.4% SiO_2, 28% FeO, 11.9% MgO and 9.1% Al_2O_3. It was placed in a crucible made of alumina (diameter 63 mm and height 99 mm) and this crucible inside another graphite crucible. The crucible was charged with 5 gr of ULC steel and 100 g of slag. The experiments were carried out at 1600 °C. The amount of charged carbon was calculated based on the amount of FeO in the slag. The slag height was measured with a Mo rod. The foaming process took 6 min in all experiments, after that the alumina crucible was quenched with liquid nitrogen. Foaminess was measured with the ratio of foam height to the initial height or relative foaming height. The relative foaming height for all the conventional carbon sources was in the range from 7 to 9 and for biochar in the range from 5.5 to 6.2. The relevant result is that a mixture made of 50% biochar-50% coke gave values in the range from 8 to 8.5, suggesting that can be a good replacement of conventional carbon sources. In addition to this, they also evaluated the composition along the quenched slag. They found differences in the vertical direction from top to bottom. At the top they found Fe_2O_3 attributed to slag oxidation and at the bottom the presence of iron particles produced by the reduction of FeO.

The research group from Prof. G. Irons at McMaster University in Canada reported several works dealing with the injection of carbon particles and slag foaming [75, 94, 95, 138–140, 202, 203]. They developed a model to predict the carbon gasification rate, bubble diameter, film thickness and bubble life time as a function of the carbon injection rate. They assumed each particle formed one bubble. They suggested the maximum injection rate of carbon particles at which foaming becomes inefficient. Ji et al. [138–140] injected coal particles into CaO (22–40%)-SiO_2 (7–15%)-FeO (31–43%)-MgO (7–20%)-Al_2O_3 (2–9%) slags. The experiments were carried out at

1600 °C in a 75 kW induction furnace with a capacity 65 kg, using MgO crucibles with an interior diameter of 191 mm and height of 295 mm, a graphite susceptor was placed outside the crucible to enhance the slag melting rate. The charge consisted on average of 17 kg of ULC carbon steel and 3–4 kg of slag powder. The height of steel was about 90 mm and the initial slag height was about 40–70 mm. When the steel temperature reached 1600 °C, 90 g of copper were added to identify iron droplets from steel and slag, then the slag was added, in three batches and fully melted.

The carrier gas to inject the carbon particles was nitrogen. The gas was passed through a flowrate transducer and then injected to the slag and another part of the gas was introduced into the feeder for coal injection. The feeder was suspended from a load cell to monitor the solids flow rate. The pressure in the feeder was controlled by a pressure transducer. The load cell, pressure transducer and flow rate were controlled by PLC.

Metal and slag samples were taken before and after each experiment. The crucible was covered with a stainless-steel lid with a port for injecting the carbon particles, a port for gas extraction and a port to insert a steel rod to measure foam height. The outlet gas was passed through a filter to collect the dust. Gas samples were taken during the experiments using gas tight syringes. The lance was preheated 5 min using only nitrogen then the injection of carbon particles started.

The coal particles had a small size, 75% was in the range from 100 to 300 μm with a fixed carbon content of 92% and 3.92% of volatile matter. They were injected into the slag using a lance and a carrier gas (nitrogen) with a gas flow rate of 7.75 Nl/min. The tip of the lance was 2.0 cm above the metal surface. The injection time was from 30 to 122 s and the mass injected was from 13 to 82 g. The maximum foam height varied from 11 to 20 cm.

The dust produced varied from 1.2 to 3.5 g, containing 3.2 to 12.9%C, equivalent to about 1% of the total injected carbon. In the experiments, the CO produced reacted with oxygen on the free board and part of the CO transformed into CO_2. The gas composition includes CO, CO_2, N_2 and O_2. The CO_2 includes CO_2 from post-combustion and CO_2 produced by oxidation of the carbon particles. To define the degree of coal gasification, a model was developed, with a number of assumptions which include the definition of a rate parameter, χ, by trial and error. Figure 7.136 shows the carbon balance using their model. It shows the carbon distribution; fraction of carbon gasified, carbon in the slag and carbon dissolved in liquid steel. After 3 min most of the carbon is gasified. An important finding from this work is the definition of the maximum injection rate of coal particles. Below this point gasification depends mainly on the injection rate. The saturation injection rate for their experimental conditions was found to be 0.014 molC/s kg slag, equivalent to 10 gC/min kg slag, which is also equivalent to a maximum gasification of 0.011 molC/s kg slag, as shown in Fig. 7.137.

Wu et al. [188] compared the effect of the coke particle size on the foam height. They measured the foam height using a metallic wire and the gas injected was nitrogen. The results shown in Fig. 7.138 correspond to a slag containing 41%CaO-26%SiO_2-8%MgO-15%Al_2O_3. It can be seen that as the particle size decreases, the foam height increases, reaching a maximum value with particles of 76 μm. Note the

Fig. 7.136 Carbon balance of injected coal particles. After [140]

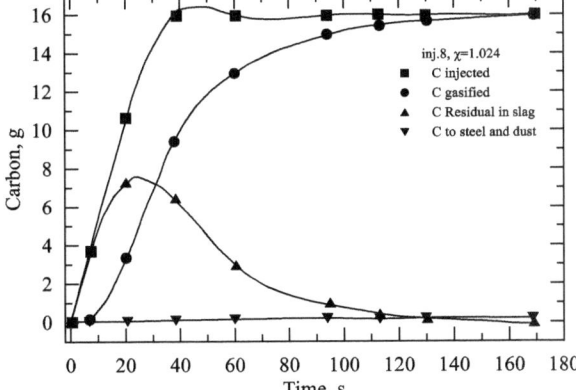

Fig. 7.137 Gasification of injected coal particles as a function of the injection rate. After [140]

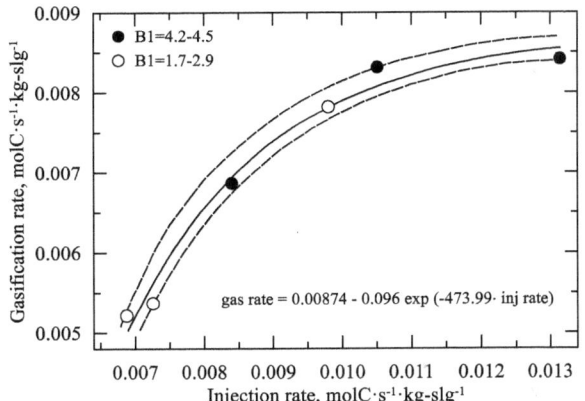

error in the units for the gas flow rate in this figure, it is possible that the variable is actually the superficial gas velocity because the way they define the foaming index in this report $\left(\Sigma = h_{foam}^{max}/V\right)$.

Ogawa et al. [3] in 1992 reported additional observations in situ using carbonaceous materials similar to bath smelting. The size of the carbon particles was from 10 to 25 mm. Their results indicated a decrease in the residence time of CO bubbles as the carbon/slag mass ratio increased from 0.1 to 0.5. Increasing the mass of the carbon particles increased bubble coalescence. In the conclusions they reported that smaller particles decreased the foaming height due to a larger surface area and higher bubble coalescence. This behavior is contrary to what has been reported that small particles improve foaming due to higher effective viscosity but this is due to the carbon particles they employed which are big particles. Katayama et al. [3] subsequently, citing the paper from Ogawa, indicated that through X-ray fluoroscopic analysis they found that the bubbles that promoted foaming had a bubble size lower than 1 mm, furthermore, a foamy slag could be formed with two layers when the rate

Fig. 7.138 Effect of has
flow rate on the foam height.
After [188]

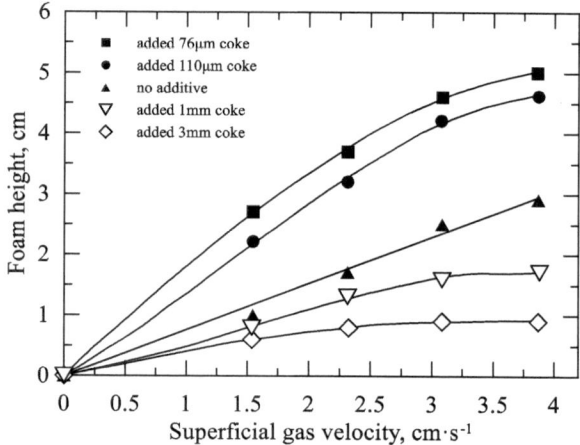

of production of CO was low (or its equivalent of rate of production of FeO). At high
rates of production of CO, the foam height increased. In bath smelting it is desired
only one slag layer of foamy slag. Katayama et al. calculated the slag density from
the measurements on the slag height, as shown in Fig. 7.139. Foamy slags have a
density from 0.6 to 1.2 ton/m^3.

The specific consumption of carbon particles for slag foaming ranges from 2 to
15 kg/ton [204]. Increasing oxygen consumption and arc length, both increase the
requirements of carbon particles. DC EAF's operate with longer arcs. Nekrasov [77]
reported that carbon injection is about 0.3 kg C/min ton and oxygen injection from
25 to 40 Nm^3/ton with a ratio of C/O from 0.3 to 0.8 kg C/Nm^3 O$_2$.

Svensson [205] observed foaming by addition of iron particles containing carbon
(with carbon ranging from 0.3 to 3.9%C) into 35%FeO-35%SiO$_2$-40%CaO slags at
1600 °C. He observed an incubation time of a few seconds, 4–8 s, for the system to
start foaming, following by strong foaming and finally further decrease in intensity.

Fig. 7.139 Effect of gas
evolution rate on the average
slag density. After [3]

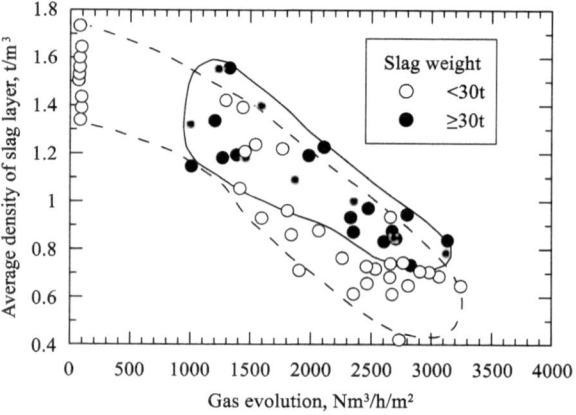

The total foaming time increased from 1 to 10 min as the carbon content in the particles increased.

Kipepe and Pan [206] reported one of the few experiments on slag foaming in induction furnaces. They employed a small furnace of 5 kg. Once the scrap was melted, they injected oxygen at a constant flow rate and simultaneously also injected carbon particles. The oxygen flow rate was kept constant at 100 ml/s and the injection rate of carbon particles was increased from 21 to 55 g. The foam height was measured with an alumina rod. They measured the melting time, electric energy consumption, the temperature of the slag surface and also calculated the heat losses. Their results indicated a minimum in energy consumption with the addition of 40 g of C, below or above this value energy consumption increased. The foam height also decreased when the carbon addition exceeded 40 g, attributed to an excess of carbon particles.

7.10 The Isothermal Saturation Diagrams (ISD)

In 1992, Burstrom et al. [207] reported pilot plant and industrial trials that supported the strategy to pay attention to two aspects; a high effective slag's viscosity and a high generation rate of CO. Their work is discussed below in a another section. Pretorius and Carlisle [194] in 1999 applied the model suggested by Burstrom et al. They divided the main slag components into two groups; refractory oxides (CaO, MgO) and fluxing oxides (SiO_2, Al_2O_3, FeO, MnO, CaF_2). As shown below, refractory oxides have much higher melting temperature in comparison with fluxing oxides.

	MgO	CaO	Al_2O_3	MnO	SiO_2	CaF_2	FeO
T_m, °C	2800	2600	2030	1850	1720	1420	1370

Pretorius and Carlisle [194] suggested the need to achieve both CaO and MgO saturation, to improve foaming which also reduces refractory wear due to chemical attack. Actually, a slag should be first designed to guarantee both CaO and MgO saturation to minimize refractory consumption. As a secondary benefit is the promotion of foaming conditions.

The net production rate of CO depends on the oxidation rate due to oxygen injection which produces FeO and its rate of reduction rate which consumes FeO. If there is enough amount of carbon particles, FeO can be controlled between a certain concentration to control slag fluidity.

From a practical viewpoint, one part of the success to reach good foaming conditions depends on the timing to add fluxes and the rate of oxygen and carbon particles. If the furnace is charged in batches the concentration of CaO and MgO in the beginning will be very high resulting in a highly viscous slag. Oxygen injection increases the formation of FeO and then decreases the slag's viscosity. Using a dynamic mass balance helps to provide the adequate amount of additions.

There are several models which can be employed to calculate the slag composition to achieve CaO and MgO saturation, or both, for example; Bergman [208] in 1989 proposed the following relationship between MgO_{sat} and optical basicity (Λ) at 1600 °C:

$$\log(\%MgO)_{sat} = 7.499 - 9.121\,\Lambda$$

Park [209] carried out experimental measurements with CaO-FeO-Fe_2O_3-SiO_2-MgO-(MnO)-(Al_2O_3) slags, at 1600 °C, deriving the following expression:

$$\%MgO_{sat} = 0.00816X^2 - 1.404X + 62.31$$
$$X = (\%CaO) + 0.45(\%FeO + \%Fe_2O_3) + 0.55(\%MnO)$$

The experimental MgO saturation values at 1600 °C reported by Tromel et al. [210] are shown in the ternary diagram shown in Fig. 7.140.

Pretorius and Carlisle used the ternary phase diagrams for the CaO-SiO_2-FeO-MgO systems at 1600 °C to define the composition for dual saturation. In this analysis TiO_2, VO_2, Cr_2O_3 were neglected because its concentration is usually below 2%. MnO has a similar effect to FeO, therefore it was grouped together. Figure 7.141 summarizes the MgO saturation concentration at 1600 °C as a function of basicity (CaO + MgO/SiO₂). This value is the MgO saturation concentration at 1600 °C without Al_2O_3.

Since Al_2O_3 decreases the solubility of MgO, Selin proposed the following correction factor:

$$MgO_{sat} = 0.615 \times \left(\frac{\%Al_2O_3}{\%Al_2O_3 + SiO_2}\right)(\%MgO_{ref} - 6)$$

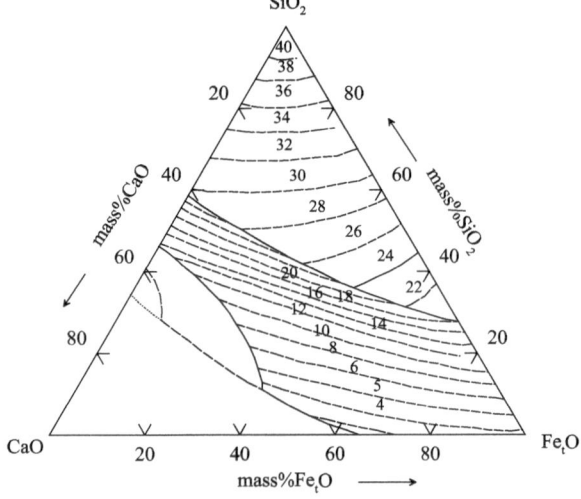

Fig. 7.140 MgO Saturation limits (wt%), given by the numbers inside the diagram. After [210]

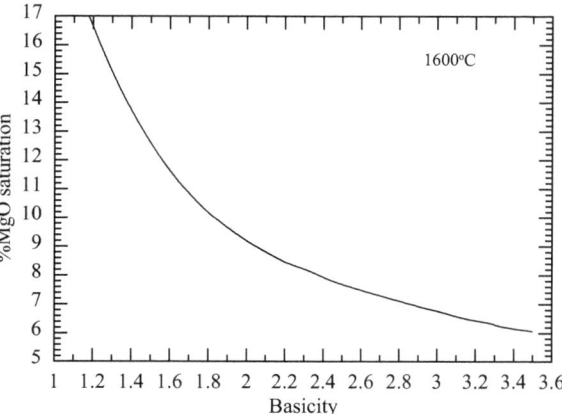

Fig. 7.141 MgO_{ref} at 1600 °C as a function of basicity. After [194]

The correction factor for temperature is:

$$\%MgO_{sat} = \%MgO_{ref} + 0.0175 \times (T - 1600)$$

The advantage of the graphical representation suggested by Pretorius and Carlisle is that is not only quite simple to calculate and visualize in a two-dimensional plot but more important, the MgO saturation is defined as a function of FeO in the slag. The slag basicity remains with minor changes throughout the heat in comparison with the changes on FeO, therefore is more convenient to describe the saturation limits at constant basicity values. The two-dimensional plots defining the MgO saturation concentration as a function of FeO at constant basicity values were called Isothermal Saturation Diagrams (ISD), also known as isothermal solubility diagrams or isothermal stability diagrams. Figure 7.142 shows the case for a basicity of 1.5. According with the previous discussion, the objective is to design a slag chemistry that reaches dual saturation and targets a composition which is slightly above this value to have solid particles precipitated. In this figure, point (a) represents the chemical composition for dual saturation. The dashed line around the liquidus region represents slags with suspended particles. It can be observed that as the concentration of FeO increases from point (a), the region of liquid increases. Line a-b represents the saturation line for MgO, it is observed that the MgO_{sat} decreases as FeO increases. In the diagrams there are three points, X, Y and Z which represent three different slags, showing what happens as they move in a direction of higher FeO. Slag X moves to point X_1 due to oxygen injection which causes an increment of FeO, at this point the effective viscosity is high due to the presence of solid particles. When the concentration of FeO increases to point X_2 it reaches the melting point for that slag. Further increment of FeO makes the slag fully liquid and with lower viscosity. Liquid slag X' is non-saturated and will attack the refractory. Slag Y is better than slag X because even if FeO increases is never fully liquid from Y to Y'. Slag Z has always a very high viscosity. From the three cases, slag Y_1 is the best because is closer to the dual saturation point and has a good effective viscosity. They suggest to achieve a slag which is about 1–1.5% in excess of the MgO saturation value. The authors reported additional diagrams for basicities of 2.0, 2.5 and 3.0. From these diagrams it was observed the following: (i) the liquid region decreases as basicity increases, (ii) to reach the liquidus region it is necessary to increase FeO, and (iii) MgO solubility decreases as basicity increases. As a corollary of the previous analysis, the target should be to reach point (a) at each basicity. Figure 7.143 summarizes three cases with increasing basicity.

The previous diagrams were designed for the typical steelmaking temperature of 1600 °C. If the temperature increases, which is the common case at the end of a heat, the foaming conditions will deteriorate due to a lower slag viscosity. Figure 7.144 shows the ISD for a basicity of 2.0 and two temperatures, 1600 and 1700 °C. As temperature increases the region of liquid phase increases. The authors suggest additions of fluxes to increase the effective viscosity to counteract the effect

Fig. 7.142 Isothermal saturation diagrams proposed by Pretorius and Carlisle. Ref. [194]

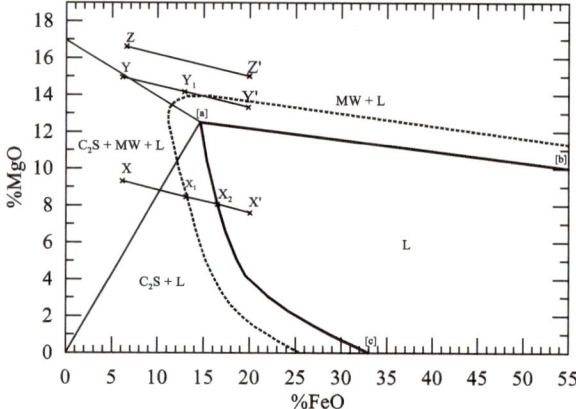

Fig. 7.143 Target of MgO saturation as a function of basicity. After [194]

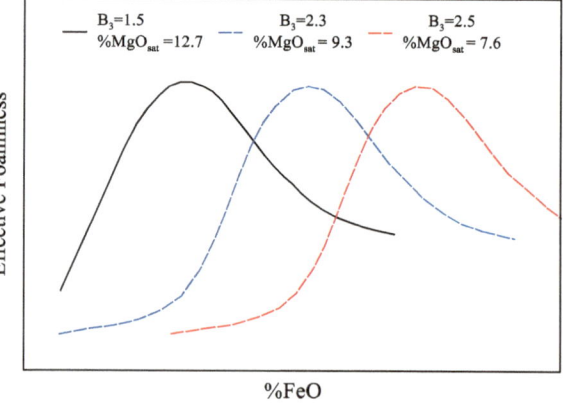

Fig. 7.144 ISD comparing two temperatures at a basicity of 2.0. After [194]

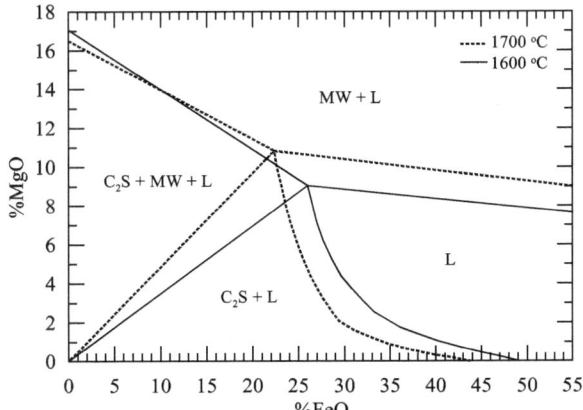

of temperature, in particular the use of dolomitic stone because its calcination is endothermic and also produces a gas.

Although the previous diagrams are quite practical to use they are limited to the conditions they were developed (slag basicity, temperature and FeO). De Almeida et al. [211] indicates that their analysis excludes the presence of other secondary metallurgical species that can affect MgO solubility. Luz et al. [74] proved significant differences when the slag basicity was lower than 1, for example for $B_2 = 0.9$ there was no precipitation of C_2S and saturation of the liquid slag was not possible, suggesting the application of thermodynamic software to carry out a more detailed calculation for each specific case. Bennett and Kwong [212] and Luz et al. [213] reported the application of FactSage. However even if thermodynamic software is more accurate, such as Factsage, especially for slags with low basicity, is still not as reliable as the direct measured values. Tayeb et al. [174] compared experimental values with values predicted with FactSage, as shown in Fig. 7.145. For basicity lower than 2.0 the agreement is quite satisfactory but for more basic slags the MgO solubility calculated from FactSage is about 1 to 2wt% lower than their experimental data.

Heo and Park [165] made an important observation about the concept of MgO saturation. This term is usually applied to define the saturation concentration of MgO, however a more proper definition is magnesio-wüstite saturation (MO) because is difficult to precipitate MgO as a single oxide in the EAF process. The same authors also stressed the importance to work under MO saturation conditions to improve not only foaming but also dephosphorization [214]. Figure 7.146 shows the use of ISD to describe slag compositions corresponding to liquid slags and saturated slags, suggesting about 5% of solids.

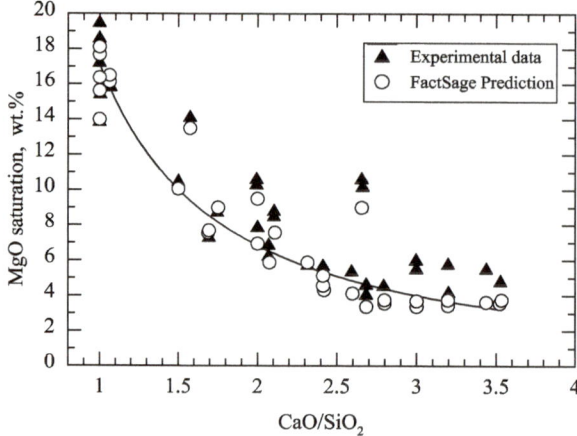

Fig. 7.145 Comparison of MgO$_{sat}$ between experimental values and those predicted using Factsage. After [174]

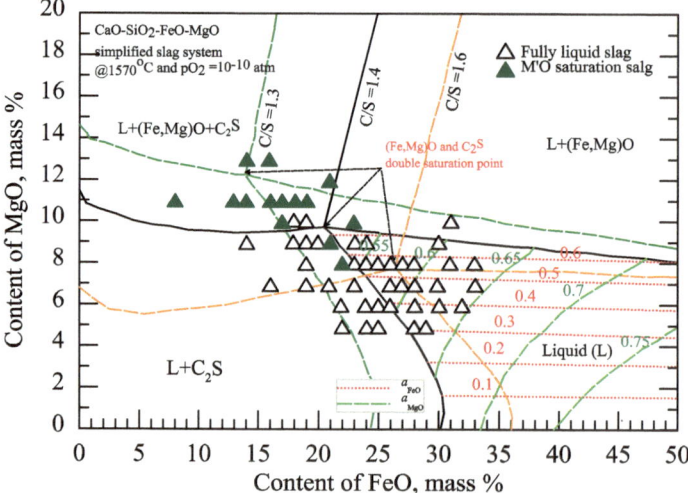

Fig. 7.146 ISD describing fully liquid and saturated slags. After [214]

7.11 Effect of Bubble Size

Ogawa and Tokumitsu [2, 132, 215], in 1990, using X-ray equipment, observed slag foaming in situ at 1500 °C. Sulphur was added to liquid slag as FeS. They found that slag foaming was possible when the bubble size was 1 mm in diameter, as shown in Fig. 7.147. This figure shows a peak in the foam height when the bubble size was 1 mm and started to decrease when the bubbles increased from 1 to

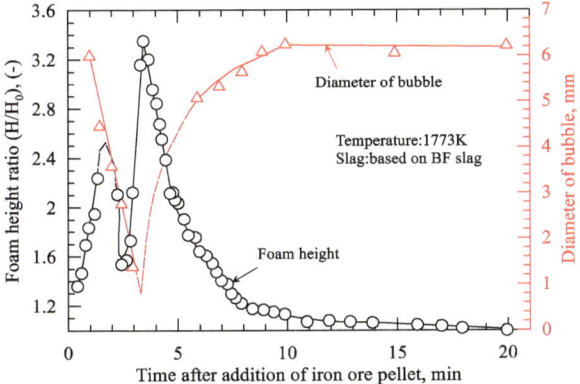

Fig. 7.147 Change in foam height ratio and bubble size evolved at slag/metal interface using BF slag and iron ore pellets. After [132]

6 mm. The slag composition was; 33.8%CaO-37.5%SiO₂-13.1%Al₂O₃-3.7%MgO-1%MnO-1%FeO-10%BaO. At the peak of the foaming height, the slag height was more than 3 times the original slag height. When the bubbles are large, about 6 mm, the foam fully collapsed. Figure 7.148 is a schematic representation of the foaming height increasing due to a larger generation of small size bubbles. The bubbles are generated due to additions of iron oxide pellets into the slag. The foam height increased with the decrease in bubble size. The bubble size decreases under the following conditions; (1) the slag is wetted by the liquid metal, (2) FeO increases and (3) sulphur decreases (Fig. 7.149). Figure 7.150 shows the relationship between iron oxide and bubble size. At low concentrations of iron oxide in the slag, below 6.5% FeO, the bubble size is larger, about 6 mm. Above 20% FeO in the slag, the figure predicts a bubble size of 1 mm. Figure 7.151 shows that small bubbles are formed at the slag/metal interface but its size increases when sulphur is present. Notice that these results are valid for a hot metal containing a large concentration of sulphur.

In a previous work by Jiang and Fruehan [216] in 1991, they found that injecting gas through a single orifice of 1.75 mm, the bubbles produced ranged from about 10 to 15 mm. In order to produce smaller bubbles, Zhang and Fruehan [128] in 1995 employed two methods; in the first method they employed a closed alumina tube with an internal diameter 4.7 mm, containing four orifices with diameters less than 1 mm around the perimeter near its end. The slag contained CaO-SiO₂-FeO-(Al₂O₃) with a basicity equal to 1, 15%FeO and about 5% Al₂O₃ due to dissolution of the crucible made of alumina. The gas injected was argon. In the second method the gas was produced by reduction of FeO at the slag/metal interface. The slag, employed, 60 g, was formed by 40%CaO-40%SiO₂-5%FeO-15%Al₂O₃. The melt was formed by 40 g of carbon-saturated Fe-C-S alloy. Sulphur was modified from 0.002 to 0.17% in order to control the bubble size. At low concentrations of sulphur, the bubble size was smaller. They observed that foams formed by gas produced at the slag/metal interface were similar to beer foams, with fine bubbles. Figure 7.152 reports the

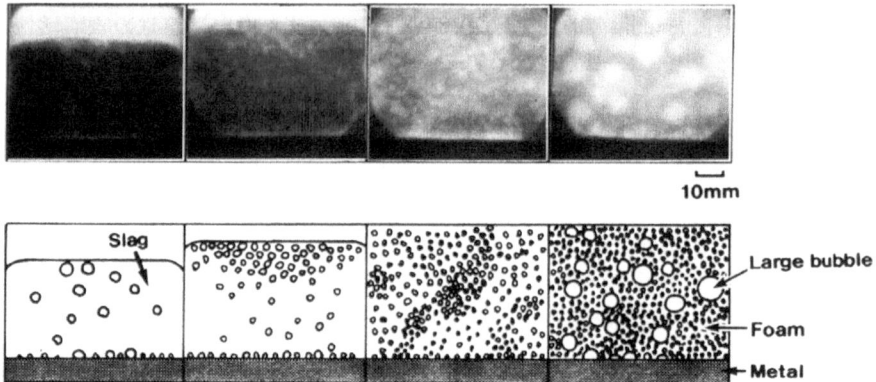

Fig. 7.148 XRF images of slag foaming after the addition of iron ore pellets. After [132]

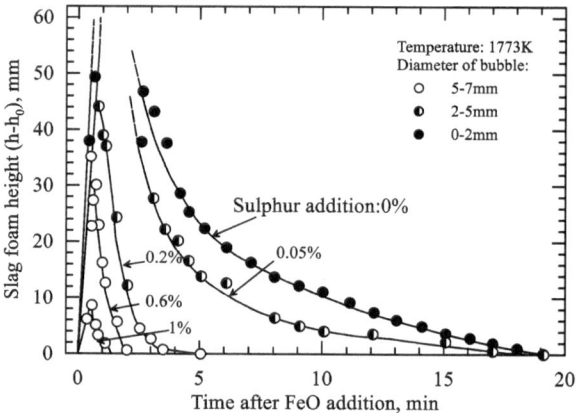

Fig. 7.149 Effect of sulphur and bubble size on slag foaming. After [132]

fluctuation in the bubble size for an experiment with high concentration of S; 0.11%. The value employed in the calculations is the average value. Table 7.19 summarizes the results. In this table, the measured values of gas generation rate in cm^3/min were transformed to mol/cm^2 s. The gas generation rate is the value observed at steady state. The foam height was the difference between the foamed height and the initial height, which was estimated as 3.5 cm. From this table, it can be observed that decreasing sulphur also decreases the bubble size and decreasing the bubble size the foaming index increases. Figure 7.153 shows the effect of the bubble size on the foaming index. Bubbles generated at the slag/metal interface by chemical reactions are smaller than the bubbles created due to argon injection [76]. Figure 7.154 shows the effect of sulphur on the bubble size. Lower sulphur produces smaller bubbles. It should be noticed that due to desulphurization, sulphur from the metal diffuses to the slag phase. Sulphur in Fig. 7.154 corresponds to sulphur in the metal phase.

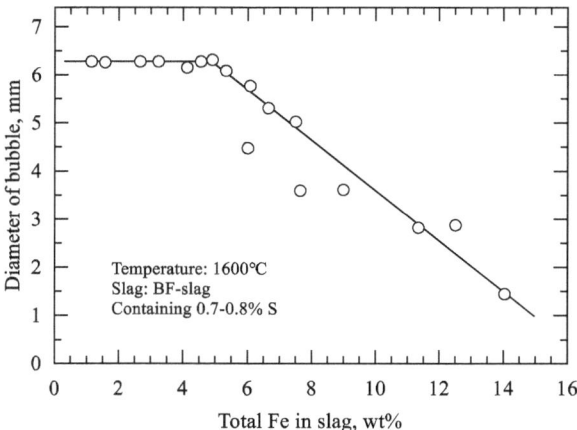

Fig. 7.150 Effect of iron oxide in the slag on the bubble size. After [2]

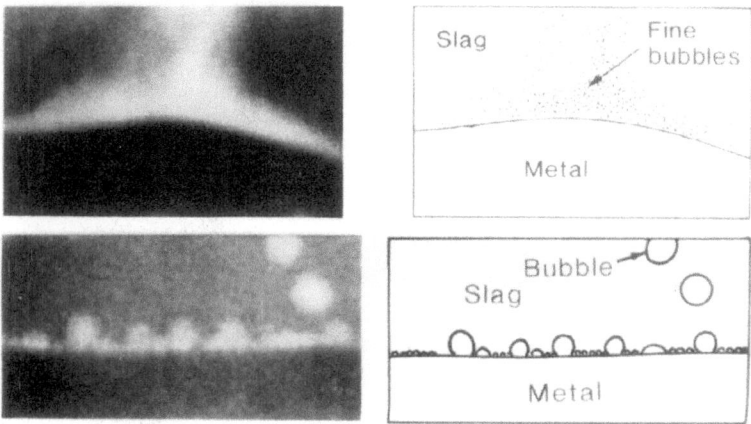

Fig. 7.151 XRF images of bubbles evolved at the slag/metal interface. Without sulphur (Upper) and with sulphur (lower). After [132]

Zhang and Fruehan argued that it is the sulphur in the metal phase which controls the bubble size, not the sulphur in the slag phase. It has been previously discussed that increasing sulphur in the metal phase increases the contact angle, this in turn promotes the formation of larger bubbles.

Kim et al. [126] citing the work carried by Yi and Rhee, reported an average bubble size from 7 to 9 mm in $CaO\text{-}SiO_2\text{-}FeO$ slags.

Jung and Fruehan [127] measured the bubble size, as shown in Fig. 7.155. In this plot the predicted values using Sano and Mori's correlation is included. It is observed that the experimental values are higher than those predicted by Sano and Mori's

Fig. 7.152 Fluctuation in
the bubble size and average
bubble size. After [128]

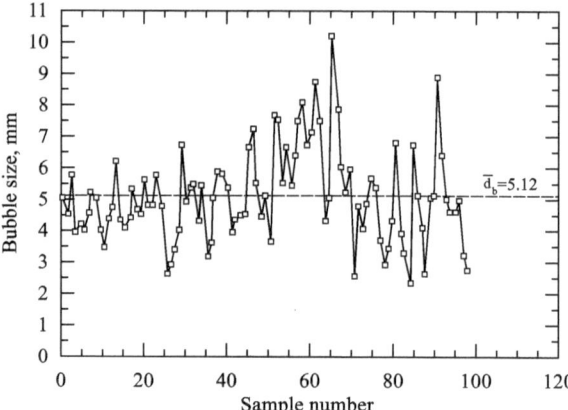

Table 7.19 Foams generated by gas/metal reaction and foaming index. After [128]

%S_o	%S_f	dCO/dt, mol/cm^2·s	$\bar{d}_b$, mm	h, mm	Σ, s
0.002	0.002	3.07×10^{-6}	0.70	16	56
0.064	0.032	2.54×10^{-6}	1.7	8.8	25
0.054	0.015	2.70×10^{-6}	3.0	2.5	7.5
0.11	0.056	2.07×10^{-6}	5.1	1.5	5.7
0.17	0.071	1.50×10^{-6}	5.4	1.1	5.75

Fig. 7.153 Effect of bubble
size on the foaming index.
After [128]

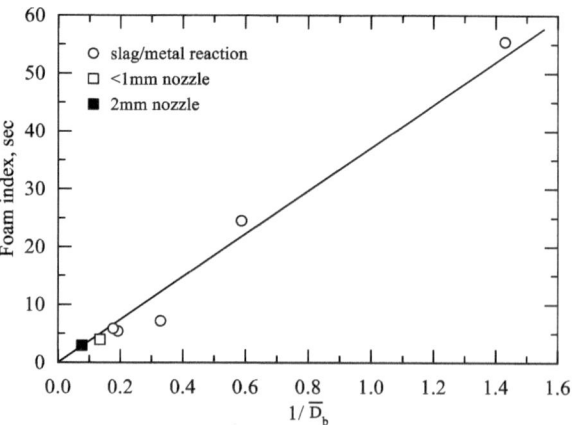

equation, attributed to the lack of a viscosity term in this equation. The predicted
value is about 7 mm, compared with 12 and 17 mm for the experimental values.

Zhu and Sichen [97] reported a strong effect of the superficial gas velocity on
the bubble size using silicon oil and injected nitrogen, as shown in Fig. 7.156. It
can be observed that as the superficial velocity increases, also increases the foam

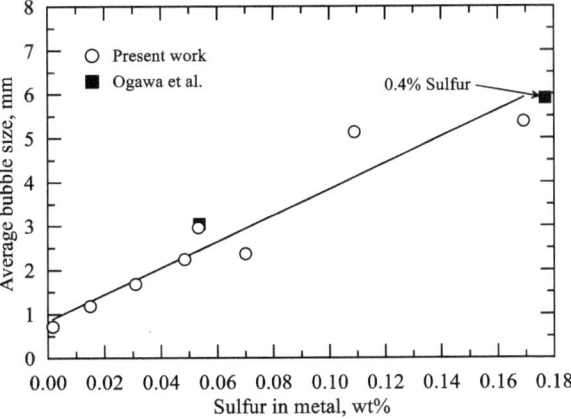

Fig. 7.154 Effect of sulphur on the bubble size. After [128]

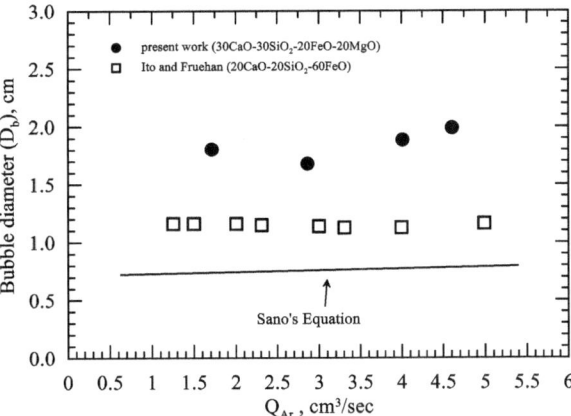

Fig. 7.155 Bubble size as a function of gas flow rate. After [127]

height, however the slope is different, indicating that in addition to the superficial gas velocity also the chemical composition has effect on the bubble size.

Hong et al. [88] reported that an increase in silica concentration decreased the size of CO gas bubbles due to lower surface tension, on the contrary sulfur tended to increase the size of CO gas bubbles.

Kapilashrami et al. [109] reported that the bubble size decreased by decreasing the viscosity of the lower (metal) phase based on bubble formation below the interface. Once they were formed, they were retained for a short period of time, depending on the viscosity of the lower phase, and allowed to grow. Their bubbles were formed by chemical reaction between oleic acid and bicarbonate in an aqueous solution, releasing CO_2 bubbles. Once the bubbles formed, they ascended to the free surface with a velocity depending on its diameter. For a bubble of 1 mm the velocity was 30 mm/s. In Fig. 7.157 the authors schematically represented all different stages of bubble formation.

Fig. 7.156 Effect of the superficial gas velocity on the foam height. After [97]

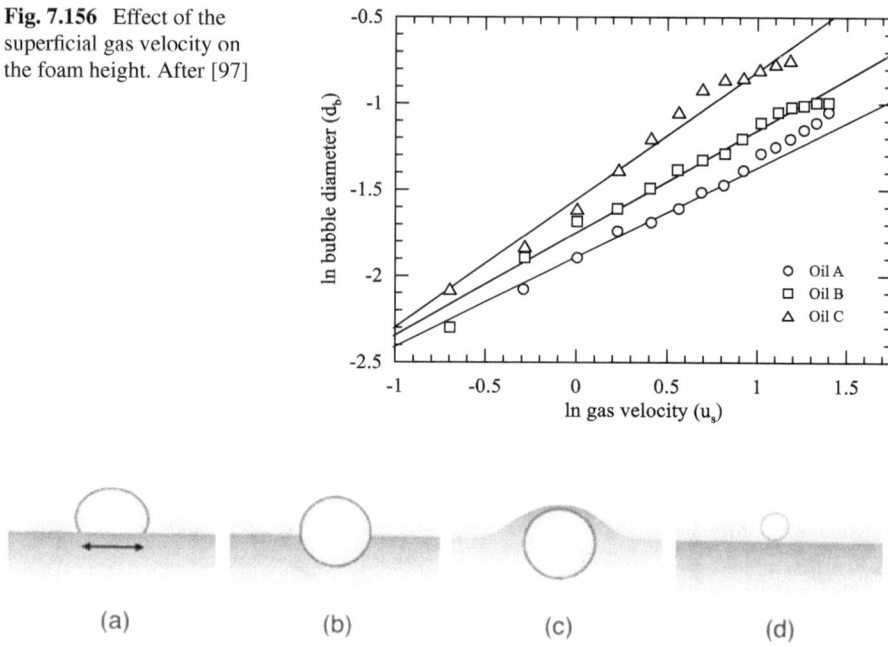

Fig. 7.157 Possible modes of bubble generation; **a** above interface, **b** at the interface, **c** below the interface, **d** in the bulk of the upper interface

Stadler et al. [96] concluded that a foam with small spherical bubbles is more stable than a foam with larger size however their measurements on bubble size were inaccurate and lacked experimental evidence to reach this conclusion.

7.12 Effect of Slag Volume

Previously it has been indicated that Bikerman [53] reported that increasing the initial volume of liquid also increased the foaming index until a critical value was reached. Their results were obtained using organic substances, indicating the criterion that the minimum slag height should be at least half of the crucible's diameter.

Ozturk and Fruehan [93] investigated the effect of the initial slag volume on slag foaming, from 1450 to 1600 °C. They employed a molybdenum disilicide resistance furnace, an alumina crucible with 41 mm diameter and 300 mm height, argon introduced with an alumina pipe of 1.57 mm diameter placed 5 mm above the bottom of the crucible. The electric probe consisted of two molybdenum wires with a diameter of 0.76 mm protected by alumina tubes. The slag was prepared using reagent grade chemicals (CaO-SiO_2-Fe_3O_4-Al_2O_3). The amount of slag was 160 g, equivalent to a slag height of 42 mm. The foam height as a function of the superficial velocity was measured with an electric probe and at the end it was quenched. The experiments were conducted with three different slag amounts; 80, 140 and 205 g, their corresponding foaming index were 2.10, 2.23 and 2.36, respectively, indicating a small increment in the foaming index. These results, on a practical basis, would indicate that the foaming index is independent of the slag volume as long as there is enough slag to form the foam. This finding justifies to exclude the slag volume on the foaming index, but they also suggested further confirmation.

Lin and Irons [217] reported the foam height as a function of the initial height of the liquid, using a water model, as shown in Fig. 7.158. Air was injected through the bottom using a porous metal disc with a pore size of 2 μm and a water column of 7.5 cm in diameter. In these experiments the gas bubbles crossed the initial height of the liquid and the foam height. It can be observed that the foam height increases as both the initial liquid's height and superficial gas velocity linearly increases up to a point where bubbles coalesce and then the foam is suppressed when the churn-turbulent flow is produced at high superficial velocities. In the last flow regime, the bubbles are large with spherical-cap shape. The foaming index applies to the first regime, in this example for superficial gas velocities below 12 cm/s.

Zhu and Sichen [97] carried out experiments with three different oil heights; 2.5, 3.75 and 5.0 cm. The results are shown in Fig. 7.159. It can be observed a linear relationship between the initial oil height and foam height.

Further experiments by Zhu and Sichen [108] producing gas by chemical reaction in a water model indicate that the initial height has a small effect in increasing the maximum foam height but the reaction lasted longer and also the foam remained at a higher height, as shown in Fig. 7.160. This case corresponds to addition of 10 g of oxalic acid. Figure 7.161 shows in more detail the effect of increasing the initial liquid height on the foam height at different superficial gas velocities. Increasing the initial slag height also increases the foam height.

Further research in the same laboratory by Wu et al. [98] confirmed the increment in foam height when the initial height increased, as shown in Fig. 7.162.

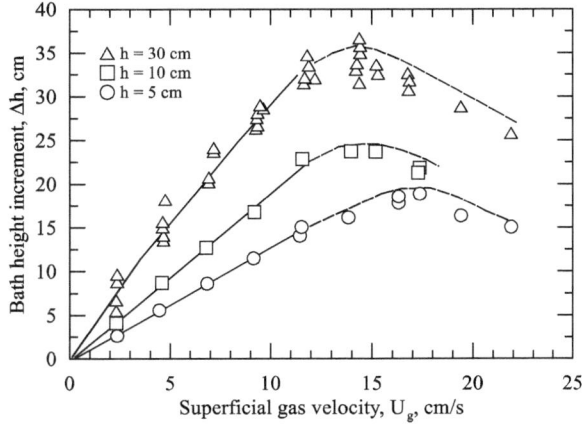

Fig. 7.158 Effect of initial liquid's height and superficial gas velocity on foam height. After [217]

Fig. 7.159 Effect of the
initial oil height on the foam
height; **a** $U_s = 0.5$ cm/s,
b $U_s = 1.0$ cm/s and **c** $U_s = 1.5$ cm/s

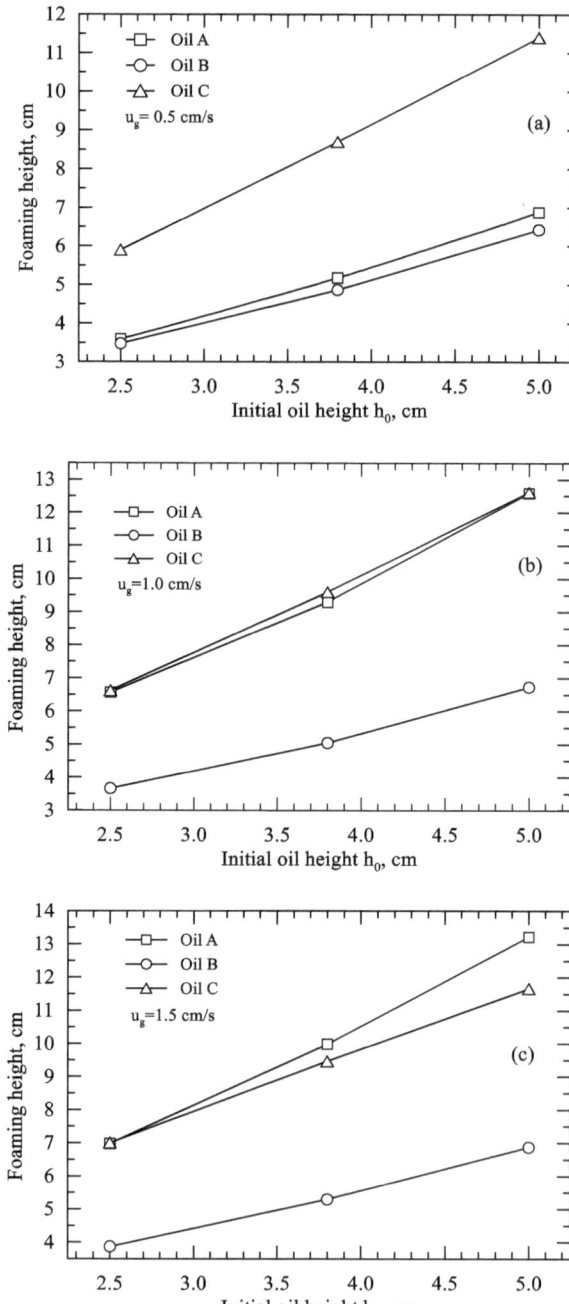

Fig. 7.160 Effect of the initial oil height on foaming. After [108]

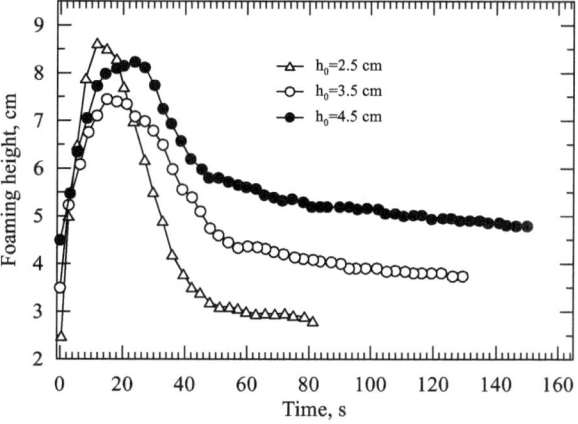

Fig. 7.161 Effect of the initial height on foam height at different superficial gas velocities. After [108]

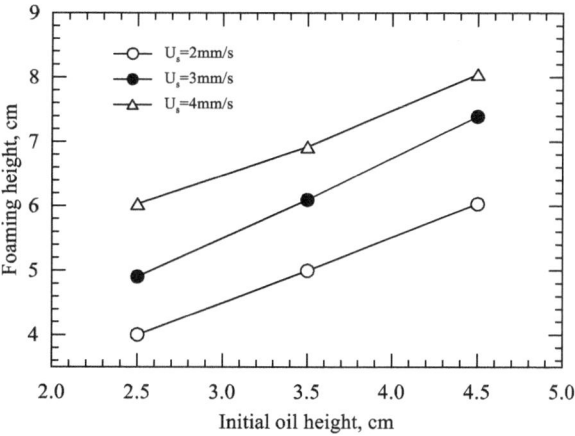

Fig. 7.162 Effect of the initial height of liquid on foam height. After [98]

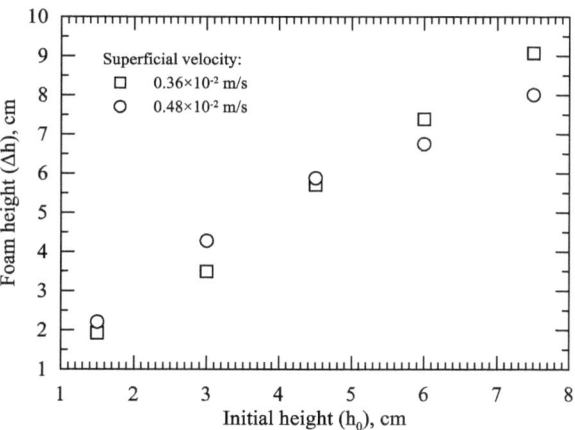

7.13   Reduction Rate of FeO and CO Gas Generation Rate

The rate of reduction of FeO by solid carbon has been extensively investigated. In a report to the 54th Committee of the Japan Society for Promotion of Science in 1966, Mori suggested the reduction of FeO by solid carbon is carried out through the following steps [218]:

- Diffusion of FeO to the reaction interface
- Interfacial chemical reactions

$$(FeO) + C_{(s)} = \underline{Fe} + CO_{(g)}$$
$$(FeO) + CO_{(g)} = \underline{Fe} + CO_{2(g)}$$
$$C_{(s)} + CO_{2(g)} = 2CO_{(g)}$$

- Separation of the reaction products from the reaction surface.

Sugata et al. [218] in 1974 carried out experiments with a carbon rod at low and high rotating speeds. At low rotating speeds both diffusion and chemical reaction were controlling mechanisms. At high rotating speeds they carried out experiments injecting CO and found that the reduction rate was much lower in comparison with a carbon rod which indicates the direct chemical reaction between CO and FeO is not relevant. They suggested the reaction between solid carbon and FeO as the controlling mechanism.

A review paper by Fruehan [219] from 1977 described that the following overall reaction;

$$(FeO) + C_{(s)} = Fe + CO_{(g)}$$

Is not a direct reaction but takes place through gaseous intermediate products. Due to non-wetting between slag and carbon particles, solid carbon is consumed by oxidation with CO_2 through the Boudouard reaction, the CO produced reacts with FeO dissolved in liquid slag:

$$CO_{2(g)} + C_{(s)} = 2CO_{(g)}$$
$$CO_{(g)} + (FeO)_{(l)} = Fe_{(l)} + CO_{2(g)}$$

Five steps are involved: two simultaneous reactions, FeO with CO at the gas/liquid interface and $C_{(s)}$ with CO_2 at the gas/solid interface. FeO should diffuse from the bulk to the gas/liquid interface, the product CO_2 should diffuse out of the liquid/gas interface inside the gas halo to the solid/gas interface, and diffusion of CO out of the gas halo surrounding each carbon particle. The rate limiting step is the gas–solid chemical reaction. Figure 7.163 illustrates the two reactions.

Ozawa et al. [220] in 1986 suggested a different reaction mechanism which involves direct reaction between the carbon particles and FeO from the slag, where high volatile matter can form a gas halo slowing down the chemical reaction. This

Fig. 7.163 Reduction of iron oxide with solid carbon. After [140]

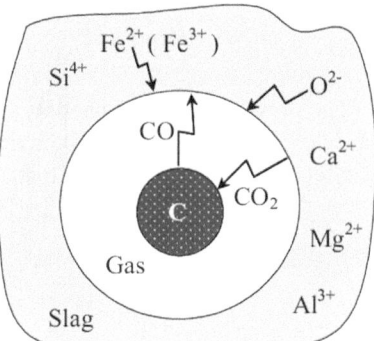

model is unrealistic because doesn't consider that carbon particles are non-wetting with the slag.

The expression derived by Story and Fruehan [221] for the rate of production of CO is the following:

$$\frac{dCO}{dt} = \frac{(D_e \rho Sk)^{\frac{1}{2}}}{RT}\left(p_{CO_2} - p_{CO_2}^{eq}\right)$$

where dCO/dt is rate of production of CO per unit external surface area of the particle (mol/cm^3 s), ρ is the molar density of carbon (moles/m^3), S is the specific surface area (m^2/mole), k is the intrinsic rate constant (m/s), p_{CO_2} is the partial pressure of CO_2 for the slag/gas reaction (atm) and $p_{CO_2}^{eq}$ is the equilibrium partial pressure of CO_2 for the gas/solid reaction (atm).

Hara and Ogino [222] measured and observed the rate of reduction of FeO in FeO-CaO-SiO$_2$ slags using a carbon rod immersed in liquid slag. They found a linear relationship between the carbon consumed and the volume of CO generated. After an incubation period the rate of evolution of CO increased, reached a peak and then decreased.

Hong et al. [88] estimated the evolution of CO from the reduction of iron oxide by solid carbon. The evolution rate of CO was calculated assuming first order reaction:

$$\frac{d(\%FeO)}{dt} = -k_A(\%FeO)$$
$$Q_{CO} = \left(\frac{RT}{p}\right)(\%FeO)_0 W_s \frac{\{1-\exp(-k_A t)\}}{100 M_{FeO}}$$

where k_A is the apparent reaction rate constant (1/s), t is time (s), Q_{CO} is the gas evolution rate at time t (cm^3/s), $(\%FeO)_0$ is the initial FeO in the slag (mass %), W_s is the mass of molten slag (g), T is temperature (K), R is the gas constant (J/mol K) and p is the pressure (MPa). The average gas evolution rate from time 0 to time t is expressed as follows:

$$\overline{Q}_{CO} = \left(\frac{RT}{p}\right)(\%FeO)_0 W_s \frac{\{1 - \exp(-k_A t)\}}{100 M_{FeO} t}$$

The previous expressions only consider FeO in the slag and injection of carbon particles. In practice, the production of CO comes from two chemical reactions; decarburization due to oxygen injection and reduction of FeO by the carbon particles injected.

$$-\frac{d(\%FeO)}{dt} = k_A\left[(\%FeO) - (\%FeO)_{eq}\right]$$

The generation rate of CO is controlled by the following factors:

- Injection rate and quality of solid carbon
- Activity of FeO in the slag
- Injection rate of oxygen
- Temperature.

Molloseau and Fruehan [223] studied the reduction rate of FeO in CaO-SiO$_2$-(3–35)FeO-Fe$_2$O$_3$ slags by iron droplets containing 2.8%C and 0.01%S using X-ray fluoroscopy. The CO evolved was measured using a constant volume pressure increase (CVPI) technique. They found that increasing the FeO concentration, increased the generation rate of CO but for each concentration there was a critical temperature at which this rate was higher, for example, Fig. 7.164 shows the results with 20% FeO, it shows that the temperature that produces more CO is 1400 °C. When FeO is 10 and 30% the maximum volume of CO was produced at a temperature of 1500 °C. Additions of hematite increased the generation rate of CO due to a higher oxygen potential.

Matsuura et al. [224] developed a decarburization model which takes into account the reaction between oxygen injected and carbon dissolved in liquid steel and also the reduction of iron oxide due to the injection of carbon particles. For the slag/metal part, the DeC rate is divided into two stages. In the first stage the carbon concentration

Fig. 7.164 Generation rate of CO as a function of temperature. After [223]

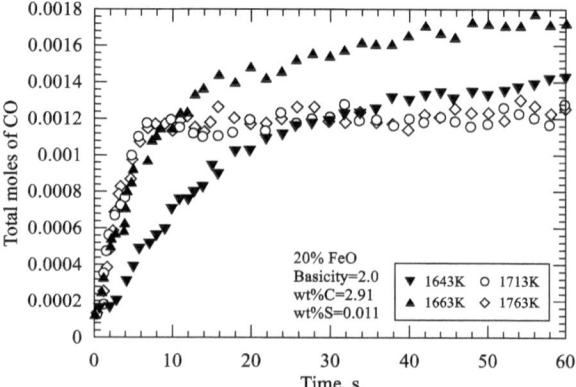

is high and the reaction is limited by diffusion of oxygen. In the second stage carbon is low, not all FeO is reduced, in this case the reaction rate is limited by diffusion of carbon to the reaction interface. The DeC rate can be expressed by the following two expressions, describing the first and second stage, respectively:

$$\frac{d[\%C]_t}{dt} = -\frac{Q_{O_2}^t \times (1000/22.4) \times (2 - R_{PC}) \times 0.012}{W_m^t} \times 100$$

$$\frac{d[\%C]_t}{dt} = -\frac{k_m A\rho}{W_m^t}\left([\%C]_t - [\%C]_t^{eq}\right)$$

where $Q_{O_2}^t$ is oxygen flow rate at time t, R_{PC} is the post-combustion ratio, k_m is the mass transfer coefficient for C in liquid steel, A is the reaction interfacial area, ρ is the density of liquid steel, W_m^t is the weight of liquid steel.

The FeO concentration results from the difference between the initial FeO and a balance between the iron oxidized and FeO reduced.

$$W_{FeO}^{t'} = W_{FeO}^t + \left(W_{Fe-Oxi}^t - W_{FeO-red}^t\right) \times \frac{72}{56}$$

The reduction of FeO by solid carbon was defined by the following expression which assumes control due to mass transfer of FeO in the slag and also assumes the interfacial area is proportional to its weight in the slag.

$$\frac{d(\%FeO)}{dt} = k_{red} W_{CM}^t \frac{(\%FeO)_t}{100}$$

where k_{red} is the reaction rate constant between carbon and iron oxide in the slag, W_{CM}^t is the weight of carbon particles at time t. They assumed k_{red} is a constant with a value of 0.57 min^{-1}. The model was applied to an EAF of 120 ton of nominal capacity, charged with pig iron (48 ton) and scrap (67 ton). The hot heel was 15 ton of liquid steel and 6 ton of slag from the previous heat. The total coke injected was 877 kg and oxygen 4212 Nm^3. The model results predicted very high values of FeO, in the range from 40 to 60%.

Zhu and Sichen [108] carried out experiments with a physical model at room temperature, with gas generation by chemical reaction between oxalic acid and carbonate dissolved in water. Their results clearly show the relationship between the gas generation rate and the foam height, as shown in Fig. 7.165 with an initial oil height of 2.5 cm. The results indicate a foam height increase from 4 to 6 times the initial liquid height.

7.14 Gas Generation Rate with Limestone and Dolomite

Raw dolomite and limestone can be used in the EAF to replace metallurgical lime and burnt dolomite, with the additional benefit of gas generation which can improve the slag foaming conditions. Although this is a real possibility, these materials not

Fig. 7.165 Foam height and gas generation rate as a function of time using different amounts of oxalic acid. After [108]

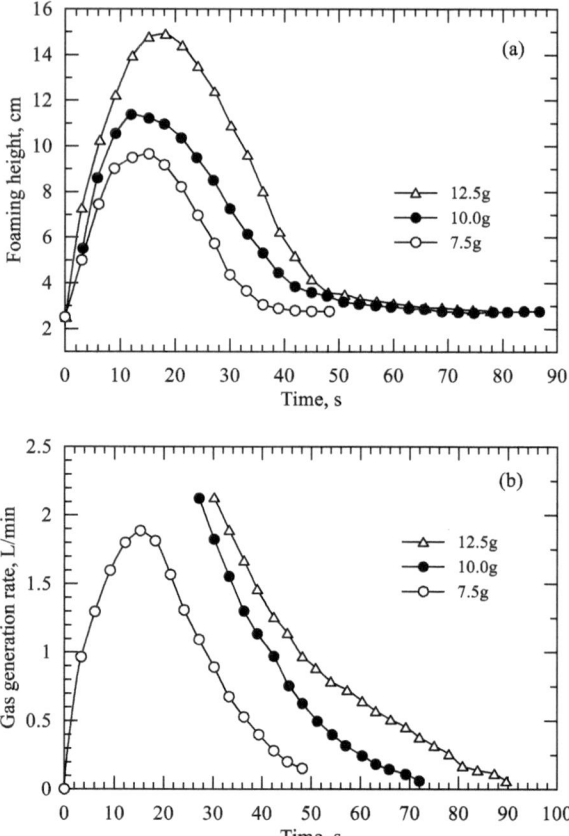

only increase the consumption of energy but also increase the tap-to-tap time and are not a common practice.

Chu et al. [225] studied foaming due to the thermal decomposition of $CaCO_3$ spheres of 4 g in $Na_2B_4O_7$ slag at 925 °C and developed a model to predict their experimental data. The foam height was measured with a molybdenum wire every 30 or 60 s. Figure 7.166 reports their results. It can be shown that decreasing the particle size from 5 to 3 mm, the foam height increases but on the other hand its foam life is shorter, an indication that the particles are consumed at a faster rate. Further work by the same group reported that dolomite spheres had a longer foam life than limestone [226].

Limestone and dolomite have been employed in slag foaming of stainless steels. Chychko et al. [227, 228] carried out fundamental investigations on slag foaming. They employed especial iron crucibles with an inverted U shape at the bottom that allowed a thermocouple to be located outside the crucible but close to the region where the pieces of carbonate are located. The slag, 150 g, containing $31CaO$-$31SiO_2$-$7MgO$-$6Al_2O_3$-$25FeO$ had an estimated melting temperature of 1200 °C, viscosity

Fig. 7.166 Foaming behavior due to CaCO₃. After [225]

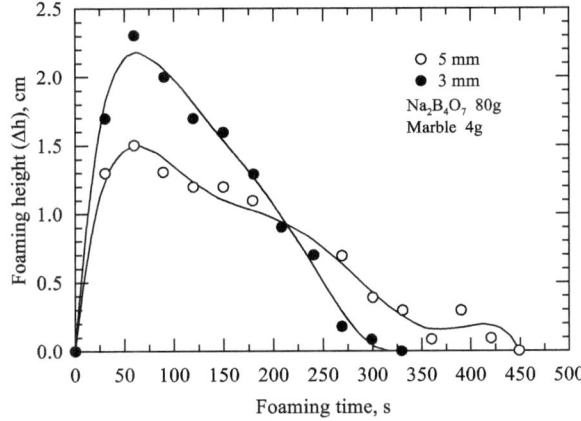

of 690 cP at 1350 °C and 390 cP at 1450 °C. Dolomite contained 21MgO-29CaO-2.3SiO$_2$-0.44Al$_2$O$_3$ + FeO and limestone about 98.4%CaCO$_3$. About 1.3–1.8 g of carbonates at room temperature were added in each experiment, with different shapes; spherical samples about 10 mm diameter, rectangular wedge samples about 12 × 20 mm. To avoid flotation, each sample was attached to iron tablets of 3.5 g. Results were satisfactory for 1350 and 1400 °C. At 1450 °C they reported violent foaming. After comparison of the calculated energy consumed using carbonates and the actual measured values, they found that the measured values were about 56–79% of the theoretical values, the difference due to the improved foaming conditions, with similar results for both types of carbonates and no effect of the carbonate shape. The calculated foaming index with carbonates was 0.35 s and compared with the "same system" reported by Ito and Fruehan, without specific details on the slag chemistries employed, indicating a value of 24 s, this discrepancy was taken as a reason to reject the application of the foaming index to dynamic conditions.

In another part of their work Chychko et al. reported results of carbonate additions into the EAF. Using from 0 to 20 kg/ton, energy consumption increased about 1.2–1.3 kWh/kg if the addition was made in the initial charge but was lower if the addition was made under flat bath conditions.

7.15 Prediction Models of Slag Foaming

Most of the foaming indexes that have been proposed describe the foaming behavior based only on the physical properties of the slag. Their models indicate that foaming increases by increasing the slag's viscosity and decreased if the density and surface tension is increased. A high value of viscous forces tend to retain the bubbles in the slag meanwhile a high value of surface tension forces provide low surface elasticity and lower foam stability. This section discusses in detail the models proposed and the resulting equations to predict slag foaming.

Fig. 7.167 Relationship
between the void fraction
and the superficial gas
velocity. After [85, 86]

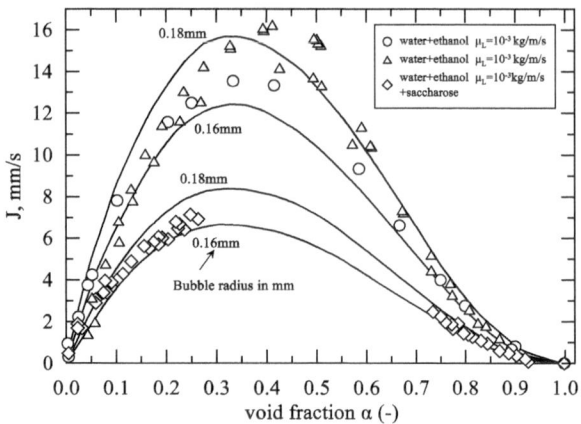

7.15.1   Ogawa et al. Model

Ogawa et al. [85, 86] in 1992 applied the drift-flux model developed by Graham
Wallis [229] in 1961. It was the first attempt at a description of slag foaming from
a fundamental viewpoint. Wallis reported the following expression to estimate the
void fraction in a system with two layers; foam layer (α_1, top) and dispersed gas
layer (α_2, bottom). The dispersed gas slayer is the layer where bubbles ascend freely.

$$\alpha_1 = \tfrac{1}{2}\left(1 + \sqrt{1 - 4J/U_\infty}\right)$$
$$\alpha_2 = 1 - \alpha_1$$

where J is the superficial gas velocity and U_∞ is the rise velocity, defined by the
modified Stokes law:

$$U_\infty = \frac{(\rho_L - \rho_G)gR^2}{3\mu_L}$$

where ρ_L and ρ_G are densities of liquid slag and gas phases, μ_L is the slag's viscosity,
α_1 and α_2 are the void fractions. The experimental data from Ogawa et al. confirmed
the validation of Wallis model, as shown in Fig. 7.167.

7.15.2   Lin and Guthrie's Model

Lin and Guthrie [217] in 1995 reported results on slag foaming using water modeling,
as described in a previous section. They focused their work to bath smelting conditions. Under those conditions they found the conventional foaming index proposed
by Fruehan cannot be applied because the superficial gas velocity is extremely high,

above the regime of bubbly flow. For a bubbly flow regime, the volume of gas in the foam linearly increases with the gas flow rate, giving the foaming index suggested by Bikerman;

$$V_g = \Sigma Q_g$$
$$\Sigma = \frac{\Delta h}{U_g}$$

where V_g is the volume of gas in the foam, Δh is the foam height and U_g the superficial gas velocity.

The model they proposed for bath smelting was based on the general definition for the volume of gas in the foam, assuming the gas bubbles rise through the liquid and through the foam, as follows:

$V_g =$ (number of bubbles in foam) $\times$ (V_b)

$N_b =$ (frequency of bubble formation) $\times$ (mean residence time of bubbles in foam bed)

$$N_b = \frac{Q_g}{V_b} \frac{(h + \Delta h)}{U_s}$$

The volume of gas becomes:

$$V_g = Q_g \frac{(h + \Delta h)}{U_s}$$

where h is the height of liquid, U_s is the bubble rise (slip) velocity, V_b is the volume of a bubble, N_b is the number of bubbles. Replacing equivalent values for V_g and Q_g in the previous expression, and separating the value of the foam height:

$$V_g = \Delta h\, A$$
$$Q_g = U_g A$$
$$\Delta h = \frac{U_g}{(U_s - U_g)} h$$

This expression indicates the foam height depends on the bubble rise velocity, gas superficial velocity and initial liquid's height.

From the previous expression, the rise velocity can be re-cast in the following form:

$$U_s = \frac{U_g}{\left(\frac{\Delta h}{\Delta h + h}\right)} = \frac{U_g}{\varepsilon}$$

where ε is the average hold up of gas in the foam. From experimental data they found a constant value for $\varepsilon = 0.8$, therefore, the rise velocity can be estimated. Furthermore, the authors also derived the following relationship for the bubble size as a function of the superficial gas velocity.

$$d_b = \frac{2}{g}\left[\frac{1-\varepsilon}{\varepsilon}U_g\right]^2$$

Its application to industrial data suggested a bubble size from 100–250 mm, extremely high values which they assumed it can be formed due to coalescence. The previous expression can be rearranged in terms of the definition of the foaming index, as follows:

$$\Sigma = \frac{\Delta h + h}{U_g + (0.5gd_b)^{0.5}}$$

Surprisingly, in this expression the foaming index is not directly a function of the physicochemical properties of the slag, however the authors argue that its effect is given through the definition of the value for the bubble size.

7.15.3   Iron's Model of Void Fraction

Gou and Irons [169] in 1996 and Kapoor and Irons [104] in 1997 derived mathematical models to describe the gas fraction due to slag foaming. These models are based on the drift-flux model from Wallis who proposed the idea that a relative motion exists between the gas and the liquid phases, therefore the total gas flux is the sum of the average superficial gas velocity times the gas fraction and a flux due to the relative motion between the two phases. Their first model was derived ignoring the initial gas momentum and the presence of surfactants on gas-slag flow but in the second model they took them into account as a lump parameter, as follows:

$$\frac{U_s}{\theta} = C_0 U_s + L_p\left(\frac{\sigma g \Delta \rho}{\rho_l^2}\right)^{\frac{1}{4}}$$

where L_p is a lump parameter which accounts for the initial height of the liquid, presence of dissolved surfactants, density of the gas and boundary conditions imposed on the flow by bubble deformation characteristics, θ is the gas fraction and C_0 is a distribution parameter. The previous equation can be re-arranged to estimate the gas fraction as a function of the other variables. C_0 and L_p are obtained from a drift-flux analysis of a gas-slag system. Figure 7.168 shows a drift-data analysis from a 5-ton smelter.

Derived from their analysis, the definition of the foaming index was given by the following expression:

$$\Sigma = \frac{\Delta h + h_0}{C_0 U_s + \overline{U}_{Gj}}$$

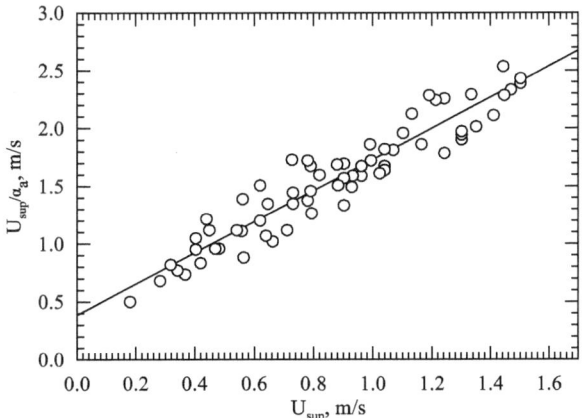

Fig. 7.168 Drift-flux analysis with data from a 5-ton smelter. After [104]

where: $\overline{U}_{Gj}$ is the average gas drift velocity. The expression indicates that the foaming index does not depend just on the physical properties of the slag but also on the superficial gas velocity and the initial slag height. Therefore, "the adoption of a slag foaming index, measured under classical foaming regime to study the gas-slag system under entirely different flow conditions is fundamentally incorrect". The foaming regime is produced at low superficial gas velocities.

7.15.4 Lahiri-Seetharaman Equation

Lahiri and Seetharaman [230] in 2002 derived an expression, from first principles, for the foaming index, starting with the following definition:

$$\text{Rate of change of foam volume} = \text{rate of gas generation} - \text{rate of bubble break} - \text{up}$$
$$\frac{dV}{dt} = Q - k_D N V_b$$

where dV/dt is the rate of change of foam volume, Q is the rate of gas generation, k_D is the rate constant for bubble break-up, N is the number of bubbles and V_b is the average volume of a bubble. The void fraction (θ) is equal to ratio between the volume of bubbles and the total foam volume, i.e., $\theta = NV_b/V$. If the previous expression is defined in terms of θ, simplified for steady state conditions ($dV/dt = 0$), the foam volume defined in terms of the foam height, and the vessel has uniform cross sectional area, they obtained the following expression:

$$\Sigma = \frac{h}{U_s} = \frac{1}{k_D\theta}$$

The rate constant for bubble break-up (k_D) is inversely proportional to the time required for drainage (τ_D). The drainage time results from a ratio of the volume of liquid entrapped in a unit area between two consecutive layers of bubbles and drainage rate per cross-sectional area. The drainage rate was defined based on a fundamental work carried out by Narsimhan and Ruckenstein, who defined this parameter as a function of the drainage velocity. Another equation developed by Dasai and Kumar was employed to define the drainage velocity, assuming that surface tension remains constant. In this way Lahiri and Seetharaman derived the following relationship for k_D:

$$k_D = 8.82 \times 10^{-6} \frac{A n_p \theta c_v d_b \rho g}{(1 - \theta)\mu}$$

where A is a proportionality constant, n_p is the number of Plateau border channels per bubble, c_v is a polynomial of the inverse of the dimensionless surface viscosity and d_b is the bubble diameter. Replacing this value in the previous expression, the foaming index can be re-cast as follows:

$$\Sigma = 11560 \left(\frac{A n_p c_v}{U_s} \right) \left(\frac{1 - \theta}{\theta^2} \right) \left(\frac{\mu}{\rho d_b} \right)$$

For pentagonal dodecahedra shaped bubbles n_p is 10, assuming $\theta = 0.9$, $A = 1$, the value of the three terms on the right-hand side is 143. The authors obtained 150 by least-squares.

$$11560 \left(\frac{A n_p c_v}{U_s} \right) \left(\frac{1 - \theta}{\theta^2} \right) = 150$$

Then, the final expression became:

$$\Sigma = 150 \left(\frac{\mu}{\rho d_b} \right)$$

It is to be noticed that this expression excludes the effect of surface tension. Figure 7.169 compares the foaming index predicted with the previous equation and compares with the equation obtained by Zhang and Fruehan. Both equations provide similar predicted results. On the basis of their analysis, they questioned if surface tension has the role previously reported by different researchers. In previous works it has been reported that surface active elements largely increase the foaming index by a factor of 3–5, this would require to include surface tension by a factor of 27 since the exponent reported was 1/3. The actual decrease in surface tension was only about 10%. The slags considered in the figure have surface tensions in the range from 463 to 525 mN/m. These arguments support their conclusion that the effect of surface tension cannot be defined from data analysis.

Although the previous expression does not include surface tension, it is still involved in the definition of the bubble size. One equation that describes de bubble size is given below, which depends on the liquid's surface tension. This expression

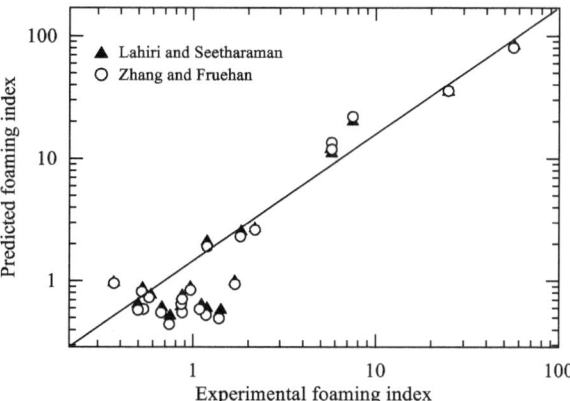

Fig. 7.169 Comparison of two equations that predict the foaming index. After

is valid for low flow rates (Re < 500).

$$d_b = \left(\frac{6\sigma d_n}{\rho g}\right)^{\frac{1}{3}}$$

If this expression is replaced, the foaming index can be described as a function of the surface tensions, as follows:

$$\Sigma = C\left(\frac{\mu}{\rho^{\frac{2}{3}}\sigma^{\frac{1}{3}}}\right)$$

7.15.5 Fruehan's Models

The optimum slag physicochemical properties to promote slag foaming have been studied extensively, particularly by Fruehan and co-workers (Ito and Fruehan [125, 166], Jiang and Fruehan [216] and Zhang and Fruehan [128]) and developed the, now classical expressions, to estimate the foaming index. Since these expressions were derived using dimensional analysis, they are further discussed in Sect. 7.15.12. These researchers defined the effect of slag density (ρ), slag viscosity (μ), slag surface tension (σ) and bubble diameter (d) on the foaming index (Σ). The foaming index is a measure of the time it takes for the gas phase to go through the slag and escape. Its experimental value can be computed from the ratio between slag height and the superficial gas velocity. One of the expressions reported is the following [128]:

$$\Sigma = 115\frac{\mu^{1.2}}{d^{0.9}\rho\sigma^{0.2}}$$

The equation indicates that in order to improve slag foaming the slag viscosity should be increased and its surface tension decreased. Acid slags and lower FeO increase the slag viscosity, therefore have a higher foaming index. This type of equations provides information on the slag physical properties involved and their relative importance, however is not enough to have the proper slag properties. The equation is only useful to define the potential foaming conditions. Our own work, described in Sect. 7.15.7 emphasize the need to include the instantaneous generation rate of CO in order to define the final conditions for slag foaming which resulted in a new expression called dynamic foaming index.

7.15.6 Zhu and Sichen Model

Zhu and Sichen [97] in 2000 developed a model based on a one-dimensional analysis of mass and momentum conservation which predicts the foam height as a function of a dimensionless number. The model was developed considering the following assumptions:

- The gas velocity and gas fraction only change in the vertical direction
- Buoyancy and drag forces on the bubbles are neglected
- The work done for deformation of a viscous material, i.e. viscous dissipation, is neglected
- The liquid velocity is zero. Mass and momentum conservation, only for the gas phase
- Steady-state conditions.

The momentum balance involves the upward motion due to buoyancy forces and its opposite force or drag force. Combining both conservation equations defines an expression for the gas velocity. The velocity at the origin, $z = 0$, is the superficial gas velocity and the void fraction (θ) is obtained from the ratio of these velocities. Mass conservation for the liquid slag is defined by the following expression:

$$\int_0^{h_{max}} (1 - \theta)dz = h_0$$

where h_{max} is the maximum foaming height, h_0 is the initial foam height.

Replacing the expression derived for the void fraction, after integration, simplification of terms and separate h from the final expression, the following relationship was derived:

$$h_{max} = \frac{a}{b(1-2a)} \ln\left(\frac{2}{1+a}\right) + \frac{h_0}{1-2a}$$

$$a = \frac{U_s}{c}$$

$$b = \frac{3}{4} \frac{C_D}{d_b} \frac{\rho_l}{\rho_g}$$

$$c = \left(\frac{4d_b g}{3C_D}\right)^{\frac{1}{2}}$$

The expression for h_{max} can be further simplified since the first term on the right side is much smaller with the second term. Then, the final expression becomes:

$$h_{max} = \frac{h_0}{1-2a} = \left(\frac{1}{1-2N_{foam}}\right) \cdot h_0$$

$$N_{foam} = U_s \left(\frac{3C_D}{4d_b g}\right)^{\frac{1}{2}}$$

The model indicates that the maximum foam height increases with the initial slag height and by increasing the dimensionless number $N_{foam.}$ up to 0.5.

The drag coefficient can be estimated based on the equation proposed by Kuo and Wallis:

Re_b	≤ 0.49	$0.49 \leq Re_b \leq 100$	$Re_b > 100$	$Re_b = \frac{\rho_l g}{12\mu_l} - d_b^3 - (\rho_l - \rho_g)$
C_D	$\frac{16.0}{Re_b^1}$	$\frac{20.68}{Re_b^{0.643}}$	$\frac{6.3}{Re_b^{0.385}}$	

The bubble diameter can be estimated using a modified expression based on equations proposed by Sano and Mori (1980) and Davidson and Schuler (1960), valid for liquids with viscosity from 19.2 to 97 mPa s:

$$d_b = 0.03\left(\frac{\sigma}{\rho_l}\right)^{0.5}\left(\frac{\mu_l}{\rho_l}\right)^{0.25} U_s^{0.65}$$

Figure 7.170 compares the experimental data and the predicted values. The authors indicated that the model is very sensitive to the bubble size. In order to compare foams with only one layer the results shown correspond to superficial velocities above 1.75 cm/s. It can be observed that the model predictions are better when the foam height increases. The model always predicts an increment in foam height as the superficial velocity increases. The authors attribute the deviation from the model's predictions due to turbulence which is not accounted in the model.

Further sensitivity analysis with their model is shown in Fig. 7.171. It is observed that as the dimensionless number increases, the foam height also increases. The dimensionless number can be increased by increasing the drag coefficient or decreasing the bubble size. Increasing the drag coefficient or decreasing the bubble size decreases the Re number. The authors suggest the dimensionless number is a good indicator of the foaming capacity. According with this model, the foaming height will be two times the initial height if N_{foam} is 0.25, five times if N_{foam} is 0.40 and ten times if N_{foam} is 0.45. Above N_{foam} of about 0.4 the increment in the foam height is exponential.

Fig. 7.170 Measurements and predicted values of foam height as a function of the superficial gas velocity; **a** slag A, **b** slag C. After [97]

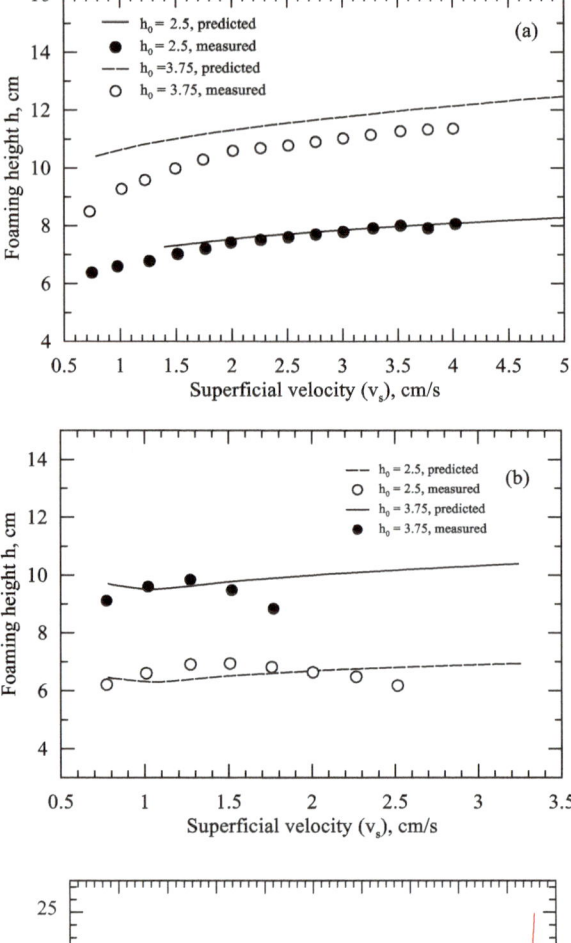

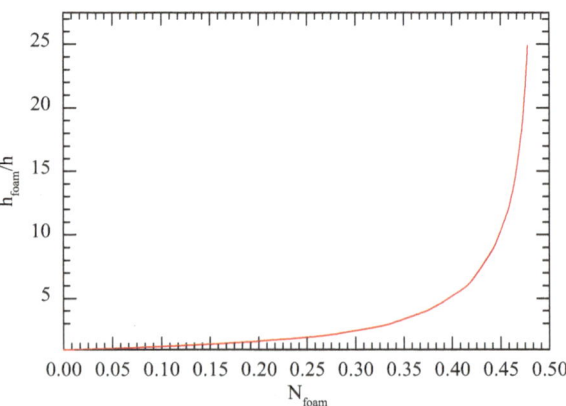

Fig. 7.171 Effect of the dimensionless number on the foam height ratio. After [97]

Fig. 7.172 Model predictions for two types of oils with small changes in the bubble size. After [108]

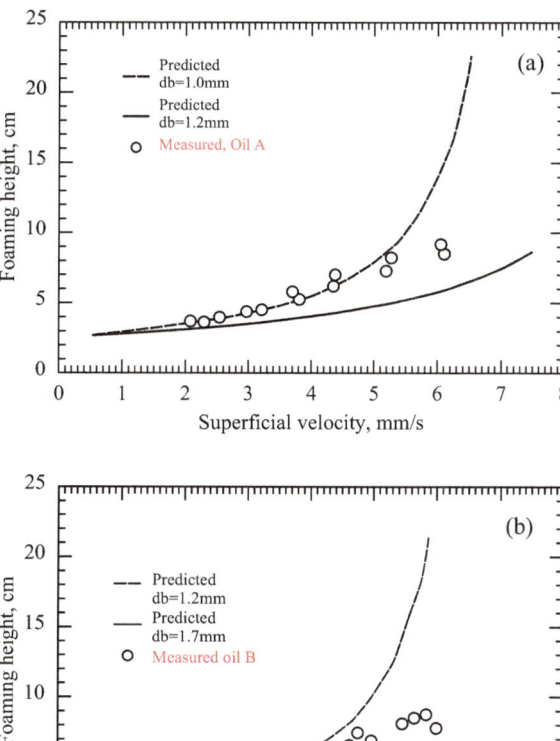

Further work by Zhu and Sichen [108] producing the gas by chemical reaction between oxalic acid and carbonate dissolved in water show the large sensitivity of this model to the bubble size. Figure 7.172 shows the results for two types of oils and the model predictions with small changes in the bubble size. Figure 7.172a shows a good prediction with a bubble size of 1.0 mm but a small change to 1.2 mm produces a poor reproducibility of the experimental data.

7.15.7 Morales's Model

7.15.7.1 Model Description

Conejo et al. [231] in 1999 provided a comprehensive description of slag foaming at the industrial practice using 100% DRI, describing the equipment for injection of carbon particles, characterization of carbon particles, measurements of carbon and FeO in the slag, etc. This work was the origin of the model further developed in collaboration with Morales and Hernandez. DRI contains both FeO and carbon. Low quality DRI, below 90% metallization, introduces a high volume of oxygen, which can even be equivalent to 100% of the total oxygen injection requirements. For the particular metallic charge and volume of oxygen employed in one steel plant these authors suggested the addition of a minimum of 5 kg C/ton, this value is achieved charging DRI with 2.3% C. In addition to this, Conejo et al. suggested to control dissolved carbon in steel above 0.06% which is obtained with an initial addition of coke, about 10 kg/ton, a continuous addition of carbon particles, about 7 kg/ton. Energy consumption and FeO were decreased by increasing the injection of carbon particles in the range from 4 to 14 kg/ton. The decrease in energy consumption was in the range from 40 to 50 kWh/ton. Using a high amount of carbon particles, above 14 kg/ton, can be detrimental to the erosion of the delta due to combustion of very fine particles. They also reported a relationship between C dissolved in the slag and FeO. To decrease FeO below 30% the carbon dissolved in the slag should be in the range from 0.10 to 0.15%. 70% of the particles had a particle size from 0.074 to 0.6 mm and 30% were carbon fines (less than 0.074 mm). They indicated that the zone of injection should guarantee enough residence time for each particle.

Morales et al. [232–238] between 1997 and 2002 reported in detail one of the first comprehensive mathematical models in the EAF using DRI. This work was primarily carried out by H. Hernandez, forming the basis of his PhD thesis in 2000 [233]. The model computes the optimum amount of materials to be charged into the furnace which minimize the cost of steel and the total energy requirements, subsequently a kinetic model computes the rate of melting of DRI and the rate of change in slag composition. The kinetic model employs thermodynamic activities which were computed using the modified quasi-chemical slag's model. The main contribution of this work was a detailed kinetic analysis on the reduction of iron oxides with the injected carbon particles and the subsequent generation rate of CO. With this information the authors proposed a new way to estimate a dynamic foaming index. Below is a summary of the main elements of the reduction of iron oxide with carbon particles and the proposed dynamic foaming index.

The formation of iron oxide has two sources, FeO from DRI and FeO from oxygen injection. The melting rate of DRI was defined as a function of its metallization based on statistical analysis of plant data. The reduction rate of FeO has two sources, at the slag/metal interface and by the injected carbon particles, as shown by the following expression:

$$\frac{d\left(W_{FeO}^{sl}X_{FeO}^{sl}\right)}{dt} = \sum X_{FeO}^{DRI}\frac{dW_{DRI}}{dt} + \frac{M_{FeO}}{M_O}\sum K_1 Q_{O_2} - M_{FeO}AN_{CO} - r_{FeO}^{C_s}$$

where; The LHS term is the change in FeO concentration, the first two terms on the RHS refer to the production of FeO (FeO in DRI and FeO produced by oxygen injection), the third term is the rate of decarburization at the slag/metal interface and the last term is the rate of reduction of FeO due to injection of carbon particles.

The kinetic model on the reduction of iron oxide by carbon particles first computes the fraction of particles that penetrate into the slag followed by two simultaneous reactions which consume the carbon particles. The rate of consumption of these particles is based on the shrinking core model and it is defined its residence time in the slag based on the particle size. Each particle reacts until its residence time is reached and reaches the slag surface or its radius is zero (fully consumed particles). The sub-model developed to estimate the fraction of particles that enter the liquid slag was based on a previous momentum analysis equation reported by Farias and Robertson [239] in 1982 for the injection of powders into liquid steel. Their energy balance includes the energy of the particles in the gas phase, energy to overcome an interfacial resistance, energy to displace a portion of the liquid in front of the interface due to penetration of the solids and to maintain its own energy within the liquid. The expression developed was the following:

$$\frac{1}{2}\rho_g\theta_{p(t)}^g\left(v_p^g\right)^2=\frac{1}{2}\rho_g\theta_{p(t)}^l\left(v_p^l\right)^2+\frac{1}{2}\rho_l\theta_{l(t)}^p\left(v_l^p\right)^2+\sigma_{lg}(f-\cos\beta)\cdot\frac{A_p\theta_{p(t)}^g}{V_p}$$

where: $\theta_{p(t)}^g$ is the volume fraction of particles that cross the gas–liquid interface, $\theta_{p(t)}^l$ is the volume fraction of particles in the liquid phase, $\theta_{l(t)}^p$ is the fraction of liquid displaced due to the penetration of solid particles, v_p^g, v_p^l, ρ_g, ρ_l, σ_{lg}, V_p and A_p are the particle velocity in the gas phase, particle velocity in the liquid phase, gas density, liquid density, surface tension at the gas–liquid interface, particle volume and surface area of the particle, respectively. This equation contains two unknowns, to solve it is necessary to fix one of them. The fact that the two volume fractions, $\theta_{p(t)}^g$ and $\theta_{p(t)}^l$ is not equal to one is not intuitive as it might appears at first sight because their control volumes are different:

$$\theta_{p(t)}^g + \theta_{p(t)}^l \neq 1$$

The control volume "carrier gas–solid particles" defines the volume fraction of particles that cross the gas–liquid interface. The control volume "liquid slag-solid particles" defines the volume fraction of particles in the liquid. Further work by Farias and Irons found an important displacement of the liquid (they used water), however the slag viscosity is much larger than water and it is not expected any displacement, then $\theta_{l(t)}^p$ can be taken as zero. On the other hand, a plot between $\theta_{p(t)}^g$ and $\theta_{p(t)}^l$ shows that the largest values for $\theta_{p(t)}^g$ are found when $\theta_{p(t)}^l = 1$. Under these conditions the previous energy balance can be reduced to the following expression:

$$\theta_{p(t)}^g = 1 - \frac{12\sigma_{lg}(f-\cos\beta)}{\left(v_p^g\right)^2 d_p\rho_p} = 1 - R_{gl}$$

where: $\theta_{p(t)}^g$ represents the fraction of particles that enter into the slag, f is the difference between the gas–liquid interface area while the particle is passing through the interface and the area of the undisturbed surface, σ_{lg} is the gas–liquid surface tension,

β is the solid–liquid contact angle, v_p^g is the particle velocity in the gas phase, d_p is the particle size, ρ_p is the particle density.

This equation is very important because in one single equation most of the variables that control slag foaming are combined together.

Figure 7.173a shows graphically the reduction of iron oxide by solid particles. This model was based on the work by Richards and Brimacombe [240] for zinc fuming in 1985. The jet with the fraction of particles that enter the slag reaches a maximum penetration depth (z_{max}). At that stage the jet is dissipated and the entrained particles will ascend towards the free surface. The time that it takes for the particles to exit the nozzle and reach the maximum penetration depth is much shorter compared with the time for rising to the free surface. It was estimated a time of 0.8 s. The displacement of the particles in 2D is obtained by integration of the two following differential equations:

$$\int_{Z_{mx}}^{0} dz = \int_{0}^{t} U_{pz}dt$$

$$\int_{x_1}^{x_f} dr = \int_{0}^{t} U_{px}dt$$

The integration limits correspond to two possible conditions; large particles reach the free surface without full conversion or small particles can be consumed before reaching the free surface. The residence time of the particles, 7–10 s, was calculated for a slag height of 30 cm [233]. The position of the lance was estimated to be 10 cm above the slag/metal interface. The kinetic model for the reduction of iron oxides with solid particles according to Fig. 7.173b involves two simultaneous reactions; solid carbon reacts with CO_2 (Boudouard reaction) producing CO and then FeO is reduced by CO at the slag/gas bubble interface. The two reactions involve CO, CO_2 and the carbon particles. The instantaneous particle radius is defined with a set of six differential equations which define mass and volume balances for the three previous species. The mass balance involves the gas species which have the following form:

$$\frac{d\left(C_i V_g^b\right)}{dt} = \pm A_b k_{FeO}\left(1 - \theta_{SiO_2}\right)\left(C_{FeO} - C_{FeO}^{eq}\right) + nr_B$$

where "i" refers to CO, CO_2. On the RHS a positive sign for CO and negative sign for CO_2, n is $= 1$ for CO and ½ for CO_2. k_{FeO} is the mass transfer coefficient obtained from a relationship between the Pe and Nu numbers. The term $\left(1 - \theta_{SiO_2}\right)$ describes the available sites for chemical reaction. r_B is the reaction rate for the Boudouard reaction, which is defined by the following expression:

$$r_B = A_0 \cdot \exp(-E_a/RT) \cdot p_{CO_2}$$

Fig. 7.173 Mechanism of iron oxide reduction by carbon particles; **a** fraction of particles that enter the slag phase, **b** reactions consuming the carbon particles. After [232]

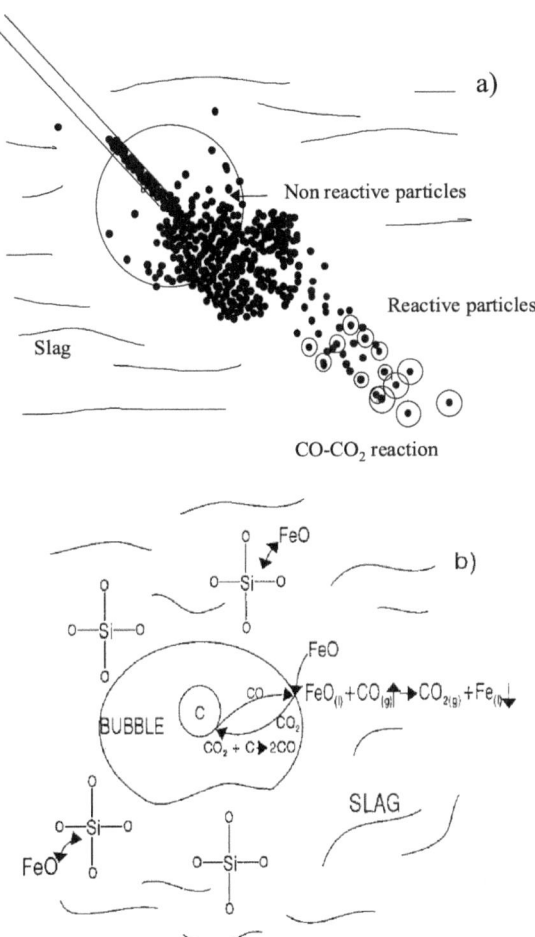

where r_B is the reaction rate of the Boudouard reaction, taken from the work carried out by Tarby and Philbrook in 1967. From a work carried out by Skiner and Smooth for bituminous carbon with 20% impurities, the values of A_0 and E_a were taken as $3.13 \cdot 10^6$ and 196.2 kJ/mol.

Richards and Brimacombe [240] assumed the whole gas–liquid interface is available for chemical reaction, which is not true, as shown in Fig. 7.174. This figure was made with data from Mills and Keene for $CaO\text{-}SiO_2\text{-}FeO$ slags, computing the activity using the quasi-chemical model and the fraction of sites occupied by SiO_2 on the gas surface, the excess surface concentration for silica, was calculated with the Gibbs adsorption equation divided by the SiO_2 saturation concentration. In the figure it is shown that SiO_2 decreases the available sites for the reduction of iron oxides at the bubble surface.

The final expression for the reduction of iron oxide becomes:

Fig. 7.174 Surface area occupied by SiO_2 at different FeO concentrations. After [232]

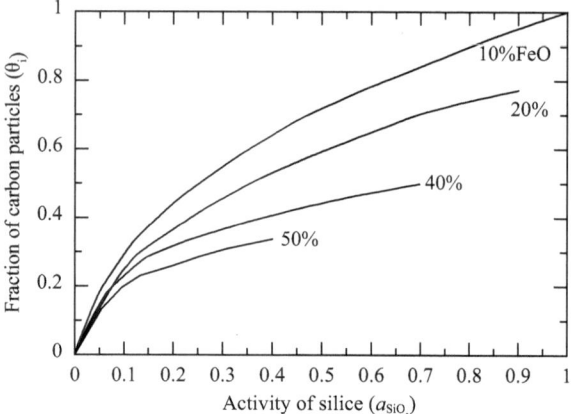

$$\frac{dW_{FeO}^{C_s}}{dt} = Q_s \theta_{p(t)}^g \frac{M_{FeO}}{M_C} \cdot \left[1 - \left(\frac{r_{p(t)}^f}{r_p^0} \right)^3 \right]$$

where $r_{p(t)}^f$ and r_p^0 are the instantaneous particle size and initial particle size, respectively, and Q_s is the mass flow rate of solids. This equation assumes the initial particle size is constant. In order to account for a particle size distribution, the Montecarlo method was applied. The particle size distribution is arranged in a plot of distribution frequency (%) versus particle size, then the method randomly chooses between 0 and 1 corresponding to certain particle size.

The previous expression was the most important contribution of this model, it is extremely important because it provides in a single equation the influence of the injection variables; the computation of $\theta_{p(t)}^g$ involves the carrier gas flow rate, slag properties (surface tension), injection angle, wetting slag/particle and particle properties; the computation of the particle size involves the effect of the initial particle size and its reactivity, the final expression also involves the flow rate of carbon particles. Figure 7.175 shows the degree of consumption of the carbon particles after they complete its residence time. It can be observed that for a time of 10 s, only the smaller particles of 100 μm are fully consumed and for the same time the larger particles are consumed less than 20%.

Figure 7.176 shows the effect of SiO_2 on the rate of reduction of iron oxide due to injection of carbon particles. It can be observed that silica decreases the rate of reduction due to its effect to decrease the fraction of sites available for chemical reaction.

Figure 7.177 shows model's results indicating the effect of the injection rate of carbon particles on the rate of reduction of iron oxide for different flow rates of the carrier gas and slag basicity with particles of 200 μm. The reduction rate of iron oxide increases by increasing the flow rate of particles from 0 to 70 kg/min and by

Fig. 7.175 Degree of consumption of carbon particles of different size as a function of its residence time. After [234]

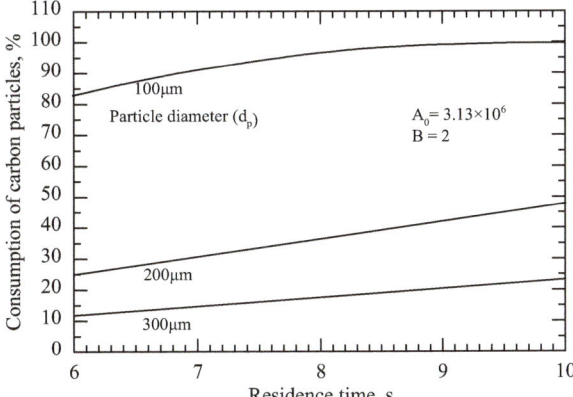

Fig. 7.176 Effect of SiO$_2$ on the rate of reduction of FeO with injection of carbon particles. After [234]

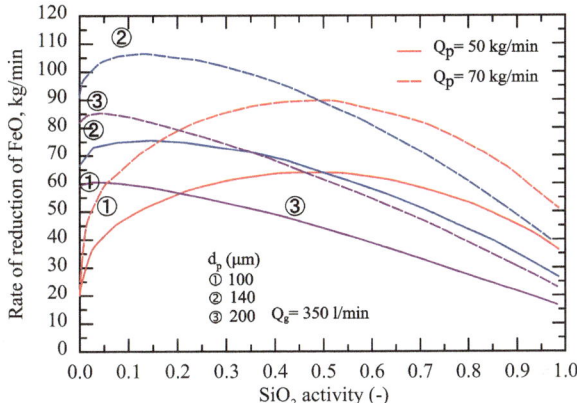

increasing the carrier gas flow rate from 250 to 400 Nl/min. The effect of slag basicity is more complex as explained below:

- At very low flow rates of the carrier gas, 200 Nl/min, the reduction rate for acid slags is higher than basic slags
- As the gas flow rate increases to 250 l/min, the difference in reduction rates for acid and basic is smaller. The slag basicity makes no difference, the reduction rate is controlled by the injection rate of solids
- At high gas flow rates of the carrier gas, above 300 Nl/min, the reduction rate is higher for basic slags
- The plot also compares two slags with different basicity and the same concentration of FeO (CaO/SiO$_2$, 15%FeO, 6%MgO, 2%Al$_2$O$_3$). In this case the rate of reduction of FeO is lower for acid slags.

The carbon particle size has a strong effect on the reduction rate of FeO, as shown in Fig. 7.178. This figure shows also the effect of slag basicity and carrier gas flow

Fig. 7.177 Effect of the
flow rate of carbon particles,
carrier gas flow rate and
basicity on the rate of
reduction of iron oxides.
After [232]

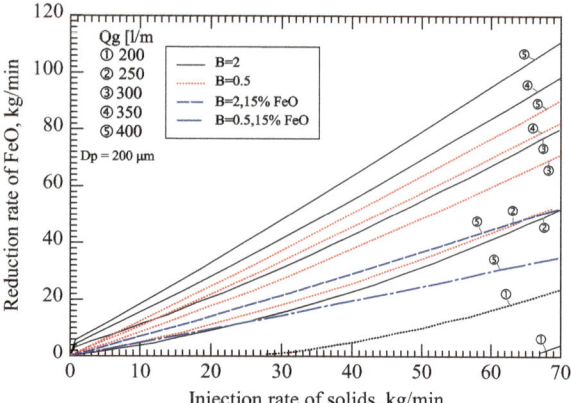

rate, keeping constant the flow rate of carbon particles. In all cases, increasing the
flow rate of the carrier gas, the reduction rate of FeO also increases. The effect of
the particle size changes as its size increases, as follows:

- Increasing the particle size has two opposing effects, for a constant gas flow rate
  and slag basicity; increasing the particle size increases the rate of reduction of
  FeO until it reaches a maximum value and then it decreases.
- The maximum rate of reduction of FeO is achieved by decreasing the particle size,
  i.e. from 100 to 140 μm, highest gas flow rate, 400 Nl/min and lower basicity
  (0.5).
- The effect of particle size and slag basicity on the reduction rate of FeO has a
  complex behavior; in acid slags the reduction rate is increased by increasing the
  particle size but once the peak in the rate of reduction of FeO is achieved and the
  particle size is further increased, there is a new mechanism that promotes a higher
  rate of reduction with basic slags.

Fig. 7.178 Effect of the
carbon particle size on the
rate of reduction of iron
oxide. After [232]

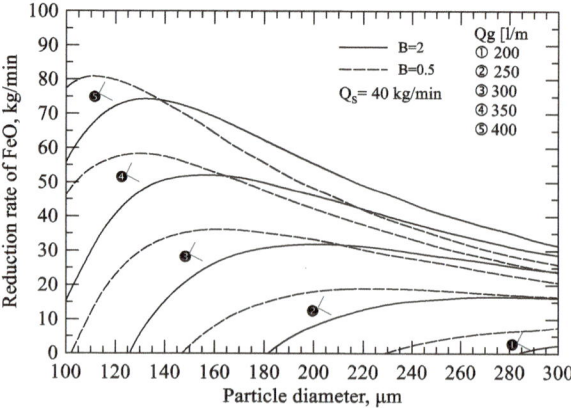

Fig. 7.179 Effect of the mass flow of carbon particles on the rate of reduction of FeO; predicted and experimental data. After [232]

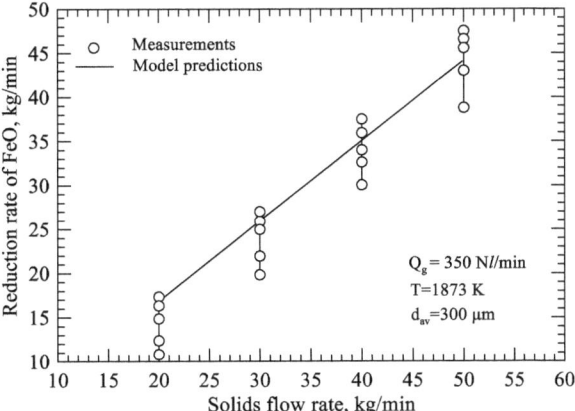

The previous findings were original in the scientific community in 1997, although the work carried out keeps simplifications that deviate from the real foaming practice at the industrial scale. This model has shown that more variables should be included to properly predict slag foaming under dynamic conditions in the industry.

Figure 7.179 compares experimental and numerical data predicted by the model. It can be observed that the model proposed is capable to provide a good estimation of the real values, considering that the industrial data have fluctuations in all the variables and in the model average values were employed.

The model just described provides the fundamental relationships among the relevant process variables to optimize the carbon injection practice. From experience, it should never be assumed that because there is a large amount of injected carbon particles, there will be automatically good foaming conditions, sustained for a long time. This is not true. There are many variables involved and this requires optimization. The location of injection of carbon particles should be designed to allow the longest residence time and guarantee its consumption. In the previous results it has been shown that smaller carbon particles can be fully consumed but this can also be misleading because the smaller the particles the fraction of them which enter the slag decreases due to a lack of momentum and then are rapidly transferred to the top slag surface and then to the gas atmosphere. These smaller particles then are combusted at the delta region causing overheating and erosion, affecting its life service.

7.15.7.2   Dynamic Foaming Index (DFI)

The work carried out by Fruehan and co-workers became the standard practice to study slag foaming since 1989. The use of a foaming index has been disputed as conceptually wrong by several researchers, for example Gou and Irons [169], however, it has also been proved to be useful to describe foaming at the industrial scale [241].

Our work [93, 99, 127, 128, 141, 144, 166, 216, 242] criticized the foaming index developed by Fruehan because mainly depends on the properties of the slag. The gas phase is represented only by the bubble size. As has been discussed previously, the real foaming practice with carbon particles involves many more variables, however, in addition to the slag properties the instantaneous generation rate of CO should be included. On this basis, a new expression was defined, called dynamic foaming index (DFI):

$$\Sigma_{DFI} = f_{CO(t)} \cdot \Sigma$$
$$f_{CO(t)} = \frac{Q^{sl}_{g(t)}}{Q_{sl(t)}}$$
$$Q^{sl}_{g(t)} = \frac{dV_{CO}}{dt} + \frac{1}{[2460 + 18(\%FeO)]} \frac{dW_{sl}}{dt}$$

where Σ_{DFI} is the dynamic foaming index, $f_{CO(t)}$ computes the ratio of slag volumes with $\left(Q^{sl}_{g(t)}\right)$ and without CO $\left(Q_{sl(t)}\right)$ and Σ is a static foaming index which depends only on the physical properties of the slag. The kinetic model previously described is capable to estimate the generation rate of CO due to the reduction of FeO both at the slag/metal interface and by the carbon particles. In both cases the reduction rate of FeO is assumed to take part by two simultaneous reactions which involve the Boudouard reaction with solid carbon and carbon dissolved in the slag. For further details on the equations, see Ref. [233]. The concept on the DFI involves two terms, the gas generation term and the slag properties term. If the slag properties are adequate for slag foaming but there is not enough gas, slag foaming will be poor and vice versa, a large generation of gas CO without the adequate slag properties foaming will not happen or will be limited to the extent of the imbalance. The foaming index will be high if both the gas generation rate is high and with the proper slag properties. The dynamic foaming index and the proposed control with the distortion of the electric arc to control slag foaming were both novel ideas when were originally proposed in 2001. The arc distortion is a measure of the noise intensity, therefore using this parameter there is no need for additional equipment to measure slag foaming. The inverse of the distortion implies lower noise and higher foaming conditions. We reported this original idea in 2001 [234–238], unfortunately it was not transformed into a patent. The same idea was later developed into hardware by other companies [243]. Other technologies include the measurement of vibrations to detect the foaming index [155] or a combination of sound and electric parameters [162]. The DFI was validated under industrial conditions as shown in Figs. 7.180, 7.181 and 7.182 for different heats. The stage at which the concept was validated was only qualitative but promising.

Kapilashrami et al. [134] in 2006 disproved the DFI concept, arguing that is not a measure of slag foaminess under dynamic conditions because it is reduced to "the ratio of the foam volume to the rate of change of the slag volume". Their criticism is based on the assumption that the volume of CO generated is much larger than the volume of slag, and therefore $f_{CO(t)}$ becomes a ratio between the rate of CO generated to the rate of change of the slag volume. In their words, the DFI "becomes the ratio of the foam volume to the rate of change of the slag volume and is neither a

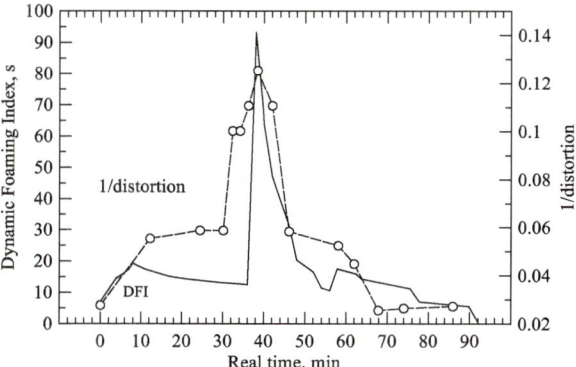

Fig. 7.180 Validation of DFI. After [235]

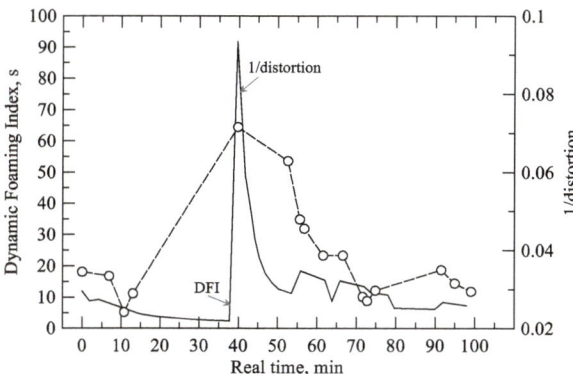

Fig. 7.181 Validation of DFI. After [235]

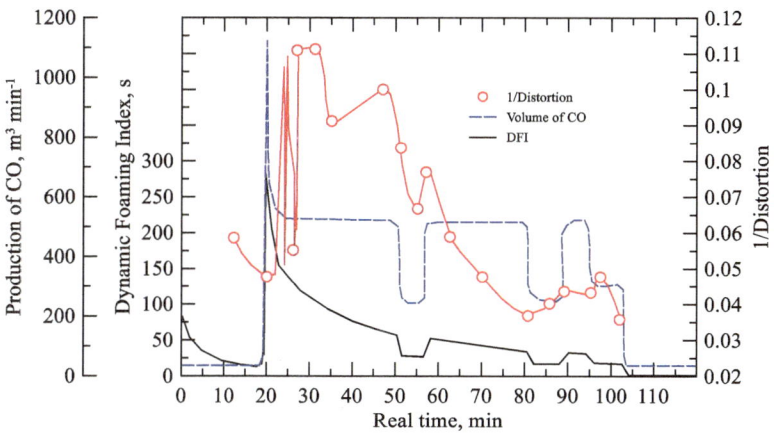

Fig. 7.182 Validation of DFI. After [238]

measure of the foam stability nor the foaminess of slag under dynamic conditions".
This criticism is wrong because, in spite that the volume of CO is much higher than
the slag volume, the equation involves an instantaneous generation rate of CO and
therefore adding this term was necessary, proved with the experimental data shown
in Figs. 7.180, 7.181 and 7.182.

Perhaps a better argument to keep the foaming index only as a function of the slag
physical properties is to stick to the original definition. The foaming index is a ratio
between the foam volume and the gas generation rate which can also be defined as
a ratio of foam height to the superficial gas velocity. According with this definition,
the variable that depends on the gas generation rate is the foam volume (or foam
height using an equivalent expression):

$$\overline{V}_{foam} = \Sigma \cdot Q_g$$
$$h_{fs} = \Sigma \cdot U_s$$

The slag foaming capacity (L_{fs}) is represented not by the foaming index but by
the foam height. The purpose of the proposed DFI was to define a new index which
also includes the instantaneous gas generation rate.

7.15.8 Oosthuizen et al. Model

Oosthuizen et al. [241] in 2001 derived a simplified approach to estimate the foaming
index in accordance with the definition by Bikerman. In Bikerman's equation the
foam height can be estimated if the foaming index and the superficial gas velocities
are known. They employed a linear relationship between the foaming index and
FeO according to Jones et al. and the superficial gas velocity of the gas produced.
The molar flow rate of gas produced derives from three sources; carbon injection,
decarburization at the slag/metal interphase and nitrogen in the carrier gas. In the
following analysis it is assumed that 1 kg of nitrogen is injected for every 150 kg of
graphite

$$G_{gas} = G_1 + G_2 + G_3$$
$$G_1 = \frac{M_{CO}}{M_C} w_C$$
$$G_2 = \frac{M_{CO}}{M_C} k_{DeC}\left(X_C - X_C^{eq}\right)$$
$$G_3 = \frac{1}{150} w_C$$

Then,

$$G_{gas} = \frac{1}{M_{CO}}(G_1 + G_2) + \frac{1}{M_{N_2}} G_3$$

Using the ideal gas law to define the volume of gas produced by the cross-sectional area of the EAF, it is obtained the superficial gas velocity:

$$U_s = \frac{G_{gas}RT}{PA}$$

where G_{gas} is the molar flow rate of gas produced, G_1 molar flow rate of CO produced due to carbon injection, G_2 is the molar flow rate of CO produced by decarburization, G_3 is the molar flow rate of nitrogen in the carrier gas, w_C is the injection rate of carbon particles and M represents molecular mass. The authors derived the following relationship, valid for 20–40% FeO:

$$\Sigma = 20\ 172.58\ (\%FeO) - 2.07$$

In this expression the foaming index linearly increases as FeO increases, which is un-realistic and the magnitude is also probably wrong, probably a typo in the constant. The foam height is calculated from the foaming index and the superficial gas velocity:

$$h_{fs} = \Sigma \cdot U_s = (20\ 172.58\ (\%FeO) - 2.07) \times \frac{G_{gas}RT}{PA}$$

Figure 7.183 reports the results from this model. The change in foam height is directly related to changes in FeO and gas production rate. In this figure the lowest foam depth is obtained when FeO is maximum at 44 min, in this moment oxygen injection is stopped.

Fig. 7.183 Evolution of foam height based on model by Oosthuizen et al. [241]

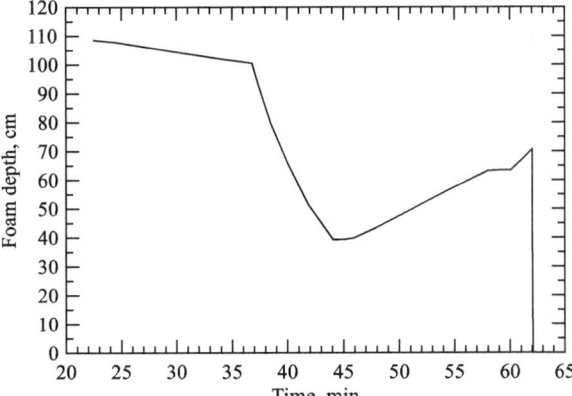

7.15.9   Kim et al. Model

A similar model to Ito and Fruehan was also reported by Kim et al. [126] in 2001. They predicted the foaming index using their own relationship (similar to Ito and Fruehan) but with different constants, 214 for CaO-based slags and 999 for MgO_{sat}. The gas generation rate was estimated considering the combustion of carbon, the reduction of FeO and the Boudouard reactions:

$$C_{(s)} + O_{2(g)} = CO_{(g)}$$
$$CO_{(g)} + O_{2(g)} = CO_{2(g)}$$
$$dC_{comb}/dt = (-4.84 \times 10^{-2} + 3.80 \times 10^{-5}T) \cdot p_{O_2}$$
$$(FeO) + C_{(s)} = Fe + CO_{(g)}$$
$$dC_{red}/dt = 1.67 \times 10^{-7}(\%FeO)$$
$$CO_{2(g)} + C_{(s)} = 2CO_{(g)}$$
$$dC_{gasif}/dt = 6.35 \times exp(-19500/T)$$

where dC/dt represents the rate of production of gas in mol/cm^2 s. The gas generation rate is transformed to superficial velocity and the foam height computed as follows:

$$h = h_0 + \Sigma U_s^g$$

The net CO generation rate at each foam height was transformed to energy savings considering a certain degree of post-combustion, taking the exothermic heat of reaction as 67.1 kcal/mol. The predicted foam height in Fig. 7.184 is based on an EAF of 120 ton of nominal capacity, 5 m diameter, 55 min tap-to-tap time and a temperature of 1510 °C. The slag composition consisted of $CaO\text{-}SiO_2\text{-}20FeO\text{-}MgO_{sat}\text{-}5Al_2O_3$ with a binary basicity ratio of 1.2, 12 ton of slag with a density of 3.2 ton/m^3 and void fraction of 0.8–0.9. The carbon injection rate was 18 kg/min assuming a combustion degree from 30 to 50%. As an example, the model predicts a foam height of 71.9 cm with a superficial velocity of 35.5 cm/s and a foaming index of 2.03. The foam height value seems too high but comparison with industrial data was not reported.

7.15.10   Pilon et al. Model

Pilon et al. [244, 245] in his PhD thesis from 2001 developed a model to predict the foam height as a function of the slag thermophysical properties, the bubble size and the superficial gas velocity. His approach follows previous suggestions from Hrma and others indicating that foams start after a minimum superficial gas velocity, if $U_g < U_g^m$ the gas in the liquid is not enough to create a foam layer, when the superficial gas velocity increases and $U_g = U_g^m$ it is formed a monolayer of bubbles holding a linear relationship with the foam height if the superficial gas velocity is in the range $U_g^m < U_g < U_g^c$. Once the superficial gas velocity reaches a critical value,

Fig. 7.184 Foam height based on Kim et al. [126] and its effect on energy savings

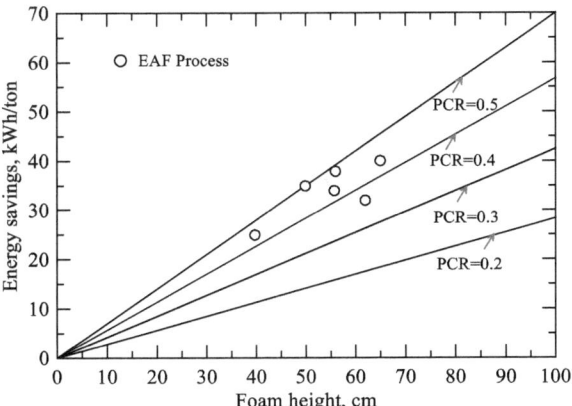

$U_g > U_g^c$ an excess of gas accumulates, then the foam volume grows continuously. The conventional foaming index is valid for the second stage.

Pilon developed an expression from first principles starting from two sub-models developed by Bhakta and Ruckenstein [246] and Narsimhan and Ruckenstein [247, 248] which define the vertical distance for the top and bottom of a foam as a function of the void fraction and the volumetric drainage rate through the Plateau borders, with the following assumptions:

- Foam consists of dodecahedron bubbles of the same size
- The Plateau borders are randomly oriented
- Drainage through the Plateau borders due to thinning is negligible compared to that due to gravity
- Coalescence of bubbles and Ostwald ripening within the foam are absent
- Surface tension is constant
- The wall effects are negligible
- The foam is under isothermal conditions
- The void fraction at the bottom is assumed to be constant $\theta(z_2,t) = 0.74$.

The model is shown in Fig. 7.185, the rate of change of heights z_1 and z_2 are defined by the following expressions:

$$\frac{dz_1}{dt} = \frac{\theta(z_1,t) \cdot q_{PB}^D(z_1,t)}{1 - \theta(z_1,t)}$$

$$\frac{dz_2}{dt} = \frac{U_g}{\theta(z_2,t)} - q_{PB}^D(z_2,t)$$

where U_g is the superficial gas velocity, $\theta(z, t)$ is the gas void fraction, $q_{PB}^D(z, t)$ is the volumetric flow rate of liquid through the Plateau border at location z and time t. Since the foam height, $h = z_2 - z_1$:

Fig. 7.185 Schematic
representation of foam
formation. After [244, 245]

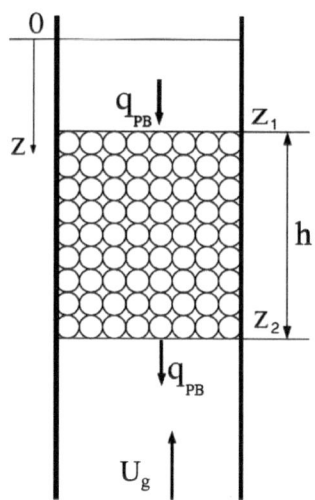

$$\frac{dh}{dt} = \frac{d(z_2 - z_1)}{dt} = \frac{U_g}{\theta(z_2,t)} - q_{PB}^D(z_2,t) - \frac{\theta(z_1,t) \cdot q_{PB}^D(z_1,t)}{1 - \theta(z_1,t)}$$

The term on the RHS $\frac{U_g}{\theta(z_2,t)}$ represents the increase in foam height due to the incoming gas and the other terms on the RHS of the equation represent a decrease in foam height due to liquid drainage and bubble breakup at the top of the foam.

The volumetric flow rate of liquid due to drainage is based on the expressions developed by Narsimhan and Ruckenstein [247, 248].

$$q_{PB}^D(z, t) = 3.6 \times 10^{-3} c_v \frac{[1 - \theta(z, t)]^2}{\theta(z, t)} \left\{ \frac{\rho g r^2}{\mu} + \frac{1.39}{\alpha} \frac{\sigma r^2}{\mu} \frac{\partial}{\partial z} \left[\left(\frac{\theta(z, t)}{[1 - \theta(z, t)]^2} \right)^{\frac{1}{2}} \right] \right\}$$

The foam height as a function of time is computed by solving the differential equation, however Pilon found this procedure complicated and sensitive the initial boundary conditions, instead, he applied dimensional analysis to the previous equations to simplify the expressions in terms of dimensionless numbers. The procedure is explained in detail in refs [244, 245]. The new dimensionless expression contained only two dimensionless numbers:

$$\left(\frac{\mu \left(U_g - U_g^m \right) r_0}{\sigma r_0} \right) \times \left(\frac{h}{r_0} \right) = C \left(\frac{\rho g r_0^3}{\mu \left(U_g - U_g^m \right) r_0} \right)$$

The first term on the LHS is the capillary number, a ratio of viscous forces over surface tension forces. The term on the RHS is a ratio of the Re and Fr numbers. r_0 is the average bubble radius. Therefore, the final expression becomes:

$$Ca\left(\frac{h}{r_0}\right) = C\left(\frac{Re}{Fr}\right)^n$$

where

$$Ca = \frac{\mu\left(U_g - {}^m_g\right)}{\sigma}$$
$$Re = \frac{\rho(U_g - U_g^m)r_0}{\mu}$$
$$Fr = \frac{(U_g - U_g^m)^2}{gr_0}$$

A particular solution for the dimensionless relationship was derived using foaming data from the literature. The results are expressed by the following equation and Fig. 7.186.

$$\left(\frac{h}{r_0}\right) = \frac{2905}{Ca}\left(\frac{Fr}{Re}\right)^{1.8}$$

or,

$$h = \frac{2905}{Ca}\frac{\sigma}{r_0^{2.6}}\frac{\left[\mu\left(U_g - U_g^m\right)\right]^{0.8}}{(\rho g)^{1.8}}$$

This equation was found to be valid for the following conditions:

- Slag viscosity: 46–12,100 mPa s
- Slag density: 1200–3000 kg/m^3
- Slag surface tension: 69.5–478 mN/m

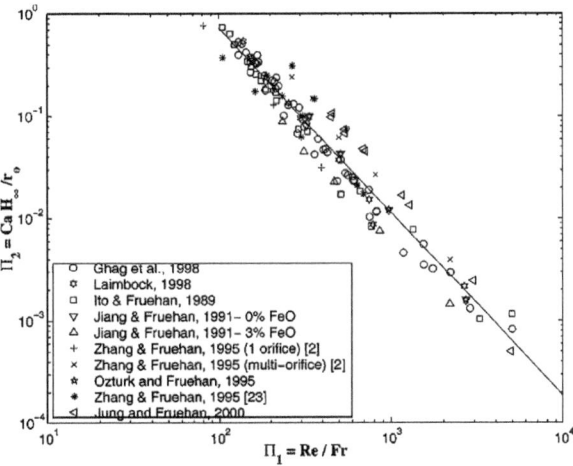

Fig. 7.186 Pilon's model to predict foam height. After [244, 245]

- Superficial gas velocity: 0–40 mm/s
- Average bubble radius: 0.7–20 mm.

The minimum superficial gas velocity, U_g^m, was estimated assuming a linear relationship between foam height and the superficial gas velocity.

The equation reported by Pilon contains the surface tension term on the numerator indicating that increasing surface tension increases the foam height, however the term in the equation involves a ratio of surface tension to bubble radius and they are not independent from each other. Based on the available information, as the surface tension decreases the bubble radius decreases by the same order and the ratio $\frac{\sigma}{r_0^{2.6}}$ increases.

7.15.11 Wu et al. Model

Wu et al. [98] developed a semi-empirical model and was validated with results from a physical model. The results from the physical model gave the idea to formulate the mathematical model. The oils selected had the following physical properties; kinematic viscosity from 50×10^{-6} m^2/s to 515×10^{-6} m^2/s, surface tension 21 mN/m and density 970 kg/m^3. The experiments were carried out at different superficial velocities, from 0.18 to 2.16 cm/s and an initial height of the liquid of 4.5 cm in most of the cases. The results show that increasing the superficial gas velocity, the foam height initially increases, rapidly reaches a maximum value and then decreases, similar to a Gaussian distribution function. Figure 7.187 shows the effect of the superficial gas velocity on the foam height. It also shows that above a kinematic viscosity of 280×10^{-6} m^2/s, the foam is suppressed. When the foam height decreases, it is attributed to the formation of larger bubbles, either because of a liquid too viscous or excessive turbulence.

Fig. 7.187 Effect of the superficial gas velocity on foam height at different kinematic viscosities. After [98]

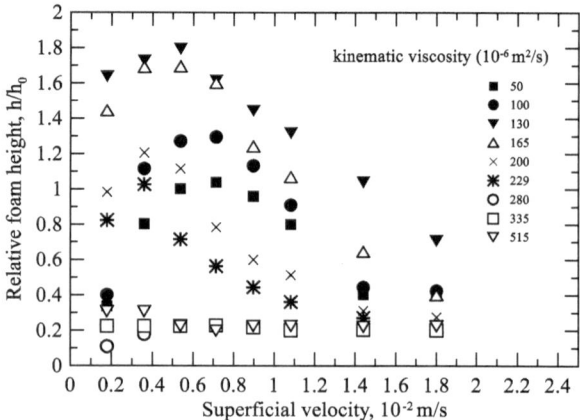

On the basis of their experimental data, a Gaussian relationship between the kinematic viscosity, the superficial gas velocity and foam height, the authors choose a Gaussian distribution function to describe this behavior.

$$\frac{\Delta h}{h_0} = f_1 \exp[-C_1(\nu - \nu_0)]^2$$
$$\frac{\Delta h}{h_0} = f_2 \exp\left\{-C_2\left[\left(U_g - \left(\tfrac{\nu_0}{\nu}\right)^{\frac{1}{2}} U_{g0}\right)\right]^2\right\}$$

where f_1 is a function of the superficial gas velocity, ν is the kinematic viscosity of the liquid, ν_0 is the kinematic viscosity at which the foaming of the liquid at a given superficial velocity shows a maximum, C_1 is a C_2 are constants, U_g is the superficial gas velocity, U_{g0} is the superficial gas velocity at which the foaming of the liquid at a given kinematic viscosity shows a maximum, f_2 is a function of the physical properties. The term $\left(\frac{\nu_0}{\nu}\right)^{\frac{1}{2}}$ takes into account the shift of the maximum foaming height due to the kinematic viscosity effect. Figure 7.188 compares the experimental data and model predictions of foam height as a function of the kinematic viscosity and superficial gas velocity.

Fig. 7.188 Experimental data and model predictions. After [98]

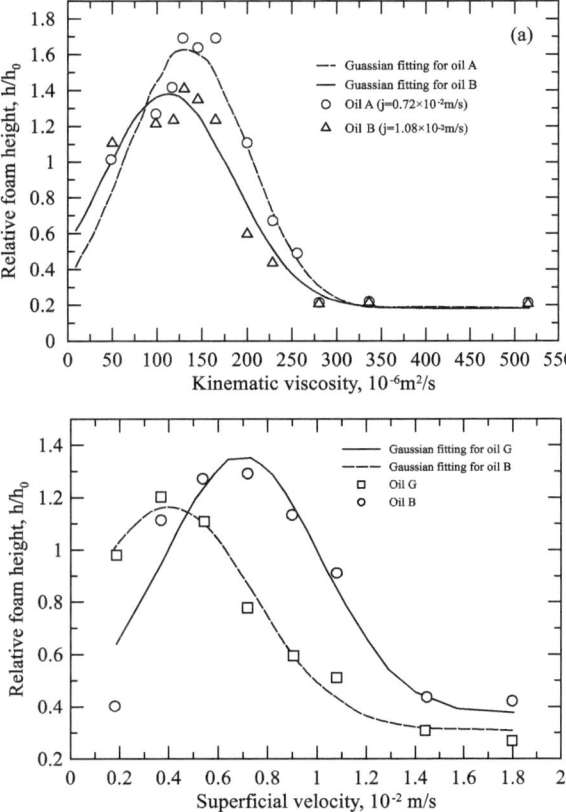

The final expression assumes the kinematic viscosity and superficial gas velocity control the foam height, therefore both expressions are coupled, inserting the Morton number in the final expression because they indicate it defines the residence time of the bubbles (the higher Mo, the higher residence time and larger foam height), as follows:

$$\frac{\Delta h}{h_0} = C_3 \left(\frac{v \rho^3 g}{\sigma^3} \right)^{C_4} \exp\left\{ -C_2 \left[\left(U_g - \left(\frac{v_0}{v} \right)^{\frac{1}{2}} U_{g0} \right) \right]^2 \right\} \exp[-C_1(v - v_0)]^2$$

The values of the constants were optimized to fit their experimental data:

$$\frac{\Delta h}{h_0} = 2.238 \left(\frac{v \rho^3 g}{\sigma^3} \right)^{0.045} \exp\left\{ -8150 \left[\left(U_g - \left(\frac{v_0}{v} \right)^{\frac{1}{2}} U_{g0} \right) \right]^2 \right\}$$
$$\times \exp\left[-1.04 \times 10^8 (v - v_0) \right]^2$$

Since the model was developed to fit their experimental data, the prediction is good, as would be expected, however this model has a large limitation for its application to other systems.

7.15.12 Slag Foaming Models by Dimensional Analysis

Dimensional analysis is a powerful tool employed to describe the relationship between multiple variables involved in any natural phenomena. Another book by the author is focused only on the fundamentals and application of dimensional analysis in metallurgy [249]. Chapter 11 in that book describes the derivation of equations of different cases involving slag foaming, step by step.

Ito and Fruehan [125, 166] in 1989 reported the first dimensional analysis on slag foaming. In this analysis the foaming index was defined as a function of the following variables:

$$\Sigma = f(\rho, \sigma, \mu, g)$$

The general relationship obtained by dimensional analysis was expressed with the following two dimensionless numbers:

$$\left(\frac{\Sigma \rho^{\frac{1}{4}} g^{\frac{3}{4}}}{\sigma^{\frac{1}{4}}} \right) = f\left(\frac{\sigma^{\frac{3}{4}} \rho^{\frac{1}{4}}}{\mu g^{\frac{1}{4}}} \right)$$

Using physical properties for $CaO\text{-}SiO_2\text{-}FeO\text{-}Al_2O_3$ slags, without solids precipitated, they reported the following particular solution:

$$\Sigma = 570\frac{\mu}{\sqrt{\rho\sigma}}$$

This equation suggests the strongest effect on the foaming index is the slag viscosity.

Jiang and Fruehan [216] studied bath smelting slags. In this work they suggested the calculation of the slag physical properties using additive models to estimate the slag density and surface tension, and Urbain's model to estimate the viscosity:

$$V = x_1\overline{V}_1 + x_2\overline{V}_2 + x_3\overline{V}_3 + \cdots$$
$$\sigma = x_1\overline{\sigma}_1 + x_2\overline{\sigma}_2 + x_3\overline{\sigma}_3 + \cdots$$
$$\mu = A \cdot T \cdot \exp^{(B/T)}$$

where x_i is the molar fraction, $\overline{V}_i$ is the partial molar volume (assumed to be equal to the molar volume of the pure components), $\overline{\sigma}_i$ is the surface tension of the pure components, μ represents viscosity in Ns/m^2, A and B are viscosity parameters.

The general expression with the same set of variables previously considered by Ito and Fruehan results in the following alternate general relationship:

$$\left(\frac{\Sigma g\mu}{\sigma}\right) = f\left(\frac{\rho\sigma^3}{\mu^4 g}\right)$$

The dimensionless group on the right side is the Morton number. This number is used to characterize the bubble shape, which in turn define the rising velocity and represents a ratio of gravitational-viscous forces and surface tension forces. Using physical properties for bath smelting slags, they reported the following particular solution:

$$\Sigma = 115\frac{\mu}{\sqrt{\rho\sigma}}$$

In a further work, Zhang and Fruehan [128] included the effect of the bubble size, increasing from two to three the dimensionless numbers.

$$\Sigma = f(\rho, \sigma, \mu, d_b, g)$$

The general expression derived by dimensional analysis is the following:

$$\left(\frac{\sigma g\mu}{\sigma}\right) = f\left(\frac{\mu^4 g}{\rho\sigma^3}\right)\left(\frac{\rho^2 d_b^3 g}{\mu^2}\right)$$
$$N_\sigma = C \times Mo^a \times Ar^b$$

where N_Σ is a dimensionless number which includes the depending variable, Mo is the Morton number and Ar is the Archimedes number. Using the experimental data, the values obtained for C, a and b were; 4.7, 0.39 and $-$ 0.28, respectively. Figure 7.189 shows the predicted values with the particular solution, including

Fig. 7.189 Prediction of foaming index by dimensional analysis. After [128]

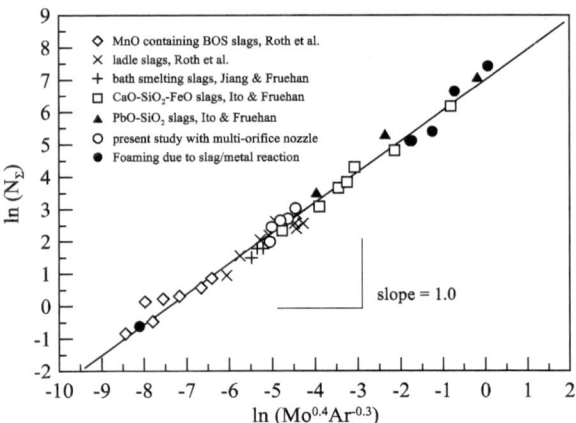

previous results from the same research group. In this plot, collecting the available data, the value of the constant is 2.7. The previous equation can be recast, as follows:

$$\Sigma = 115 \frac{\mu^{1.2}}{\sigma^{0.2} \rho d_b^{0.9}}$$

These authors reported that once the bubble diameter is included the foaming index is more sensitive to changes in the slag viscosity and less sensitive to changes in the surface tension.

Jouhari et al. [250] and Bhoi et al. [172] carried a similar work to Ito and Fruehan however their measurements on foam height were carried out with a steel rod, which is a very inaccurate method. By the time they reported their results it was well established the importance of the bubble size. Their results were reported to be similar to those by Ito and Fruehan. Their results reported the value of the constant to be 500 and 470, respectively.

Ghagh et al. [251] studied the foaming index in water-glycerol solutions. Their dimensional analysis included (ρg) as one variable and instead of the absolute value of surface tension they employed the value of the surface tension depression ($\Delta\sigma$) because this value is the one that directly affects slag foaming. In this regard, Nekrasov et al. [77] indicated that it is not the absolute value of surface tension which affects foaming because in aqueous systems with surface tension in the range from 1 to 10 mJ/m^2 a decrease of 1 to 2 mJ/m^2 significantly improves foaming but in high temperature systems with slags having surface tension in the range from 400 to 600 mJ/m^2, a decrease of 100 mJ/m^2 is sometimes not enough to improve foaming. The surface tension depression is equal to the difference in the surface tension of the solvent and the solution. Their general and particular expressions derived by dimensional analysis were the following:

$$\left(\frac{\Sigma \Delta \sigma}{\mu d_b}\right) = f\left(\frac{\Delta \sigma}{(\rho g) d_b^4}\right)$$

$$\Sigma = 2 \times 10^6 \frac{\mu \Delta \sigma}{(\rho g)^{2.32} d_b^{3.64}}$$

The particular solution adequately reproduced their experimental data for water-glycerol solutions but failed to reproduce high-temperature data. They argued that the surface tension depression in high temperature systems was much higher, about an order of magnitude and suggested that instead of the surface tension depression, it was better to use the surface elasticity, according with the following expression:

$$E = \frac{A d\sigma}{dA} = \frac{d\sigma}{d \ln A} = -\frac{d\Delta\sigma}{d \ln \Gamma_i}\frac{d \ln \Gamma_i}{d \ln A} = -E_M \frac{d \ln \Gamma_i}{d \ln A}$$

where A is the surface area, $\Delta\sigma$ is the surface tension depression, Γ_i is the Gibbs relative surface excess of the surface-active species and E_M is the Marangoni dilational modulus equal to $\frac{d\Delta\sigma}{d \ln \Gamma_i}$ and represents the maximum elasticity of the film, also has the same units as $\Delta\sigma$, therefore in a revised dimensional analysis, $\Delta\sigma$ was replaced by E_M, obtaining the following particular solution:

$$\Sigma = 5.43 \times 10^5 \frac{\mu E_M^{0.89}}{(\rho g)^{1.89} d_b^{2.78}}$$

They found out that the correlation coefficient was not better than the previous expression but slightly lower, 0.926 and 0.906, respectively. In its last attempt to improve the correlation coefficient they argued that the assumption of achieving maximum surface elasticity was inaccurate and they replaced its value for an effective surface elasticity, calculated in accordance with the equations suggested by Malysa et al. [252]. Replacement of E_M by E_{eff} increased the correlation coefficient to 0.958. The final expression is indicated in Table 7.20.

Skupien and Gaskell [184] repeated the calculations from Jiang and Fruehan [216] employing their own experimental measurements on surface tension and extrapolated the experimental measurements from Lee and Gaskell. Their results also indicate a higher influence of slag viscosity on the foaming index in comparison with surface tension. The final expression is indicated in Table 7.20.

Kim et al. [126] conducted an extensive investigation on slag foaming for CaO-SiO$_2$-FeO-MgO$_{sat}$-X slags where X represents MnO, Al$_2$O$_3$, CaF$_2$ and P$_2$O$_5$. They carried out a dimensional analysis similar to Jiang and Fruehan [216]. The results evaluated correspond to two types of slags, with and without MgO; CaO-SiO$_2$-FeO-Al$_2$O$_3$ (CSFA) and CaO-SiO$_2$-30FeO-MgO$_{sat}$-X (CSFAM-X) where X corresponds to MnO, Al$_2$O$_3$, CaF$_2$ and P$_2$O$_5$. It should be noticed that they exclude the effect of the bubble diameter. Figure 7.190 shows that MgO gives a slightly different slope, but the effect of the variables involved is identical.

Stadler et al. [96] carried out experiments at low temperature with three columns of different diameters. Air was injected through porous sinter disks with two different pore sizes. The surface tension was modified with additions of ethanol and methyl

Table 7.20 Summary of foaming indexes reported by different investigations

	Year	First author	Variables	System		Σ
1	1989	Ito [125, 166]	$\Sigma = f(\rho, \sigma, \mu, g)$	FeO-CaO-SiO$_2$-(Al$_2$O$_3$), 1250–1400 °C	5	$\Sigma = 570\,\dfrac{\mu}{\sqrt{\rho\sigma}}$
2	1991	Jiang [216]	$\Sigma = f(\rho, \sigma, \mu, g)$	FeO-CaO-SiO$_2$-MgO-Al$_2$O$_3$, 1500 °C	5	$\Sigma = 115\,\dfrac{\mu}{\sqrt{\rho\sigma}}$
3	1995	Zhang [128]	$\Sigma = f(\rho, \sigma, \mu, d_b, g)$	FeO-CaO-SiO$_2$-Al$_2$O$_3$	6	$\Sigma = 115\,\dfrac{\mu^{1.2}}{\sigma^{0.2}\rho d_b^{0.9}}$
4	1995	Lin [217]	$\Sigma = f(\Delta h, d_b, g, U_g)$	Air/water/silicone oil	5	$\Sigma = \dfrac{\Delta h + h}{U_g + (0.5 g d_b)^{0.5}}$
5	1996 1997	Gou [169] Kapoor [104]	–	5-ton smelter	–	$\Sigma = \dfrac{\Delta h + h_0}{C_0 U_s + U_{Gj}}$
6	1998	Ghagh [251]	$\Sigma = f(d_b, E_{eff}, \mu, \rho g)$	Water-glycerol	5	$\Sigma = 1 \times 10^6\,\dfrac{\mu E_{eff}}{(\rho g)^2 d_b^3}$
7	2000	Jouhari [250]	$\Sigma = f(\rho, \sigma, \mu, g)$	FeO-CaO-SiO$_2$	5	$\Sigma = 500\,\dfrac{\mu}{\sqrt{\rho\sigma}}$
8	2000	Skupien [184]	$\Sigma = f(\rho, \sigma, \mu, g)$	FeO-CaO-SiO$_2$, 1300 °C, 1400 °C	5	$\Sigma = 100\,\dfrac{\mu^{0.54}}{\rho^{0.39}\sigma^{0.15}}$
9	2000	Zhu	–	–	–	$h_{max} = \left(\dfrac{1}{1 - 2N_{foam}}\right)\cdot h_0$
10	2001	Kim [126]	$\Sigma = f(\rho, \sigma, \mu, g)$	FeO-CaO-SiO$_2$-MgO-Al$_2$O$_3$ (CaO-based)	5	$\Sigma = 214\,\dfrac{\mu}{\sqrt{\rho\sigma}}$
11	2001	Oosthuizen [241]	–	FeO-CaO-SiO$_2$-MgO-Al$_2$O$_3$-CaF$_2$-MnO-P$_2$O$_5$ (MgO$_{sat}$)	–	$\Sigma = 999\,\dfrac{\mu}{\sqrt{\rho\sigma}}$
			–	–	–	$\Sigma = \dfrac{h}{U_g} = h\dfrac{PA}{G_{gas}RT}$

(continued)

Table 7.20 (continued)

	Year	First author	Variables	System	Σ
12	2002	Lahiri [230]	–	–	$\Sigma = 150\left(\dfrac{\mu}{\rho d_b}\right)$
13	2002	Pilon [244, 245]	–	Aqueous and slag systems	$h = \dfrac{2905}{Ca}\dfrac{\sigma}{r_0^{2.6}}\left[\dfrac{\mu\left(U_g-U_g^m\right)}{(\rho g)^{1.8}}\right]^{0.8}$
14	2005	Pilon [253]	$h = f\left(\begin{array}{c}\rho,\, r_b,\, \sigma,\, \mu,\\ g,\, U_r\end{array}\right)$ 7	Aqueous and slag systems	$\dfrac{h}{r} = 2617 \times \dfrac{\mu^{0.73}U^{0.79}\sigma^{1.01}}{\rho^{1.74}g^{1.77}r^{3.51}}$
15	2010	Wu [98]	$\dfrac{\Delta h}{h_0} = 2.24\left(\dfrac{\nu\rho^3 g}{\sigma^3}\right)^{0.045}\exp\left\{-8150\left[\left(U_g - \left(\dfrac{\nu_0}{\nu}\right)^{\frac{1}{2}}U_{g0}\right)\right]^2\right\}\exp\left[-1.04\times10^8(\nu-\nu_0)\right]^2$		
16	2013	Attia [254]	$h = f\left(\begin{array}{c}\rho,\, r_b,\, \sigma,\, \mu,\\ g,\, U_r,\, D,\, S_0\end{array}\right)$ 9	Aqueous systems	$h = 118 \times \dfrac{\sigma}{r_0^{1.64}}\dfrac{\mu^{0.8}\left(U_g-U_g^m\right)^{1.76}}{(\rho g)^{1.8}(DS_0)^{0.96}}$

Fig. 7.190 Dimensional analysis by Kim et al. Taken from Ref. [78, 78]

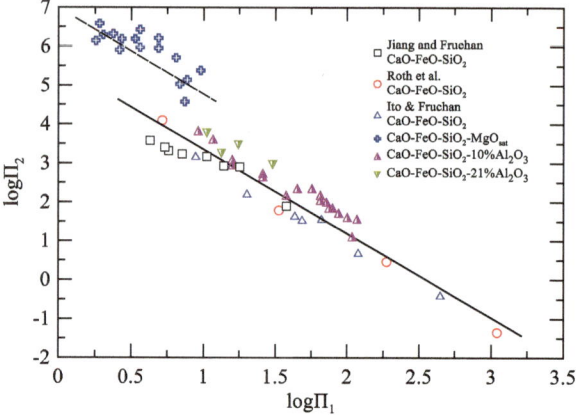

isobutyl carbinol (MIBC) and changes in viscosity using glycerol and sucrose. In addition to this, supersaturated solutions of $MgSO_4$ were employed to evaluate the effect of solid particles. The bubble size was measured extracting samples with a capillary tube, this procedure was limited to bubbles smaller than 5 mm. Their dimensional analysis was carried out including the physical properties of both the slag and the gas phase.

$$\Sigma = f(\rho, \rho_g, \sigma, \mu, \mu_g, g, U_g, h, D)$$

Since they employed only air, it was unnecessary to derive an expression involving the physical properties of air. The general relationship obtained by dimensional analysis was expressed with the following four dimensionless numbers:

$$\left(\frac{\Sigma g\mu}{\sigma}\right) = f\left(\frac{\rho\sigma^3}{\mu^4 g}\right)\left(\frac{\rho_g U_g D}{\mu_g}\right)\left(\frac{D}{h}\right)$$

The particular solution using their experimental data was the following:

$$\Sigma = C\left(\frac{\mu^{0.704} h^{0.25}}{g^{0.574}\sigma^{0.278}\rho^{0.426}D^{0.603}U_g^{0.353}}\right)$$

This equation was unable to satisfactorily reproduce their experimental data indicating possible errors in its derivation.

Until now, the most accurate model to predict slag foaming is due to Pilon et al. [244, 245]. They initially developed the following expression by scaling a governing differential equation:

$$h = \frac{2905}{Ca}\frac{\sigma}{r_0^{2.6}}\frac{\left[\mu\left(U_g - U_g^m\right)\right]^{0.8}}{(\rho g)^{1.8}}$$

However, it was also indicated that this equation needed further improvements because of the assumption of a constant bubble size in its derivation. A new expression should consider Ostwald ripening, bubble coalescence and a bubble size distribution. The effect of the initial height of liquid is still not clear as well as the surrounding atmosphere and the gas injected. In regard to temperature, since the model predictions are accurate enough in spite of the large differences in temperature it was considered that the equation takes into account its effect on the value of the thermophysical properties. In 2005, Lotun and Pilon [253] carried out a dimensional analysis including 7 variables:

$$h = f\left(\rho, r_b, \sigma, \mu, g, U_r\right)$$

where: U_r is called reduced superficial gas velocity $\left(U_g - U_g^m\right)$. The general equation was the following:

$$\frac{h}{r} = C \times Ca^a Re^b Fr^c$$

Using a vast number of different experimental data reported in the literature, the particular solution was the following:

$$\frac{h}{r} = 2617 \times \frac{\mu^{0.73} U_r^{0.79} \sigma^{1.01}}{\rho^{1.74} g^{1.77} r^{3.51}}$$

In a more recent paper, Attia et al. [254] expanded the number of variables to 8 to include the effect of Ostwald ripening; D, the diffusion coefficient of the gas in the liquid phase and S_0, the Ostwald coefficient of solubility, which defines the volume of saturated gas absorbed by unit volume of pure liquid at a given temperature and pressure. A new dimensionless relationship was developed and evaluated with aqueous solutions, resulting in the following particular solution:

$$h = 118 \times \frac{\sigma}{r_0^{1.64}} \frac{\mu^{0.8} \left(U_g - U_g^m\right)^{1.76}}{(\rho g)^{1.8} (DS_0)^{0.96}}$$

Table 7.20 summarizes the previous equations obtained by dimensional analysis and other equations derived from analytical, water models and numerical models.

7.15.13 Slag Foaming Models by CFD

A foam is a colloidal system which involves multi-phase flow due to the presence of a liquid, a gas and carbon particles. Under these conditions is no longer a Newtonian fluid and the Navier–Stokes equations cannot longer apply. The Navier–stokes equation is part of the foundations of CFD. A Newtonian fluid obeys Newton's law of viscosity, in this law viscosity is a constant. Sichen et al. [81] indicate they have made measurements that show the foam does not actually flow at all, instead, the gas moves like a plume and only the area very close to this plume is affected. These arguments can probably explain the poor number of investigations reporting CFD models on slag foaming in the past. However, the work carried out recently in Australia since 2013 at Swinburne University of Technology (SUT) has proven that it is possible to succeed in developing robust CFD models as will be shown below.

Campolo et al. [255] in 2007 reported a mathematical model by co-injecting oxygen and carbon particles but very limited in terms of slag foaming. Wei et al. [256] in 2020 reported the injection of carbon particles using coherent jets indicating that in this way the particles can be accelerated inside the slag a longer distance. They also reported improvements in terms of a decrease in FeO in the slag but many details are missing.

Sattar et al. from SUT, in 2013 [257, 258] and 2014 [259] started the most complete CFD modelling work on slag foaming until now. Their work is based on previous models developed for aqueous systems that predict bubble coalescence and bubble break up and is limited to validate the experimental work carried out by Fruehan's group in the past injecting argon into a high temperature system. They employed an Euler-Euler multi-phase flow model, the liquid phase was treated as a continuum and the gas phase (bubbles) was considered as a dispersed phase and also treated as interpenetrating continua. The model assumes the bubbles are pentagonal dodecahedrons but in reality, differ in shape and size. The foam density and viscosity were computed treating the foam as a mixture, as follows:

$$\rho_f = \rho_l\theta_l + \rho_g(1 - \theta_l)$$
$$\mu_f = \mu_l\theta_l + \mu_g(1 - \theta_l)$$

The momentum interfacial exchange is significant, due to bubble coalescence and bubble break up. A population balance model was included which results from bubble formation, coalescence and break-up. Coalescence occurs due to bubble interactions and liquid drainage.

The coalescence model developed by Prince and Blanch was employed to evaluate bubble interactions. This coalescence rate is the product of a collision rate and a collision efficiency. Coalescence due to liquid drainage was evaluated using the model from Tong et al. The coalescence rate depends on the number of films per bubble, the failure rate of films separating bubbles and a probability term. In turn the failure rate depends on the time required to drain the liquid. The model from Luo and Svendsen was used to describe bubble break-up which results from the product of a collision frequency and the probability of bubble break up. This probability is defined by the energy and size of the eddys. Only eddys which have enough energy can promote bubble break up. A number of other additional sub-models were also employed to define all the parameters involved. The system of equations was solved using the commercial CFD software AVL Fire 2009.2. Figure 7.191 shows the crucible dimensions and meshing based on the experimental work carried out by Jiang and Fruehan [216]. Figure 7.192 compares the experimental data and predicted results with the CFD model. The agreement is quite satisfactory.

In 2014, Sattar et al. [260] applied their model to predict slag foaming in the BOF, as can be seen in Fig. 7.193 however they found some disagreements. Hewage et al.

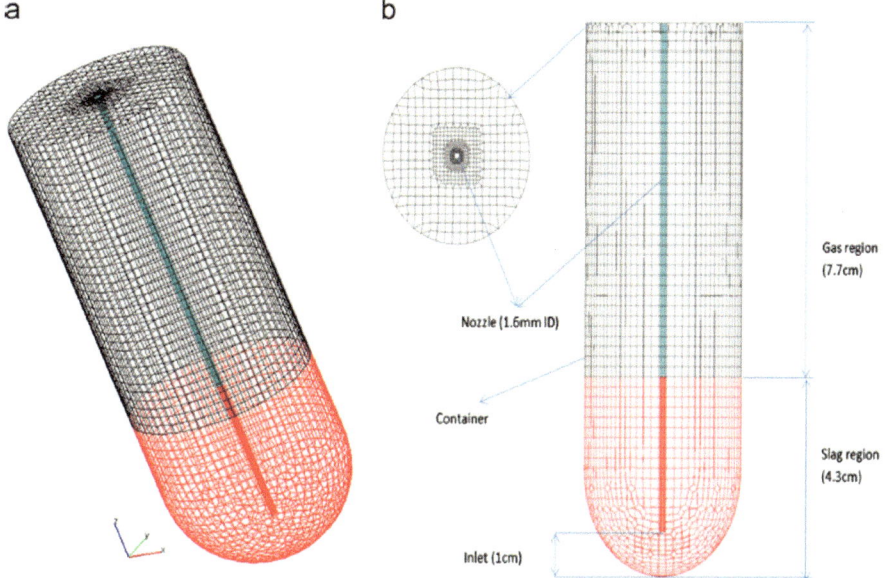

Fig. 7.191 Meshing. **a** 3D mesh, **b** outline. After [259]

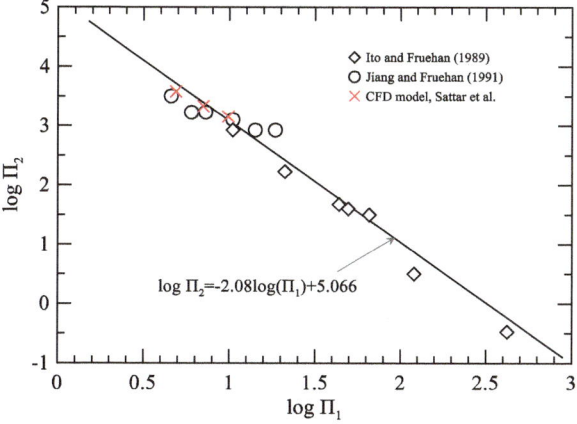

Fig. 7.192 Experimental and predicted results, in terms of dimensionless numbers. After [259]

Fig. 7.193 Predicted versus experimental on foam height using Sattar's model. After [260]

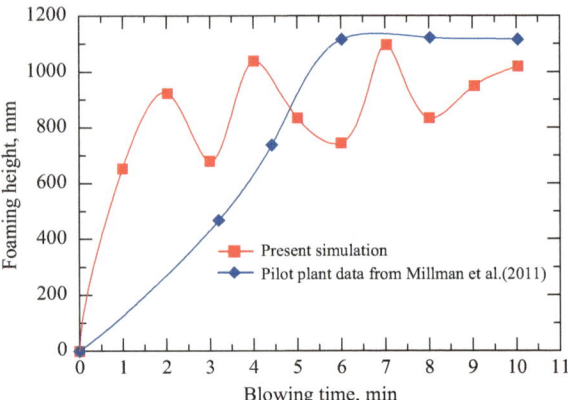

[261] in 2015 attempted to improve Sattar's model, with limited success, indicating the need for major improvements that include a more accurate kinetic model which describes the changes in slag composition and a review of the whole set of sub-models originally included in Sattar's model.

7.16 Slag Foaming in the Industrial Practice

7.16.1 Experimental Reports on Slag Foaming

In 1992, Burstrom et al. [207] investigated the effect of slag composition, slag temperature, saturation degree of the slag, methods of gas production, and coal/coke quality on slag foaming in the EAF. The conclusion of this work is that slag foaming can be controlled with two main parameters; a high effective viscosity and a high decarburization rate. The investigation included heats in a pilot plant and industrial tests. The pilot plant tests were carried out at MEFOS, with an EAF charged with 8 ton of scrap and 850 kg of synthetic slag. The bath surface area was 4.2 m^2. The melt temperature varied from 1580 to 1650 °C. The foam height was measured with a video camera and the gas produced was controlled with their own process control systems. The foam height was evaluated as a function of the production rate of CO, resulting in a linear relationship for different concentrations of FeO:

$$h = h_0 + k \cdot \frac{dCO}{dt}$$

where h is the foam height in cm, h_0 is initial slag height in cm, dCO/dt is the production rate of CO in kg/min, and k is a constant, called foam stability in min cm/kg.

Two types of experiments were carried out, first, to evaluate the slag composition, slag temperature and coal/no-coal injection of carbon particles. The results changing temperature, slag composition and injection/non-injection of coal particles are shown in Table 7.21, which summarizes the foam stability factors for different conditions. The lowest values for the foam stability factor are obtained when only oxygen is injected without carbon particles. Injecting only oxygen leads to only exothermic reactions which increase the slag temperature (above of liquid steel) and therefore the slag viscosity decreases. The authors also suggested a higher MgO saturation degree and lower slag temperate to get higher foam stability factors. This conclusion is not clear from the table because the case with a higher saturation degree and lowest slag temperature doesn't have the highest foam stability factor.

Figure 7.194 confirms the strong relationship between the gas generation rate and foam height. Increasing the volume of CO from decarburization increases the foam height. It can also be observed that the foam stability factor is higher if FeO is increased above 10%, noticing that the FeO was always lower than 21%.

The second group of experiments on the pilot plant evaluated the effect of carbon reactivity on foam height, as shown in Fig. 7.195. Table 7.22 summarizes their results. The largest k values are obtained using powdered graphite and petroleum coke, on

Table 7.21 Effect of slag composition, slag temperature and coal/non-coal injection on the foam stability factor. After [207]

Injection	FeO	CaO	SiO$_2$	MgO$_{sat}$	T steel, °C	T slag, °C	h$_0$	k
O$_2$	13	49	6		1595	1620	15	12.5
O$_2$	15	45	7	2.8	1650		15	8
O$_2$ + coal	21	45	6	2.4	1580	< 1300	15	20
O$_2$ + coal	17	46	11	2.9	1585	1400	15	33
O$_2$ + coal	18	46	11	2.1	1650	1420	15	24
O$_2$ + coal + lime	14	48	12	4.9	1650	1390	15	14

Fig. 7.194 Foam height as a function of gas production, using different carbon particles. After [207]

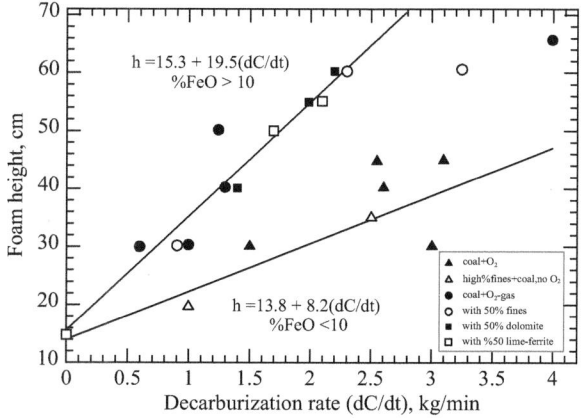

Fig. 7.195 Foam height as a function of gas production, for industrial trials. After [207]

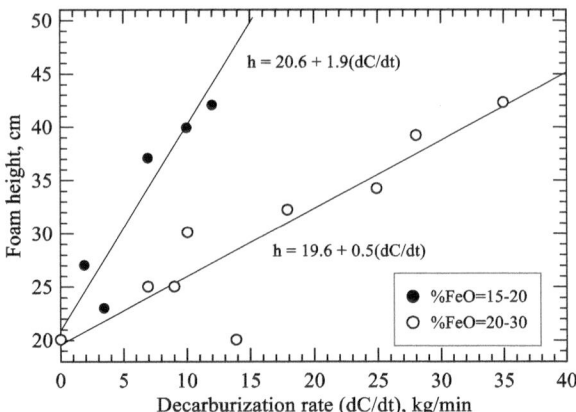

the other hand, using highly volatile coal or coke fines yields the worst results. They observed that graphite was rapidly consumed and then the foam collapsed, suggesting petroleum coke as the best option.

The industrial scale experiments confirmed the linear relationship between gas generation and foam height. Figure 7.195 shows the results, noticing that in this case the concentration of FeO ranged from 15 to 30%. These results indicate that FeO higher than 20% decrease the foam stability factor.

The same authors evaluated the effect of the ratio $(\mu_e/\mu) \times 100$ on the foam stability factor and confirmed a positive trend, as shown in Fig. 7.196. This result would suggest to add particles to increase the slag's effective viscosity.

Boemer and Rodl [262] reported results on physical and numerical modeling on foaming with an injection lance. Although this publication omits details of the research work, they reported important information on the optimum lancing conditions. They simulated the co-injection of oxygen and carbon particles in a DC-EAF of 190 ton. The results are indicated in Fig. 7.197. The foaming index was reported as the ratio of the $V_{foam}/V_{foam\ reference}$. It is shown that the foaming index increases by increasing the oxygen flow rate, increasing the vertical angle and by decreasing the distance from the surface to the lance.

Table 7.22 Effect of reactivity of carbon particles on the foam stability factor. After [207]

Exp.	Source of carbon	Lowest measured T in the slag, °C	h_0	k
1	Powdered graphite	1360	15	18
2	Coke fines	1690	15	7
3	Brown coal coke	1500	15	10
4	Petroleum coke	1380 low T	15	27
		High T	15	17
5	Highly volatile coal	1540	15	8

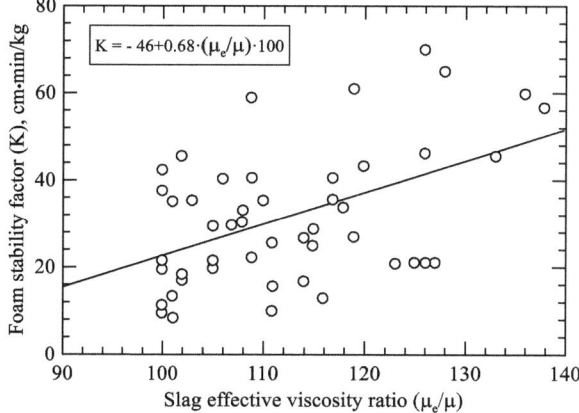

Fig. 7.196 Effect of the slag effective viscosity ratio on foam height. After [207]

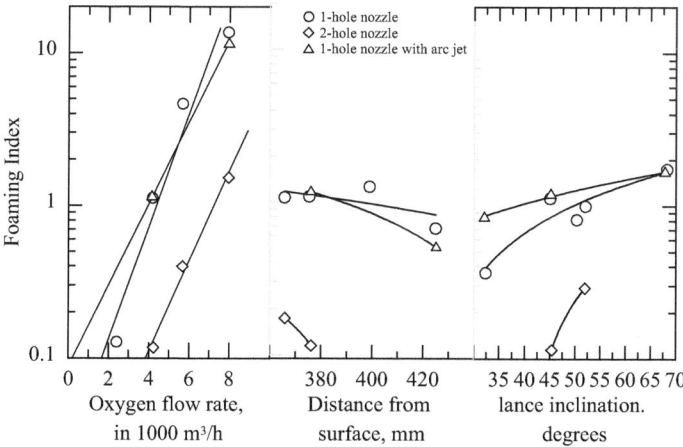

Fig. 7.197 Effect of lance injection characteristics on slag foaming. After [262]

Morales et al. [263] in 1995 reported foaming practices in the EAF using DRI. One important feature of DRI operations is a larger volume of slag in comparison with scrap-based operations. The furnace had a nominal capacity of 220-ton, hot heel of 30 ton and a maximum electric power of 98 MW during most part of the heat. The DRI feeding rate was 2800 kg/min with 3000 Nm^3 O$_2$/heat. In their work they compared 4 different practices; A, B, C and D. In practice A, a total of 1500 kg of coke was added during the heat. Practice B, in addition to 1550 kg of coke charged continuously, graphite was injected. In practice C, coke was increased to 2000 kg and graphite injection. In this practice 1000 kg of coke were added in the beginning. Practice D was a slight modification of practice C. When carbon was lower than 0.1% the flow rate of carbon was increased to 30 kg/min and before tapping 2 ton of lime were added. The carbonaceous material was graphite with more than 80% particles below 74 μm (mesh 200), injected at a rate of 25 kg/min, using a SIT-2000 equipment with air as the carrier gas. Practice D was chosen because they found a lower energy consumption. They also suggested optimum FeO from 18 to 20% and bath temperature lower than 1580 °C, however these values were suggestions based on theoretical considerations and observations but not derived from a control of those variables.

Conejo et al. [231] in 1999 reported detailed information on the slag foaming practice in the EAF. Unfortunately, this document was reported only in Spanish. The following is a summary of this publication. Much of the metallurgical practices have changed but the analysis is still useful, in particular this work suggests to control the injection of carbon particles based on the carbon dissolved in the slag, the operation with a balanced C/O ratio, and also defined a way to define the optimum C in DRI to achieve a balanced C/O operation and an optimum specific consumption of carbon particles for the conditions investigated. The EAF had the old design with tapping spout which required higher tapping temperatures, carbon injection was made using both disposable lances located at the slag door and an automatic system of oxygen injection system with an attached lance. The melt-shop consisted of 4 AC-EAF of 220 ton of nominal capacity with transformers of 125 MVA, shell diameter of 7.2 m and electrodes of 711 mm in diameter. The metallic charge was 100% DRI in most heats. The metallization of DRI ranged from 87 to 95% and carbon in the DRI from 1.8 to 2.0%. The specific consumption of oxygen was in general from 10 to 15 Nm^3/ton at flow rates from 2250 Nm^3/h with one lance and 4500 Nm^3/h with two lances, at a pressure of 13 kg/cm^2. The carbon injection systems could deliver particles at flow rates from 30 to 120 kg/min, at a pressure of 3.5 bar using dry carbon particles. Operating with a low flow rate of solids gave clogging problems in the lance. The upper limit was also limited due to violent foaming as well as production of flames. This system operates continuously. The lance diameters were 38.1 and 31 mm for the consumable lance and automatic lance, respectively. The lances should be immersed in the slag to avoid back splashing and create steel scabs on the lances which impede its movement. It has been recommended to immerse about 10 times the nozzle diameter. Flames can also be produced if the lance is not immersed properly. The type of carbon was petroleum coke with the particle size and composition indicated in Table 7.23. It can be observed that more than 80% of the particles are ≤ mesh 100. The measured specific surface area was 4.31 m^2/g. The particle density is lower than the slag, about 1.5–2 g/cm^3. Its thermal conductivity is much larger than liquid steel, from 27 to 39 cal/m °C s.

The way to define an optimum carbon injection practice from a practical viewpoint was based on the consumption of electric energy and the achievement of a balanced C/O practice that controlled the amount of FeO in the slag. The following correlations considering a single variable shows a large dispersion which is common because multiple variables change at the same time not only the one analyzed, for example DRI metallization and carbon, temperature, oxygen injection rates, etc. In spite of this limitation, it was possible to suggest the optimum carbon injection conditions as will be discussed further. Figure 7.198 shows the relationship between the

Table 7.23 Chemical composition and particle size of carbon particles. After [231]

	Chemical composition, wt%					Particle size distribution							
	Volatile matter	Ash content	Fixed carbon	H$_2$O	S		18	20	30	70	100	200	– 200
Spec, %	≤ 3	≤ 3	≥ 93	≤ 1	≤ 0.5	mm	1	0.8	0.59	0.21	0.15	0.07	– 0.07
Aimcor, %	3.06	11.3	85.6	2.1	1.6	%	8	1.2	3.6	6.8	13.2	38.0	29.2

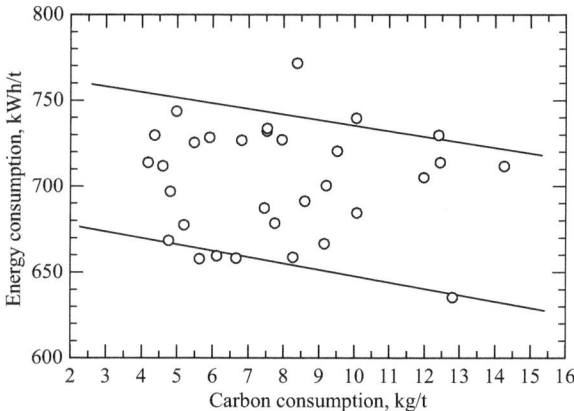

Fig. 7.198 Effect of carbon injection on energy consumption. After [231]

specific consumption of carbon particles and electric energy consumption. The large energy consumption is due to a metallurgical practice with 100%DRI. On average, increasing the consumption of carbon particles from 5 to 10 kg/ton decreases energy consumption by about 20 kWh/ton. This effect includes the exothermic contribution from higher consumption of oxygen and improved slag foaming conditions.

The optimum consumption of carbon particles should be defined based on controlling the C/O balance which in turn depends on DRI metallization, oxygen flow rate and coke additions. A proper addition keeps under control the concentration of FeO, as shown in Fig. 7.199. On average, FeO was decreased from 33 to 27% by increasing carbon injection from 5 to 10 kg/ton.

In their report, Conejo et al. proposed a novel analysis to describe carbon injection using a C$_{slag}$-FeO relationship. From the previous figure it can be estimated the average carbon injection consumption to control FeO. If FeO should be controlled below 30%, it is required about 7 kg/ton of carbon injection, then according with Figs. 7.200 and 7.201 this value is equivalent to an average concentration of carbon in the slag of 0.10%. Jones et al. [173] suggest a carbon consumption of 2–5 kg/ton for medium powered furnaces and 5–10 kg/ton for DC furnaces because they operate with longer arcs.

Another important aspect investigated was the C-O balance. In order to reach C-O balance, according with stoichiometry, the ratio O/C should be 1.33. Figure 7.202

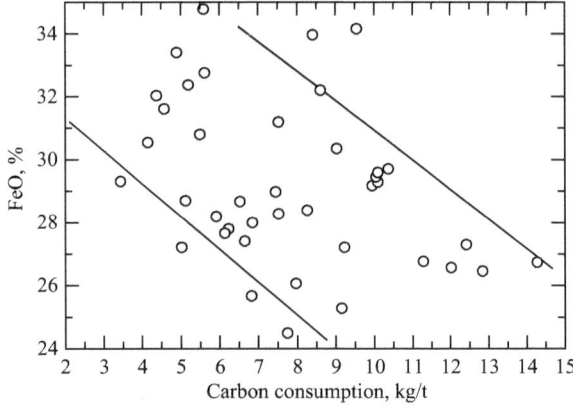

Fig. 7.199 Effect of carbon injection on FeO in the slag. After [231]

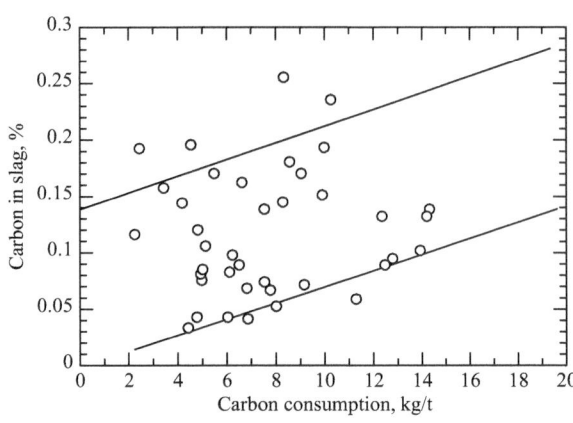

Fig. 7.200 Relationship between carbon injection and C dissolved in the slag. After [231]

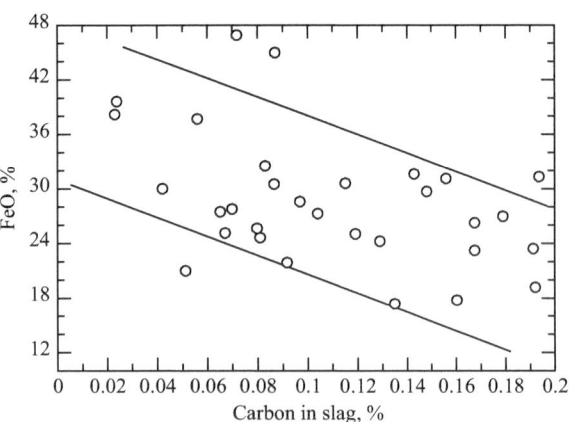

Fig. 7.201 Relationship C_{slag}-FeO. After [231]

Fig. 7.202 Evaluation of C-O balance. After [231]

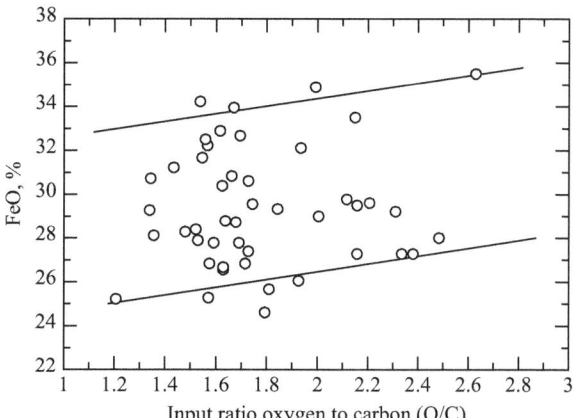

shows that the plant was operating above this value, indicating an excess of oxygen. In their work, the authors also proposed a practical approach to estimate the amount of carbon in DRI to reach C-O balance as a function the specific consumption of oxygen.

Conejo et al. [264] also reported the use of recycled MgO refractory at the beginning of a heat to obtain an early slag foaming. The hot heel is a standard practice, however the slag left in the EAF from the previous heat has a high concentration of FeO and is highly fluid. Under those conditions is unable to foam. The additions were based on the target of achieving MgO saturation. Schroeder [83] also suggested an early formation of foamy slag by putting lime or mill scale in the bottom of the first charge bucket. Mill scale should be ruled out because of the high oxidizing conditions of the previous slag.

Aminorroaya and Edris [180, 181] reported slag foaming under industrial conditions. They analyzed the optimum values in the rate of carbon injection and slag chemistry which decrease energy consumption. The experimental information was taken from an industrial EAF with 200 ton of liquid steel, 5.5 m in diameter, power transformer of 90 MVA, 5–10 ton of hot heel, electrodes of 60 cm. The metallic charge consists of DRI/scrap in a ratio about 81/19. DRI had a metallization of 91% and 1.93%C. A triple supersonic lance injects oxygen, methane and carbon particles. At minute 18, the oxygen flow rate was 1150 Nm^3/h and at minute 30 was increased to 2300 Nm^3/h. The carbon employed was petroleum coke with a particle size from 0 to 3 mm, injected below the slag surface at an angle of 43°.

The foaming index was computed estimating the physical properties, density and surface tension with data from Mills and Keene and Urbain's model to estimate the viscosity;

$$\rho = 2460 + 18(\%FeO + \%MnO)$$
$$\sigma = 754.24 - 569.4(\%SiO_2/100) - 137.13(\%FeO/100)$$
$$\mu = AT\exp(B/T)$$

The viscosity was estimated at 1550 °C assuming a ternary system FeO-(CaO + MgO + MnO)-(Al$_2$O$_3$ + SiO$_2$ + P$_2$O$_5$). The final value employed was the effective viscosity which considers the presence of solid particles. The authors reported that the foam height was measured but no further information was provided. The first parameter to optimize was the carbon injection rate. Figure 7.203 reports the effect of the carbon injection rate on the foam height. It can be observed a maximum foam height for an injection rate of 9 kg/ton. An additional analysis between the carbon injection rate and energy consumption confirmed that for this value it is obtained the minimum energy consumption. The values on foam height seem unrealistic, reaching values close to 1 m. If the information was true, an injection rate of 1 kg/min would be enough to protect the electric arcs because it corresponds to a foam height of 60 cm; the values of the arc length in the industrial practice are below 50 cm.

Figure 7.204 shows the optimum range of FeO. It shows that FeO in the range from 20 to 24% reaches the highest foam height, decreasing after this value.

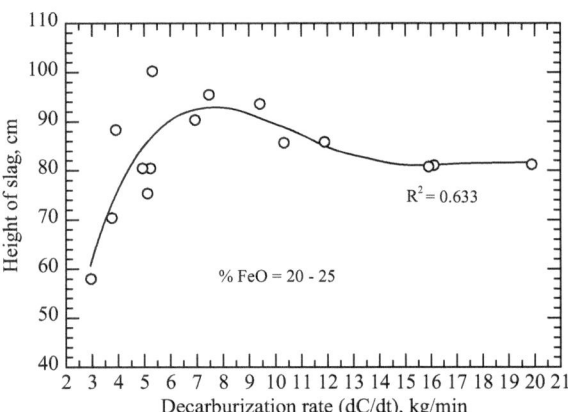

Fig. 7.203 Effect the carbon injection the on the foam height. After [180, 181]

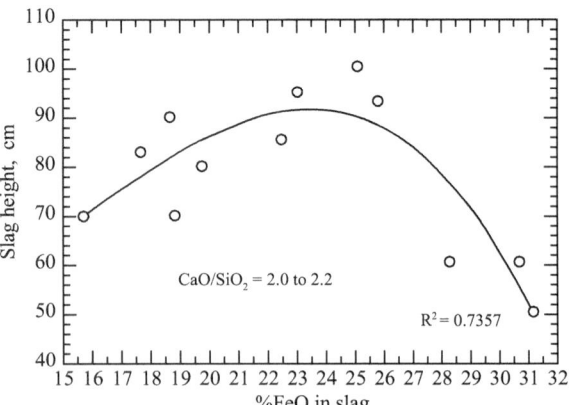

Fig. 7.204 Effect FeO on the foam height. After [180, 181]

Nucor Yamato Steel's strategy to improve foaming conditions was based on a high generation rate of CO and higher slag viscosity [265]. High rates of production of CO were possible charging 14.5 kg C/ton and injecting 6.8 kg C/ton with oxygen injection of 47–52 Nm^3/ton. Slag viscosity was controlled with a binary basicity ratio from 2.0 to 2.2 and avoiding high temperatures. This practice corresponds to approx. 40% electric energy and 60% chemical energy.

Jung and Fruehan [127] in 2000 applied a previous correlation reported by Zhang and Fruehan to estimate the foaming index during 50% of the blow in a BOF of 200 ton with a diameter of 6 m. The appendix in this publication describes the models employed to estimate the density, surface tension and viscosity of the slags. The bubble diameter was also estimated. With the calculated foaming index, the foaming height was estimated, as follows:

$$h = \Sigma \left(U_g^s - U_g^o \right) + h^0$$

where: U_g^o and h^0 are the superficial gas velocity and slag height before foaming begins.

The decarburization rate was calculated from which the superficial gas velocity in the reactor was also calculated. Figure 7.205 shows the calculated values for both foam height and decarburization rate. It can be shown the intense foaming conditions at high decarburization rates, especially during the first third part of the blow, which anticipated potential slopping problems.

Heo and Park [165] carried out industrial trails on EAF's of 70 and 150 tons. Slags contained CaO-SiO_2-MgO-Fe_tO-MnO-Al_2O_3-P_2O_5. The average slag chemistry is shown in Table 7.24. They separated the slags into two types; type A were MgO saturated and type B, were MO saturated. MO saturation refers to MgO/FeO saturation as explained before in this chapter. The amount of slag for type A was 7 ton and for type B 14 ton. The foam height was measured with three vibration sensors installed in the outer wall. The value reported is the average value.

Fig. 7.205 Evolution of the calculated foam height and decarburization rate in a BOF during the first 50% of the blow. After [127]

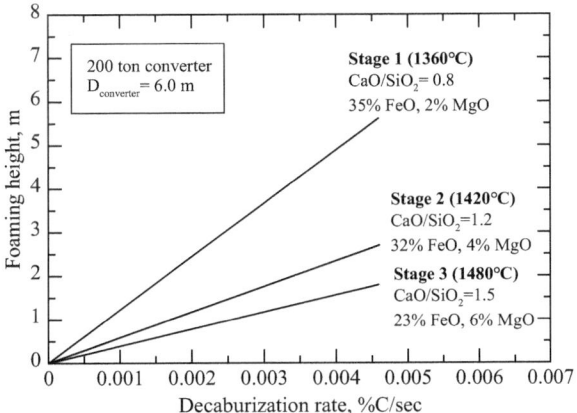

Table 7.24 Average slag chemistry of plant trials. After [165]

Type	SiO$_2$	Al$_2$O$_3$	Fe$_t$O	CaO	MgO	MnO	P$_2$O$_5$	C/S	Condition
A	17.4	15.7	24.6	24.7	7.5	9.9	0.18	1.42	100% liquid
B	20.3	8.4	17.0	32.1	10.9	11.0	0.38	1.99	MO sat

The foaming index employed was the one reported by Ito and Fruehan $\left(\Sigma = C\mu/\sqrt{\rho\sigma}\right)$ with C = 999. The physical slag properties were estimated. The effective slag viscosity was calculated using the Einstein-Roscoe equation. The superficial gas velocity was estimated from the rate of reduction of FeO;

$$U_s^g = \frac{Tk_s}{273} \cdot \frac{S}{A}$$
$$k_s = 2.88 \exp\left(-\frac{39700}{RT}\right) \cdot a_{FeO}$$

where; k_s is the reaction rate constant (mol FeO/cm^2·s), based on results reported by Sugata et al. [167], S is the interfacial area between slag and solid carbon, A is the cross-section area. The S/A ratio was assumed as 20 (After Ito and Fruehan). The initial slag height was estimated considering an EAF diameter of 5 m, slag density of 3.63 ton/m^3 and total slag mass of 7 tons. Figure 7.206 compares the calculated and measured values. Measured values range from 25 to 50 cm but calculated values predict values from 18 to 22 cm, underpredicting the real values. It seems clear that the authors oversimplified a dynamic process because the slag chemistry and gas generation rate changes throughout the heat.

Nal et al. [266] reported the importance of independent flow control for carbon injection. In their original design the carbon particles were supplied by one injection machine but only the lance with the shortest connection consumed a much higher fraction of particles in comparison with those located at a longer distance. They solved

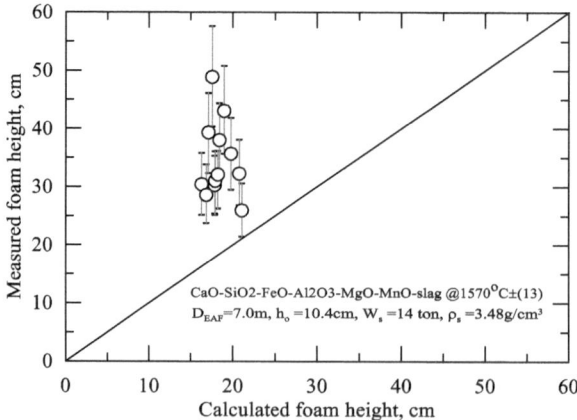

Fig. 7.206 Comparison between measured and calculated foam height. After [165]

this problem by individual flow control devices and separate injection machines, at three injection points. The injection rate for each lance was in the range from 0 to 80 kg/min.

7.16.2   Summary on Experimental Investigations on Slag Foaming

Mukai [267] in 1991 summarized in a small review the effect of different variables on slag foaming, shown in Table 7.25. Mukai reported slag foaming for laboratory and industrial practice indicating huge differences in foam height. Those differences were attributed to the way the foam is created. The foam height created by gas injection is different to a foam height where gas is produced by chemical reaction at high temperature, other differences are the bubble size, the rate of gas supply to the slag, bubble expansion, crucible diameter, etc. Mukai then concluded that because of these differences the foam heights cannot be compared.

Table 7.25 Summary on the effect of different variables on slag foaming

Properties of slag phase	Foam height	Foaming system	First author
$CaO/SiO_2 > 1$	Maximum around C/S = 1–1.5	Slag + metal	Kitamura [122]
	Minimum around C/S = 1.2	Slag	Ito
$CaO/SiO_2 < 1$	Decrease with decreasing C/S	Slag + metal	Kitamura [122]
	Increase with decreasing C/S	Slag	Ito
	Increase with decreasing C/S	Slag	Hara
Viscosity	Maximum at the lowest viscosity	Slag + metal	Kitamura [122]
	Increase with increasing viscosity	Slag	Ito
	Increase with increasing viscosity	Slag + graphite	Sugata
Surface tension	Decrease with increasing surface tension	Slag	Hara
	Increase with increasing surface tension	Slag + graphite	Sugata
Solid phase	Decrease due to the precipitation of solid phase	Slag + metal	Kitamura [122]
	Increase due to the presence of solid particles	Slag	Ito
	Increase due to the presence of solid suspension	BOF process	Kozakevitch

7.17 Developments on Slag Foaming

7.17.1 Foaming of Stainless Steels

Stainless steel slags are primary composed of CaO, SiO_2, MgO, Al_2O_3 and Cr_2O_3. FeO is present in small concentrations, below 2%. The main source of CO derives from the reduction of Cr_2O_3 by either C or CaC_2:

$$Cr_2O_{3(l)} + 3C_{(s)} = 2Cr_{(l)} + 3CO_{(g)}$$
$$Cr_2O_{3(l)} + 3C_{(s)} = 2Cr_{(l)} + 2CO_{(g)} + CaO_{(l)}$$

Cr_2O_3 can also be reduced with Si or Al:

$$Cr_2O_{3(l)} + 1.5Si_{(l)} = 2Cr_{(l)} + 1.5SiO_{2(l)}$$
$$Cr_2O_{3(l)} + 2Al_{(l)} = 2Cr_{(l)} + Al_2O_{3(l)}$$

The reduction reaction with carbon is promoted with high C and Cr_2O_3 activities, low pressure, high temperature and low Cr activity in the metal. Cr_2O_3 has a high melting point, 2435 °C, and produces highly viscous slags. The slag melting point can be decreased by addition of silica and FeO containing materials. During oxygen injection, Cr is oxidized prior to iron oxidation, however Si prevents Cr oxidation and also Cr_2O_3 is simultaneously reduced, decreasing its concentration below 5%.

Kozakevitch [268] pointed out that chromium oxide produced foamy slags in the open-hearth and Bessemer processes and cited the work carried by Leyba and Komar from 1941 who reported that C_2O_3 had a very small solubility in open hearth slags, maximum 3–4%. The excess chromium oxide remains in the slag as suspended solid.

Swisher and McCabe [269] in 1964 reported the effect of Cr_2O_3 on slag foaming in the temperature range from 1580 to 1640 °C indicating that it contributed to foam stability only for concentrations below 0.4%, with a slag's basicity of 0.64, as shown in Fig. 7.207. Below 0.4%, Cr_2O_3 dissolves in liquid slag but exceeding this concentration it precipitates. Contrary to this result, Hara et al. were able to produce foaming with slags containing 2% Cr_2O_3 controlling the gas flow rate, as shown in Fig. 7.208.

Görnerup and Jacobsson [270] suggested an idealized practice to control the concentration of Cr_2O_3 at low values, as shown in Fig. 7.209. It can be observed an early addition of iron oxide followed by oxygen and carbon injection. This example shows an initial concentration of about 15% Cr_2O_3. Chromium and iron oxides are reduced, the reduction rate of Cr_2O_3 increases when oxygen injection stops and the final reduction is achieved during tapping. Is important to control the reduction rate of FeO with injected carbon otherwise if FeO is not reduced it can oxidize Cr. These authors suggest to limit Cr_2O_3 to a maximum of 15% in the beginning, start oxygen injection at an early stage to produce heat, add mill scale to promote CO generation and aim for high silicon, about 0.1% during tapping.

Fig. 7.207 Effect of Cr_2O_3 on foam life. After [269]

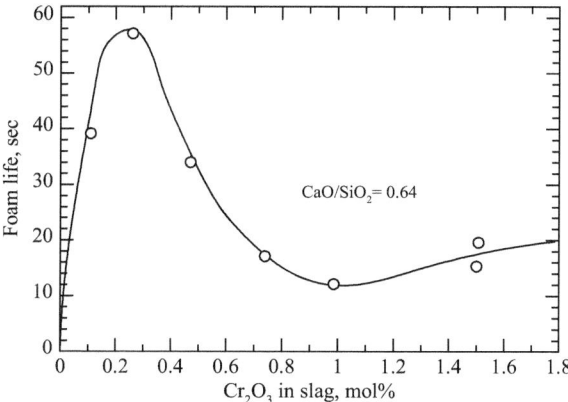

Fig. 7.208 Effect of gas flow rate and Cr_2O_3 on slag foaming. After [124]

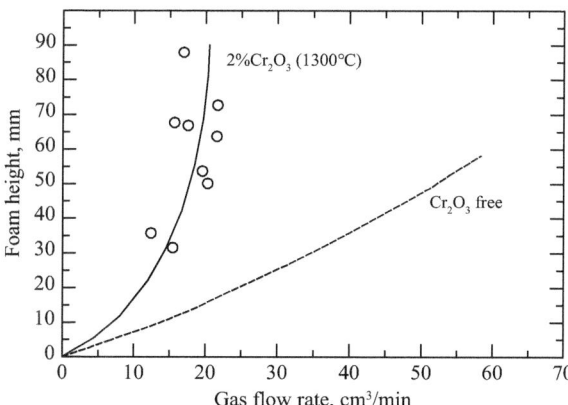

Fig. 7.209 Stainless steels slag foaming practice. After [270]

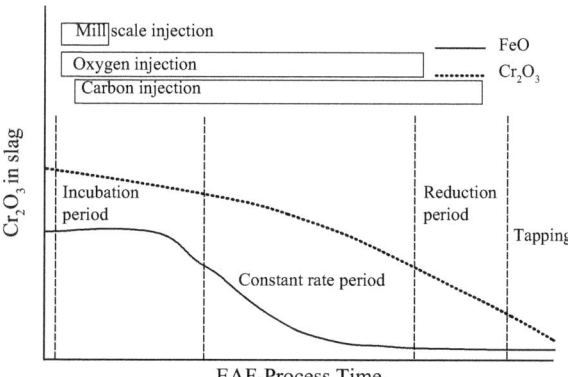

Divakar et al. [271] studied the reduction of oxides in stainless steelmaking and also investigated slag foaming. They employed a resistance furnace with a graphite crucible of 45 mm diameter and 90 mm height, an alumina tube was placed inside so that only the bottom surface reacts with the slag. The furnace was sealed and opened to take samples. X-ray was employed to define the moment of melting and its images video-recorded during a maximum of 2 min. The slag height was measured manually from the video recording. The gas generation rate was calculated from the measured reduction rates of all oxides present in the slag. The slag mass was 50 g, containing CaO-SiO_2-$(0$–$10)FeO$-$(2$–$20)Cr_2O_3$-MnO-Al_2O_3-V_2O_5. The main temperatures investigated were 1550 and 1600 °C. The results for two slags are shown in Fig. 7.210. Figure 7.210a shows a maximum gas generation rate at 10 min with the lowest foam height, below 10 mm. On the contrary when the gas generation rate decreases, the foam height increased. These results clearly indicate that there is no relationship between foam height and gas generation rate. Such a conclusion was not further discussed because it is against a well-established foaming practice. The possible reason was inadequate physical slag properties due to very low FeO and high Cr_2O_3, therefore very high viscosities.

Fig. 7.210 Foam height as a function of time for **a** $5MgO$-$7Al_2O_3$-$2MnO$-$6FreO$-$10Cr_2O_3$ at 1550 °C and **b** $5MgO$-$7Al_2O_3$-$2MnO$-$6FeO$-$5Cr_2O_3$-$5V_2O_5$ at 1600 °C. After [271]

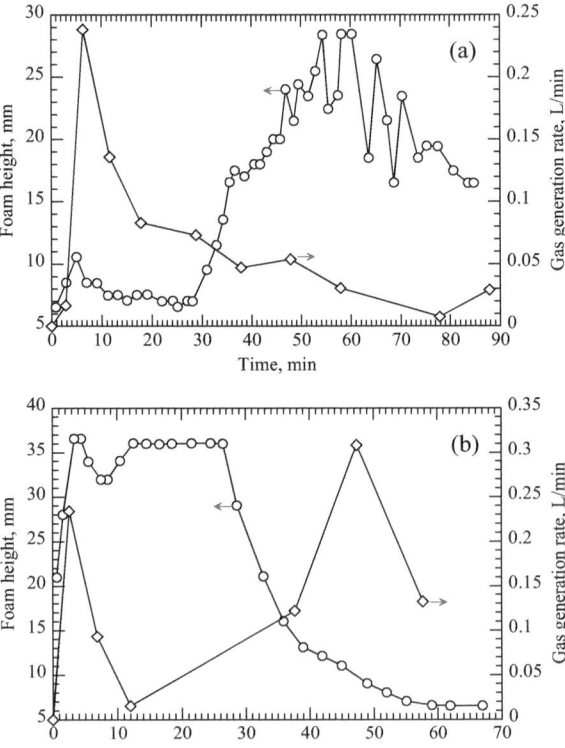

Kapilashrami et al. [143] expanded the previous work, including the calculated viscosity and bubble size. They found no relationship between the foaming index and slag viscosity, bubble size and gas generation rate, attributed to the dynamic conditions of their experimental work. They found an increment in the viscosity during each experiment due to the reduction of both FeO and Cr_2O_3, from 70 cP to a maximum value of about 130 cP. The increment in viscosity was enhanced due to Al_2O_3 dissolution from the refractory. The bubble size also changed during the experiments. The bubble size for different experiments ranged from 4 to 11 mm. Figure 7.211 shows the change in the foaming index during each experiment. All slags contained $7\%Al_2O_3$-$5\%FeO$-$5\%MgO$. Slag A and B had the same composition; $10\%Cr_2O_3$-$2\%MnO$, Slag C; $5\%Cr_2O_3$-$2\%MnO$-$2\%V_2O_5$ and slag D; $5\%Cr_2O_3$-$2\%MnO$-$5\%V_2O_5$. It can be observed a peak in the foaming index and then decreases. Their results on foam height and gas generation didn't show a correlation because when the gas generation rate shows a minimum value, foam still prevails. Slag D with V_2O_5 produced the largest foam height, about 35 mm in comparison with the peaks for slags A and B with 20 and 15 mm, respectively. The bubble size increased from slag A to D. Slag D with the largest bubble size, 10–11 mm, had the highest foam height. Based on these results, they argue that the foaming index is not a suitable parameter to describe slag foaming under dynamic conditions.

In two additional publications Kapilasharmi et al. [134, 272] added a few more experiments confirming their previous conclusions, discrediting the use of the conventional foaming index to describe slag foaming under dynamic conditions. The classical foaming index is based on steady state conditions where the gas generation rate is equal to the gas bubble break up. In principle it would seem they are correct in their conclusion because they found that the foaming index from their experimental work could not be correlated with changes in slag viscosity, gas generation rate and bubble size, variables that have been confirmed in all the existing literature as having a dominant role. Their work has several particularities; (1) The foaming index was based on calculated viscosities, rather than measured values, (2) The actual gas

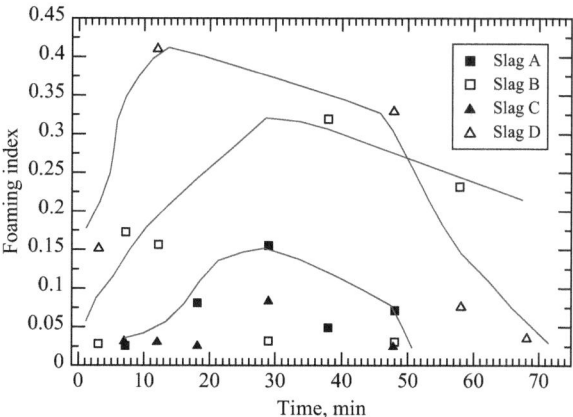

Fig. 7.211 Foaming index as a function of time. After [143]

generation rate was estimated from the reduction rate rather measured. The volume flow rate of gas escaped was estimated subtracting the volume flow rate of gas generated from the volume flow rate of gas hold up in the slag ($\dot{V}_{ge} = \dot{V}_{gg} - \dot{V}_{g\text{-slag}}$), (3) The foaming index was calculated from the ratio h/U_s however they do not report the values of the superficial velocities employed, (4) their results have contradictory results because in one report they claim there is no effect of the bubble size on the foaming index [143] but in another one they conclude there is a relationship between foam height and bubble size [272], (5) their experimental work is focused on one particular type of steels, stainless steels, while the majority of research done in this area is for carbon steels. Some of these differences can explain the different results but more work on stainless is still needed to clarify its behavior.

Figure 7.212 shows their experimental results for the foaming index and foam height. The bubble size is given below. Slag 6, the one with the largest bubble size, reported the largest foam height during the first 60 min. It also can be observed that there is a general trend between the foaming index and the foam height. Comparison is possible only with 4 slags because the authors omitted to report the foam height of slags 2 and 5. For slag 1 the foam height increases, reaches a maximum and then decreases, the foaming index also increases as the foam height increases but once the foam height decreases, the foaming index remains. There is a mismatch at the end. For slag 3 the foaming index increases in line with an increment in the foam height up to about 60 min and then both variables change in opposite directions. There is a mismatch at the end. In slags 4 and 6, the foaming index changes in proportion to the changes in the foam height.

Slag	1	2	3	5	6
D_b, mm	< 2	3–6	2–8	11–15	14–25

Kerr and Fruehan [273] compared the following additions on stainless steel's slags foaming; NiO, $CaCO_3$, $Ca(NO_3)_2(H_2O)_2$ and other oxide briquettes. Except for NiO, the other additions improved slag foaming. The foaming efficiency was evaluated in terms of the CO produced, reaction rates and amount required. Calcination of limestone releases CO_2 and is an endothermic reaction which also lowers the slag viscosity. Both factors improve slag foaming. The authors estimated that if 1-cm diameter particles are added, 80 kg of limestone would be required with an injection rate of 2.2 kg/s. Hydrated calcium nitrate can both dissociate and react with carbon:

$$Ca(NO_3)_2(H_2O)_2 = CaO + N_2 + 2H_2O + 2.5O_2$$
$$Ca(NO_3)_2(H_2O)_2 + 5C = CaO + N_2 + 2H_2O + 5CO$$

They found that calcium nitrate produces 5.5 mol of gas per mole and $CaCO_3$ only one mol. The higher production rate of CO reduces the mass of calcium nitrate required, about 0.8 kg/s.

Briquettes containing Fe_2O_3 also gave good results. Fe_2O_3 in comparison with FeO has a faster rate of reduction to FeO than FeO to Fe.

Fig. 7.212 Dynamic changes of; **a** foaming index and (b) foam height. After [134]

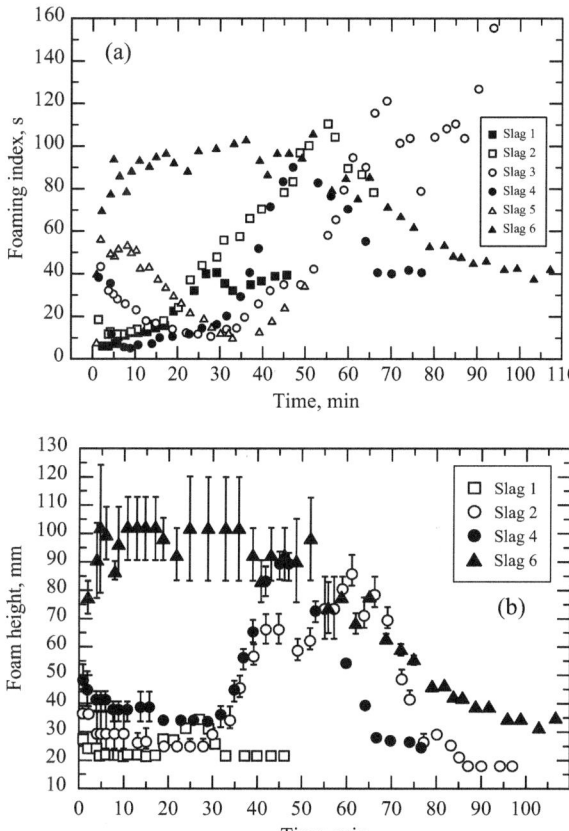

Karbowniczek et al. [274] also proposed the use of calcium nitrate, in particular calcium ammonium nitrate, $Ca(NO_3)_2 (NH_4NO_3)_{0.2}(H_2O)_2$ a by-product from fertilizer production. The total moles of gas produced are 8 and 10 kg of this material can produce about 8 Nm^3 of gas equivalent to 56 m^3 at 1600 °C. They carried out foaming experiments confirming that Ca-ammonium nitrate is beneficial for slag foaming but this effect can be affected if Cr_2O_3 in the slag increases, in particular above 10%.

There are a large number of patents dealing with addition of mixtures to improve slag foaming of stainless-steel slags [275–277]. US Patent 8673047 indicates that the conventional way to foam slags of carbon steels, injecting carbon particles, cannot be applied to stainless steels. The method suggests the addition of briquettes containing Fe_2O_3 and carbon, 81 and 18%, respectively, additionally assisted with FeCr HC and limestone. In US Patent 8403402 it is suggested to use $CaCO_3$ and 40–70% of FeCr HC. The foaming mixture is added from the roof at four points, about 20–30 kg of mixture per m^2 of surface area. Reichel et al. [278] reported the application of briquettes containing iron oxide, limestone and FeCr HC with a controlled density (higher than 3.5 ton/m^3) to promote the reaction at the slag/metal interface.

7.17.2   Slag Foaming with Waste Materials

Slag foaming requires oxygen and carbon and both components are present in different waste materials produced in iron and steelmaking operations. Recycling would be a positive effect on the environment and from an economic view point.

In an internal report made by the author to ArcelorMittal-Aviles in 2011 [279] it is described in detail an early idea to recycle by-products to improve the foaming conditions. At that time the oily-mill sludges (OMS) had to be landfilled at a high cost, about 120 euros/ton. This material contains almost 95–97% iron oxide. Briquettes were prepared mixing OMS with carbon and lime, using cement as a binder. Trails with an induction furnace proved the feasibility to use these briquettes as a foaming agent. Limiting the amount to less than 3% was suggested to avoid any negative effects on energy consumption.

Davydenko et al. [280, 281] in 2015 proposed the use of by-products as a source of oxygen, mixed with carbon and then briquetted. Fe, Ni and Cr could be recovered. This approach helped simultaneously to recover metals and also to improve the foaming conditions during the production of stainless steels. The mill-scale has iron oxide and chromium oxide as the two main components. The production of 25–42 Mton/year of stainless steels in Sweden produces 1.2–2.1 Mton/year of mill scale, about 30–70 kg mill-scale/ton of rolled products. The evaluation of the foaming conditions in industrial trials included visual observations and measurements on slag density. The changes in slag density from 1 to 20% were used as an indication of the volume of CO generated. Their results showed a satisfactory correlation with a visual index for the foaming conditions.

7.17.3   Slag Foaming with Hydrogen

The near future corresponds to hydrogen metallurgy. Fossil fuels will be employed to a minimum extent in order to control global warming. Iron and steelmaking processes are under development to use hydrogen at a much higher scale than in the past. There are specific areas that will need further developments. One of them is the role of carbon as a reducing element of wüstite in liquid slags. This reaction is fundamental to control slag foaming, also contributes to improved stirring conditions and removal of nitrogen. If carbon is almost eliminated (zero is impossible because steel needs carbon), the iron oxide in DRI or the iron oxide formed by oxygen injection (if chemical energy is required to partially replace the expensive electric energy) it will require a reducing element. In principle, hydrogen can be injected and reduce FeO. This reaction has been investigated in the past. Ban-Ya et al. [282] studied the reduction rate of FeO with $Ar-H_2$ gas mixtures in the temperature range from 1400 to 1450 °C. They reported a higher reduction rate by increasing hydrogen in the gas phase and also a much faster rate, about 20 times, compared to the reduction of solid wüstite. The gas formed can in principle replace the role of CO as foaming agent but until today this specific role has never been investigated.

7.18 Effect of Slag Foaming on Energy Consumption

The major benefit with a foamy slag is the minimization of heat losses. Figures 7.213 and 7.214 shows the larger heat losses when there is no foamy slag.

Sanchez et al. [283] reported a numerical model that predicts the size and temperature of the hot spots as a function of slag foaming height. Table 7.26 summarizers their results. The model predictions indicate a huge temperature on the walls when there is no slag, about 1827 °C, decreasing to 1387 °C when the arcs are fully covered.

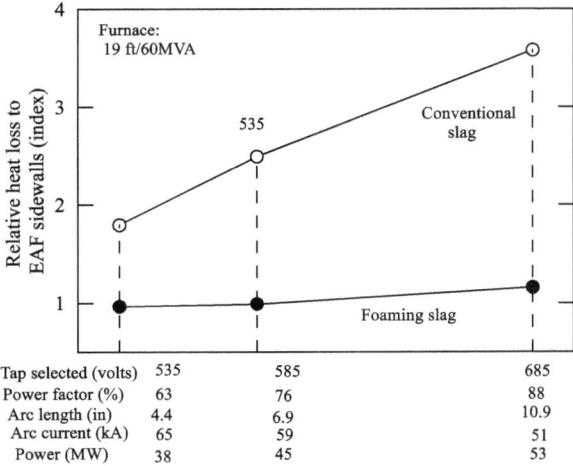

Fig. 7.213 Effect of electric parameters on energy losses with and without foamy slag. After [173]

Tap selected (volts)	535	585	685
Power factor (%)	63	76	88
Arc length (in)	4.4	6.9	10.9
Arc current (kA)	65	59	51
Power (MW)	38	45	53

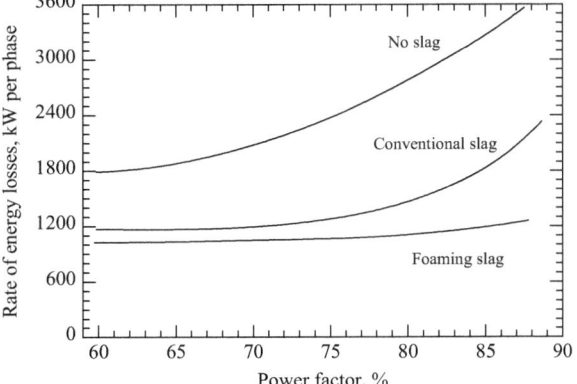

Fig. 7.214 Effect of power factor on energy losses with and without foamy slag. After [173]

The diameter of the hot spots increases from 0 for full arc coverage to 1.4 m and the incident radiation increases from 1.84 to 3.15 hMW/m^2 for the same conditions. It is quite clear the importance of slag foaming on the heat losses.

Fedina et al. cited by Chycko et al. [228] reported energy savings from 10 to 30 kW/ton in an EAF of 150 ton.

Aminorroaya and Edris [180] reported energy savings for an industrial EAF of 200 ton charged with 80% DRI and 20% scrap, in the order of 90 kWh/ton due to improved foaming conditions. They reported the slag height but didn't provide details about its measurement. The injected carbon particles had a particle size from 0 to 3 mm. They reported an optimum value of 9 kg C/ton, for this value energy consumption was the lowest. A maximum foaming height of about 90 cm was obtained with FeO in the range from 20 to 24% and a binary basicity ratio from 2 to 2.2.

Agnihotri et al. [284] carried investigated the effect of foaming on energy consumption in a EAF of 40 ton of nominal capacity charged with 72% scrap, 8%DRI and 18% pig iron, varying the coke injection rate from 5 to 25 kg/ton. They assumed the changes in energy consumption were due to improved foaming conditions, resulting in a decrease in energy consumption from 50 to 60 kWh/ton equivalent to 10–12% and a decrease on the total melting time of 11 min, equivalent to 12%. They suggested an optimum FeO in the range from 17 to 21% and slag basicity from 1.8 to 2.2.

Table 7.27 summarizes reported values on energy consumption applying slag foaming. There is a large scattering in those results. Industrial reports suggest savings in energy consumption from 50 to 100 kWh/ton.

Table 7.26 Evolution of hot spots as a function of slag height. After [283]

h, cm	Area, m^2	Avg. diameter, m	Avg. incident radiation, MW/m^2	Maximum temperature, °C
45	0	0	1.84	1387
33.75	0	0	2.07	1477
22.5	0.24	0.55	2.39	1647
11.25	0.47	0.7	2.77	1727

Table 7.27 Summary on savings on energy consumption using slag foaming

First author	kWh/ton	Notes
Buydens [155]	10–15	
Fedina [228]	10–30	EAF 150 ton
Eissa et al. [285]	100	Decrease from 680 to 580 kWh/ton with FeO from 22 to 26%
Aminorroaya [180]	90	
Morales et al. [263]	90	EAF 210 ton, 100% DRI, from 730 to 640 kWh/ton
Agnihotri [284]	50–60	EAF 40 ton, 510–450 kWh/ton with FeO from 17 to 21%

7.19 Challenges on Slag Foaming

Comparison of foaming height and foam life from different investigations is not possible because of different definitions of the foaming parameters and experimental conditions. In most cases the physical slag properties have been calculated with different models instead of using exact experimental values. These limitations have been cited by numerous researchers in the past [184]. As a consequence, the foaming capacity has been reported to be different for similar slags and the relative weight assigned to each variable is different with different activation energies, etc. Several reviews have reported inconsistencies in the experimental results, however in some cases, a more rigorous analysis shows that there is no such inconsistency. For example, Nekrasov et al. [77] report large discrepancies in optimum FeO concentrations. The reason of the large discrepancies results from different experimental conditions and in particular the source of the gas phase, i.e., different behavior if the gas is injected or internally produced. Since 1991 Mukai [267] reported these differences.

The same slag component can show opposite results depending on its concentration and the composition of the other components, for example CaF_2 has a large effect to decrease the slag viscosity (which decreases foaming) but also decreases the surface tension (which increases foaming). It has also been discussed the amphoteric nature of some of the oxides in the slag which under acid conditions act as basic, decreasing polymerization and slag viscosity but in basic slags act as acids, increasing the slag viscosity. In other cases, increasing the surfactants concentration, which decrease the surface tension and improve foaming, reach the point of excess surface concentration which depletes all the available surface sites and then no longer improve foaming. Therefore, it is important not to assume that a result obtained for a given set of experimental conditions will hold systematically for other set of conditions, even if the changes seem to be small.

Iron oxide plays a critical role on slag foaming. It is commonly defined as wüstite but in reality, there is a mixture of FeO and Fe_2O_3 in liquid slags. An accurate description of the role of iron oxide should involve these two types of oxides which have different chemical properties.

The concept of foaming index or unit of foaminess as first described by Bickerman in 1938 and extensively promoted from the work by Fruehan's group in the 1990s has been proved to be a useful but limited concept. These limitations arise from the following factors:

- The original concept assumes a linear relationship between the superficial gas velocity and the foam height but in reality, there are three different regions; non-foaming below a minimum superficial gas velocity, linear region and non-linear region above a critical superficial gas velocity.
- Foaming does not only depend on the physico-chemical properties of the system but also on the gas generation rate, type of gas, minimum superficial gas velocity, bubble size, bubble size distribution, etc. It would be convenient to develop a unified foaming index which includes the gas injected or gas generated and the

slag physical properties. Pilon et al. disregarded the foaming index concept and instead focused on the value of the foam height as a function the system properties and gas velocity.

- The foaming index assumes it continuously increases as viscosity increases but is not true, at a critical value starts to decrease

One of the main challenges on slag foaming is the production of steel without fossil fuels. Slag foaming is based on the reduction of FeO by solid carbon. One solution is the use of biomass and another solution is to minimize the amount of carbon. Fruehan [286] has suggested a minimum decarburization of 0.3%C in order to achieve good foaming conditions. Another possibility is to inject a gas through the bottom to improve the stirring conditions in such a way that the bubbles remain a longer time in the slag. The real challenge in the near future is to produce a stable slag foaming without carbon, using hydrogen to reduce the iron oxide, in addition to this, the development of slag foaming models by CFD is one of the fields less developed.

References

1. Kozakevitch P (1969) Foams and emulsions in steelmaking. J Met 21:57–68
2. Hara S, Ogino K (1992) Slag-foaming phenomenon in pyrometallurgical processes. ISIJ Int 32:81–86
3. Ogawa Y, Katayama H, Hirata H, Tokumitsu N, Yamauchi M (1992) Slag foaming in smelting reduction and its control with carbonaceous materials. ISIJ Int 32:87–94. https://doi.org/10.2355/isijinternational.32.87
4. Ameling D, Ullrich W, Sittard M, Petry J, Wolf J (1986) Investigations on the formation of slag in electric arc furnaces (in German). Stahl Eisen 106:625–630
5. Wunsche E, Simcoe R (1984) Electric arc furnace steelmaking with quasi-submerged arcs and foamy slags. Iron Steel Eng 61:35–42
6. Adams W, Alameddine S, Bowman B, Lugo N, Paege S, Stafford P (2001) Factors influencing the total energy consumption in arc furnaces. In: 59th electric arc furnace conference, Phoenix, AZ, USA, 11–14 Nov 2001, pp 691–702
7. Bowman B, Krüger K (2009) Arc furnace physics. StahlEisen
8. Zhang L, Taniguchi S (2000) Fundamentals of inclusion removal from liquid steel by attachment to rising bubbles. Int Mater Rev 45:59–82
9. Karbowniczek M (2002) Model of slag foaming in electrical arc furnaces. In: 7th European electric steelmaking conference, Venice, Italy, 26–29 May 2002
10. Kitchener JA (1964) Ch 2. Foams and free liquid films. In: Danielli JF, Pankhurst KGA, Riddiford AC (eds) Recent progress in surface science, vol 1, pp 51–93
11. Millman S, Malmberg D, Baath L (2001) Radio-wave interferometry for BOS slag control (EC 19473)
12. Shinotake A, Takamoto Y (1993) Combustion and heat transfer mechanism in iron bath smelting reduction furnace. Rev Metall Cah D'Informations Tech 90:965–973. https://doi.org/10.1051/metal/199390070965
13. Bikerman JJ (1973) Foams. Springer
14. Stevenson P (2012) Foam engineering: fundamentals and applications. https://doi.org/10.1002/9781119954620
15. Pugh R (2016) Ch. 11. Bubble size measurements and foam test methods. Bubble and foam chemistry. Cambridge University Press, pp 372–404

16. de Vries AJ (1958) Foam stability. Part 1. Structure and stability of foams. Recl Des Trav Chim Des Pays-Bas 77:81–91
17. de Vries AJ (1958) Foam stability. Part II. Gas diffusion in foams. Recl Des Trav Chim Des Pays-Bas 77:209–223. https://doi.org/10.1021/j150568a024
18. de Vries AJ (1958) Foam stability. Part III. Spontaneous foam destabilization resulting from gas diffusion. Recl Des Trav Chim Des Pays-Bas 77:283–296
19. de Vries AJ (1958) Foam stability: Part IV. Kinetics and activation energy of film rupture. Recl Des Trav Chim Des Pays-Bas 77:383–399. https://doi.org/10.1002/recl.19580770412
20. De Vries AJ (1958) Foam stability: Part V. Mechanism of film rupture. Recl Des Trav Chim Des Pays-Bas 77:441–461. https://doi.org/10.1002/recl.19580770510
21. Kitchener JA, Cooper CF (1959) Current concepts in the theory of foaming. Q Rev Chem Soc 13:71–97
22. Walstra P (1989) Ch 1. Principles of foam formation and stability. In: Wilson A (ed) Foams: physics, chemistry and structure. Springer, pp 1–15
23. Schramm LL, Wassmuth F (1994) Foams: basic principles. In: Schramm LL (ed) Foams: fundamentals and application in the petroleum industry. American Chemical Society, pp 3–45
24. Nexhip C, Sun S, Jahanshahi S (2004) Physicochemical properties of foaming slags. Int Mater Rev 49:286–298. https://doi.org/10.1179/095066004225021945
25. Wang J, Nguyen AV, Farrokhpay S (2016) A critical review of the growth, drainage and collapse of foams. Adv Colloid Interface Sci 228:55–70. https://doi.org/10.1016/j.cis.2015.11.009
26. Calhoun SGK, Chandran Suja V, Fuller GG (2022) Foaming and antifoaming in non-aqueous liquids. Curr Opin Colloid Interface Sci 57:101558. https://doi.org/10.1016/j.cocis.2021.101558
27. Gochev G, Platikanov D, Miller R (2016) Chronicles of foam films. Adv Colloid Interface Sci 233:115–125. https://doi.org/10.1016/j.cis.2015.08.009
28. Scriven L, Sternling C (1960) The Marangoni effects. Nature 187:186–188
29. Volpe C Della, Siboni S (2018) The Wilhelmy method: a critical and practical review. Surf Innov 6:120–132. https://doi.org/10.1680/jsuin.17.00059
30. Wikipedia. J.J. Bikerman. https://de.wikipedia.org/wiki/Jacob_Joseph_Bikerman
31. Wilde PJ (2000) Interfaces: their role in foam and emulsion behaviour. Curr Opin Colloid Interface Sci 5:176–181. https://doi.org/10.1016/S1359-0294(00)00056-X
32. Van der Mensbrugghe MG (1867) XXXVI. On the tension of liquid films. Dublin Philos Mag J Sci 33:270–282. https://doi.org/10.1080/14786446708639785
33. Drelich J, FAng C, White C (2002) Measurement of interfacial tension in fluid-fluid systems. Encycl Surf Colloid Sci 3152–3166
34. Boussinesq MJ (1913) Sur l'existence d'une viscosité superficielle, dans la mince couche de transition séparant un liquide d'un autre fluide contigu. Ann Chim Phys 29:349–357
35. Joly M (1964) Surface viscosity. In: Danielli JF, Pankhurst KGA, Riddiford A (eds) Recent progress in surface science, vol 1, pp 1–50
36. El Omari Y, Yousfi M, Duchet-Rumeau J, Maazouz A (2022) Recent advances in the interfacial shear and dilational rheology of polymer systems: from fundamentals to applications. Polymers (Basel) 14. https://doi.org/10.3390/polym14142844
37. Miller R, Wustneck R, Kragel J, Kretzshmar G (1996) Dilational and shear rheology of adsorption layers at liquid interfaces. Colloids Surfaces A 111:75–118
38. Wasan DT, Nikolov AD, Lobo LA, Koczo K, Edwards DA (1992) Foams, thin films and surface rheological properties. Prog Surf Sci 39:119–154. https://doi.org/10.1016/0079-6816(92)90021-9
39. DeVries AJ (1958) Foam stability. A fundamental investigation of the factors controlling the stability of foams. Rubber Chem Technol 31:1142–1205. https://doi.org/10.5254/1.3542363
40. Wilde PJ, Clark DC (1996) Chapter 5. Foam formation and stability. In: Hall GM (ed) Methods of testing protein functionality. Blackie Academic and Professional, London, England, pp 110–152
41. Young T III (1805) An essay on the cohesion of fluids. Philos Trans R Soc London 95:65–87

42. Cramb AW, Jimbo I (1989) Calculation of the interfacial properties of liquid steel. Slag systems. Steel Res 60:157–165. https://doi.org/10.1002/srin.198900893
43. Loth E (2008) Quasi-steady shape and drag of deformable bubbles and drops. Int J Multiph Flow 34:523–546. https://doi.org/10.1016/j.ijmultiphaseflow.2007.08.010
44. Denkov N, Tcholakova S, Politova-Brinkova N (2020) Physicochemical control of foam properties. Curr Opin Colloid Interface Sci 50:101376. https://doi.org/10.1016/j.cocis.2020.08.001
45. Mysels KJ, Frankel S, Shinoda K (1959) Soap films: studies of their thinning and a bibliography. Pergamon Press
46. Anazadehsayed A, Rezaee N, Naser J, Nguyen AV (2018) A review of aqueous foam in microscale. Adv Colloid Interface Sci 256:203–229. https://doi.org/10.1016/j.cis.2018.04.004
47. Sun Q, Tan L, Wang G (2008) Liquid foam drainage: an overview. Int J Mod Phys B 22:2333–2354. https://doi.org/10.1142/S0217979208039514
48. Nexhip C (1997) Fundamentals of foaming in molten slag systems, PhD thesis, Dept. Chem. Eng., Melbourne University, Australia
49. Nexhip C, Sun S, Jahanshahi S (2000) Some observations on the draining of CaO-SiO$_2$-Al$_2$O$_3$ slag bubble films. Metall Mater Trans B Process Metall Mater Process Sci 31:1105–1115. https://doi.org/10.1007/s11663-000-0086-z
50. Nexhip C, Sun S, Jahanshahi S (1998) Studies on bubble films of molten slags. Philos Trans R Soc A Math Phys Eng Sci 356:1003–1012. https://doi.org/10.1098/rsta.1998.0204
51. Ghosh P (2009) Coalescence of bubbles in liquid. Bubble Sci Eng Technol 1:75–87. https://doi.org/10.1179/175889709X446543
52. Swisher JH, McCabe C (1964) Cr$_2$O$_3$ as a foaming agent in CaO-SiO$_2$ slags. AIME Trans 230:1669–1675
53. Bikerman JJ (1938) The unit of foaminess. Trans Faraday Soc 34:634–638
54. Foulk CW, Miller JN (1931) Experimental evidence in support of the balanced-layer theory of liquid film formation. Ind Eng Chem 23:1283–1288
55. Rudin BAD (1957) Measurement of the foam stability of beers. J Inst Brew 63:506–509
56. Rusanov AI, Krotov VV, Nekrasov AG (1998) New methods for studying foams: foaminess and foam stability. J Colloid Interface Sci 206:392–396. https://doi.org/10.1006/jcis.1998.5697
57. Jeelani SK, Ramaswami S, Hartland S (1990) Effect of binary coalescence on steady-state height of semi-batch foams. Chem Eng Res Des 68:271–277
58. Hartland S, Bourne JR, Ramąswami S (1993) A study of disproportionation effects in semi-batch foams-II. Comparison between experiment and theory. Chem Eng Sci 48:1723–1733. https://doi.org/10.1016/0009-2509(93)80131-9
59. Hartland S, Barber AD (1974) A model for cellular foam. Trans Inst Chem Eng 52:43–52
60. Hrma P (1990) Model for a steady state foam blanket. J Colloid Interface Sci 134:161–168. https://doi.org/10.1016/0021-9797(90)90262-M
61. Hutzler S, Lösch D, Carey E, Weaire D, Hloucha M, Stubenrauch C (2011) Evaluation of a steady-state test of foam stability. Philos Mag 91:537–552. https://doi.org/10.1080/14786435.2010.526646
62. Barbian N, Ventura-Medina E, Cilliers JJ (2003) Dynamic froth stability in froth flotation. Miner Eng 16:1111–1116. https://doi.org/10.1016/j.mineng.2003.06.010
63. Barbian N, Hadler K, Ventura-Medina E, Cilliers JJ (2005) The froth stability column: linking froth stability and flotation performance. Miner Eng 18:317–324. https://doi.org/10.1016/j.mineng.2004.06.010
64. Barbian N, Hadler K, Cilliers JJ (2006) The froth stability column: measuring froth stability at an industrial scale. Miner Eng 19:713–718. https://doi.org/10.1016/j.mineng.2005.09.021
65. Iglesias E, Anderez J, Forgiarini A, Salager J (1995) A new method to estimate the stability of short-life foams. Colloids Surf A 98:167–174
66. Garret P (1993) Recent developments in the understanding of foam generation and stability. Chem Eng Sci 48:367–392

67. Garrett PR (2014) The science of defoaming: theory, experiment and applications. CRC Press, Boca Raton, FL, USA
68. Garrett PR (2015) Defoaming: antifoams and mechanical methods. Curr Opin Colloid Interface Sci 20:81–91. https://doi.org/10.1016/j.cocis.2015.03.007
69. Yi SH, Rhee CH (1997) Effects of additives on the foaming behaviour of the FeO-SiO$_2$ based slag. Steel Res 68:429–433. https://doi.org/10.1002/srin.199700578
70. Komarov SV, Kuwabara M, Sano M (2000) Suppression of slag foaming by a sound wave. ISIJ Int 40:431–437. https://doi.org/10.1016/S1350-4177(00)00051-1
71. Cooper CF, Kitchener JA (1959) Foaming of silicate melts. J Iron Steel Inst 193:48
72. Kitchener JA (1962) Confirmation of the Gibbs theory of elasticity of soap films. Nature 194:676–677
73. Ralston J (2020) The scientific legacy of Joseph Kitchener—its impact in colloid science and flotation. Miner Eng 149. https://doi.org/10.1016/j.mineng.2020.106230
74. Luz AP, Ávila TA, Bonadia P, Pandolfelli VC (2011) Slag foaming: fundamentals, experimental evaluation and application in the steelmaking industry. Refract Worldforum 3:91–98
75. Zhu TX, King MP, Coley KS, Irons GA (2012) Progress in slag foaming in metallurgical processes. Metall Mater Trans B Process Metall Mater Process Sci 43:751–757. https://doi.org/10.1007/s11663-012-9654-2
76. Liukkonen M, Penttila K, Koukkari P (2014) A compilation of slag foaming phenomenon research. Theoretical studies, industrial experiments and modelling. Espoo, Finland
77. Nekrasov IV, Sheshukov OY, Metelkin AA, Sivtsov AV, Tsymbalist MM (2016) Slag conditions in electrosmelting: a review. Steel Transl 46:435–442. https://doi.org/10.3103/S09670 91216060097
78. Min DJ, Tsukihashi F (2017) Recent advances in understanding physical properties of metallurgical slags. Met Mater Int 23:14–18. https://doi.org/10.1007/s12540-017-6750-5
79. Luz AP, Tomba Martinez AG, López F, Bonadia P, Pandolfelli VC (2018) Slag foaming practice in the steelmaking process. Ceram Int 44:8727–8741. https://doi.org/10.1016/j.cer amint.2018.02.186
80. Panchal H, Chadha K, Jariwala F, Gadkari U, Gohil R (2020) Effect of slag foaming in EAF & factors affecting it: a review. Int J Res Appl Sci Eng Technol 8:1466–1470. https://doi.org/10.22214/ijraset.2020.31200
81. Sichen D, Huss J, Vickerfält A, Berg M, Martinsson J, Allertz C et al (2022) The laboratory study of metallurgical slags and the reality. Steel Res Int 93:2100132. https://doi.org/10.1002/srin.202100132
82. Bowman B (1990) Effects on furnace arcs of submerging by slag. Ironmak Steelmak 17:123–129
83. Schroeder DL (2000) The advantages of foaming slag control in EAF operation. Steel Times 228:368–369
84. Pujadas A, McCauley J, Tada Y, Mathis G, Iacuzzi M (2002) Electric arc furnace energy optimization at Nucor Yamato Steel. In: 7th European electric steelmaking conference, Venice, Italy, 26–29 May 2002, pp 1.399–1.411
85. Ogawa Y, Huin D, Gaye H, Tokumitsu N (1992) Physical model of slag foaming. In: 4th international conference on molten slags and fluxes, vol 1, pp 374–379. https://doi.org/10.2355/isijinternational.33.224
86. Huin D, Ogawa Y, Gaye H, Tokumitsu N (1993) Physical model of slag foaming. ISIJ Int 33:224–232. https://doi.org/10.2355/isijinternational.33.224
87. Terashima H, Nakamura T, Mukai K, Izu D (1992) Analysis of bubble shapes generated at mercury-aqueous solution interface. J Jpn Inst Met 56:422–429. https://doi.org/10.2320/jin stmet1952.56.4_422
88. Hong L, Hirasawa M, Sano M (1998) Behavior of slag foaming with reduction of iron oxide in molten slags by graphite. ISIJ Int 38:1339–1345. https://doi.org/10.2355/isijinternational.38.1339
89. Hong L, Hirasawa M, Sano M (2000) Effect of gas injection on the behavior of slag foaming with smelting reduction of iron oxide by graphite. High Temp Mater Process 19:165–175. https://doi.org/10.2355/isijinternational.38.1339

90. Yi SH, Kim SM (2002) Effect of surface-active species on the foam stability in the EAF slag system. Scand J Metall 31:148–152. https://doi.org/10.1034/j.1600-0692.2002.310209.x

91. Yoshida F, Akita K (1965) Performance of gas bubble columns: volumetric liquid-phase mass transfer coefficient and gas holdup. AIChE J 11:9–13. https://doi.org/10.1002/aic.690110106

92. Zhang B, Wang R, Hu C, Liu C, Jiang M (2021) Effect of viscosity on dynamic evolution of metallurgy slag foaming. ISIJ Int 61:1348–1356. https://doi.org/10.2355/ISIJINTERNATIONAL.ISIJINT-2020-404

93. Ozturk B, Fruehan RJ (1995) Effect of temperature on slag foaming. Metall Mater Trans B 26:1086–1088. https://doi.org/10.1007/BF02654111

94. King MP (2009) Carbon injection into electric arc furnace slags. MSc thesis, McMaster University

95. Zhu TX (2011) Carbon injection into electric arc furnace slags. MSc thesis, McMaster University

96. Stadler SAC, Eksteen JJ, Aldrich C (2006) Physical modelling of slag foaming in two-phase and three-phase systems in the churn-flow regime. Miner Eng 19:237–245. https://doi.org/10.1016/j.mineng.2005.05.018

97. Zhu MY, Sichen D (2000) Modelling study of slag foaming phenomenon. Steel Res 71:76–82. https://doi.org/10.1002/srin.200005693

98. Wu LS, Albertsson GJ, Sichen D (2010) Modelling of slag foaming. Ironmak Steelmak 37:612–619. https://doi.org/10.1179/030192310X12690127076550

99. Ito K, Fruehan RJ (1989) Study on the foaming of CaO-SiO$_2$-FeO slags: Part I. Foaming parameters and experimental results. Metall Trans B 20:509–514. https://doi.org/10.1007/BF02654600

100. Kleppe W, Oeters F (1977) Model experiments on foaming in the top-blown BOP converter (in German). Arch Fur Das Eisenhuttenwes 48:193–197

101. Pahl MH, Franke D (1995) Schaum und Schaumzerstörung - ein Überblick. Chemie Ing Tech 67:300–312. https://doi.org/10.1002/cite.330670306

102. Mirsandi H, Baltussen MW, Peters EAJF, van Odyck DEA, van Oord J, van der Plas D et al (2020) Numerical simulations of bubble formation in liquid metal. Int J Multiph Flow 131. https://doi.org/10.1016/j.ijmultiphaseflow.2020.103363

103. Gudenau HW, Wu K, Nys S, Rosenbaum H (1992) Formation and effects of slag foaming in smelting reduction. Steel Res 63:521–525. https://doi.org/10.1002/srin.199201753

104. Kapoor A, Irons GA (1997) A physical modeling study of fluid-flow phenomena in gas-slag systems. ISIJ Int 37:829–838

105. Urquhart RC, Davenport WG (1970) Stabilization of metal-in-slag emulsions. JOM

106. Ghag SS (1996) Physical modelling of slag foaming. PhD Thesis, The University of Queensland

107. Warczok A, Utigard TA (1994) Low temperature physical modelling of slag foaming. Can Metall Q 33:205–215. https://doi.org/10.1179/cmq.1994.33.3.205

108. Zhu MY, Jones T, Sichen D (2001) Modelling study of slag foaming by chemical reaction. Scand J Metall 30:51–56. https://doi.org/10.1034/j.1600-0692.2001.d01-37.x

109. Kapilashrami A, Lahiri AK, Seetharaman S (2005) Bubble formation through reaction at liquid-liquid interfaces. Steel Res Int 76:616–623. https://doi.org/10.1002/srin.200506066

110. Zhang X, Qiu G, Lv X (2015) Simulation study on solution foaming by controlling gas generation reaction in water-glycerol system. In: Jiang T (ed) 6th international symposium on high-temperature metallurgical processing, TMS, pp 413–420

111. Qiu G, Shan C, Zhang X, Lv X (2017) Physical modelling of slag-foaming phenomenon resulted from inside-origin gas formation reaction. Ironmak Steelmak 44:246–254. https://doi.org/10.1080/03019233.2016.1210360

112. Stadler SAC (2002) An experimental study of slag foaming. MSc Dissertation, Department of Process Engineering, University of Stellenbosch

113. Stadler SAC, Eksteen JJ, Aldrich C (2007) An experimental investigation of foaming in acidic, high FexO slags. Miner Eng 20:1121–1128. https://doi.org/10.1016/j.mineng.2007.01.013

114. Tucker JP, Deglon DA, Franzidis JP, Harris MC, O'Connor CT (1994) An evaluation of a direct method of bubble size distribution measurement in a laboratory batch flotation cell. Miner Eng 7:667–680. https://doi.org/10.1016/0892-6875(94)90098-1

115. Wang R, Zhang B, Liu C, Jiang M (2020) A novel modeling method of slag foaming in BOF. Metall Mater Trans B Process Metall Mater Process Sci 51:1941–1946. https://doi.org/10.1007/s11663-020-01921-w

116. Wang R, Zhang B, Liu C, Jiang M (2020) Physical modelling of dynamic evolution of metallurgical slag foaming. Exp Therm Fluid Sci 113. https://doi.org/10.1016/j.expthermflusci.2020.110041

117. Wang R, Zhang B, Hu C, Liu C, Jiang M (2021) Modeling study of metallurgical slag foaming via dimensional analysis. Metall Mater Trans B Process Metall Mater Process Sci 52:1805–1817. https://doi.org/10.1007/s11663-021-02147-0

118. Zhang B, Wang R, Liu C, Shi P, Jiang M (2022) Physical modeling of metallurgical slag foaming induced by chemical reaction. JOM 74:4930–4937. https://doi.org/10.1007/s11837-022-05483-x

119. Wang R, Zhang B, Hu C, Liu C, Jiang M (2022) Physical modeling of slag foaming in combined top and bottom blowing converter. JOM 74:151–158. https://doi.org/10.1007/s11837-021-04984-5

120. Birk W, Arvanitidis I, Jonsson P, Medvedev A (2000) Physical modelling and control of dynamic foaming in an LD-converter process. In: IEEE industry applications conference, vol 4, pp 2584–2590

121. Harada Y, Ishihara M, Saito N, Nakashima K (2017) Impedance measurement of simulated foaming slag for evaluation of gas phase fraction. ISIJ Int 57:1733–1741. https://doi.org/10.2355/isijinternational.ISIJINT-2017-265

122. Kitamura S, Okohira K (1992) Influence of slag composition and temperature on slag foaming. ISIJ Int 32:741–746

123. Paramguru RK, Galgali RK, Ray HS (1997) Influence of slag and foam characteristics on reduction of FeO-containing slags by solid carbon. Metall Mater Trans B Process Metall Mater Process Sci 28:805–810. https://doi.org/10.1007/s11663-997-0007-5

124. Hara S, Ikuta M, Kitamura M, Ogino K (1983) Foaming of molten slags containing iron oxide. Tetsu-To-Hagane 69:1152–1159

125. Ito K, Fruehan RJ (1989) Slag foaming in smelting reduction processes. Steel Res 60:151–156. https://doi.org/10.1002/srin.198900892

126. Kim H-S, Min D-M, Park J-H (2001) Foaming behavior of CaO-SiO$_2$-FeO-MgO sat-X (X=Al$_2$O$_3$, MnO, P$_2$O$_5$, and CaF$_2$) slags at high temperature. ISIJ Int 41:317–324

127. Jung SM, Fruehan RJ (2000) Foaming characteristics of BOF slags. ISIJ Int 40:348–355. https://doi.org/10.2355/isijinternational.40.348

128. Zhang Y, Fruehan RJ (1995) Effect of the bubble size and chemical reactions on slag foaming. Metall Mater Trans B 26:803–812. https://doi.org/10.1007/BF02651727

129. Corbari R, Matsuura H, Halder S, Walker M, Fruehan RJ (2009) Foaming and the rate of the carbon-iron oxide reaction in slag. Metall Mater Trans B Process Metall Mater Process Sci 40:940–948. https://doi.org/10.1007/s11663-009-9270-y

130. Chang YE, Lin CM, Shen JM, Chang WT, Wu W (2021) Effect of MgO content on the viscosity, foaming life, and bonding in liquid for the CaO-SiO$_2$-MgO-5Al$_2$O$_3$-30FeO slag system. Metals (Basel) 11:1–12. https://doi.org/10.3390/met11020249

131. Xiang J, Wang X, Yang M, Wang J, Shan C, Fan G et al (2021) Slag-foaming phenomenon originating from reaction of titanium-bearing blast furnace slag: continuous monitoring of foaming height and calibration. J Mater Res Technol 11:1184–1192. https://doi.org/10.1016/j.jmrt.2021.01.111

132. Ogawa Y, Tokumitsu N (2001) X-ray fluoroscopic observation of slag foaming. Tetsu-To-Hagane 87:14–20

133. Kapilashrami A (2004) Interfacial phenomena in two-phase systems: emulsions and slag foaming. PhD Thesis, Royal Institute of Technology

134. Kapilashrami A, Görnerup M, Lahiri AK, Seetharaman S (2006) Foaming of slags under dynamic conditions. Metall Mater Trans B Process Metall Mater Process Sci 37:109–117. https://doi.org/10.1007/s11663-006-0090-z

135. Koch K, Ren J (1994) Photoelectric measurement of the foaming process in the reduction of slags containing iron oxide. Steel Res 65:3–7. https://doi.org/10.1002/srin.199400918

136. Ren J, Westholf M, Koch K (1994) The influence of MgO, K_2O, Na_2O and gas pressure on slag foaming behaviour under reducing conditions. Steel Res 65:213–218

137. Khanna R, Rahman M, Leow R, Sahajwalla V (2007) Novel sessile drop software for quantitative estimation of slag foaming in carbon/slag interactions. Metall Mater Trans B Process Metall Mater Process Sci 38:719–723. https://doi.org/10.1007/s11663-007-9078-6

138. Ji F, Barati M, Coley KS, Irons GA (2002) Some experimental studies of coal injection into slags. In: 60th EAF conference proceedings, pp 511–524

139. Ji F, Barati M, Coley K, Irons GA (2004) A kinetic study of carbon injection into electric arc furnace slags. In: 7th international conference on molten slags fluxes and salts, vol 1

140. Ji FZ, Barati M, Coley K, Irons GA (2005) Kinetics of coal injection into iron oxide containing slags. Can Metall Q 44:85–94. https://doi.org/10.1179/cmq.2005.44.1.85

141. Zhang Y, Fruehan RJ (1995) Effect of carbonaceous particles on slag foaming. Metall Mater Trans B 26:813–819. https://doi.org/10.1007/BF02651728

142. Chang YE, Lin CM, Shen JM, Luo SF, Yu KW, Wu W (2023) The effect of the addition of alumina on the viscosity, surface tension, and foaming efficiency of $2.5(CaO/SiO_2)$-xAl_2O_3-$yFeO$-MgO melts. Ceram Int 49:21994–212003. https://doi.org/10.1016/j.ceramint.2023.04.024

143. Kapilashrami A, Gornerup M, Lahiri AK, Seetharaman S (2004) Foaming of model slags containing V_2O_5 under dynamic conditions. In: VII international conference on molten slags fluxes and salts, pp 479–482. https://doi.org/10.1142/9789814434447_0027

144. Zhang Y, Fruehan RJ (1995) Effect of gas type and pressure on slag foaming. Metall Mater Trans B 26:1088–1091. https://doi.org/10.1007/BF02654112

145. Malmberg D, Hahlin P, Nilsson E (2007) Microwave technology in steel and metal industry, an overview. ISIJ Int 47:533–538. https://doi.org/10.2355/isijinternational.47.533

146. Kobayashi S, Hatono A, Katogi K, Kuriyama A, Ichihara K (1983) Slag level gauge using microwaves in BOF. Tetsu-To-Hagane/J Iron Steel Inst Japan 69:51–59. https://doi.org/10.2355/tetsutohagane1955.69.1_51

147. Maki Y, Sakimura H, Sawada H, Iwamura T, Akimoto K (1984) Measurement of hot metal level in a torpedo car by microwave. Tetsu-To-Hagane/J Iron Steel Inst Japan 70:1103–1109. https://doi.org/10.2355/tetsutohagane1955.70.9_1103

148. Ruuska J, Ollila S, Baath L, Leiviska K (2006) Possibilities to use new measurements to control LD-KG converter. In: 5th European oxygen steelmaking conference, Aachen, Germany, 26–28 June 2006, pp 1–8

149. Sakamoto Y, Sumio K, Kobayashi Y, Hatono A (1980) Method and apparatus for measuring slag foaming using microwave lever meter. US 4210023

150. Sakamoto Y, Kobayashi S, Hatono T (1977) Acoustic measurement of slag forming. JPS Patent 5433790A

151. Heenatimulla J, Brooks GA, Dunn M, Sly D, Snashall R, Leung W (2022) Acoustic analysis of slag foaming in the BOF. Metals (Basel) 12. https://doi.org/10.3390/met12071142

152. De Vos L, Cnockaert V, Bellemans I, Vercruyssen C, Verbeken K (2021) Critical assessment of the applicability of the foaming index to the industrial basic oxygen steelmaking process. Steel Res Int 92:1–9. https://doi.org/10.1002/srin.202000282

153. Deo B, Overbosch A, Snoeijer B, Das D, Srinivas K (2013) Control of slag formation, foaming, slopping, and chaos in BOF. Trans Indian Inst Met 66:543–554. https://doi.org/10.1007/s12666-013-0306-2

154. Baumert J (1978) Method and apparatus for monitoring slag thickness in refining crucible. US Patent 4098128

155. Buydens JM, Nyssen P, Marique C, Salamone P (1998) The dynamic control of the slag foaming operation in the electric arc furnace. Rev Metall 95:501–509. https://doi.org/10.1051/metal/199895040501

156. Dicker J (2014) Monitoring of slag foaming and other performance indicators in an electric arc furnace. MSc thesis, University of New South Wales
157. Bardernheuer V, Oberhauser P (1970) Comparison of the vibration and sound measurement to control the formation of slag in the Basic Oxygen Converter. Stahl Eisen 90:789–795
158. Kawakami M (1976) Desulphurization and dephosphorization by bottom injection of alkali bearing lag powder. Scand J Metall 5:113–123
159. Mucciardi F, Barrera G (1984) The use of vibrations for dynamic control in primary steelmaking operations. In: Steelmaking conference proceedings, pp 221–229
160. Mucciardi F (1987) Monitoring liquid-gas interactions with an accelerometer. Can Metall Q 26:351–357. https://doi.org/10.1179/cmq.1987.26.4.351
161. Jeong JJ, Ban SJ, Kim SW (2010) Estimation of slag foaming height from vibration signals in electric arc furnaces. In: ECTI-CON2010: the 2010 ECTI international conference on electrical engineering/electronics, computer, telecommunications and information technology, pp 866–869
162. Matschullat T, Rieger D, Krüger K, Döbbeler A (2008) Foaming slag and scrap melting behavior in electric arc furnace—a new and very precise detection method with automatic carbon control. Arch Metall Mater 53:399–403
163. Matschullat T, Rieger D, Dittmer B, Krüger K, Döbbeler A, Mees H (2012) Results of foaming slag and scrap meltdown control SIMELT CSM/FSM based on structure-borne sound in electric arc furnace operation. In: AISTech 2023 iron & steel technology conference proceedings, pp 2577–2590
164. Humber R, Dobbeler A (2010) Improved performance with modern automation solutions for electric steelmaking. In: 10th European electrical conference, pp 614–623
165. Heo JH, Park JH (2019) Assessment of physicochemical properties of electrical arc furnace slag and their effects on foamability. Metall Mater Trans B Process Metall Mater Process Sci 50:2959–2968. https://doi.org/10.1007/s11663-019-01671-4
166. Ito K, Fruehan RJ (1989) Study on the foaming of CaO-SiO_2-FeO slags: Part II. Dimensional analysis and foaming in iron and steelmaking processes. Metall Trans B 20:515–521. https://doi.org/10.1007/BF02654601
167. Sugata M, Sugiyama T, Kondo S (1972) Reduction of FeO in molten slags with solid carbon. Tetsu-To-Hagane 58:1363–1375
168. Lee J, Kim J, Hwang H, Son K, Jeon W, Kim Y et al (2020) Modeling of slag foaming height of electric arc furnace using stepwise regression analysis. Metall Res Technol 117. https://doi.org/10.1051/metal/2020008
169. Gou H, Irons GA, Lu WK (1996) A multiphase fluid mechanics approach to gas holdup in bath smelting processes. Metall Mater Trans B Process Metall Mater Process Sci 27:195–201. https://doi.org/10.1007/BF02915045
170. Pilon L, Viskanta R (2004) Minimum superficial gas velocity for onset of foaming. Chem Eng Process Process Intensif 43:149–160. https://doi.org/10.1016/S0255-2701(03)00012-6
171. Kitamura N, Ogahira K, Tanaka S, Hirai M (1983) Fundamental study on foaming suppression conditions for hot metal slag (in Japanese). Tetsu-To-Hagane 69:S135
172. Bhoi B, Jouhari AK, Ray HS, Misra VN (2006) Smelting reduction reactions by solid carbon using induction furnace: foaming behaviour and kinetics of FeO reduction in CaO-SiO_2-FeO slag. Ironmak Steelmak 33:245–252. https://doi.org/10.1179/174328106X79994
173. Jones JAT, Bowman B, Lefrank PA (1998) Electric arc furnace steelmaking. In: Fruehan RJ (ed) The making, shaping, and treating of steel, 11th edition, steelmaking and refining, Pittsburgh, PA, USA, pp 525–660
174. Tayeb MA, Assis AN, Sridhar S, Fruehan RJ (2015) MgO solubility in steelmaking slags. Metall Mater Trans B Process Metall Mater Process Sci 46:1112–1114. https://doi.org/10.1007/s11663-015-0352-8
175. Hellbrugge VH, Endell KE (1941) Connections between chemical composition and the viscosity of slags and their technical importance. Arch Fur Das Eisenhuttenwes 14:307–315
176. Bockris JO, Mackenzie JD, Kitchener JA (1955) Viscous flow in silica and binary liquid silicates. Trans Faraday Soc 51:1734–1748

177. Yokoyama S, Takeda M, Ito K, Kawakami M (1992) Rate of reduction of chromite ore by the dissolved carbon in molten iron and slag foaming during the reduction. Tetsu-To-Hagane 78:29–36

178. Capodilupo D, Masucci P, Brascugli G, Angelis VD (1990) Operating improvements in electric steel production on the Terni EAF after introduction of slag foaming practice. In: Proceedings of the 6th international iron steel congress, ISIJ, Nagoya, Japan, 21–26 Oct, pp 98–104

179. Wu L (2011) Study on some phenomena of slag in steelmaking process. PhD Thesis, Royal Institute of Technology

180. Aminorroaya S, Edris H (2002) The effect of foamy slag in the Electric Arc Furnaces on energy consumption. In: 7th European electric conference, Venice Italy, 26–29 May, pp 2447–2456

181. Aminorroaya S, Edris H (2002) The effect of foamy slag on power consumption in Electric Arc Furnaces (in Arabic). Comput Methods Eng 21:195–206

182. Xiang J, Wang J, Li Q, Shan C, Qiu G, Yu W et al (2020) Slag-foaming phenomenon originating from reaction of titanium-bearing blast furnace slag: effects of TiO_2 content and basicity. Can Metall Q 59:151–158. https://doi.org/10.1080/00084433.2020.1715696

183. Hara S, Ogino K, Kitamura M (1990) The surface viscosities and the foamines of molten oxides. ISIJ Int 30:714–721. https://doi.org/10.2355/isijinternational.30.714

184. Skupien D, Gaskell DR (2000) The surface tensions and foaming behavior of melts in the system CaO-FeO-SiO_2. Metall Mater Trans B Process Metall Mater Process Sci 31:921–925. https://doi.org/10.1007/s11663-000-0068-1

185. Gaye H, Lucas LD, Olette M, Riboud PV (1984) Metal-slag interfacial properties: equilibrium values and "dynamic" phenomena. Can Metall Q 23:179–191. https://doi.org/10.1179/cmq.1984.23.2.179

186. Yamashita K, Sukenaga S, Matsuo M, Saito N, Nakashima K (2014) Rheological behavior and empirical model of simulated foaming slag. ISIJ Int 54:2064–2070. https://doi.org/10.2355/isijinternational.54.2064

187. Chang WT, Lin CM, Su YL, Li CC, Chang YE, Shen JM et al (2021) Effect of FeO content on foaming and viscosity properties in FeO-CaO-SiO_2-MgO-Al_2O_3 slag system. Metals (Basel) 11:1–10. https://doi.org/10.3390/met11020289

188. Wu K, Qian W, Chu S, Niu Q, Luo H (2000) Behavior of slag foaming caused by blowing gas in molten slags. ISIJ Int 40:954–957. https://doi.org/10.2355/isijinternational.40.954

189. Lahiri AK, Yogambha R, Dayal P, Seetharaman S (2003) Foam in iron and steelmaking. High Temp Mater Process 22:345–352

190. Liu Z, Pandelaers L, Blanpain B, Guo M (2018) Viscosity of heterogeneous silicate melts: a review. Metall Mater Trans B Process Metall Mater Process Sci 49:2469–2486. https://doi.org/10.1007/s11663-018-1374-9

191. Kondratiev A, Jak E (2001) Review of experimental data and modeling of the viscosities of fully liquid slags in the Al_2O_3-CaO-'FeO'-SiO_2 system. Metall Mater Trans B Process Metall Mater Process Sci 32:1015–1025. https://doi.org/10.1007/s11663-001-0090-y

192. Wright S, Zhang L, Sun S, Jahanshahi S (2000) Viscosity of a CaO-MgO-Al_2O_3-SiO_2 melt containing spinel particles at 1646 K. Metall Mater Trans B Process Metall Mater Process Sci 31:97–104. https://doi.org/10.1007/s11663-000-0134-8

193. Wright S, Zhang L, Sun S, Jahanshahi S (2001) Viscosities of calcium ferrite slags and calcium alumino-silicate slags containing spinel particles. J Non Cryst Solids 282:15–23. https://doi.org/10.1016/S0022-3093(01)00324-6

194. Pretorius EB, Carlisle RC (1999) Foamy slag fundamentals and their application to Electric Furnace Steelmaking. Iron Steelmak 26:79–88

195. Martinsson J, Sichen D (2016) Study on apparent viscosity of foam and droplet movement using a cold model. Steel Res Int 87:712–719. https://doi.org/10.1002/srin.201500200

196. Bindal S, Nikolov A, Wasan S, Lambert D, Koopman D (2001) Foaming in simulated radioactive waste. Environ Sci Technol 35:3941–3947

197. Mombelli D, Dall'Osto G, Villa G, Mapelli C, Barella S, Gruttadauria A et al (2021) Study on the pneumatic lime injection in the electric arc furnace process: an evaluation on the performance benefits. Steel Res Int 92. https://doi.org/10.1002/srin.202100083

198. Bianco L, Baracchini G, Cirilli F, Moriconi A, Moriconi E, Marcos M et al (2013) Sustainable EAF steel production (GreenEAF) 26208
199. Bianco L, Baracchini G, Cirilli F, Di SL, Moriconi A, Moriconi E et al (2013) Sustainable electric arc furnace steel production: GREENEAF (Nachhaltigkeit der Stahlproduktion mit dem Elektro-Lichtbogenofen: GREENEAF). BHM Berg- Huettenmaenn Monatsh 158:17–23. https://doi.org/10.1007/s00501-012-0101-0
200. DiGiovanni C, Li D, Ng KW, Huang X (2023) Ranking of injection biochar for slag foaming applications in steelmaking. Metals (Basel) 13. https://doi.org/10.3390/met13061003
201. Kieush L, Schenk J, Koveria A, Hrubiak A, Hopfinger H, Zheng H (2023) Evaluation of slag foaming behavior using renewable carbon sources in electric arc furnace-based steel production. Energies 16. https://doi.org/10.3390/en16124673
202. King MP, Ji F, Irons GA, Coley KS (2009) Kinetics of carbon reaction with electric arc furnace slags during slag foaming. In: AISTech 2009 proceedings, vol I, pp 617–626
203. Zhu T, Coley KS, Irons GA, Ray SK (2011) Carbon injection rate for optimum electric arc furnace slags foaming. In: AISTech—the iron & steel technology conference proceedings, vol I, pp 849–856
204. Vieira D, De Almeida RAM, Bielefeldt WV, Vilela ACF (2016) Slag evaluation to reduce energy consumption and EAF electrical instability. Mater Res 19:1127–1131. https://doi.org/10.1590/1980-5373-MR-2015-0720
205. Svensson J (2018) The effect of carbonaceous iron on slag foaming. Degree project KTH
206. Kipepe TM, Pan X (2014) Importance and effect of foaming slag on energy efficiency. In: 71st world foundry congress: advanced sustainable foundry, WFC 2014
207. Burström E, Ye G, von Scheele J, Selin R (1992) Mechanism of formation of foaming slags in electric arc furnaces. In: 4th international conference on molten slags and fluxes, pp 352–357
208. Bergman Å (1989) Some aspects on MgO solubility in complex slags. Steel Res 60:191–195
209. Park JM (2001) MgO solubility in BOF slag equilibrated with ambient air. Steel Res 72:141–145. https://doi.org/10.1002/srin.200100098
210. Trömel G, Koch K, Fix W, Grosskurth N (1969) The influence of magnesium oxide on the equilibrium in the system FeO-CaO-FeO-SiO$_2$ and on the sulfur distribution at 1600°C (in German). Arch Für Das Eisenhüttenwes 40:969–978
211. De Almeida RAM, Vieira D, Bielefeldt WV, Vilela ACF (2017) Slag foaming fundamentals—a critical assessment. Mater Res 20:474–480. https://doi.org/10.1590/1980-5373-MR-2016-0059
212. Bennett J, Kwong KS (2010) Thermodynamic studies of MgO saturated EAF slag. Ironmak Steelmak 37:529–535. https://doi.org/10.1179/030192310X12706364542669
213. Luz AP, Thiago AA, Bonadia P, Pandolfelli V (2011) Thermodynamic simulations and isothermal solubility diagrams as tools for slag foaming control. Ceram Int 37:2947–2950. https://doi.org/10.1016/j.ceramint.2011.04.020
214. Heo JH, Park JH (2021) Effect of slag composition on dephosphorization and foamability in the electric arc furnace steelmaking process: improvement of plant operation. Metall Mater Trans B Process Metall Mater Process Sci 52:3613–3623. https://doi.org/10.1007/s11663-021-02322-3
215. Ogawa Y, Tokumitsu N (1990) Observation of slag foaming by X-ray fluoroscopy. In: Sixth international iron and steel congress, ISIJ, Nagoya, Japan, pp 147–152
216. Jiang R, Fruehan RJ (1991) Slag foaming in bath smelting. Metall Trans B 22:481–489. https://doi.org/10.1007/BF02654286
217. Lin Z, Guthrie RIL (1995) A model for slag foaming for the in-bath smelting process. Iron Steelmak 22:67–73
218. Sugata M, Sugiyama T, Kondo S (1974) Reduction of iron oxide contained in molten slags with solid carbon. Trans Iron Steel Inst Jpn 14:88–95. https://doi.org/10.2355/isijinternational1966.14.88
219. Fruehan RJ (1977) The rate of reduction of iron oxides by carbon. Metall Trans B 8:279–286. https://doi.org/10.1007/BF02657657

220. Ozawa M, Kitagawa S, Nakayama S, Takesono Y (1986) Reduction of FeO in molten slags by solid carbon in the electric arc furnace operation. Trans Iron Steel Inst Jpn 26:621–628. https://doi.org/10.2355/isijinternational1966.26.621

221. Story SR, Fruehan RJ (2000) Kinetics of oxidation of carbonaceous materials by CO_2 and H_2O between 1300°C and 1500°C. Metall Mater Trans B 31:43–54. https://doi.org/10.1016/0008-6223(65)90013-8

222. Hara S, Ogino K (1990) Reduction of molten iron oxide-based slags by solid graphite. Tetsu-To-Hagane/J Iron Steel Inst Jpn 76:360–367. https://doi.org/10.2355/tetsutohagane1955.76.3_360

223. Molloseau CL, Fruehan RJ (2002) The reaction behavior of Fe-C-S droplets in CaO-SiO_2-MgO-FeO slags. Metall Mater Trans B Process Metall Mater Process Sci 33:335–344. https://doi.org/10.1007/s11663-002-0045-y

224. Matsuura H, Manning CP, Fortes RAFO, Fruehan RJ (2008) Development of a decarburization and slag formation model for the electric arc furnace. ISIJ Int 48:1197–1205. https://doi.org/10.2355/isijinternational.48.1197

225. Chu S, Niu Q, Wu K, Wang Y (2000) Decomposition of $CaCO_3$ in molten borate and its effect on slag foaming behavior. ISIJ Int 40:549–553. https://doi.org/10.2355/isijinternational.40.549

226. Chu S, Wu K, Wang Y (2001) Foaming behavior in molten slag caused by decomposition of carbonate minerals. Chin J Process Eng 1:262–267

227. Chychko A, Seetharaman S (2011) Foaming in electric arc furnace. Part I: Laboratory studies of enthalpy changes of carbonate additions to slag melts. Metall Mater Trans B Process Metall Mater Process Sci 42:20–29. https://doi.org/10.1007/s11663-010-9447-4

228. Chychko A, Teng L, Seetharaman S (2012) Foaming in electric arc furnace—Part II: Foaming visualization and comparison with plant trials. Metall Mater Trans B Process Metall Mater Process Sci 43:1078–1085. https://doi.org/10.1007/s11663-012-9690-y

229. Wallis GB (1961) Some hydrodynamics aspects of two-phase flow and boiling (paper 38). Int Dev Heat Transf 2:319–340

230. Lahiri AK, Seetharaman S (2002) Foaming behavior of slags. Metall Mater Trans B Process Metall Mater Process Sci 33:499–502. https://doi.org/10.1007/s11663-002-0060-z

231. Conejo AN, Torres R, Cuellar E (1999) Industrial analysis on the reduction of iron oxide through injection of carbon particles into the electric arc furnace (in Spanish). Rev Metal 35:111–125. https://doi.org/10.3989/revmetalm.1999.v35.i2.613

232. Morales RD, Rodriguez-Hernandez JH, Garnica P, Romero JA (1997) A mathematical model for the reduction kinetics of iron oxide in electric furnace slags by graphite injection. ISIJ Int 37:1072–1080

233. Rodriguez-Hernandez JH (2000) Multiphase flow in steelmaking operations (in Spanish). PhD thesis, IPN/ESIQIE

234. Conejo A, Morales R, Rodriguez H (2001) Mathematical modeling of the EAF process using direct reduced iron. 59th electric furnace conference, Phoenix, AZ, USA, 11–14 Nov 2001, pp 797–810

235. Morales RD, Rodríguez-Hernández H, Conejo AN (2001) A mathematical simulator for the EAF steelmaking process using direct reduced iron. ISIJ Int 41

236. Rodriguez HH, Conejo AN, Morales RD (2001) Theoretical analysis of the interfacial phenomena during the injection of carbon particles into EAF slags. Steel Res 72

237. Morales RD, Conejo AN, Rodríguez HH (2002) Process dynamics of electric arc furnace during direct reduced iron melting. Metall Mater Trans B Process Metall Mater Process Sci 33

238. Morales RD, Rodríguez-Hernández H, Vargas-Zamora A, Conejo AN (2002) Concept of dynamic foaming index and its application to control of slag foaming in electric arc furnace steelmaking. Ironmak Steelmak 29:445–453

239. Farias L, Robertson DG (1982) Physical modelling of gas-powder injection into liquid metals. In: Process technology conference proceedings, Iron and Steel Society, pp 206–222

240. Richards GG, Brimacombe JK (1985) Kinetics of the zinc slag-fuming process: Part II. Mathematical model. Metall Trans B 16:529–540. https://doi.org/10.1007/BF02654851
241. Oosthuizen DJ, Viljoen JH, Craig IK, Pistorius PC (2001) Modelling of the off-gas exit temperature and slag foam depth of an electric arc furnace. ISIJ Int 41:399–401. https://doi.org/10.2355/isijinternational.41.399
242. Matsuura H, Fruehan RJ (2009) Slag foaming in an electric arc furnace. ISIJ Int 49:1530–1535. https://doi.org/10.2355/isijinternational.49.1530
243. Sedivy C, Krump R (2008) Tools for foaming slag operation at EAF steelmaking. Arch Metall Mater 53:409–413
244. Pilon L (n.d.) Interfacial and transport phenomena in closed-cell foams. PhD Thesis, Purdue University
245. Pilon L, Fedorov AG, Viskanta R (2001) Steady-state thickness of liquid-gas foams. J Colloid Interface Sci 242:425–436. https://doi.org/10.1006/jcis.2001.7802
246. Bhakta A, Ruckenstein E (1997) Decay of standing foams: drainage, coalescence and collapse. Adv Colloid Interface Sci 70:1–124. https://doi.org/10.1016/s0001-8686(97)00031-6
247. Narsimhan G, Ruckenstein E (1986) Hydrodynamics, enrichment, and collapse in foams. Langmuir 2:230–238. https://doi.org/10.1021/la00068a021
248. Narsimhan G, Ruckenstein E (1986) Effect of bubble size distribution on the enrichment and collapse in foams. Langmuir 2:494–508. https://doi.org/10.1021/la00070a020
249. Conejo AN (2021) Fundamentals of dimensional analysis: theory and applications in metallurgy. Springer. ISBN: 978-981-16-1601-3
250. Jouhari AK, Galgali RK, Datta P, Bhattacharjee S, Gupta RC, Ray HS (2000) Foaming during reduction of iron oxide in molten slag. Ironmak Steelmak 27:27–31. https://doi.org/10.1179/030192300677345
251. Ghag SS, Hayes PC, Lee HG (1998) Model development of slag foaming. ISIJ Int 38:1208–1215. https://doi.org/10.2355/isijinternational.38.1208
252. Małysa K, Miller R, Lunkenheimer K (1991) Relationship between foam stability and surface elasticity forces: fatty acid solutions. Colloids Surf 53:47–62. https://doi.org/10.1016/0166-6622(91)80035-M
253. Lotun D, Pilon L (2005) Physical modeling of slag foaming for various operating conditions and slag compositions. ISIJ Int 45:835–840. https://doi.org/10.2355/isijinternational.45.835
254. Attia JA, Kholi S, Pilon L (2013) Scaling laws in steady-state aqueous foams including Ostwald ripening. Colloids Surf A 436:1000–1006
255. Campolo M, Andreoli M, Tognotti L, Soldati A (2007) Modelling of a multiphase reacting turbulent jet: application to supersonic carbon injection in siderurgic furnaces. Chem Eng Sci 62:4439–4458. https://doi.org/10.1016/j.ces.2007.05.019
256. Wei G, Zhu R, Yang S, Hu S, Yang L, Chen F (2020) Carbon powder mixed injection with a shrouding supersonic oxygen jet in electric arc furnace steelmaking. Metall Mater Trans B Process Metall Mater Process Sci 51:2298–2308. https://doi.org/10.1007/s11663-020-01920-x
257. Sattar MA, Naser J, Brooks G (2013) Numerical simulation of two-phase flow with bubble break-up and coalescence coupled with population balance modeling. Chem Eng Process Process Intensif 70:66–76. https://doi.org/10.1016/j.cep.2013.05.006
258. Sattar MA, Naser J, Brooks G (2013) Numerical simulation of creaming and foam formation in aerated liquid with population balance modeling. Chem Eng Sci 94:69–78. https://doi.org/10.1016/j.ces.2013.01.064
259. Sattar MA, Naser J, Brooks G (2014) Numerical simulation of slag foaming on bath smelting slag (CaO-SiO$_2$-Al$_2$O$_3$-FeO) with population balance modeling. Chem Eng Sci 107:165–180. https://doi.org/10.1016/j.ces.2013.11.037
260. Sattar MA, Naser J, Brooks G (2014) Modelling of slag foaming coupled with decarburisation. In: TMS David G.C. Robertson symposium on pyrometallurgy, San Diego, CA, 16–20 Feb, pp 401–408
261. Hewage AK, Naser J, Brooks G (2015) Numerical simulation of slag foaming in oxygen steelmaking. In: 11th international conference on CFD in the minerals and process industries, CSIRO, Melbourne, Australia, 7–9 Dec, pp 1–6

262. Boemer A, Rodl S (2000) Optimisation of the electric arc furnace lancing strategy by physical and numerical simulation. Steel Res 71:197–203

263. Morales RD, Lule R, Lopez F, Camacho J, Romero JA (1995) The slag foaming practice in EAF and its influence on steelmaking shop productivity. ISIJ Int 35:1054–1062

264. Conejo AN, Lule RG, Lopéz F, Rodriguez R (2006) Recycling MgO-C refractory in electric arc furnaces. Resour Conserv Recycl 49:14–31. https://doi.org/10.1016/j.resconrec.2006.03.002

265. Jones J, Safe P, Wiggins B (1999) Optimization of EAF operations through offgas system analysis. In: Electric furnace conference, Pittsburgh, PA, USA, 14–16 Nov 1999, pp 459–480

266. Nal Ö, Dolapçioğlu S, Gottardi R, Partyka A, Miani S (2017) Efficiency of the EAF with Telescope roof. In: AISTech 2017 conference, vol 1, pp 1149–1160

267. Mukai K (1991) Some views on the slag foaming in iron and steelmaking processes. Tetsu-to-Hagane 77:856–858

268. Kozakevitch P (1949) Tension superficielle et viscosite des scories synthetiques. Rev Metall 66:572–582

269. Echterhof T (2021) Review on the use of alternative carbon sources in EAF steelmaking. Metals (Basel) 11:1–16. https://doi.org/10.3390/met11020222

270. Görnerup M, Jacobsson H (1998) Foaming slag practice in electric stainless steelmaking. Iron Steelmak 25:59–66

271. Divakar M, Görnerup M, Lahiri AK, Divakar M, Go M (2002) Simultaneous reduction of oxides and dissolution of alumina in stainless steelmaking slag. Ironmak Steelmak 29:297–302. https://doi.org/10.1179/030192302225005150

272. Kapilashrami A, Lahiri AK, Görnerup M, Seetharaman S (2006) The fluctuations in slag foam under dynamic conditions. Metall Mater Trans B Process Metall Mater Process Sci 37:145–148. https://doi.org/10.1007/s11663-006-0095-7

273. Kerr JJ, Fruehan RJ (2004) Additions to generate foam in stainless steelmaking. Metall Mater Trans B Process Metall Mater Process Sci 35:643–650. https://doi.org/10.1007/s11663-004-0005-9

274. Karbowniczek M, Tuvnes P, Engh TA (2008) Foaming of stainless steel slags with Cr_2O_3 using Steelcal. Metall Foundry Eng 34:93–104. https://doi.org/10.7494/mafe.2008.34.1.93

275. Reichel J, Rose L, Karbowniczek M (2011) Method for the production of a foamed slag in a metal bath. US Patent 8403402, 25 Oct 2011

276. Reichel J, Rose L (2013) Process for producing a foamed slag on austenitic stainless melts in an electric arc furnace. US Patent 8409320, 2 Apr 2013

277. Reichel J, Rose L (2014) Process for producing foamed slag. US Patent 8673047, 18 Mar 2014

278. Reichel J, Rose L, Kempken J, Damazio M, Carvalho R, RazLoss H, et al (2008) EAF-foamy slag in stainless steel production. New extremely efficient technology. Easy to handle and cost-efficient. Arch Metall Mater 53:391–397

279. Conejo AN (2011) Oily-mill sludge recycling in the electric arc furnace. Aviles, Spain

280. Davydenko A, Karasev A, Lindstrand G, Jönsson P (2015) Investigation of slag foaming by additions of briquettes in the EAF during stainless steel production. Steel Res Int 86:146–153. https://doi.org/10.1002/srin.201400036

281. Davydenko A, Karasev A, Glaser B, Jönsson P (2019) Direct reduction of Fe, Ni and Cr from oxides of waste products used in briquettes for slag foaming in EAF. Materials (Basel) 12. https://doi.org/10.3390/ma12203434

282. Ban-ya S, Iguchi Y, Nagasaka T (1984) Rate of reduction of liquid wustite with hydrogen. Testu Hagane 70:689–696

283. Sanchez JLG, Conejo AN, Ramirez-Argaez MA (2012) Effect of foamy slag height on hot spots formation inside the Electric Arc Furnace based on a radiation model. ISIJ Int 52

284. Agnihotri A, Singh PK, Singh D, Gupta M (2021) Foamy slag practice to enhance the energy efficiency of electric arc furnace: an industrial scale validation. J Mater Today Proc 46AD:1537–1542

285. Eissa M, Megahed G, El-Deab M, Fathy A, Gawad S, Farahat R (2017) Effect of slag chemistry on energy consumption in electric arc furnace. In: 4th international conference on energy engineering, Aswan, Egypt, 26–28 Dec, p 4

286. Fruehan RJ (1985) Scrap in iron and steelmaking: technologies to improve the use of scrap. Iron Steelmak 12:31–36

Chapter 8
Scrap Quality and Scrap Pre-heating

8.1 Steel Scrap Quality

Scrap quality is a property which can have different meanings depending on its purpose;

- Scrap quality that yields a high melting rate; in this case physical properties are more important, in particular its bulk density
- Scrap with low concentration of residual elements because of its negative effect on the properties of the final steel products; in this case chemical properties are more important.

Evidently both, physical and chemical properties, are always important in terms of defining scrap quality. Non-ferrous content in the scrap increases the consumption of energy. However, scrap quality and both price and availability move in opposite directions. A practical way to define high-quality steel scrap is a material that has the physical and chemical properties which allow higher melting rates in the steelmaking reactor, the production of steel with a certain chemical composition that eliminates casting problems and gives the desired mechanical properties.

A valued property of steel scrap is its infinite recyclability, which inevitably leads to accumulation of residual elements, a problem which has not yet been solved economically with existing technologies. Steel is the world's most recycled material. The recycling rate of common steel products ranges from 70 to 106% [1], for example, for automobiles is about 106%, ships and bridges 98%, appliances 90%, concrete rebar 70%, and aluminum cans 67%. The recycling rate above 100% indicates that there is a higher steel recovery than the steel required to manufacture new steel products.

The main raw material for the Electric Arc Furnace (EAF) is steel scrap and is also the main reason that the specific energy consumption to produce steel is the lowest, in comparison with integrated steel plants. A widely cited reference from the world steel association is the savings obtained by recycling steel scrap which

A. N. Conejo, *Electric Arc Furnace: Methods to Decrease Energy Consumption*,
https://doi.org/10.1007/978-981-97-4053-6_8

otherwise would be produced if virgin materials are employed [2]; 1,400 kg of iron ore, 740 kg of coal, and 120 kg of limestone are saved for every 1,000 kg of steel scrap. In addition to this, about 975 Mton CO_2 are saved because of the consumption of 650 Mton of steel scrap.

8.1.1 Steel Scrap Classification

Ferrous scrap classification based on its origin:

- Obsolete scrap: reclaimed scrap after the end of life of steel products. It is highly heterogeneous because its origin is very diverse.
- Prompt scrap: Prompt scrap is ferrous metal that has not yet reached the marketplace. These are the residuals of large-scale industrial manufacturing processes.
- Home scrap: internal scrap.

The Institute of Scrap Recycling Industries (ISRI) [3] classified commercial steel scrap in a large number of types and its English terminology prevails worldwide. From this classification the most common types are the following:

- Heavy melting scrap (HMS). HMS-1 and HMS-2. It is heavy scrap made of obsolete scrap and wrought iron characterized by a large thickness and high density. HMS-1 has a thickness higher than 6 mm, doesn't include blackened and galvanized steel, it has a higher density than HMS1. HMS-2 has a minimum thickness of 3.1 mm and is lighter than HMS-1, contains blackened and galvanized steel. The sources of HMS are very diverse; structural steel from demolished buildings, bolsters from rail cars, vintage ship, heavy gauge large diameter torched tubing and pipes.

  HMS is the most commercialized type of scrap due to its high density. Its share of the market is about 80% of the obsolete scrap.
- Bushelings: Clean industrial sheet steel scrap usually with cutting edges, clippings, stampings, punchings, or skeleton scrap. Type 1 is more homogeneous and doesn't include black, galvanized and zinc-coated materials.
- Shredded: steel scrap processed through a shredder and magnetically separated originating from automobiles, appliances, heavy melting steel, and miscellaneous sheet steel.
- Bundles or pressed scrap: Hydraulically compressed bundles (with a baling machine) consisting of steel sheet scrap, beverage cans, clippings, or skeleton scrap that may include black, galvanized and zinc coated materials. Bundles are made in two sizes; small bundles (914 × 610 × 610 mm) and large bundles (914–1219 × 822 × 822 mm). Since this is a high-density scrap, it should be placed at the bottom of the scrap bucket to ensure contact with the hot heel and reach faster melting. Smaller size bundles can be heated faster. Large bundles in front

of a burner create overheating due to blow-back of oxygen. It is important for the production of steels with low content of residual elements

- Turnings: Scrap generated during turnings, with a typical spiral appearance
- Pig iron: Cast Iron or Blast Furnace Iron that has been reduced from iron ore. Liquid iron is cast into individual small pigs, grouped pigs that have been broken into smaller pieces, or pyramids. Residual elements is extremely low.

High density scrap helps to decrease the number of scrap charges but also takes longer time for melting. The scrap apparent density given below is partially based on the specifications of one steel plant [4]. Due to the large heterogeneous nature of steel scrap these values are only an estimated reference.

Scrap apparent density. After [4]	
	Density, ton/m^3
Turnings	0.4–0.6
HMS-1	0.7–1.0
Bushelings-2	0.8
Shredded-2	0.97–1.23
Bundles-1	1.1–1.6
Pig iron	> 4.4

Figure 8.1 illustrates the main types of ferrous scrap.

The final and most important classification of steel scrap is carried out at the scrap yard, to define the scrap charging process into the baskets. An optimum scrap charging practice can contribute with energy savings in the order of 20 kWh/ton [5]. This practice aims to achieve the following objectives:

- Provides a cushion of light scrap at the bottom to protect the refractory
- Guarantee a defined liquid steel chemical composition
- Its size distribution and density allow the rapid formation of a molten pool to immerse the electrodes and reach an overall high melting rate
- Avoids cave-ins
- Avoids high density scrap in front of the burners
- Provide protection of the sidewalls and roof from arc radiation
- Reaches an overall high density to minimize recharges.

In order to avoid human errors, the charging process is automatized. Due to the large amount of information to sort-out steel scrap, one of the most common methods is to use machine learning methods. Baumert et al. described the implementation of image analysis using a probabilistic neural network classifier (PNN) in baskets of 100 ton. The system included a color digital camera, a precision distance laser scanner and a light projector. The classifier processes images and weight of about 20 layers and finally reports the average density of steel scrap in the basket, which ranges from 0.5 to 1.5 ton/m^3, with an average of 0.75 ton/m^3.

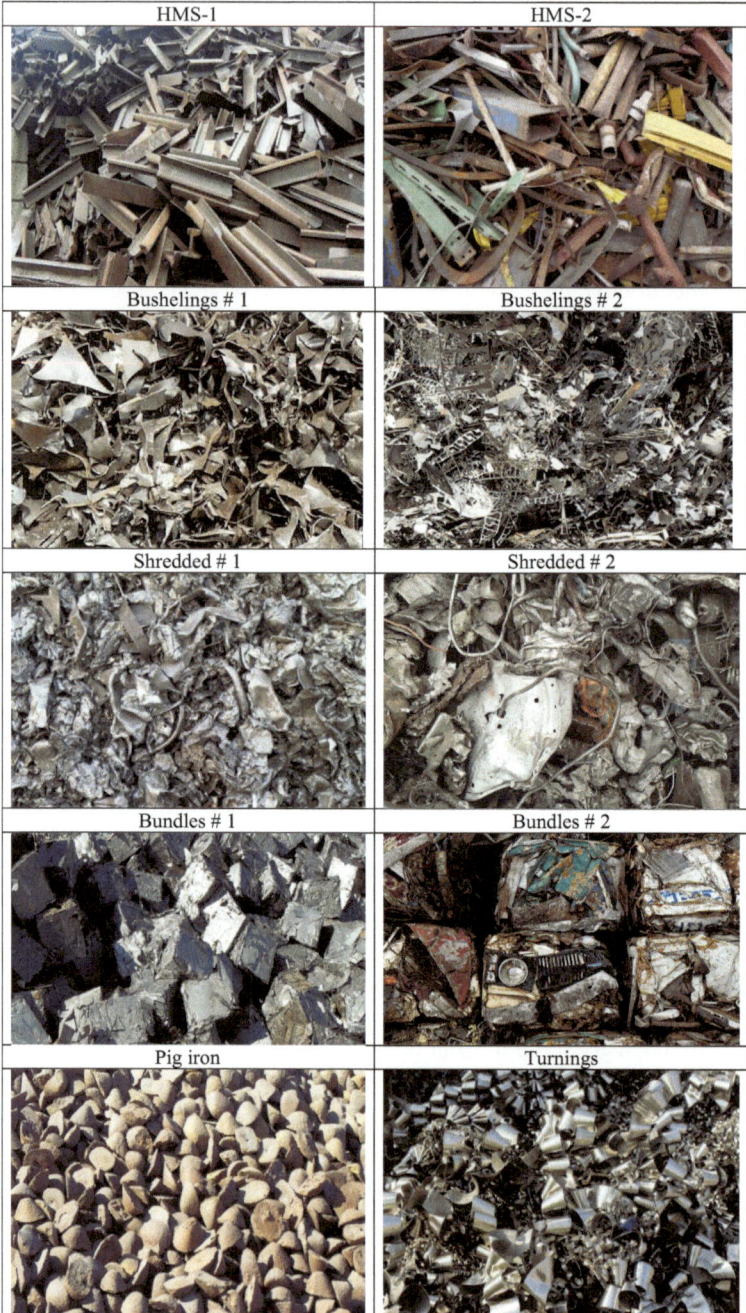

Fig. 8.1 Types of steel scrap

Scrap density in the Consteel process is not an issue in comparison with charging in scrap buckets. The main requirement is to match energy input with the scrap feed-rate [5]. Light scrap is charged at the bottom of the conveyor to avoid melting or oxidation in the pre-heating tunnel, also bushelings because can act a barrier to heat transfer to the scrap at the bottom. High-density scrap is charged on the top to improve its pre-heating. Shredded scrap fills the gaps and increases the overall scrap density.

8.1.2 Scrap Melting Stages in a 100% Scrap Operation with Batch Charging

Batch charging is the dominant system for scrap charging in the EAF using scrap buckets. There are two types of scrap buckets; "orange peel" and "clamshell". The orange-peel is an old design and the clamshell is the dominant design. The orange-peel design has the advantage that opening is faster and the scrap drops straight but on the other hand the clamshell has an automatic closure system and can be sat on a flat surface.

Charging operation: The electrodes are lifted and the roof is turned, a crane places the scrap bucket on top of the furnace and the first bucket of scrap is charged, then the roof is closed and the electrodes descend until the electric arcs are formed. Each charge can take up to 3 min with fast moving systems. To reach higher melting rates the scrap should be evenly distributed and avoid the presence of non-conductive materials. Gaps should be avoided because promote the collapse of scrap (cave ins). Adding a small amount of heavy scrap on the top can improve the bulk density of the scrap inside the furnace but at the same time it should be located away from the pitch circle to avoid that falls and break the electrodes (cave-ins). Using light scrap on the top accelerates the initial step called bore-in.

Boring: Boring represents the initial melting of scrap around the electrodes, mainly in the vertical direction. During this stage is possible to apply maximum power transformer (higher voltage) because the intense arc radiation is shielded by the steel scrap. Boring is a stage of maximum noise. It can be reduced when light scrap is charged on the top and melts faster, thus covering the arcs. About 15% of the scrap is melted during the bore-in period.

Flat bath: Stage of complete scrap melting. In old metallurgical practices, before the slag foaming practice was dominated, once the scrap was melted, the arc length was decreased to avoid intense radiation to the refractory walls. In current practices is possible to keep maximum power transformer and high voltage if the arc is immersed in a foamy slag.

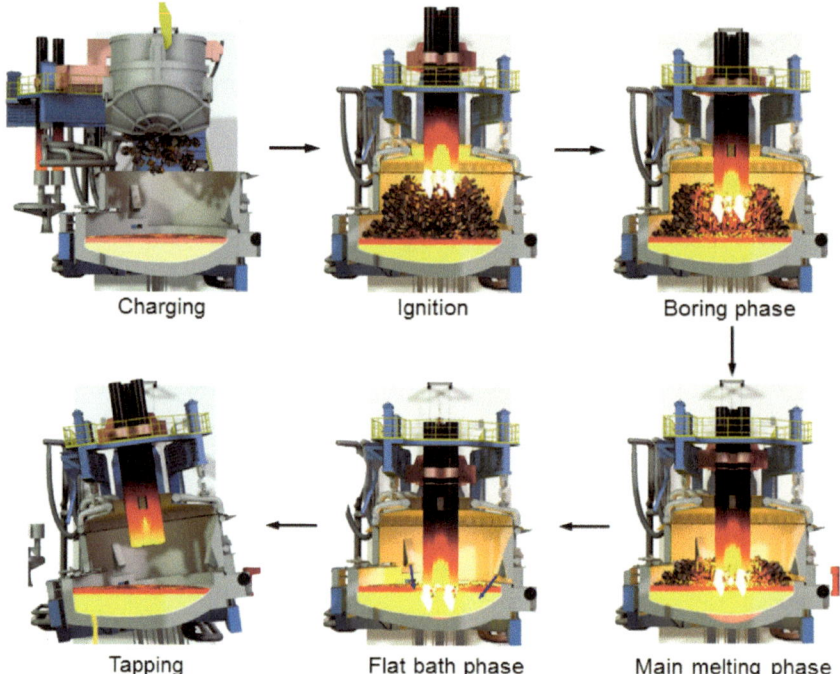

Fig. 8.2 EAF steps with 100% scrap operation. Pictures taken from SMS company [6]

Furnace turn-around: When steel reaches its final carbon and phosphorous targets, as well as temperature of liquid steel the next step is tapping. Electric power is turned-off. The final stage, turnaround covers the end of tapping until the next heat starts. It is also an important part of the power-off time but necessary to inspect the furnace, make refractory repairs to the hearth and slag line, if necessary. Figure 8.2 illustrates schematically the main stages during scrap melting.

8.1.3 Steel Scrap Size, Density and Non-metallic Matter

Scrap size: small pieces of scrap decrease the number of recharges but if the size is too small can weld together and behave like scrap of high density, then it will take much longer time to melt. Adams et al. [7] suggest a size in the range from 100 to 1000 mm with a bulk density from 0.6 to 0.9 ton/m^3.

The *scrap bulk density* or apparent density results from dividing its weight over the volume of the container. The optimum value of the bulk density should minimize the number of re-charges and provide the highest melting rate. Light scrap can have a higher melting rate but it increases the number or recharges, on the other hand, heavy scrap decreases the number of recharges but it increases the melting rate. Heavy scrap should be properly placed; big pieces of heavy scrap in front of the oxyfuel burners region create overheating and also because it takes longer time to melt, heavy scrap should reach the hot heel as early as possible. Bowman and Kruger [8] indicates that the optimum density is hard to define, suggesting a range from 0.4 to 0.8 ton/m^3. Classification of scrap based on density is complex because of the large range of scrap density within the same group. The following classification is based on one Swedish steel plant, as reported by Carlsson et al. [9].

- Light: 0.3–1.0 ton/m^3
- Light-Mid: 0.56–1.0 ton/m^3
- Mid: 0.7–1.4 ton/m^3
- Heavy: > 1.4 ton/m^3.

One reason why heavy scrap has a longer melting time is because the liquid cannot drain to the bottom. The limiting case with the highest density equivalent to pure iron (7.8 ton/m^3), can be obtained due to a major delay causing the molten metal to freeze, in this case remelting can be achieved by tilting the furnace to drain the liquid metal.

Non-ferrous matter in Scrap: The presence of non-ferrous matter such as dust, concrete, etc. affects the metallic yield on one side and on the other side it increases the amount of fluxes, slag and final dust produced. Increasing the amount of slag decreases the metallic yield for the same oxidation degree. Eddy currents can be applied to separate non-ferrous, non-magnetic materials.

Humidity: Scrap humidity is a problem not only because it increases energy consumption but also because it involves the potential for explosions. Humidity artificially increases the weight of scrap. Snow and ice have the same effects.

8.1.4 Residual Elements

One of the main problems of scrap quality is the presence of residual elements, also called tramp elements. Residual elements cannot be removed under the oxidizing conditions that prevail in steelmaking. Residual elements are alloying elements which affect the steel properties for a given application. Copper is one of the principal residual elements. Since residual or tramp elements cannot be removed during steelmaking or the available technologies are not scalable to a massive production of residuals free-scrap, efforts have been mainly focused on sorting and separation. Once a tramp element is found during steelmaking and exceeds the chemical specifications the only solution is to dilute the heat with liquid steel from another heat, this is not only risky but also drastically increases the tap-to-tap time and energy consumption for the heats involved.

In general, the higher the quality of the final steel product the concentration of residual elements should be lower. This was in fact the main limitation for the EAF process to produce high quality steels for more than five decades since this technology was developed. Their original market was on long products which tolerate higher concentration of residual elements but eventually with the use of DRI and hot metal in minimills, which for practical purposes have no-residual elements (in the order of ppm), it was possible to cover the production of the flat products market. One of the typical applications of flat products is the production of steel for automobiles. Automobile steel requires < 0.06% Cu but the typical concentration in obsolete steel scrap is about 0.2–0.4%.

The big problem of residual elements is that not only affects the final mechanical properties but also during continuous casting, due to segregation, they can promote cracks that turn the steel product into scrap. Kapoor et al. [10] reviewed the effect of residual elements during continuous casting. The segregation ratio defines the partition ratio of one element between the solid and the liquid, when this partition ratio is the unity there is no segregation, when it is lower than 1 it means the concentration of the element is higher in the liquid with respect to the solid. The higher the deviation from 1 it means the element has a greater tendency for segregation. The segregation ratio of copper and tin in delta iron is 0.7 and 0.27, respectively.

As mentioned before, due to the complexity to remove tramp elements from steel scrap, the best solution today is to optimize its collection, sorting, classification and preparation. Activities which are carried out by scrap dealers. Their daily operation is usually out of control of the steel companies; however, it has been pointed out the need to organize this work. An example of a good organization is the Swedish steel industry. Sweden produces a relatively small amount of steel, about 4.4 Mton/year, but has created a supply network with only one procurement intermediary, owned and controlled by five steel producers [11].

8.1.5 Scrap Mix Optimization

Scrap is a very heterogeneous material. The order in which scrap is placed in the buckets, represented by layers of scrap, affects boredown arc stability, collapse to the molten bath and electrode breakage.

There is no one single criteria to define the selection of scrap. The scrap mixture is based on several factors, as follows:

1. Residual elements
2. Bulk density
3. Price
4. Melting rate
5. Availability.

An ideal scrap mix should meet the constraints on residual elements, minimize the number of charges, minimize the melting time and reach the target steel composition at the lowest possible cost.

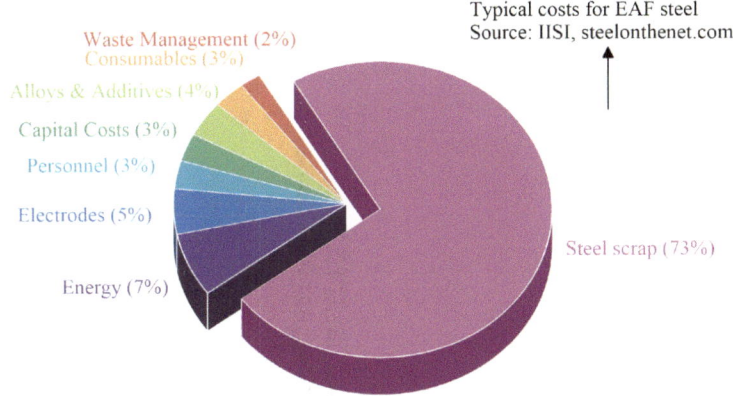

Fig. 8.3 Production cost of steel in scrap-based operations. Taken from [12]

All steel plants apply strategies for scrap mix optimization. Malfa et al. [12] report that scrap mix optimization can decrease the production costs by 2–3% simply because scrap represents the largest share in the cost to produce steel in scrap-based operations, from 60 to 80%. Figure 8.3 shows the results suggested by the International Iron and Steel Institute.

Whipp and Martin [13] in 2000 gave a clear picture about scrap mix optimization in the real practice; the first issue starts by defining which department makes the decision on the purchased scrap, is it production, quality or purchasing? Each one of those departments will make the decision on different priorities, which will move the decision to different scrap quality, production requesting scrap that can be melted faster, quality requesting scrap with the lower amount of impurities and procurement requesting the cheaper scrap. Cheap scrap will require additional fluxes, longer melting time, decrease the metallic yield and demand higher energy consumption. The application of optimization models solves the problem because such models make the decision based on the minimization of the cost of steel subject to different constraints, such as residual elements, alloying elements, scrap availability, changes in scrap prices, etc. Maiolo and Evenson[14] reported an optimization model to define the scrap mix using LSRA. Each type of scrap was characterized in terms of their energy requirements, bulk density, residual elements and price. Linear correlations to estimate energy consumption, residuals, volume and price were developed using the mass fraction of the various scrap materials. Since they reported that energy consumption varies linearly with power-on time, the impact on productivity could also be evaluated. They claimed that by introducing the optimization model it was possible to decrease the amount of scrap, power-on time and electric energy consumption. Further advantages could be added by including the benefits of metallic yield on productivity. Gyllenram and Westerberg [15] applied optimization tools to study the final cost of steel using cheap scrap. They found that the initial savings with cheap scrap are outweighed due to higher energy consumption, longer melting times, etc., suggesting prime scrap as a better choice. Further work [16] analyzing the gangue content in DRI also concluded that removing SiO_2 from the iron ore has an overall lower cost in comparison with the benefits of high quality DRI. The economic benefits of high quality DRI, decreasing SiO_2 from 4.0 to 2.0% were estimated in the order of 20 USD/ton DRI (15 USD/ton pellet), while the estimated cost for iron ore beneficiation was about 2.5 USD/ton pellet.

8.1.6 Steel Scrap Availability and Export Bans

According with the world steel association, in 2021, it is estimated that the global steel industry consumed about 2.3 billion tonnes of iron ore, 1.1 billion tonnes of metallurgical coal and 680 million tonnes (Mton) of recycled steel to produce 1.95 billion tonnes of crude steel. The consumption of steel scrap grew on average 12% per year from 1950 to 2008 from about 60 Mton per year to 450 Mton per year, within this period the rate was even higher during the last decade [17]. China, as the main producer of steel through the EAF is also the largest consumer of steel scrap; it consumed 40% of the world total in 2021 [18] equivalent to 245 Mton, in spite of having only a small fraction of the EAF share, 10.6%. The production of steel by the EAF was only 89.4 Mton, indicating that the primary steel sector (BOF) is also a large consumer. The small share for the EAF is in part due to the low availability of scrap and a national policy that has restricted imported scrap in the past. In 2021 only 0.6 Mton of scrap were imported.

In 2014, Wübbeke and Heroth [19] evaluated the situation of steel scrap recycling in China. They reported 800 take-back stations with 502 companies. Steel scrap recycling in those days was dominated by two large companies (CRD and FG). Baowu had its own affiliated company to supply steel scrap. Taylor [20] and other sources in 2020 reported the growth in the number of car shredders, increasing from 70 in 2017 to 317 in 2020.

The low availability of higher amount of steel scrap was attributed to the following problems:

- Lack of a comprehensive strategy for steel scrap recycling which includes the lack of large processing and distributions bases, contradictory policies that complicate import licensing and a fragmented supply chain
- Short domestic supply
- High import prices associated with a high turbulence in the iron and steel industry
- Outdated recycling technologies
- Low share of steel production by the EAF.

Wang et al. [21] evaluated the scrap produced in China from 1949 to 2017 reporting a total of 946 Mton of accumulated obsolete scrap in almost 70 years, by contrast, their forecast for the availability of scrap from 2018 to 2025 predicts a much larger growth, from 250 to 380 Mton of scrap. The accumulated amount of steel stock in China, according to Wübbeke and Heroth [19], from 1949 to 2010 was 4.56 billion tons with an annual growth in the range from 10 to 13%, estimating 5.75 billon tons for 2012. These values have the same order of magnitude of Japan and the USA. Most of the steel stock in China was produced from 2000 and has not reached its end-of-life.

The car industry is an important source of obsolete steel scrap. In 2002 car production was 3.2 million, increased to 7.1 million in 2006, and in 2009 China became the largest car manufacturer in the world with 13.79 million cars. The growth rate was exponential from 2008 until 2017, reaching 29 million cars per year. China had 90.86 million cars on the roads in 2012 [19]. Liu et al. [22] described in detail the steps for car dismantling according with the regulations issued by the Chinese government to guarantee a sustainable car dismantling industry. Zhao and Chen [23], reported the existence of a large number of illegal dismantlers and its negative effects not only on an efficient way to recycle steel scrap but also on bad practices which affect the environment, for example leaking waste oil into the ground, damping of other toxic substances to groundwater, burning plastic to recover copper wire, etc. Illegal dismantlers lack of technology that affects the output and quality of steel scrap since it is processed manually. The black market exists due to higher profits by the car owners. Zhang et al. [24] reported data from the Ministry of Commerce of China, indicating a very low rate of end-of-life vehicles (ELV's), about 16.5% in 2019. Their forecast for 2050 is that the accumulated amount of ELV will reach 1.48 billion, 86% coming from urban areas.

Several suggestions to increase the supply of steel scrap in China include the following:

- Optimization of recycling management,
- Development of industry standards
- Implement price subsidies to promote the use of steel scrap.

Table 8.1 shows the countries with the largest consumption of scrap according with the Bureau of International Recycling [25]. In 2021, the largest scrap importers were in order of imported tonnes; Turkey (25.0 Mton), EU (5.3 Mton), USA (5.2 Mton), India (5.1 Mton), S Korea (4.7 Mton), Pakistan (4.1 Mton), Taiwan (3.1 Mton). The largest exporters were: EU (19.4 Mton), USA (17.9 Mton), UK (8.2 Mton), Japan (4.8 Mton), Canada (4.8 Mton) and Russia (4.1 Mton). A country with a high dependance on imported scrap is more vulnerable, for example Turkey had to decrease steel production in 2007 due to disruptions in its supply chain [19].

Table 8.2 shows the consumption of steel scrap in China from 2001 to 2012 according to Chen et al. [26]. The value for 2012 represented 14% of the world total.

The forecasted worldwide growth of the EAF from its share in 2021 of 30%, according with a study conducted by Mackenzie [18], will be about 48% in 2030 and 60% in 2050. In China the expected growth is an increase to 20% in 2030, adding new EAF plants with a capacity of 80 Mton, therefore the consumption of steel scrap will be double in less than a decade. However, the gradual switch to the EAF will be global in the next decade demanding more scrap. The USA is one of the largest exporters of steel scrap with a volume of 18 Mton (16% of the global market) [18] but it is also a large consumer. In 2024 it will require 10.5 Mton of scrap. India is also expected to radically increase its imports of steel scrap. As of 2019 India had

Table 8.1 Consumption of steel scrap by country in Mton (after [25])

	2017	2018	2019	2020	2021
China	147.7	187.8	215.9	232.62	226.21
EU 28/27	93.6	90.39	86.4	75.2	87.8
USA	58.8	60.1	60.7	50.2	59.4
Turkey	30.27	31.3	27.9	30.0	34.8
Japan	35.77	36.5	33.6	29.17	34.7
Russia	29.34	31.7	30.1	30.0	32.1
S. Korea	30.67	29.9	28.6	25.8	28.2

Table 8.2 Consumption and supply of steel scrap in China, in Mton (after [26])

	2001	2002	2003	2004	2005	2006	2007	2008	2009	2010	2011	2012
Total consumption	34.4	39.2	48.2	54.0	63.3	67.2	68.5	72.0	80.0	86.7	91.0	84.0
Supply industry	13.3	13.4	15.3	17.0	22.2	27.5	26.8	28.6	30.4	33.0	35.6	36.5
Supply society	19.0	22.8	32.2	33.0	36.8	38.0	43.3	42.2	45.8	51.9	50.8	44.2
Import	9.8	7.9	5.6	7.5	7.1	4.4	1.2	2.6	10.2	4.4	5.1	3.7

47 EAF's and 1128 induction furnaces. It is the second largest producer of steel, with 118 Mton in 2021 and plans to grow to 300 Mton in 2030, with 35–40% by EAF and induction furnaces [27]. The consumption of scrap will increase from a level of 30 Mton in 2020 (27 Mton from domestic source and 7 Mton imported) to about 70 Mton in 2030. This scenario clearly illustrates why many countries have imposed restrictions to export scrap and guarantee its domestic supply. According to the GMK center [18] in 2022, 43 countries imposed restrictions to export scrap and would be expected to rise to 71 in 2023. To face the issue of drastic changes in the price of steel scrap, Yellishetty et al. [17] proposed the establishment of "scrap stabilization funds" to which steel companies/countries can contribute when scrap prices are low and withdraw from when prices are high, in order to make scrap recycling a profitable venture at all times. They also indicated that such a program has already been implemented in Canada. In addition to this they also propose new policies to organize the informal sector of peddlers/waste-pickers, which represent about 1–2% of the population in developing countries.

   An important analysis, country-wise, is to know the scrap market in order to make improvements in terms of achieving both a higher domestic scrap availability and scrap quality. For example, a study [28] about the scrap market in Sweden reported a consumption of 45% of domestic scrap by 10 steel plants (six companies). The scrap is provided by six major scrap dealers and one scrap broker (co-owned by the six companies). The typical composition of the metallic charge includes 20–30% of internal scrap, 50–60% purchased scrap and 15–20% scrap substitutes. In their study they found that trust between the steel plant and the supplier was very important, grounded on the consistency of the scrap quality provided.

The scrap availability will not be enough to cover its demand to produce steel in the EAF in the next 50 years, this is in part because steel products reach their end of life over a wide range of time, on average from 15 to 19 years [17], therefore, DRI as an alternative material is also expected to grow. The GMK center [18] reported that 23 DRI new facilities were announced for the period 2024–2026 with a total capacity of 40 Mton. The production of DRI worldwide was 114 Mton.

Pauliuk et al. [29] in 2013 reported an interesting futuristic analysis on the limits to the growth of steel production in the world and its implications on scrap demand for the next 80 years. The analysis is based on the in-use stock of steel products. They indicated that most developed countries have reached a per capita value from 10 to 16 tonnes and OECD countries from 6 to 16 tonnes. Müller et al. [30] in 2011 estimated that the global average was 2.7 ton per capita. Saturation will be achieved when the country completes urbanization and infrastructure development, estimated when the steel consumption per capita is about 14 tonnes. At this point, the demand for steel will decrease. Figure 8.4 shows the USA would reach the saturation point in 2020, western Europe in 2030, China in 2050, India in 2075, Latin-America in 2100 and Africa in 2150. As of today, the predictions are proved to be wrong but the idea of updating this analysis seems relevant to estimate the future needs of steel scrap in the world. Eventually, according with the previous analysis, the demand for steel will remain constant or decrease.

In 2013, Oda et al. [31] reported forecasts on the consumption of steel scrap and production of steel. Figure 8.5 shows historical data on scrap consumption from 1870 to 2010. From 1900 to 1940 there was a slow but gradual increment in scrap consumption reaching about 100 Mton/year. The period from 1950 to 1980 shows a sharp increment in scrap consumption, from 100 to 400 Mton/year, followed by a stagnation period from 1980 to 2000, below 400 Mton/year. The rise in steel production in China in 2000 marked a second sharp increment in scrap consumption reaching almost 600 Mton/year. If we simplify the weight of one car as 1 ton, this represents

Fig. 8.4 Final steel demand by region. After [29]

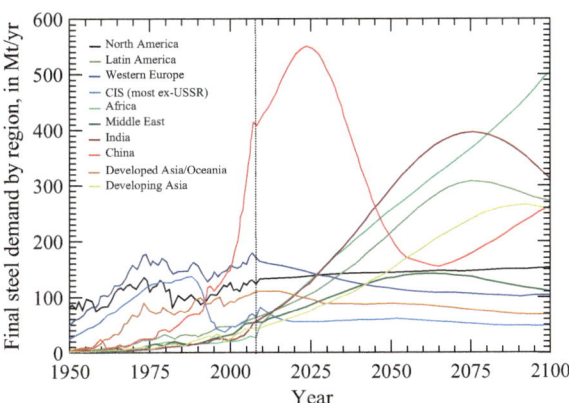

recycling 600 million cars per year. Currently, the mass of iron and steel in a car is 1058 kg (80%) and other metals 20% [32]. The forecast from Oda et al. on scrap consumption up to 2050 is shown in Fig. 8.6. According with their forecast, they estimate that 764 Mton of old scrap will be available by 2050. The total contribution from home scrap and new scrap remains stagnant at about 220 Mton, with old scrap as the main source of scrap available. By 2050 the scrap available will be about 1.1 billion ton. Their forecast for the production of steel by 2050 is about 2.3 billion ton of steel, similar to that predicted by the International Energy Agency [33] considering their two scenarios, SPS and SDS (stated policies scenario and sustainable development scenario). These scenarios clearly indicate limitations to supply the demand of scrap to produce steel and consequently they anticipate the blast furnace process will still remain as the main route to produce steel by 2050, 56% by the BOF and 44% by the EAF, as shown in Fig. 8.7. However, their analysis excludes the contribution of DRI and also another forecast predicts different values on scrap availability. Predictions from the World Steel Association [34] in 2021 indicate the end of life of steel products will increase by 600 Mton in 2030 and 900 Mton in 2050, the contribution of EU, USA and Japan will remain at about 180 Mton with the largest growth from China. In this scenario the availability of scrap is about 150 Mton higher.

In the previous forecasts the role of DRI has not been included. The growth rate of DRI has had a very dynamic growth in the last few years, from 73 Mton in 2016 to 135 Mton in 2023, a growth of 62 Mton in only 7 years. The international energy agency predicts a total production of DRI close to 411 Mton by 2050. World steel dynamics gives a more conservative value at 272 Mton.

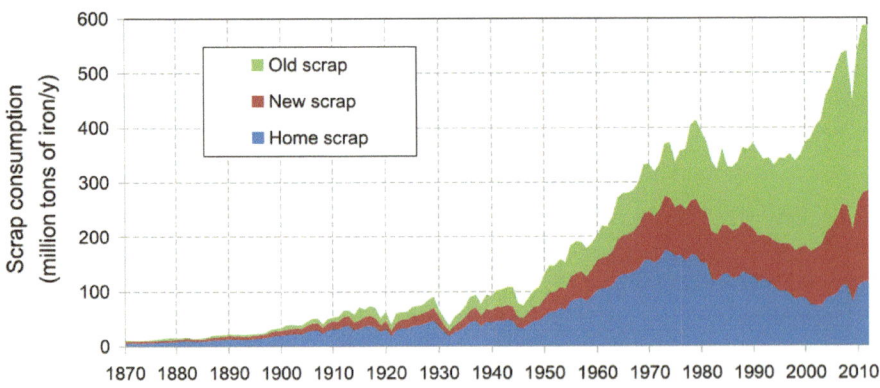

Fig. 8.5 Historical consumption of steel scrap from 1870 to 2010. After [31]

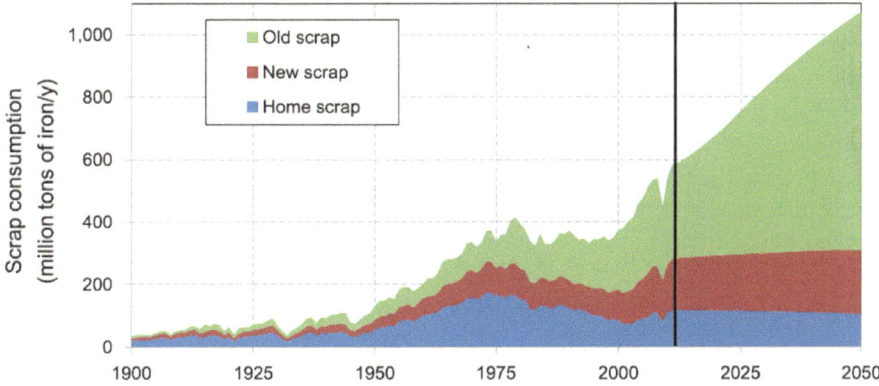

Fig. 8.6   Forecast to 2050 on consumption of steel scrap. After [31]

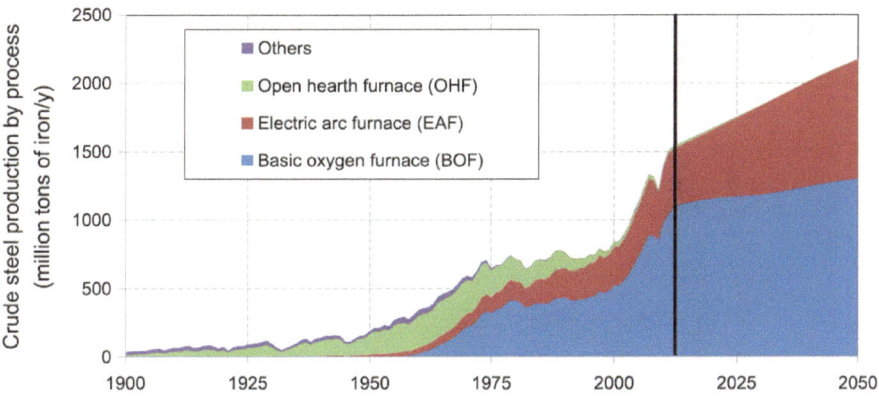

Fig. 8.7   Forecast steel production by process up to 2050. After [31]

8.1.7   *Effect of Scrap Recharges on Energy Consumption*

This is one of the most important factors to control energy consumption in scrap-based operations when scrap is charged in batches. Each time the roof is swung a large amount of heat is radiated from the furnace and heat is lost to the surroundings. Scrap charging in batches is one of the most important elements of the power-off time and affects not only energy consumption but also productivity. Jones et al. [35] suggest heat losses from 10 to 20 kWh/ton each time the roof is opened for scrap charging. Adams et al. [7] reported measurements with 1–6 scrap baskets, obtaining an average value on the heat losses of 10 kWh/ton per basket. The number of baskets depend on the bulk density of the scrap. The number of baskets or recharges is in general from 2 to 3 and power-off time of 3 min per charge. Bosley et al. [36] indicate a loss of productivity of 5% for every minute of power-off time.

8.1.7.1   Large Shell Volume EAF

Single bucket charging decreases energy consumption and also increases productivity. Laurenti et al. [37] in 2005 indicated that a medium size EAF of 82 ton of nominal capacity, operated with single bucket, it was possible to reach a power-on time less than 30 min, a productivity of 140 ton/h and annual production of 1 Mton. They employed large buckets with a volume of 140 m^3 and given the bulk density was above 0.65 ton/m^3, it was possible to get single-bucket charging. The volume of the EAF was increased to 150 m^3 (6.7 m diameter and 3.1 m height of WC panels). A standard EAF with the same capacity has a shell volume of about 80 m^3.

VAI Fuchs (Primetals) developed the model Ultimate, characterized not only by a larger shell volume but also an intense melting and refining rate capacity. The first ultimate EAF started operations in 2005 in Russia and later on in Saudi Arabia and Turkey in 2007, with capacities of 180, 160 and 250 ton, respectively. The supplier reported the performance of the EAF installed at Colakoglu Turkey [38], commissioned in 2007, with the following characteristics: charged with 100% scrap, 250 ton of nominal capacity, 45 ton hot heel, average oxygen injection 32 Nm^3/ton, bottom gas injection, transformer of 240 + 20% MVA, upper shell diameter 9 m, upper shell height 3.15 m, roof height 1.4 m, electrode diameter 711 mm and pitch circle diameter 1450 mm, shown in Fig. 8.8. The real furnace capacity was 356 ton on average with a productivity of 320 ton/h. Due its large size and scrap density, the number of charges ranges from 2 to 3, with an average electric energy consumption of 324 kWh/ton.

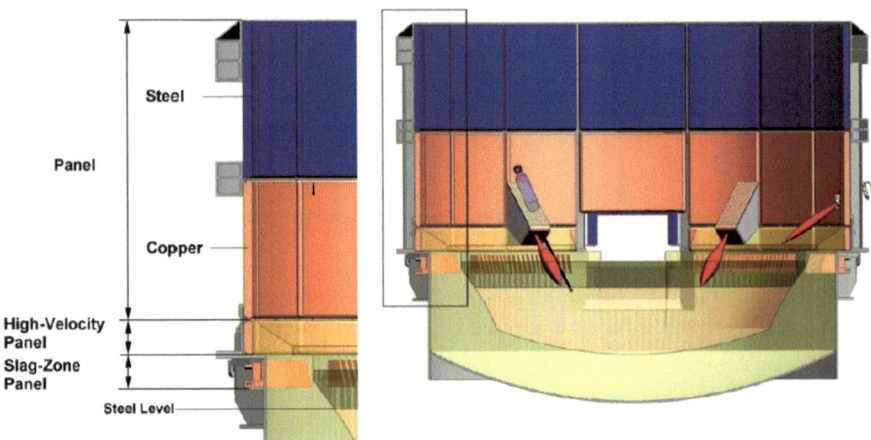

Fig. 8.8 EAF with a higher shell volume. After [38]

8.1.7.2 Telescopic EAF

Fuchs (Inteco) in 2009 [39] invented a design to avoid scrap recharges using a movable shell, as shown in Fig. 8.9. This design solves a previous limitation on the maximum shell height. The maximum height is defined by the allowable length of the electrodes submerged in the furnace, if this length exceeds 3–3.5 m, according with Fuchs, electrode breakages often occur, interrupting furnace operation. In this design the shell is formed by two cylinders, the lower one is fixed and the upper one is movable, the roof and the electrodes both have separate lifting mechanisms, also the roof is not upwardly curved but flat to which increases the immersion depth of the electrodes. When the scrap melts the level of the charge descends and with it, it is possible to maintain under control the immersion depth. The upper cylinder is liftable from 0.3 to 0.8 m.

The larger volume shell of the telescopic EAF increases the residence time of the off-gas in contact with the scrap. Its main purpose is to operate without recharges, only one bucket of scrap, however this goal has limitations depending on the scrap density available. Nal et al. [40] indicate that this is possible if the bulk scrap density is above 0.75–0.8 ton/m^3, if this is not possible the benefits of the telescopic EAF are vanished. Charging materials that increase the scrap density like pig iron or DRI are not a solution because of its lower thermal conductivity. High quality scrap of higher density is more expensive. The first telescopic EAF was installed in Bastug Turkey in 2010 with a nominal capacity of 140 ton and additional hot heel of 25 ton. Equipped with an Areva transformer of 150 MVA, secondary voltage from 951 to

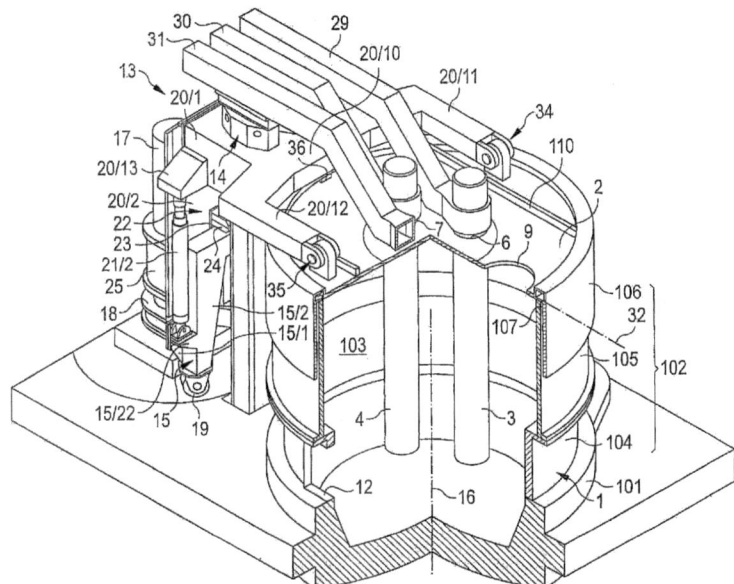

Fig. 8.9 Telescopic EAF. After [39]

1455 V, 45 MVA reactor and 2.4 ohms reactance at the highest tap, capable to get 120 MW of active power. A useful shell volume of 300 m^3 (8.5 m diameter, 5.4 m height). Electrodes of 710 mm diameter and 1400 mm pitch circle. The scrap mix has an average minimum density of 0.59 ton/m^3. With this scrap mix (using low density scrap from 0.3 to 0.4 ton/m^3) only 40% of the heats achieved the target of one single charge. The difference between one single charge and two charges results in savings on energy consumption from 40 to 70 kWh/ton. To achieve a uniform melting rate the EAF was modified with 9 burners with a maximum capacity of 5 MW using 500 Nm^3/h of natural gas and 1000 Nm^3/h of oxygen. In oxygen mode it can inject up to 2000 Nm^3/h. In 2015 the EAF was upgraded, increasing its diameter to 8.7 m. The number of single charging heats increased by 57% but still below the target of a single scrap practice. They attributed this to the lack of higher density scrap.

8.2 Fundamental Studies on Scrap Preheating

8.2.1 Models on Scrap Preheating

Zhang and Oeters [41] in 1999 reported a design for scrap pre-heating, shown in Fig. 8.10. Scrap is in a shaft EAF with a height of 10 and 4.5 m diameter, preheating time was 1 h, the reactor's capacity 50–60 ton. The authors addressed a number of reports in the 1990's which indicated a large scrap oxidation due to scrap pre-heating, therefore the objective of this work was to predict the temperature of solid scrap and the off-gas as well as the extent of scrap oxidation. The proposed preheating system involves gradual post-combustion with oxygen injection at different levels along the shaft. Scrap descends continuously at a rate of 167 ton/h, the initial scrap temperature was 25 °C and the off-gas temperature was 400 °C. The model involves heat and mass transfer analysis. Heat transfer by convection between the gas and the scrap, also heat transfer by radiation from gas to scrap and between scrap pieces. To simplify the model, it was assumed that voids are uniform and spherical, also heat transfer by conduction within the scrap pieces was neglected because it is only relevant for large pieces of scrap. Iron oxidation was neglected if the temperature was below 600 °C. The rate of iron oxidation was defined based on the two following reactions:

$$Fe_{(s)} + CO_{2(g)} = FeO_{(s)} + CO_{(g)}$$
$$Fe_{(s)} + H_2O_{(g)} = FeO_{(s)} + H_{2(g)}$$

In Fig. 8.11a it can be observed that the maximum scrap temperature reaches the melting temperature at a height of 10 m, with an average value of 1200 °C. The maximum iron oxidation is about 250 kg/h which represents only 0.2% of the production rate of 167 ton/h. this result is quite different to other reported data indicating values of 10%, however this is due to the continuous movement of the scrap and a short residence time with oxidizing gases at the highest temperatures.

Fig. 8.10 Shaft design employed in the model developed by Zhang and Oeters. After [41]

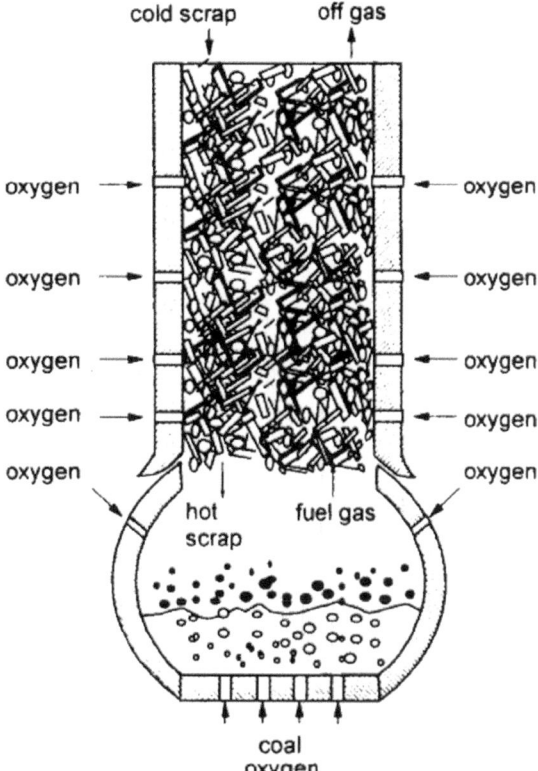

Scrap in shaft EAF's remain static until it is charged and therefore a higher oxidation rate is possible, however the same approach can be used to define the oxidation behavior in continuously charged processes.

The fist experimental studies on scrap preheating with burners were carried out by Mandal and Irons [42–45] from 2008 to 2013. The experimental furnace has a volume of 1 m³ and the dimensions are shown in Fig. 8.12. The side wall where the burner is inserted has two brick layers. The fuel employed was propane. The experimental work was designed to study the effect of power burner and scrap size on the temperature distribution and thermal burner efficiency. Four types of scrap were investigated; small shredded, large shredded, bushelings and heavy scrap. The burner power employed were 8.3, 12.8 and 17.3 kW. The temperature distribution was measured introducing thermocouples at one side of the furnace, as shown in Fig. 8.13. A data acquisition system captured data every 30 s and to interpolate/extrapolate the temperature distribution on the whole furnace, the software TECPLOT was employed. The burner thermal efficiency was computed comparing the temperature of the exhaust gas with the adiabatic flame temperature.

Figure 8.14 shows the results of temperature distribution in °C after 60 min with a top cover. It can be observed that the volume of preheated scrap increases as the

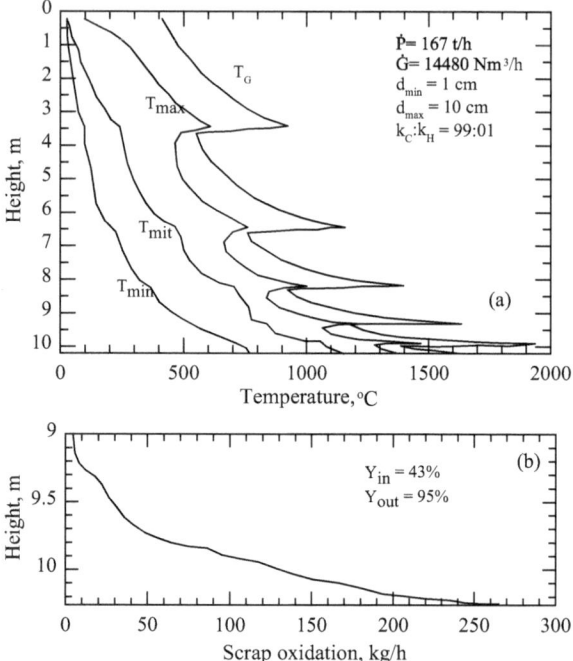

Fig. 8.11 Model results on scrap preheating. After [41]

size of the scrap decreases, this is because the A/V ratio increases. Figure 8.15 shows the decrease in thermal efficiency as a function of time. It can be observed that the temperature of the off-gas is higher using heavy scrap which indicates lower heat transfer to the scrap, consistent with a decrease in thermal efficiency. At 60 min, the temperature difference between small shredded scrap and heavy scrap is about 80 °C.

At the time of developing its numerical model, Mandal and Irons [45] argued that commercial packages were unable to handle the complexity of the problem, defined in the following terms:

- Particles are non-uniform and much larger than in other applications
- Gas velocities are also higher, flow is non-Darcian ($Q \neq U/A$) and turbulent
- Porosity is much higher than experienced with regular or spherical shaped particles
- Porosity is spatially non-uniform and changes with time.

Initially, they found that the measured temperatures of the thermocouples were lower than that of the predicted gas. Thermocouples receive heat from the gas and radiate heat to the scrap and were included as part of the computational domain. According with the model, for the same particle size but different porosity, they found that decreasing the porosity improved the heat transfer efficiency, for example for a porosity decrease from 94 to 86%, the heat transfer coefficient increased from 265 to 651 W/m³K, suggesting some advantages of scrap of higher density.

Nal et al. [40] observed the following practical aspects on scrap pre-heating:

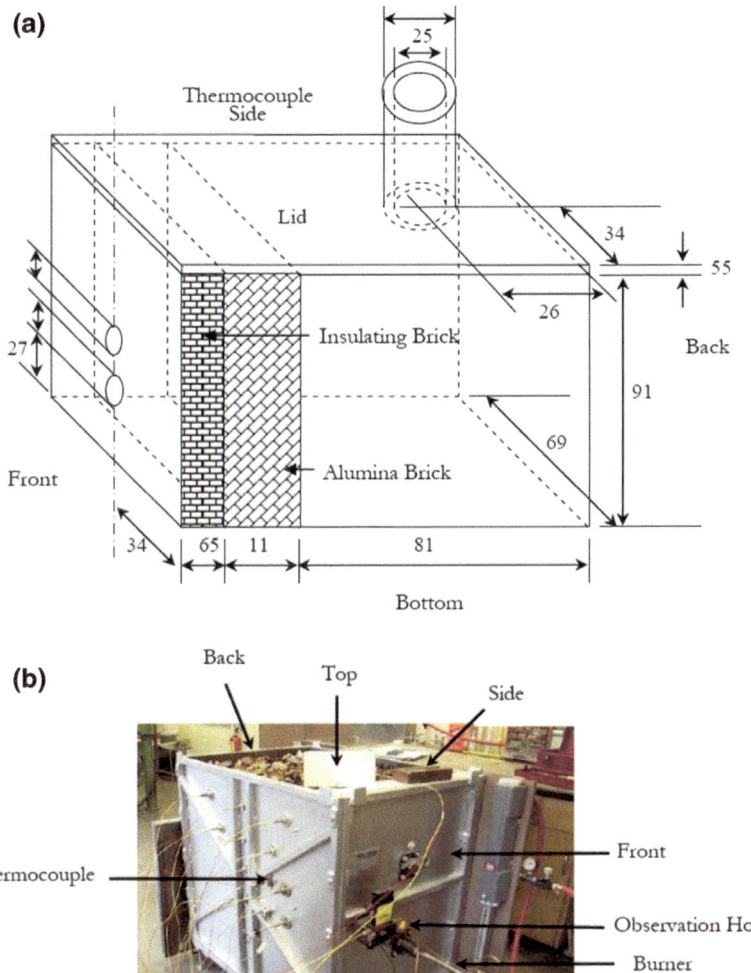

Fig. 8.12 Experimental set up to study scrap preheating; **a** top and **b** bottom. After [42]

- Scrap temperature reaches a maximum value about 8 min of burner operating time, corresponding to maximum burner efficiency of 85–90%.
- Smaller scrap size is heated faster because of its larger surface/volume ratio
- If more heat is transferred to the scrap, the temperature of the off-gas will be lower
- More burner power produces more heating.

Scrap preheating and further melting is controlled by four primary driving forces; temperature gradients, concentration gradients, frozen-shell formation, further

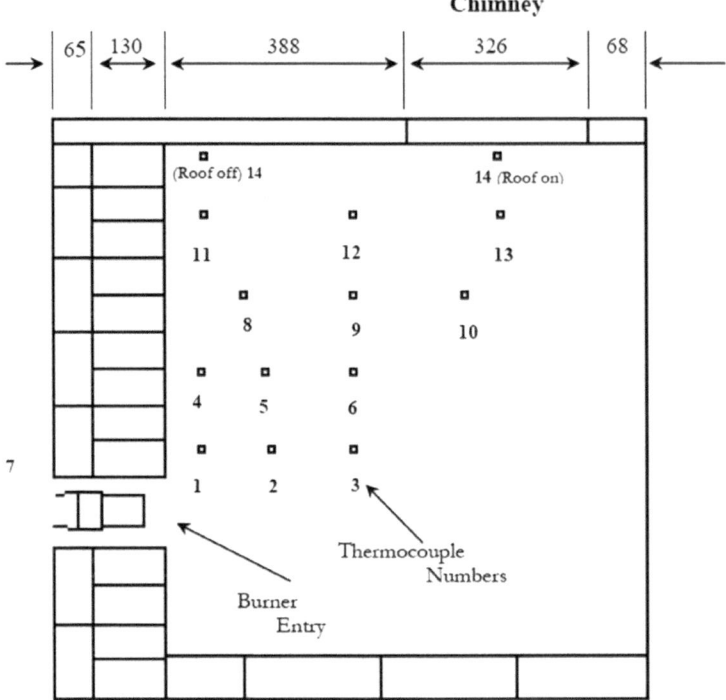

Fig. 8.13 Location of thermocouples. After [42]

melting when scrap pieces reach the liquid bath, and the velocity field in the molten bath (stirring conditions). The temperature gradient is related to the temperature of the scrap and its melting temperature. Li et al. [46] developed a model in which the scrap pieces were immersed in the hot heel, scrap is heated from the top and cooled from the bottom. This model suggested that in order to increase the melting rate the scrap should have lower scrap bulk density, increase the hot heel depth, and stirring drastically decreases the melting time.

8.2.2 Effect of Scrap Oxidation on Energy Consumption

Tang et al. [47] investigated the oxidation of scrap due to oxyfuel burners. Scrap oxidation has different effects; during scrap preheating the oxidation reaction increases the preheating temperature and decreases the scrap melting point, however during melting the oxide has to be reduced increasing energy consumption and if it is not fully reduced it will decrease the metallic yield. They validated their model using the experimental data from Mandal and Irons [44]. Iron oxidation was based on the following reaction:

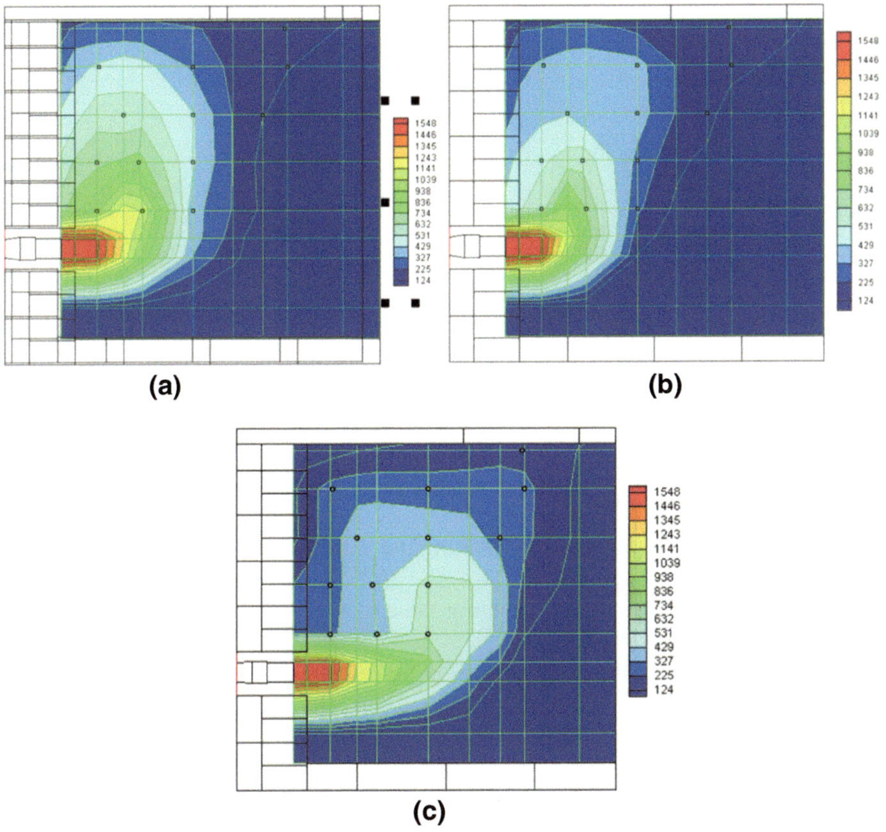

Fig. 8.14 Temperature distribution during scrap preheating with a burner of 12.8 kW. **a** small shredded, **b** large shredded and **c** heavy scrap. After [42]

Fig. 8.15 Exhaust gas temperature and thermal efficiency during scrap preheating. After [42]

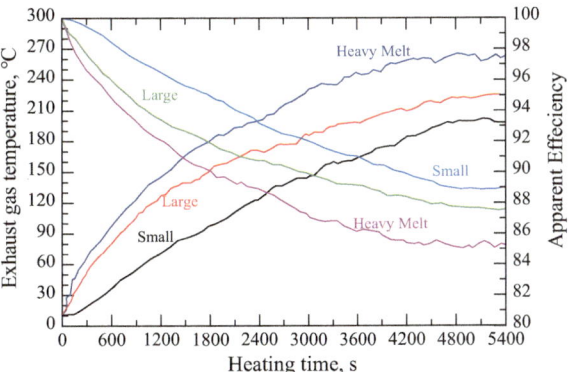

Fig. 8.16 Degree of scrap
oxidation using oxyfuel
burners. After [47]

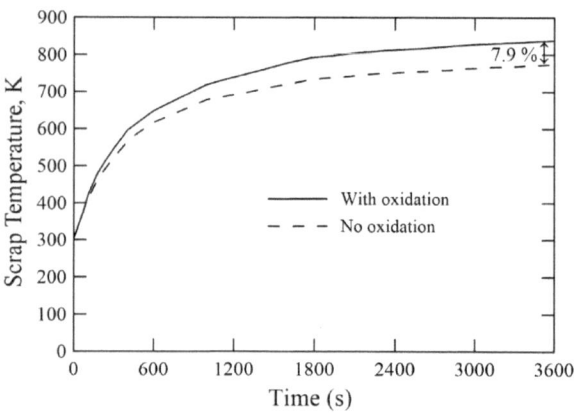

$$Fe_{(s)} + 1/2\,O_{2(g)} = FeO_{(s)};$$

The oxidation rate was calculated, taking the value of the rate constant equal to 4 $\times$ 10^{-7} g^2/cm^4 s.

Figure 8.16 shows the results showing the extent of iron oxidation after 60 min, which corresponds to 7.9%.

According to Adams et al. [7] the reduction of 1% Fe_3O_4 (either original rust or oxidized during preheating) consumes 13 kWh/ton. The effect of water, snow and ice is similar because it also oxidizes the scrap. The removal of water vapor at 1000 °C requires additional energy, about 30 kW/ton.

8.3 Scrap Pre-heating Technologies

8.3.1 Sensible Heat and Chemical Energy in the Off-Gas

Scrap preheating can be carried out simultaneously in several ways:

- Using oxy-fuel burners
- Using post-combustion
- Sensible heat from the off-gas.

The off-gas system is composed of the primary off-gas extraction system using the fourth hole and a secondary system that captures the fugitive emissions to the meltshop. Jones et al. [48] report the following values for a 120–150 ton EAF:

1. Off-gas flowrate: 850–1130 Nm^3/min
2. Off-gas temperature: 1700–1930 °C
3. Off-gas composition: 5–15% H_2, 10–30% CO
4. Rate of heat extraction: 0.879–1.465 kWh/min.

The off-gas from the primary system is water cooled to temperatures below 650 °C, then using heat exchangers its temperature can be below 315 °C. It reaches the baghouse at temperatures below 130 °C [48].

The sensible heat in the off-gas can be in the range from 10 to 30% from the total energy input. This energy is lost, particularly during flat bath operation because the heat is not transferred to the scrap. This energy can be employed for preheating or other applications such as steam production. EAF operations that involve more chemical energy (higher C-DRI, hot metal) produce a higher amount of energy in the off-gas.

In addition to the sensible heat in the off-gas there is another component that should be considered, its chemical energy as a fuel, as shown in Fig. 8.17, which indicates that both the sensible heat and the chemical energy in the off-gas ranges from 25 to 35% [49]. These results correspond to EAF's from the USA. An estimate, assuming a conservative value of energy losses in the off-gas of 30%, indicates that it corresponds to about 3% of the total energy employed in the US steel industry and its economic value about 182 million US dollars per year [49].

The amount of energy required to produce steel at a tapping temperature of 1620–1640 °C is about 390 kWh/ton. This energy is divided in energy to bring the metallic charge from room temperature to its melting point (1530–1540 °C), energy to provide the heat for the melting step and the rest is the heat to bring it to the tapping temperature, their corresponding values are; 294 kWh/ton for heating, 75 kWh/ton for melting and 25 kWh/ton for superheat [50]. It is clear that the largest amount of heat is involved during the heating process.

Yang et al. [51] describes in detail a larger number of applications for the energy in the off-gas of EAF's. One of the applications of this energy is in scrap pre-heating. These authors estimate energy savings in the order of 15 kWh/ton for every increment of 100 °C in the temperature of scrap, in addition to more advantages resulting from an increment in the melting rate. This value would yield energy savings in the order of 75

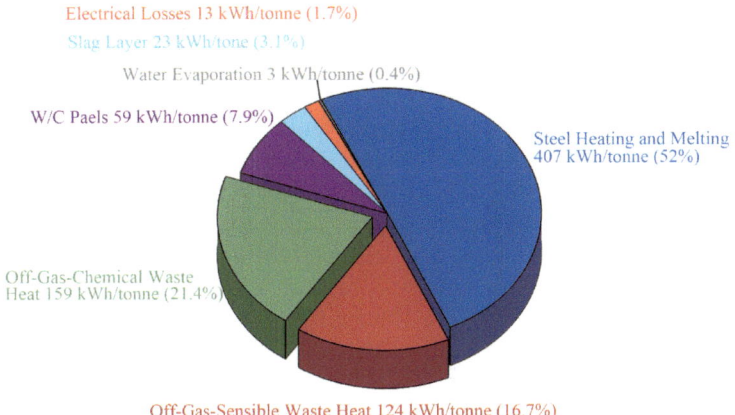

Fig. 8.17 Energy output from EAF's in the USA. After [49]

kWh/ton if the scrap is preheated to 500 °C. UCAR compared energy consumption using preheated scrap and found energy savings in the order of 40 kWh/ton. The systems included Shaft, Consteel and bucket preheaters, the range on energy savings was from 30 to 50 kWh/ton [7]. Considering both references, most probably an average value on energy savings lies in the order of 10 kWh/ton for every increment of 100 °C.

Scrap preheating decreases the thermal efficiency of both post-combustion and burners because the temperature gradient is decreased. Scrap preheating has rapidly evolved in the last 40 years. The first attempts at recovering the energy from the off-gas gas for scrap preheating was using scrap buckets. Today, the number of available technologies is quite extensive. There are several reviews on this subject [35, 36, 50, 52–55]. A summary of these technologies is given below.

8.3.2 Scrap Bucket Preheating Systems

Due to high electricity costs, plants in Europe and Japan started to carry-out scrap preheating using scrap buckets, since the late 1960's. The EAF off-gas and combinations of off-gas and natural gas were employed. Natural gas scrap preheating can increase scrap temperature to 540–650 °C. Scrap oxidation is severe if the scrap preheating temperature is above 650 °C [35], subsequently, scrap oxidation decreases metallic yield.

Scrap bucket preheating is a simple way to recover the sensible heat from the off-gas, however it has many limitations: (1) Tap-to-tap time is the first element to consider to evaluate the feasibility of scrap bucket preheating. If this time to preheat exceeds the tap -to tap time it will be inefficient. (2) Shorter bucket life, (3) Sticking problems, and (4) large temperature gradients from top to bottom.

Plastics and oily scrap which ignite at about 260 °C, release hydrocarbons which contribute not only with foul smells into the environment but with toxic dioxins. Dioxins is the generic name of a group of toxic polychlorinated dioxins (PD) and polychlorinated furans (PF) which persist in the atmosphere for a long time. Some types of dioxins are extremely toxic to human health; they can cause cancer, reproductive system damage, develop mental problems, damage to the immune system, and can interfere with hormones. However, dioxins can be destroyed using post-combustion and quenching the gas. Dioxins are destroyed at high temperatures (above 850 °C) but can be formed again during cooling in the range from 200 to 600 °C [56]. Therefore, the off-gas should be quenched below 200 °C.

Typically, the off-gas leaves the furnace at 1200 °C, it enters the bucket at 815 °C and leaves the bucket at 200 °C [36]. The amount of heat-exchange depends on the time and the heat transfer coefficient which is a function of the scrap size. Scrap can be preheated in a range from 300 to 450 °C, reducing energy consumption from 40 to 50 kWh/ton. Oily scrap is not adequate to preheat in the bucket because it will ignite and create problems, unless the system is designed to handle it.

It has also been observed that FeO in the dust sticks to the surface of the scrap as a consequence the dust generation rate is decreased. Jones et al. [35] reports 20% lower dust, from 18 to 14 kg/ton and enrichment of ZnO from 22 to 30%.

Danarc process: Danarc is a design developed by Danieli and consists of a single charging scrap bucket using a tall EAF and a tall scrap bucket. The volume of both vessels contain similar amounts of scrap. The scrap bucket is located in a position similar to the height of the EAF roof and can be placed on the top to discharge the preheated scrap. The scrap bucket is water cooled. Preheating is enhanced using burners. Some performance results obtained at ABS were reported by Danieli [57, 58] in 1999 that show improvements but the values reported for the same company are different; one report indicates a decrease in energy consumption from 359 to 302 kWh/ton (savings of 57 kWh/ton) and decrease in power-on time from 34 to 30 min (4 min shorter) while in a second report the energy savings are 70 kWh/ton and 3 min shorter power-on time.

8.3.3 Shaft Pre-heating and Continuous Charging Systems

8.3.3.1 EOF

The first idea of a vertical shaft for scrap preheating was the development of the Energy Optimizing Furnace (EOF) in the early 1980's in Brazil by CSP [59] used in the production of steel from hot metal and large amounts of preheated scrap. This development is credited to the guidance of Willy Korf [60], a German entrepreneur who was the first to embrace the Minimill and who also contributed to the development of the Midrex process. The large shaft from the EOF is placed on top of the furnace and is divided into several chambers. The scrap temperature reaches 850 °C. The furnace can operate with up to 40–50% of scrap and the difference is hot metal.

Fruehan and Nassaralla [61] in 1998 provided some specific information about this process: Scrap is preheated in a series of preheat chambers, reaching 800–1200 °C. Hot metal is charged followed by scrap from the lower preheating chamber. A 50/50 hot metal/scrap operation required about 70 Nm^3 O$_2$/ton, one third to decarburize the hot metal and the rest for post-combustion using 20 kg coal/ton. The blowing time is about 27–32 min and tap to tap time less than 60 min. Two EOF's were operating in Brazil, one EOF of 80 ton was operated by Tata steel India and one EOF in Italy. The productivity is equivalent to a modern EAF however sulphur is a problem. The EOF is shown schematically in Fig. 8.18.

8.3.3.2 Fuchs Furnace

The concept of the shaft scrap-preheating furnace was developed by G. Fuchs from Germany. He had a company that built scrap buckets for scrap preheating. His patent

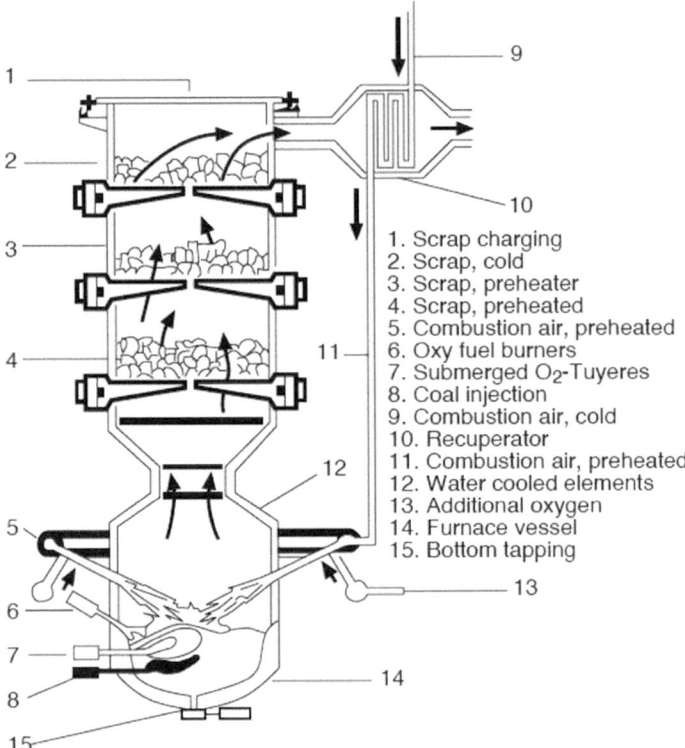

1. Scrap charging
2. Scrap, cold
3. Scrap, preheater
4. Scrap, preheated
5. Combustion air, preheated
6. Oxy fuel burners
7. Submerged O_2-Tuyeres
8. Coal injection
9. Combustion air, cold
10. Recuperator
11. Combustion air, preheated
12. Water cooled elements
13. Additional oxygen
14. Furnace vessel
15. Bottom tapping

Fig. 8.18 EOF process. After [61]

on the shaft furnace was from 1986 [62]. Most frequently shaft furnaces are DC-EAF's with one central electrode [63], as shown in Fig. 8.19.

The first AC shaft EAF was installed in 1991 in the UK. The EAF is fully charged with scrap, both the interior and the shaft regions. Scrap gradually descends depending on the melting rate. A new scrap basket is charged when there is enough space.

The first DC *finger shaft EAF* was installed in 1994 in Hylsa Mexico. The EAF has a nominal capacity of 150-ton, hot heel of 44 ton (30%), power transformer of 156 MVA and three bottom electrodes. The average metallic charge consisted of 45% scrap and 55% DRI. The finger has to be water cooled, consuming about 15 kWh/ton. Scrap-retaining "fingers" suspend the scrap within the shaft. The fingers flip down to discharge the preheated scrap. After tapping, preheated scrap from a previous heat is discharged into the EAF. As the scrap melts, it descends, leaving space for the fingers to be closed and charged again to preheat a second scrap basket. A maximum preheating temperature of 400 °C has been estimated for scrap in the finger shaft EAF [55]. Hylsa Mexico experienced severe problems with the fingers. Jones et al. [35] reported the following consumptions; electric energy consumption

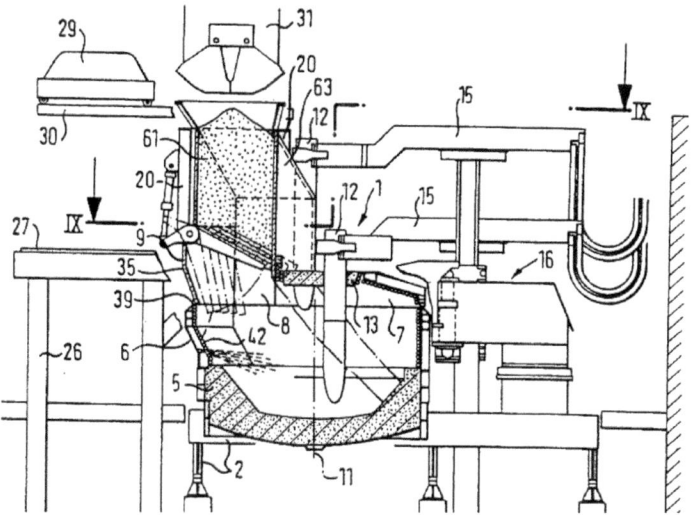

Fig. 8.19 Fuchs shaft EAF. After [62]

394 kWh/ton, electrode consumption 0.9 kg/ton, oxygen consumption 20 Nm^3/ton, carbon for slag foaming 4.5 kg/ton, power-on time 48 min.

8.3.3.3 Fuchs COSS Shaft EAF

The first prototype of the Continuous Optimized Single Shaft Furnace (COSS) was tested in 2006. The shaft is separated from the furnace shell and doesn't have water cooled panels. It uses a pusher for the continuous charging of the preheated scrap. The portion of scrap at the bottom of the shaft has the highest temperature and is the one charged in batches. A limitation reported in this process is the use of water cooling in the hydro-cylinders and pushers because as the preheating temperature increases above 450 °C there are more risks of water leakages [55]. The shaft is mounted on a transfer car, sitting on load cells. Prior to tapping the shaft doesn't have to be removed. There is a by-pass for the off-gas entering the shaft. It can be used during shaft maintenance and to increase the temperature of the gas leaving the shaft prior to post-combustion.

The first unit was installed in 2007 in the Shagang group, China. The shaft volume allows to preheat 90 ton of scrap, potentially reaching temperatures from 800 to 1000 °C. The furnace is operated with 40% hot metal and 60% pre-heated scrap. Includes 5 coherent jets for decarburization. Electric energy consumption falls to less than 100 kWh/ton with 45% hot metal and 40 Nm^3/ton of oxygen. During the scrap and hot metal charging period no electricity is used. The supplier claims a guaranteed electrode consumption less than 0.5 kg/ton This is in part due to the elimination of electrode breaks due to fallen scrap. Additionally, it is also claimed a lower investment

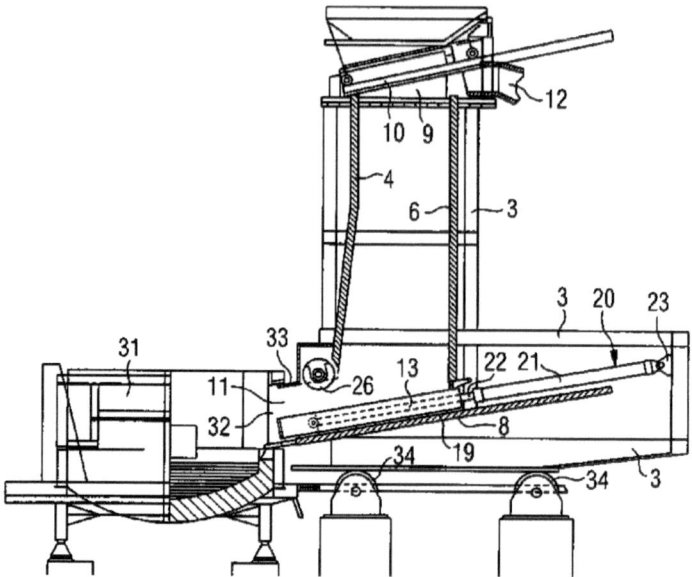

Fig. 8.20 Fuchs COSS EAF. After [65]

cost compared to Consteel [52]. INTECO acquired the rights of Fuchs systems in 2015. CVS, a Turkish company has reported its own technology which is alike to the concept of COSS [64]. The COSS shaft EAF is shown schematically in Fig. 8.20.

8.3.3.4 Quantum Finger Shaft EAF

Primetals, a merger between Mitsubishi and Siemens VAI in 2015, continued the development of the Quantum design. The first furnace started operations in Tyasa Mexico at the end of 2014. This is also a finger shaft EAF, the main difference is the design of the fingers and the gradual scrap feeding of the shaft through an automatic inclined elevator. The fingers, similar to pitchforks, are introduced into the shaft through its sidewalls, as shown in Fig. 8.21. A report on energy consumption indicates values in the range from 280 to 300 kWh/ton [66].

8.3.3.5 Double Shaft Arc Furnace (SHARC)

This design was developed by the German company SMS. A previous version, developed by their former company Mannesmann Demag, was the process *CONTIARC*. Scrap arrives to the top by conveyors, taken by magnets and discharged throughout the ring area. The electrode is protected in an enclosed case.

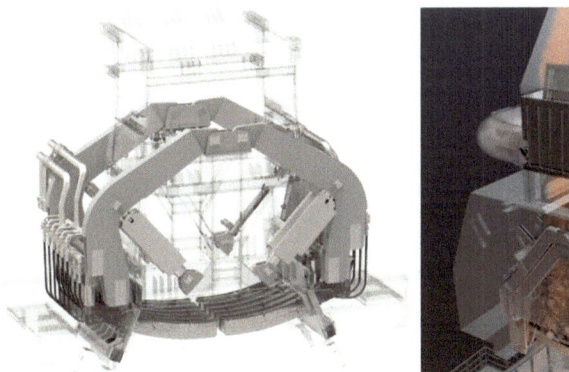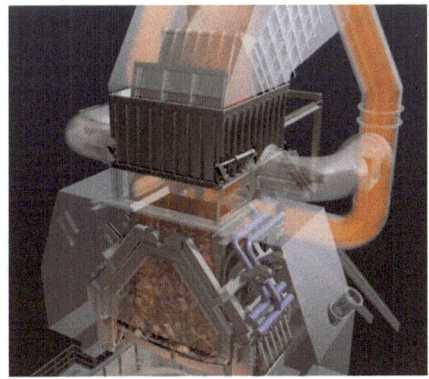

Fig. 8.21 Quantum EAF. After [66]

The main feature of a SHARC furnace is two identical preheating shafts vertically arranged which results in a symmetric and more homogeneous scrap distribution, as shown in Fig. 8.22. It contains 34 water cooled fingers that can be retracted all together or individually. It is a DC furnace with an electrode of 700 mm (28″). The scrap falls around the electrode, using a cover to protect the arm of the electrode. Scrap is charged into the two shafts and preheated by a previous heat. After tapping the preheated scrap is discharged and on top of this scrap is charged cold scrap which will be preheated during the melting period of the underlying scrap. Once the scrap starts to melt and descends, the fingers are extended again to hold and preheat the scrap for the new heat. The performance indicators of a SHARC furnace operating in Greece are the following [67]: average preheating temperature, 550 °C, energy consumption from 290 to 360 kWh/ton, electrode consumption 0.6 kg/ton. The scattering on energy consumption is attributed to an intermittent operation. The hot off-gas is removed from the furnace at both sides directly below the charge openings. Odenthal et al. [68] reported a numerical model to describe the fluid dynamics due to oxygen injection. They found that the gas temperature in both shafts is not uniform, with an average value of 1040 °C.

8.3.3.6 IHI Shaft EAF

This process was developed by the Japanese company Ishikawajima-Harima Heavy Industries (IHI) in 1996 [69]. It has some similarities with the COSS shaft EAF. The preheated scrap uses two scrap pushers. A DC EAF furnace installed in Japan operates with two electrodes. The power feeding bus is arranged in a way that the two arcs deflect towards the center of the furnace and the thermal load at the walls is lower, allowing for the use of refractory brick which decreases energy losses. Scrap falls in the center of the furnace between the two electrodes. Preheating is carried out in two

Fig. 8.22 SHARC-EAF. After [68]

chambers, a charging chamber connected to the EAF and a preheating chamber. The charging chamber has two rows of pushers. Scrap is charged continuously. There is almost no data on this process and most probably the process has been dicsontinued. Jones et al. [35] reported the following value; the hot heel is large, 110 ton, tapped steel is 140 ton, energy consumption about 236 kWh/ton. IHI transferred the EAF business unit to Paul Wurth in 2014. The IHI shaft EAF is shown schematically in Fig. 8.23.

8.3.3.7   ECOARC Shaft EAF

Developed by NKK, currently Steel Plantech an Engineering Company founded in 2001 as a result of a merger between JFE engineering, Hitachi, Kawasaki and Sumitomo. The shaft is connected directly on top of the melting chamber forming a single unit. It doesn't have fingers or pushers to transfer the preheated scrap allowing to increase oxygen consumption and obtain higher temperatures in the off-gas without deformation of any finger or pusher [70]. Scrap can be charged continuously into the preheating shaft and there is always solid scrap in the furnace in contact with the melting chamber. As the scrap melts, it descends from the shaft. Since there is solid scrap in the melting chamber the temperature of liquid steel is low, from 1500 to 1540 °C therefore to start the refining period the furnace is tilted 15°. The arc is never in contact with solid scrap. Solid scrap melts when enters in contact with superheated liquid steel [70]. It operates with a low noise, below 100 dB [71]. In

Fig. 8.23 IHI process. After
[35]

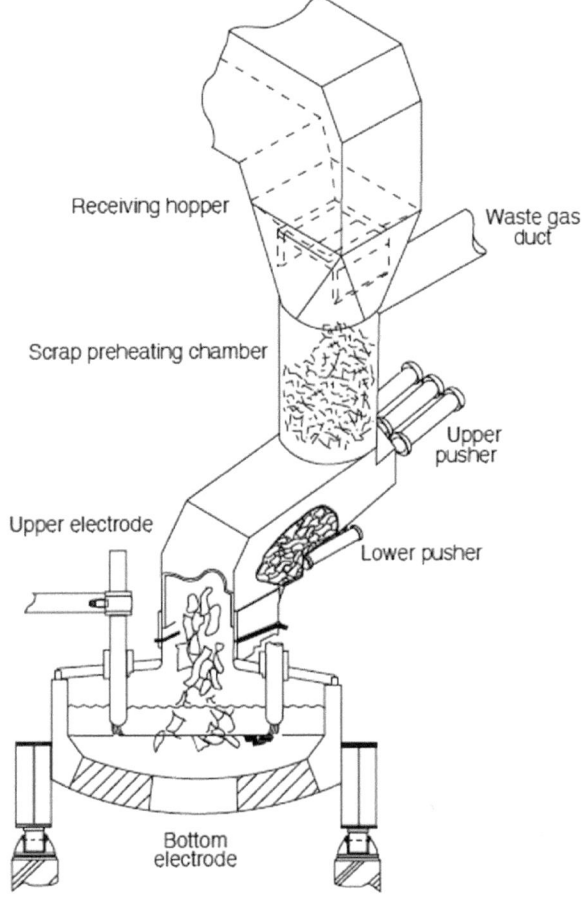

Receiving hopper

Waste gas
duct

Scrap preheating chamber

Upper
pusher

Upper electrode

Lower pusher

Bottom
electrode

2015 there were operating 6 plants, four of them in Japan, using AC-EAF's with a single electrode. The supplier claims savings in electric energy consumption from 100 to 150 kWh/ton and a target of electric energy consumption of 210 kWh/ton [72]. A plant from Korea reported a scrap pre-heating temperature of 800 °C and energy consumption below 300 kWh/ton [54]. The ECOARC shaft EAF is shown schematically in Fig. 8.24.

8.3.3.8 COMELT Process

The COMELT process was developed by VAI (currently Primetals) in the early 1990's. The main characteristic of this design is the presence of four inclined electrodes using DC current, as shown in Fig. 8.25. This design has the advantage to withdraw the electrodes to charge scrap in the center. The hot gases raise and preheat

Fig. 8.24 ECOARC
process. After [54]

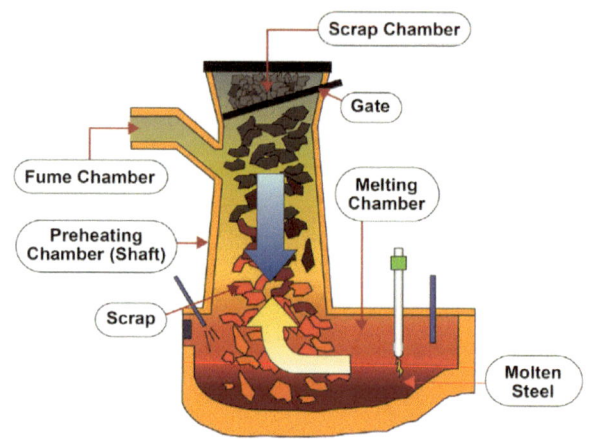

the scrap. Müller et al. [73] claimed improvements over the conventional EAF, such as 30% lower electrode consumption and lower energy consumption. The process didn't pass the pilot plant test most probably due to huge capital investments, however the idea of a different electrode layout opens the possibility to explore new ideas in this subject.

Fig. 8.25 COMELT
process. After [73]

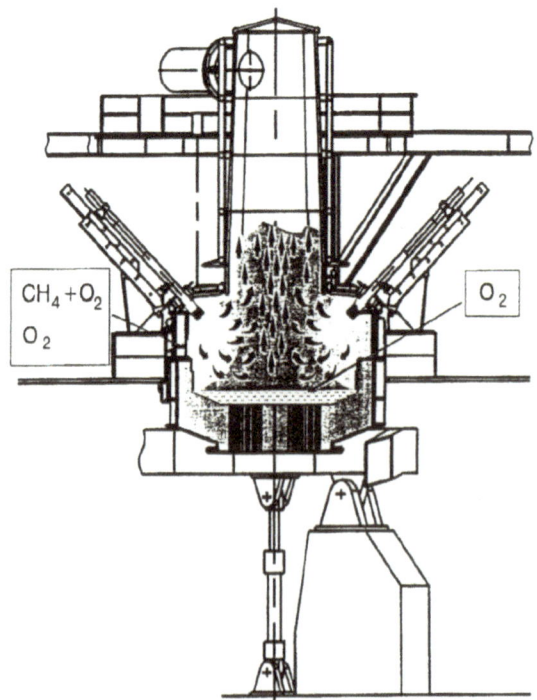

Fig. 8.26 Twin-shell process. After [35]

8.3.4 Two Furnaces Coupled Together

8.3.4.1 Twin Shell EAF

It weas developed by Nippon steel in the early 1980's. In this design one EAF is used
to preheat the scrap using the off gas from a second EAF and both have a common
transformer. The process commonly uses one charge of scrap, in this case the EAF
works below its nominal capacity. When there are two charges of scrap the second
charge is cold or preheated externally in a scrap bucket. The first trials were conducted
by SKF in Sweden and Nippon Steel in the early 1980's. Today several suppliers
offer this design. In order to improve arc stability at the beginning of scrap melting,
a vertical burner can be used to preheat the scrap (O/gas = 4), producing a hole
1500 mm deep (Köhle et al.). A DC-EAF has the advantage that only one electrode
is moved. Some limitations of this design are the higher investment cost and higher
maintenance [55]. The twin shell EAF is shown schematically in Fig. 8.26.

8.3.4.2 CONARC: The Twin BOF-EAF Process

This process was developed by Mannesmann Demag Huettentechnik now SMS and
the basic idea is to use the best of both the BOF and EAF processes; a high melting
rate of the EAF and the high decarburization rate of the BOF, using the vessels of an
EAF. The EAF employed for decarburization has one or two top lances for oxygen

Fig. 8.27 CONARC
process. After [75]

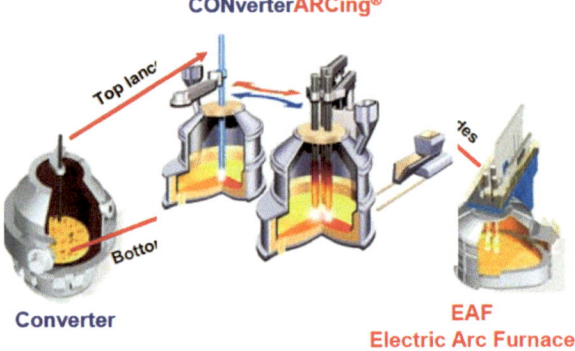

injection. The process reached commercial operation in 1998 with units operating in Japan and South Africa. Kappes [74] in 2000 claimed a successful operation at the commercial scale in South Africa using hot metal from a Corex plant, from 0 to 70%, with oxygen flow rates in the order of 130 Nm³/min. The EAF capacity was 170 ton. Kleinschmidt et al. [75] reported what appears to be the last information from this process at the commercial scale. Construction of another plant in 1999 in Dolvi India, each EAF with a nominal capacity of 180-ton, feed with hot metal and DRI. The design included two top lances to increase the amount of hot metal. Plans for erection of another plant in Orissa India in 2007 were indicated in the same report and in 2012 a steel magazine further disclosed the construction of a third plant at Essar's Hazira in India. Figure 8.27 shows a 3D perspective of the process.

8.3.5   Pre-heating and Continuous Charging Systems

8.3.5.1   Brusa Process

This process was invented by U. Brusa from Italy in 1974 [76]. It is a predecessor of the Consteel process. Scrap is continuously charged into the EAF. Scrap preheating occurs in a rotary furnace connected to the EAF. Preheating is carried out by both the off-gas and the use of natural gas. An installation for a small capacity EAF with 36 ton used a 13 m long rotary furnace, 5 burners at the bottom and the residence time was short, from 6 to 10 min. The use of natural gas was intense, equivalent to 73% of the total heat [50]. The temperature of the scrap was raised up to 450 °C with the off-gas and close to a 1000 °C using natural gas [77]. The rotation of the scrap inhibited sticking problems but it was a problem at high temperatures. The decrease in electric energy consumption was high, in the order of 220 kWh/ton using 30 Nm³/ton of natural gas, however, the concept of a rotary furnace implies several limitations [50]: First, bigger furnaces would require a large rotary furnace and in turn a large height for the meltshop and second, operates only with special fragmentized scrap, all of which increases the cost of steel.

8.3.5.2 CONTIARC

The CONTIARC process was developed by Mannesmann Demag Huettentechnik now SMS, with pilot plant tests in 1996 at Aachen university. It is a DC-EAF with the central electrode protected from falling scrap with an annular wall, scrap is continuously charged by a conveyor to the ring shaft where is picked up by a series of magnets and is distributed evenly throughout the ring shaft area. There is always solid scrap in the furnace which protects heat radiation to the walls. The scrap is preheated by the rising gases and due to the large amount of scrap it provides a large surface area which decreases the amount of dust. Hofmann et al. [78] reported that according to design the expected energy consumption would be about 200 kWh/ton, up to 40% less dust which will also decrease the capacity of the gas cleaning system and lower electrode consumption, about 0.8 kg/ton less than AC-EAF. The CONTIARC EAF is shown schematically in Fig. 8.28.

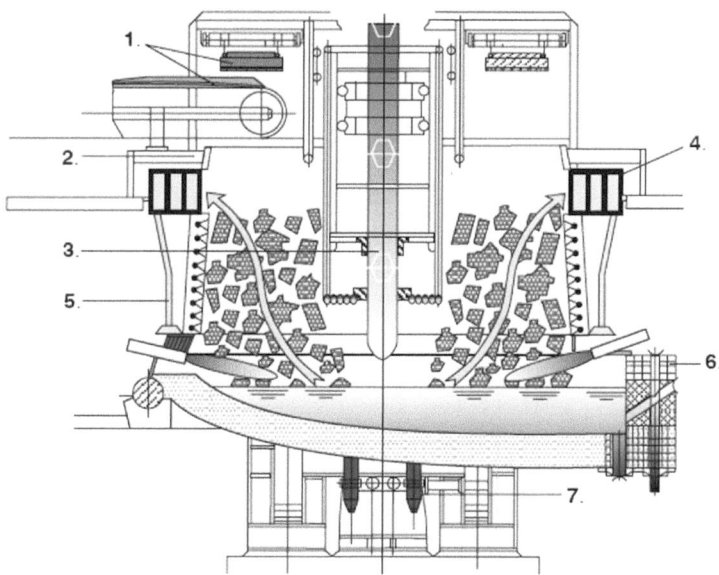

Fig. 8.28 CONTIARC process. After [35]

8.3.5.3 CONSTEEL

The Consteel process was developed by J. Vallomy in 1985 [79]. It is one of the most accepted technologies for scrap preheating with a total of 79 installations in 2020 [80]. Scrap is preheated on an oscillating conveyor in a preheating tunnel and charged continuously into a large hot heel. The EAF can be operated at full power throughout the heat. Hot heel is in the order of 40% of liquid steel in the EAF. Fluxes are added through the roof or by injection through the walls.

Due to the dynamic nature of this process, Jones et al. [35] suggest that good operating results can be achieved only if several key variables are controlled simultaneously; bath temperature, scrap feed rate and scrap chemistry, oxygen injection rate, batch carbon levels and slag composition. One target is to ensure slag foaming conditions otherwise the heat radiated to the walls will be extremely high due to the flat bath operational conditions in this process. Scrap charging in the conveyor should be arranged in a way that ensures the proper average density to enhance heat transfer and reach a higher preheating temperature before it reaches the furnace. Scrap of high density, such as bundles will have a poor heating efficiency due to its high density. Shredded scrap due to a low density and high surface area can be heated faster.

Due to scrap preheating, energy consumption is largely decreased. EAF's with scrap preheating can employ transformers of lower capacity. Figure 8.29 shows the active power as a function of the nominal capacity of 44 Consteel EAF's, according with the supplier [5]. With this process the trend of higher furnace transformers and a higher ratio MVA/ton has been reversed to values from old furnaces, about 0.5, decreasing the capital investment in the transformer. The supplier also claims a productivity higher than 2 ton/h per MW if the hot heel is above 50%.

One of the first installations was Nucor steel-Darlington in the USA in 1986 [36]. Nucor retrofitted an old EAF. The tunnel was 20.1 m in length and equipped with 16 natural gas burners. The Ameristeel plant was described by Jones et al. [35]; 75 ton EBT-EAF designed to tap 40 ton and a hot heel of 30–35 ton (40–46%). The

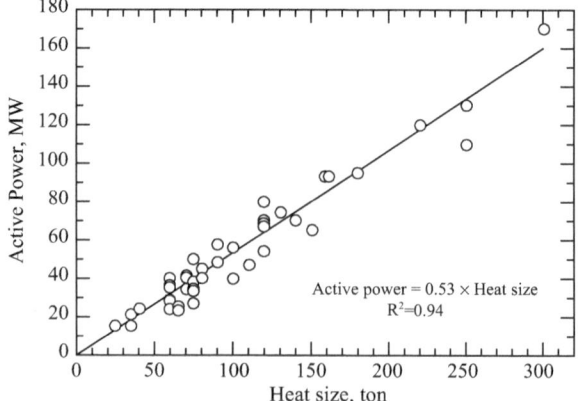

Fig. 8.29 Active power as a function of the nominal capacity of Consteel EAF's. After [5]

length of the tunnel was 24 m with 60 natural gas burners. About 70–75% of the CO generated in the EAF is available as fuel in the preheater.

Fragmented scrap is charged into a vibrating conveyor. The off-gas moves in a direction opposite to the conveyor. The section where preheating occurs is called the tunnel. The off-gas coming from the EAF is post-combusted using air injectors. The EAF is closed at all times, reducing energy losses. Full power can be applied from the beginning if there are also good foaming conditions. Full power from the beginning allows to use a transformer of smaller capacity. The operation is stable because the arcs are fully immersed in a foaming slag and also flat bath conditions are met during the entire heat. Noise levels drop by 10 dB compared with batch charging at the same power level [36]. A report from the supplier indicates a noise decrease, from 115 dB on average for top charging to 85 dB on average for the Consteel process [81].

Assuming a high preheating temperature, the estimated melting rate is about 2.2–2.5 ton/h·MW, much higher than that for a conventional EAF which is in the order of 1.2–1.5 ton/h·MW [36, 82]. There is some debate about the value of the scrap preheating temperature in Consteel. Values in the range from 150 to 300 °C have been reported [54, 83] much lower than the range initially suggested by Vallomy, from 800 to 1000 °C [79]. The CONSTEEL EAF is shown schematically in Fig. 8.30

A mathematical model was developed by De Miranda et al. [85] to estimate the temperature profile of the scrap in the tunnel for the Ori Marin plant, with a length of the tunnel of 24 m. They reported a large thermal stratification, with surface temperatures of 800 °C at a small region on the surface of the scrap layer but much lower temperatures, from 100 to 200 °C in the rest of the scrap layer. This result clearly indicates poor preheating conditions.

Toulouevski and Zinurov [50] have summarized the advantages and disadvantages of horizontal shafts, as follows.

- Advantages; (1) scrap is arranged in relatively thin layers which improves heat transfer, (2) Heat losses are decreased because the arcs are covered by a foaming

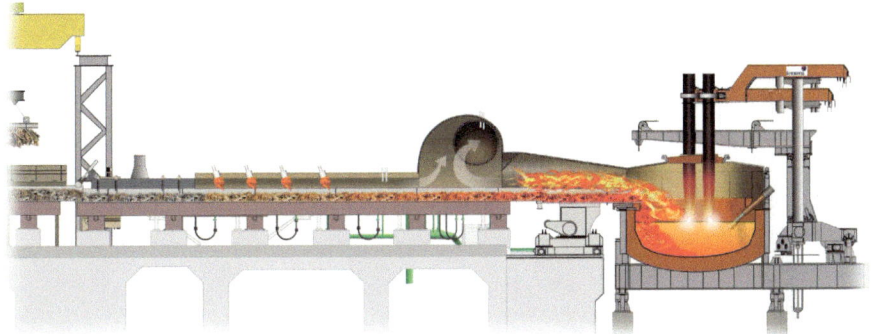

Fig. 8.30 Consteel process. Second generation. After [84]

slag, improving the life of water-cooled panels (3) covered arcs produce lower noise and increase arc stability which allows higher power factor and active power (4) lower FeO and higher yield, (5) electrode breaking is eliminated and, (6) the scrap that has reached a high temperature is immediately discharged into the molten bath.

- Disadvantages: heat transfer efficiency during scrap pre-heating is low because the gas flow moves parallel to the surface of the scrap layer with low penetration into the scrap layer, which leads to a high temperature difference between the surface and the inner layer of scrap, in addition to this, they also indicated that the productivity of Consteel furnaces is lower in comparison with modern furnaces with conventional technology. In Consteel EAF's the main heat transfer mechanism is by convection heat transfer in comparison with conventional EAF's where heat transfer involves both radiation and convection. Makarov [83] provides data from a Russian plant, AMZ, with a poor performance of the Consteel process, as shown in Table 8.3. The large energy consumption could be associated to the poor preheating temperature that is reported (150–200 °C) but also to poor foaming conditions. The author compares the heat transfer mechanisms in the conventional process and a process like Consteel where the arc is not surrounded by scrap. The fastest rate of heat transfer occurs by radiation from the arc, since that rate is proportional to T^4. Since radiation is not taking an important role because it is submerged in a foaming slag, conduction and convection heat transfer dominate the rate of heat transfer. This idea leads the author to estimate a lower arc efficiency for the Consteel process compared with a conventional one.

Memoli et al. [82] provided evidence to indicate that although convection is one of the main heat transfer mechanisms, the convection heat transfer coefficient in conventional EAF's can be low due to low stirring conditions but for the case of

Table 8.3 Comparison of performance of two Consteel plants [83]

	AMZ, 2010	Ori Martin, 2003
EAF total capacity, ton	120	115
Hot heel, ton	50	40
Conveyor speed, m/min	5	–
Total length of conveyor, m	–	48.5
Tunnel length, m	30.5	24
Tunnel width, m	–	2.0
Tunnel height, m	–	3.0
Thickness scrap layer, m	0.7	0.4
Scrap feeding rate, ton/min	4–5	–
Scrap preheating temperature, °C	150–200	–
Energy consumption, kWh/ton	416	–
Electrode consumption, kWh/ton	1.6	–

Consteel they argued that the oxygen injected and a liquid containing a fraction of solid scrap provides a higher value for the heat transfer coefficient which would explain why an increase in the hot heel decreases the power-on time.

Toulouevskii et al. [86] suggested the use of high-power burners to increase the scrap preheating temperature in Consteel. A second generation of Consteel has included an additional preheating section using natural gas, before the scrap enters the tunnel [87]. This section is also a tunnel with refractory lining. This extra energy should increase the preheating temperature. The supplier claims values of energy consumption in the order of 300 kWh/ton. Herin and Busbee [88] reported values from 370 to 390 kWh/ton.

Slag foaming is critical in the Consteel process because there is no scrap to shield the radiation from the electric arcs and the heat losses to the water-cooled panels would increase.

A big concern about all scrap preheating technologies using the off-gas has been raised by Toulouevski and Zinurov [50]. They argue that approximately 75% of the energy recovered preheating the scrap is needed to reheat the resulting off-gas at the combustion chamber. The same authors argue that the improvements on energy consumption with scrap pre-heating systems do not off-set the loss in productivity and added maintenance to complex shafts. The cited loss in productivity, however, seems to be a false claim since a preheated scrap actually increases the melting rate.

Table 8.4 summarizes some results on scrap preheating for different technologies. The preheating temperature varies from 200 to 1000 °C. Lee and Sohn [54] have suggested a limit on the maximum scrap preheating temperature at 800 °C because of partial melting and scrap sticking which would block its discharge into the EAF.

It has been reported [5] that scrap size is not a problem in the Consteel process because it can employ from very small pieces up to 1.5 m. The practical size limitation is the distance between the tip of the conveyor and the electrodes, as shown in Fig. 8.31.

Small pieces of scrap are charged at the bottom of the conveyor and large density pieces on the top. This applies for both shredded scrap and bushelings. The best performance is achieved when the scrap feed rate is controlled to achieve a higher

Table 8.4 Energy savings for various scrap pre-heating technologies (Adapted from Lee and Sohn [54])

Type	Company	Brand	°C	kWh/ton
Horizontal shaft	BBS Brusa	BBS Brusa	1000	220
Vertical shaft	JP Plantech	ECOARC	800	90
Vertical shaft	INTECO	COSS	500	60
Vertical shaft	VAI	Finger shaft	500	60
Horizontal shaft	TENOVA	Consteel	300	60
Bucket charge	DANIELI	DANARC	450	40
Twin shell	JFE/	Twin shell	200	57

Fig. 8.31 Main scrap size
limitation in the Consteel
process. After, [5]

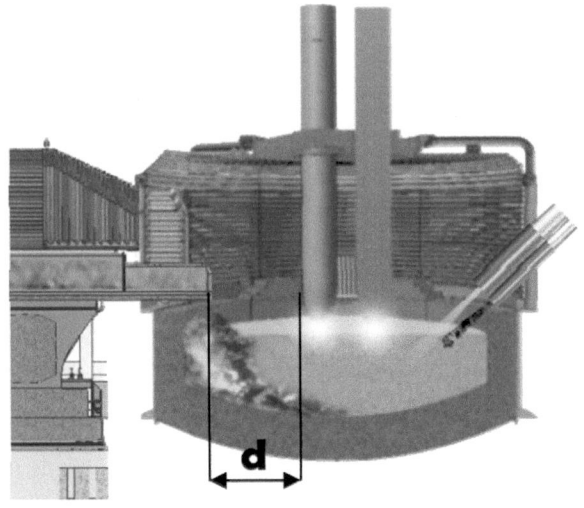

heat transfer for a given production rate of liquid steel. By increasing the residence
time of the scrap in the conveyor solid scrap also retains a higher fraction of dust.

8.3.6 Analysis of Heat Recovery from the Off-Gas

8.3.6.1 Efficiency on Heat Recovery for Scrap Preheating

Nimbalkar et al. [89] evaluated the efficiency of energy recovery from the off-gas for
scrap preheating in EAF's in the USA in 2014. To put the situation of scrap preheating
into perspective in the USA, out of 173 EAF's, 42% are old, built before 1990, only
33% had capacities above 100 ton and only 10% had scrap preheating systems, 9
furnaces using Consteel and 9 furnaces using shaft and twin-shell scrap preheating.
In Table 8.5 these authors made a comparison of scrap preheating processes. This
comparison gives a clear picture of the advantages and disadvantages of current scrap
preheating technologies, indicating that there is only partial heat recovery on all of
these processes (in the range from 25 to 120 kWh/ton), with a relatively large amount
of heat in the exhaust gas that leaves after preheating, formation of toxic gases, scrap
sintering creates problems to descend into the EAF, scrap oxidation (especially small
pieces of scrap) decreases the metallic yield, decreasing the tap to tap time decreases
the residence time of the off-gas and heat transfer to the scrap, limited preheating
in Consteel to the lower layers as well as a large hot heel which should be heated,
in addition to uneven scrap preheating, its higher productivity per MVA is due to
lower transformer but the melting rate is lower, about 1.4 times lower in comparison
with scrap heated by radiation heat transfer. A report by ESTEP in 2021 [90] in
addition to the previous limitations of scrap pre-heating technologies also includes

high investment costs (CAPEX). The report suggests a heat recovery in the order of 10–15% for scrap pre-heating and 16–20% if the sensible heat in the off-gas is employed for steam production. They suggest the standard pay-back time in the steel industry in general is about 3 years but in the case of scrap pre-heating can be from 5 to 10 years.

Figure 8.32 shows schematically the current limits on heat recovery using scrap preheating.

Considering the limitations described previously, Nimbalkar et al. [89] propose the development of new systems which include the following aspects;

Table 8.5 Comparison of scrap preheating technologies (after [89])

Method	$T_{preheat}$, °C	Advantages	Disadvantages	WHR potential
Bucket	315–450		Partial heat recovery	Max. 30–45% energy recovery, shorter t-t-t by 9–10%,
Shaft fuchs		Decrease energy up to 18%, Decrease 20% dust		
Twin shaft				
Finger shaft	Up to 800[a]	Through utilization of off-gas during heat cycle	Energy savings depend on type of scrap and post-combustion degree	Energy savings 90–110 kWh/ton
Brusa	450	Continuous preheating	Needs special fragmented scrap, sticking	Homogeneous heating
Consteel	400–600[a]	335–355 kWh/ton	Heating mostly the top layer, huge hot heel	Max. 50% energy recovery
COSS		Benefits of Shaft and continuous preheating	Partial heat recovery	80–100 kWh/ton
Telescopic		Single bucket charging	Partial heat recovery	Savings 20–30 kWh/ton

[a]Data from supplier

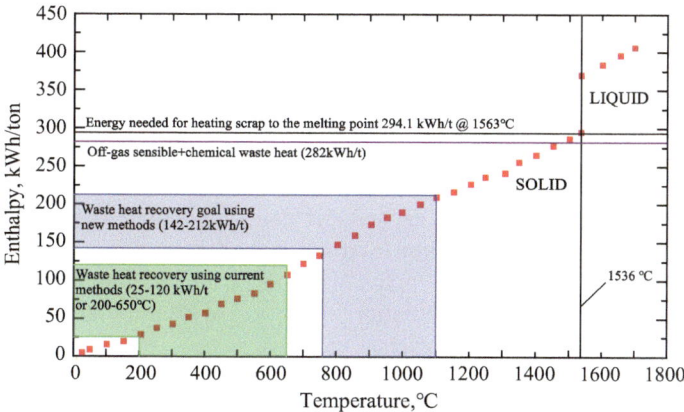

Fig. 8.32 Current heat recovery by scrap preheating. After [89]

- Exhaust gas preconditioning by removing dust
- Use a heat recovery system designed to complete the combustion of gases containing chemical heat, then heat can be stored and transferred
- Use of a heat accumulator
- Scrap preheating using preheated hot gases that contain no combustible components
- Use clean gases to produce steam and steam to produce electrical power.

A heat recovery system based on the previous ideas has not been yet developed.

8.3.6.2   Efficiency on Heat Recovery for Steam Production

Hartfuss et al. [91] reported a novel idea to produce steam using $Ca(OH)_2$ instead of water. The off-gas shows large fluctuations in temperature, mass flow rate and load of dust particles. The peak temperature in the combustion chamber during the melting period reached 1400 °C and decreased to 250 °C during scrap charging, the mass flow rate fluctuated from 28 to 37 kg/s. The dust concentration was reported as 20 kg/m³, equivalent to 580 g/s. Calcium hydroxide particles, smaller than 500 μm, are used in the quenching section. The heating process causes the formation of lime and steam. Steam is partially separated and the rest allowed to produce $Ca(OH)_2$ once again. One limitation is the mixing with EAF dust which affects the quality of the lime produced.

8.3.6.3   Efficiency on Heat Recovery for Electricity Generation

Yang et al. [51] compared four alternatives to recycle the off-gas from an EAF charged with 50% hot metal; scrap pre-heating, electricity generation, steam production and coal gas production. In all cases the available energy was 274 kWh/ton which corresponds to 18% of sensible heat in the off-gas and 19% of potential chemical heat due to the presence of CO in the off-gas. Their calculations indicate the largest recovery efficiency of 71% when the potential chemical energy is employed to produce coal gas, followed by applications that use the sensible heat in the off-gas; 65% for electricity generation, 53% for scrap preheating and 17% for steam production.

Schliephake et al. [92] described some of the challenges to produce steam because in addition to achieve a higher energy recovery it should also keep in balance the production rate and consumption rate of steam. Water at high pressure 10 bar and 184 °C provides a higher cooling capacity, in comparison with cold water, 250 kJ/kg and 83.5 kJ/kg, respectively. They used water in a temperature range from 105 to 159 °C. In the case described the demand of steam for the vacuum furnace was about 7 ton/h in comparison with the potential production of 20 ton/h. Part of the excess capacity was stored using Ruth buffers.

8.4 Airtight and Closed-Door Operation

Air infiltration is an important subject, it has multiple ramifications and scrap pre-heating is only one of them. Air infiltration has a large influence on post-combustion, scrap pre-heating, mass and energy balances, nitrogen in both liquid steel and off-gas volume, etc. Air infiltration into the furnace should be eliminated as much as possible because air is not only a source of oxygen, nitrogen and hydrogen but also consumes energy and increases the volume of the off-gas. With air infiltration the volume of the off-gas will be much higher. The excess gas which is not captured escapes from the furnace to the shop, as a result of air infiltration. Hajidavalloo et al. [93] estimated a total flow rate of the off gas about 10.4 kg/s, of which 4.0 kg/s is gas released from the EAF, indicating that air infiltrated is about 6.4 kg/s, a higher flow rate compared to the gases produced by the EAF.

Oxygen from air infiltrated reacts with carbon from the electrodes. Nitrogen consumes more energy and hydrogen can get absorbed by the slag. The main source of air infiltration is the slag door but also occurs at the electrode ports and roof ring. Air infiltration can account for 15 kWh/ton in energy losses [7]. In this sense, an airtight closed-door operation can be taken as another method to decrease energy consumption.

Liu et al. [94] developed mass and energy balances for an EAF of 100 ton charged with 100% scrap and both scrap and hot metal (70–30). The infiltrated air was estimated from measurements of the off-gas. Their results indicate the infiltrated air ranged from about 50–100 Nm^3/min and found a slight decrease in the total energy consumption, however they also pointed out that the final result will depend on the amount of CO and H_2 released from the molten bath and post-combusted by the infiltrated air.

The EAF operates under negative pressure. Bender et al. [95] indicates a common negative pressure of − 0.02 in. but they suggest the possibility to make it slightly positive, about + 0.3 in. which could reduce electric energy by 50 kWh/ton.

In spite that an air tight, closed-door operation is always recommended, in practice is not usually applied because, among other reasons, requires higher slag door maintenance that increases the power-off time. The slag door is necessary for different purposes:

- Visual inspection of the refractory inside the EAF and furnace gunning
- Manual steel and slag sampling
- Manual temperature measurements
- Manual or automatic oxygen and carbon injection.

The previous operations imply safety risks for the furnace operators who work in front of the slag door. Currently, with the best available technologies, those operations can be performed by robots. For example, Wolf [96] describes the benefits of using gunning robots. Manual gunning is carried out with long lances, 6–8 m, introduced through the slag door, gunning rate is low, from 60 to 80 kg/min. The intense heat and weight of the equipment make this operation difficult and increase the power-off time. With a gunning robot several lances can be introduced to include different materials, the gunning rate is in the range from 125 to 150 kg/min.

Gottardi et al. [97] described a new slag door that avoids air infiltration, eliminating the tunnel of the slag door an splitting it into two parts. The lower section can be used to push scrap and release slag if necessary. The new design allows a higher slag volume in the EAF which contributes to improved foaming conditions and higher yield of the carbon injected.

In a closed-door operation, the slag is not allowed to leave. It has advantages and disadvantages. Uyen [98] reported results in a 45 ton EAF using 60–65% DRI. Foaming height reached from 1.5 to 2 m and allowed to increase the active power, increase metallic yield by 2.1% and a drastic reduction in FeO. This practice was not found adequate for low P steels due to phosphorous reversion and also, they reported risks of slopping.

References

1. Mulvaney D, Richards RM, Bazilian MD, Hensley E, Clough G, Sridhar S (2021) Progress towards a circular economy in materials to decarbonize electricity and mobility. Renew Sustain Energy Rev137. https://doi.org/10.1016/j.rser.2020.110604
2. WSA (2023) Steel and raw materials. World Steel Assoc. https://worldsteel.org/wp-content/uploads/Fact-sheet-raw-materials-2023.pdf
3. Institute of Scrap Recycling Industries (ISRI) (2022) Scrap specifications circular. Washington DC
4. Steel Dynamics (2021) Iron & steel scrap specifications manual
5. Memoli F, Jones JAT, Picciolo F (2013) How changes in scrap mix affect the operation of Consteel® EAFIn: . AISTech 2013 conference proceedings, vol 1, pp 795–808
6. Odenthal HJ, Kemminger A, Krause F, Sankowski L, Uebber N, Vogl N (2018) Review on modeling and simulation of the electric arc furnace (EAF). Steel Res Int 89:1–36. https://doi.org/10.1002/srin.201700098
7. Adams W, Alameddine S, Bowman B, Lugo N, Paege S, Stafford P (2001) Factors influencing the total energy consumption in arc furnaces. In: 59th electric furnace conference, Phoenix, AZ, USA, 11–14 Nov 2001, pp 691–702
8. Bowman B, Krüger K (2009) Arc furnace physics. StahlEisen
9. Carlsson LS, Samuelsson PB, Jönsson PG (2020) Modeling the effect of scrap on the electrical energy consumption of an electric arc furnace. Processes 8. https://doi.org/10.3390/pr8091044
10. Kapoor I, Davis C, Li Z (2021) Effects of residual elements during the casting process of steel production: a critical review. Ironmak Steelmak 48:712–727. https://doi.org/10.1080/030 19233.2021.1898869
11. Berlin D, Feldmann A, Nuur C (2022) Supply network collaborations in a circular economy: a case study of Swedish steel recycling. Resour Conserv Recycl 179:106112. https://doi.org/10.1016/j.resconrec.2021.106112
12. Malfa E, Nyssen P, Filippini E, Dettmer B, Unamuno I, Gustafsson A et al (2013) Cost and energy effective management of EAF with flexible charge material mix. BHM Berg Huettenmaenn Monatsh 158:3–12. https://doi.org/10.1007/s00501-012-0103-y
13. Whipp R, Martin R (n.d.) Effective use of scrap optimization in melt shop operations. In: 2000 electric furnace conference proceedings, pp 771–780
14. Maiolo JA, Evenson EJ (2001) Statistical analysis and optimization of EAF operations. In: 59th electric furnace conference, Phoenix AZ, USA, 11–14 November, 2001, pp 105–112

15. Gyllenram R, Westerberg O (2016) The impact of scrap upgrading on EAF production cost and environmental performance. ESTAD 2015, Dusseldorf, pp 1–6
16. Arzpeyma N, Gyllenram R, Jönsson PG (2020) Development of a mass and energy balance model and its application for HBI charged eafs. Metals 10. https://doi.org/10.3390/met100 30311
17. Yellishetty M, Mudd GM, Ranjith PG, Tharumarajah A (2011) Environmental life-cycle comparisons of steel production and recycling: sustainability issues, problems and prospects. Environ Sci Policy 14:650–663. https://doi.org/10.1016/j.envsci.2011.04.008
18. Tarasenko A (2022) The material of the future: why scrap metal has already become a strategic raw material. https://gmk.center/en/posts/the-material-of-the-future-why-scrap-metal-has-alr eady-become-a-strategic-raw-material/%0D%0A
19. Wübbeke J, Heroth T (2014) Challenges and political solutions for steel recycling in China. Resour Conserv Recycl 87:1–7. https://doi.org/10.1016/j.resconrec.2014.03.004
20. Taylor B (2020) Where the shredders are. Recycl Today. https://www.recyclingtoday.com/ news/metal-shredding-recycling-bir-china-usa-mexico-india-japan/. Accessed 19 Jan 2024
21. Wang M, Tian Y, Liang Y, Zhou R, Luo Y, Li X (2020) Forecast scrap generation and emission reduction of china's steel industry. Adv Intell Syst Comput 283–292. https://doi.org/10.1007/ 978-3-030-21248-3_21
22. Liu M, Chen X, Zhang M, Lv X, Wang H, Chen Z et al (2020) End-of-life passenger vehicles recycling decision system in China based on dynamic material flow analysis and life cycle assessment. Waste Manag 117:81–92. https://doi.org/10.1016/j.wasman.2020.08.002
23. Zhao Q, Chen M (2011) A comparison of ELV recycling system in China and Japan and China's strategies. Resour Conserv Recycl 57:15–21. https://doi.org/10.1016/j.resconrec.2011.09.010
24. Zhang L, Lu Q, Yuan W, Jiang S, Wu H (2022) Characterizing end-of-life household vehicles' generations in China: spatial-temporal patterns and resource potentials. Resour Conserv Recycl 177:105979
25. BIR (2021) World steel recycling in figures 2017–2021
26. Chen W, Yin X, Ma D (2014) A bottom-up analysis of China's iron and steel industrial energy consumption and CO_2 emissions. Appl Energy 136:1174–1183. https://doi.org/10.1016/j.ape nergy.2014.06.002
27. Ministry of Steel (India) (2020) Steel scrap recycling policy
28. Compañero RJ, Feldmann A, Tilliander A (2021) Circular steel: how information and actor incentives impact the recyclability of scrap. J Sustain Metall 7:1654–1670. https://doi.org/10. 1007/s40831-021-00436-1
29. Pauliuk S, Milford RL, Müller DB, Allwood JM (2013) The steel scrap age. Environ Sci Technol 47:3448–3454. https://doi.org/10.1021/es303149z
30. Müller DB, Wang T, Duval B (2011) Patterns of iron use in societal evolution. Environ Sci Technol 45:182–188. https://doi.org/10.1021/es102273t
31. Oda J, Akimoto K, Tomoda T (2013) Long-term global availability of steel scrap. Resour Conserv Recycl 81:81–91. https://doi.org/10.1016/j.resconrec.2013.10.002
32. Statista (2023) Average weight of metal content in US and Canadian-built light vehicles between 2007 and 2017, by type. Statista 2023:1. https://www.statista.com/statistics/882580/us-and-can adian-built-vehicles-average-metal-content-weight-by-type/#:~:text=In2017%2C
33. IEA. Iron and Steel Technology Roadmap (2020) Towards more sustainable steelmaking. https://doi.org/10.1787/3dcc2a1b-en
34. World Steel Association (2021) Scrap use in the steel industry. https://worldsteel.org/wp-con tent/uploads/Fact-sheet-on-scrap_2021.pdf. Accessed 16 May 2023
35. Jones JAT, Bowman B, Lefrank PA (1998) Electric arc furnace steelmaking. In: Fruehan RJ (ed) The making, shaping and treating of steel. Steelmaking and refining, 11th edn. Pittsburgh, PA, pp 525–660
36. Bosley J, Clark J, Dancy T, Fruehan R, McIntyre E (1987) Techno-economic assessment of electric steelmaking through the year 2000. Pittsburgh, PA
37. Laurenti S, Gottardi R, Miani S, Partyka A (2005) High performance single-bucket charging EAF practice. Ironmak Steelmak 32:195–198. https://doi.org/10.1179/174328105X38099

38. Abel M, Hein M (2008) The Simetal ultimate at Colakoglu/Turkey. In: AISTech 2008 Proceedings, AISTech, PIttsburgh, PA (Paper 67)
39. Fuchs G (2009) Melting furnace, in particular electric arc furnace. US Patent 2009/0274190, 2009
40. Nal Ö, Dolapçioğlu S, Gottardi R, Partyka A, Miani S (2017) Efficiency of the EAF with telescope roof. In: AISTech 2017 Conference 2017, vol 1, pp 1149–1160
41. Zhang L, Oeters F (1999) Possibilities of counter-current scrap pre-heating with melting by use of 100% fossil energy. Steel Res 70:296–308. https://doi.org/10.1016/b978-0-12-248291-5/50008-0
42. Mandal K, Irons GA (2008) A study of scrap heating by burners. In: AISTech 2008 Proceedings. PIttsburgh, PA, USA, 5–8 May 2008
43. Mandal K, Irons GA (2010) Numerical modeling of scrap heating by burners. In: AISTech 2010 Proceedings, pp 801–810
44. Mandal K, Irons GA (2013) A study of scrap heating by burners. Part I: experiments. Metall Mater Trans B 44:184–195
45. Mandal K, Irons GA (2013) A study of scrap heating by burners: part II—numerical modeling. Metall Mater Trans B Process Metall Mater Process Sci 44:196–209. https://doi.org/10.1007/s11663-012-9752-1
46. Li J, Provatas N, Irons GA (2008) Modeling of late melting of scrap in the EAF. Iron Steel Technol 5:216–223
47. Tang G, Chen Y, Silaen AK, Krotov Y, Zhou CQ (2019). Effects of steel scrap oxidation on scrap preheating process in an electric arc furnace. In: 10th international symposium on high-temperature metallurgical processing, pp 453–465. https://doi.org/10.1007/978-3-030-05955-2_43
48. Jones J, Safe P, Wiggins B (1999) Optimization of EAF operations through offgas system analysis. In: Electric furnace conference. Pittsburgh, PA, USA, 14–16 November 1999, pp 459–480
49. Thekdi A, Nimbalkar S, Keiser J, Storey J (2015) Preliminary results from electric arc furnace off-gas enthalpy modeling. In: The iron & steel technology conference and exposition. Oak Ridge National Lab (ORNL), Oak Ridge, p 15
50. Toulouevski YN, Zinurov IY (2010) Preheating of scrap by burners and off-gases. In: Springer-Verlag (ed) Innovation in electric arc furnaces. Berlin, pp 93–113
51. Yang LZ, Zhu R, Ma GH (2016) EAF gas waste heat utilization and discussion of the energy conservation and CO_2 emissions reduction. High Temp Mater Process 35:195–200. https://doi.org/10.1515/htmp-2014-0183
52. Fuchs G, Rummler K, Haissig M (2008) New energy saving electric arc furnace designs. In: AISTech iron & steel technology conference, p 9061
53. De Beer J, Worrell E, Blok K (1998) Future technologies for energy-efficient iron and steel making. Annu Rev Energy Environ 23:123–205. https://doi.org/10.1146/annurev.energy.23.1.123
54. Lee B, Sohn I (2014) Review of innovative energy savings technology for the electric arc furnace. JOM 66:1581–1594. https://doi.org/10.1007/s11837-014-1092-y
55. Toulouevski YN, Zinurov IY (2017) Fuel arc furnace (FAF) for effective scrap melting: from EAF to FAF. Springer Nature Singapore. https://doi.org/10.1007/978-981-10-5885-1
56. Lehner J, Friedacher A, Gould L (2004) Low cost solutions for the removal of dioxin from electric arc furnace offgas. Rev Metall 2004:49–56
57. Fior A (1999) The Danieli Danarc Plus M2 furnace at ABS meltshop. METEC, pp 78–83
58. Michielan A, Lavaroni G, Fior A (2000) The Danieli Danarc™ furnace at ABS, pp 745–752
59. Weber R, Nosé D, Morsoletto L, Pfeifer HC (1994) Last achievements with the EOF process. Rev Metall 439–444
60. Ondracek J, Bauerschmidt A (1998) Willy Korf-German entrepreneur: case A and case B. Entrep Theory Pract 23:49–70. https://doi.org/10.1177/104225879802300204
61. Fruehan RJ, Nassaralla CL (1998) Alternative oxygen steelmaking processes (chap 13). The Making Shaping and Treating of Steel—Steelmaking and Refininig, vol, 1998, pp 743–759

62. Fuchs G, Ehle J (1986) Electric arc furnace having a space provided on one side of the furnace vessel for accomodating charging material. US patent 4617673. US patent 4617673
63. Madias J (2014) Electric furnace steelmaking. Treatise process. Metall 3:271–300. https://doi.org/10.1016/B978-0-08-096988-6.00013-4
64. Rummler K, Tunaboylu A, Ertas D (2012) New generation in pre-heating technology for electric steelmaking higher productivity with reduced power. In: 43rd Steelmaking seminar, Belo Horizionte, Brazil, 20–23 May 2012
65. Fuchs G (2009) Charging device, especially charging stock preheater. US patent 7497985
66. Apfel J, Mueller A, Beile H (2016) EAF quantum-results 2015. 11th European electric steelmaking conference, p 47
67. Metzen A, Germershausen T, Bader J, Bergs A (2016) SHARC-Shaft arc furnace with efficient scrap preheating concept provides low conversion costs. Metall Plant Technol 52–57
68. Odenthal HJ, Kemminger A, Krause F, Vogl N (2017) A holistic CFD approach for standard and shaft-type electric arc furnaces. AISTech Iron Steel Technol Conf Proc 1:1101–1114
69. Ogushi M, Takeuchi O, Yamamura I, Iura T, Yoshida H (1996) Electric arc melting furnace. US Patent 5590150. US Patent 5590150
70. Yamaguchi R, Mizukami H, Maki T, Ao N (2000) ECOARC technology. 58th electric furnace conference. Orlando FL, pp 325–336
71. Sugasawa T, Kato H, Nagai T (2012) The first ECOARC in kingdom of Thailand: Introduction of the high efficiency furnace. In: SEAISI conference. Bali Indonesia, 28–31 May 2012, p 10
72. Nagai T, Sato Y, Kato H, Fujimoto M, Sugasawa T (2015) The most advanced power saving technology in EAF: introduction to ECOARC. METEC/ESTAD conference. Dusseldorf, Germany, 15–19 June 2015
73. Muller HG, Hofer LP, Berger HA, Pirklbauer JF, Gould LP (1996) Advanced electric arc furnace solutions for im proved steelmaking performances (in French). Rev Metall 93:485–96
74. Kappes H (2000) A new steelmaking concept at Saldanha. Rev Métallurgie 97:897–904. https://doi.org/10.1051/metal/200097070897
75. Kleinschmidt G, Wimar M, Kempken J, Falkenreck U (2005) Competent solutions for new challenges in steelmaking. In: 15th IAS steelmaking conference, pp 101–110
76. Brusa U (1974) Electric furnace for heating and melting scrap iron and steel. US Patent 3789126. US Patent 3789126
77. EPRI Center for Materials Production (1997) Electric arc furnace scrap preheating
78. Hofmann W, Reichelt W (1996) Contiarc—a new scrap melting technology. Steel Times 224(103):105–106
79. Vallomy J (1985) Apparatus for continuous steelmaking. US Patent 4532124
80. Stagnoli P (2020) Innovation in EAF. Webinar Steelmint 2020. https://www.youtube.com/watch?v=4nkJhw4K7Wo
81. Memoli F, Ferri M (2007) New track record for Consteel due to new environment-friendly features. MPT Int 58–65
82. Memoli F, Guzzon M, Giavani C (2011) The evolution of preheating and the importance of hot heel in supersized Consteel® systems. In: AISTech conference. Indianapolis, IN, USA, 2–5 May 2011, pp 823–832
83. Makarov AN (2019) Effect of the architecture on energy efficiency of electric arc furnaces of conventional and Consteel designs. Metallurgist 62:882–891. https://doi.org/10.1007/s11015-019-00743-9
84. Memoli F, Giavani C, Malfa E (2012) Consteel® evolution^{TM}—the second generation of Consteel technology. Ind Heat 2012. https://www.industrialheating.com/articles/90594-consteel-evolution---the-second-generation-of-consteel-technology
85. De Miranda U, Di Donato A, Volponi V, Zanusso U, Argenta P, Pozzi M (2003) Scrap continuous charging to EAF, ECSC 20883
86. Toulouevskii YN, Zinurov IY, Shver VG (2012) New possibilities of Consteel furnaces. Russ Metall 2012:449–453. https://doi.org/10.1134/S0036029512060213
87. Memoli F, Rondini N, Giavani C, Malfa E (2013) Consteel evolution, the second generation of Consteel technology. In: 44th steelmaking seminar, Araxa MG

88. Herin H, Busbee T (1996) The Consteel process in operation at Florida steel. Iron Steelmak 23:43–46
89. Nimbalkar S, Thekdi A, Keiser J, Storey J (2014) Waste heat recovery from high temperature off-gases from electric arc furnaces. AISTech Iron Steel Technol Conf Proc 1:1113–1123
90. ESTEP (2021) Improve the EAF scrap route for a sustainable value chain in the EU Circular Economy scenario
91. Hartfuß G, Schmid M, Scheffknecht G (2020) Off-gas waste heat recovery for electric arc furnace steelmaking using calcium hydroxide ($Ca(OH)_2$) dehydration. Steel Res Int 91:1–8. https://doi.org/10.1002/srin.202000048
92. Schliephake H, Born C, Granderath R, Memoli F, Simmons J (2010) Heat recovery for the EAF of Georgsmarienhütte, Germany. In: AISTech conference, Pittsburgh, PA, USA, 3–6 May 2010, pp 745–752
93. Hajidavalloo E, Dashti H, Nejad MB (2013) Exergy and energy analysis of an AC steel electric arc furnace under actual conditions. Int J Exergy 12:380
94. Liu Y, Wei G, Tian B (2023) Analysis and optimisation on the energy consumption of electric arc furnace steelmaking. Ironmak Steelmak 50:1–15. https://doi.org/10.1080/03019233.2023.2172826
95. Bender M, Zemp R, Ineichen R (1996) Influence of electric arc furnace pressure on power consumption. Iron Steel Eng 73:73–77
96. Wolf C (2013) Gunning robots for the hot repair. In: International technical conference on refractories (UNITECR 2013), pp 27–32
97. Gottardi R, Partyka A, Miani S, Novak H, Klipa N (2011) Enhanced slag door for electric arc furnace. In: AISTech 2011 proceedings. Indianapolis, IN, USA, 2–5 May 2011, pp 857–861
98. Uyen J (2002) EAF "closed door process" at Corporación Aceros Arequipa S.A. In: 7th European electric steelmaking conference. Venice, Italy, 26–29 May 2002, pp 1.47–1.56

Chapter 9
Stirring in the EAF

9.1 Driving Forces for the Motion of Liquid Steel in a Conventional EAF

9.1.1 Introduction

The share of steel production by the EAF process will rapidly increase in the next decade due to several advantages over the integrated route; higher flexibility in raw materials and less production of CO_2, among several other factors, however, the EAF has also several limitations; use of expensive electric energy and poor stirring conditions. The stirring conditions of the conventional EAF are poor. Forced convection in the EAF can be carried out in several ways but the two primary techniques are; bottom gas injection (BGI) and electromagnetic stirring (EMS). EMS is an old technique but barely exploited in the EAF, until recently. Fundamental research on mixing phenomena in the EAF is poor, compared with the mixing studies in the ladle. Yang et al. [1] recently reviewed the subject of stirring technologies in EAF steelmaking, however, the subject requires not only a more critical analysis of the current developments in addition to an analysis of the current challenges to achieve higher stirring intensities in the EAF.

This chapter is divided into three main parts; (1) An analysis of the driving forces for the motion of the liquid in a conventional EAF, (2) Bottom gas injection and (3) electromagnetic stirring. The driving forces in a conventional EAF include natural convection, oxygen injection through supersonic and coherent jets, stirring due to the escape of the CO bubbles generated by decarburization and electromagnetic forces. Bottom gas injection (BGI) is currently the most extended practice to improve the stirring conditions. There is a relatively large amount of work carried out involving different arrangements with porous plugs and its effect on fluid flow phenomena, which includes mixing time, spout height, mass transfer, melting rate, slag eye formation, velocity and temperature distributions. Argon has been the typical injected gas but other gases have also been employed, including CO_2. More recently, electromagnetic stirring (EMS) has also made significant progress and promises important advantages. Both technologies, BGI and EMS, are compared, including its reported benefits on electric energy consumption.

9.1.2 Driving Forces for the Motion of Liquid Steel in a Conventional EAF

The mixing conditions in a conventional EAF are driven from four sources:

(i)   Natural convection due to buoyancy forces
(ii) Oxygen injection
(iii) CO generation
(iv) Electromagnetic forces around the pitch circle

All of these forces combined are still insufficient to eliminate thermal and chemical gradients within the liquid. Gonzalez et al. [2, 3] reported in 2010 fluid flow results from a mathematical model for an industrial AC-EAF of 210 tons. This work clearly illustrates the large thermal stratification in a conventional EAF, indicating thermal gradients higher than 100 °C between cold and hot regions. They reported that thermal stratification can be decreased by increasing arc length from 25 to 45 cm, as shown in Fig. 9.1, this is because the average speed of liquid steel was increased from 4.77 to 5.06 cm/s, for a slag-free system. Fornander and Nilsson [4] cited a report with a thermal gradient of 50 °C in one EAF. Operation with Ultra High Power (UHP)-EAF's would increase the magnitude of the thermal gradients.

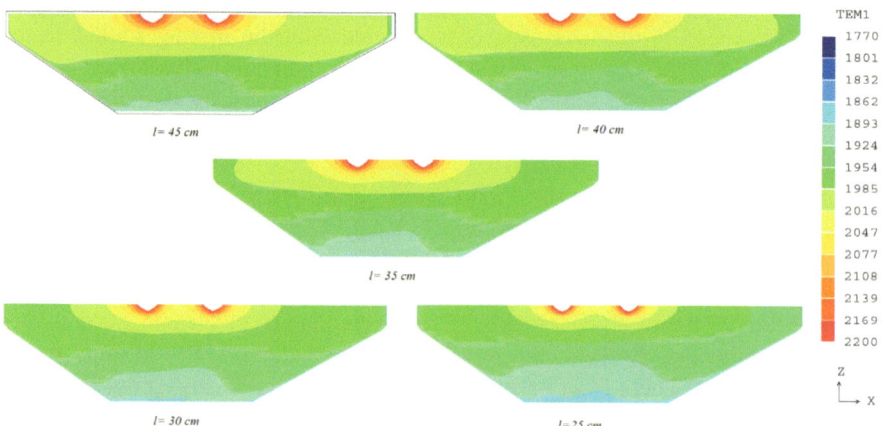

Fig. 9.1 Thermal stratification in a conventional EAF as a function of arc length (*l*). Temperature in K [3]

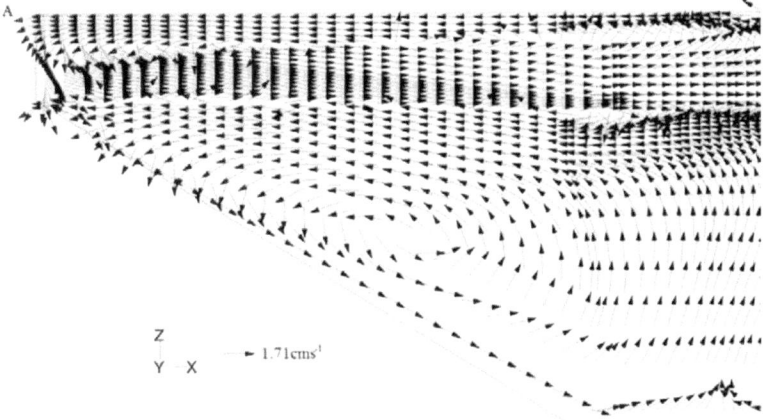

Fig. 9.2 Fluid velocities in the steel/slag system under natural convection [5]

The top slag layer has a large influence on the velocity of liquid steel. Ramirez et al. [5] reported results from a numerical model that shows the velocity profiles in liquid steel and liquid slag with a slag thickness of 35 cm, in an EAF of 210 ton, as shown in Fig. 9.2. It can be observed that, due to buoyancy forces, both liquids have the same flow patterns. Each liquid ascends to the top and then forms a recirculation loop. The momentum exchange at the slag/metal interface creates shear stresses because each liquid moves in a different direction. These authors reported a decrease in the average velocity of liquid steel from about 5 cm/s to about 2 cm/s.

More recently, Elkoumy et al. [6] compared the average velocity of liquid steel in a 225 ton EAF-EBT with and without any heating from the electrodes and found that without any heating the liquid remains stagnant.

The low velocities in liquid steel in a conventional EAF is one of its major limitations because their effects are multiple:

- Lower melting rate of the metallic charge (larger tap-to-tap time)
- Lower dissolution rate of fluxes (lower dephosphorization rates)
- Higher energy consumption
- Lower decarburization rate
- Higher consumption of oxygen
- Higher deviations from slag/metal equilibrium
- Lower metallic yield due to higher FeO in the slag
- Higher thermal and chemical stratification
- Lower productivity

Due to much higher stirring conditions, the rate of decarburization in the BOF is 4–6 times higher than in the EAF [7]. It is quite clear that one of the main advantages of the BOF process over the conventional EAF process is a higher mixing intensity.

Table 9.1 Velocity of liquid steel due to natural convection

Year	Author	Characteristics	Velocity, cm/s
2018	Elkoumy et al. [6, 8]	Natural convection	max: 0.01–0.04
		Forced convection due to local EMS	max: 0.62
2010	Gonzalez et al. [9]	Increasing arc length from 25 to 45 cm	avg: 4.7–5.0

In the following sections the four driving forces for the motion of liquid steel in a conventional EAF are described in more detail.

9.1.2.1 Motion Due to Natural Convection

Motion due to natural convection occurs due to density differences. These density differences can be due to temperature gradients or the mixing of a gas into a liquid. Elkoumy et al. [6, 8] compared the maximum velocity of liquid steel in a 225 ton EAF-EBT with natural and forced convection, changing the temperature and slag thickness. Under natural convection the maximum velocity they reported ranged from about 0.01 to 0.04 cm/s and under forced convection it increases by one order of magnitude. The authors omit to report the average velocity which would be a better parameter to describe the influence of the electric arcs. Their suggestion for an optimum slag thickness of 15 cm in order to decrease its chemical attack and especially reduce heat losses is totally unrealistic because the optimum slag thickness should be defined in principle based on the height of the electric arc. Gonzalez et al. [9] reported a higher melting rate of DRI when the arc length was increased from 25 to 45 cm, this was due to a higher velocity of the liquid. Table 9.1 compares the maximum velocity with natural convection with other cases that include the effect of electromagnetic stirring and a raise in the arc length.

9.1.2.2 Forced Convection Due to Oxygen Injection

Oxygen injection, in particular if the oxygen is injected using supersonic lances, creates a high turbulent zone during impingement with the liquid, reaching high velocities, however there is a sudden momentum dissipation away from this zone, consequently the effect of oxygen injection on the mixing conditions of the entire EAF has a limited effect. Guo et al. [10] developed a numerical model based on the VOF-DPM algorithms to compare three stirring mechanisms; CO generation, oxygen injection and bottom gas injection, in a 190 ton AC-EAF, at 1600 °C. Oxygen injection was injected vertically using a lance with a flow rate of 51 Nm^3/min producing an average velocity of liquid steel of 2 cm/s. Wang et al. [11] reported numerical model results on the motion of the liquid due to oxygen injection with supersonic lances. The EAF had a nominal capacity of 150 ton and the height of liquid steel was 1 m. The gas flow rate was 50 Nm^3/min. At the point of impact of the oxygen jet with liquid steel a depression is formed, the liquid in that region can achieve velocities in the range from 10 to 50 cm/s, decreasing in the vertical direction. For every 1 cm decrease in the vertical direction the velocity decreases about 0.15–0.2 cm/s in the middle of the molten bath and about 0.6–0.7 cm/s near the furnace bottom. Han et al. [12] reported the velocity of the jet and the velocity of the oxygen jet in a 60-ton EAF with a flow rate of 33 Nm³/min (equivalent to 0.79 kg/s) and lances with a vertical angle of 45°. They reported a decrease in the velocity of the oxygen jet from 671 to 121 m/s from the exit of the Laval nozzle to the point of impact with liquid steel. The maximum velocity of the liquid was 21 cm/s and its average velocity of 1.5 cm/s. He et al. [13] developed a numerical model for an EAF of 150 ton. with a top slag layer of 10 cm and height of liquid steel of 1 m, to describe fluid flow and decarburization rates. They found that increasing the oxygen flow rate from 8 to 33 Nm³/min, the maximum speed of molten steel increased from 1.4 to 21 cm/s. They also reported the following equations for the impact area, flow of liquid steel through the exposed area and the penetration depth of the oxygen jet, as a function of the oxygen gas flow rate:

$$S = 0.0004\,Q_{O_2} - 0.0932$$
$$Q_S = 0.1798\exp(0.003\,Q_{O_2})$$
$$H = 0.61\,Q_{O_2}d \cdot \sin\theta + 3.1$$

where S is the impact area in m², Q_{O_2} is the oxygen flow rate in Nm³/h, Q_S is the flow of liquid steel through the exposed area in Nm³/min, H is the penetration depth in cm, d is the outer diameter of the oxygen lance in cm, θ is the lance vertical angle.

Thongjitr et al. [14] reported a large increase in the distance from the furnace wall to the surface of the molten bath from 1 to 2 m due to erosion of the refractory. This increment in distance can be compensated using coherent jets. Comparing three jets for a distance of 2 m; conventional, coherent with CH_4 as shrouding gas (0.069 kg/s), coherent with both $CH_4 + O_2$ as shrouding gas (CH_4 + 0.277 kg/s), the three of them with the same central O_2 flow rate at 0.715 kg/s, resulted in the following penetration depths: 28.6 cm for the supersonic jet, 37.8 cm for the coherent-CH_4 jet and 42.7 cm for the coherent-CH_4-O_2 jet. The coherent-CH_4-O_2 jet with a distance of 1 m resulted in a very high penetration depth of 72.8 cm.

Chen et al. [15] reported that increasing the oxygen flow rate, the vortex center is moved away from the oxygen lance and deeper in the molten pool. Table 9.2 summarizes reported values on the maximum and average velocity of liquid steel due to oxygen injection.

Mixing time due to oxygen injection has been investigated by several researchers [16–20], including the effect of the oxygen flow rate (Q), number of lances (N), horizontal angle (θ_h), vertical angle (θ_v) and distance from the lance tip to the surface of the molten bath (L). The largest influence is due to the gas flow rate.

He et al. [16] measured mixing time due to oxygen injection in a water model with a geometric scale 1:10 from a prototype of 150 ton., changing the gas flow rate, the number of lances, the vertical and horizontal angles. They reported the largest

Table 9.2 Effect of oxygen jets on liquid steel's velocity

Year	Authors	Characteristics	Velocities
2000	Guo et al. [10]	190 ton, 1 vertical lance, 3060 Nm^3/h	Liquid's velocity: 2 cm/s
2010	Wang et al. [11]	Supersonic jet, 150 ton, $H_l = 1$ m, 3000 Nm^3/h	Liquid's velocity: 10–50 cm/s (impingement zone) 0.6–0.7 cm/s (bottom)
2010	Han et al. [12]	60 ton, 2000 Nm^3/h	Jet velocity: 671 m/s, exit velocity 121 m/s, impact zone Avg velocity liquid: 1.5 cm/s
2011	He et al. [13]	Increasing O_2 from 500 to 2000 Nm^3/h	Maximum velocity liquid increased from 1.4 to 21 cm/s

influence due to the gas flow rate, as shown by the following equation:

$$\tau = 309 - 77Q - 0.72S - 1.23A - 0.57B - 0.44C$$

where τ is mixing time in s, Q is the oxygen flow rate in Nm^3/h, S is the horizontal angle with 2 lances, A is the vertical angle with one lance, B is the vertical angle with two lances, C is the vertical angle with 3 lances. Mixing time due to oxygen injection was also reported by Yang et al. [17].

$$\tau = 1648Q^{-2.35}L^{0.6}\theta^{0.2}$$

where L is the distance between the oxygen blowing point and furnace wall in m, θ is the vertical angle.

Qin et al. [18] simulated oxygen injection with three lances, reporting mixing time for an EAF without EBT of 100 ton of nominal capacity using a water model with a geometric scale 1:10, varying the oxygen flow rate (43–83 Nl/min), the horizontal angle (10–30 degrees) and vertical angle (30–50 degrees). Their results indicate a strong effect of the horizontal angle, suggesting $10°$ in conjunction with a vertical angle of $50°$ to reach the shortest mixing time. Decreasing the vertical angle increases the distance from the top of the lance to the liquid. He et al. [19] studied by CFD the effect of the horizontal angle on the velocities of liquid steel, comparing two angles $0°$ and $20°$ for an EAF with 150 ton of nominal capacity. Increasing the horizontal angle from $0°$ to $20°$ promotes tangential forces which enhance the velocity of the liquid. Jia et al. [20] employed a water model with a geometric scale 1:3.5 to investigate the effect of the horizontal angle ($10°$ to $30°$), the vertical angle ($5°$–$15°$) and lance height (57–67 cm) on mixing time, using two lances. The shortest mixing time was obtained with a horizontal angle of $30°$, a vertical angle of $15°$ and a lance height of 67 cm. Ramirez and Conejo [21] reported a strong effect of the vertical angle on fluid flow and in turn on the decarburization rate, suggesting to decrease from current values in the range $40°$–$45°$ to $20°$.

Table 9.3 Mixing time due to oxygen injection

Year	Authors	$\theta_h,°$	$\theta_v,°$	L, cm	Optimum
1997	Caffery et al.	30			$\theta_h = 0$, 30 (optimum based on avg U)
2012	Qin et al.	10–30	30–50		$\theta_h = 10°$, $\theta_v = 50°$
2013	He et al.	0–20			Tangential forces increased
2016	Jia et al.	10–30	5–15	57–67	$\theta_h = 30°$, $\theta_v = 15°$, L = 67

Table 9.3 summarizes the previous results. The vertical angle in steel plants is usually in the range from 45 to 50 degrees. Three of the reports in the table agree on 30 degrees for the optimum horizontal angle.

9.1.2.3 Forced Convection Due to CO Generation

Oxygen injection has two components for the motion of liquid steel; momentum transfer due to impingement of the jet and the resulting CO due to decarburization (DeC). CO bubbles provide an additional source of stirring. Guo et al. [10] simulated two decarburization rates, low at 0.002%C/min, equivalent to 0.034 kg/s and high, equivalent to 0.136 kg/s. The generation of CO was estimated in an area close to the injection of oxygen. The average velocity due to CO generation was about 3.6 cm/s, similar in magnitude to bottom gas injection. The combined effect of bottom gas injection and CO generation increased the average velocity to about 6 cm/s. In a later work, Guo et al. [22] compared the Euler-Euler and Euler-Lagrangian models, suggesting a much better prediction capability using the Euler-Lagrangian modelling approach. Chen et al. [23] reported a mathematical model for DeC using four coherent jets with a fixed inclination angle of 45° and compared the velocity of the liquid due only to momentum stirring with the additional effect of CO stirring. The average velocity of the liquid was 1.4 cm/s due to momentum stirring, increasing to 13.4 cm/s due to both momentum and that with CO stirring. The volume average velocity of liquid steel including the effect of CO is much higher than that reported by Guo et al. [10]. The values reported by these two groups of researchers shows considerable differences, partially explained by the fact that the case with a lower velocity uses one oxygen jet and most probably the result corresponds to the lower decarburization rate (not clarified in the report) while the second case employs four coherent jets and a higher decarburization rate. However, the reported average velocity due to oxygen injection by the two groups of researchers was 2 and 1.4 cm/s, respectively, surprisingly a lower velocity for the case using 4 coherent jets. Therefore, in the first case CO from decarburization doubles the velocity but in the second case the increment is 10th fold. Clearly, this subject requires further confirmation. Motion of liquid steel due to CO increases when the concentration of carbon is higher, however it severely decreases at the end of the heat when the carbon content is lower. Table 9.4 summarizes the average velocity of liquid steel only due to CO stirring.

Table 9.4 Fluid flow due to CO bubbles from decarburization

Year	Authors	Characteristics	$\overline{U}$, cm/s
2000	Guo et al. [10]	190-ton, 1 vertical lance, 3060 Nm³/h, two DeC rates	3.6
2020	Chen et al. [23]	Four coherent jets, 0.020%C/min	13.4

He et al. [13] reported the decarburization rate as a function of the flow of liquid steel through the exposed area to the oxygen jet, as follows:

$$r_c = -0.069 \, \exp\left(-\frac{Q_s}{15.176}\right) + 0.074$$

where r_c is the decarburization rate in %/min.

9.1.2.4 Forced Convection Due to Electric Arc Electromagnetic Forces in AC-EAF and DC-EAF

Electromagnetic forces or Lorentz forces result from the product of the current density and the magnetic flux density. The Lorentz forces in the arc plasma impinge on the liquid bath and move radially, parallel to the liquid bath producing shear stresses that induce motion of the liquid. Alexis et al. [24] reported that the shear stresses are proportional to the radial jet velocity. The velocity of the arc jet is higher at higher arc currents [25]. The direction of motion of liquid steel depends on the phase sequence in a three electrode-EAF. The phase sequence should be adjusted counterclockwise and the rotation of the magnetic field will also be in the same direction, which avoids electrode loosening. Elkoumy et al. [8] evaluated stirring under forced convection due to the electromagnetic forces from an electric arc power of 120 MW, indicating an increase in the maximum velocity of liquid steel by one order of magnitude from 0.07 to 0.62 cm/s by increasing the slag thickness from 5 to 30 cm. This result is unexpected because the top slag layer plays a role to decrease the motion of liquid steel as demonstrated by Amaro et al. [26].

AC-EAF: The magnitude of the Lorentz forces with AC are low and limited to the region around the zones of impingement of the electric arcs. Current can flow from one electrode to the other, through the slag and the metal, however since the electric conductivity of the metal is much higher than the slag, current flows through the metal [27]. Eunny and Lahiri [27] reported that the peak in current density is through the corner of the electrode tip. Zhu and Wei [28] reported large velocities around the pitch circle, about 3 cm/s, however the velocity of the liquid at the walls decreased to 0.6 cm/s.

DC-EAF: Lorenz forces are stronger in DC-EAF compared to AC-EAF because electron transfer is not limited to the region of liquid steel around the electrodes. In DC-EAF electron transfer occurs from the top cathode to the bottom anode, the passage of electric current through liquid steel with its own magnetic fields causes electrically induced flow [29].

Wang et al. [30] reported that the diameter of the bottom electrode influences the circulation patterns and also that a higher electrode diameter at the bottom, 2 m versus 0.2 m, produces better mixing conditions. Similar results have been reported by other researchers [29, 31]. Yang and Wu [31] reported an increment in the average velocity from 3.64 to 4.51 cm/s by increasing the bottom electrode radius from 0.1 to 0.48 m. On the contrary, in a previous report by Wang et al. [32] they compared a small and a larger bottom electrode, 0.1 and 1 m respectively, in a small EAF with a depth of the molten pool of 0.5 m and reported higher electromagnetic forces using the small bottom electrode.

Liu et al. [33] reported the recirculation rate of liquid steel and mixing time for two DC-EAF's with a capacity of 30 and 75 tons. The recirculation rate increased by increasing the electric current. For the larger EAF with a current of 45 kA, with a depth of liquid steel of 0.7 m and cathode diameter of 0.5 m, it was 3.2 ton/s, each recirculating cycle took 23 s and mixing time was 69 s.

Kazak investigated fluid flow due to electromagnetic forces in DC-EAF's [34, 35]. One problem in DC-EAF is a larger erosion of the refractory bottom due to intense fluid flow caused by electromagnetic forces, in the range of 0.01–8659 N for a 410-ton EAF with four bottom electrodes and two top electrodes. Flow of current from the bottom (anode) to the top (cathode) electrode and its associated magnetic field creates a vortex flow of liquid metal, as shown in Fig. 9.3. Liquid metal rises from the bottom and forms recirculation loops. A column of liquid metal of high velocity forms between the two electrodes, reaching values close to 50 cm/s at the bottom electrode and about 10–20 cm/s below the top electrode, drastically decreasing away from this column with values about 5 cm/s [36]. Overall, thermal stratification is less than 50 °C. Zhao et al. [37] also reported similar values, in the order of 30 °C for an EAF of 60 ton with a current of 20 kA and a bottom electrode with a radius of 0.7 m.

Raising the position of the bottom electrode was found to decrease the maximum velocities and a decrease in the shear stresses by about 30%. Further improvements were reported cooling down the bottom electrode temperature [38]. Simulations injecting argon suggested that bottom gas injection accelerates the shear stresses and therefore a higher refractory erosion rate [39]. Kazak et al. [39] cited the following relationship to define the velocity of the liquid due to electromagnetic forces:

$$U_o = j_o L \sqrt{\mu_o / \rho}$$

where U_o is the velocity of the liquid, j_o is current density (I/S), L is the diameter of the bottom electrode, μ_o is the permeability of free space, and ρ is the density of liquid steel

Ramirez et al. [40] reported an Eulerian-Eulerian mathematical model for a DC-EAF with and without bottom gas injection. Without gas injection the electromagnetic forces are the main driving forces for the motion of the fluid. An increase in the arc length increased the heat input but decreased the velocity of the liquid, in other words, a short arc operation improves the mixing conditions of the bath. The top slag

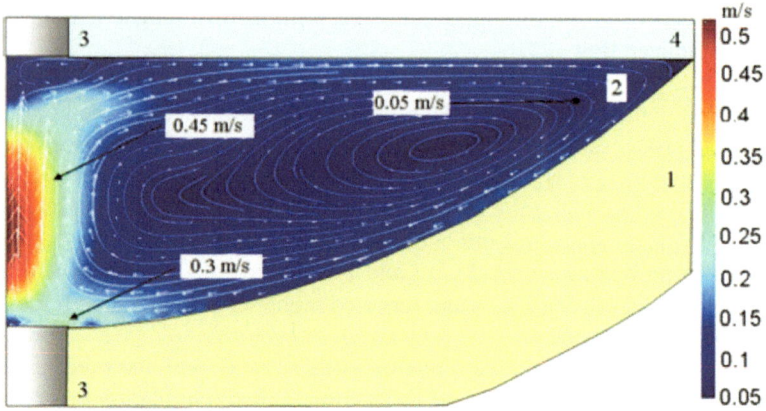

Fig. 9.3 Streamlines of vortex flow velocities in a DC-EAF (1—refractory, 2—liquid metal, 3—electrodes, 4—slag). After [35]

layer also contributes to decrease the velocity of liquid steel. Combined electromagnetic forces from the arcs and gas injection improved thermal homogenization but the area under the influence of electromagnetic forces was decreased which in turn increased the temperature of that region.

In order to enhance electromagnetic forces and reach a higher mixing intensity in the DC-EAF, Kukharev et al. [41] proposed to increase the number of electrodes, three on the top and three on the bottom. Smirnov et al. [42] studied the effect of the number of bottom electrodes and its position on the mixing intensity. They reported higher stirring intensities with two bottom electrodes giving 805 s for the shortest mixing time. Mixing time was decreased from 805 to 300 s. by increasing the arc current from 10 to 20 kA. Bowman and Krüger [43] have reported maximum velocities of liquid steel in DC-EAF in the range from 20 to 60 cm/s, for currents in the range from 40 to 100 kA. Table 9.5 summarizes the results on liquid steel velocities reported for AC and DC-EAF's.

9.2   Bottom Gas Stirring

9.2.1   Introduction

Bottom gas injection (BGI) in the ladle furnace is a proven technique and is used widely, however EAF bottom gas injection has the following significant differences with bottom gas injection in ladles; (1) During the melting process the furnace is filled with solids and the liquid bath is shallow, (2) large spouts can create arc instability, (3) The vessel is not a cylinder. It could be visualized like a cylinder but with a very steep angle. The ratio of height of liquid steel to shell diameter is in the order of

Table 9.5 Fluid flow velocities due to electromagnetic stirring in AC and DC-EAF's

AC	U, cm/s	U_{max}, cm/s
Zhu and Wei [28]	3 (pitch circle) 0.6 (walls)	
DC	$\overline{U}$, cm/s	U_{max}, cm/s
Yang and Wu [31]	3.64 ($r_b = 0.1$)	
	4.51 ($r_b = 0.48$)	
Ramirez et al. [40]		150 ($L_{arc} = 15$ cm) 100 ($L_{arc} = 35$ cm)
Wang et al. [32]	3.6–4.1	
Kazak et al. [36]		50 (bottom electrode) 10–20 (top electrode) 5 (away from the column)
Bowman and Krüger [43]		20 (40 kA) 60 (100 kA)

0.15–0.35 much lower than a ladle which has values in the order of 0.8–1.2. In spite of these differences, BGI has proven to enhance the stirring conditions of the EAF.

Bottom gas stirring was first applied in the late 1980's in several countries. The interest on BGI decreased for almost 20 years from the early 1990's probably due to operational problems with porous plug blockage and maintenance issues, however from about 2010 onwards the interest in this technology raised again and numerous investigations have been reported to identify the optimum injection conditions to improve mixing phenomena.

Lazcano et al. [44, 45] from the former Metallurgical Mexican Laboratories (IMIS) in the late 1980's installed porous plugs to inject natural gas in two Mexican steel plants but the systems were eventually cancelled due to severe wearing of the porous plugs and enormous difficulties to change them before the end of life of the hearth's crucible. Similar problems were reported by Nisshin Steel in Japan [46] in 1990. In this case they designed a new system consisting of a removable bottom section.

9.2.2   Bottom Gas Injection Systems

Bottom gas injection can be done in three ways, schematically shown in Fig. 9.4:

- Direct system
- Indirect system
- Hybrid system.

In the direct system the porous plug is in contact with liquid steel, stirring intensity is higher, gas consumption is lower, the plume is narrow and forms a slag eye, the

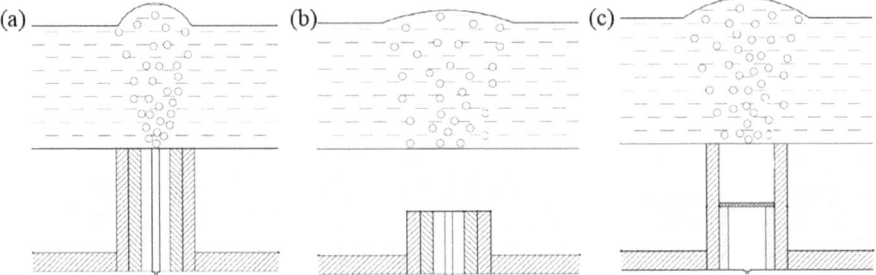

Fig. 9.4 Bottom gas injection systems. **a** Direct, **b** indirect and **c** hybrid

stirring conditions are carefully controlled but porous plugs have a short life span requiring more maintenance. Another disadvantage of the direct system is the shallow bath. The principal mixing mechanism due to gas injection is the dragging force of the liquid by the ascending bubbles. With shallow baths the bubble residence time is small. Increasing the gas flow rate in a direct system has the following limitations;

- The height of the spout increases, increasing steel reoxidation
- Heat losses increase due to direct exposure of liquid steel to the atmosphere
- Risk to induce short-circuit current when liquid steel gets in contact with the electrodes
- Electrode washing when liquid steel gets in contact with the electrodes

Dong et al. [47] suggested a maximum spout height of 5 cm. By 2013, Kirschen et al. [48] argued that the direct systems were the most common system in the steel industry worldwide. Ma et al. [49] reported a lifespan of 700 heats for the direct system. The injection elements for ladles and EAF's are different. According to Ishihara from Shinagawa refractories [50], the plugs for EAF's are not porous but consist of small diameter pipes, as shown in Table 9.6.

In the indirect system the porous plug is immersed in the hearth's crucible, covers a wider area avoiding the slag eye and refractory life is much longer but the stirring conditions cannot be accurately controlled and gas consumption is higher. In this case the permeability of the refractory above the porous plug should be carefully selected to avoid infiltration of liquid steel and at the same time promote diffusion of the stirring gas.

Table 9.6 Comparison of injection devices (after [50])

Application	EAF	BOF	Ladles
Injection device	Pipes	Pipes	Porous plug
Diameter of pores/pipes	1–2 mm	1–2 mm	100–300 μm
Number of pipes	3–20	50–200	–
Refractory type	MgO-C	MgO-C	High alumina

Mimura and Ususaka [51] described disadvantages for the two previous systems; a short life span of about 150–300 heats for the direct system and then demanding its replacement, and about 3000 heats for the indirect system but several limitations due to clogging of the ramming mass and dispersion of the stirring gas. They proposed a hybrid combination with a porous plug at the bottom and on the top a cylinder containing ramming mass of equal or higher permeability than the furnace bottom. An intermediate section provides conditions to ensure a homogeneous supply of gas.

Tonghua Steel (part of Shougang group) installed an indirect system in a Consteel EAF with 70 ton of nominal capacity equipped with three porous plugs. Dong et al. [47, 52] reported the following benefits in this plant; a higher decarburization rate, from 0.04%C/min to 0.10%C/min, about 10 Nm^3/ton lower oxygen consumption and lower iron oxide, about 2.5% and shorter tap to tap time. The slag/metal system was closer to equilibrium, for example for 0.015% C reported 18%FeO. Iscor (currently ArcelorMittal South Africa) installed an indirect system designed by Thyssen-Stahl AG, called TLS, in one EAF with a capacity of 55 tons consisting of three porous plugs located between the electrodes at a circle diameter of 1600 mm, outside of the electrode pitch circle diameter of 1200 mm to avoid instability of the electric arcs [53]. They choose the indirect type because of a longer life of the hearth's crucible. The system installed came also with a deflector plate and a cylinder. The deflector plate avoids porous plug clogging due to metal infiltration and the cylinder covers a region to avoid the gas to escape towards the sidewalls of the furnace. They initially reported problems due to blockage of the porous plugs which were solved maintaining a thickness of the ramming material above the plugs at least 250 mm thick, otherwise there is the formation of a top layer from 150 to 200 mm consisting of infiltrated slag that contains cracks and allows the penetration of liquid steel, furthermore, at all times it was ensured that the gas flow rate should not exceed a maximum value, in this case 5 Nm^3/h because excessive uncontrolled gas flow rate leads to disruption of the refractory material. The benefits observed in this plant using bottom gas stirring were: shorter tap-to-tap time (3 min), lower electrode consumption (0.15 kg/ton), lower FeO (5–10%), higher metallic yield (1.2%), however energy consumption was not decreased, which the authors attributed to extra energy to reduce FeO.

Gulyaev et al. [54] compared the direct and indirect system in a conventional EAF of 100 ton, using three porous plugs. In the direct system two porous plugs are located between the electrodes slightly away from the pitch circle and the third one is located closer to the EBT region. In the indirect system three plugs are located below the pitch circle and a fourth plug is located along the centerline of EAF, close to the EBT. The gas flow rate is adjusted in three stages, increasing from 20 to 60 Nl/min. The indirect system provided better results than the direct system, comparing P, S and lime consumption. Overall, energy consumption was decreased 15–20 kWh/ton, lime also decreased 8–10 kg/ton and power-on time decreased by 3 min.

Due to the cyclic nature of the operation of the EAF which involves melting the metallic charge, refining and tapping, bottom gas stirring is carried out in three stages [55]. Moderate gas flow rate during the melting process, then a higher flow rate during flat bath operation and finally a third stage during the tapping process.

Directional Porosity Porous plugs (DPP) can be made of MgO-C with small stainless-steel tubes of 1 mm in diameter. The gas flow rate is controlled separately and the typical range varies from 10 to 100 Nl/min.

9.2.3   Injection Gases

Ar, N_2, natural gas, CO_2, CO and air have been employed as stirring gases in the EAF [56]. Argon is the dominant source for gas stirring due to its inert properties.

Lazcano et al. [44] replaced argon by natural gas. They claimed that the decomposition of natural gas into carbon and hydrogen was negligible due to the short residence time, therefore it can be employed in the production of low carbon steels. This technology was applied in the early 1990's in Mexico but problems with high frequency to change the porous plugs and difficulties to change them forced to eliminate its use. The same authors suggested a device that fully closed the slag door avoiding ejections of steel out of the furnace during BGI [44].

Zhu et al. have extensively investigated the use of CO_2 in steelmaking [57–59]. The stirring intensity with CO_2 is higher because it reacts with C, Si, Mn and Fe. During decarburization, one mole of CO_2 can produce two moles of CO, therefore the stirring intensity of CO_2 increases with carbon concentration but on the other hand, when carbon is below 0.2% the dissociation ratio of CO_2 falls below 40% or is almost suppressed depending on the concentration of sulphur [60]. In addition to this, CO_2 injection involves endothermic reactions with C and Fe contributing to a cooling effect around the porous plugs and therefore contributing not only to higher dephosphorization rates but also increasing the life of the porous plugs. Zhu et al. [58] also reported higher decarburization rates using CO_2 because of its oxidizing nature. The concentration of nitrogen was lower injecting N_2-CO_2 mixtures in comparison with Ar-N_2 mixtures. In spite of these advantages CO_2 has not replaced argon as the main stirring gas. Wang et al. [59] addressed some of its current limitations in comparison with argon: (1) It can increase C, (2) it can oxidize Cr, (2) there is no agreement on the optimum amount, (3) it requires a preparation process which increases its cost, and compared to oxygen the decarburization rate is lower, increasing the tap-to-tap time.

9.2.4   Optimum Injection Layout Based on Mixing Time Measurements

Mixing time has been commonly employed as a criterion to define the optimum gas injection layout. Most of the water and mathematical modeling studies on mixing time in metallurgical reactors have been carried out under isothermal conditions. Banerjee and Irons [61] demonstrated that mixing time is higher in non-isothermal systems, especially at low gas flow rates, indicating that thermal stratification opposes mixing and that mixing time in isothermal systems cannot be applied to top heated systems, like the EAF. In spite of this isolated report, isothermal studies constitute one of the basis for decisions on mixing efficiency.

Conventional EAF without EBT: Rui et al. [62] investigated mixing time for 5 different layouts in a water model with a geometric scale 1:7 and a flow rate in the range from 3 to 11 Nl/min. The layouts included 1, 3 and 4 porous plugs. In the layout with one porous plug, the plug was placed in the center. With 3 and 4 porous plugs, each case placed the plugs in the pitch circle and at 0.6R, as shown in Fig. 9.5. Their results on mixing time indicate that the number of porous plugs is not the dominant factor. Both layouts with 3 and 4 porous plugs gave maximum mixing efficiency if the plugs were placed on the pitch circle, between the electrodes. Li et al. [63] reported isolines for mixing time that clearly indicate the position of the dead zones on the regions closer to the walls and top volumes of liquid.

He et al. [64] compared mixing time in a water model with a geometric scale 1:4 from a small EAF with 10 ton of nominal capacity for central and off-center gas injection (0.6R). They reported greater mixing efficiency for central gas injection in comparison with off-central gas injection, as shown in Fig. 9.6. This behavior is in contrast to what it is found in ladles, indicating very different flow patterns in both reactors due to a different H/D ratio.

Liu and Wen [65] reported mixing time in a water model with a geometric scale 1:4.3 from a small EAF with 20 ton of nominal capacity for different nozzles, from 1 to 4, located around the pitch circle (0.18R). The slope of the hearth was 45°. Their results indicate that increasing the number of porous plugs doesn't necessarily improve the mixing conditions, obtaining shorter mixing time with one porous plug in the center (case a) as shown in Fig. 9.7. The case with three porous plugs gives the

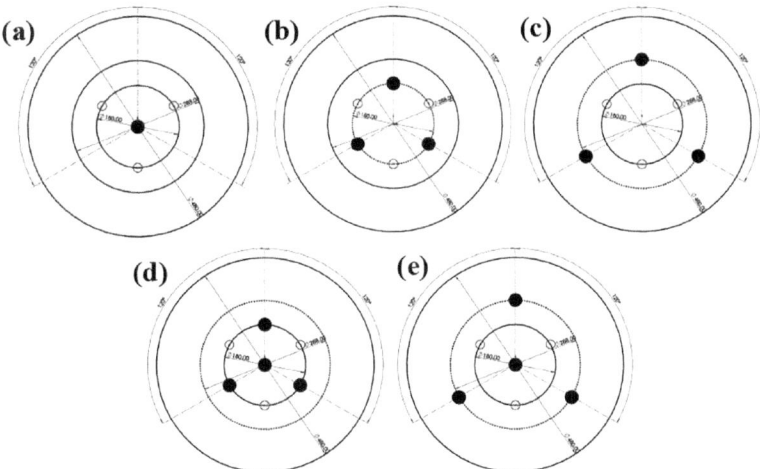

Fig. 9.5 Number and position of porous plugs investigated by Rui et al. [62]

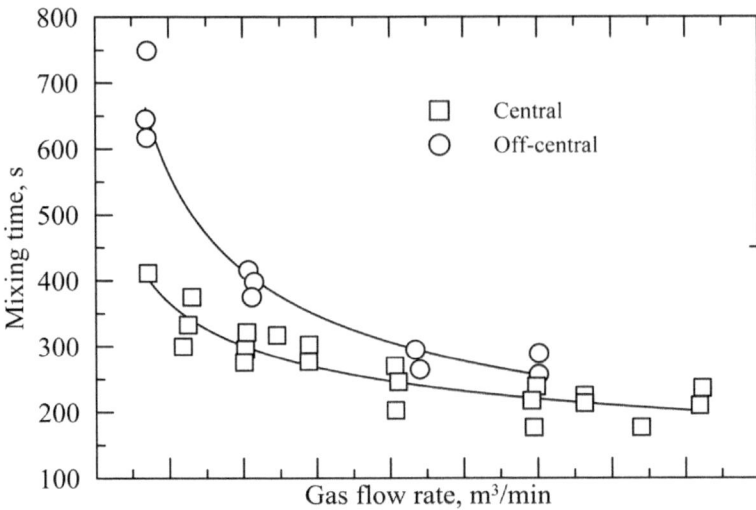

Fig. 9.6 Mixing time for central and off-central gas injection in the EAF [64]

shortest mixing time but is not far from the results obtained for case a. In addition to this, they also reported that increasing the gas flow rate decreases mixing time up to a critical value above which the mixing conditions are reversed and mixing time increases. For their model this critical value was about 8.5 Nl/min.

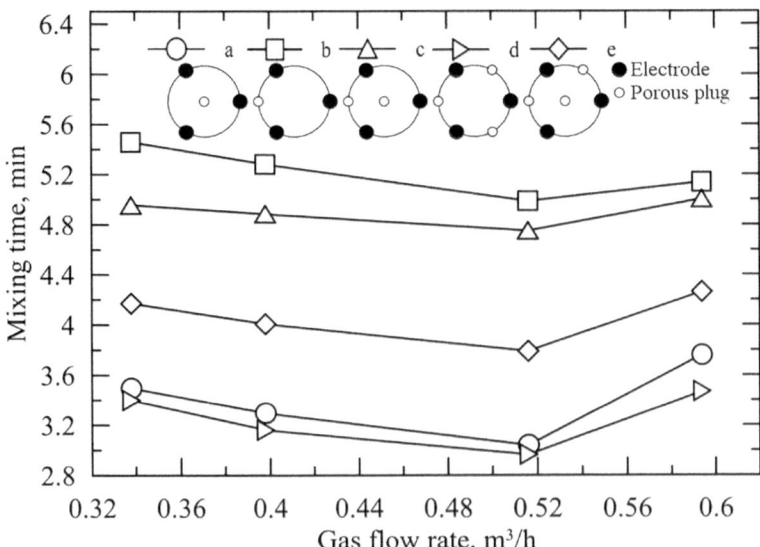

Fig. 9.7 Mixing time for different gas injection layouts, reported by Liu and Wen [65]

On the contrary, Jiang and Li [66] reported that mixing time decreases by increasing the number of porous plugs, as shown in Fig. 9.8, suggesting an optimum layout with one in the center and three on the electrode circle with a separation angle of 120°. Their results are based on a conventional EAF without EBT. They compared 4 cases; one in the center, three along the electrode circle, three on the electrode circle and one central, and one plug off center. Furthermore, they also investigated the effect of the nozzle diameter in the range from 1 to 3 mm, suggesting a diameter from 1 to 1.5 mm decreases mixing time. The experimental results about the effect of the height of the liquid on mixing time indicate that increasing this height, mixing time increases, which is wrong, however their conclusion state the opposite. Increasing the height of liquid steel allows a higher exchange of potential energy from the bubbles and stirring gets improved.

Liu et al. [67] also reported better mixing conditions with three porous plugs along the pitch circle in comparison with one centrally located. They carried out velocity measurements of the liquid with a laser doppler anemometer technique in a water model with a geometric scale 1:7 and a constant gas flow rate of 3 Nl/min. Kazakov et al. [68] used a water model for a 100 ton EAF with a geometric scale 1:25 to study mixing time changing the number of porous plugs from 1 to 4. Details on the position was not reported. The shortest mixing time was reported using 2 porous plugs; 8.9 s, compared with 14.9 and 9.7 s with 3 and 4 porous plugs, respectively. They also reported a minimum gas flow rate of 60 Nl/min to reach thermal homogenization based on thermal calculations.

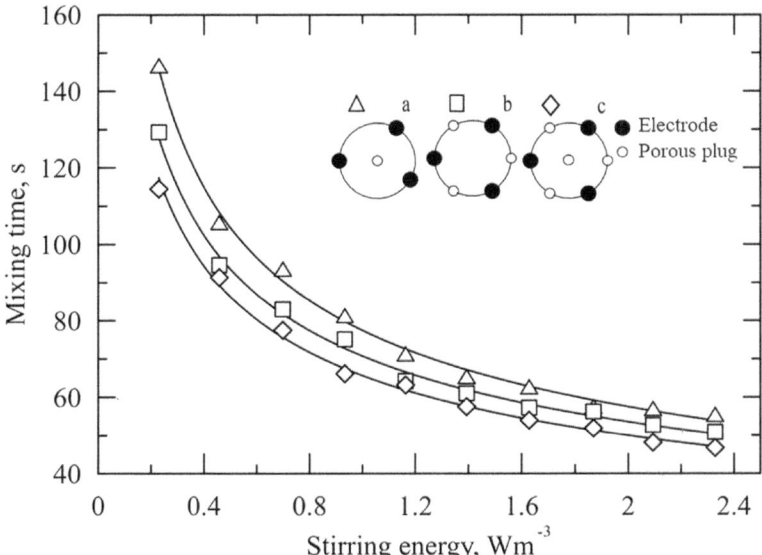

Fig. 9.8 Effect of stirring energy on mixing time for the layouts investigated by Jiang and Li [66]

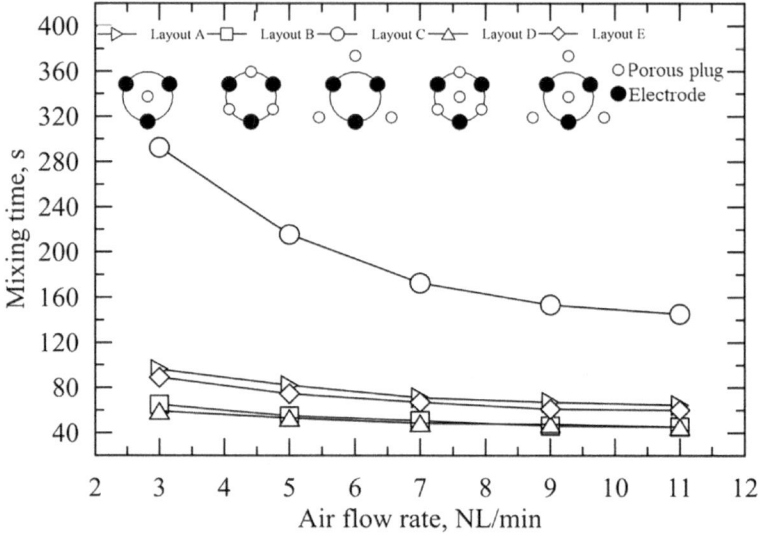

Fig. 9.9 Mixing time fort different layouts investigated by Li [69]

Li [69] reported mixing time in a water model from a conventional EAF without EBT. He investigated five different layouts containing 1, 3 and 4 porous plugs at two radial positions 0.4R and 0.65R. He reported an optimum layout consisting of 3 porous plugs, located at a radial position 0.4R, separated 120° with or without a 4th one located in the center (layouts B or C) as shown in Fig. 9.9. They developed a mathematical model which accurately predicted the experimental data.

Modern EAF with EBT: The previous results are useful to get a better understanding of mixing phenomena in the EAF due to BGI, however cannot be applied to modern EAF's with EBT. The EAF with EBT has the potential to produce a higher volume of dead zones close to the EBT and requires different layouts. Guo et al. [10] simulated bottom gas injection with three porous plugs and flow rates of 20 and 40 Nl/min per plug. The average velocity of liquid steel was 3.8 cm/s. Zhu et al. carried out physical and mathematical modeling work [70, 71]. The water model had a geometric scale 1:4 from a 75-ton EAF/EBT. They analyzed 2 different layouts, each at 3 different radial positions. Each layout consisted of two fixed positions with a separation angle of 150° and third plug with two variable positions that correspond to two different sets of separation angles (112°, 98°) and (90°, 120°), respectively, as shown in Fig. 9.10. The radial positions studied (r/R) were 0.4, 0.5 and 0.6 with reference to the bottom radius. Kerosene, about 7% by volume, was placed on the top. Mixing time was measured at three points and then the average value reported as mixing time. The results from this work clearly indicate that placing the plugs close to the crucible walls negatively affects mixing time, therefore the best layout was at a radial position of 0.4R and the following set of separation angles (150°,112°, 98°) which indicates that the location of the third plug should be closer to the EBT region. The mathematical model was developed applying the VOF algorithm and the standard k-ε turbulence model, using the dimensions of the prototype. The results show two main regions with high and low velocities, respectively. The middle region has lower velocities and is surrounded by top and bottom regions of higher velocities. The EBT region is another small zone with lower velocities than the rest of the liquid steel. For a constant gas flow rate, the average velocity for the optimum layout was 0.78 cm/s in comparison to 0.54 cm/s for the plugs located close to the walls, furthermore, using as a reference the volume of liquid with a velocity lower than 0.06 cm/s, the dead volumes were defined as 0.49 and 1.31 m³, for the best and worst layouts, respectively. The gas flow rate has a large effect on the average velocity and dead volumes. Increasing the gas flow rate from 50 to 200 Nl/min increased the average velocity from 0.5 to 1.2 cm/s and the dead volume decreased from 1.4 to 0.55 m³. Industrial trials comparing a conventional EAF without stirring and stirring with 3 porous plugs in accordance with the best layout from their physical and mathematical modeling work, indicated three main advantages; higher dephosphorization rates, shorter tap-to-tap times by two minutes and lower FeO, from an average of 23.4 to 19.3%. The solubility product %C × %O was decreased by about 25%. The temperature gradient was also decreased from 34 to 11 K using bottom gas injection. In another study using a cylindrical vessel representing bottom gas injection in a 3 ton EAF that produces high nitrogen FeV it was reported that the radial position closer to the wall (0.7R) produced higher mixing conditions [72]. A water model with a geometric scale ratio 1:2 and kerosene were employed to simulate the prototype. The ratio height of liquid to diameter was 0.15, three nozzles symmetrically located with a separation angle of 120° and an oil thickness of 26% were employed in the cold model. The radial positions investigated varied from 0.3 to 0.7R. Although this study is cited to describe the effect of the nozzle radial position on the stirring conditions in the EAF is far from representing the geometry of modern EAF's and therefore their results cannot be compared with an EAF equipped with EBT.

Yang et al. [73, 74] developed a mathematical model using the VOF algorithm and the k-ε turbulence model to study the effect of the gas flow rate, in particular changing the gas flow rate for the porous plug located closer to the EBT. They used a constant radial position of 0.5R (R is the radius of the shell diameter) and a constant separation angle (135°, 112° and 113°), as shown in Fig. 9.11a. Their numerical model results indicate that the dead area decreases by increasing the gas flow rate at the porous plug closer to the EBT. It decreased 37% from 0.66 to 0.41 m³ by increasing the gas flow rate from 100 to 200 Nl/min. Yang et al. [75, 76] additionally reported numerical model predictions using the VOF algorithm and the Large Eddy Simulation (LES) turbulence model indicating a maximum velocity of liquid steel of 3.0 cm/s. Keeping the gas flow rate of two porous plugs constant at 133 N/min and increasing the gas flow rate of the third porous plug closer to the EBT from 100 to 267 Nl/min, they reported an increase in the average velocity from 0.28 to 0.45 cm/s, as shown in Fig. 9.11b.

Yang et al. [77] carried out physical and mathematical modelling to compare uniform versus non-uniform gas injection. The results from the water model indicated a shorter mixing time using a linear non-uniform gas injection, however the

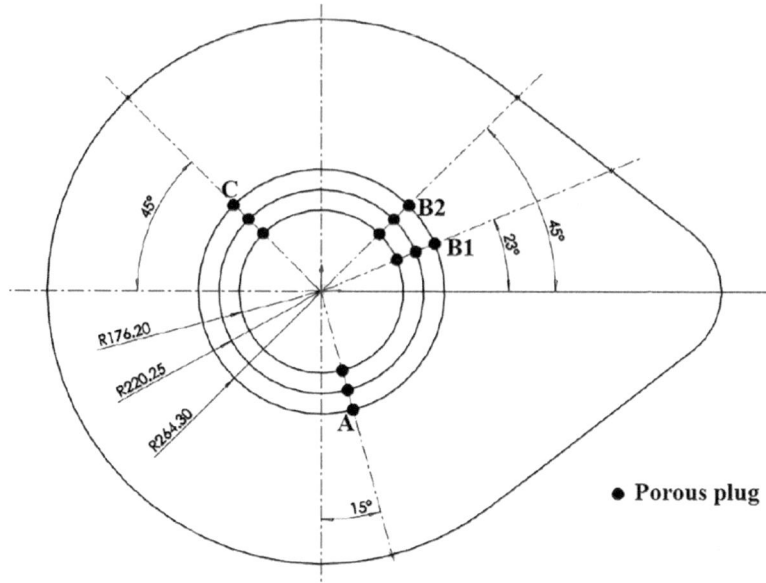

Fig. 9.10   Position of porous plugs investigated by Zhu et al. [70, 71]

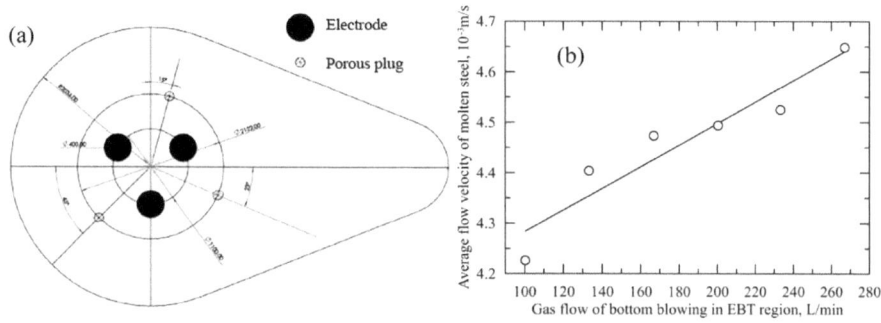

Fig. 9.11   (a) Position of porous plugs and (b) Average velocity in the study carried out by Yang et al. [73, 74]

comparison was made using different total gas flow rates. The numerical model results for a 100-ton EAF were carried out with a total gas flow rate of 400 Nl/min. Their results also indicate that a linear non-uniform gas flow rate gives higher stirring intensity The average velocity of liquid steel increased from 1.09 to 2.07 cm/s from uniform (133–133–133) to linear non-uniform gas injection (80–133–186). Increasing the gas flow rate in the porous plug closer to the EBT zone didn't produce better mixing conditions. Industrial results using non-uniform gas injection indicate conditions for the metal/slag system closer to equilibrium with a decrease in the C-O solubility product to 0.0030 and a decrease in FeO of 8%.

Table 9.7 summarizes reported relationships that predict mixing time as a function of gas flow rate (or its equivalent, stirring energy). From this table it can be observed that only one report included the effect of the height of liquid steel. In reference [66] they reported a plot and one equation indicating that increasing the height also increases mixing time, which is wrong. The sign was changed in the table, in agreement with the conclusion by the same authors.

In spite of the relatively large amount of research work studying the effect of the injection layout on mixing time, the optimum layouts show large differences in terms of the optimum radial position, number of porous plugs and their separation angle. One of the reasons is that different experiments consider different set of variables. Table 9.8 summarizes the large variation in geometric scale ratios (from 1:1 to 1:7), with and without EBT, height of liquid/diameter, etc.

Table 9.9 summarizes the optimum layouts reported in previous investigations using mixing time as the decision criterion. The values suggested for the number of porous plugs, its radial position and separation angle is quite variable. It is interesting to observe that two cases suggest central gas injection, the worst layout in the case of ladles, which suggests a completely different flow structure between ladles and the EAF. Few investigations have carried out an extensive analysis of each of the previous variables. Li [69] extensively studied the radial position but using a conventional EAF without EBT. Zhu et al. [70, 71] also compared three radial positions and two separation angles, suggesting a radial position at 0.4R.

Velocities of liquid steel due to bottom gas injection are summarized in Table 9.10. Basically two groups have reported results; Guo et al. [10] and Zhu et al. [70, 71, 73–77] in Canada and China, respectively. Guo et al. reported an average velocity of 3.8 cm/s using a total flow rate of 60 Nl/min. On the other hand, Zhu et al. have reported lower average velocities in spite of larger gas flow rates (100 to 400 Nl/min), from 0.28 to 2.07 cm/s. Yang et al. [77] have reported that a non-uniform pattern of gas flow rate increases the average velocity.

The effect of the steel physical properties on fluid flow in most of the previous investigations has not been addressed in detail. Elkoumy et al. [6, 8] reported that

Table 9.7 Mixing time relationships for the EAF

Year	Equation	Notes	References
1994	$\tau_{center} = 2162Q^{-0.277}$ $\tau_{off\text{-}center} = 4294Q^{-0.377}$		[64]
1995	$\tau = 434\varepsilon^{-0.35}$	$\varepsilon = \frac{0.285QT}{W}\log(1 + H/148P)$	[78]
1996	$\tau = 65.72\varepsilon^{-0.39}$ $\tau = 3.29Q^{-0.43}H^{0.65}$	$\varepsilon = \frac{6.1825\times10^{-3}QT}{V}\log(1 + \rho_l gH/P_o)$ 4-porous plugs	[66]
1998	$\tau = 80.79\varepsilon^{-0.48}$	1-porous plug	[63]
2002	$\tau = 1648Q^{-2.35}L^{0.6}\theta^{0.2}$	Effect of O_2 injection	[17]
2011	$\tau = 366\varepsilon^{-0.277}$	Three porous plugs	[52]

Table 9.8 Dimensions of EAF models in previous investigations

		Authors	EAF	EBT	Scale	D_b, mm	D_t, mm	H_l, mm	d_n, mm	h_s, mm	H_l/D_t
1	1994	He et al. [64]	10	No	1:4	420	660	150		-	0.22
2	2000	Guo et al. [10]	190	Yes	1:3			1260		150	
3	2000	Li [69]	30	No	1:7	30	50	10	3	0	0.20
4	2013	He et al. [19]	150	Yes	1:1		6200	1000	66	62	0.16
5	2015	Zhu et al. [62]	75	Yes	1:4	881	1175	325	2.5	25	0.27
6	2016	Ma et al. [49]	70	Yes	1:1		5700	1000			0.17
7	2017	Liu et al. [72]	3	No	1:2	870	870	135	2.5	35	0.15
8	2017	Yang et al. [65]	100	Yes	1:1	2122	4072	1430	11	165	0.35

the steel physical properties do not affect the velocity of the liquid. They studied small changes on the viscosity of liquid steel (0.0039–0.0047 Pa·s). On the contrary, Hu et al. [80] considered changes in steel chemical composition from the start of melting with a molten metal containing high carbon, about 2.5% and melts at low temperature, under these conditions the changes in viscosity are in the range from 0.0049 to 0.0113 Pa·s. The large change in viscosity has a large impact on the stirring conditions throughout the heat. They reported, for a constant gas flow rate of 100 Nl/min and three porous plugs, a decrease in the average velocity of the liquid from 0.8 to 0.5 cm/s and this effect was attributed to a decrease in viscosity. Physical modeling work confirmed that the residence time of bubbles decreases as the temperature of the liquid increases (corresponding to a decrease in liquid's viscosity).

Wei et al. [81] compared the continuous gas injection mode with a pulse gas injection mode by physical and mathematical modeling. The water model had a geometric scale ratio 1:4. KCl was added as a tracer, 50 ml in each one of the three nozzles. They choose a step-wise pulse instead of a sinusoidal method of injection using a programmable logic controller (PLC). Mode 1 was the conventional continuous way of gas injection with a constant gas flow rate of 2.12 Nl/min in the water model. Gas injection in the pulse mode was made during 1.2 s. In mode 2, the first pulse was injected at low gas flow rate of 1.59 Nl/min for 0.6 s and then switched to a higher value of 2.65 N/min for another 0.6 s. In mode 3 the gas flow rate in the second pulse was increased to 3.18 Nl/min but for a shorter period of time

Table 9.9 Optimum injection layouts

	Year	Authors	r/R	N	θ	Notes
1	1994	He et al. [64]	0	–	–	Comparing 0 and 0.6R, Better central gas injection
2	1995	Rui et al. [62]	Pitch circle	3	120°	5 layouts (N = 1 to 4, r/R = 0 and 0.6R). N is not a dominant factor
			0, pc	4	0°,120°	
3	1995	Liu et al. [67]	Pitch circle	3	–	r/R = 0, pitch circle
4	1996	Liu and Wen [65]	0	1	–	r/R = 0, 0.18, N = 1 to 4. Above 8 Nl/min τ increases
			Pitch circle	3	120°	
5	1996	Jiang and Li [66]	Pitch circle	4	120°	r/R = 0, pitch circle, N = 1,3, 4
6	1997	Caffery et al. [79]	Pitch circle	4	0°,120°	Single layout, not optimum
7	2000	Li [69]	0.6	3	120°	N = 1 to 4. No effect of d_n
			0,0.6	4	0°,120°	
8	2014	Kazakov et al. [68]	–	2	–	N = 1 to 4. Minimum 60 Nl/min thermal homogenization
9	2015	Zhu et al. [70, 71]	0.4	3	150°,112°, 98°	r/R = 0.4, 0.5 and 0.6 (150°, 112°, 98°) (150°, 90°, 120°)
10	2017	Zhu et al. [72]	0.7	–	–	N = 3, (120°,120°,120°) r/R = 03 to 0.7. Cylinder

Table 9.10 Velocities and dead volumes due to bottom gas injection

	Year	Authors	$\overline{U}$, cm/s	V_{dead}, m³	Notes
1	2000	Guo et al. [10]	3.8	–	190-ton EAF, H = 1.26 m, h_s = 0.15–0.3 m, Q = 20 Nl/min, N = 3
2	2015 2016	Zhu et al. [70, 71]	0.78 (r/R = 0.4) 0.54 (r/R = 0.6)	0.49 (r/R = 0.4) 1.31 (r/R = 0.6)	0.5 to 1.2 cm/s (Q from 50 to 200 Nl/min)
3	2017 2019	Yang et al. [73, 74]	–	0.66 to 0.41 (Q from 100 to 200 Nl/min)	r/R = 0.5 (135°, 112°, 113°)
4	2018	Yang et al. [75, 76]	0.43[1] 0.46[2]	–	[1]133–133–100 Nl/min [2]133–133–267 Nl/min
5	2020	Yang et al. [77]	1.09 (133–133–133) 2.07 (80–133–186)	–	Q_{total} = 400 Nl/min

of 0.4 s. For the mathematical model they used the VOF algorithm and the standard k-ε turbulence model, with a scale 1:1. They found shorter mixing times using the pulse injection mode compared with the conventional mode, 115.2, 108.9, 105.8 s for modes 1, 2 and 3, respectively. The numerical model reported higher velocities and lower volume of dead zones for mode 3. The average velocities and dead zones for modes 1, 2 and 3 were 0.0048, 0.072 and 0.0086 m/s and 1.42, 1.13 and 0.96 m^3, respectively. Mode 2 was applied at a steel plant and the reported benefits were a decrease of FeO by 2% and a decrease in the C-O solubility product, from 0.0030 to 0.0028. A disadvantage with this mode of operation is a shorter refractory life, from 612 heats for the conventional mode to 566 heats with the pulse mode.

Kirschen et al. [82–84] reported the number of porous plugs and gas flow rates for a number installations. EAF's with a capacity ranging from 45 to 140 ton have been designed with 3 porous plugs and gas flow rates ranging from 25 to 110 Nl/min. The number of porous plugs was increased up to 5 in EAF's with a capacity higher than 150 ton. The performance indicators were highly variable; yield increased from 0.5 to 4.5%, oxygen consumption decreased from 0.25 to 10 Nm^3/ton, electric energy consumption decreased from 5 to 17 kWh/ton and power-on time decreased by 1.5 to 9 min.

Ishihara [50] summarized gas flow rates employed in different steel plants producing different types of steels. For carbon and special steels the gas flow rate ranged from 25 to 75 Nl/min and for stainless steels the range was from about 25 to 150 Nl/min, as shown in Fig. 9.12.

He et al. [56] summarized some of the industrial bottom gas injection conditions in the early 1980's, as shown in Table 9.11, which was updated with more recent

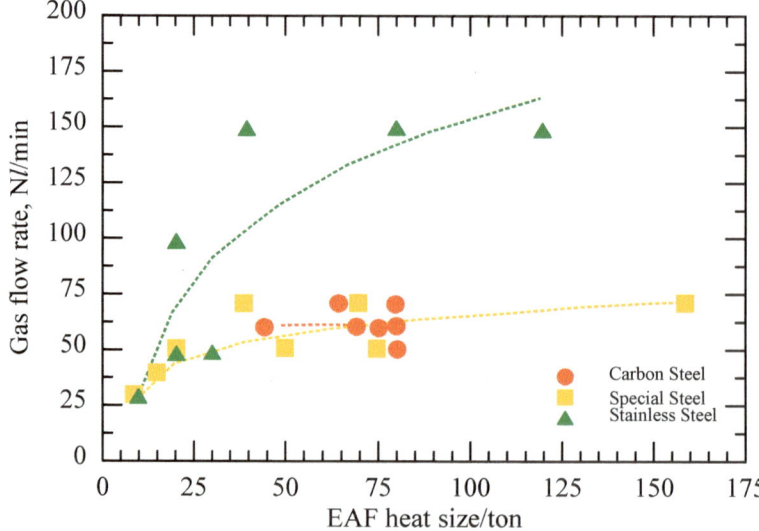

Fig. 9.12 Relationship between EAF capacity and gas flow rate for BGI. After [50]

Table 9.11 Industrial gas injection conditions

Ref	Country	Year	Ton	Gas	Q, Nlmin	N
[56]	Japan	1981	98	Air, O_2, N_2	10,000–30,000	6
[56]	Japan	1987	50	Ar, N_2	210–900	3
[56]	Japan	1986	50	N_2	300–900	3
[56]	Mexico	1986	5	Nat gas + Ar	100–3600	2
[56]	Mexico	1989	45	Nat gas + N_2	70–100	2
[56]	USA	1985	20	Ar + N_2	190	3
[56]	France	1987	85	Ar	150–250	3
[47]	China	2012	50	Ar + N_2	$13 + 30 = 43$	3
[49]	China	2016	65	Air, Ar, N_2	20	3
–	China	2023	110	Ar	$40 \times 3 = 120$	3
[85]	World	2013	6–250	Ar	10–110	1–5

information. The range in gas flow rate is highly variable but predominates a range from 10 to 110 Nl/min. The total gas flow rate depends on the number of porous plugs and capacity of the EAF. The most typical layout includes 3 porous plugs.

9.2.5 Slag Eye and Spout Height Due to Bottom Gas Injection

Due to the relatively shallow depth of liquid steel in the EAF, the gas flow rate cannot be excessive, otherwise the following problems arise as the gas flow rate is increased:

- The height of the spout increases, increasing steel reoxidation
- Heat losses of liquid steel increase
- Risk to induce short-circuit current when liquid steel gets in contact with the electrodes
- electrode washing

Dong et al. [47, 52] suggested a maximum spout height of 5 cm to control the previous problems.

Li et al. [86] studied the limits on gas flow rate to prevent electrode washing for a given arc length using a 1:7 geometric scale EAF model. The arc length was estimated using the following equation, which is derived from a linear relationship between arc voltage and arc length:

$$L_{arc} = \frac{U_{arc}^{rms} - U_{ak}}{E}$$

where L_{arc} is the arc length in mm, U_{arc}^{rms} is the rms arc voltage, U_{ak} is the voltage drop at the plasma column (taken from 35 to 40 V), and E is electric field intensity in the plasma column which varies from 0.5 to 2 V/mm but is commonly taken from 1 to 1.15 V/mm for good foaming conditions and about 0.6 V/mm for poor foaming conditions [87, 88]. The arc length in a 30-ton EAF with an arc voltage of 320 V was estimated at about 288 mm, equivalent to 41 mm in the water model. They obtained the following relationship for the spout height (Δh) as a function of stirring energy (ϵ) and height of the liquid (H):

$$\Delta h(m) = 0.0815\epsilon^{0.664}H^{-0.65}$$

This equation indicates that for the same gas flow rate the spout height is inversely proportional to the height of the liquid, therefore at the beginning of a heat the spout height is higher because the molten bath is shallow. Similar results were reported by Krishnapisharody and Irons, using a ladle [89]. Combining the results from the two previous equations, a maximum gas flow rate in each porous plug can be defined to control the height of the spout so that it remains lower than the arc length, if the location of the porous plugs is in the pitch circle.

He et al. [64] reported the following relationship, including only the effect of the gas flow rate:

$$\Delta h = 0.299Q^{0.539}$$

where Δh is the height of the spout in mm and Q is the gas flow rate in cm^3/min.

9.2.6   Melting Rate and Mass Transfer Due to Bottom Gas Injection

Cheng et al. [90] investigated the effect of bottom gas injection on the melting rate and mass transfer in a water model using three porous plugs. They employed a geometric scale 1:4.8 from an EAF with a capacity of 25 ton and gas flow rates corresponding to 0–100 W/ton in terms of stirring energy. The melting rate of scrap was simulated with ice and benzoic acid was added to simulate mass transfer from oil to water. They reported higher melting rates and higher rates of mass transfer by increasing the stirring energy and their results were summarized by the two following equations:

$$t_m = 1720\epsilon^{-0.67}$$
$$k_v = 0.0144\epsilon^{0.872}$$

where t_m is the melting time in s, k_v is the term for the mass transfer coefficient (kA/V) in s^{-1} and ϵ is the stirring energy.

Yang et al. [17] studied the effect of stirring due to oxygen injection with two lances on mass transfer using a water model with a geometric scale 1:10 and the following expression for the volumetric mass transfer coefficient was derived:

$$k_b A = 9.6 Q^{0.4}$$

where k_b is the mass transfer coefficient in m/min and A is the interfacial area in cm^2.

9.2.7 Effect of EAF Geometry (H/D Ratio) on Mixing Efficiency

The major challenge to reach better stirring conditions in the EAF with bottom gas injection is the shallow depth of liquid steel. This condition imposes limits on the maximum gas flow rate. Wei et al. [91] suggest a maximum gas flow rate of 150 Nl/min. Cao [92] reported higher velocities in the liquid and lower energy consumption by increasing the ratio H/D. A 10% increment in the ratio H/D decreased energy consumption by 3.4%.

Yang and Wu [31] reported an increment in the average velocity of liquid steel from 3.64 to 4.82 cm/s by increasing the H/D ratio from 0.3 to 0.4, using a bottom electrode diameter of 0.1 m in a DC-EAF of 12 ton with an electric current of 14 kA and arc length of 0.33 m.

Ramirez et al. [40] reported the effect of the H/D ratio for a DC-EAF, indicating that increasing this ratio from 0.11 to 0.23 increased by about 3 degrees the temperature of liquid steel.

Timoshenko et al. [93–95] has reported numerical modelling results using bottom gas injection, changing the height/diameter ratio of the EAF. Their results indicate that increasing the H/D ratio from 0.18 to 0.33, the time to heat the liquid from 1500 to 1650 °C decreases from 12 to 16%. Belkovskii and Kats [96] reported an optimum value of 0.38 at which stirring was maximum, considering a decarburization rate of 0.02%/min. They suggested a range from 0.22 to 0.38. All the previous results suggest a change in the current design of the EAF, increasing the H/D ratio, in order to improve the mixing conditions.

In ladles the optimum mixing conditions have been reported for an H/D ratio of about 1–1.5 [97]. It should be noticed that an increase in the H/D ratio decreases the slag/metal interfacial area which would decrease the rate of removal of impurities.

9.2.8 Combined Stirring: Oxygen Injection and Bottom Gas Injection

Oxygen injection is a basic component of the EAF process. Its effect on the overall mixing efficiency should always be taken into account. One of the first studies using CFD to describe the stirring conditions with both oxygen injection and bottom gas injection was reported by Caffery et al. in 1997 [79]. They compared two layouts

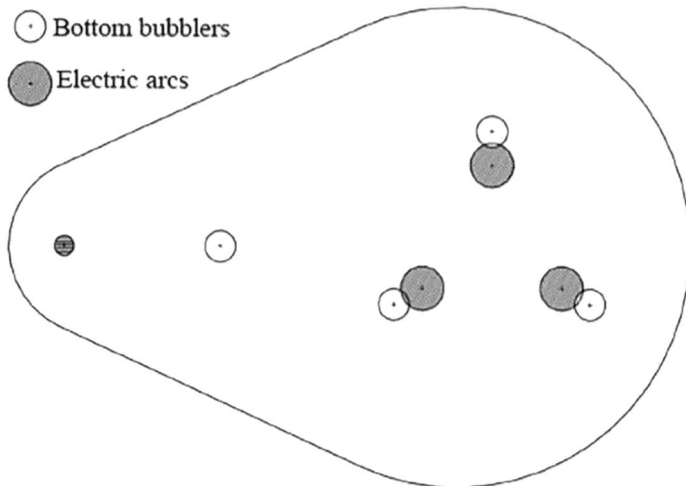

Fig. 9.13 Layout of porous plugs in the study conducted by Caffery et al. [79]

with 3 and 4 porous plugs, as shown in Fig. 9.13, two different lance angles (0, ±
30) and three gas flow rates (2.5, 5 and 10 Nm^3/h). Higher stirring intensities were
found when the lance angle was placed at ± 30 degrees. This angle is with respect
to the furnace center line. The average velocity of the liquid was increased from 6–7
to 14–15 cm/s changing the lance angle from 0 to ± 30 degrees. To obtain higher
velocities is important to avoid direct impact of the streamlines between the velocity
fields created by bottom gas injection and the supersonic lances. The fourth porous
plug close to the EBT was considered useful to increase the stirring condition in
the balcony area. A maximum gas flow rate of 5 Nm^3/h (83 Nl/min) was suggested,
probably because its location is just beneath the electrodes. A final suggestion was
made to consider replacing the 3 porous plugs for one central position between the
electrodes.

Chen et al. [98] conducted mixing time measurements in a water model and
reported improved mixing conditions using two oxygen injection lances with a horizontal angle of 30° and two porous plugs located along the line between slag door
and EBT. Mixing is also improved if the distance from the tip of the lance to the
liquid, is decreased.

Ma et al. [49] reported a numerical model comparing a conventional EAF equipped
only with supersonic lances and the case equipped with coherent jets as well as
bottom gas injection (3 porous plugs), called combined blowing. They applied the
VOF model and the standard k-ε turbulence model. They reported an increase in
the velocity of the liquid of about 98% and a decreased volume of dead zones of
about 79% changing from conventional to combined blowing. The average velocity
increased from 1.31 to 2.6 cm/s and the dead zones, defined for regions with
an average velocity lower than 1.0 cm/s decreased from 92.5 to 13.3%. Higher

stirring conditions led to enhanced decarburization (from 01.2 to 0.138%/min), higher dephosphorization rates (from 70 to 85%) and lower FeO (from 17 to 14%). They also reported a life of the porous plugs of about 700 heats with an erosion rate of about 0.5 mm/heat. The bottom gas flow rate was 20 Nl/min.

Wei et al. [99–103] developed a new technology for submerged oxygen injection called COherent jets with CO_2 and O_2 MIxed injection (COMI). Supersonic lances for gas injection cannot be submerged because are water cooled and any water leakage can create a serious explosion. With this limitation the gas jet has a short penetration into the liquid. Coherent jets have a much higher penetration depth and its stirring intensity is higher. The COMI system uses a shrouding gas containing C_3H_8 or CH_4 which absorbs heat upon dissociation and then forms a mushroom at the tip, protecting the lance. The proportion of CO_2 on the CO_2-O_2 mixture varies from 15 to 25%. The fire spot temperature with pure oxygen is about 2500 °C and with 15% CO_2 decreases to about 1700 °C. Submerged CO_2-O_2 injection increases the average velocity of liquid steel compared with a conventional supersonic lance, from 1.3 to 2.6 cm/s, two times higher. Further analysis by physical and mathematical modelling included the effect of the horizontal arrangement, gas flow rate, dip angle and depth of immersion. Two submerged lances and three horizontal arrangements were tested. A horizontal arrangement where one jet improved the momentum of the second one contributed to a higher average velocity of the liquid. The horizontal angle was 20°. The dip angle was reported to have a low influence however the evaluated range was small, from 5 to 15 degrees. The submerged lances affect the middle and lower volumes of the liquid and the upper lances affect the upper volume. Considering an EAF with 75-ton capacity, the gas flow rate in the submerged lances was increased from 200 to 600 Nm^3/h and for the coherent jets the gas flow rate was 2000 Nm^3/h with an inclination angle of 44 degrees. Further analysis by Wei et al. [104] comparing pure gas and a mixture of gas and powders indicate that the presence of solids enhance the momentum of the immersed jet. Increasing the vertical angle from 0° to 15°, the penetration depth in the vertical direction increases from about 5 to 21 cm but the penetration depth in the horizontal direction decreases from about 45 to 26 cm, with a gas flow rate of 600 Nm^3/h. Both the vertical and horizontal components of the penetration depth increase by increasing the load of particles.

Zhu and Wei [28] developed a numerical model for an AC-EAF with a nominal capacity of 100 ton to describe the velocity of the liquid due to oxygen injection, bottom gas injection using three porous plugs, and electromagnetic forces. The average velocity increases up to 7 cm/s with the combined effects of these three components. The following expression was derived to describe the average velocity:

$$\overline{U} = 0.217 \times \log\left(\frac{Q_{O_2}}{1094}\right) + 0.1039 \times \log\left(\frac{Q_{BGI}}{1.073}\right) + 0.0013 \times e^{\left(\frac{S}{4539}\right)}$$

where $\overline{U}$ is the average velocity of the molten bath in m/s, Q_{O_2} is the injected oxygen in Nm^3/h, Q_{BGI} represents bottom gas injection in Nm^3/h and S is the apparent power in KVA.

Table 9.12 Average velocity of liquid steel due to several stirring methods

Year	Authors	Characteristics	layout	$\overline{U}$, cm/s
1997	Caffery et al. [79]	$O_{2,inj}$ 3 and 4 porous plugs, 1 lance changes in horizontal angle $\pm$ 30°	$O_{2,inj.}$ 0° + BGI	6–7
			$O_{2,inj.}$ 30° + BGI	14–15
2000	Guo et al. [10]	EAF 190 ton, 3060 Nm³/ h, 1 vertical O_2 lance, DeC 0.002%C/min, 3 plugs (20/40 Nl/min)	$O_{2,inj}$	2
			CO	4
			BGI	4
			CO + BGI	6
2016	Zhu et al. [28]	100 ton EAF,	$O_{2,inj.}$ 0° + BGI + EMS	7
2016	Ma et al. [49]	Supersonic vs Coherent jets + BGI (20 Nl/min)	Supersonic jet	1.31
			Cojet + BGI	2.6
2018	Wei et al. [99–103]	Coherent jets + submerged (O_2 + 15–25% CO_2)	Cojet + BGI	2.6
2020	Chen et al. [23]	4 Coherent jets, vertical angle 45°	$O_{2,inj}$	1.4
			$O_{2,inj.}$ + CO	14

BGI bottom gas injection, *Cojet* Coherent jet, *CO* CO due to decarburization

Guo et al. [10] compared stirring due to CO and BGI. Bottom gas injection consisted of three porous plugs with flow rates of 20 and 40 Nl/min per plug. The combination of both methods gave an average velocity of 6 cm/s.

Table 9.12 summarizes reported velocities of liquid steel using oxygen injection, bottom gas stirring, CO from the decarburization reaction, electromagnetic stirring and combinations of these methods.

9.3   EAF Electromagnetic Stirring

9.3.1   Fluid Flow Due Electromagnetic Stirring Devices

Electromagnetic stirring (EMS) is an old technique but barely exploited in the EAF, until recently. EMS was first commercially applied in 1947 at Uddeholm Sweden [105]. EMS caused by the electric arcs was noticed since the early development of the EAF, however it was also noticed its weak effect. Several methods were proposed since the early 1920's to enhance this effect [106, 107]. Dreyfus, a German engineer working for ASEA (ABB) proposed the use of coils at the bottom of the EAF. The flow rate of liquid steel can be controlled depending on the intensity of the low frequency current through the stirrer windings. The stirring direction can be reversed by changing the current direction. The stirring intensity is controlled in accordance

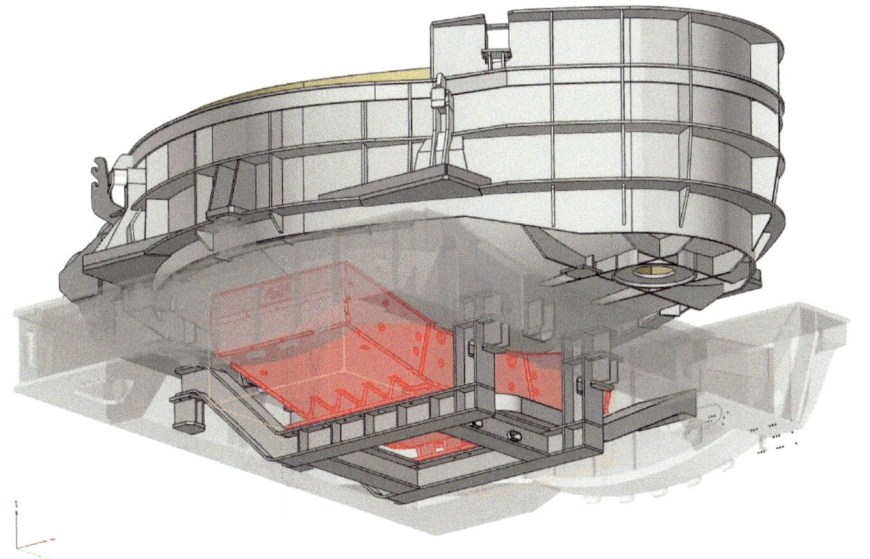

Fig. 9.14 EAF with electromagnetic stirring. After [109]

with the different stages in the EAF operation. A non-magnetic furnace shell bottom is required or welded over the previous shell.

EMS was the first technology applied to improve the mixing conditions in the EAF according to ABB [105], however in spite of its benefits (no slag eye) until now this technology has been less commercially successful in comparison with bottom gas stirring, perhaps due to limitations much larger investment costs. More than 150 units have been sold by ABB since 1947 [108]. Figure 9.14 illustrates EMS at the bottom of an EAF.

Fornander and Nilsson [4] made early reports in 1950 about the benefits of electromagnetic stirring. They reported lower improved deoxidation (lower FeO and O) resulting in higher desulphurization rates. The average S concentration was 200 ppm without stirring, compared with 160 ppm with stirring. Additionally, the temperature difference between the slag and the molten liquid steel was decreased from 102 to 26 °C. Walther [110] described in detail the design of the first EMS installation in the USA, which included an extra weight of 23 ton for an EMS installed in a 90 ton EAF and indicating the EMS had an average electric energy consumption of 10 kWh/ ton. He reported several technical benefits such as negligible thermal stratification, higher melting rate, lower dissolved oxygen and improved de-slagging. In addition to the technical aspects he also reported a safer operation that improved job attitudes.

Widlund et al. [111] through mathematical modeling reported that electromagnetic stirring increases ten times the velocity of the liquid (one order of magnitude) in comparison with natural convection in the EAF, as shown in Fig. 9.15. Arzpeyma et al. also reported a reduced melting time and an increased heat transfer coefficient by a factor of four with EMS [112].

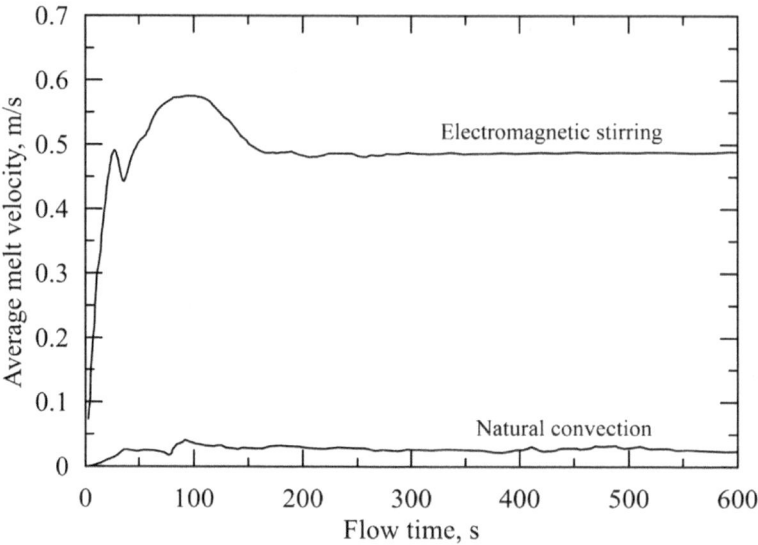

Fig. 9.15 Computed results comparing natural convection and electromagnetic stirring [111]

Teng et al. [108, 113–115] has reported results from recent improvements in electromagnetic stirring by ABB. The average designed velocity was reported in the range from 0.2 to 0.5 m/s. Results from a 90 ton AC-EAF indicate multiple benefits such as [108, 113, 114]: higher arc power, a small thermal stratification, in the order of 5 °C, superheat was decreased, higher melting rate, shorter power on time (5%), lower energy consumption by about 14 kWh/ton, lower electrode consumption (4–6%), higher yield (0.2–0.5%), decarburization rate increased from 0.03 to 0.06%C/min, tap oxygen and FeO closer to equilibrium values, FeO was decreased by 3% and oxygen was decreased from 618 to 504 ppm. In stainless steelmaking they reported to have solved problems of un-melted FeCr at the bottom of the crucible using EM stirring, due to higher mixing intensities, Cr_2O_3 was decreased 3% and energy consumption was decreased by 26 kWh/ton. The orientation of the magnetic field and current intensity were investigated in a Consteel EAF [116, 117]. The melting rate in this process is controlled by convection heat transfer with typical thermal gradients of about 150 °C. Using a water model, the authors reported a higher melting rate if the magnetic field was in the direction of the scrap charged continuously rather than in the directions of either the slag door or the EBT area. In such direction the melting rate was 75% higher compared with the case of no stirring conditions. They also reported that increasing the electric power intensity above 30% made a minor effect on further decreasing the melting rate.

de Santis et al. [118] studied mixing time with and without EMS in water models. Froude similarity was initially employed to define the gas flow rate using 4 submerged water pumps but the final criterion was based on achieving similar stirring energies.

The water pumps reproduced the flow induced by a travelling magnetic field generated by EMS. Additionally, they made industrial measurements using Cu as a tracer. Mixing time with and without EMS was reported as 62 and 260 s, respectively.

9.3.2 EMS Versus Bottom Gas Injection

De Santis et al. [118] compared EMS and bottom gas injection. Bottom gas injection was carried out with three porous plugs, 100 l/min in each plug. Details on the location of the plugs and EMS intensity was not provided. According with this comparison and from water models, EMS gave a shorter mixing time in comparison with BGI, with a mixing time of 155 s and 240 s, respectively. The advantage for EMS remains if stirring due to oxygen injection is included, however the added effect of oxygen injection has less impact on mixing time compared to the case using BGI, as shown in Fig. 9.16.

Electromagnetic stirring and bottom gas stirring in ladles have also been compared. Chung et al. [119] reported a numerical model, indicating that gas injection provides better mixing conditions but electromagnetic stirring higher removal capacity of non-metallic inclusions, in particular below 20 microns, attributed to more homogeneous stirring conditions.

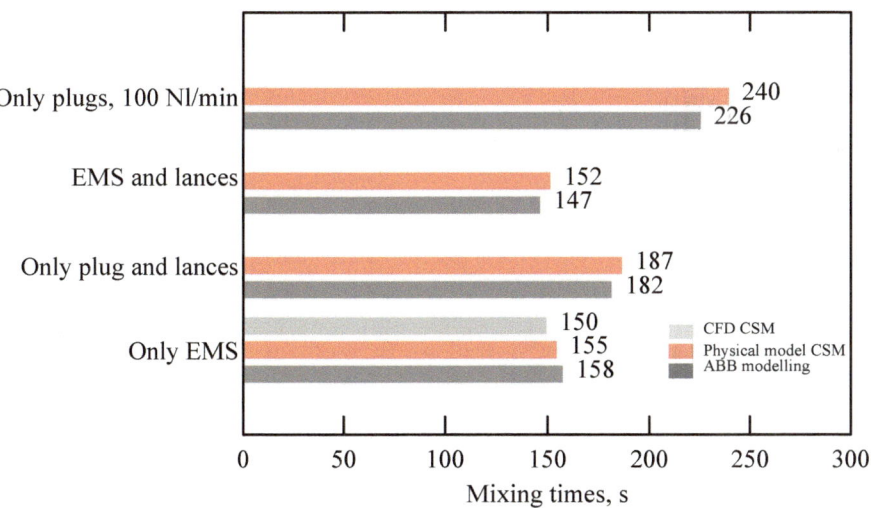

Fig. 9.16 Comparison of mixing times for BGI and EMS. After [118]

9.4 Benefits from EAF Stirring

Fagionato et al. [120] reported an early application of bottom gas injection in 1989, indicating a decrease in energy consumption in the range from 10 to 25 kWh/ton. Schade [121] in 1991 reported a decrease in energy consumption in the range from 11 to 22 kWh/ton and increment in metallic yield of 0.5%. Cao [92] reported a decrease in energy consumption of 15 kWh/ton and lower tap-to-tap time, about 5–6%. Kirschen et al. [48] reported at least 5 kWh/ton in electric energy savings, 0.5% increment in metallic yield and a decrease from 0.5 to 1 min in power-on time.

Makoto et al. [122] in 1987 reported drastic improvements injecting nitrogen from 30 to 60 Nm^3/h through one porous plug in one 80 ton EAF: A decrease of electric energy consumption of about 12 kwh/ton, decrease of tap-to-tap time of 7–10 min and about 25 °C lower tapping temperature. Nitrogen removal due to gas stirring is a common phenomenon in DRI steelmaking operations [123, 124]. Denitrogenization with common DRI has low efficiency because of the low DRI density, affecting only the upper volumes of liquid steel in the EAF. Injection of DRI fines has a higher efficiency. Anghelina et al. [124] reported that due to high reaction rates, bubbles become saturated with nitrogen and there is no need to inject DRI fines too deep. Wei et al. [58] compared the denitrogenization rate with argon and CO_2. The higher stirring intensity using CO_2 decreased nitrogen to 12 ppm, in contrast to 34 ppm N using argon.

EAF stirring has proved to give important benefits to improve the EAF process, as shown in Table 9.13 from different steel plants. Most of these results correspond to steel plants with BGI. Most of the data from EMS have been provided by suppliers.

Final remarks: The conventional EAF has very low stirring conditions of liquid steel. The driving forces in a conventional EAF derive from buoyancy forces, momentum transfer by the oxygen jets, decarburization and local electromagnetic forces at the electrode regions. In general, these forces are weak or operate in a narrow region causing not only a low average velocity but also thermal stratification of liquid steel. This limitation of a conventional EAF is a major drawback, especially if this reactor intends to increase its share in production of steel worldwide.

Bottom gas stirring and electromagnetic stirring have proved to give important benefits to improve steelmaking operations. By increasing the velocity profiles of liquid steel and decreasing the dead zones, EAF stirring also has increased the rate of decarburization, increased the melting rate, decreased the power-on time, decreased FeO and increased metallic yield, decreased electric energy consumption and overall, a higher productivity. However, our fundamental knowledge on the hydrodynamics and reaction rates under higher turbulent conditions still needs additional research work. The contribution from decarburization to the velocity fields and stirring intensity requires further investigation to confirm the few reports available. Most measurements on mixing time in the EAF have been limited to define the effect of the gas flow rate for different injection layouts. The effect of EAF dimensions (diameter and height of liquid) and slag layer properties (thickness and physical properties) is missing and should be incorporated in the analysis. The optimum layouts that

Table 9.13 Benefits using BGI in the EAF

	Before	After	Change	Ref	Notes
Average velocity, cm/s	1.31	2.6	+ 1.3	[49]	O$_2$ inj + BGI
	0.48	0.86	+ 0.38	[81]	Pulse mode
	1.3	2.6	+ 1.3	[99]	COMI
Dead volume, %	92.5	13.3	− 79.2	[49]	U < 0.01 m/s
Dead volume, m^3	1.42	0.96	− 0.46	[49]	U < 0.0008 m/s
De C rate, %/min	0.04	0.10	+ 0.06	[47]	
	0.127	0.142	+ 0.015	[71]	
			+ 0.022	[99]	COMI
O$_2$, Nm^3/ton	42.5	32.4	− 10.1	[47]	
			− 2.1	[99]	COMI
FeO, %	18.6	16.1	− 2.5	[47]	
	23.4	19.3	− 4.1	[71]	
	21.2	19.2	− 2	[49]	Pulse mode
			− 6.3	[99]	COMI
[C]-[O] equilibrium	0.0079	0.0056	− 0.0023	[49]	
	0.0056	0.0032	− 0.0024	[71]	
	0.0030	0.0028	− 0.0002	[81]	
	0.0036	0.0030	− 0.0006	[99]	COMI
DeP rate, %	64	76	+ 12	[71]	
Met yield, %			+ 0.5 to + 3	[85]	
Energy, kWh/ton	136.23	128	− 8.23	[47]	60% hot metal
			− 5 to − 8	[85]	
Power-on time, min	59	55	− 4	[47]	
			− 0.5 to − 2	[85]	
Tap-to-tap time, min			− 5	[99]	COMI
Lime, kg/ton	61.5	51.1	− 10.4	[47]	
			− 7.8	[99]	COMI
Gases, Nl/min	0	43	+ 43	[47]	
Savings, USD ($)/ton			− 4.8	[49]	30.83 ¥/ton
			− 1.7	[85]	1.5 €/ton
	Net savings			[125]	1.4 $/ton

have been suggested show large disagreements in terms of the number of porous plugs, its radial position and its separation angle. This is because of the large differences in the experimental conditions. There is still need of a more complete physical modelling work that includes all these variables in the presence of a slag layer, complemented with numerical calculations of the velocity profiles. The proposed work should include all the driving forces: natural convection, oxygen injection, number and type of lances for oxygen injection, horizontal angle and vertical angles of oxygen lances, lance distance tip to liquid steel, rate of decarburization, initial temperature, H/D ratio, number and position of porous plugs, gas flow rate per porous plug, electromagnetic forces, heat input from the electric arcs, etc. Mass transfer and bottom gas stirring has been poorly investigated. It is necessary to evaluate if the conditions which improve mixing efficiency of liquid steel also correspond with those for the removal of impurities.

The information provided by EMS suppliers suggest an outstanding performance, however, it is necessary to confirm the claimed benefits with more available data reported by steel plants, comparing its return of investment and metallurgical benefits.

References

1. Yang L, Hu H, Yang Z, Xue B, Guo Y, Wang S (2021) A review on bath fluid flow stirring technologies in EAF steelmaking. J Iron Steel Res Int 28:1341–1351. https://doi.org/10.1007/s42243-021-00650-x
2. Gonzalez OJP, Ramírez-Argáez MA, Conejo AN (2010) Effect of arc length on fluid flow and mixing phenomena in AC Electric Arc Furnaces. ISIJ Int 50:1–8. https://doi.org/10.2355/isijinternational.50.1
3. Gonzalez OJP (2008) Mathematical modelling of DRI melting in the EAF. Morelia Technological Institute. MSc Thesis (in Spanish)
4. Fornander S, Nilsson F Inductive stirring in arc furnaces. J Met 1950:256–9
5. Ramírez-Argáez MA, Conejo AN, Guzmán YIC, Trápaga G (2010) Influence of the top slag layer on the flow dynamics in AC-electric arc furnaces. Int J Eng Syst Model Simul 2010:2. https://doi.org/10.1504/IJESMS.2010.038141
6. Elkoumy M, El-Anwar M, Fathy AM, Megahed GM, El-Mahallawi I, Ahmed H (2018) Simulation of EAF refining stage. Ain Shams Eng J 9:2781–2793. https://doi.org/10.1016/j.asej.2017.10.002
7. Fritz E, Gebert W (2005) Milestones and challenges in oxygen steelmaking. Can Metall Q 44:249–260. https://doi.org/10.1179/cmq.2005.44.2.249
8. Elkoumy M, El-Anwar M, Fathy A, Megahed G, El-Mahallawi I, Ahmed H (2018) Computational simulation model for metallurgical effects during EAF refining stage: waiting and arcing time. ISIJ Int 58:1669–1678. https://doi.org/10.2355/isijinternational.ISIJINT-2018-224
9. González OJP, Guzmán YIC, Ramírez-Argaez MA, Conejo AN (2008) Melting behavior of simulated DRI in liquid steel. Arch Metall Mater 53:359–364
10. Guo D, Gu L, Irons GA (2000) Evaluation of stirring in electric arc furnaces. In: 58th electric furnaces conference, Orlando, FL, USA, 12–15November 2000, pp 223–233
11. Wang L, Zhu R, He C (2010) Two-phase mathematical model of oxygen jet impinging on molten steel bath surface in EAF (in Chinese). Chinese J Process Eng 10:625–631
12. Han J, Chen Y, Li S (2010) Study of molten stirred by oxygen jet in EAF (in Chinese). Ind Heat 39:30–33
13. He C, Zhu R, Liu C, Li J (2011) Analysis on decarburization rate based on numerical simulation in EAF Steelmaking (in Chinese). Steelmaking 27:41–45
14. Thongjitr P, Kowitwarangkul P, Pratumwal Y, Otarawanna S (2023) Optimization of oxygen injection conditions with different molten steel levels in the EAF refining process by CFD simulation. Metals 13. https://doi.org/10.3390/met13091507
15. Chen S, Li J, Zhu R, He C (2013) Three-phase numerical simulation of oxygen jet impinging on molten steel bath in EAF (in Chinese). J Graph 34:66–70

16. He Q, Guo Z, Yan H, Ding D, Zhou Y, Lin P (2003) Experimental study on stirring characteristic of EAF's wall lances (in Chinese). Iron Steel 38:13–16
17. Yang W, Zhou J, Wang M, Ding Y, Yu P, Xu D (2002) Simulative experiment of oxygen utilization for large EAF steelmaking and its industrial application (in Chinese). J Iron Steel Res 14:1–5
18. Qin X, Cang D, Xu D, Duan J (2012) Experimental study on the stirring characteristics of oxygen lances and bath mixing in EAF (in Chinese). Energy Metall Ind 31:32–36
19. He C, Huang Q, Yang N, Liu C (2013) Research on oxygen lances optimized layout for EAF steelmaking (in Chinese). Ind Heat 42:59–63
20. Jia N, Dong K, Ma G, Wei G, Cheng T, Yue K (2016) Water model study of 45t EAF embedded type oxygen lance (in Chinese). Ind Heat 45:44–47
21. Ramirez-Argaez MA, Conejo AN (2021) CFD study on the effect of the oxygen lance inclination angle on the decarburization kinetics of liquid steel in the EAF. Metall Res Technol 118. https://doi.org/10.1051/metal/2021069
22. Guo DC, Gu L, Irons GA (2002) Developments in modelling of gas injection and slag foaming. Appl Math Model 26:263–280. https://doi.org/10.1016/S0307-904X(01)00060-9
23. Chen Y, Silaen AK, Zhou CQ (2020) 3D Integrated modeling of supersonic coherent jet. Penetration and decarburization in EAF refining process. Processes 8:1–18
24. Alexis J, Ramirez M, Trapaga G, Jönsson P (2000) Modeling of a DC Electric Arc Furnace—heat transfer from the arc. ISIJ Int 40:1089–1097. https://doi.org/10.2355/isijinternational.40.1089
25. Qian F, Farouk B, Mutharasan R (1995) Modeling of fluid flow and heat transfer in the plasma region of the dc electric arc furnace. Metall Mater Trans B 26:1057–1067. https://doi.org/10.1007/BF02654108
26. Amaro-Villeda AM, Ramirez-Argaez MA, Conejo AN (2014) Effect of slag properties on mixing phenomena in gas-stirred ladles by physical modeling. ISIJ Int 54:1–8. https://doi.org/10.2355/isijinternational.54.1
27. Sridhar E, Lahiri AK (1994) Steady state model for current and temperature distributions in an electric smelting furnace. Steel Res 65:433–437. https://doi.org/10.1002/srin.199401189
28. Zhu R, Wei G (2016) Study on the composite blowing stirring intensity in EAF steelmaking process. In: 2016 19th national steelmaking academy conference. Changsha, China, pp 7–12
29. Kurimoto H, Mondal H, Morisue T (1996) Analysis of velocity and temperature fields of molten metal in DC electric arc furnace. J Chem Eng Japan 29:75–81. https://doi.org/10.1252/jcej.29.75
30. Wang F, Jin Z, Zhu Z (2006) Fluid flow modeling of arc plasma and bath circulation in DC electric arc furnace. J Iron Steel Res Int 13:7–13. https://doi.org/10.1016/S1006-706X(06)60086-1
31. Yang J, Wu Z (1995) Effect of the bath parameters on the stirring in DC arc furnace (in Chinese). Iron Steel 30:20–23
32. Wang F, Jin Z, Zhu Z (2005) Numerical simulation of flow and temperature fields of molten bath in the refining stage of DC electric arc furnace (in Chinese). Ind Heat 34:4–9
33. Liu Y, Guo H, Liu X (1998) Numerical calculation of electromagnetic stirring, heating transfer and circulating flow in UHP DC EAF Bath (in Chinese). Iron Steel 33:14–17
34. Kazak O, Semko O (2011) Electrovortex field in DC arc steelmaking furnaces with bottom electrode. Ironmak Steelmak 38:273–278. https://doi.org/10.1179/1743281210Y.0000000006
35. Kazak O (2013) Modeling of vortex flows in direct current (DC) electric arc furnace with different bottom electrode positions. Metall Mater Trans B Process Metall Mater Process Sci 44:1243–1250. https://doi.org/10.1007/s11663-013-9899-4
36. Kazak O, Semko O (2011) Modelling magnetohydrodynamic processes in DC arc steelmaking furnace with bottom electrodes. Ironmak Steelmak 38:353–358. https://doi.org/10.1179/1743281211Y.0000000004
37. Zhao P, Wang K, Fu J, Ma T (1995) Numerical calculation on electromagnetic stirring and heat transfer in DC-EAF bath (in Chinese). J Univ Sci Technol Beijing 17:284–288

38. Kazak O (2014) Numerical modelling of electrovortex and heat flows in dc electric arc furnace with cooling bottom electrode. Heat Mass Transf Und Stoffuebertragung 50:685–692. https://doi.org/10.1007/s00231-013-1265-1
39. Kazak O, Starodumov I (2020) Comparison of argon injection and other methods of electrovortex flow control in metallurgical furnaces. Eur Phys J Spec Top 229:475–483. https://doi.org/10.1140/epjst/e2019-900141-8
40. Ramírez M, Alexis J, Trapaga G, Jönsson P, Mckelliget J (2001) Mathematical modeling of iron and steel making processes. Modeling of a DC electric arc furnace. Mixing in the bath. ISIJ Int 41:1146–1155. https://doi.org/10.2355/isijinternational.41.1146
41. Kukharev A, Bilousov V, Bilousov E, Bondarenko V (2019) The peculiarities of convective heat transfer in melt of a multiple-electrode arc furnace. Metals 9. https://doi.org/10.3390/met9111174
42. Smirnov SA, Kalaev VV, Hekhamin SM, Krutyanskii MM, Kolgatin SN, Nekhamin IS (2010) Mathematical simulation of electromagnetic stirring of liquid steel in a DC arc furnace. High Temp 48:68–76. https://doi.org/10.1134/S0018151X10010116
43. Bowman B, Krüger K (2009) Arc furnace physics. StahlEisen
44. Lazcano A, Vargas G, Maroto C (1988) Process to improve the refining of liquid metals by natural gas injection. US Pat. 4780133
45. Lazcano A, Vargas G (1990) Startup and preliminary results of bottom blowing natural gas into an EBT arc furnace at DEACERO S.A. Iron Steelmak 17:58–62
46. Sugiura M, Nukushina N (2008) Bottom-Tuyere-changing of EAF for stainless steel. In: Proceedings ICS 2008: the 4th international congress on the science and technology of steelmaking, pp 174–177
47. Dong K, Zhu R, Liu W (2012) Bottom-blown stirring technology application in Consteel EAF. Adv Mater Res 361–363:639–643. https://doi.org/10.4028/www.scientific.net/AMR.361-363.639
48. Kirschen M, Hanna A, Zettl KM (2013) Benefits of EAF bottom gas purging systems. AISTech Iron & steel technology conference proceeding, vol 1. Pittsburgh, PA, USA, pp 761–767
49. Ma G, Zhu R, Dong K, Li Z, Liu R, Yang L et al (2016) Development and application of electric arc furnace combined blowing technology. Ironmak Steelmak 43:594–599. https://doi.org/10.1080/03019233.2016.1144547
50. Ishihara E (2023) Introduction to the technology of gas purging plug refractories for steelmaking EAF. Shinagawa Technical Report, p 66
51. Mimura N, Ususaka M (1992) Gas introducing and stirring apparatus of metallurgical vessel. GB2254682A
52. Dong K, Li J, Zhu C (2011) The Application of bottom blowing stirring to Consteel furnace (in Chinese). Ind Heat 40:60–62
53. Wijngaarden MV, Pieterse AT (1994) Bottom-stirring in an electric-arc furnace: performance results at Iscor Vereeniging works. J South African Inst Min Metall 27–34
54. Gulyaev MP, Filipov VV, Enders VV, Shumaher E, Shumaher E, Frantski R, et al (2002) First in CIS systems of steel bottom blowing with inert gases in electric arc furnace. In: 7th European electronic conference, vol 1. Venice, Italy, 26–29 May 2002
55. Badr K, Kirschen M, Cappel J (2011) Chemical energy and bottom stirring systems—cost effective solutions for a better performing EAF. METEC 2011, June 27–July 1, Dusseldorf, pp 12.1–12.8
56. He P, Zhang DM, Deng KW, Zhang RS (1992) Stirring technology of bottom blowing gas inEAF (in Chinese). Iron Steel 27:65–70
57. Zhu R, Han B, Dong K, Wei G (2020) A review of carbon dioxide disposal technology in the converter steelmaking process. Int J Miner Metall Mater 27:1421–1429. https://doi.org/10.1007/s12613-020-2065-5
58. Wei G, Zhu R, Dong K, Li Z, Yang L, Wu X (2018) Influence of bottom-blowing gas species on the nitrogen content in molten steel during the EAF steelmaking process. Ironmak Steelmak 45:839–846. https://doi.org/10.1080/03019233.2017.1410949

59. Wang H, Zhu R, Wang X, Li Z (2017) Utilization of CO_2 in metallurgical processes in China. Miner Process Extr Metall 126:47–53
60. Mannion FJ, Fruehan RJ (1989) Decarburization kinetics of liquid Fe-Csat alloys by CO_2. Metall Trans B 20:853–861. https://doi.org/10.1007/BF02670190
61. Banerjee SK, Irons GA (1992) Physical modelling of thermal stratification, bottom build-up and mixing in submerged arc electric smelting. Can Metall Q 31:31–40. https://doi.org/10.1179/cmq.1992.31.1.31
62. Rui S, Liu J, Wu Z, Jiang M, Li B, He J (1995) Characteristics of uniform mixing in bottom-blown electric arc furnace (in Chinese). Iron Steel 30:32–35
63. Li L, Shi D, Jiang M, Jin C (1998) Effects of stirring location and energy on mixing characteristics of bath in bottom blowing EAF (in Chinese). Steelmaking 5:30–33
64. He P, Zhang LP, Mei HS (1994) Study of hydraulic model of electric arc furnace with bottom blowing (in Chinese). Eng Chem Metall 15:110–114
65. Liu CQ, Wen QY (1996) Cold state simulation tests of bottom-blowing gas stirring of arc furnace. Ind Heat 5:17–20
66. Jiang MF, Li LF (1996) Water model research of homogenous mixing characteristics in bottom stirring EAF (in Chinese). J Northeast Univ 17:352–355
67. Liu J, Rui S, Wu Z, Jiang M, Li B, He J (1995) Flow state analysis in bath of bottom-blown arc furnace (in Chinese). J Northeast Univ 16:298–301
68. Kazakov SV, Gulyaev MP, Filipov VV (2014) Hydrodynamics of electric arc furnace bath at stirring with inert gases. In: 23rd international conference on metallurgy and materials. Brno, , Czech Republic, 21–23 May 2014, pp 1–12
69. Li B (2000) Fluid flow and mixing process in a bottom stirring electrical arc furnace with multi-plug. ISIJ Int 40:863–869. https://doi.org/10.2355/isijinternational.40.863
70. Liu F, Zhu R, Dong K, Bao X, Fan S (2015) Simulation and application of bottom-blowing in electrical arc furnace steelmaking process. ISIJ Int 55:2365–2373. https://doi.org/10.2355/isijinternational.ISIJINT-2015-352
71. Wei G, Zhu R, Dong K, Ma G, Cheng T (2016) Research and analysis on the physical and chemical properties of molten bath with bottom-blowing in EAF steelmaking process. Metall Mater Trans B Process Metall Mater Process Sci 47:3066–3079. https://doi.org/10.1007/s11663-016-0737-3
72. Liu F, Zhu R, Dong K, Bai R (2017) The preparation process of nitrogenous ferrovanadium with bottom-blowing nitriding in electrical arc furnace. Ironmak Steelmak 44:159–167. https://doi.org/10.1080/03019233.2016.1192827
73. Yang L, Jiang T, Li G, Guo Y, Chen F, Yang Z (2017) Research on the molten bath of EBT area with bottom-blowing in EAF steelmaking process. www.preprints.org. https://doi.org/10.20944/preprints201704.0077.v1
74. Lu M, Li H, Yang Z, Li X, Xing G, Yang S (2019) Effect of bottom blowing flow rate near EBT area on EAF steelmaking (in Chinese). Iron Steel 54:38–44
75. Yang ZS, Yang LZ, Guo YF, Wei GS, Cheng T (2018) Simulation of velocity field of molten steel in electric arc furnace steelmaking. In: 9th international symposium on high-temperature metallurgical processing. Phoenix AZ, pp 69–79. https://doi.org/10.1007/978-3-319-72138-5_8
76. Yang Z, Yang L, Song J, Guo Y, Wei G, Wang B et al (2018) Numerical simulation research on molten steel flow under different gas flow rates in the eccentric bottom tapping zone of electric arc furnace (in Chinese). Chinese J Eng 40:123–129
77. Yang Z, Yang L, Cheng T, Chen F, Zheng F, Wang S et al (2020) Fluid flow characteristic of EAF molten steel with different bottom-blowing gas flow rate distributions. ISIJ Int 60:1957–1967. https://doi.org/10.2355/isijinternational.ISIJINT-2019-794
78. Zhao X (1995) Development of foreign EAF bottom blowing technology (in Chinese). Steel Pipe 5:12–15
79. Caffery G, Warnica D, Molloy N, Lee M (1997) Temperature homogenisation in an electric arc furnace steelmaking bath. In: 1st international conference on CFD in the mineral & metal processing and power generation, Melbourne, Australia, 3–4 July 1997, pp 87–99

80. Hu H, Yang L, Guo Y, Chen F, Wang S, Zheng F, et al (2021) Numerical simulation of bottom-blowing stirring in different smelting stages of electric arc furnace steelmaking. Metals 11. https://doi.org/10.3390/met11050799

81. Wei G, Zhu R, Wang Y, Dong K, Wu X, Liu R et al (2018) Simulation and application of pulsating bottom-blowing in EAF steelmaking. Ironmak Steelmak 45:847–856. https://doi.org/10.1080/03019233.2018.1498759

82. Kirschen M, Ehrengruber R, Hanna A, Zettl KM (2015) Latest developments in gas purging systems for EAF. In: AISTech 2015 Conference, Cleveland, OH, USA. May 4–7 2015, pp 1974–1983

83. Kirschen M, Ehrengruber R, Zettl K-M (2016) Benefits from improved bath agitation with the RADEX DPP gas purging system during EAF high alloyed and stainless steel production. RHI Bull 2016:8–13. https://doi.org/10.17073/1683-4518-2019-11-3-10

84. Kirschen M, Hanna A, Zettl KM (2013) Benefits of EAF bottom gas purging systems. AISTech Iron Steel Technol Conf Proc 1:761–767

85. Kirschen M, Hanna A, Zettl KM (2013) Benefits of EAF bottom gas purging systems. AISTech Conf 1:761–767

86. Li L, Jiang M, Duan Z (1996) Simulation research on the rising height of the bath in bottom stirring EAF. Steelmaking 2:49–52

87. Kruger K. Electric arc furnaces. In: von Starck A, Muhlbauer A, Kramer C (eds) Handbook of thermoprocessing technology. Vulkan-Verlag, Germany, pp 207–222

88. Timm K (2006) Physics of AC arcs. In: Electric engineering of arc furnaces, Dusseldorf, Germany, 23–26 October 2006, pp 1–10

89. Krishnapisharody K, Irons GA (2007) A study of spouts on bath surfaces from gas bubbling: part I. Experimental investigation. Metall Mater Trans B Process Metall Mater Process Sci 38:367–375. https://doi.org/10.1007/s11663-007-9024-7

90. Cheng GG, Qin ZZ, Fan T (1994) Cold modeling of bottom gas stirring in electric arc furnace (in Chinese). Steelmaking 3:17–20

91. Wei G, Zhu R, Wang Y, Wu X, Dong K (2019) Technological innovations of electric arc furnace bottom-blowing in China. J Iron Steel Res Int 26:909–916. https://doi.org/10.1007/s42243-018-0163-7

92. Cao M (1998) Bottom gas stirring and smelting technology of deep bath for DC EAF. Shanghai Met 20:34–37

93. Timoshenko SN (2012) Improving bath geometry as a way of increasing EAF thermal efficiency. Donetsk DonNTU, pp 36–43

94. Timoshenko SN (2016) Computer modeling bath geometry to improve energy efficiency of electric arc furnace, system technologies. Reg Collect Sci Work Dnipro 104:33–39

95. Timoshenko SN, Stopvchenko AP, Kostetski YV, Gubinski MV (2018) Energy efficient solutions for EAF steelmaking. J Achiev Mater Manuf Eng 88:18–24

96. Belkovskii AG, Kats YL (2013) Effect of the geometric parameters of the EAF bath on the main characteristics of furnace operation. Russ Metall 2013:410–413. https://doi.org/10.1134/S0036029513060037

97. Turkoglu H, Farouk B (1991) Mixing time and liquid circulation rate in steelmaking ladles with vertical gas injection. ISIJ Int 31:1371–1380. https://doi.org/10.2355/isijinternational.31.1371

98. Chen Y, Chen S, Liao Y, Liu Y (2013) Side-bottom stirring process technology of the EAF and its application effect (in Chinese). Ind Heat 42:30–33

99. Wei G, Zhu R, Wu X, Dong K, Yang L, Liu R (2018) Technological innovations of carbon dioxide injection in EAF-LF steelmaking. J Met 70:969–976. https://doi.org/10.1007/s11837-018-2814-3

100. Wei G, Zhu R, Tang T, Dong K, Wu X (2019) Study on the impact characteristics of submerged CO_2 and O_2 mixed injection (S-COMI) in EAF steelmaking. Metall Mater Trans B Process Metall Mater Process Sci 50:1077–1090. https://doi.org/10.1007/s11663-018-1482-6

101. Tang T, Zhu R, Wei G, Tian B (2019) Simulation and optimization of embedded type oxygen lance in EAF (in Chinese). Ind Heat 48:49–53

102. Wei G, Zhu R, Han B, Yang S, Dong K, Wu X (2020) Simulation and application of submerged CO_2-O_2 injection in electric arc furnace steelmaking: modeling and arrangement of submerged nozzles. Metall Mater Trans B Process Metall Mater Process Sci 51:1101–1112. https://doi.org/10.1007/s11663-020-01816-w
103. Wei G, Zhu R, Yang S, Wu X, Dong K (2021) Simulation and application of submerged CO_2–O_2 injection in EAF steelmaking: combined blowing equipment arrangement and industrial application. Ironmak Steelmak 48:703–711. https://doi.org/10.1080/03019233.2021.1896068
104. Wei G, Zhu R, Tian B, Dong K, Yang L (2020) Impact characteristics of submerged gas–solid injection in the manufacturing process of steel (in Chinese). Chinese J Eng 42:47–53
105. Fdhila RB, Sand U, Eriksson JE, Yang H. A stirring history. ABB Rev 3:45–48
106. Clifford HE (1924). lectric furnace. US Pat. 1496299. https://doi.org/10.1145/178951.178972
107. Dreyfus L (1941) Electric furnace. US Pat. 2256518. https://doi.org/10.1088/0950-7671/20/2/406
108. Teng L, Meador M, Ljungqvist P (2017) Application of new generation electromagnetic stirring in electric arc furnace. Steel Res Int 88:1–8. https://doi.org/10.1002/srin.201600202
109. Tenova (n.d.) Innovation (EMS). https://tenova.com/node/561
110. Walther HF (1954) Induction stirring provides better control of operating techniques. J Met 6:21–23. https://doi.org/10.1007/bf03397973
111. Widlund O, Sand U, Hjortstam O, Zhang X (2011) Modeling of electric arc furnaces (EAF) with electromagnetic stirring. SteelSim-Metec
112. Arzpeyma N, Widlund O, Ersson M, Jönsson P (2013) Mathematical modeling of scrap melting in an EAF using electromagnetic stirring. ISIJ Int 53:48–55. https://doi.org/10.2355/isijinternational.53.48
113. Teng L, Hackl H, de Rezende LM (2015) The effect of arcsave on the EAF process with eccentric bottom tapping. In: 46 Seminário de Aciaria, Rio de Janeiro Brazil, 17–21 August 2015, pp 244–251. https://doi.org/10.5151/1982-9345-26378
114. Teng L, Jones A, Hackl H, Meador M (2015) Arcsave®, innovative solution for higher productivity and lower cost in the EAF. AISTech Iron Steel Technol Conf Proc 2:1965–1973
115. Teng L, Ljungqvist P, Hackl H, Andersson J (2016) Process improvement with EMS. Steel Times Int 40:59–62
116. Memoli F, Grasselli A, Andersson JZ, Lehman AF, Teng L (2019) Performance of the electromagnetic stirring system applied to the flat bath consteel® operation. In: AISTech conference, vol 2019, May, Pittsburgh, PA USA, pp 775–780. https://doi.org/10.33313/377/081
117. Pan H, Teng L, Yang H, Chen J, Grasselli A (2019) Water and numerical modelling of flat bath operation consteel furnace with electromagnetic stirring. China Steelmak
118. de Santis M, Marx K, Pierre R, Kleimt B, de Miranda U, Schrader T-F et al (2020) Improvement of electrical arc furnace operations with support of advanced multiphysics modeling simulations of the EAF process. EC-RFCS 30444
119. Chung S-K, Shin Y-H, Yoon J-K (1992) Flow characteristics by induction and gas stirring in ASEA-SKF ladle. ISIJ Int 32:1287–1296
120. Fagionato A, Galenda A, Mario F, Pawliska V, Cappelli G (1989) Results of arc furnace stirring special regard to the application at Beltrame Steelworks in Italy. In: 47th electric furnace conference proceedings., Orlando, FL USA, 29 Oct–1 Nov 1989, pp 299–304
121. Schade RJ (1991) Bottom stirring in an electric arc furnace. CMP, Pittsburgh, PA, USA
122. Makoto T, Hayashi A, Ishiguro T, Ieda K (1987) Test of bottom blowing in electric arc furnace for improvement of operation. Tetsu to Hagane 73:226
123. Hornby SA, Trotter D, Varcoe D, Reeves R (2002) Use of DRI and HBI for nitrogen control of steel products. In: Electric furnace conference, pp 687–702
124. Anghelina D, Brooks G, Irons G (2004) DOE Contract 97ID1355. Nitrogen control in EAF steelmaking by DRI fines injection
125. Worrell E, Martin N, Price L (1999) Energy efficiency and carbon dioxide emissions reduction opportunities in the U. S. Iron and Steel Sector. NBNL report

Chapter 10
Hot Metal and Hot DRI

10.1 Hot Metal

The EAF was not designed to use hot metal however due to the following conditions, charging hot metal has been found a solution;

1. Low scrap availability
2. Low scrap quality (high concentration of residual elements)
3. High scrap price
4. High price of electric energy and the need to replace it with chemical energy.

Jones et al. [1] accurately predicted in 1998 a higher share of hot metal in the EAF. This situation was especially important in China in the last decade due to abundant availability of hot metal and lack of domestic scrap. Since increasing hot metal increases the concentration of CO and CO_2, they also suggested the need to recover the energy in the off-gas, for example with higher levels of post-combustion. Hot metal will not replace scrap or DRI in the long term in the EAF because of its high production of CO_2. Most probably, no more extra blast furnaces will be built in the world, however, due to its long life, about 40 years [2], there will be plenty of hot metal available. Wang et al. [3] listed 25 furnaces of high capacity in China, with an effective volume from 4000 to 6000 m^3, commissioned between 2009 and 2019, therefore EAF operation with high ratios of hot metal will keep common at least in China. In addition to the blast furnace, hot metal can also be produced by smelting processes like COREX and through mini blast furnaces. Using hot metal in the EAF decreases the consumption of electric energy which decreases CO_2 emissions from electricity production using fossil fuels but at the same time produces CO_2, therefore its balance in terms of net savings of CO_2 is not relevant.

Yang et al. [4] reported an energy balance with 50% hot metal. For this case the amount of chemical energy was 77% and only 23% electric energy. The calculated power consumption becomes zero as the hot metal ratio reaches 80% [5].

© The Author(s), under exclusive license to Springer Nature Singapore Pte Ltd. 2024   555
A. N. Conejo, *Electric Arc Furnace: Methods to Decrease Energy Consumption*,
https://doi.org/10.1007/978-981-97-4053-6_10

10.1.1 Benefits of Hot Metal in the EAF

Charging hot metal offers the following benefits:

1. Decreases energy consumption
2. Increases the melting rate (higher productivity) as long as the EAF has a high DeC capacity
3. Decreases tap-to-tap time
4. Use of cheaper-lower quality scrap with higher concentration of residual elements can be tolerated
5. Is not necessary to add extra carbon due to the large amount of C in hot metal (about 4–5%)
6. Transformer capacity can be lower because melting requirements are minimized.

However, to reach those benefits, the following aspects should be taken into account:

- Additional oxygen requirements
- Extra capacity of the off-gas system to cool down a higher volume of gas
- Additional fluxes to control slag basicity due to higher input of silicon, consequently higher volume of slag
- Control of explosions due to intense production of CO
- Productivity losses due to delays to charge hot metal, in particular if the EAF is not equipped to charge hot metal
- Relative price of hot metal with respect to scrap or DRI and the final production cost
- Higher volume of CO_2 emissions.

The energy contribution of hot metal comes from two sources; its sensible heat and the exothermic heat of reaction due to oxidation of its elements. The tapping temperature and chemical composition of hot metal is highly variable depending on the burden composition and grades of steel to be produced. The chemical composition and temperature of hot metal from a Mexican steel plant charged with 100% pellets indicate a tapping temperature from 1430 to 1490 °C and the following chemical composition; 4.2–4.8%C, 0.5–0.9%Si, 0.4–0.7%Mn. According to Sampaio et al. [6] one ton of hot metal at 1430 °C has a sensible heat of 250 kWh, based only on the iron content.

Adams et al. [7] estimated a contribution of 4.5 kWh/ton per 1% replacement of scrap by hot metal, based on the following analysis: if hot metal is charged from 1150 to 1350 °C, the sensible heat varies from 223 to 272 kWh/ton (on average 2.48 kWh/ton per 1% replacement), oxidation of Si and Mn add 1.4 kWh/ton per 1% replacement of scrap, oxidation and dissolution of C contribute with 0.5 and 0.6 kWh/ kg, respectively (1.1 kWh/kg or about 0.6 kWh/ton per 1% hot metal addition). Pfeifer et al. [8] estimated a contribution from 3.19 to 5.02 kWh/ton per 1% replacement of scrap. Zinurov et al. [9] reported a range from 3.5 to 4.5 kWh/ton per 1% replacement of scrap.

Duan et al. [10] reported a decrease in energy consumption from 35 to 40 kWh/ ton, from 171 to 132 kWh/ton, employing 55–57% hot metal in an EAF of 50 ton,

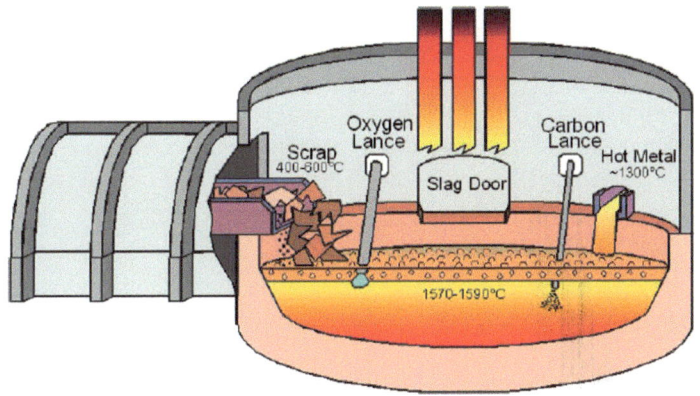

Fig. 10.1 Consteel EAF charged with hot metal. After [12]

also the tap-to-tap time was decreased from 5 to 10 min. They initially charged 20 ton of scrap, then the first part of hot metal previously pretreated was added, 20–25 ton and finally a second part was added, about 20 ton. The hot metal pretreated was decarburized from 4 to 1.5–2.2%C.

The Consteel process uses continuous scrap preheating but can also be charged with hot metal. The largest number of Consteel furnaces has been built in China, about one third of the approximately 80 units worldwide. The first unit was built in Shaoguan in 2000 and in Wuxi I&S the second, in 2001. The electric energy consumption reported for Shaoguan was 260 kWh/ton using 30% hot metal [11]. Figure 10.1 shows a Consteel charged with hot metal.

Table 10.1 summarizes reported or estimated values on energy consumption due to addition of hot metal into the EAF. 1% of hot metal contributes with savings from 3 to 5 kWh/ton. Notice the differences in charging hot metal + DRI with hot metal + scrap, using the same HMR, with DRI the electric energy consumption was 450 kWh/ton and 260 kWh/ton with scrap.

10.1.2 EAF Practice with Hot Metal

Hot metal should be added gradually to avoid violent decarburization with the initial highly oxidized slag, also avoid large fluctuations in the steel chemistry which might result in explosions. Hot metal can be charged in different ways: top charging onto a liquid hot heel using a ladle, boring a hole in the scrap and top charging with a ladle, using a launder through the slag door or in another zone of the EAF. Figure 10.2 shows schematically different ways to charge hot metal into the EAF. Using a launder is the safer and more efficient way to charge hot metal. Zinurov [9] reported a charging rate from 2.5 to 5 ton/min depending on the tap-to-tap time.

Table 10.1 Energy savings with hot metal

	Reference	Decrease in EEC, kWh/ton per 1%	EEC, kWh/ton	% hot metal
[7]	Adams	4.5		
[8]	Pfeifer	3.19–5.02		
[13]	AM Mexico	4–6		
[9]	Zinurov	3.5–4.5		
[6]	AM South-Africa		450	30 (+ DRI)
			200	75 (+ DRI)
[10]	Duan		132	55–57
[14]	Solver		225	35
			30	70
[15]	Gottardi		180	40
			0	70
[16]	Xu		200–250	35–40
[6]	Wuxi I&S China		250–265	25
[6]	Shaoguan China		20–30	70
[11]			260	30

Solver et al. [14] described the methods to charge hot metal into the EAF. The advantage by charging hot metal through the open roof is that the EAF is not modified but this method results in fumes and splashes that can affect the water-cooled panels, the crane has to be used and oxygen injection cannot get started until the roof is closed. Also, the delay time and heat losses increase, affecting productivity. A twin-shell DC-EAF with a capacity of 155 ton was modified to install one side runner in each shell. They used two hot metal ratios, 35% (60 ton) and 70% (120 ton) obtaining a consumption of electric energy of 225 and 30 kWh/ton, respectively.

Gottardi et al. [15] reported results from two plants in China, Shagang Zhangjiagang works and Tianjin Iron and Steel. Both plants had similar AC-EAF's with a capacity of 110 ton of liquid steel, using 6 burners/supersonic lances with a maximum capacity for oxygen injection of 15,000 Nm^3/h, equivalent to 2.27 Nm^3 O_2/ton min, similar in magnitude to the conventional BOF. The process was designed for 40% hot metal and a decarburization rate of 0.12%C/min using 4 oxygen lances operated at 2000 Nm^3/h each, equivalent to 300 Nm^3/h per m^2 of liquid batch surface. These authors reported two important findings. First, the hot charging method has an important effect on tap-to-tap time. They found that placing the runner on top of the EBT balcony gives better performance in comparison with the runner located in front of the slag door due to early oxygen gas injection in the first case, as shown in Fig. 10.3, but this modification was possible only at the EAF from Shagang. First, scrap is charged, then, after the first minute of power-on, hot metal is charged. The pouring rate varies from 5 to 8 ton/min. When half of the hot metal is charged, some lances operate in supersonic mode. Once all Si is oxidized, all lances operate in supersonic mode together with two carbon injectors. Second, the maximum productivity, about

Fig. 10.2 Different ways to add hot metal into the EAF: **a** top charging, **b** launder through the slag door and **c** launder lateral wall. After [9]

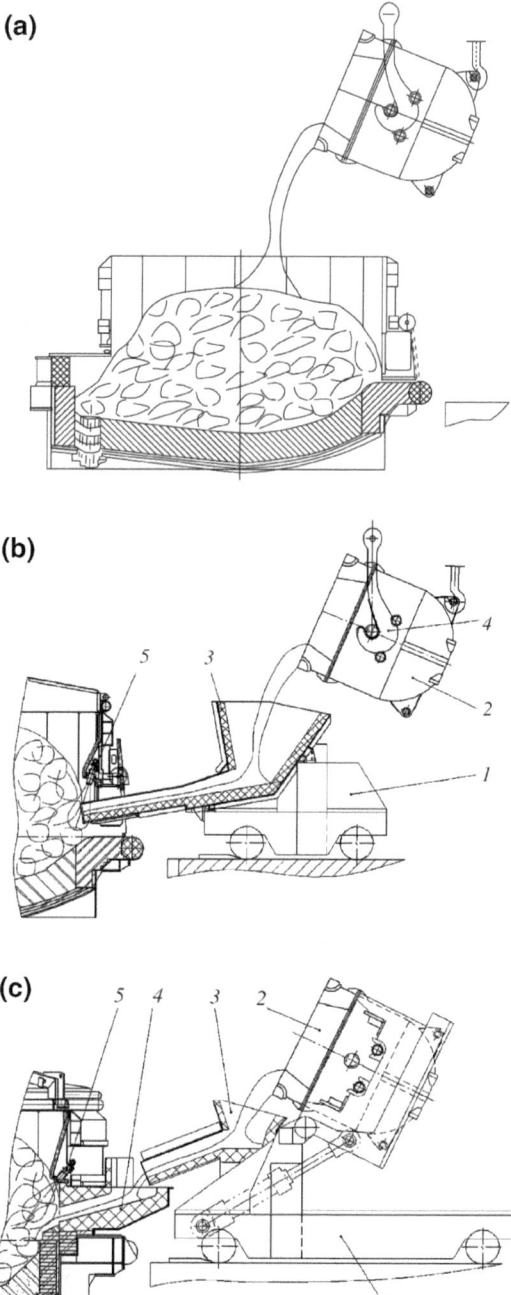

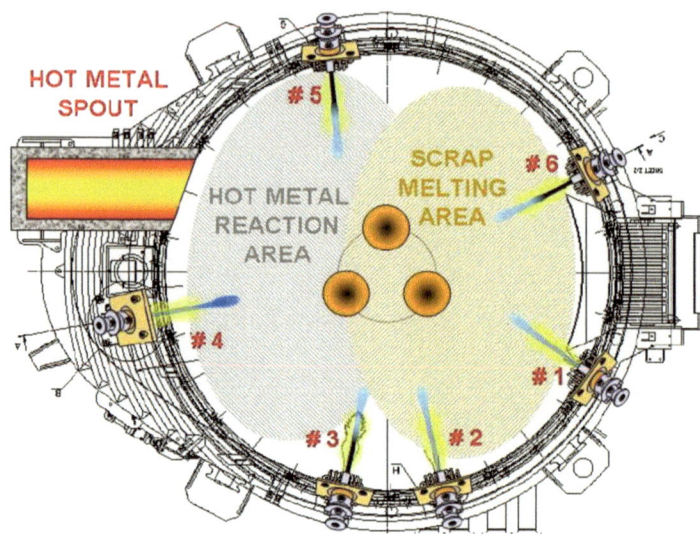

Fig. 10.3 Optimum position for hot metal charging. After [15]

220 ton/h, was reported for a hot metal ratio of 40%, as can be seen in Fig. 10.4. Electric energy consumption linearly decreased with the HMR, from 350 kWh/ton using 10% hot metal to 220 kWh/ton with 40% hot metal and zero electric energy consumption with 70% hot metal, as shown in Fig. 10.5. The decarburization rate increased with the HMR, from 0.05%C/min and 10% hot metal to about 0.10%C/min and 40% hot metal. They calculated the decarburization rate for each heat based on an estimated initial value based on the composition of the charge and defining the beginning of decarburization when intense flames could be seen, the final time was defined for the first steel sample. The total initial carbon was about 2% charging 40% hot metal (1.7% hot metal and 0.3%C from scrap), in this case the oxygen injection in supersonic mode took about 16 min.

Xu et al. [16] reported the optimum value of hot metal from 35 to 40%, for these values energy consumption was about 200–250 kWh/ton and productivity reached a maximum value of 140 ton/h. They employed a DC-EAF charged with hot metal with an average composition of 4.3% C and 0.65% Si and a temperature from 1250 to 1400 °C. The oxygen injection system consisted of three injectors, each one with a capacity of 2000 Nm³/h, in addition to this, two burners with oxygen injection of 900 Nm³/h each one. The specific consumption of oxygen was 47 Nm³/ton and the tap-to-tap time was 44 min.

Mandal et al. [17] reported results on the optimum HMR for two different size EAF's. The furnace with a higher capacity and a diameter of 6.1 m got a peak in productivity of 350 ton/h for 32% hot metal and the EAF's with a smaller diameter of 5.8 m had a peak in productivity of 300 ton/h with 40% hot metal. The reason for this difference was not explained, it can be due to different factors such as differences in oxygen injection capacity.

Fig. 10.4 Productivity as a function of HMR. After [15]

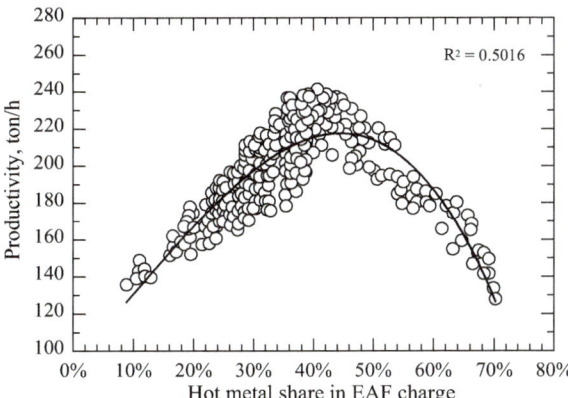

Fig. 10.5 Electric energy consumption using hot metal in the EAF. After [15]

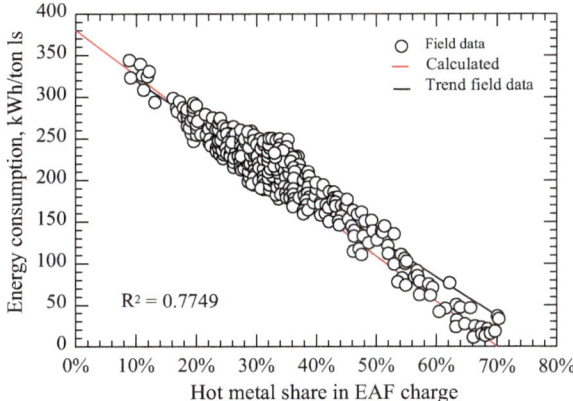

It has also been recommended to avoid decreasing C below 0.3% until the final stage because decarburizing from high carbon levels can yield high FeO and iron losses to the fuming system. The FeO should be controlled to promote enough oxidizing conditions to remove phosphorous.

Grant et. al. [18, 19] explained why carbon injection is even more critical when hot metal is used. This is due to the poor mixing conditions in conventional EAF without stirring mechanisms, then, the bath charged with hot metal exhibits higher concentration gradients, in particular if there is only one oxygen lance. Due to oxygen injection, carbon gradients are developed in particular in the region close to the injection point where the concentration of FeO is much higher. Sudden mixing due to cave-ins, fall of ferrobergs, tilting or other reasons promote a violent reaction between FeO and C that produces CO:

$$FeO_{(l)} + \underline{C} = Fe_{(l)} + CO_{(g)}$$

The reaction is endothermic and decreases the temperature of the molten steel. By injecting carbon in the zone of oxygen injection, the oxygen concentration gradients are decreased.

Pig iron is similar to hot metal in chemical composition, except that it is solid. It is usually produced when the BOF is under maintenance and the BF is kept in operation. It is produced in small ingots, from 3 to 50 kg and more commonly about 10 kg each. It has a bulk density much higher than scrap, about 3.3–3.5 ton/m^3. Due to its high carbon content, its melting point is much lower than that of scrap, therefore it can agglomerate and form ferrobergs during melting, at the same time due to formation of large pieces, it takes longer time for its melting. Because of these problems, it is limited to less than 20% of the total weight of the metallic charge and should be charged to avoid contact with the furnace walls. Grant et al. [19] suggest to charge it in the middle of the EAF to make sure it gets closer to the electric arcs.

Charging hot metal and DRI involves also high risk of explosions in particular if the metallization of DRI is low, below 90%.

The additional amount of fluxes in the EAF due to Si in hot metal can be double in comparison with a scrap practice, in the order of 35 kg lime/ton of hot metal [9].

10.1.3 Limits on Hot Metal Addition into the EAF

Arcelor Mittal Mexico reported energy savings of about 4–6 kWh/ton per 1% replacement of DRI [13]. The maximum amount of hot metal was 23%, about 60 tons. The maximum amount of hot metal was due to limitations on the capacity of injection of oxygen, about 8600 Nm^3/h. Once hot metal was charged, oxygen was injected for about 20 min without connecting the EAF. The average composition of the metallic charge was 4.05% C, 0.37% Si, 0.063% S and 0.042%P for hot metal, 94% metallization, 2.5% C, 6% gangue, 0.0032%S and 0.04% P for the DRI. Additional benefits using hot metal were lower FeO, from 36 to 30%, and shorter power-on time, from 60 to 44 min. On the other hand, they also reported a decrease in productivity, from 70 to 50 ton/h due to the lack of adequate facilities to charge hot metal into the EAF.

Lee and Sohn [20] reported a limit of 40% on the hot metal ratio (HMR) in a DC EAF of 150 ton, with a capacity to inject 4000–8000 Nm^3 O$_2$/h, otherwise the decarburization rate decreases due to oxygen supply, increasing the tap-to-tap time. Increasing the HMR ratio from 0.4 to 0.75 increased the oxygen requirements from about 4500–7500 Nm^3/h. 50–65 ton of hot metal were charged, using a ladle, once the electrodes completed the boring phase on the initial charged scrap, which was 90–105 ton. In a typical heat, hot metal was added at minute 5 and completed the addition at minute 9. Oxygen injection started at minute 16 with a flow rate of 4000 Nm^3/h. The oxygen flow rate was progressively increased in 20 min to 8000 Nm^3/h. The maximum flow rate of oxygen, at 9000 Nm^3/h takes place during 5 min and then it is decreased to 5000 Nm^3/h, finally at minute 47 oxygen injection stops. The normal DeC rate with scrap was from 0.02 to 0.04%/min and with hot metal increased from 0.04 to 0.08%C/min depending on the carbon content in the molten metal, as can be seen in Fig. 10.6. The savings on energy consumptions were not reported but the benefits on the tap-to-tap time were reported as 0.34 min per each 1% of hot metal. They also reported closer equilibrium conditions, defined in terms of lower dissolved O and lower FeO.

Lee et al. [21] also compared the dephosphorization (DeP) degree with scrap and with hot metal, as shown in Fig. 10.7. The DeP degree was higher in the case of the steel produced with hot metal, on average 50 and 75%, respectively.

Lee et al. [22] extended the analysis to higher HM ratios which shows the optimum hot metal ratio is 40%, in terms of tap-to-tap time, as shown in Fig. 10.8. Above 40% hot metal and for the steel pants evaluated, the found oxygen supply limitations to achieve higher decarburization rates.

Analysis for heats with 36% HM reported a higher P partition ratio and C-O values closer to equilibrium in comparison with no additions of HM, attributed to

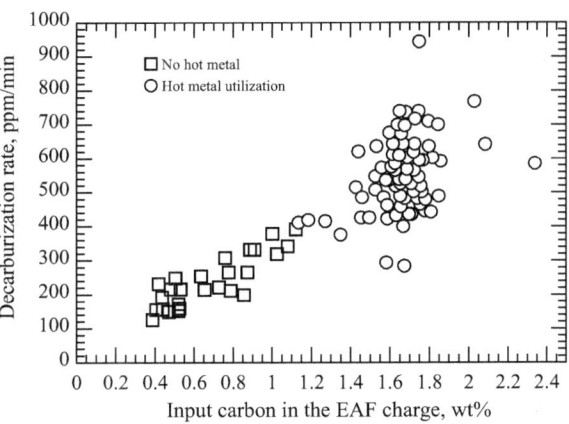

Fig. 10.6 DeC rates with and without hot metal. After [20]

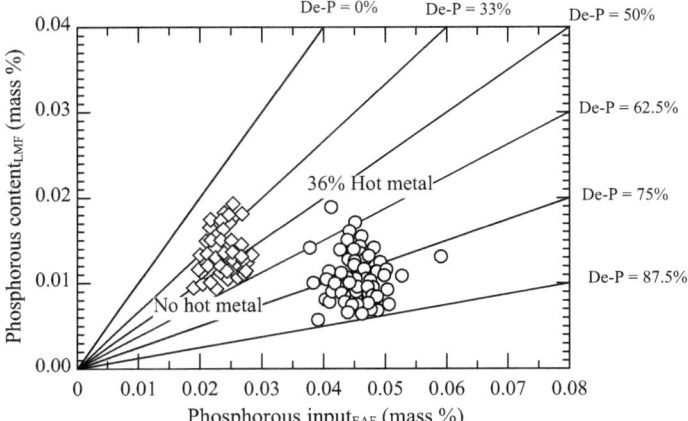

Fig. 10.7 Comparison of dephosphorization degree using hot metal. After [21]

Fig. 10.8 Effect of hot metal ratio on tap-to-tap time. After [22]

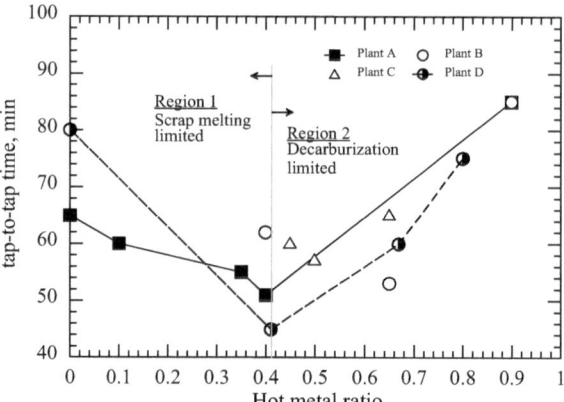

enhanced stirring conditions due to formation of CO [21]. Nitrogen decreases when hot metal is charged because of the higher generation of CO bubbles. Values from 50 to 70 ppm N have been reported [9]

Since increasing hot metal, increases the requirements of oxygen injection, this can be a major limitation for old EAF's in particular. Tian et al. [2] compared oxygen consumption in two plants with different HMR, as shown in Fig. 10.9. It can be observed that the specific oxygen consumption for 40% hot metal is about 30–40 Nm^3/ton and increases to 40–50 Nm^3/ton when the hot metal increases from 60 to 80%.

In a conventional BOF, the specific gas flow rate ranges from 2.5 to 3.8 Nm^3 O_2/ton min and using combined blowing from 3.5 to 5.0 Nm^3 O_2/ton min [23]. Table 10.2 shows actual oxygen injection capabilities in EAF plants.

Since high C and high Si in hot metal is at the same time an advantage in terms of supply of chemical energy but also a disadvantage in terms of large requirements on oxygen consumption, time for removal of those elements and its waste, it was

Fig. 10.9 Oxygen consumption as a function of HMR. After [2]

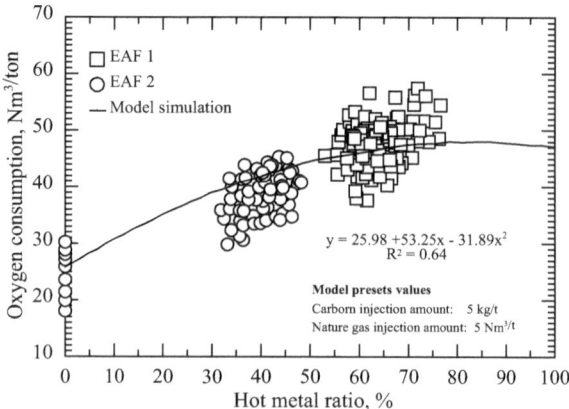

Table 10.2 Oxygen injection capacity in some EAF plants

		Max O_2, Nm^3/h	O_2, Nm^3/ton	O_2, Nm^3/ min ton	Maximum hot metal, %	EAF, ton	t-t-t, min
[13]	AM MX	8600	43	0.65	23	220	
[20]	Lee	9000		1.00	40	150	47
[16]	Xu		47.1	1.07	35	100	44
[15]	Gottardi	15,000	68	2.27	40	110	< 35

suggested to produce special hot metal with lower concentrations in C and Si using a cupola furnace. This idea was from an Indian inventor [24] because India has a large number of cupolas and small EAF's. The suggested chemistry of hot metal was C lower than 3%, silicon lower than 0.2% and manganese lower than 0.2%, together with a temperature above 1375 °C.

10.1.4   Increased Capacities on Post-combustion and the Off-Gas System

Using hot metal will produce a higher volume of CO and in order to take advantage of its chemical energy it would be desirable to implement post-combustion which also would require a higher capacity by the off-gas emissions system. Irawan et al. [25] applied the integral EAF model developed by Bekker to study the effect of hot metal on energy consumption and CO generation and also developed a CFD model to describe the temperature distribution along the dedusting system. Increasing hot metal from 0 to 30% (50% DRI and the rest was scrap) decreased energy consumption from 672 to 624 kWh/ton and also increased the production of CO. At its highest peak of CO generation, the increment was from about 4700–5400 kg CO using 30% hot metal. Post-combustion of the CO increased the temperature of liquid steel from 1628 to 1899 °C if the furnace power increases from low to high power.

10.1.5   Carbon Footprint with Hot Metal in the EAF

Hot metal in comparison with scrap has a much higher carbon concentration and therefore, increasing the hot metal ratio (HMR) in the EAF will increase the carbon footprint, however the extent of this influence is reduced by its effect on the decrease on electric energy. Liu et al. [26] reported two practices, one with 100% scrap and another with 60% scrap-30% hot metal, obtaining a total of 360 kg CO_2/ton and 759 kg CO_2/ton, respectively, an increment of 399 CO_2/ton with 30% hot metal, more than 100% compared to 100% scrap. The electric energy consumption using

Fig. 10.10 Carbon emissions in the EAF with different ratios of hot metal. After [2]

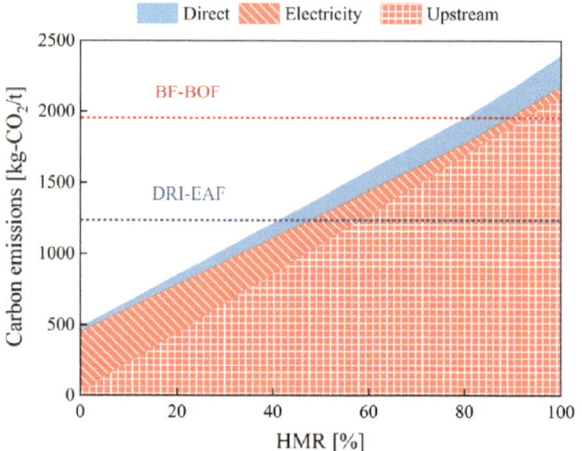

hot metal was decreased from 3.3 to 2.3 GJ/ton but the total energy consumption was increased from 4.7 to 8.4 GJ/ton due to higher input of chemical energy.

Na et al. [27] argues that the EAF in China has not contributed to decrease CO_2 emissions because of the widespread use of hot metal to replace steel scrap. The emissions of CO_2 for the EAF were reported in the order of 1419 ton CO_2/ton steel. Tian et al. [2] reported the carbon foot print for the EAF with increasing amounts of hot metal, as shown in Fig. 10.10. Their calculation estimate about 500 kg CO_2/ton steel for 100% steel scrap and about 2100 kg CO_2/ton steel for hot metal produced in the BOF. The figure shows that HM increases CO emissions taking into the upstream emissions (CO_2 emissions to produce the hot metal in the BF). In this plot the lines are references using 100%DRI in the EAF and the emissions in the BF. When the amount of DRI is 41.3%, it reaches the value of emissions obtained with 100% DRI in the EAF.

Yang et al. [28] reported a decrease in carbon emissions when the HMR ratio was increased. This particular result was obtained assuming that electricity was produced with coal and even if carbon emissions in the EAF were increased due to higher HMR ratio, the net result, according with their analysis, gave lower net emissions, for example for scrap-based EAF the total carbon emission were 146.9 kg/ton and with 50% hot metal, it decreased to 137 kg/ton.

10.2 Hot DRI

10.2.1 Operation with 100% DRI

Direct Reduced Iron and steel scrap have important differences, both chemical and physical, as shown in a simplified way in Table 10.3:

Table 10.3 Property comparison between scrap and DRI

	Scrap	DRI
Residual elements, wt%	0.1–0.9	0
Gangue content, wt%	0	3.9–8.4
Iron oxide, wt%	0	6–14
Carbon, wt%	0.001–0.3	1–4
Fe, wt%	99–99.9	81–86
Bulk density, ton/m^3	0.3–1	1.6–3.5
Size, mm	Variable	6–15

One of the main advantages of DRI over scrap is the low concentration of residual elements. Using DRI allows more flexibility to charge cheaper low-quality scrap which contains a higher concentration of residual elements. On the other hand, its main disadvantage is the presence of oxides of high melting point such as Al_2O_3, CaO and MgO. FeO varies depending on DRI metallization, from about 14 to 6% for a metallization from 88 to 94%, respectively. These differences are the main reason for a different EAF operation with scrap and DRI. An operation with 100% DRI should consider the following:

- Higher gangue content in DRI produces a higher volume of slag.
- Higher acid gangue content demands an increment in the amount of fluxes otherwise potentially increases refractory wear and slows down steel refining
- Higher gangue content will severely increase the consumption of electric energy
- Lower metallization increases iron oxide which in turn decreases the requirements of oxygen injection but increases energy consumption because the reduction of FeO is endothermic. Reduction of FeO by carbon contributes to the formation of CO which improves the stirring conditions and the removal of nitrogen from liquid steel.
- Unreduced iron oxide from DRI contributes to a lower metallic yield.
- The gangue content and additional fluxes will increase the tap-to-tap time
- Scrap contains low carbon, on the contrary, DRI has carbon contents from 1 to 4%. Higher carbon contributes with chemical energy and production of CO. This gas improves slag foaming and nitrogen removal. Higher carbon will require additional oxygen injection and conditions to improve the decarburization rate, such as earlier oxygen injection, to offset the difference with respect to the case of lower C in the DRI.
- Post-combustion is not required because there are no solids above the free surface to absorb that heat
- Can be added continuously, requiring only a pipe to connect the DRI silo with the EAF roof. The EAF can operate without opening the roof. If the fraction of DRI charged is below 20% it can be added in the scrap bucket.

Table 10.4 Consumption factors as a function of %DRI

Share of DRI/HIB, %	0–5 (HBI)	60–95 (DRI)
Energy consumption, kWh/ton	340–390	530–680
Natural gas, Nm^3/ton	5–10	0–2
Oxygen injection, Nm^3/ton	25–37	2–35
Coal and carbon fines, kg/ton	2–9	8–17
Fluxes, kg/ton	25–35	27–60
Tap-to-tap time, min	50–60	60–100
Metallic yield, %	90–94	87–92

Kirschen et al. [29] compared 16 industrial EAF's with scrap and using more than 50% DRI. The results are shown in Table 10.4. These results summarize the effect of DRI on EAF steelmaking.

Feeding rate: One of the most important parameters in a metallurgical practice with DRI is the feeding rate. This value depends on several factors but is essentially controlled by the temperature of liquid steel. Variations in chemical composition produce fluctuations in the temperature of liquid steel. For the same power input, increasing the feeding rate of DRI with high metallization will increase the temperature of liquid steel and on the other hand, increasing the feeding rate of DRI with low metallization will decrease the temperature of liquid steel. DRI of low metallization will contain a higher concentration of FeO that consumes more energy. On a practical basis, the EAF operator measures the temperature of liquid steel and if it has increased then, increases the feeding rate. The target is to maintain the temperature of liquid steel in the range from 1570 to 1590 °C. Hornby et al. [30] gave an example to compute the feeding rate, assuming an energy consumption of 500 kWh/ton DRI and an active power of 80 MW, the feeding rate would be:

$$kg\frac{DRI}{min} = 80\,MW \times \frac{1000\,kW}{MW} \times \frac{h}{60\,min} \times \frac{1000\,kg\,DRI}{500\,kWh} = 2666\frac{kg\,DRI}{min}$$

Equivalent to 33.3 kg/min·MW. The feeding rate can be in the range from 24 to 36 kg/min·MW. Perwaja steel [31] reported a feeding rate of 38.8 kg/min·MW using 87.7%DRI for a heat size of 75.2 ton, 60 MVA transformer and active power of 57.9 MW. The higher feeding rate was obtained increasing the intensity of oxygen and carbon injection, using 4 KT oxygen lances and 2 KT carbon injectors. Charging hot DRI, as will be discussed below, allows a much higher feeding rate in the order of 55 kg/min·MW.

Electric power profile: The operation with 100% DRI allows flat bath operation during the whole heat, therefore the furnace can be operated with maximum electric power if there is enough hot heel (10–20%).

10.2.2 Benefits of Hot DRI on Energy Consumption

The idea to recover the sensible heat of DRI was initially applied by the Purofer process in the 1970's in a limited scale [32]. Hot DRI can be transported to the EAF in two ways:

- Buckets
- Conveyor.

Transporting hot DRI in buckets produces larger heat losses, reoxidation and DRI physical degradation, therefore is not a recommended method, however it is employed sometimes when the conveyor is out of operation.

One of the requirements to use hot DRI is the proximity of the reduction reactor with the meltshop in order to decrease heat losses and transportation distance. The concept to produce hot DRI is to avoid the cooling stage and pneumatically/ mechanically transport hot DRI from the reactor's discharge to the EAF for its direct and continuous feeding. To avoid reoxidation, hot DRI is transported under a protective atmosphere. The carrier gas is separated from the hot DRI in the pressurized interface bins, then is quenched, scrubbed and recycled to the pneumatic transport circuit.

Currently, both technologies MIDREX and HYL offer the option of hot DRI delivery to the EAF. According with Abel and Hein [33], the plant in Hadeed Saudi Arabia can deliver DRI at a temperature of 650 °C using an inclined belt to an intermediate hopper above the furnace and discharged by gravity into the EAF, as shown in Fig. 10.11. During transportation nitrogen is employed to avoid carbon oxidation. Comparing the reported energy consumption with 75% cold DRI + Scrap and 100% hot DRI, it was 492 and 430 kWh/ton, respectively.

There is no doubt that hot DRI offers multiple benefits and new plants should consider this option, however most of the available information on its benefits come from the main suppliers, which offer the best face of this practice and lack a rigorous analysis of this technology.

Transportation of hot DRI involves two challenges; transporting hot materials to an elevated height and avoid any exposure with the atmosphere. Charging hot DRI can become suitable if the DRI plant and the EAF furnace are close to each other. Degradation due to handling was reported to be in the order of 1% [34]. The consumption of nitrogen is about 5–8 Nm^3/ton. It has been argued [35] that pneumatic transport is only suitable for small capacities and is more expensive, suggesting instead mechanical transport, with conveying capacities of about 210 ton/h with an inclination angle of 60° up to an elevation of 100 m and 400 ton/h up to 80 m. The lower the inclination angle and elevation the higher the conveying capacity.

HyL has reported that HyLSA was using hot DRI since 1998 using pneumatic transport and a DRI temperature from 250 to 350 °C [36]. In 2004 they experienced charging DRI at 600 °C. In their system DRI is transported pneumatically through a vertical pipe to the top using an inert gas, then through an elevated platform is moved to the EAF shop where it is discharged by gravity. Figures 10.12 and 10.13 illustrate

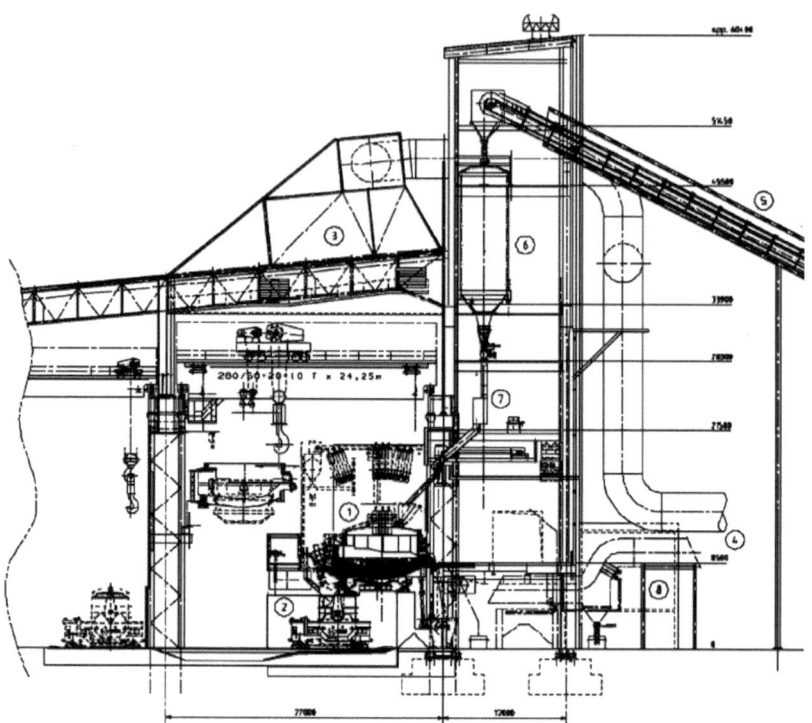

Fig. 10.11   Typical installation for hot DRI. After [33]

this system. Figure 10.12 correspond to the Monterrey plant of HYL [37], showing on the left side an insulated storage bin from which DRI is discharged by gravity to the EAF. The storage bin has the purpose to balance differences in the instantaneous throughput of the DRI and EAF plants, usually the time availability of the DRI plant is higher than the EAF plant because is a continuous process in comparison with the batch processes for the EAF. In addition to this, an alternative path allows DRI cooling. Cold DRI is employed when the DRI plant is under maintenance. Midrex [38] argues that discharging from a DRI reactor at high pressure, 5–6 barg like HyL, requires additional equipment; 20 mechanical valves and five lock hoppers.

Patrizio et al. [40] described in detail results of charging hot DRI at Emirates Steel Arkan (ESA), United Arab Emirates, with a metallic charge that included 10% cold DRI and 90% hot DRI. Energy consumption was decreased to 412 kWh/ton using hot DRI, further optimization of the process decreased electric energy consumption even further to 378 kWh/ton. The steelshop had the following characteristics [41]: EAF with a tapping capacity of 150 ton, 50 ton hot heel, 7 m lower shell diameter, electrodes 710 mm, pitch circle of 1400 mm, transformer 130 MVA + 20%, reactance 1.3 Ohm, secondary current 74 kA, secondary voltage from 650 to 1250 V, maximum active power 112–114 MW. The hot DRI injection point was located far from the 4th hole and its diameter modified to decrease the velocity of the off-gases below 30 m/

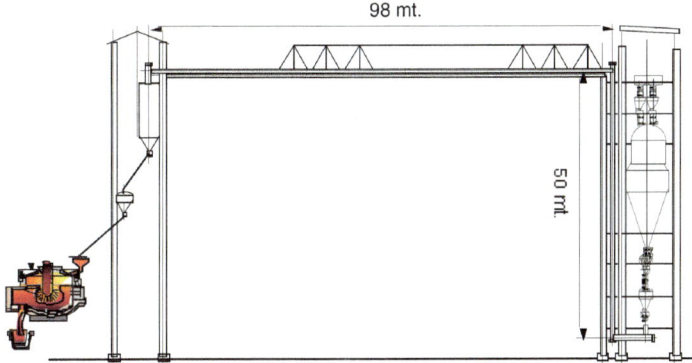

Fig. 10.12   Pneumatic hot DRI transport. Monterrey plant HyL. After [37]

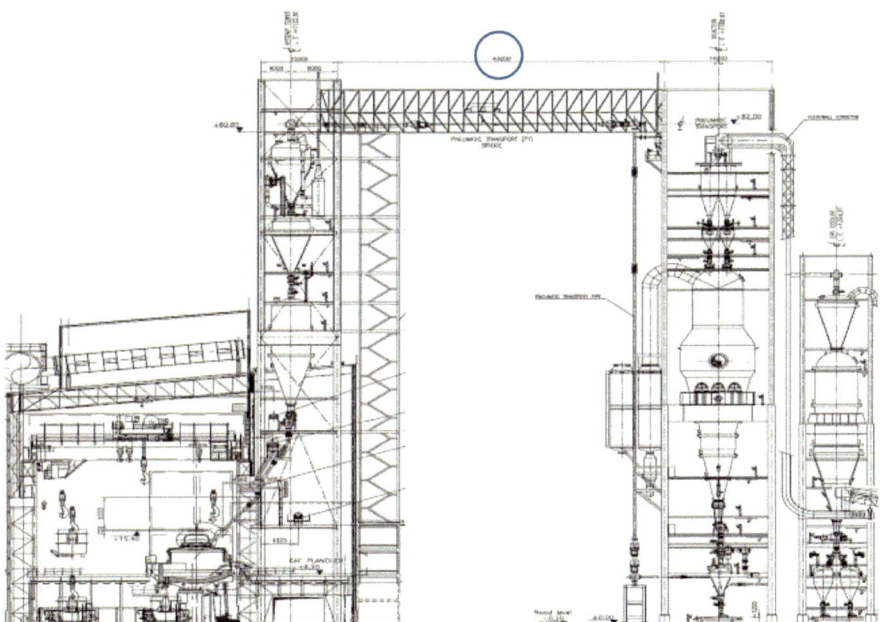

Fig. 10.13   Pneumatic transport. HyL. After [39]

s, both factors contributing to decrease the loss of DRI fines, as shown in Fig. 10.14. The EAF has five oxygen lances with a capacity of 2200 Nm^3/h and three carbon lances with a capacity of 15–30 kg/min. The average chemical composition of the DRI had a metallization of 94.9%, 2.09%C and 90 ppm S.

The melting profile using 10% cold DRI and 90% hot DRI is shown in Fig. 10.15. The EAF is connected at minute 6. Electric power, oxygen and DRI are increased progressively in the 10 following minutes; maximum power from 60 to 100–108 MW,

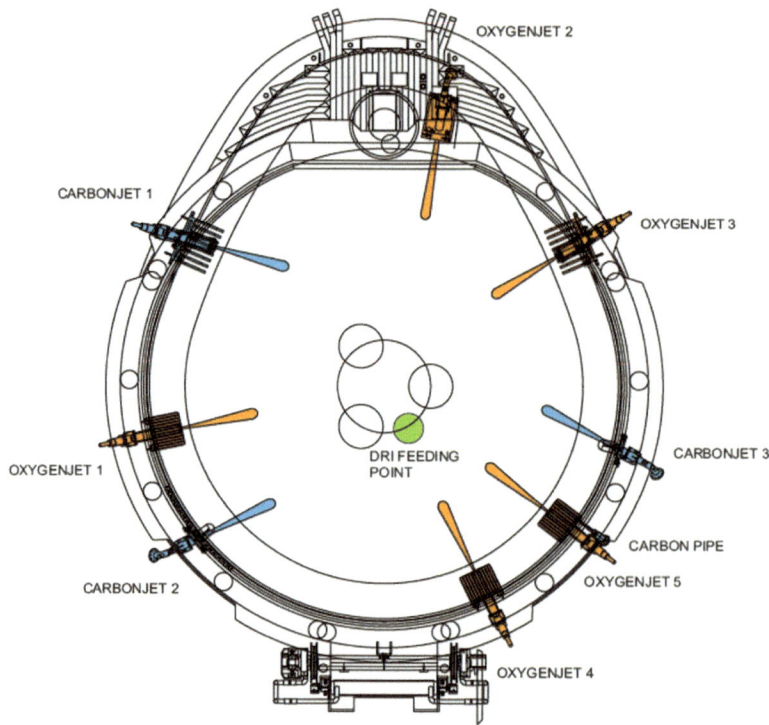

Fig. 10.14 AC-EAF at Emirates steel. After [41]

oxygen from 4000 to 10,000 Nm^3/h and DRI from 2000 to 5800 kg/min. The equivalent specific DRI feeding rate increases from 40 to 55 kg/min MW. The refractory wear index (RWI) reached values from 230 to 235 KWV/cm^2 when the maximum power was in the range from 100 to 108 MW. It has been reported that the feeding rate should be decreased when charging hot DRI to prevent carbon boil [42].

A comparison of the consumption factors with the previous metallurgical practice with and without hot DRI is shown in Table 10.5. The decrease in electric energy consumption and power-on time using hot DRI were 141 kWh/ton and 13.9 min, respectively. The decrease in energy consumption resulted from the sensible heat in the DRI (105 kWh/ton), additional chemical energy from extra oxygen injected (4 kWh/ton), higher metallization with hot DRI (23 kWh/ton) and lower power-on time (9 kWh/ton). The decarburization rate was in the order of 320 kg C/m^2 h. The yield was reported as 92% based on the ratio of liquid steel to the mass of iron in the DRI or 87.7% if the yield was defined as the ratio between liquid steel to the total DRI charged, which is the conventional definition of metallic yield. These values resulted from iron losses to the slag (28%FeO) and dust, 31 and 18 kg/ton, respectively.

Midrex claims they started using hot DRI since 1984 [43]. Midrex [44] in 2000 reported some trial results from Essar India operations using DRI at 650 °C. They

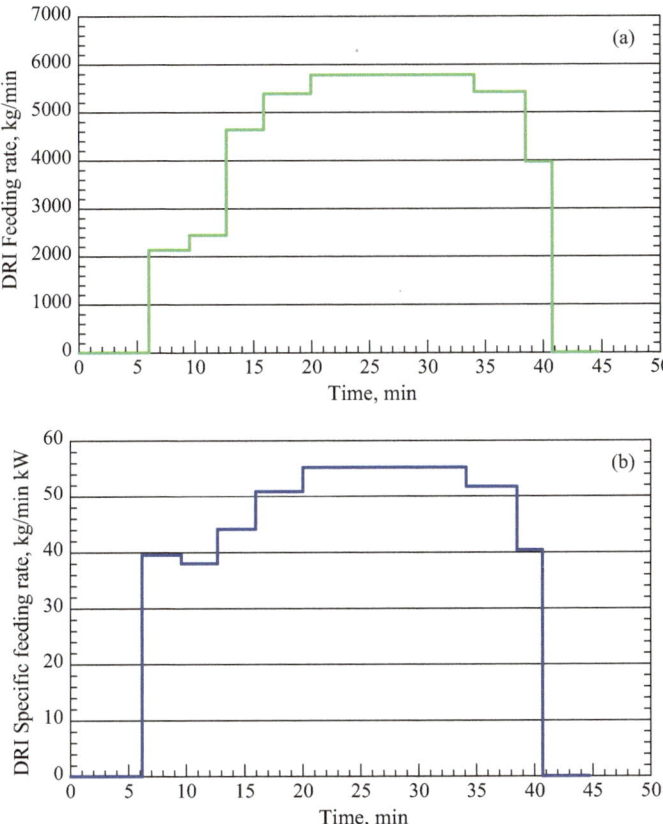

Fig. 10.15 Melting profile with 100% DRI; **a** Total DRI feeding rate and **b** specific DRI feeding rate After [41]

Table 10.5 Comparison of consumption factors. After [41]

DRI	t-t-t, min	p-on, min	EE kWh/ ton	O₂, Nm³/ ton	C inj, kg/ ton	Avg MW	Taping T, °C
90% hot + 10% cold	45.5	35	392	34.9	9.7	102	1640
100% cold	58.5	48.9	533	33.3	17.3	98	1640

employed buckets, with 35 ton and 90-ton capacity, to transport the DRI. One advantage of DRI over scrap it its high bulk density and charging the EAF with one single bucket. Due to transport limitations by bucket, the DRI charged ranged from 20 to 70%. The reported temperature loss was about 50 °C. They also reported 20–24 kWh/ton for every 100 °C increment in DRI temperature, as well as a shorter power

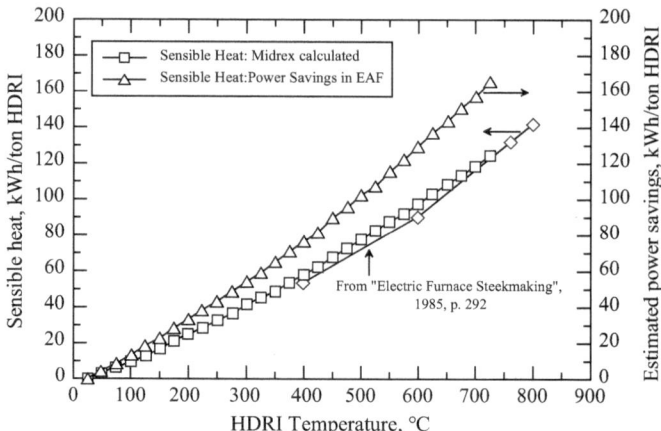

Fig. 10.16 Effect of DRI temperature on its heat content

on time. According with their calculations, using both hot and higher C DRI (2.6%) it would be possible to decrease the tap-to-time by 14 min.

Tavano et al. [32] estimated the sensible heat of DRI assuming a heat capacity for the DRI in the range from 0.17 to 0.20 kcal/kg°C, then for a mass of 1 ton and an increment of temperature of 100 °C, hot DRI would decrease energy consumption in the range from 19.8 to 23.3 kWh/ton.

Midrex reported the expected energy savings in the order of 17–20 kWh/ton per 100 °C and total savings from 120 to 140 kWh/ton with 95% hot DRI in the metallic charge, at an average temperature of 700 °C [45]. Figure 10.16 shows theoretical calculation on the sensible heat of DRI as a function of its temperature.

Martinis et al. [46] from Tenova HyL reported energy savings equivalent to 26 kWh/ton for every increment of 100 °C, as shown in Fig. 10.17. These results correspond to Emirates steel using 90% hot DRI and 10% cold DRI. The lower energy consumption corresponds to DRI with higher metallization. According with these results, using hot DRI at 600 °C would consume from 370 to 450 kWh/ton for a metallization of 96 and 93%, respectively.

In addition to this it was also reported a decrease in the melting time, about 5% per every increment of 100 °C, which also increases the productivity. For example, for a DRI with a metallization of 94%, the tap-to-tap time decreases from about 65 min using cold DRI to 45 min with hot DRI at 600 °C, a decrease in tap-to-tap time of 20 min, as shown in Fig. 10.18. These authors also reported improvements in the reactor cone in order to decrease heat losses during the discharge of hot DRI. They improved the flow of solids and decreased heat losses adding a refractory material with a friction coefficient similar to steel at low temperature.

Memoli et al. [47] from Tenova HyL further reported the predicted electric energy consumption using 100% hot DRI with different carbon content, as shown in Fig. 10.19. According with these predictions using hot DRI at 600 °C would further decrease the electric energy consumption from 440 to 375 kWh/ton by increasing

Fig. 10.17 Effect of hot DRI temperature on energy consumption. After [46]

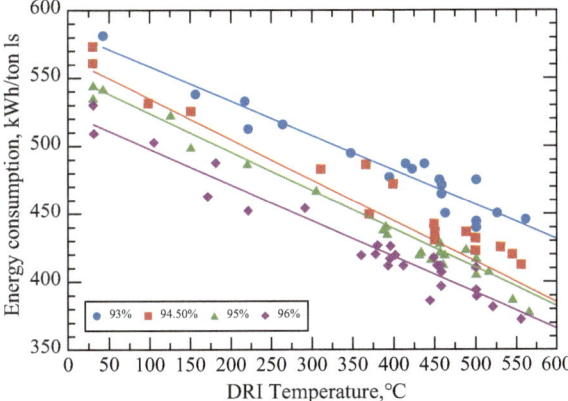

Fig. 10.18 Effect of hot DRI temperature on tap-to-tap time. After [46]

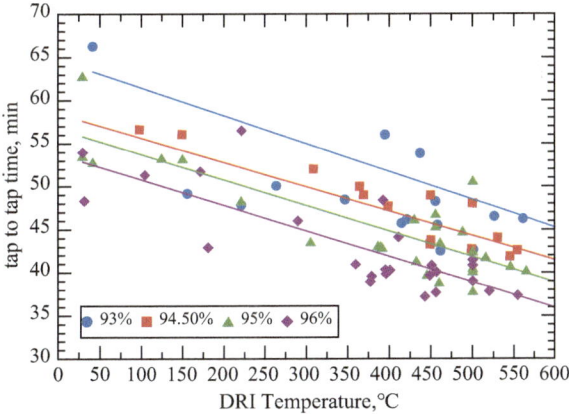

the carbon content from 2.2 to 4.0%. For a fixed carbon concentration, increasing the temperature of DRI up to 600 °C would decrease electric energy consumption by about 25%. An economic evaluation using hot DRI was reported by Duarte [48] from the same company, in 2004. In this report they claim economic benefits using hot DRI with higher carbon content, as shown in Table 10.6. From this table, it is observed that producing hot DRI is slightly more expensive than producing cold DRI, in particular if the C concentration is increased, however the decrease in electric energy consumption in the EAF has a strong impact on the final production costs and higher productivity. For the case reported, productivity can be increased from 0.8 to 1.1 Mton/year using hot DRI at 600 °C with 4.0%C. Midrex [49] reported slightly higher production costs because of a higher consumption of natural gas, about 0.1 Gcal/ton DRI, equivalent to 1 USD/ton DRI, on the basis of a price of natural gas of 9.92 USD/Gcal (2.50 USD/mmBTU). Figure 10.20 shows Midrex savings using hot DRI on the basis of a cost of electricity of 0.035 USD/kWh, electrode consumption equal to 0.004 kg/kWh and a cost of 2.3 USD/kg. For 600 °C the estimated combined savings indicate 6 USD/ton.

A review from Midrex in 2022 [38] on their hot DRI plants shows that most of their plants use a conveyor (hot transport conveyor, HTC). Their suggestions for the hot transport method are shown in Table 10.7. Direct charging by gravity is adequate for green field projects when the DRI is located above or close to the EAF but the height of the DRI reactor is raised "significantly", as can be seen in Fig. 10.21. Using hot transport vessels (HTV) is an alternative for plants where the distance between the DRI reactor and the EAF is larger than 200 m.

The first Midrex plant using direct charging by gravity was Jindal Hadeed in Sohar Oman, the project started in 2005 and commissioning until 2011, with an estimated investment of 700 million USD [50]. It was designed for a production capacity of 1.5 Mton DRI with a metallization higher than 93% and 1.5%C, the

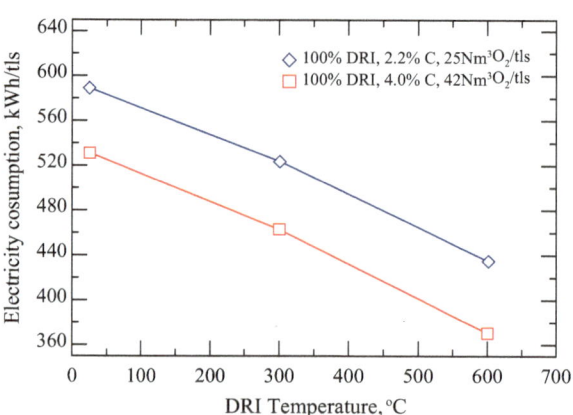

Fig. 10.19 Predicted energy consumption using hot DRI with 2.2 and 4.4%C. After [47]

Table 10.6 Effect of DRI carbon and temperature on liq. steel production cost [48]

DRI plant	Metallization	%	94		94		94	
	C_{DRI}	%	2.2		4		2.2	
	Hot DRI	°C	25		25		600	
			Unit/tls	USD/tls	Unit/tls	USD/tls	Unit/tls	USD/tls
EAF plant	Natural gas	Gcal	2.31	18.3	2.41	19.1	2.33	18.5
	Total DRI cost			80.2		80.6		80.6
	Electricity	kWh	589.6	20.6	531.4	18.6	435.4	15.2
	Total cost LS			134.2		132.9		127.4
Annual capacity		ktls/year	839		1031		1140	

Unit prices: 7.9 USD/unit, electricity 0.035 USD/kWh

Fig. 10.20 Combined
savings using hot DRI. After
[49]

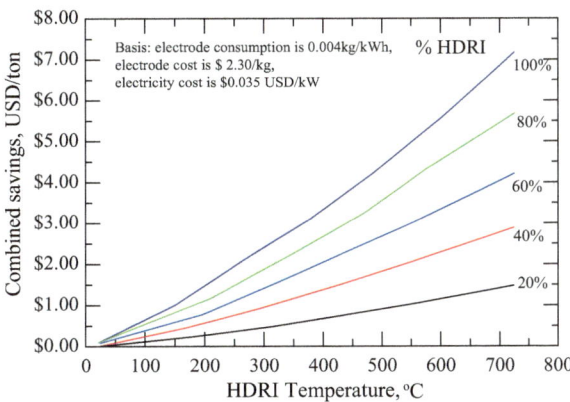

Basis: electrode consumption is 0.004kg/kWh, % HDRI
electrode cost is $ 2.30/kg,
electricity cost is $0.035 USD/kW

(y-axis: Combined savings, USD/ton — $0.00 to $8.00; x-axis: HDRI Temperature, °C — 0 to 800; curves labeled 100%, 80%, 60%, 40%, 20%)

Table 10.7 Suggested hot
transport method. After [38]

Distance DRI plant-EAF, m	0–40	40–200	> 200
Hot transport method	Gravity	HTC	HTV

discharge temperature of DRI was about 650 °C and charged into the EAF at 622 °C,
a temperature decrease of about 25 °C. In this method, if the length of the DRI chute
is large it causes DRI physical degradation. In a second design [38] Midrex used an
intermediate bin and then transported horizontally to the EAF, decreasing the height
by 12 m but with this design the discharge flow rate of the DR furnace is unknown
because the HDRI feed bin is emptying and filling at the same time.

Harada and Tanaka [52] reported a decrease in CO_2 emissions using hot DRI,
1022 kg CO_2/ton steel using 80% hot DRI.

A recent development on hot DRI is the Continuous Reduced Iron Steelmaking
Process (CRISP) developed by Gordon and Wheeler (Hatch) in 2005 [53, 54]. The
concept of charging hot DRI in this process is installing the DRI plant at a significant
height above the EAF shop, as shown in Fig. 10.22, which is not a relevant new
concept, however some of their additional features include a large rectangular EAF
with six electrodes, significant radiation from the arcs using a long arc operation,
instead of oxygen injection they employ iron ore pellets and it is proposed a continuous operation. This process anticipates the use of low quality iron ores to produce
DRI. The pilot plant test were carried out in a conventional EAF [55]. Barati [56],
on the basis of theoretical estimations concludes that this process has slightly lower
energy intensity in comparison with the conventional HDRI-EAF or CDRI-EAF
processes; 20.91, 21.99 and 22.76 GJ/ton, respectively.

10.2.3 DRI Preheating

Scrap pre-heating with the sensible heat from the off-gas is a mature technology in
the EAF, however DRI preheating with the off-gas have never been implemented on

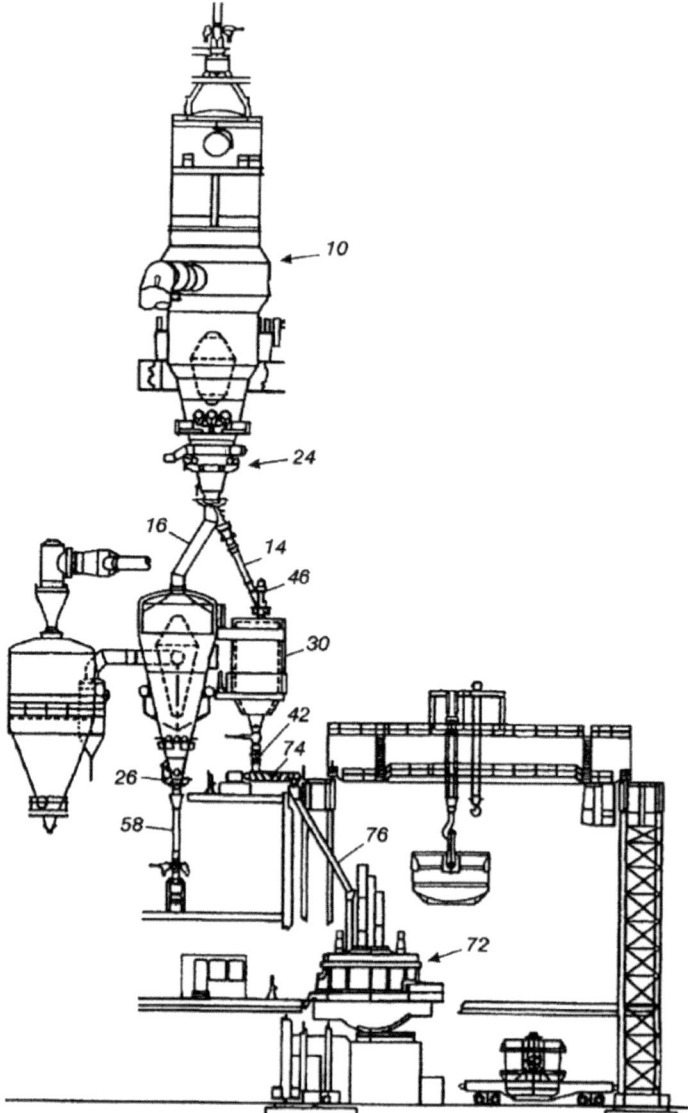

Fig. 10.21 Hot charging method by Midrex (Hotlink). After [51]

a commercial scale. The idea, however, has been suggested since the early 1970's. Tress and Hunter [58] in 1975 suggested DRI preheating with the off-gas using a special container which rotates promoting a higher gas–solid interaction. The patent suggests a short residence time of the solid with the gas in the container, about 30 s. It lacks of any information on the oxidation of DRI. A similar patent was disclosed by Kim [59] in 1987 with a chute divided into two sections, one which rotates. The

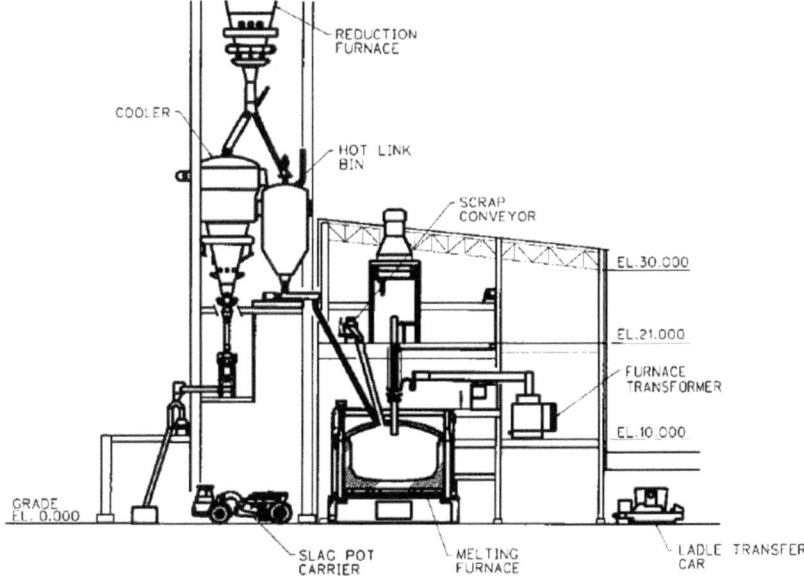

Fig. 10.22 CRISP process. After [57]

chute bifurcates to separate the final off-gas, as shown in Fig. 10.23. Details on DRI reoxidation also omitted in this patent. Scrap and DRI are very different materials both physically and chemically. DRI is porous and highly reactive, therefore reoxidation is much higher. The major concern with DRI preheating using the off gas is excessive reoxidation which the previous patents ignore.

Sensis et al. [60] in 1981 proposed a method to preheat DRI focused on the oxidation of carbon contained in the DRI using the off-gas from the EAF and burners to achieve high temperatures and high velocities, in the range from 1450 to 1550 °C and higher than 1 m/s, respectively. DRI is preheated using a vibrating conveyor to promote a loose bed of particles which are exposed to the hot gas for a short period of time, from 1 to 1.5 min. The thickness of the sponge iron bed should be from 7 to 15 cm. The basis of this method is oxidation of carbon in the DRI, therefore it should be at least 1.5% but preferably in the range from 1.7 to 4%. The authors reported one example with an initial C_{DRI} of 2.5% and after preheating in the conveyor C_{DRI} decreased to 1.35% without changes in the metallization. Another option to add carbon was saturating with a hydrocarbon oil.

Pavlicevic et al. [61] from Danieli in 2002 filed a patent to preheat DRI using the off gas from the EAF with substantial conditioning in terms of temperature and final chemical composition to promote DRI carburization. The process, shown in Fig. 10.24 shows the off gas leaving the furnace above 1200 °C, this gas is cooled and its composition adjusted using burners with natural gas. The CO/CO_2 ratio of the preheating gas should be above 1.6 and its temperature in the range from 450 to 700 °C. Subsequently, this patent was abandoned.

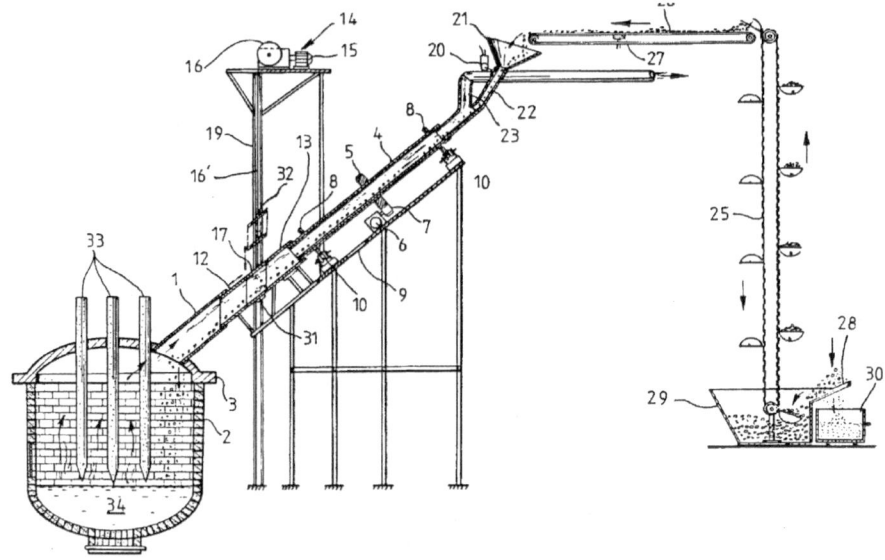

Fig. 10.23 DRI preheating using the off gas. After [59]

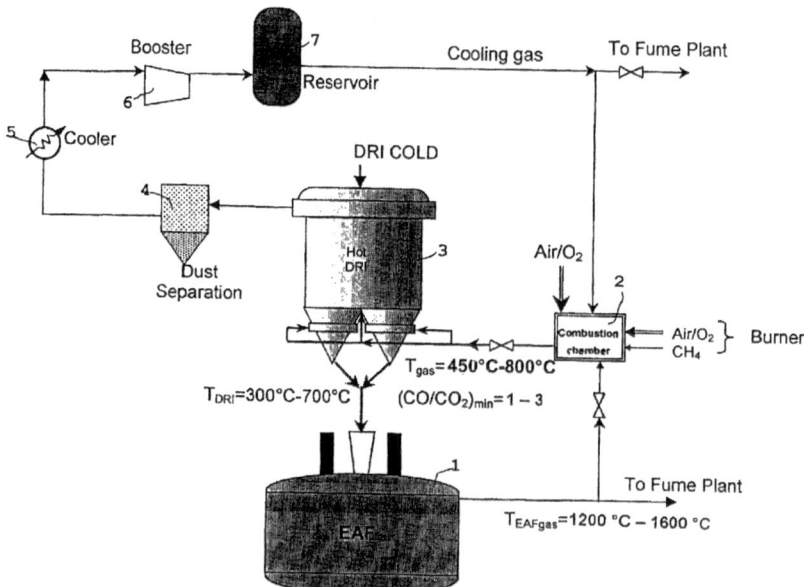

Fig. 10.24 DRI preheating using the off-gas. After [61]

Hylsa suggests that the off-gas from the EAF is unsuitable for DRI preheating. Instead, in a patent from 2000, Villarreal [62] suggests using inert gases (nitrogen), carburizing gases (natural gas) or a reducing gas (from natural gas reforming), using two preheating bins. DRI would be preheated from 400 to 800 °C. The patent does not describe its economic feasibility.

Hajidavalloo et al. [63] proposed a new method for DRI preheating using the sensible heat in the off gas, as shown in Fig. 10.25. In this method a heat exchanger inside the duct of the off-gas system heats nitrogen. Heated nitrogen is introduced into a silo containing cold DRI. The results from a model developed by these authors using as a reference an industrial size silo with a cross-section of 2.5 m × 5.1 m and height of 4.5 m containing 75 ton of DRI and an initial temperature of the gas of 600 °C, are shown in Fig. 10.26. The figure shows a rapid increment of temperature of the DRI at the bottom of the silo, about 600 °C, but the top remains cold. It takes about two hours for the material at the top to be heated to about 230 °C. The average temperature was 480 °C. A further exergy analysis [64] based on a mass flow rate of the off-gas of 10.4 kg/s, suggested a potential decrease in electric energy consumption of 21%.

More recently, in 2021 Sane et al. [65] from Air products reported some preliminary results using oxy-fuel burners to preheat DRI before it is fed into the EAF, as

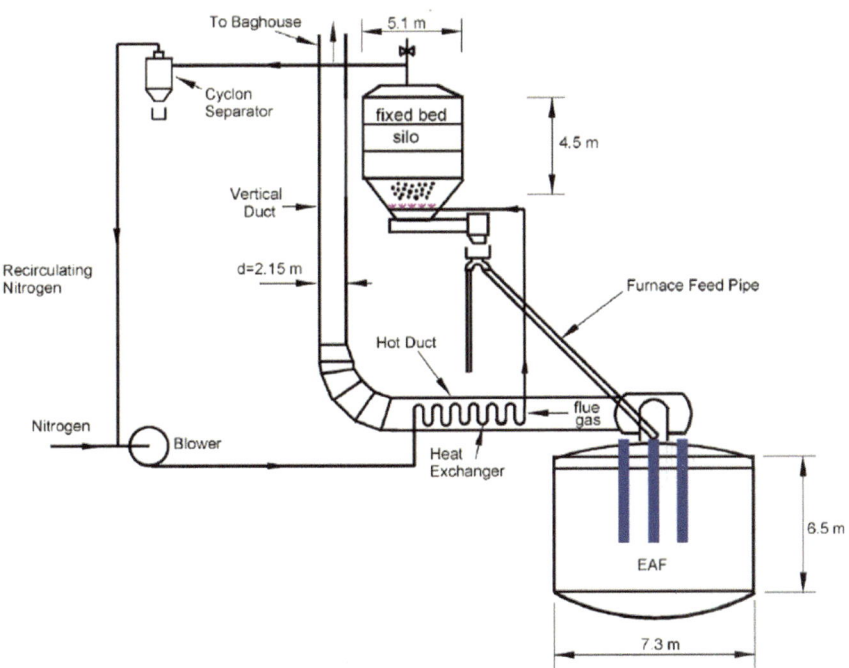

Fig. 10.25 DRI preheating using the off-gas. After [63]

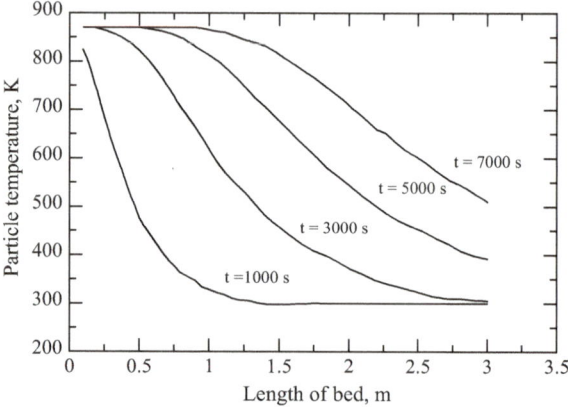

Fig. 10.26   Temperature distribution of DRI inside the silo at different times. After [63]

shown in Fig. 10.27. They claim a small surface oxidation, as shown in Fig. 10.28 using a monolayer of pellets.

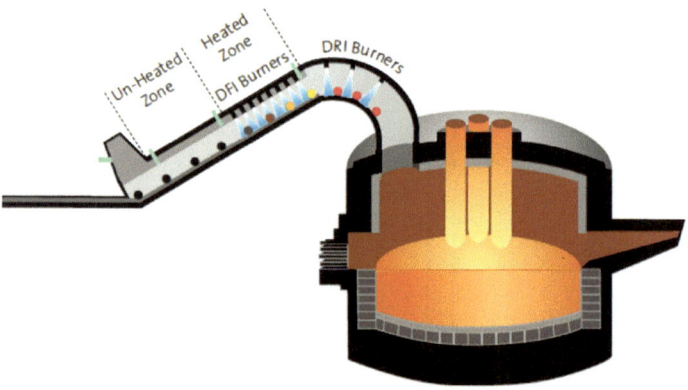

Fig. 10.27   DRI preheating using oxy-fuel burners. After [65]

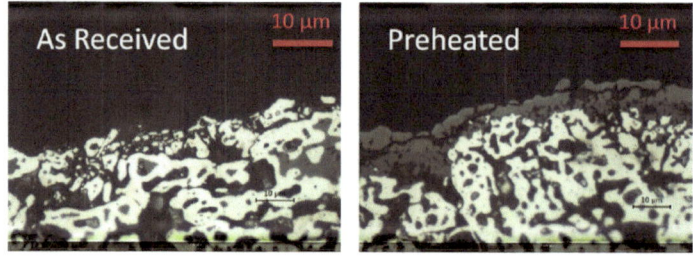

Fig. 10.28   Single pellet preheating. After [65]

10.3 Final Remarks

Hot metal and hot DRI are two important methods to decrease electric energy consumption. The largest decrease in electric energy consumption can be achieved with hot metal, to the extent that it can result in no electric energy consumption at all using more than 70% hot metal, however the feasibility of this method depends on several factors, such as oxygen injection capacity (number of lances, maximum gas flow rate per lance), off-gas system capacity and charging equipment of hot metal. Due to these limitations, optimum values are in the order of 40%. For this hot metal ratio, the decrease in electric energy consumption is in the order of 50–60%, for scrap-based operations. The critical problem with hot metal in the EAF is that eliminates one of its strongest advantages, which is low CO_2 emissions.

On the other hand, using hot DRI is also a feature for greenfield projects, promoting electric energy savings in the order of 25% when the temperature of DRI is about 600 °C. Due to the expected acceleration in DRI production in the next decade, it expected that all new projects will include charging hot DRI.

The challenge remains how to preheat DRI using the huge amount of sensible heat in the off-gas, an area that remains poorly investigated.

References

1. Jones JAT, Bowman B, Lefrank PA (1998) Electric arc furnace steelmaking. In: Fruehan RJ (ed) The making, shaping and treating of steel. Steelmaking and refining, 11th edn. Pittsburgh, PA, USA,pp 525–660
2. Tian B, Wei G, Li X, Zhu R, Bai H, Tian W et al (2022) Effect of hot metal charging on economic and environmental indices of electric arc furnace steelmaking in China. J Clean Prod 379:134597. https://doi.org/10.1016/j.jclepro.2022.134597
3. Wang Y, Zuo H, Zhao J (2020) Recent progress and development of ironmaking in China as of 2019: an overview. Ironmak Steelmak 47:640–649. https://doi.org/10.1080/03019233.2020.1794471
4. Yang LZ, Jiang T, Li GH, Guo YF, Chen F (2018) Present situation and prospect of EAF gas waste heat utilization technology. High Temp Mater Process 37:357–363. https://doi.org/10.1515/htmp-2016-0218
5. Emi T (2015) Steelmaking technology for the last 100 years: toward highly efficient mass production systems for high quality steels. ISIJ Int 55:36–66. https://doi.org/10.2355/isijinternational.55.36
6. Sampaio RS, Jones J, Vieira JB (2009) Hot metal strategies for the EAF industry. Iron Steel Technol 6:31–37
7. Adams W, Alameddine S, Bowman B, Lugo N, Paege S, Stafford P (2001) Factors influencing the total energy consumption in arc furnaces. In: 59th electric furnace conference, Phoenix, AZ, USA, 11–14 Nov 2001, pp 691–702
8. Pfeifer H, Kirschen M, Simoes JP (2005) Thermodynamic analysis of EAF electrical energy demand. In: 8th European electric conference, Birmingham UK, 9–11 May, 2005
9. Zinurov IY, Shumakov AM, Ovchinnikov SG, Pigin SN, Mamenko YF (2009) Utilization of hot metal in arc furnaces. Steel Transl 39:576–578. https://doi.org/10.3103/S0967091209070146

10. Duan J, Zhang Y, Yang X (2009) EAF steelmaking process with increasing hot metal charging ratio and improving slagging regime. Int J Miner Metall Mater 16:375–382. https://doi.org/10. 1016/S1674-4799(09)60067-4
11. Jiemin T, Fern MB, Argenta P (2005) EAF technology evolution by continuous charging. Ironmak Steelmak 32:191–194. https://doi.org/10.1179/174328105X38080
12. Jiemin T, Xuefeng M, Argenta P (2006) Charging hot metal to the EAF using Consteel Millenium Steel 79–86
13. Lopez F, Lule F, Espinoza J (2011) Use of hot metal in the EAF; Experience of AM-LZC flat carbon Mexico (In Spanish). Hierro y Acero 12:5–11
14. Solver C, Roth JL, Hoffmann M, Stoltz R, Houbart M (2011) Dragon steel corp. In: Taiwan boosted 150-ton EAF twin shell by continuous hot metal charging process. AISTech Iron Steel Technol Conf Proc I:799–809
15. Gottardi R, Miani S, Partyka A, Suber M (2011) Decarburization efficiency in EAF with hot metal charge. In: AISTech conference, vol I, Indianapolis IN, USA, 2–5 May 2011, pp 811–21
16. Xu X, Ruan X, Zhang G, Deng X, Grant M, Chen T (2006) High efficiency production practice of a 100 t DC EBT EAF at Xing Cheng steel works. AISTech Iron Steel Technol Conf Proc 2:413–422
17. Mandal T, Maity A, Chatterjee S, Mukherjee A (2017) Charging hot metal in electric arc furnaces (EAF's): reducing cost structure and expanding grade horizons. In: SEAISI conference
18. Grant MGK, Kaiser KC, Cantacuzene SM, Tao C (2005) Optimization of oxygen steelmaking in non-conventional EAF operations. In: AISTech 2005 conference, vol I, pp 545–553
19. Grant MGK, Blostein P, But S, Kaufman C (2014) Adaptation of EAF operations for unconventional raw materials. Metallurgist 58:95–104
20. Lee B, Sohn I (2014) Review of innovative energy savings technology for the electric arc furnace. JOM 66:1581–1594. https://doi.org/10.1007/s11837-014-1092-y
21. Lee B, Sohn IL (2015) Effect of hot metal on decarburization in the EAF and dissolved sulfur, phosphorous, and nitrogen content in the steel. ISIJ Int 55:491–499. https://doi.org/10.2355/ isijinternational.55.491
22. Lee B, Ryu JW, Sohn I (2015) Effect of hot metal utilization on the steelmaking process parameters in the electric arc furnace. Steel Res Int 86:302–309. https://doi.org/10.1002/srin. 201400157
23. Fritz E, Gebert W (2005) Milestones and challenges in oxygen steelmaking. Can Metall Q 44:249–260. https://doi.org/10.1179/cmq.2005.44.2.249
24. Banerjee S (2002) Process for making steel. US Patent 6424671B1
25. Irawan A, Kurniawan T, Alwan H, Muslim ZA, Akhmal H, Firdaus MA et al (2022) An energy optimization study of the electric arc furnace from the steelmaking process with hot metal charging. Heliyon 8. https://doi.org/10.1016/j.heliyon.2022.e11448
26. Liu Y, Wei G, Tian B (2023) Analysis and optimisation on the energy consumption of electric arc furnace steelmaking. Ironmak Steelmak 1–15. https://doi.org/10.1080/03019233.2023.217 2826
27. Na H, Gao C, Guo Y, Tian F (2019) CO2 emissions characteristics and source analysis of "Chinse style" electric arc furnace stellmaking route. J Northeast Univ 40:212–217
28. Yang LZ, Jiang T, Li GH, Guo YF (2017) Discussion of carbon emissions for charging hot metal in EAF steelmaking process. High Temp Mater Process 36:615–621. https://doi.org/10. 1515/htmp-2015-0292
29. Kirschen M, Hay T, Echterhof T (2021) Process improvements for direct reduced iron melting in the electric arc furnace with emphasis on slag operation. Processes 9:1–10. https://doi.org/ 10.3390/pr9020402
30. Hornby S, Madias J, Torre F (2016) DRI/HBI—exploding the myths. Steel Times Int 24–9
31. Omar AM, Appasamy T, Memoli F (2004) DC EAF with high DRI feeding rates through multipoint injection. MPT Int Int 27:58–67
32. Tavano A, Franco B, Martinis A, Taylor D (2014) Hot charging of DRI in the electric arc furnace: is it always a good deal? In: AISTech 2014 proceedings, pp 525–533

33. Abel M, Hein M (2008) The use of scrap subtitutes like cold/hot DRI and hot metal in electric arc furnaces. Arch Metall Mater 53:353–357
34. HyL (1999) 2nd HYL HYTEMP seminar
35. Moritz M, Reddemann F (2021) AUMUND: leading technology specialist in hot conveying and cooling of direct reduction products. Millenium Steel 18–22
36. Viramontes-Brown R, Lizcano-Zulaica C, De La Peña RG, Herrera-García MA (2004) An economic and productivity analysis charging hot DRI for EAF steelmaking. AISTech Iron Steel Technol Conf Proc 1:535–543
37. Yañez D (1996) The impact of hot charged, high carbon DRI on EAF mix and steelmaking practice. In: International conference on recycling iron steel industry, Las Vegas, NV, USA, 27–30 Oct 1996. ISS, p 15
38. Voelker B, Boyle S. Operational results of hot charging DRI. Direct MIDREX 3–11
39. Poodi AN (2013) DRI in use—how is it being utilized? Different plants, different solutions. World DRI pellet congress, Abu Dhabi, UAE, p 28
40. Patrizio D, Razza P, Pesamosca A (2015) Capacity enhancement at emirates steel: continuous improvement in EAF performance with hot DRI charge. AISTech Iron Steel Technol Conf Proc 2:1954–1964
41. Razza P, Patrizio D (2010) Operating results with hot DRI charge at Emirates steel industries. Millenn Steel 39–44
42. Hornby-Anderson S (2002) Educated use of DRI/HBI improves EAF energy efficiency and yield and downstream operating results. In: Associatione Italiana di Metallurgia (ed) 7th European electric steelmaking conference, Venice Italy, 26–29 May 2002, p 26–29
43. Klawonn RM, Hoffman GE (2005) Optimum utilization of hot DRI: an innovative approach to integration with steelmaking processes. AISTech Iron Steel Technol Conf Proc 1:383–393
44. Anderson SH (2000) DRI: the EAF energy source of the future? https://www.metallics.org/
45. Tennies W, Metius G, Kopfle J (2000) Breakthrough technologies for the new millenium. In: 4th European coke ironmaking congress, Paris, 19–21 June 2000, pp 256–264
46. Martinis A, Nogare D, Volpatti A, Morales J (2015) Packing energy and iron to serve the meltshop requirements. In: 45° Semin. reducao minerio ferro e Mater. primas, Rio de Janeiro Brazil, 17–21 August 2015, pp 1–13
47. Memoli F, Jones JAT, Picciolo F (2015) The use of DRI in a Consteel® EAF process. Iron Steel Technol 72–80
48. Duarte P (2004) Latest advances in direct reduction to serve mini-mills and integrated mills. In: SEAISI (ed) South East Asia iron steel conference and exhibition, Kuala Lumpur, Malaysia, May 2004, p 15
49. Montague S, Häusler W (1999) Hot charging DRI for lower cost and higher productivity. Direct Midrex 3–7
50. SIS (2005) The first hotlink steel plant. https://www.mesteel.com/countries/oman/Shadeed_I ron_Steel.pdf. Accessed 15 July 2023
51. Montague S, Voelker B (2001) Direct reduced iron discharge system and method. US patent 6214086
52. Harada T, Tanaka H (2011) Future steelmaking model by direct reduction technologies. ISIJ Int 51:1301–1307. https://doi.org/10.2355/isijinternational.51.1301
53. Gordon I, Wheeler FM (2005) Continuous steelmaking process. US Patent 6875251B2
54. Gordon I, Wheeler FM (2008) Plant for use in continuous steelmaking process. US patent 7499142B2
55. Wheeler F, Gordon Y, Broek S, Cameron I (2010) The successful piloting of CRISP, the innovative continuous steelmaking technology. La Metal Ital 27–33
56. Barati M (2010) Energy intensity and greenhouse gases footprint of metallurgical processes: a continuous steelmaking case study. Energy 35:3731–3737. https://doi.org/10.1016/j.energy.2010.05.022
57. Wheeler FM, Gordon Y, Wheeler JG (2003) Latest developments in the Hatch continuous steelmaking process. In: AISE annual convention proceedings, Pittsburgh, PA, 28 September–1 October 2003, p 1–7

58. Tress J, Hunter W (1975) Charging an electric furnace. US Patent 3929459
59. Kim Y (1987) Apparatus for continuously preheating and charging raw materials for electric furnace. US Patent 4642048. US Patent 4642048
60. Sensis S, Schwerdtfeger J, Walden K, Ameling D (1981) Process for feeding iron sponge into electric arc furnace. US Patent 4290800
61. Pavlicevic M, Burba G, Primavera A, Guastini F (2002) Process to preheat and carburate directly reduced iron (DRI) to be fed to an electric arc furnace (EAF). US Patent 2002/0011132A1. US Patent 2002/0011132A1
62. Villarreal-Treviño JA (2000) Method and apparatus for preheating of direct reduced iron used as feed to an electric arc furnace. WIPO Patent 47780
63. Hajidavalloo E, Alagheband A (2008) Thermal analysis of sponge iron preheating using waste energy of EAF. J Mater Process Technol 208:336–341. https://doi.org/10.1016/j.jmatprotec.2007.12.140
64. Hajidavalloo E, Dashti H (2010) Exergy analysis of steel electric arc furnace. In: Engineering system design and analysis, Istanbul Turkey, 12–14 July 2010, pp 1–7
65. Sane A, Buragino G, Makwana A., He X (2021) Enhancing Direct Reduced Iron (DRI) for use in electric steelmaking. Millenn Steel 14–17

Chapter 11
Hot Heel, Tapping and Energy Recovery

11.1 Hot Heel

11.1.1 Benefits and Recommended Values

Hot heel is the amount of remaining liquid steel after tapping. As a percentage it can be defined in two ways, with respect to the weight of tapped steel or the total liquid steel capacity.

$$\text{hot heel}(\%) = \frac{\text{weight of hot heel, tons}}{\text{weight of tapped steel (tons)}} \times 100$$

$$\text{hot heel}(\%) = \frac{\text{weight of hot heel, tons}}{\text{capacity of liquid steel (tons)}} \times 100$$

There is a difference when it is reported as a percentage. Many times hot heel is reported without a clear definition of the reference value. The total capacity of liquid steel is the tapped steel and the hot heel:

$$\text{Capacity of liquid steel} = \text{tapped steel} + \text{hot heel}$$

For example, an EAF with 200 ton of tapped steel and 30 ton of hot heel:

$$\text{Capacity of liquid steel} = 200 + 30 = 230 \text{ ton}$$

Hot heel in terms of tapped steel is 15% but only 13% taking as reference the capacity of liquid steel. It is important to make sure hot heel on a % basis is made using the same reference. In the following, if not indicated, % hot heel will be defined in terms of the total capacity of liquid steel.

The hot heel practice was probably introduced since the early 1970s. Post [1] in 1973 suggest a hot heel to accelerate the melting rate due to continuous addition of DRI. Pearce [2] in 1986 points out an example with the tapping weight decreased from 131 to 115 ton to leave space for the hot heel. Bosley et al. [3] in 1987 suggested values for the hot heel in the range from 10 to 20%. The hot heel practice became more popular in the 1990's as a tool to decrease slag carry over. Millman and Thornton from British steel in 1998 still considered this practice as costly and unproductive but currently is a widespread practice on EAF steelmaking.

It should be noticed that hot heel does not refer only to liquid steel because the remaining liquid includes molten slag. The fraction of slag in the hot heel is usually unknown, but as an example, it was estimated that in a hot heel of 50 ton, the amount of slag was about 40% [4].

In DC-EAF's the hot heel is mandatory in order to provide an electrical path to the return electrode. Nucor reported a longer life of the bottom electrode increasing the hot heel [5] in a DC-EAF Consteel process.

The technique remains largely empirical. It is currently a common practice because it has many benefits, such as the following:

1. Decreases carry-over slag: Abraham and Chen [6] reported a relationship between hot heel and carry over slag, shown in Fig. 11.1. In this figure, when there is not hot heel, the negative values of carry over slag indicate tapping involves only liquid slag. The capacity of liquid steel of the EAF is 136 ton [7]. From this figure, increasing hot heel 11% (15/136) decreases carry over slag from about 13.2 to 6.6 kg/ton, a total decrease of about 6.6 kg/ton. Their results would suggest a minimum of 27% hot heel to suppress carry-over slag.

2. Increases the melting rate of the metallic charge: Fig. 11.2, from an unknown source, shows productivity and melting time for an EAF with a capacity of liquid

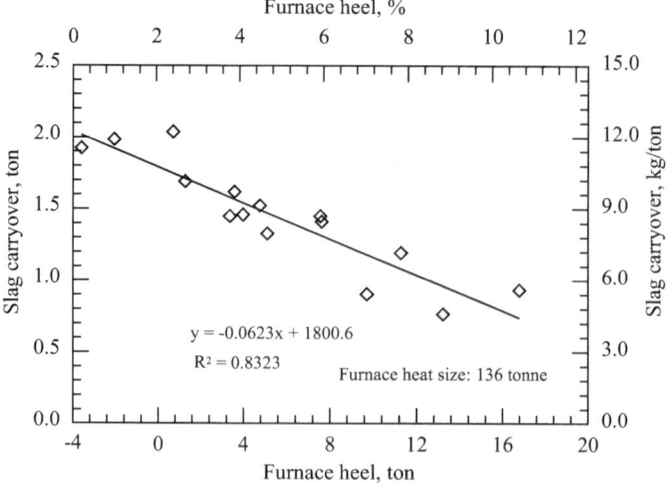

Fig. 11.1 Relationship between hot heel and carry over slag. After [6]

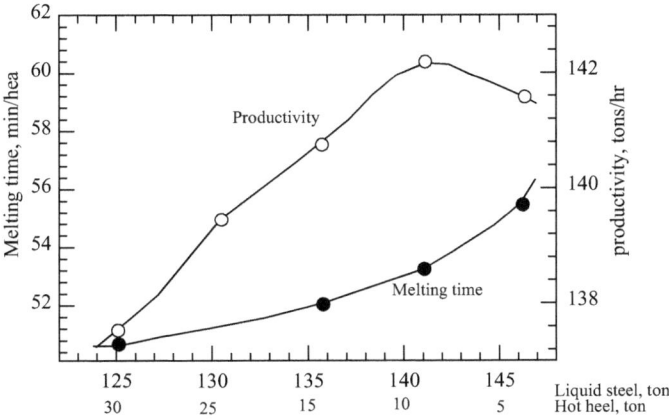

Fig. 11.2 Effect of hot heel on the melting rate and productivity. *Source* Unknown

steel of 155 ton. Hot heel varied from 6.45% to 19.3%. It can be observed that productivity increases when hot heel increases from 6.4% to 9.7% but further increments in the hot heel affect productivity in ton/h.

3. Decreases the melting time: Fig. 11.2 shows that contrary to productivity which decreases after a corresponding optimum value, the melting time continuously decreases but the rate decreases when % hot heel is above 13%.
4. Promotes an earlier slag foaming: In addition to heel remining from a previous heat there is also a remaining amount of slag. Both liquids contain sensible heat. The initial slag, however should be conditioned to promote slag foaming.
5. Earlier injection of oxygen: The beginning of oxygen injection requires the presence of a volume of liquid steel which depends on the type of oxygen injectors and distance from the tip of the lance to the molten surface of liquid steel. Coherent jets with a longer coherent distance can start earlier.
6. Improves arc stability and decreases flicker.
7. If optimized, it contributes to decrease energy consumption: The result of improved foaming conditions, higher melting rate and improvements to arc stability, all of them decrease energy consumption.

Some of its disadvantages are:

1. The main disadvantage is that consumes energy, it is a volume of steel that is continuously heated and this heat is a heat loss that decreases the overall thermal efficiency. However, in the proper amount accelerates the melting rate of the metallic charge which in the end decreases the total electric energy consumption
2. Excessive hot heel increases the total energy consumption: the optimum value for a particular EAF should be defined, however there is a lack of information on the subject

3. Excessive hot heel decreases productivity: The solution is to define the optimum value
4. Is difficult to define the initial mass of hot heel. Most EAF's do not measure the mass of hot heel due to the lack of sensors
5. It decreases the tapping volume.

According to Graftech, hot heel can save about 15 kWh/ton [8], however they suggest that the magnitude of the hot heel is not important. The hot heel in a giant 420 ton DC-EAF in Japan was reported as 120 tons, equivalent to 22% [9], injecting 33 N m³ O$_2$/ton.

Lugo [10] cited a report that increasing the hot heel from 6 to 25% obtained energy savings of 11%. Based on a private communication with Siemens, Li et al. [11] reported that there was a trend to increase the hot heel in newer EAF's, to values about 25% of the tap weight.

Hot heel in the Consteel process is the largest for EAF's, this is necessary because the main heat transfer mechanism is not radiation but convection and then accelerate the melting rate, in the order of 50% [12]. Scrap is fully immersed in the hot heel. Memoli et al. [13] reported that hot heel has a positive influence on furnace productivity, increasing from 1.5 to 2.4 ton/h MW when the hot heel increased from 37 to 60%, the highest value, 2.5 ton/h MW was reached using a hot heel of about 55%, as shown in Fig. 11.3. The reported results in this figure correspond to 20 AC-EAF Consteel plants. A higher productivity involves lower energy consumption. The effect of hot heel on the total energy consumption is shown in Fig. 11.4. The normalized value was defined for a reference value of hot heel of 46.5%.

Li et al. [14] reported the lowest energy consumption, about 250 kWh/ton in a 90 ton Consteel EAF, using a hot heel from 32 to 39%. The large hot heel in the Consteel process is the reason of achieving high melting rates in spite that convection heat transfer is the main heat transfer mechanism [13].

Table 11.1 summarizes reported values on savings in energy consumption with a hot heel practice. Table 11.2 also summarizes reported values of % hot heel for

Fig. 11.3 Effect of hot heel on productivity in the Consteel process. After [13]

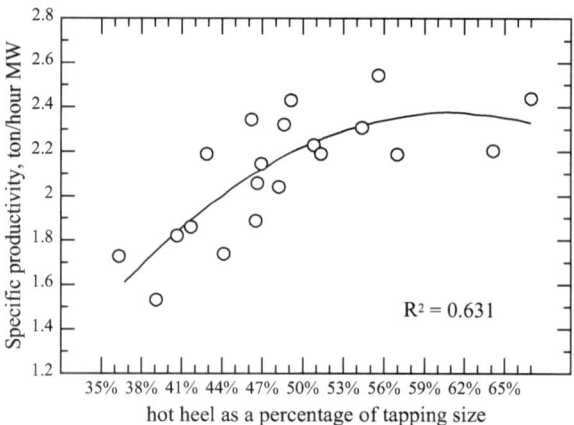

$R^2 = 0.631$

hot heel as a percentage of tapping size

Fig. 11.4 Effect of hot heel on energy consumption in the Consteel process. After [13]

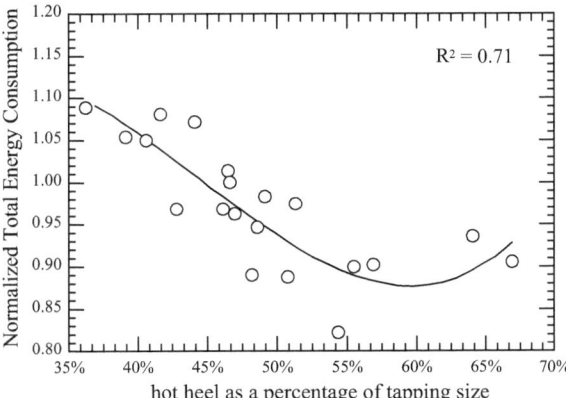

different EAF's. Jones et al. [15] gave as a rule of thumb a hot heel size in ton is equal to 1.4 ton times the power input in MW. The few data in Table 11.2 indicate current values are much higher, from 2.2 to 4.8.

INTECO [19] reported a change in the tapping weight at the expense of hot heel for an EAF of 165 ton of nominal capacity and 140 ton tapping weight. Originally the hot heel was 15% but, in the interest to increase the tapping weight to 155 ton, the shell volume was slightly increased from 300 to 330 m^3 and the hot heel decreased, without giving further details to the final value. In a way to justify this change, they argued that "the influence of hot heel on performance is not clearly visible in case of a XXL furnace size. Large hearth of the furnace and insufficient hot heel depth causes freezing of liquid metal on the bottom after scrap charge. This frozen metal has to be remelted during superheating extending the superheating time during which more oxygen is consumed causing simultaneous decrease of the metallic yield". These two arguments are weak because, on one side they are not supported with evidence and only if hot heel is not optimized and is too low, it would cause the problems described. Belkovskii and Kats [20] reported model calculations which showed a decrease in productivity and an increase in electric energy consumption as the hot heel increased, which is clearly in opposition to real observations with an adequate volume of hot heel. Their calculations didn't make a detailed analysis of the melting rate.

Table 11.1 Summary of savings on energy consumption due to hot heel

First author	EAF, ton	Hot heel, ton (%)	EE cons. kWh/ton	Savings, kWh/ton or %
Adams [8]				15 kWh/ton
Lugo [10]		6–25%		11%
Li [14]	90	43–53 (32–39)	250	

Table 11.2 Summary of reported values of hot heel for different EAF's

Reference	EAF type	MVA	MW	Tapping, ton	Hot heel, ton	Total, ton	Times, MW	Hot heel, %[a]	Metallic charge
[9]					120	420		22	
[16]	DC-twin shell			155	30	185		16.2	Scrap + hot metal
[17]	DC-twin shell			135	60	195		30.8	Scrap + hot metal
[18]	AC-EAF	130	112	150	50	200	2.2	25.0	DRI
[10]	Various			100	10–20	110–120		9.1–16.7	–
	AC-EAF	110	75	82	25–35	107–117	2.5	23.4–29.9	Scrap
[19]	Telescopic	150	120	165	25	190	4.8	13.2	Scrap
	AC-EAF			200	30	230		13.0	DRI
[13]	AC-Consteel			187	100	287		34.8	Scrap
[15]	AC-Consteel	30		70	30–35	100–105		30–33.3	Scrap
[15]	IHI			140	110	250		44	

[a]Hot heel/steel capacity

The hot heel usually contains a highly oxidizing slag. The iron oxide can be reduced using coke additions at the beginning of the new heat. In DRI-based operations it is more suitable to add coke together with DRI to improve its immersion in the molten bath.

11.1.2 Hot Heel Measurement

An accurate determination of the hot heel is important because it affects the initial slag foaming due to high concentrations of FeO from the previous heat, a low value could cause freezing due to the cold metallic charge and on the other hand, an excessive value will also increase the consumption of energy. An accurate measurement of the hot heel is extremely important in order to define its optimum value. There are different ways to measure the hot heel:

- Its value is usually reported based only on visual observations and experience by the EAF operator but this method is highly inaccurate
- An accurate method is to tap and weigh the hot heel but is time consuming
- Bar immersed manually from the top of the EAF [21] which has a graduated rule attached. This method involves risks for the operators

- Use of dynamic mass and energy balance models using the tapping weight from the preceding heat [22], however an accurate dynamic model requires multiple measurements of the different mass outlets
- Estimation of the hot heel height using the distance from the surface of hot heel edge to the tap hole, measured with binoculars that include a scale inscribed on the lens [23]
- The most recommended way is to employ sensors.

Several sensors are available to measure the level of the molten bath, for example;

- Optical fiber [5]. Heraeus electro-nite developed a technique originally employed to measure the temperature of liquid using optical fiber to measure the hot heel height.
- Radar sensors. Sagasti et al. [24, 25] described its fundamentals and challenges of operation. Radar waves are similar to microwaves. The distance to be measured by a radar is related to the time lapse between sending and receiving the signal. Meszaros et al. [24] disclosed a patent describing the measurement of the slag thickness. They indicate that differences in the electrical conductivity give different reflection times, EAF slags have values in the range from 0.5 to 1.5 Ω/cm and liquid steel 7140 Ω/cm at steelmaking temperatures. Abraham and Chen's patent [6] employs a radar to measure the hot heel, they found a relationship between hot heel and carry-over slag.
- Laser sensors [26]. Agellis developed a sensor attached to the temperature manipulator. An initial calibration is required to correlate lance position and lance height.
- Thermographic camera coupled with displacement of the electrode [27].
- Electromagnetic sensors mounted behind the safety lining [28]. A patent application indicates the effective measurement of liquid steel and refractory thickness using electromagnetic sensors around a vessel, particularly suitable for ladles.

It should be taken into account that measuring the height of the hot heel is not enough to know the mass of steel, due to a higher volume of steel as a function of the refractory life, therefore an additional sensor to provide the actual profile of the refractory or another way to estimate this profile, is required.

11.2 Tapping

11.2.1 Importance of Tapping

It is very important to understand that primary and secondary metallurgy are not fully separate processes but they are connected through an interface and this interface is tapping. It is important this concept because usually tapping is a no-man's land area. This vacuum is created because the BOF or EAF supervisor wants to immediately start a new heat after tapping and the LMF supervisor cannot control the tapping conditions. In fact, both supervisors should pay attention and understand that the production of high-quality steel during secondary refining depends on the following tapping conditions:

- Minimum possible tapping temperature: A lower tapping temperature not only saves energy consumption but also decreases the solubility of oxygen.
- Minimum possible FeO concentration: As close as possible to the equilibrium conditions guarantees the lowest FeO concentration.
- Minimize carry-over slag: related to the opposite nature of slag chemistry in primary and secondary metallurgy.
- Maximum possible use of the C deoxidation practice.

The philosophy to produce high quality steels is not only removing the largest amount of non-metallic inclusions once they are formed, which seems to be the conventional practice for a very long time, but even more important is to minimize the formation of non-metallic inclusions and this primarily depends on the concentration of FeO in the slag during tapping, the tapping temperature and the carry-over slag. Figure 11.5 illustrates this approach. If this approach is optimized and strictly supervised the benefit is not only savings in the consumption of typical deoxidizers but the production of higher quality steels. In opposition to this idea, Melville and Brinkmeyer cited in Ref. [29] suggested that there is no relationship between the degree of oxidation of steel before tapping and the total oxygen at the tundish due to rapid flotation of about 85% of alumina clusters, leaving only inclusions smaller than 30 μm. Reaching such a conclusion seems to be dangerous because it promotes the idea that the oxidizing conditions in terms of high quality steel can be neglected. It also surprises that some researchers from the steel industry state that "clean steel practices start in the ladle" [30]. Against this position, the concept of secondary metallurgy I propose to follow and the beginning of clean steel practices should start during tapping. This is a concept I have shared with students and engineers for more than 25 years. The fact that there is a poor knowledge on the subject is an indication on the lack of attention into tapping. In this context is better to follow the well-established criteria that the lower the amount of oxygen in liquid steel, the lower the amount of non-metallic inclusions [31].

11.2.2   FeO and Carry Over Slag

Iron oxide in the slag is a collateral damage due to the need to oxidize carbon in steelmaking. Each gram of iron that is oxidized decreases the metallic yield and represents a waste of money. That is not the intended purpose of oxygen injection, however, the trend to increase chemical energy and reduce expensive electrical energy has shifted the practice towards high inputs in oxygen injection. Increasing oxygen injection without overoxidation of liquid iron is possible if the following points are considered:

Fig. 11.5 Proposed paradigm to prevent the formation of non-metallic inclusions during tapping

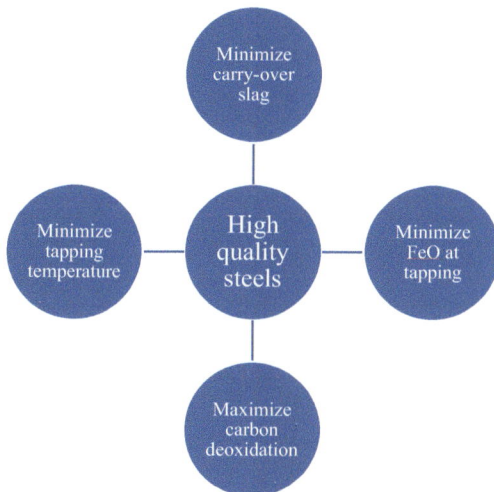

- Balance C–O: the optimum amount of oxygen injection depends in the amount of carbon in the metallic charge. Oxygen injection should be limited to the amount of carbon in the metallic charge. If there is a need to replace more electric energy by chemical energy, then the amount of carbon should also be increased simultaneously. It should always be remembered that heat transfer efficiency is higher using electric energy in comparison with chemical energy.
- Optimize the foamy slag practice: Currently, foaming is a standard practice. This practice is based on the reduction of iron oxide by addition of carbon particles into the slag. A good practice should consider that the addition of carbon particles is linked to the rate of reduction of FeO and full consumption of each particle.
- Stirring: Due to low stirring conditions in the EAF the iron oxide increases, however both in the BOF and EAF is possible to use bottom gas injection to improve the stirring conditions and approach equilibrium which decreases FeO.
- Decarburization control: Decarburization at atmospheric pressure should be limited to values above 0.05%C, otherwise the oxygen concentration rises exponentially, up to levels in the order of 1000–1500 ppm O. Deep decarburization should be left for vacuum conditions.

A good sign that the oxygen injection practice is under control is to look at the final FeO concentration before tapping. This value should be as low as possible, in the order of 18–25%, depending on the C-FeO equilibrium. In general, if this value is higher than 30% FeO the metallurgical practice is wrong.

In any case, primary slags are oxidizing slags and ideally, from a chemical view point, no carry over slag should be allowed to be transferred to the ladle during tapping, however this is almost impossible and also not wanted because some carry over slag is necessary to cover the steel from the atmosphere.

One of the main reasons to prevent carry over slag is to decrease FeO in the slag carried to the ladle to values less than 2%FeO, however it is not only FeO which is being limited but also P_2O_5. The role of phosphorous is becoming more important due to the use of low-quality iron ores, high in phosphorous. If phosphorous passes to the ladle there is a risk risk of P reversal to liquid steel and then its removal gets complicated because the slags are reducing, unfavorable for the removal of P.

The available technologies to prevent and control carry over slag in BOF and EAF have been described in a general way by Abraham and Chen [6] and Pistorius [7] and in more detail by Kapusuz et al. [32]. A report from Di Napoli cited by Pistorius compared the use of a slag stopper in the BOF; without the stopper, carry over slag was 10–15 kg/ton and P reversion was 30 ppm but with the stopper the carry over slag was reduced to 3–5 kg/ton and P reversion was about 10 ppm. The comparison made by Kapusuz suggests that the best methods involve electromagnetic and vibration signals. They described the advantages and limitations of each method. Abraham and Chen reported a direct relationship between the hot heel size and carry-over slag, suggesting a minimum hot heel of 10% to decrease the carry-over slag below 7 kg/ton.

There are two aspects that make carry-over slag a complex problem. The first aspect is the entrainment of liquid slag into liquid slag during tapping due to vortex formation. At the end of the tapping process, the stream of liquid steel carries on the inside liquid slag and therefore there is not a clear-cut separation of the two phases. The second aspect is that once the slag is detected, tapping cannot be stopped instantaneously. Depending on the diameter of the tapping hole, the tapping rate can be in the order of 1.5 ton/s. The tapping hole diameter increases every heat, leading to higher amounts of carry-over slag.

Pistorius [33] estimated the amount of carry-over slag considering several measurements; the amount of aluminum employed for deoxidation during tapping, the decrease in oxygen in liquid steel, the remaining amount of aluminum and the slag composition, By stoichiometry is possible to estimate the amount of carry-over slag.

11.2.3 Tapping Temperature

Increasing the tapping temperature increases the solubility not only of oxygen but also nitrogen and hydrogen because the stream of liquid steel is exposed to the atmosphere. An increment in oxygen shifts the equilibrium to higher concentrations of FeO. In addition to this, increasing the tapping temperature increases energy consumption. The sensible heat of liquid steel is 0.24 kWh/ton per °C. Pearce's estimation [2] reported an increment in energy consumption of 5 kWh/ton per every 10 °C increment in the tapping temperature. Similar to the values reported by Conejo et al. [34, 35] from 4 to 4.4 kWh/ton, as shown in Fig. 11.6. In this figure efficiency means the thermal efficiency of the EAF. Consequently, the tapping temperature should be as low as possible. This value depends on the additions of ferroalloys during tapping, tapping time, initial refractory temperature and requirements from secondary metallurgy.

Fig. 11.6 Effect of tapping temperature on energy consumption. After [35]

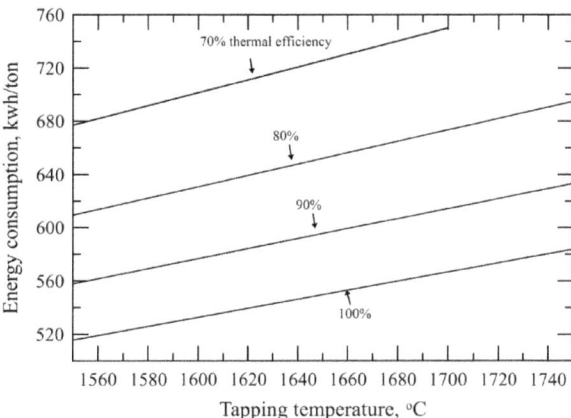

The tapping temperature is defined based on delays in the following stage, type of steel, tapping system, etc. It is commonly accepted 1600 °C as a standard reference, therefore, superheat above this temperature causes an increase on energy consumption. Superheat can represent extra energy up to 70 kWh/ton [8]. The energy to bring liquid steel from its melting temperature to 1600 °C is 25 kWh/ton, however sometimes the tapping temperature can reach up to 1700 °C due to long a waiting time after tapping. Increasing the taping temperature brings several problems:

- Energy consumption increases
- The solubility of oxygen increases, therefore there is more steel reoxidation, which in turn will require higher amount of deoxidizers and this will increase the concentration of non-metallic inclusions
- Efficiency of ferroalloys decreases.

Therefore, the tapping temperature should be kept with the minimum superheat possible.

11.2.4 Concept of NMI's Capacity

The author has proposed for a long time, both with people at steel plants and academia, to pay more attention to the tapping conditions to produce clean steels. In the author's opinion, all the efforts are placed on the removal of inclusions once they are formed. To explain clean steel in a different way is like the saying "The person who cleans the most is not the cleanest, but the person who produces less trash".

In the same line of ideas, I have proposed to develop a similar index to those employed for the removal of other impurities. The limits on steel cleanliness are imposed by the refining capacity of slags. The slag refining capacity on sulphur and phosphorus have been defined as C_S and C_P, which are directly related to the definition of partition ratios, sulphur partition ratio and phosphorous partition ratio,

L_S and L_P, respectively. The same idea can be applied to the removal non-metallic inclusions (NMI). Why is that a similar approach has not been proposed in the past? Most probably this is because the research work that relates slag chemistry and NMI is recent and the knowledge on the physicochemical variables is still under development.

Researchers from the former IRSID in 1994 [36] developed the basic methodology to estimate the type of formed inclusions in terms of the steel chemistry and the corresponding slag chemistry in contact with that type of steel, then ultimately controlling the slag chemistry is possible to design the final chemical composition of NMI.

The concept of a slag NMI capacity, C_{NMI}, can be defined similar to other capacities since the tools to estimate the concentration of particles removed from liquid steel are known.

$$C_{NMI} = \frac{\text{mass NMI removed to slag}}{\text{mass NMI in liquid steel}}$$

The removal of NMI varies depending on many variables; fraction of liquid and solid NMI, NMI chemistry, particle size, stirring conditions, interfacial tension, wettability, concentration gradients between inclusion and slag, etc., however a simpler approach would be to assume that all inclusions capable to reach the slag/metal interface and with the energy requirements to break the energy barrier to separate them from liquid steel are finally removed. The removal of NMI from liquid steel has three stages, the first one is its transport to the slag/liquid interface, then at the interface it should overcome surface tension forces to enter into liquid steel and finally, if the particle is solid, undergo a final dissolution process. Each stage is subject to a different set of known variables. Transport of NMI to the slag/metal interface depends on the stirring conditions. It has been suggested an optimum bubble size for the removal of NMI and also the effect of the injection layout on the removal efficiency. Zhang and Taniguchi [37] suggested a bubble size from 1 to 5 mm. The lowest bubble size is better in terms of improving the amount of NMI adhered to the bubbles but the rising time increases and the terminal velocity is low, below $0.1~\text{ms}^{-1}$ and then re-entrained back to the recirculating liquid. Once the inclusion reaches the slag/metal interface, wettability is the dominant variable that defines the entrainment of NMI into liquid steel. Solid inclusions have a larger contact angle in comparison with liquid inclusions and therefore better wettability, then are absorbed by liquid slags faster [36]. Choi et al. [38] and Valdez et al. [39] in the early 2000s reported equations to describe the dissolution process of solid inclusions, providing iso-dissolution lines in ternary diagrams. These equations define the dissolution time as a function of a concentration gradient (NMI-slag) and slag viscosity. A higher concentration gradient and lower viscosity enhance the dissolution rate.

The amount of research work on the field of NMI is huge, both experimentally and numerically. It would be useful to summarize that work in a more unified approach that define the process variables that control the refining capacity of slags to remove NMI, using dimensionless parameters that define the following general relationship, taking into consideration that the chemical composition finally results in specific physical properties of the phases involved.

$$C_{NMI} = f(Re, We, \dots)$$

Strandh et al. [40] studied the formation of a steel film around liquid inclusions at the slag/liquid steel interface. They found that only in large inclusions, above 180 mm, the steel film is formed. The separation of inclusions to the slag was possible if the overall wettability was positive, defined as $\cos \theta_{IMS} = \sigma_{IM} - \sigma_{IS}/\sigma_{MS}$, where σ is the interfacial tension, I is inclusion, M is metal and S is slag.

I shared the concept of slag inclusion capacity to Prof. Lifeng Zhang in 2014. Recently in 2023 his research group reported the following expression to define the slag inclusion capacity [41]:

$$Zh = \frac{g \cdot \rho_{slag}^2 \cdot (C_{sat} - C) \cdot d_{p,0}^3}{\eta_{slag}^2}$$

where g is the gravitational acceleration, ρ is the density of the slag, C_{sat} of the % of saturated mass of the inclusions phase in the slag, C is the mass % of the inclusion contained in the refining slag, η is the viscosity of the slag, d is the initial particle of the inclusion. They found that as Zh increases, also increases the dissolution rate of inclusions. This is one step forward but there is still more work to do to prove that increasing the slag inclusion capacity, there is a decrease in the concentration of NMI in liquid steel.

Prof. Holappa from Aalto University didn't agree with the concept of slag inclusion capacity when the author presented the idea [42]. In his opinion, the saturation point, which can be defined for other impurities such as S and P, but cannot be properly defined for NMI because the slag cannot get saturated with NMI.

11.2.5 Eccentric Bottom Tapping (EBT)

Eccentric Bottom Tapping (EBT), according to Teoh [43], was introduced around 1983. This was an important innovation that helped to decrease the tapping time and a decrease in steel reoxidation due to a liquid stream more uniform. The decrease in tapping time and a more uniform stream help to decrease the tap-to-tap time, reduce electrode consumption and decrease electric energy consumption. The Center for materials production [44] reported a decrease of 15 kWh/ton. Tapping time with an EBT system is shorter because the EAF inclination is shorter, decreasing from 40° for a conventional EAF with spout to 12–20° with EBT [45]. An EAF with EBT is shown in Fig. 11.7.

Free opening of the EBT is important to avoid delays. It has been reported that a good sealing and free opening can be achieved with a proper control of the sand

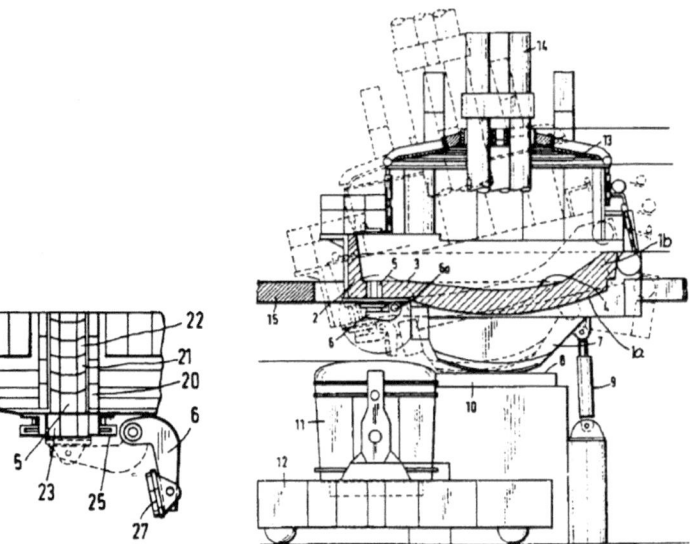

Fig. 11.7 EAF with EBT. After [45]

particle size (1–6 mm), which depends on the operation conditions [46]. To obtain good sealing conditions the sand employed should form a molten layer on the top of the tap hole. The rest of the sand remains at low temperatures and falls down when the slide gate is opened. The ferrostatic pressure breaks the top sealing discharging the molten metal. Dunite mineral can be employed, which is formed by more than 90% olivine (Mg_2SiO_4). The life of the EBT is less than 200 heats.

Kirschen et al. [47] carried out a full CFD investigation on the fluid flow characteristics between a conventional EBT (cylindrical shape) and a new conical shape, considering data from four EBT-EAF's, shown in Table 11.3. It was assumed a slag layer of 150 mm. the conical shape gave better results; lower velocities, lower turbulence and higher tapping rates. Comparing an EAF with 170 ton of steel, at the same tapping angle and sill level, EBT diameter of 170 mm, changing from cylindrical to comical; velocity decreased from 4–6 to 3–5 m/s, the Re number was about 30% lower and the tapping rate increased from about 0.8 to 1.0 ton/s. Increasing the tap diameter to 250 mm increased the tapping rate to 2 ton/s. Although increasing tap diameter increases the tapping rate, it was not recommended high values because carry over slag increases. The lower Re number and decreased turbulence helped to decrease refractory wear. The internal pressure was also decreased with the conical shape which decreased the potential of MgO-C decomposition. This can happen when the internal pressure is above 10^3 Pa at 1600 °C (0.01 bar). In the conical shape all pressures were below that level. The cylindrical shape has a small zone at the inlet region with pressures higher than 10^3 Pa. Figure 11.8 shows the cylindrical and conical shapes.

Table 11.3 EAF's investigated, comparing cylindrical and conical EBT

	EAF 1	EAF 2	EAF 3	EAF 4
Type	DC	AC	AC	AC
Capacity, ton	170	85	80	250
EBT channel length, mm	1230	950	1180	1200
Sill level above tap inlet (at 10° tilt angle), mm	730	350	530	600
Fe-pressure from melt level, 10^5 Pa	1.30	0.88	1.19	1.23

After [47]

Fig. 11.8 EBT.
a Cylindrical shape,
b conical shape. After [48]

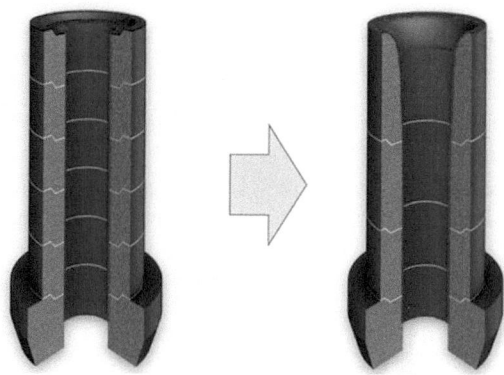

11.3 Energy Recovery

Schliephake et al. [49] made an important observation on the subject of energy recovery; It should be "a secondary option behind reducing the energy input. It is better to use one (1) kWh less than to recover one (1) kWh".

Due to the low thermal efficiency of the EAF (50–60%) the energy in the resulting by-products, off-gas and slag can be recovered. The following is a brief discussion.

11.3.1 Energy Recovery from the Off-Gas

The energy in the off-gas is about 10–30% of the input energy. Chapter 8 discussed in detail the energy recovery from the off-gas for scrap preheating, however energy recovery from the off-gas has more applications. Yang et al. [50] made a review on the recovery technologies of the off-gas from the EAF. For China in particular, due to the large use of hot metal in the EAF the off-gas temperature raises to 1400 °C and its net output is about 13–20% of the total energy consumed. There is a large difference in CO produced with hot metal and scrap, 82.3 kg/ton and 2.1 kg/ton, respectively. The amount of gas generated in the EAF, according to Yang et al. is given in Table 11.4.

For a large steel group in China producing 105 Mton/year, the total heat consumed to produce steel is equivalent to 12.4 GJ/ton and the energy lost in the off gas is about 1.7×10^8 GJ/year (about 170 PJ/year) [51]. Nimbalkar et al. [52] provided an estimate of the total amount of energy lost in the off-gas in the US steel industry per year, about 14 PJ/year. This value was based on the following assumptions; 30% heat losses in the off gas (the total heat in the off gas is about 16% sensible heat and 21% chemical heat), as shown in Fig. 11.9, annual production of 80.5 Mton of steel/year with 61.3% by the EAF, and an average energy consumption of 606 kWh/ton.

Yang et al. [50, 53] carried out a mass and energy balance for a case with 50% hot metal, introducing a total energy of 738 kW/ton, out of which 170 kWh/ton was electrical energy. The sensible heat from the outlet streams were; 397 kWh/ton in liquid steel, 43 kWh/ton in the slag, 140 kWh/ton in the off-gas, 134 kWh/ton chemical heat from CO and 34 kWh/ton by the WCP. The sensible heat in the off-gas results from 200 to 400 N m³/ton of gas produced with a temperature of 1200 °C and specific heat capacity of 1.137 kJ/kg °C. The total heat in the off-gas for the reference case is $140 + 134 = 274$ kWh/ton. These authors evaluated different scenarios to recover the sensible heat (140 kWh/ton) in 4 applications: scrap preheating, power

Table 11.4 Specific gas generated in EAF's and its chemical composition

	Small EAF	HP-EAF	UHP-EAF	SUHP-EAF
N m³/h ton	500–700	700–800	800–1000	1000–1200
	Gas composition			
	CO	CO₂	N₂	O₂
wt%	1–34	12–20	45–74	5–14

After [50]

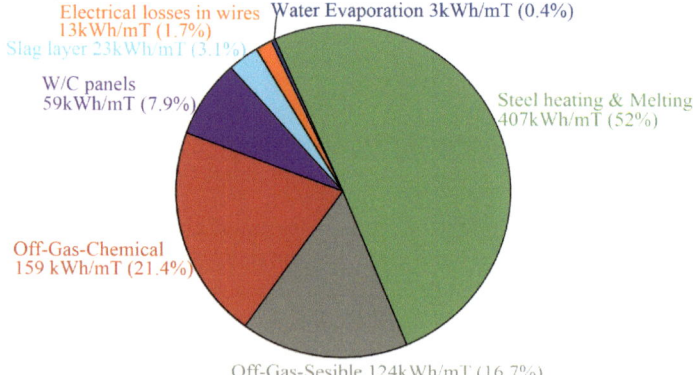

Fig. 11.9 Outputs from an energy balance for the EAF. After [52]

generation, steam production and coal gas production. The results are summarized in Fig. 11.10. The energy recovered from scrap preheating was estimated based on the assumptions of an average temperature reached by the steel scrap (500 °C) and the average decrease in energy per 100 °C (15 kWh/ton); 5 × 15 = 75 kWh/ton. The energy recovery for electricity generation was based on an estimation that reported a value of 91 kWh/ton. The amount of steam produced was estimated as 334 × 10^3 ton of saturated steam/year from a 100 ton EAF, the value was compared with the amount of coal required to produce steam and its energy contribution, equivalent to 24 kWh/ton. If the chemical heat in the off-gas is used to produce coal gas the estimated energy recovery was 134 kWh/ton. According with the previous estimated values, the energy recovery in % for scrap preheating, electricity generation, steam production and production of coal gas were; 53, 65, 17 and 70%, respectively. The previous study concluded that scrap preheating is the most mature option with a significant energy recovery.

The technology to produce steam from the off-gas from the EAF is called evaporative cooling system (ECS). Schliephake et al. [49] described in detail this system and in particular the ECS installed at one plant in Germany, as shown in Fig. 11.11. Since radiation heat transfer is higher at higher temperatures, the ECS is placed at the beginning of the off-gas duct, resulting in a temperature decrease in the off-gas to about 600 °C. The heat in the off-gas from 600 to about 180–250 °C is taken away by a heat boiler.

An evaporative cooling system, contrary to a conventional cooling system is that ECS employs pressurized water at the boiling point with typical values in the range from 13 bar/192 °C to 28 bar/230 °C. Since evaporation involves more heat compared to water heating, the ECS consumes 35% less water in comparison with conventional cooling. In the ECS not all the water is evaporated, only a fraction, about 5–12%.

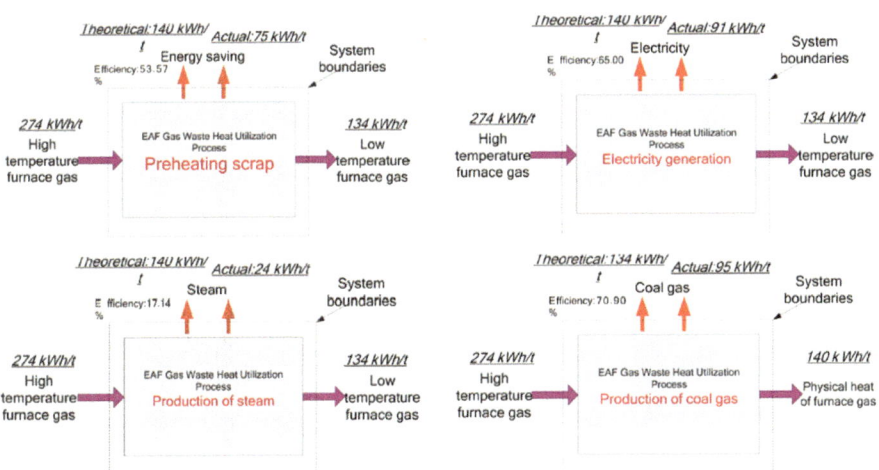

Fig. 11.10 Case studies on energy recovery from the EAF off-gas. After [53]

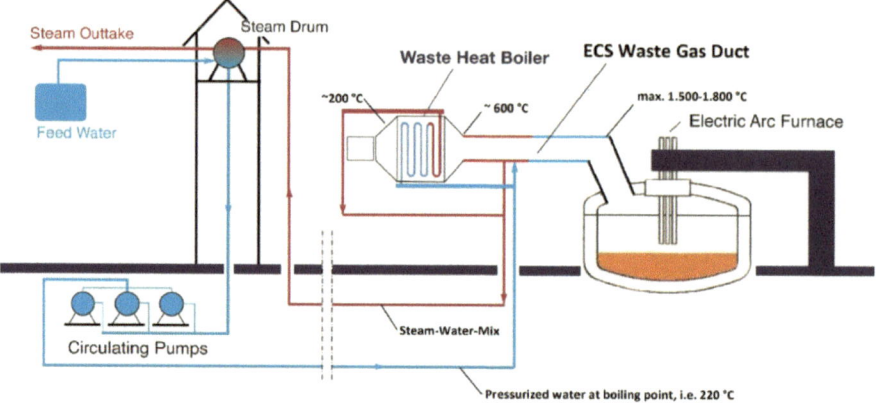

Fig. 11.11 Evaporative cooling system at Georgsmarienhütte. After [49]

Figure 11.12 compares the cooling capacity, which clearly shows the much higher cooling intensity for ECS in comparison with water cooling, 83 and 250 kJ/kg, respectively.

The dynamic nature of gas generation from an EAF and the resulting steam production can be observed in Fig. 11.13 for four heats. In each heat there are instantaneous peaks generation of steam but on average the value was about 20 ton/h. In this case the steam was employed in a vacuum degassing process, however the average demand is much smaller, about 7 ton/h.

Steam can also be employed for power generation but in this case, it has to be superheated steam, in addition to this the fluctuating nature of the off-gas produced in terms of flow rate, chemical composition and temperature adds complexity to this option. Zhang et al. [54] reviewed and also proposed a new thermodynamic cycle

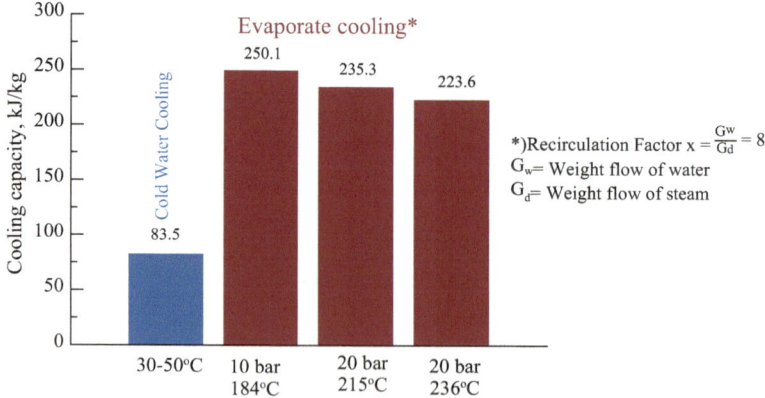

Fig. 11.12 Comparison of the cooling capacity between water cooling and ECS. After [49]

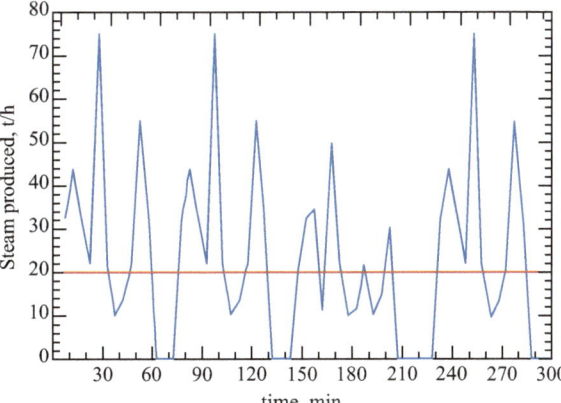

Fig. 11.13 Steam produced by ECS for 4-heats. After [49]

which improves the heat recovery efficiency from the off-gas. They indicated that one part of the solution to the intermittent temperature of the off-gas is to employ a heat storage process, also they indicated that the conventional water Rankine cycle (RC) has a low thermal efficiency. By modelling work they found that a combined cycle, Brayton-Rankine cycle, increased the heat recovery efficiency from 22 to 27%, corresponding to a net output power of 162 and 198 kW, respectively.

11.3.2 Energy Recovery from Iron and Steelmaking Slags

The amount of energy in iron and steelmaking slags correspond to approximately 30 TWh/ton of liquid steel per year, 17 TWh/ton from ironmaking slags and 13 TWh/ton from steelmaking slags. These values were estimated for the hot metal and liquid steel produced in 2019; 1250 Mton and 1867 Mton, respectively. The specific production of slag by process reported by the World Steel Association [55] was: 275 kg slag/ton of hot metal, 126 kg slag/ton of liquid steel produced in the BOF and 167 kg slag/ton steel produced in the EAF. Additionally, it was estimated an average energy in liquid slag of 50 kWh/ton. The total value is similar to that reported by Lee and Sohn of 35 TWh/ton [56]. Barati et al. [57] reported much higher slag enthalpies, for example 500 kWh/ton for EAF slags, but probably his reference is per ton of slag.

Barati et al. [57] reported a full review on energy recovery methods from high temperature slags and discussed their current challenges. Those challenges mainly derive from the low thermal conductivity of slags; 0.1–0.3 W/m K for liquid slags and 1–3 W/m K for solid slags. The solution has been to pulverize the slag and increase the surface area. The methods based on the recovery of the thermal energy are more mature, however they have not reached the commercial stage. Energy recovery is reported in the range from 40 to 65% with most cases cooling the slag in the range from 200 to 300 °C. The market for reducing slags requires a glassy phase which can be achieved at higher cooling rates, then increasing the air flow rate but this decreases the final temperature of the hot air.

References

1. Post G (1973) The optimum method for converting sponge iron to steel in arc furnaces. In: Symposium on science and technology of sponge iron its conversion to steel, Jamshedpur, India, 19–21 Feb. National Metallurgical Laboratory
2. Pearce J (1986) Development trends in EAF steelmaking. JOM 38:38–45. https://doi.org/10.1007/BF03257895
3. Bosley J, Clark J, Dancy T, Fruehan R, McIntyre E (1987) Techno-economic assessment of electric steelmaking through the year 2000. Pittsburgh, PA USA
4. Lule R, Lopez F, Lowry M, Kundrat D, Wyatt A, Kuntze J et al (2018) Improvements in yield in an all-DRI-fed EAF from minimization of FeO generation during melting as well as post-reduction of FeO from residual slag. Iron Steel Technol 15:36–41
5. Quiney P, Fitch K, Turner P, Meads L (2021) Bath level management in the Consteel DC electric arc furnace. In: AISTech 2021 proceedings, Nashville, TN, USA, June 29–July 1, pp 317–329
6. Abraham S, Chen S (2011) Process for optimizing steel fabrication. US patent 2011/0174457A1
7. Pistorius PC (2019) Slag carry-over and the production of clean steel. J S Afr Inst Min Metall 119:557–561. https://doi.org/10.17159/2411-9717/kn01/2019
8. Adams W, Alameddine S, Bowman B, Lugo N, Paege S, Stafford P (2001) Factors influencing the total energy consumption in arc furnaces. In: 59th Electric arc furnace conference, 11–14 Nov, Phoenix, AZ, USA, pp 691–702
9. Emi T (2015) Steelmaking technology for the last 100 years: toward highly efficient mass production systems for high quality steels. ISIJ Int 55:36–66. https://doi.org/10.2355/isijinternational.55.36
10. Lugo N (2014) Electric arc furnace best operation practices. In: 45th ABM steelmaking seminar, Porto Alegre, Brazil, 25–28 May, pp 44–53. https://doi.org/10.5151/1982-9345-24178
11. Li J, Provatas N, Irons GA (2008) Modeling of late melting of scrap in the EAF. Iron Steel Technol 5:216–223
12. Madias J (2014) Electric furnace steelmaking. Treatise Process Metall 3:271–300. https://doi.org/10.1016/B978-0-08-096988-6.00013-4
13. Memoli F, Guzzon M, Giavani C (2011) The evolution of preheating and the importance of hot heel in supersized Consteel® systems. In: AISTech conference, Indianapolis, IN, USA, 2–5 May, pp 823–832
14. Li J, Zhao R, Fu J, Jin Y, Feng B, Tu C et al (2002) Optimizing amount of metal charge and hot heel for a 90 ton Consteel EAF. Spec Steel 23:50–51
15. Jones JAT, Bowman B, Lefrank PA (1998) Electric arc furnace steelmaking. In: Fruehan RJ (ed) Making, shaping, and treating of steel, 11th edn. Steelmaking and refining volume, Pittsburgh, PA, USA, pp 525–660
16. Solver C, Roth JL, Hoffmann M, Stoltz R, Houbart M (2011) Dragon steel corp. In: Taiwan boosted 150-ton EAF twin shell by continuous hot metal charging process. AISTech—iron and steel technology conference proceedings, vol I, pp 799–809
17. Lee B, Sohn IL (2015) Effect of hot metal on decarburization in the EAF and dissolved sulfur, phosphorous, and nitrogen content in the steel. ISIJ Int 55:491–499. https://doi.org/10.2355/isijinternational.55.491
18. Razza P, Patrizio D (2010) Operating results with hot DRI charge at Emirates Steel Industries. Millenn Steel 39–44
19. Nal Ö, Dolapçioğlu S, Gottardi R, Partyka A, Miani S (2017) Efficiency of the EAF with telescope roof. In: AISTech 2017 conference, vol 1, pp 1149–1160

20. Belkovskii AG, Kats YL (2015) Effect of the mass of the liquid residue on the performance characteristics of an EAF. Metallurgist 58:950–958. https://doi.org/10.1007/s11015-015-0023-7

21. Hundermark R, van Rooyen Q, van Manen P, Steyn C, Sadri A, Chataway D (2018) Development of continuous radar level measurement for improved furnace feed control. Extr 2018:287–302. https://doi.org/10.1007/978-3-319-95022-8_23

22. Kleimt B, Köhle S, Kühn R, Zisser S (2005) Application of models for electrical energy consumption to improve EAF operation and dynamic control. In: 8th European electric steelmaking congress, Birmingham, UK, pp 183–197

23. Kundrat D, Wyatt A (2006) Procedure for measuring and managing the hot heel at Mittal steel Indiana Harbor's EAF

24. Meszaros G, Marquardt R, Walker D, Estocin J, Kemeny F (2000) Measuring the thickness of materials. US patent 6166681

25. Sagasti J, Ahualli F, Schonhofer M (n.d.) Use of radar technology to determine freeboard in steel ladles. AustralTek

26. Bloemer P, Nilsson JP, Lyons A, Filipovic I, Guarise M (2014) Electromagnetic bath level measurement system improves EAF melting process control. In: 45th ABM steelmaking seminar, vol 1, Porto Alegre, Brazil, 25–28 May

27. Vicente A, Gutierrez JA, Arteche JA, Macaya I (2016) Liquid steel level measurement at electric arc furnaces without increasing the power off time. https://computervision.tecnalia.com

28. TataSteel (2013) Apparatus and method for measuring the liquid metal level in a metallurgical vessel. European patent application 2568264A1

29. Zhang L, Thomas BG (2003) State of the art in evaluation and control of steel cleanliness. ISIJ Int 43:271–291. https://doi.org/10.2355/isijinternational.43.271

30. Pretorius EB, Oltmann HG, Schart BT (2013) An overview of steel cleanliness from an industry perspective. In: AISTech 2013 proceedings, Pittsburgh, PA, USA. 6–9 May, pp 993–1026

31. Kiessling R (1980) Clean steel—a debatable concept. Met Sci 14:161–172. https://doi.org/10.1080/02670836.2020.12097372

32. Kapusuz H, Güvenc M, Mistikoglu S (2019) A review study on ladle slag detection technologies in continuous casting process. Int Adv Res Eng J 03:144–149. https://doi.org/10.35860/iarej.421657

33. Pistorius PC (2022) Data analysis to asses carry-over slag. In: Steenkamp JD et al (eds) Furn. tapping 2022, pp 51–58

34. Conejo AN, Cardenas JGG (2006) Energy consumption in the EAF with 100% DRI. In: AISTech conference, vol 1, pp 529–535

35. Cárdenas JGG, Conejo AN, Gnechi GG (2007) Optimization of energy consumption in electric arc furnaces operated with 100% DRI. In: METAL, Hradec Nad Moravici, Czechia, 22–24 May, pp 1–7

36. da Costa e Silva ALV (2018) Non-metallic inclusions in steels—origin. J Mater Res Technol 7:283–299

37. Zhang L, Taniguchi S (2000) Fundamentals of inclusion removal from liquid steel by attachment to rising bubbles. Int Mater Rev 45:59–82

38. Choi JY, Lee HG, Kim JS (2002) Dissolution rate of Al_2O_3 into molten CaO-SiO_2-Al_2O_3 slags. ISIJ Int 42:852–860. https://doi.org/10.2355/isijinternational.42.852

39. Valdez M, Shannon GS, Sridhar S (2006) The ability of slags to absorb solid oxide inclusions. ISIJ Int 46:450–457. https://doi.org/10.2355/isijinternational.46.450

40. Strandh J, Nakajima K, Eriksson R, Jönsson P (2005) A mathematical model to study liquid inclusion behavior at the steel-slag interface. ISIJ Int 45:1838–1847. https://doi.org/10.2355/isijinternational.45.1838

41. Gou L, Liu H, Ren Y, Zhang L (2023) Concept of inclusion capacity of slag and its application on the dissolution of Al_2O_3, ZrO_2 and SiO_2 Inclusions in CaO–Al_2O_3–SiO_2 Slag. Metall Mater Trans B Process Metall Mater Process Sci 54:1314–1325. https://doi.org/10.1007/s11663-023-02763-y

42. Holappa L (2022) Personal discussion on the concept of slag's non-metallic inclusion capacity

43. Teoh LL (1989) Electric arc furnace technology: recent developments and future trends. Ironmak Steelmak 16:303–313
44. CMP (1992) Electric arc furnace efficiency. Center for Materials Production. Report 92-10, Pittsburgh, PA USA
45. Wronka B, Schnitzer H, Baare R, Bauer H, Otto J (1991) Tiltable arc furnace. US patent 5054033
46. Ruisanchez E, Martinez J (2015) Taphole free opening optimization in the EAF through monitorized grain size distribution control of the EBT filler sand: laboratory testing and industrial application. In: 70 ABM annual congress, pp 1123–1132
47. Kirschen M, Rahm C, Jettler J, Hackl G (2008) Steel flow characteristics in CFD improved EAF bottom tapping systems. Arch Metall Mater 53:365–371
48. Zettl KM, Zottler P, Kirschen M (2013) State of the art tapping solutions for bottom tapping EAFs. BHM Berg-Huettenmaenn Monatsh 158:13–16. https://doi.org/10.1007/s00501-012-0104-x
49. Schliephake H, Granderath R (2010) Heat recovery for the EAF of Georgsmarienhütte. In: AISTech 2010 proceedings, vol I, Pittsburgh, PA, USA, 3–6 May, pp 745–752
50. Yang LZ, Jiang T, Li GH, Guo YF, Chen F (2018) Present situation and prospect of EAF gas waste heat utilization technology. High Temp Mater Process 37:357–363. https://doi.org/10.1515/htmp-2016-0218
51. He L (2013) Waste heat power generation technology of steelmaking EAF. Ind Furn 35:16–18 (in Chinese)
52. Nimbalkar S, Thekdi A, Keiser J, Storey J (2014) Waste heat recovery from high temperature off-gases from electric arc furnaces. In: AISTech—iron and steel technology conference, proceedings, vol 1, pp 1113–1123
53. Yang LZ, Zhu R, Ma GH (2016) EAF gas waste heat utilization and discussion of the energy conservation and CO_2 emissions reduction. High Temp Mater Process 35:195–200. https://doi.org/10.1515/htmp-2014-0183
54. Zhang F, Xing M, Tang W, Wang R (2020) A compound cycle for power generation by utilizing residual heat of flue gas in electric steelmaking process. Energy Power Eng 12:45–58. https://doi.org/10.4236/epe.2020.122004
55. World Steel Association (2018) Steel industry co-products. Worldsteel position paper
56. Lee B, Sohn I (2014) Review of innovative energy savings technology for the electric arc furnace. JOM 66:1581–1594. https://doi.org/10.1007/s11837-014-1092-y
57. Barati M, Esfahani S, Utigard TA (2011) Energy recovery from high temperature slags. Energy 36:5440–5449. https://doi.org/10.1016/j.energy.2011.07.007

Chapter 12
EAF-Water Cooling

12.1 Development of Water-Cooled Panels (WCP)

12.1.1 Origin of WCP

In the late 1950s due in part to low power transformers the tap-to-tap time lasted a long time, from 4 to 6 h. The concept of UHP-EAF was developed by Schwabe and Robinson in the early 1960s [1, 2]. C. Robinson was in operations at Northwestern Steel and Wire Co. (NWSW) and W. Schwabe was director of research at Union Carbide. The basic idea was to increase the transformer capacity from about 25–40 to 74–82 MVA and correspondingly get a higher specific power input, from 0.3–0.35 to 0.46–0.51 MVA/ton. The challenges at that time were the changes associated with a higher input of electric power, for example much higher radiation to the walls, electric current through the electrodes, etc. The transformer replaced had a capacity from 40 to 50 MVA with electric current in the range from 52 to 56 kA, the new transformer had a capacity from 74 to 82 MVA and operating currents from 75 to 80 kA, with an estimated active power of 60 MW. The electrodes diameter (24 in.) was not changed but the supplier improved its quality to handle higher currents. Their results confirmed the importance to employ UHP. The plant increased its productivity, 1980 ton ingot/h and electric energy consumption decreased to 480–490 kWh/ton ingot, the best records ever in the world at that time. They recognized the importance to decrease the power-off time and sustain a high active power during the heat in order to benefit from the change in higher capacity transformer but also suggested the need to work with short arcs to decrease arc radiation. Slag foaming was mentioned superficially indicating that the electrodes were only partially submerged in a foamy slag. Although UHP was a breakthrough concept because it decreased the tap-to-tap time by half, to about 2–3 h its full achievements were limited due to the short life of the refractory, in particular at the hot spots, from 250 to 300 heats for the refractory walls to 30–100 heats for the hot spots [3]. The authors also mentioned the use of water jets on the roof intended to offer a shield at the hot spot regions. NWSW had started the production of steel through the EAF since 1936 when they installed EAF's of 10-ton capacity, in 1940 were replaced by 50-ton EAF's, in 1963 by 150-ton ingot, in 1968 the EAF capacity increased to 250 ton and in 1971 the first 400-ton EAF being the largest in the world at that time.

Although the name UHP, coined by Robinbson and Schwabe in 1963 implies a super-large transformer, they were aware that in the future new developments would make the term obsolete. In the 1980s the International Iron and Steel Institute (IISI) defined as Super UHP, (SUHP) those transformers with a capacity higher than 70 MVA/ton.

The solution to achieve the full potential of high-capacity transformers was given by Japanese steelmakers who replaced ordinary refractory bricks by water cooled panels in the early 1970s, first as water cooled panels that externally cool the refractory from the hot spots regions and subsequently as WCP directly exposed to radiation, as will be described subsequently in more detail. This innovation allowed to operate with longer arcs and higher active powers.

12.1.2 Advantages and Disadvantages of Water-Cooled Panels (WCP)

Advantages:

- Water cooled panels allow a long arc operation which otherwise would be unacceptable by common refractories
- Higher electric power which leads to faster melting rates
- Increase productivity
- Reduced time to erect a new refractory lining and its subsequent maintenance.

Disadvantages:

- The biggest disadvantage of WCP is the increment in heat losses. Bosley [4] in 1987 reported that heat losses increased from 10 to 25 kWh/ton after the installation of WCP. Heat losses account for about 10–20% of the total energy input. Heat losses involve a substantial cost. It has been estimated that each 1% of energy consumption has a cost of 0.14–0.15 USD/ton [5].
- Increasing heat losses decreases the thermal efficiency
- WCP are more expensive than refractories
- Increases the consumption of water in steelmaking: The consumption of water in the EAF is not high because water is recycled. Colla et al. [6] indicate an average intake of 28.2 m^3/ton and an average discharge of 26.5 m^3/ton, which gives a consumption of 1.6–3.3 m^3/ton, mainly due to evaporation.

RHI Magnesita [7], the largest supplier of refractories in the world, carried out simulations on heat losses using copper cooling blocks in 8 EAF's of different capacity, dimensions and tap-to-tap time, concluding that installing Cu cooling blocks is not economical, even if they increase furnace availability and save refractory because the cost on energy, gunning mixes and Cu blocks are even more expensive. This conclusion suits the goals of a refractory producer, however, it is clear that refractories at the sidewalls are not suitable for SUHP furnaces because of the extreme radiation conditions at the hot spots. On the other hand, Simon [8] summarized a more broad scenario, indicating that the specific energy consumption using WCP increases in some cases but in others decreases, however, if the operation is adequate (good foaming conditions), energy consumption should not increase.

12.2 Heat Losses Through Water Cooling Panels (WCP)

12.2.1 Radiation Intensity

The heat losses through water cooled panels depends on the thickness of the foaming slag. Guo and Irons [9] reported a numerical model to predict the radiation intensity as a function of the slag thickness and temperature with a bath of molten steel of 1.26 m high. Figure 12.1 show their results using an arc length of 0.452 m. For the normal operating range of conditions, the heat extracted by cooling water is in the range from 12 to 15 MW according with the model predictions. If the heat losses by convection are included, 3–6 MW, the result agrees with the reported experimental values from 17 to 22 MW.

Sanchez et al. [10] reported the effect of arc coverage due to slag foaming on the formation of hot spots. It was shown that a long arc operation without foamy slag present a huge risk to melt the water-cooled panels (WCP). Their results, shown in Table 12.1 indicate the area of the hot spots increases when the arc coverage decreases due to poor slag foaming conditions. When the arc coverage decreases

Fig. 12.1 Predicted values of radiation intensity to the WCP. After [9]

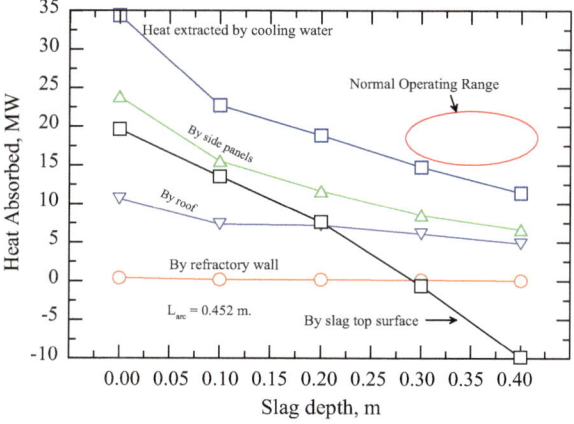

Table 12.1 Radiation intensity as a function of slag height

Slag height, cm	Arc covered, %	Total radiation, MW	Hot spots area, m^2	Max T hot spots, °C
0	0	30.33	1.7	1827
11.25	25	22.74	0.47	1727
22.5	50	15.15	0.24	1647
33.75	75	7.56	0	1477
45	100	0	0	1387

After [10]

below 50%, the temperature on the hot spots exceeds the melting temperature for the water-cooled panels.

12.2.2   Heat Losses Through Water Cooling Panels (WCP)

The amount of heat losses through water cooling is largely influenced by radiation coming from the electric arc covered by foamy slag and the hot electrodes, representing from 10 to 20% of the total input energy, at least twice the sensible heat in the slag. A summary of the heat losses using WCP is given in Table 12.2, according to Kirschen et al. [7]. The large range of heat losses for WCP, 50–330 kW/m^2, is due to different foaming conditions, slag thickness on the WCP, furnace diameter, active power, arc length, furnace capacity in tons and tap-to-tap time. The same authors compared this value with the heat losses using refractory lining, which are much lower, about 4–10 kW/m^2.

Toulouevski and Zinurov [11] reported heat losses from 200 to 300 kW/m^2 with a heavy layer of slag and from 600 to 700 kW/m^2 with a bare surface, depending on arc coverage. For oxygen tuyeres in contact with liquid steel the heat flux density can be as high as 4000–6000 kW/m^2.

Table 12.2 Summary of heat losses due to WCP

Reference	EAF, ton	Heat losses, kWh/ton	Heat losses, %	Reference	Heat losses, kW/m^2
Kirschen (2004)	100	83	10.6	Kruger (1998)	35–50
Kirschen (2007)	145	143	18.9	Pavlicevic (2002)	200–600
Pujadas (2002)	125	134	17.3	Kirschen (2007)	100–330

After [7]

Schwabe [1] in 1962 was the first one who investigated arc radiation from the electric arc to the furnace walls. The deflection of the arc was recorded by high-speed photography, this deflection was the reason that arc radiation intensified to three spots on the sidewalls, called hot spots. His findings were finally expressed in a refractory index, which indicates that refractory wear is proportional to the cube of the arc length. According to Howe, cited in Ref. [12], the effect is even more intense, refractory wear is proportional to the fifth power of arc length. Later on, Bowman and Fitzgerald [13] in 1973 confirmed Schwabe's results and also concluded that the heat flux at the hot spots is almost twice at that for the cold spots during melting. In addition to this, they also measured the slag splashing by the electrodes, estimating a mass transfer of about 330 kg/h at a velocity from 3 to 10 m/s.

Bisio et al. [14] investigated the temperature distribution in tubular WCP, comparing two construction materials, steel and copper, as a function of the heat flux radiated. The following equation from a heat balance relates the heat flux with the slag melting temperature, slag thickness, inlet–outlet water temperatures, pipe thickness and thermal conductivities:

$$Q = \frac{T_{ms} - \frac{1}{2}(T_i - T_0)}{(1 + \alpha) + \frac{h_r}{k_r} + \frac{h_s}{k_s}}$$

where Q is the heat flux per unit area in MW/m^2, T_{ms} is the slag melting temperature, T_i and T_0 are the inlet and outlet water temperatures in K, α is the water heat transfer coefficient in W/m^2 K (assumed to be 10,000), k_r and k_s are the thermal conductivities of the WCP and solid slag in W/m K, respectively (30 for steel and 320 for copper), h_r and h_s are the thickness of the cooling pipe and solid slag, respectively. From the previous equation the slag thickness is estimated when it reaches its melting point at a given heat flux. In Fig. 12.2 the authors reported the maximum temperature of the pipe, where T_{ri} and T_{re} are the temperatures across the thickness of the pipe, internal at the water/pipe interface and external at the pipe/slag interface, respectively. The pipe thickness was equal to 10 mm. It can be observed that the steel pipe reaches very high temperatures when it is exposed to high heat fluxes. They suggest a normal heat flux value of 0.2 MW/m^2 in zones not exposed to high radiation intensity and 2 MW/m^2 for the hot spot regions. At the hot spot regions the steel pipes can reach above 1000 °C. This is the reason why copper panels should be placed at hot spots.

In Fig. 12.3 it is shown the effect of the slag melting temperature on the heat extracted by the WCP. It is shown that using slags with higher melting temperatures helps to decrease the energy losses.

Fallah et al. [15] also reported a positive effect by increasing the slag melting temperature. Figures 12.4 and 12.5 show that increasing the slag melting temperature from 1370 to 1450 °C decreases the increment of water temperature at the walls from 11 to 8 °C and energy losses from 15 to 10%, respectively. In these simulations the slag layer thickness at the WCP was 3.5 cm with a thermal conductivity of 2.2 W/m K and a water flow rate of 19 m^3/h.

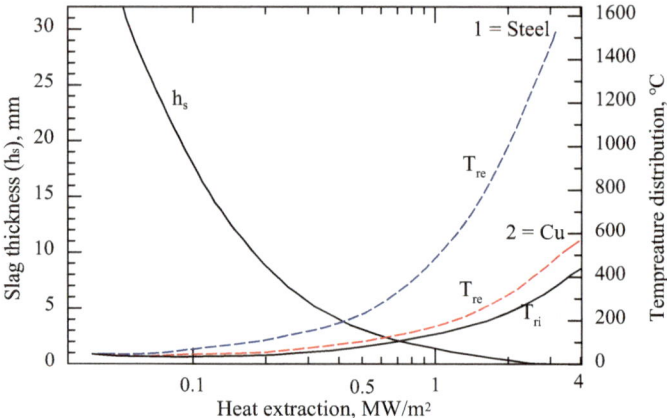

Fig. 12.2 Temperature distribution in pipes made of steel (1) and copper (2). After [14]

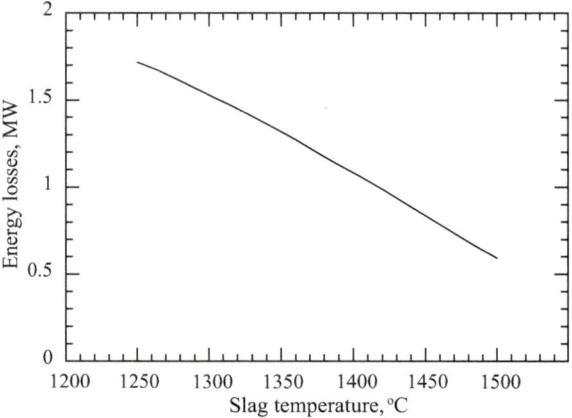

Fig. 12.3 Effect of slag temperature on energy losses. After [14]

Bowman and Krüger [16] estimated the heat losses as a function of slag thickness, indicating a range from 50 to 250 kW/m^2 for a slag thickness of 70 and 20 mm, respectively.

Gharib Mombeni et al. [17] investigated the temperature distribution in tubular WCP from the roof of an EAF. These panels are exposed to a higher cyclic thermal load due to scrap charging. These fluctuations induce thermal fatigue. Figure 12.6 shows a conventional arrangement of the tubular WCP. As can be observed, this arrangement has multiple bends, a sensitive region subject to both stagnant fluid and dead zones, which in turn can lead to cracks, as shown in Fig. 12.7. These authors developed a CFD model considering as boundary conditions; water inlet and outlet temperatures of 35 and 55 °C, respectively, water flow rate 8.6 kg/s, ambient

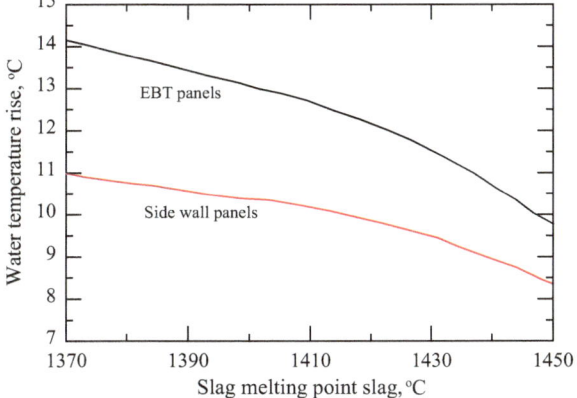

Fig. 12.4 Effect of the slag melting temperature on the increment in water temperature of the WCP. After [15]

Fig. 12.5 Effect of the slag melting temperature on energy losses to the WCP. After [15]

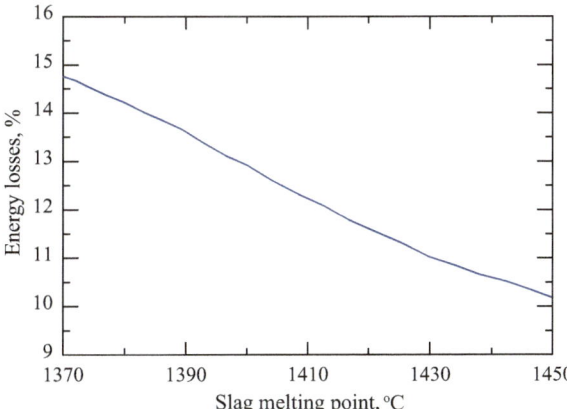

temperature of 50 °C. When the roof is opened for scrap charging it takes 20 s to open and rotate the roof, then the lower surface is exposed to the ambient atmosphere, it remains open for 120 s until the scrap is charged, then another 20 s to close the roof. Figure 12.8 shows the model results indicating an average temperature decrease in the hot face of about 45 °C, however the bends are sensitive regions because the water velocity decreases by one order of magnitude, from about 2 to 0.2 m/s, promoting higher temperatures and different thermal expansion with the rest of the pipe. In order to overcome this problem, the authors suggested a circular tubular geometry for the WCP, as shown in Fig. 12.9. With this geometry they found a more uniform fluid flow which decreased the mechanical stresses due to thermal fatigue.

Fig. 12.6 Arrangement of
tubular WCP from the roof
of an EAF. After [17]

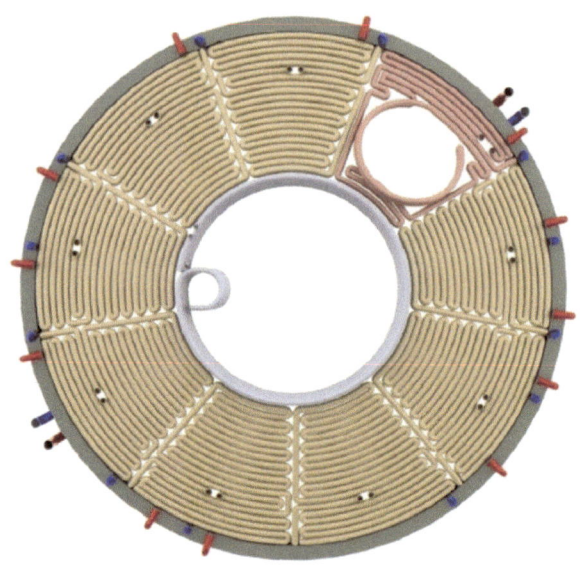

Fig. 12.7 Tubular WCP
from the roof of an EAF,
showing cracks. After [17]

12.3   Types of WCP

12.3.1   Box-Type WCP

The use of water cooling in the EAF has a long history. A design proposed by Blagg
in 1935 considered water cooling on the roof [18]. Kudrin et al. [19] also reported
a design that included water cooling in the roof since 1959 in Russia, as shown in
Fig. 12.10. They reported indirect water cooling, using tubular rings, on the external
surface of the roof of an EAF.

Fig. 12.8 Transient temperature distribution in the hot face of tubular WCOP from the roof of an EAF. After [17]

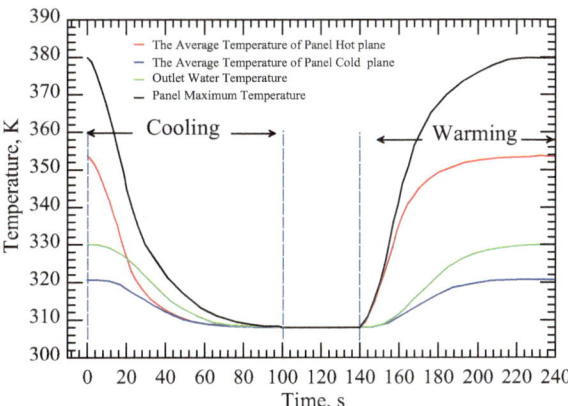

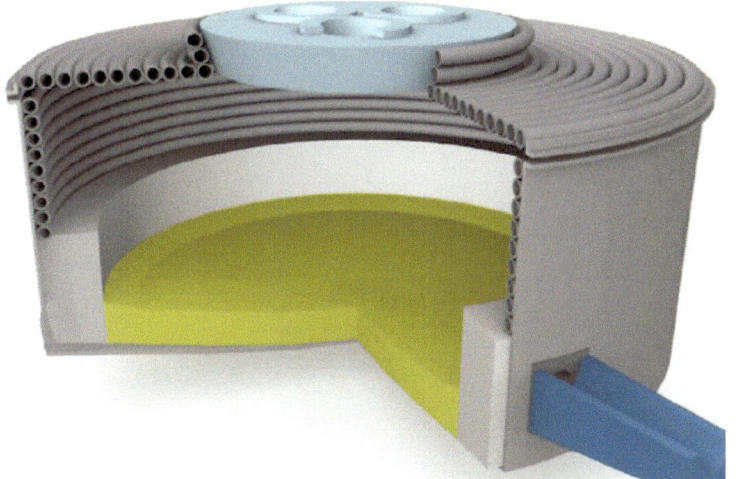

Fig. 12.9 Suggested circular tubular WCP. After [17]

In 1965, Franzen [20] described a similar design using water cooling in the EAF roof. Steel pipes are embedded in the bricks that allow the water to enter and leave, as shown in Fig. 12.11. The roof is a special part of the EAF subject to extreme transient heating conditions. The author mentions a life of 92 heats, with the life of the delta section having the shortest life. The author estimated that increasing 25% the refractory life would balance the extra costs related with labor and water cooling.

In 1974 Nanjyo from IHI [3] disclosed in US patent 3829595 a back-up secondary layer of metallic boxes, water cooled, as shown in Fig. 12.12. The cooled boxes are placed 200–500 mm above the level of molten steel or slag. Considering Fig. 12.12a, water is introduced at the bottom pipe.

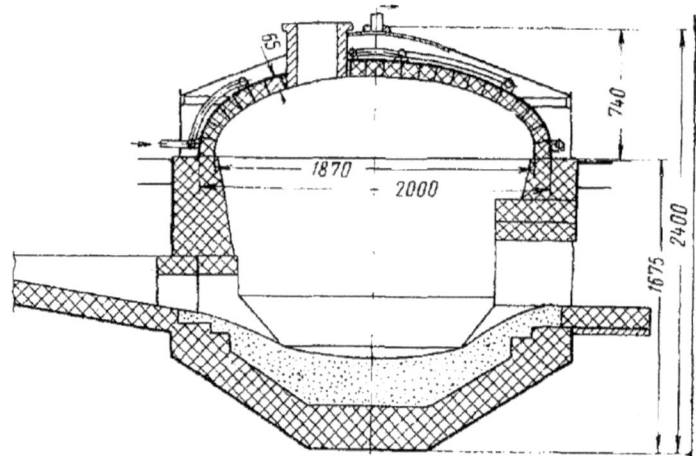

Fig. 12.10 Earliest application of water cooling of the roof of an EAF. After [19]

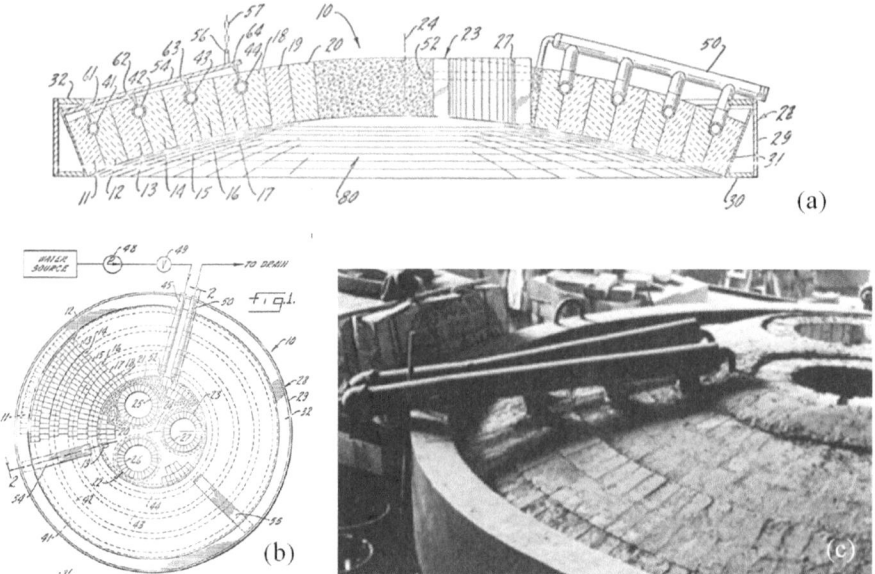

Fig. 12.11 Application of water cooling of the roof of an EAF. **a** Roof cross sectional view, **b** roof, top view, **c** photo. After [20]

In 1976, Mizuno [21] from Daido steel in US Patent 3940552 disclosed the final concept of a water cooled panel (box panel), directly exposed to arc radiation from the electrodes, as shown in Fig. 12.13. Mizuno pointed out the limitations of a previous design by IHI, indicating that the high temperature gradient in the refractory exposed to the arcs created severe problems of spalling, ending with the cooled box exposed

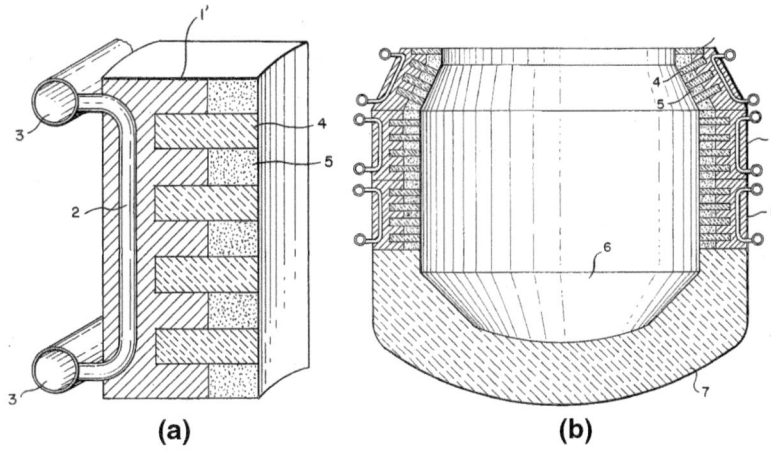

Fig. 12.12 First design of water-cooled panels from 1974; **a** section, **b** whole EAF. After [3]

to the arcs. In a new design, the WCP was designed with multiple fins on the hot face which allow adherence of liquid slag, with a thickness from 15 to 20 mm. This layer of slag insulates the WCP, the increment in water temperature is from 3 to 5 °C. The fins are welded to the panel, upwardly inclined and form a grid with different designs.

Copper box panels which include internal pipes is shown in Fig. 12.14. The copper panel is casted. This panel is designed for high heat loads, higher than 126 kW/m^2.

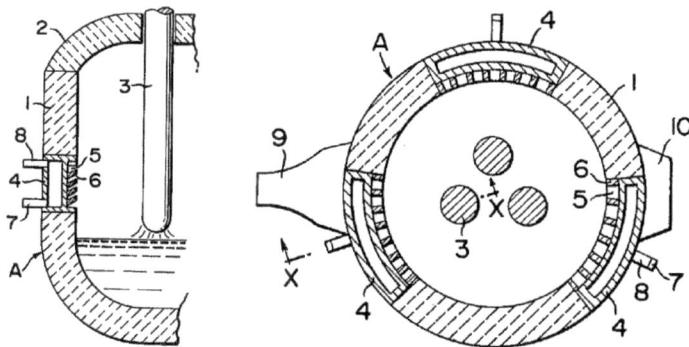

Fig. 12.13 Water cooled box panels proposed by Daido steel in 1976. After [21]

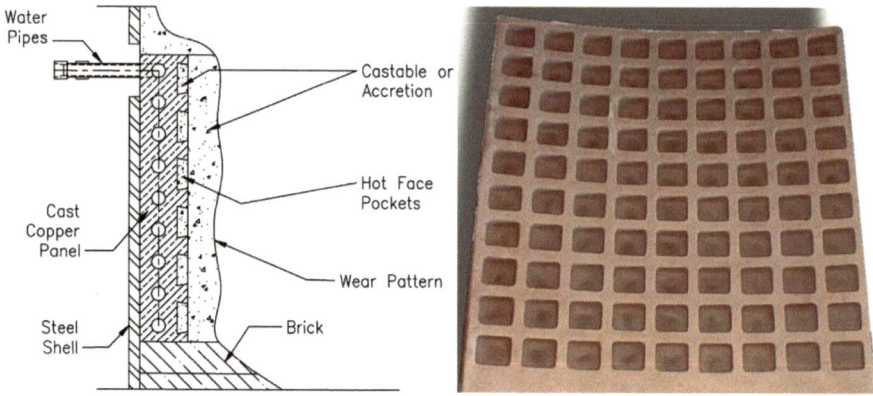

Fig. 12.14   Water cooled box panels proposed by Daido steel in 1976. After [21]

12.3.2   Tubular WCP

Zangs [22] from Demag (currently SMS) in 1980 proposed the first idea for tubular WCP, shown in Fig. 12.15. He noticed that WCP made in square or rectangular boxes have the disadvantage of dead flow zones and the risk of formation of air bubbles which largely decrease heat transfer and promote local overheating. The design proposed involves forced flow through pipes. The pipes include welded fins for slag adherence with a thickness from 10 to 20 mm. The inlet temperature of the cooling water varies from 15 to 30 °C and the outlet temperature from 60 to 65 °C at a water pressure of 4–5 bar, with a flow velocity of 2–4 m/s. The specific water consumption was 7–9 m^3 per m^2 of supplied surface. Figure 12.16 shows an example of an industrial tubular WCP.

The tubular design can result in dead zones of the liquid which in turn causes overheating producing gas bubbles which reduce heat transfer. Overheating is a cause for crack formation and can lead to water explosion with leaked water. Grajeda [23] suggested a tubular design that decreases the dead zones as shown in Fig. 12.17. In this design he suggests the use of baffles, marked as number 22 which include a hole, through this hole a small volume of liquid enter at the common dead zone in a tubular design, destroying the dead zone.

Tubular WCP are typically installed 350 mm above the liquid steel level [24].

Meysami et al. [25] have proposed that square pipes yield higher transfer rates in comparison with circular pipes. Figure 12.18 compares both cases, indicating that heat extraction is higher using square pipes.

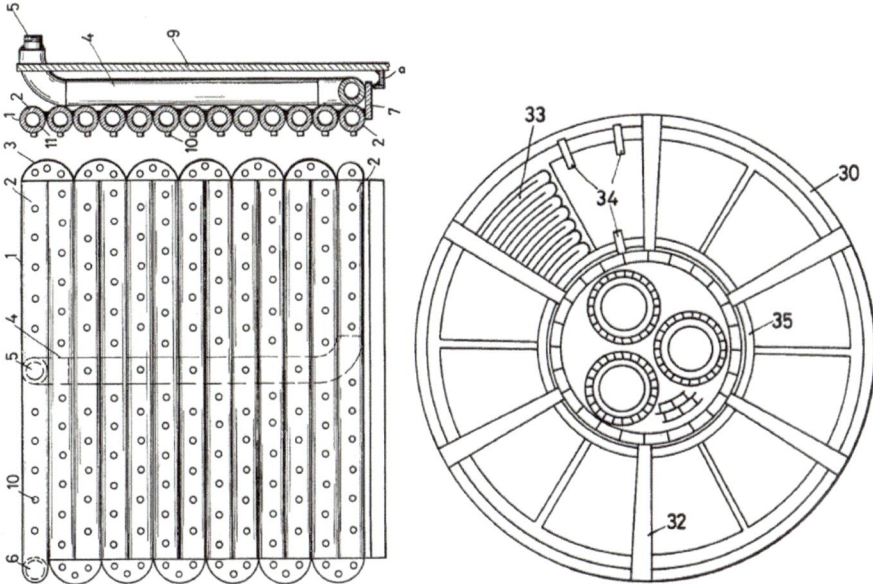

Fig. 12.15 Tubular design of WCP. After [22]

Fig. 12.16 Tubular water-cooled panels

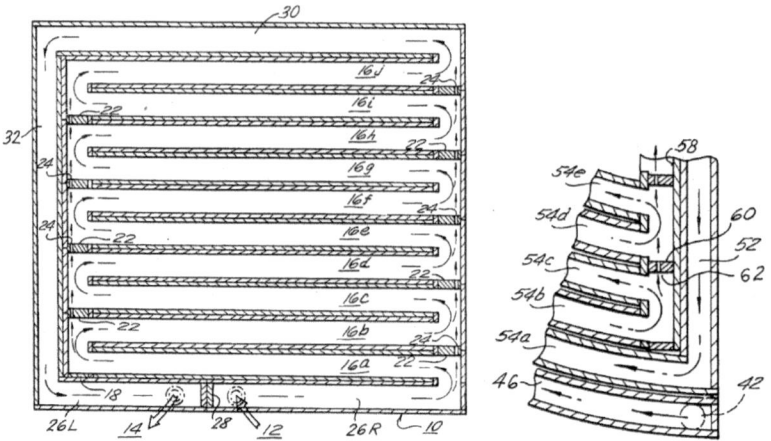

Fig. 12.17 Tubular design of WCP with baffles to eliminate dead zones. After [23]

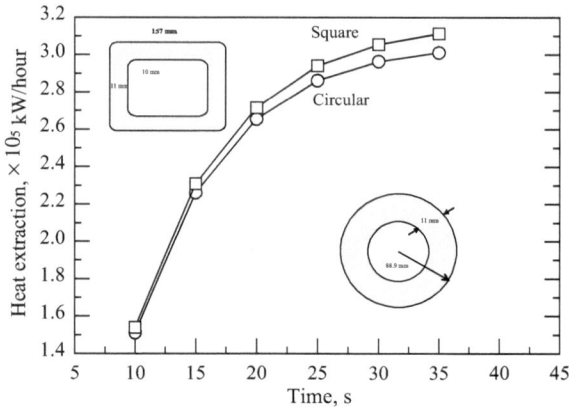

Fig. 12.18 Left: Heat extraction rates, right: circular and square pipes. After [25]

12.3.3   Spray WCP

Several EAF spray water cooling designs were disclosed by Union Carbide [26, 27] in the late 1980s, focused primarily on the roof, as shown in Fig. 12.19. These patents criticize some of the limitations of tubular WCP; cracks lead to water leakages and the risk of explosions, water is pressurized and due to high temperatures, the formation of precipitates is promoted which cause buildup and reduce heat transfer, if bubble are formed due to hot spots, the bubbles not only reduce the heat transfer capacity but also can create explosions, a loss of pressure enhances not only steam formation but also solids tend to settle, individual panels must be removed for maintenance and

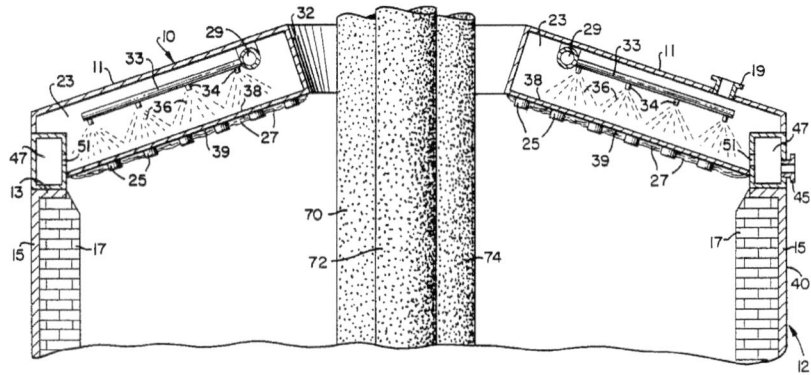

Fig. 12.19 EAF roof spray cooling. After [27]

in general the tubular system is complex due to the need of hoses, pipes, valves, etc. Spray cooling, according with these patents, eliminates all those previous limitations.

US Patent 5561685 from Union Carbide in 1996 [28] discloses details of spray cooling for the EAF walls, as shown in Fig. 12.20. Its inventors claim that spray cooling has many advantages over the conventional tubular WCP, for example; avoids water leakages, in case that steam is formed due to hot spots the vaporization heat contributes to additional cooling, cooling is more uniform, structure is lighter, is cheaper, etc.

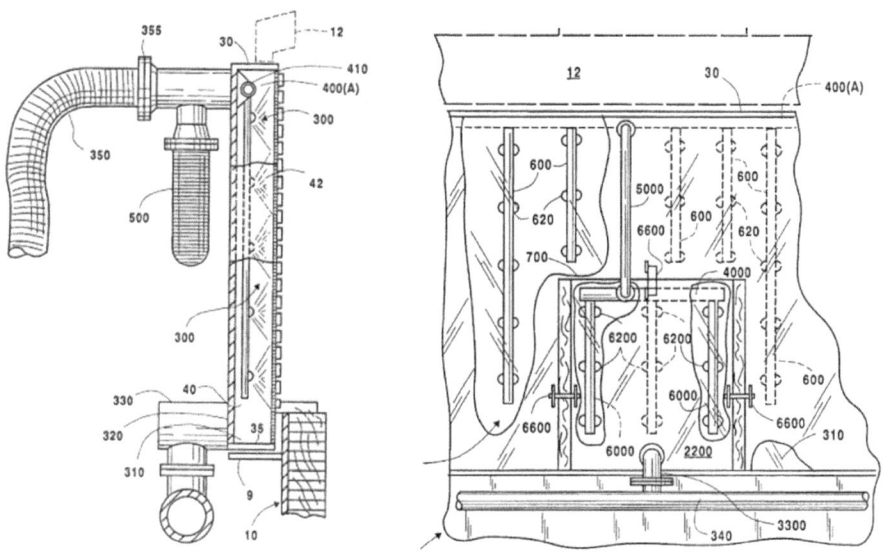

Fig. 12.20 Spray cooling of EAF walls. After [28]

Contreras et al. [29, 30] compared tubular and spray WCP, indicating that the tubular design has a higher cooling capacity but the spray system not only provides a more uniform heat extraction but is also safer. They reported lower energy loses with an optimum slag thickness on the walls of 45 mm and maximum arc coverage. These authors described the spray cooling WCP of one steel plant in Mexico which consisted of a DC-EAF charged with 30% scrap and 90% DRI. These panels are shown in Figs. 12.21 and 12.22.

Fig. 12.21   EAF with spray water cooling. Outer surface. After [29]

Fig. 12.22   EAF with spray water cooling. Inner surface. After [29]

12.4 Slag Coating of WCP

Water cooled panels should always be covered by a slag layer formed by splashing or by gunned materials. Coating WCP with a slag layer is a protection from heat radiation from the arcs due to its low thermal conductivity, about 0.12–0.13 W/ m K. The slag not only forms a refractory layer but also decreases arcing possibilities during scrap melting. The thickness of the slag layer varies considerable according to Simon [8], who summarizes different reports with the following values; 5–15 mm, 10–50 mm and as high as 30–120 mm. The thermal resistance of the slag layer is proportional to the slag thickness and inversely proportional to its thermal conductivity.

Simon [8] investigated the effect of the slag layer on the WCP on heat losses. He reported that the slag layer is not homogeneous but is formed by alternate layers of metal and slag with a broad range of thickness, as shown in Fig. 12.23, depending on the zone of the panels, its average value was 17.2 mm. The average content of metal was 22.4%. The phase composition included C_2S, FeO and complex oxides. Its density was on average 3.15 kg/m^3 at room temperature, with a heat capacity of 724 J/kg K at room temperature and a thermal conductivity from 1.45 W/m K at 300 °C to 1.54 W/m K at 800 °C. The average porosity of the slag was 22.6%. Contreras [29] reported an average thermal conductivity of 2.4 W/m K.

Trejo et al. [31] estimated the heat losses in an industrial 120 MVA AC-EAF of 150 ton charged with 70 ton of scrap and using an arc length of 45 cm. Figure 12.24 shows the heat losses as a function of the slag layer thickness. The figure shows that

Fig. 12.23 Morphology of slag and metal layers. Top surface is the hot face. After [8]

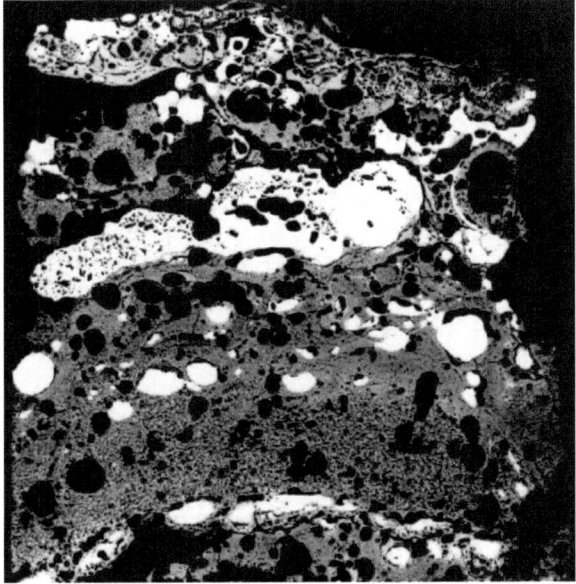

Fig. 12.24 Heat losses as a
function of slag layer
thickness in the WCP. After
[31]

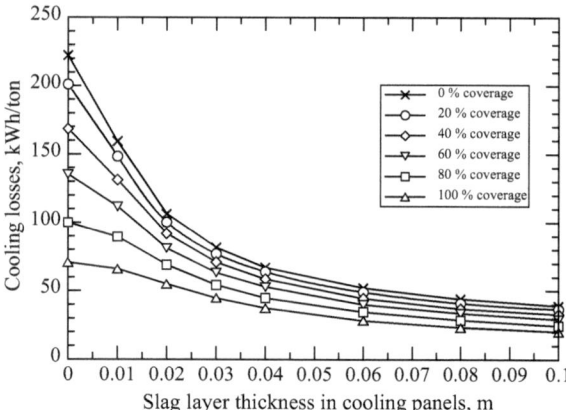

heat losses severely increase when the slag thickness becomes thinner, especially below 4 cm.

Radiation heat flux can be decreased from 700 kW/m^2 without a protective layer to 220 kW/m^2 for slag covered surfaces [32]. Another report [33] indicated that the energy extracted was about 100 kWh/m^2 for the first heat without a slag coating and decreased to 50 kWh/m^2 with the slag coating.

Many strategies have been developed to improve the slag coating on the WCP. A common technique is to provide anchors welded to the surface of the WCP that assist in the retention of slag splashed and the slag gunned on the walls. The slag layer thickness is higher in Cu-WCP due to higher heat transfer.

Borlée et al. [33] found that metallic slag holders are inefficient because the coating easily breaks due to the harsh environment and also they can melt when placed in the hot spot areas. To overcome these problems, they suggested a refractory coating which includes an insulating layer inserted between the metallic part of the panels and the refractory lining. With this modification they reported that energy losses decreased from an average of 150–85 kWh/heat, in an EAF of 105 ton with a transformer of 100 MVA and a tap-to-tap time of 42 min. A CFD model was developed which showed intense thermal gradients of the refractory in contact with the cooled tubes which can result into cracks. Kirschen [34] proposed to replace the gunning material by a refractory panel.

To avoid pipe's permanent deformation, the temperature difference across the pipe should not be too high. The water temperature rise is in the range from 8 to 17 °C and peak values of 28 °C [35].

Heat transfer to the WCP is both by convection and radiation, however convection is much lower than radiation and can be neglected, according to Bowman and Krüger [16]. In their calculations they employed a bath emissivity of 0.7, view factors in the order of 0.3 and a convection heat transfer coefficient of 3.8 W/m K. They reported that radiation heat transfer represented more than 95% of the total heat transfer to the walls.

12.5 WCP Design

12.5.1 Tubular WCP

The following is a summarized list of some of the variables that are involved in the design of WCP.

- *Water flow rate*: In AC-EAF's, the specific water flow rate for side wall panels is about 150 l/min m^2 (9 N m^3/h m^2) and 170 l/min m^2 (10.2 N m^3/h m^2) for roof panels [36]. Similar values have been reported in other references [24, 37, 38]. Zuliani et al. [38] gives a range from 163 to 183 l/min m^2 (9.8–11.0 N m^3/ h m^2) if the area is in the range from 102 to 130 m^2 and a total flow rate from 16.6 to 23.8 N m^3/min. Wandekoken et al. [24] suggests a range from 10 to 15 N m^3/h m^2. The highest water flow rate should be for WCP at the hot spots. Braverman et al. [39] reported values for WCP at the hot spots for one of the oldest designs, refractory bricks with a frontal section using a block of WCP, indicting an extremely large water flow rate, 338 l/min m^2 (20.3 N m^3/h m^2) for summer conditions but much lower for winter conditions. In summer the water temperature ranged from 30 to 32 °C but in winter was much lower, from 10 to 18 °C. The lower the water temperature, provides a higher extraction heat capacity as shown in Fig. 12.25 reported by Kuz'min et al. [40].

  An extremely high-water flow rate makes a water leak very dangerous. Calculation of the water flow rate is based on a thermal balance between the heat supplied to the WCP and the heat extracted by cooling water, as follows:

$$\text{Heat transfer to the WCP} = \text{Heat extracted by cooling water}$$

$$\dot{q}_r A_{ep} = \dot{m} C_p \Delta T$$

$$\dot{q}_r \left(\pi D_{ep} \frac{L}{2} \right) = \left(A_{ip} U_{wp} \rho_w \right) C_p \Delta T$$

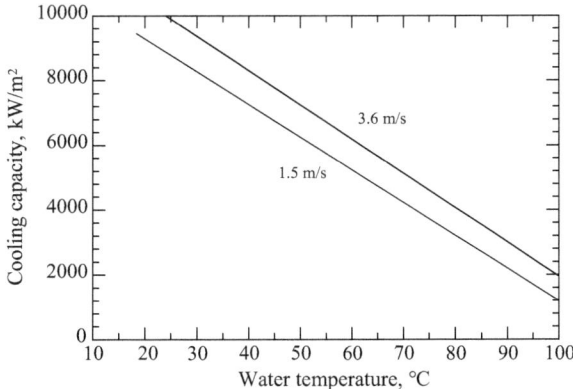

Fig. 12.25 Cooling capacity under forced convection as a function of water temperature for two water velocities; (1) 1.5 m/s and (2) 3.6 m/s. Taken from [40]

$$\dot{q}_r\left(\pi D_{ep}\frac{L}{2}\right) = \left(\frac{\pi D_{ip}^2}{4}\right)U_{wp}\rho_w C_p \Delta T$$

where $\dot{q}_r$ is the heat transferred to the WPC from the bath and the electric arcs, in kW/m^2, D_{ep} is the external pipe diameter in m, $L/2$ indicates that only half of the area of the pipes is exposed to radiation, D_{ip} is the internal pipe diameter in m, U_{wp} is the water velocity in m/s, ρ_w is the water density, $1000 \ kg/m^3$ at room temperature, C_p is the water heat capacity, $4.184 \ kJ/kg \ K$ and ΔT is the temperature increment in cooling water. $1 \ kW = 3600 \ kJ/h$.

The length of the pipe is estimated from the previous expression, given the values of $\dot{q}_r$ and U_{wp}:

$$L = \frac{D_{ip}^2 U_{wp}\rho_w C_p \Delta T}{2\dot{q}_r D_{ep}}$$

With this length, the total external surface area is:

$$A_{ep} = \pi D_{ep}\frac{L}{2}$$

Them, the required flow rate in m^3/h is:

$$Q_w = A_{ip}U_{wp} = \left(\frac{\pi D_{ip}^2}{4}\right)U_{wp}$$

- *Water velocity*: the minimum water velocity should be enough to avoid steam formation. The water velocity depends on the pipe diameter. The diameter of boiler tube grade ranges from 70 to 90 mm [35]. The water velocity is in the range from 1.5 to 3 m/s but can be up to 5 m/s for regions with high radiation intensity [35].
- *Water quality*: In order to minimize the deposit of accretion on the pipes which reduce its heat transfer capacity, the hardness and suspended particles should be minimized. The solubility of ions increases with temperature with the exception of calcium [41], therefore when the water is heated Ca ions precipitate as $CaCO_3$. This phenomenon is enhanced with the presence of particulate matter which can be present when water is later-on cooled down exposed to the atmosphere. Water treatment by reverse osmosis removes dissolved solutes, decreasing the hardness of water.

  Toulouevski and Zinurov [11] suggest to avoid critical temperatures for the precipitation of solids, for untreated water they recommend a maximum outlet temperature from 40 to 50 °C and for chemically treated water below 70 °C.
- *Water pressure (inlet, pressure drop)*: The outlet water pressure should be at least 20 psi [35]. The pressure drop can be estimated with the following equation:

Table 12.3 Comparison of thermal properties [35]

Property	Steel	Cu
Thermal conductivity, W m^{-1} K^{-1}	50	383
Thermal load, GJ m^{-2} h^{-1}	7	21

$$\Delta P_L = f \frac{L}{D} \frac{\rho U_{avg}^2}{2}$$

where ΔP_L is the pressure drop, f is the friction factor, L is length, D is the diameter of the pipe and the last term is the dynamic pressure.

The friction factor can be estimated using different correlations, for example, Colebrook's correlation:

$$\frac{1}{f^{0.5}} = -2 \log_{10} \left[\left(\frac{2.51}{Re \cdot f^{0.5}} \right) + \left(\frac{\varepsilon/d_h}{3.7} \right) \right]$$

where f is the friction factor, Re is the Reynolds number, ε is the roughness of the pipe in m and d_h is the hydraulic diameter in m.

- *Panel construction material*: Panels can be made of boiler plate steel or copper. Copper has seven times higher thermal conductivity in comparison with steel, as shown in Table 12.3.
- *Pipe thickness*: Pipe thickness should be as large as possible to withstand strikes during scrap charging but also as low as possible to maximize heat transfer. Wall thickness is in the range from 8 to 10 mm [35]

Contreras et al. [29, 30] developed a model which predicts pressure and heat losses in the WCP of a DC-EAF of 145 ton, with an electric power supply of 140 MW and 14 WCP. The following is a short summary of this work: Table 12.4 summarizes the length (L_{pn}), loss coefficients of valves and curvatures (K_{LVpn}, K_{LCpn}), surface area of each panel (A_{pn}) and total wall surface area (A).

Pressure losses or head losses are due to the frictional resistance of the piping system, commonly expressed as an equivalent fluid column height (h_L), defined as follows:

$$h_L = \frac{\Delta P_L}{\rho g} = \left(\frac{f_{pn} L_{pn}}{d_{pipe}} + K_{Lpn} \right) \left(\frac{U_{pn}^2}{2g} \right)$$

where h_L are the head losses, f_{pn} is the friction factor, U_{pn} is the average water velocity in each panel (= Q_{pn}/A_{pn}) and K_{Lpn} are the minor loss coefficient for each panel. Once the head losses are defined, the pumping power to overcome the pressure loss can also be defined.

The arrangement of WCP is in parallel with one common inlet and one common outlet.

Heat losses were estimated assuming the heat radiated to the WCP derives only from the arc and bath surface, the heat radiated is then transferred through the panel

Table 12.4 Characteristics of WCP

	1	2	3	4	5	6	7	8	9	10	11	12	13	14
L_{pn}, m	47.16	37.55	35.39	37.18	37.18	39.42	26.64	27.19	29.66	37.18	37.18	39.42	35.39	37.18
K_{LCpn}	9.2	10.7	7.1	6	6	12	2.8	2.8	2.8	6	6	12	7.1	6
A_{pn}/A, m^2	0.045	0.062	0.062	0.062	0.062	0.062	0.046	0.048	0.058	0.062	0.062	0.062	0.062	0.062

After [29, 30]

$A = 56.5\ m^2$, int. pipe diameter (d_{pipe}) $= 0.0667$ m, pipe roughness $\in\ = 0.045 \times 10^{-3}$ m, loss coefficient valve (K_{LVpm}) $= 0.2$, $\dot{Q}_o = 0.3194\ m^3/s$

layers (slag, panel, water) and finally extracted by the water. The three previous processes are defined by the following expressions:

$$\dot{Q}_{pn} = A_{s1}F_{12pn}\sigma\varepsilon_1\left(T_{s1}^4 - T_{s2pn}^4\right) + A_{sarc}F_{arcpn}\sigma\varepsilon_{arc}\left(T_{arc}^4 - T_{s2pn}^4\right)$$

$$\dot{Q}_{pn} = H_{pn}A_{pn}\left(T_{s2pn} - T_{s5}\right)$$

$$\dot{Q}_{pn} = \dot{m}_{pn}C_p\left(Tf_{outpn} - Tf_{in}\right)$$

where $\dot{Q}_{pn}$ is the heat transfer in each panel, A_{s1} and A_{sarc} are the surface of the molten bath and arc, respectively, F_{12pn} and F_{arcpn} are the view factors for molten bath-panel and arc-panel, respectively, σ si the Stefan-Boltzmann constant, ε_1 and ε_{arc} are emissivities for the slag and arc, respectively, T_{s1} is the temperature of the surface of the molten bath, T_{s2pn} is the outer slag layer temperature on each panel, T_{arc} is the arc temperature, H_{pn} is a global heat transfer coefficient for each panel, A_{pn} is the surface area of each panel, T_{s5} is the water average temperature $\left(\frac{Tf_{in}+Tf_{out}}{2}\right)$, Tf_{in} is the inlet water temperature, Tf_{out} is the outlet water temperature, $\dot{m}_{pn}$ is the water mass flow rate in each panel $\left(= \rho_{water}U_{pn}A_{intpipe}\right)$ where U_{pn} is the water velocity in each panel. The global heat transfer coefficient results from three resistances; conduction heat transfer through both the slag layer and the steel pipe as well as convection heat transfer. The convective heat transfer coefficient was calculated based on Gnielinski's correlation for internal flow, as follows:

$$H_{pn} = \frac{1}{A_{pn}R_{totalpn}} = \frac{1}{A_{pn}\left(Rcond_{slagpn} + Rcond_{pipepn} + Rconv_{waterpn}\right)}$$

$$H_{pn} = \frac{1}{\left(\frac{L_{slagpn}}{K_{slag}} + \frac{L_{pipe}}{K_{steel}} + \frac{1}{h_{pn}d_{pipe}L_{pn}}\right)}$$

where L_{slagpn} is the slag layer thickness in each panel, L_{pipe} is the pipe thickness, K_{slag} and K_{steel} are the thermal conductivities for slag and steel, respectively, $Rcond_{slagpn}$ is the conduction resistance through the slag layer, $Rcond_{pipepn}$ is the conduction resistance through the pipe, $Rconv_{waterpn}$ is the convection resistance through the cooling water, h_{pn} is the internal heat transfer coefficient (HTC) of each panel, d_{pipe} is the internal diameter of each pipe, L_{pn} is the length of each panel. The internal HTC is computed with the following expression:

$$h_{pn} = \frac{K_{water}}{d_{pipe}} \frac{\left(\frac{f_{pipe}}{8}\right)\left(Re_{pn} - 1000\right)\left(Pr_{water}\right)}{1 + 12.7\left(\frac{f_{pipe}}{8}\right)^{0.5}\left(Pr_{water}^{\frac{2}{3}} - 1\right)}$$

where K_{water} is the water thermal conductivity, f_{pipe} is the friction factor which can be computed using Colebrook's equation, Re_{pn} is the Reynolds number in each panel, Pr_{water} is the water Pr number. The slag surface temperature (T_{s2pn}) and the outer pipe surface temperature (T_{s3pn}) are related by the following expression:

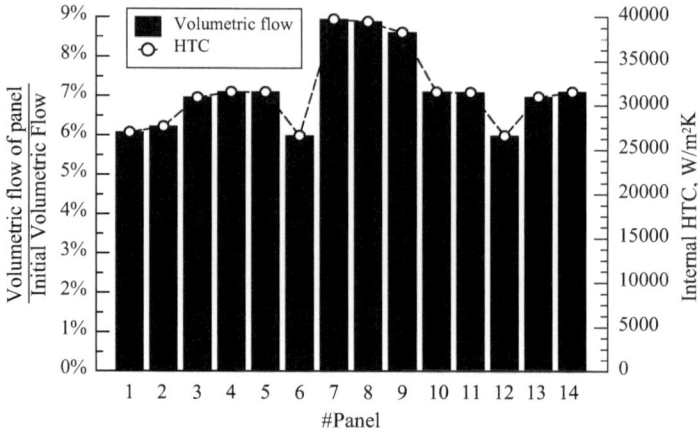

Fig. 12.26 Internal heat transfer coefficient (HTC) and fraction of water flow rate in each panel. After [29, 30]

$$\dot{Q}_{pn} = \frac{T_{s2pn} - T_{s3pn}}{Rcond_{slagpn}}$$

Some of the relevant data employed in their calculations were the following: molten bath surface area 42.74 m², arc diameter 0.15 m, emissivity of slag 0.8, emissivity of the arc 0.3, arc's temperature 10,000 K, water inlet temperature 310 K, molten bath surface temperature 1873 K, pipe thickness 1.109 cm, slag and steel thermal conductivity 2.4 and 60.5 W/m K, respectively. Figure 12.26 shows the internal HTC for each panel. The magnitude ranges from about 25 to 40 kW/m² K. The panels with a higher water flow rate have a higher HTC. They also reported that heat losses decrease from 0.7 to 0.2 MW when the thickness of the slag layer increases, from 0 to 3.4 cm. It also decreases when the arc is fully covered.

Khodabandeh et al. [42] carried a detailed CFD model that investigated the surface temperature of tubular WCP as a function of furnace diameter and the thickness of the panel's slag layer. The EAF employed in the simulations had a capacity of 120 tons, transformer of 90 MVA and roof inclination angle of 108°. Panels W1 in Fig. 12.27 have twice water-cooling flow rates in comparison with panels W2, 5.55 and 2.77 kg/s, respectively. The roof has a water flow rate of 9.72 kg/s. The water inlet and outlet temperatures were 23 and 36.5 °C, respectively.

The radiation model was divided in two parts. In the first part the temperature distribution inside the furnace was estimated, then in a second part, the temperatures at the walls were defined as boundary conditions to describe the temperature distribution inside the panels. Radiation heat transfer was computed based on the following expressions:

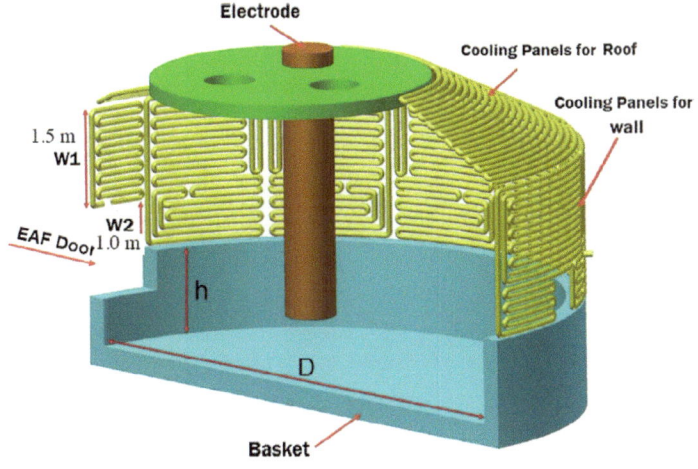

Fig. 12.27 EAF employed in the simulations. After [42]

$$J_i = \varepsilon_i \sigma T_i^4 + (1 - \varepsilon_i) \sum_{j=1}^{N} \left(VF_{ij} \cdot J_i \right)$$

$$Q_{i\text{-RAD}} = A_i \sum_{j=1}^{N} VF_{ij} \left(J_i - J_j \right)$$

$$Q_{arc\text{-RAD}} = 0.75 \cdot P_{arc}$$

where J_i represents radiosity for each surface, ε_i is surface emissivity, σ is the Stefan-Boltzmann constant, VF_i are view factors, T_i is the surface temperature, $Q_{i\text{-RAD}}$ is the energy radiated by surface i and $Q_{arc\text{-RAD}}$ is the heat radiated by the arc (based on Logar et al.). One of the important parameters are the view factors. A view factor is the fraction of radiation that leaves surface i and impact on surface j. Heat transfer inside the WCP is a case of conjugate heat transfer, involving heat transfer in solids and fluids. The model was validated using experimental measurements. The wall surface temperatures were measured using image thermography. Comparing the average temperature in one panel between experimental and numerical model predictions gave 865 and 827 °C, respectively, an error of 4%, furthermore, the predicted and experimental outlet water temperature were 39.2 and 41.5 °C, respectively, indicating, a good approximation of the model predictions.

Figure 12.28 shows the effect of furnace diameter and roof inclination angle on the surface temperature of the WCP (W1). Increasing the furnace diameter decreases the radiation intensity to the walls and therefore the temperature of the WCP decreases, for example, from 600 °C and 6 m to 580 °C and 9 m diameter. Increasing the roof inclination angle from 90° to 108° slightly increases the radiation intensity.

The effect of slag thickness, from 5 to 25 mm, on the temperature of WCP is shown in Fig. 12.29. It can be observed a decrease in the maximum temperature

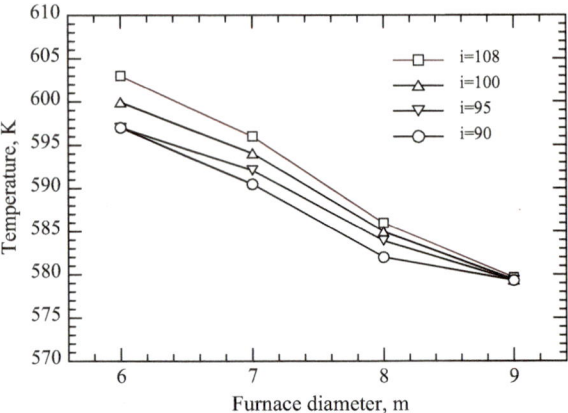

Fig. 12.28 Effect of furnace diameter on temperature, lower zone. After [42]

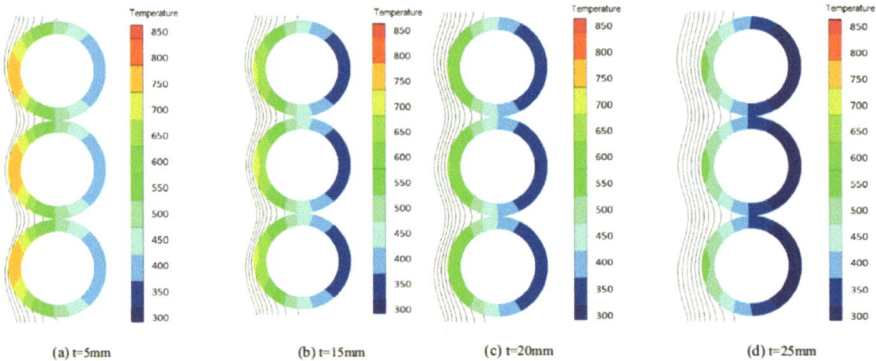

Fig. 12.29 Effect of slag layer thickness on the temperature of WCP. After [42]

from 747 to 495 °C when the slag layer thickness increases from 5 to 25 mm, a significant temperature decrease of 252 °C.

12.5.2 Spray WCP

Spray water cooling occurs when the liquid is forced through a small orifice nozzle and shatters into a dispersion of fine droplets which impact the heated surface, then the droplets spread on the surface and evaporate or form a thin liquid film. Spray cooling is characterized by a uniform heat removal. The heat removal capacity of water spray cooling is about 10 MW/m^2 [43]. The spray can be generated in two ways; air-atomized nozzles assisted and pressure nozzles. Air atomized nozzles use an air jet to help break up the liquid into finer droplets at higher velocity. The droplet

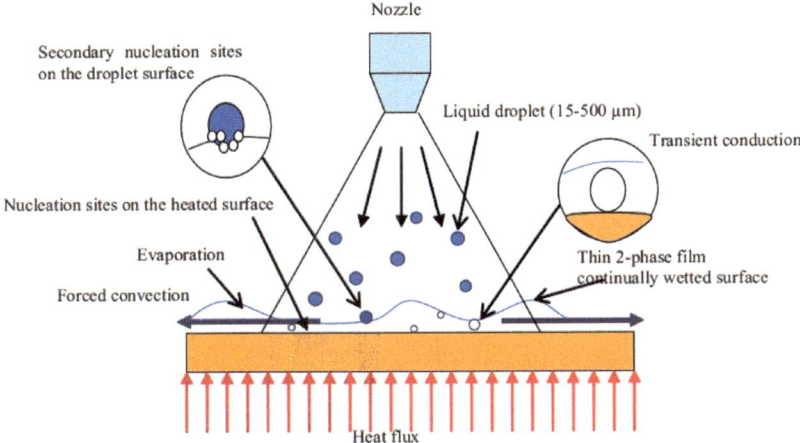

Fig. 12.30 Spray cooling and heat transfer mechanisms. After [45]

size has a wide range which is usually reported by the Sauter mean diameter (d_{32}). According to Kim [44] the mechanisms of spray cooling are still poorly understood due to the large number of variables involved and its complex interaction; different droplet size distributions, droplet number density, velocities, nozzle geometry, impact angle, surface roughness, gas content, interaction nozzle-wall and nozzle-nozzle, spray volumetric flux, etc. As a result, current investigations show opposing results, for example, some works indicating that the inclination angle affects heat transfer and other reports with opposite results. Figure 12.30 shows four heat transfer mechanisms; evaporation, forced convection due to droplet impingement on the heated surface, nucleation sites both on heated surface and droplet surfaces.

The heat transfer coefficient is a critical parameter to quantify the rate of heat extraction, as shown in the following equation:

$$q = h_s(T_o - T_s)$$

where q is the rate of heat extraction, T_o and T_s are the WCP surface temperature and spray water, respectively, h_s is the heat transfer coefficient (HTC), which should be obtained experimentally. Mzad and Khelif [43] developed a numerical model to investigate the effect of water pressure on the HTC. The HTC increased with pressure, as shown in Fig. 12.31, within a range from 9 to 39 kW/m^2 K

Abbasi et al. [46] carried experimental measurements on the HTC, reporting the following relationship which describes the HTC as a function of the water pressure. Figure 12.32 describes this relationship.

$$h = \left(C\rho^{\frac{1}{2}}C_p \overset{a}{Pr}\right)P^{\frac{1}{2}} = C_1 P^{\frac{1}{2}}$$

Fig. 12.31 Predicted HTC as a function of water pressure and surface temperature. After [43]

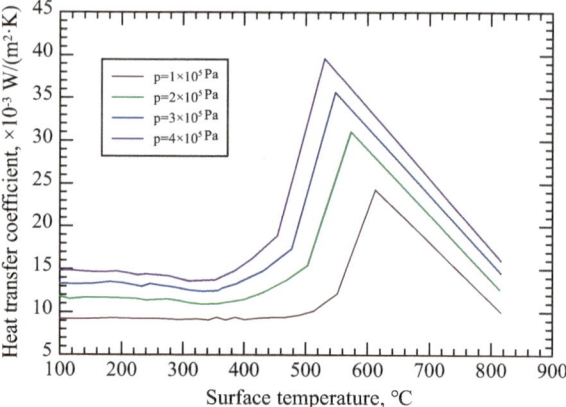

Fig. 12.32 Linear relationship between the HTC and pressure. After [46]

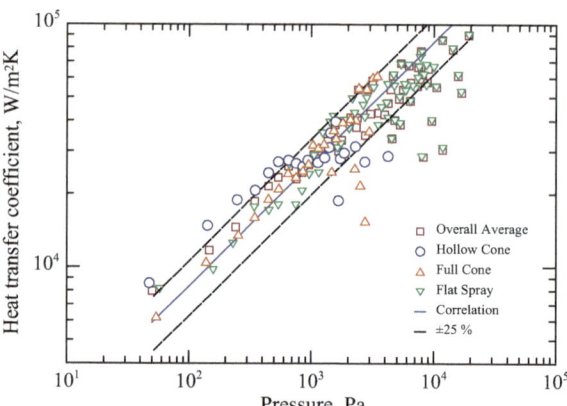

12.6   Maintenance of WCP

The lifespan of WCP is usually large, at least 10,000 heats and can reach as much as 20,000 heats. However, its life span depends on different factors, such as design, construction material, location in the EAF, slag coating practice, slag foaming conditions, scrap charging practice, oxygen injection practice, etc. In principle, WCP are subject to cyclic mechanical and thermal stresses which promote cracks.

Water leaks detection is a very important aspect using WCP. The methods based on controlling the flow rate and off-gas composition work well for large water leaks, from 90 to 180 l/min but not for small water leaks in the order to 30 l/min. Alshawarghi et al. [37] have proposed predictive models based on machine learning to control water leaks.

Liquid water can be placed on the top of liquid steel and will boil immediately. The reactions products involve both water vapor and hydrogen. Hydrogen can be formed as follows:

$$\underline{Fe} + H_2O_{(g)} = (FeO) + H_{2(g)}$$
$$\underline{C} + H_2O_{(g)} = CO_{(g)} + H_{2(g)}$$

The risk of explosions is due to entrapment of liquid water due to sudden fall of scrap or other reasons, then the trapped water will boil and yield the explosion.

Detection of water leaks by off-gas analysis is complicated due to variable concentrations of water vapor and hydrogen, which is due to the following sources: electrodes water cooling, moisture in the metallic charge, moisture in the air ingress, combustion of oily materials in the scrap and combustion products from the burner. These processes yield a large variable concentration of H_2O and H_2 during the heat, which depend on the prevailing oxidizing or reducing conditions of the gas in the freeboard, more H_2O under oxidizing conditions and more H_2 under reducing conditions. In order to evaluate a water leak it is necessary to measure not only H_2O but also H_2 simultaneously. A system designed by Tenova [38] in addition to measuring H_2O and H_2, also includes a statistical fingerprint model to evaluate different risk levels of water leaks. It has also been reported that a large water leak can be detected due to the presence of a blue flame between the electrodes [24]. Wandekoken et al. [24] proposed the application of ELD (electronic leak detection) technology.

A potential case of water leaks is in the copper WCP that contain burners [47]. In this case, if the scrap blocks the passage of the flame, it produces a "flash back", then the temperature of the WCP can be excessively high in particular at the nose section of the burner closer to the scrap. The jet of oxygen can also contribute to the same phenomena when oxygen injection starts with pieces of un-melted scrap close to the burners. The over-heating is not easily detected when the zone overheated is small because the effect of the water temperature is not high.

12.7 Water Cooling of Electrodes and Oxygen Lances

12.7.1 Water Cooling of Electrodes

This practice was first developed in Japan by Nippon Steel Corporation [48]. One form of electrodes consumption is due to oxidation. The rate of oxidation increases with temperature. To decrease the rate of oxidation, about 1 m^3 of water/h per electrode is discharged from the top of the electrodes [32]. This water is turned into steam. The energy involved corresponds to 600 kWh/m^3, assuming a temperature of 1000 °C. It is important to check the benefits on electrode consumption and the cost of energy because the energy cost is normally higher than electrode savings [32].

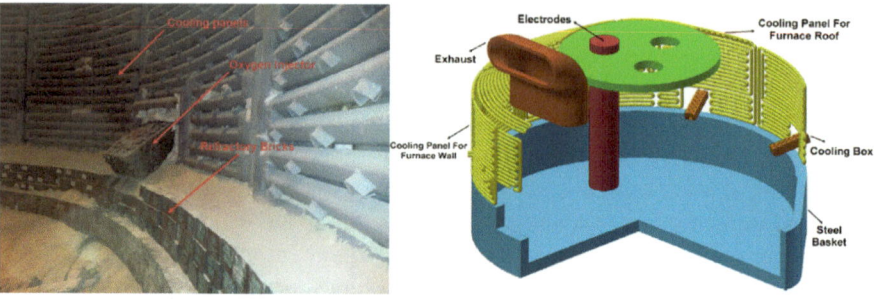

Fig. 12.33 Location of cooling boxes for the oxygen lances. After [49]

12.7.2 Water Cooling of Oxygen Lances

Oxygen injection is a routine practice in steelmaking operations. Currently, oxygen injection is carried out with lances using supersonic or coherent jets. In both cases, the high temperatures in the lances require water cooling. Modern EAF's employ coherent jets fixed at the bottom of the furnace shell, as shown in Fig. 12.33. Khoda-bandeh et al. [49] developed a numerical model to improve the water cooling capacity of oxygen lances, illustrating the importance of a proper design to avoid high temperatures. They proposed a new design that eliminated 180° pipe bends, obtaining a maximum lance temperature of 170 °C which extended the life of the lances.

12.8 Water Cooling of Flexible Cables

The water flow rate for the flexible cables ranges from 11 to 95 l/min according to Ref. [35] or from 10 to 140 l/min according to Ref. [50]. Figure 12.34 is based on Ref. [50], these results are valid for a cable strip length of 10 m, a rise in water temperature of 5 °C and a pH of water from 6.5 to 10.

Inadequate water cooling of the flexible cables can increase the electrical resistance of the conductors and also become embrittled and in extreme cases the copper wire inductors can catastrophically melt, as reported by Pinney [51] and shown in Fig. 12.35.

12.9 Replacement of Water as Cooling Medium

One of the major concerns of water as a cooling medium is the risk of explosions when the WCP's get cracks and water comes into contact with liquid steel. Ionic liquids have been proposed as a solution to this problem [52]. The ionic liquids proposed are stable at temperatures below 200 °C and for short periods of time can

Fig. 12.34 Estimated flow rate of cooling water for flexible cables. After [50]

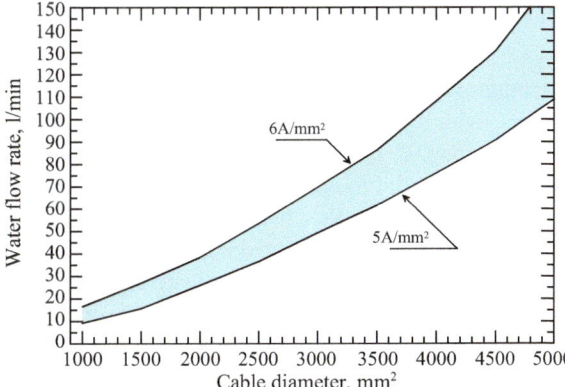

Fig. 12.35 Catastrophic failure due to improper cooling resulting in the solder melting. After [51]

be used up to 250 °C, also they are non-corrosive, non-flammable and have a dynamic viscosity similar to water. A much higher cost using these fluids would be expected but the reference didn't provide information. Kennedy et al. [53] reviewed in more detail a variety of fluids to replace water, from organic and fluorinated fluids to the use of a Pb-Bi eutectic alloy.

Soultanzadeh et al. [54] reported the use of nanofluids to partially replace water. Water nanofluids have a small fraction of nano-particles, below 100 nm, in the order of 1–10%. The presence of these particles increases its heat extraction capacity. In their work the authors reported the addition from 3 to 5% alumina nano-particles. The numerical model developed reported the heat radiated to the walls was 103 kW/m^2 K, using 9 kg/s of water the maximum temperate in the hot spots was 96 and 76 °C using 5% alumina nano-particles.

12.10   EAF Operation to Decrease Heat Losses to the WCP

The main way to decrease heat losses to the WCP is by improving the slag foaming practice which minimizes heat radiation to the WCP. This subject is covered in detail in Chap. 7. Here, only a brief discussion is presented. Treppschuh et al. [55] developed a model to calculate the thermal losses to the wall and roof of the WCP, then, according with the thermal load the electric power is adjusted accordingly. The model was based on statistical correlations and Monte-Carlo simulations. The results from this model clearly indicate an increment in the thermal load when the scrap bucket is melted and the operation was under flat bath conditions, for example peaks in the thermal load of about 6 MW for the walls and 12 MW for the roof. When the real values exceed the model predictions the electric power is decreased. Applying this model, the authors reported energy savings of 6%. Another model implemented was based on monitoring the sound and correlated with the voltage. When the normalized sound level exceeded a critical value, 0.7, the voltage was decreased, decreasing the arc length.

12.11   Recovery of Radiated Energy to the WCP

One of the latest concepts on energy recovery is the recovery of the energy radiated to the WCP [56–58]. The total amount of heat in the cooling water is large but its exergy is low due to the low temperature, therefore energy recycling had not been attempted before. The new concept involves application of Thermophoto Voltaic (TPV) electricity generation, first proposed in the late 1950's. A review on the subject is included by Utlu et al. [56] who indicate the concept has not reached the commercial stage but seems to be promising. Saboohi et al. [57] estimate an increase of the thermal efficiency of 0.8% using TPV.

References

1. Schwabe WE (1962) Arc heat transfer and refractory erosion in electric steel furnaces. In: EAF conference, pp 195–206
2. Robinson CG, Schwabe WE (1965) Ultra-high power electric steel furnace operation. JOM 17:75–80. https://doi.org/10.1007/bf03398860
3. Nanjyo T, Aoshika M (1974) Electric direct-arc furnace. US patent 3829595
4. Bosley J, Clark J, Dancy T, Fruehan R, McIntyre E (1987) Techno-economic assessment of electric steelmaking through the year 2000, Pittsburgh, PA, USA
5. Tischenko P, Timoshenko SN, Tishenko A (2000) New generation of the EAF panels with renewable slag: lowering of heat loss, combining of duties for energy saving. In: Met. 2000, pp 1–13
6. Colla V, Matino I, Branca TA, Fornai B, Romaniello L, Rosito F (2017) Efficient use of water resources in the steel industry. Water (Switzerland) 9:1–15. https://doi.org/10.3390/w9110874

7. Kirschen M, Kronthaler A, Rahm C (2007) Impact of water cooled Cu-blocks in refractory linings on the energy balance of electric arc furnaces. In: UNITECR'07 10th unified international technical conference on refractories, Dresden, Germany, 18–21 Sept, pp 166–169

8. Simon MJ (1989) The thermal performance of water cooled panels in electric arc steelmaking furnaces. Sheffield Hallam University, UK

9. Guo D, Irons GA (2003) Modeling of radiation intensity in an EAF. In: Proceedings of 3rd international conference on CFD in the minerals and process industries, Melbourne, Australia, 10–12 Dec 2003, pp 223–228

10. Sanchez JLG, Conejo AN, Ramirez-Argaez MA (2012) Effect of foamy slag height on hot spots formation inside the electric arc furnace based on a radiation model. ISIJ Int 52

11. Toulouevski YN, Zinurov IY (2010) Innovation in electric arc furnaces. Scientific basis for selection. Springer, Berlin

12. Freeman ER, Medley JE (1978) Efficient use of power in electric arc furnaces. IEE J Electr Power Appl 1:17–24. https://doi.org/10.1049/ij-epa.1978.0004

13. Bowman B, Fitzgerald F (1973) Hot spots in arc furnaces. J Iron Steel Inst 211:178–186

14. Bisio G, Rubatto G, Martini R (2000) Heat transfer, energy saving and pollution control in UHP electric-arc furnaces. Energy 25:1047–1066. https://doi.org/10.1016/S0360-5442(00)00037-2

15. Fallah A, Meratian M, Edris H, Zamani B (2008) Simulation of water cooled panel and its effect on efficiency of electric arc furnace. In: 4th International congress on the science and technology of steelmaking, pp 183–186

16. Bowman B, Krüger K (2009) Arc furnace physics. StahlEisen

17. Gharib Mombeni A, Hajidavalloo E, Behbahani-Nejad M (2016) Transient simulation of conjugate heat transfer in the roof cooling panel of an electric arc furnace. Appl Therm Eng 98:80–87. https://doi.org/10.1016/j.applthermaleng.2015.12.004

18. Blagg GE (1935) Electric furnace. US patent 1992465

19. Kudrin V, Smolin A, Sosonkin O, Blinov P (1969) Operation of an electric arc furnace with water-cooled roof. Metallurgy 23–25

20. Franzen EF. New furnace innovations-2. Internally water-cooled roof. JOM 17AD; 4

21. Mizuno S (1976) Water-cooled panel for arc furnace. US patent 3940552

22. Zangs L (1980) Vessel for metal smelting furnace. US patent 4207060

23. Grajeda I (1987) Cooling panel for electric arc furnace. US patent 4637034

24. Wandekoken TP, Maia BT, Hopperdizel P (2015) Earliest leak detection (ELD): a new concept of security in water leak detection in electric arc furnaces. Iron Steel Technol 12:88–93

25. Meysami AH, Rahimzadeh HA, Najafabadi RA, Isfahani TD (2017) Comparison of heat transfer power of the cooling panel with square cross section and circular cross section in electric arc furnaces steel-making by the use of computational fluid dynamics. Int J ISSI 14:38–42

26. Heggart R, McClintock W, Engstrom R (1987) Furnace cooling system and method. US patent 4715042

27. Burwell W (1989) Cooling system and method for molten material handling vessels. US patent 4815096

28. Lehr D, Roberts G, Miner F, Burwell W, Arthur M (1996) Modular spray cooled side-wall for electric arc furnaces. US patent 5561685

29. Contreras J (2018) Study of heat transfer in tubular-panel and spray cooling systems applied to the electric arc furnace walls. MSc thesis, ITESM Mexico

30. Contreras-Serna J, Rivera-Solorio CI, Herrera-García MA (2019) Study of heat transfer in a tubular-panel cooling system in the wall of an electric arc furnace. Appl Therm Eng 148:43–56. https://doi.org/10.1016/j.applthermaleng.2018.10.134

31. Trejo E, Martell F, Micheloud O, Teng L, Llamas A, Montesinos-Castellanos A (2012) A novel estimation of electrical and cooling losses in electric arc furnaces. Energy 42:446–456. https://doi.org/10.1016/j.energy.2012.03.024

32. Adams W, Alameddine S, Bowman B, Lugo N, Paege S, Stafford P (2001) Factors influencing the total energy consumption in arc furnaces. In: 59th Electric arc furnace conference, 11–14 Nov, Phoenix, AZ, USA, pp 691–702

33. Borlée J, Wauters M, Zampetti L, Volponi V, Rondi M, Picco M et al (2006) New cooling panels for reduction of heat losses in EAF steelmaking. Report ECSC 22404
34. Kirschen M (2016) Multilayer cooling panel and electric arc furnace. EP2818816B9
35. Jones JAT, Bowman B, Lefrank PA (1998) Electric arc furnace steelmaking. In: Fruehan RJ (ed) Making, shaping and treating of steel, 11th edn. Steelmaking and refining volume, Pittsburgh, PA USA, pp 525–660
36. Haissig M (1999) Electric arc furnace that uses post combustion. US patent 5943360
37. Alshawarghi H, Elkamel A, Moshiri B, Hourfar F (2019) Predictive models and detection methods applicable in water detection framework for industrial electric arc furnaces. Comput Chem Eng 128:285–300. https://doi.org/10.1016/j.compchemeng.2019.06.005
38. Zuliani DJ, Scipolo V, Khan M, Negru O, Bilski W (2014) Real-time water detection in EAF steelmaking. Iron Steel Technol 84–95
39. Braverman EM, Zhitnik GG, Pilyushenko VL, Zvyagintsev AA, Legostaev GS, Garchenko OA et al (1983) Performance of water-cooled panels in the wall linings of large-capacity electric furnaces. Metallurgist 27:53–56
40. Kuz'min MG, Cherednichenko VS, Bikeev RA, Cherednichenko MV (2014) Water cooled units in ultrapower electric arc furnaces. Russ Metall 933–939. https://doi.org/10.1134/S00 36029514120052
41. Gleason J, Lee D (2000) Establishing peak EAF cooling water performance. In: AISTech 2000 conference, pp 339–348
42. Khodabandeh E, Ghaderi M, Afzalabadi A, Rouboa A, Salarifard A (2017) Parametric study of heat transfer in an electric arc furnace and cooling system. Appl Therm Eng 123:1190–1200. https://doi.org/10.1016/j.applthermaleng.2017.05.193
43. Mzad H, Khelif R (2016) Effect of spraying pressure on spray cooling enhancement of beryllium-copper alloy plate. Procedia Eng 157:106–113. https://doi.org/10.1016/j.proeng. 2016.08.344
44. Kim J (2007) Spray cooling heat transfer: the state of the art. Int J Heat Fluid Flow 28:753–767. https://doi.org/10.1016/j.ijheatfluidflow.2006.09.003
45. Yan Z, Zhao R, Duan F, Wong TN, Toh KC, Choo KF et al (2011) Spray cooling. In: Two phase flow, phase change and numerical modeling. InTech, p 26
46. Abbasi B, Kim J, Marshall A (2010) Dynamic pressure based prediction of spray cooling heat transfer coefficients. Int J Multiph Flow 36:491–502. https://doi.org/10.1016/j.ijmultiph aseflow.2010.01.007
47. Bowers J, Farmer C, Miani S (2015) Comparison of temperature measurement in copper elements in the EAF. In: AISTech—iron and steel technology conference, proceedings, vol 2, pp 1809–1817
48. Madias J (2014) Electric furnace steelmaking. Treatise Process Metall 3:271–300. https://doi. org/10.1016/B978-0-08-096988-6.00013-4
49. Khodabandeh E, Rahbari A, Rosen MA, Najafian Ashrafi Z, Akbari OA, Anvari AM (2017) Experimental and numerical investigations on heat transfer of a water-cooled lance for blowing oxidizing gas in an electrical arc furnace. Energy Convers Manag 148:43–56. https://doi.org/ 10.1016/j.enconman.2017.05.057
50. Furukawa (n.d.) Water cooled cable for electric furnace, 4. https://www.FurukawaCoJp/En/Pro duct/Catalogue/Pdf/Watercable_d356ePdf
51. Pinney MG (2019) Adding life to water-cooled power cables through innovation in core construction © 2019 by the Association for Iron & Steel Technology. AISTech 2019 proceedings, pp 2473–2478
52. Hanel M, Filzwieser A, Krassnig H-J, Degel R (2019) ILTEC—highlighting potentials and eliminating concerns regarding the safe and water-free cooling technology. In: BHM Berg— Und Hüttenmännische Monatshefte, vol 164, pp 281–286. https://doi.org/10.1007/s00501-019-0819-z
53. Kennedyl MW, Nos P, Bratt M, Weaver M (2013) Alternative coolants and cooling system designs for safer freeze lined furnace operation. In: TMS conference, pp 299–314

54. Soultanzadeh MB, Haratian M, Mehmandoust B, Moradi A (2023) Numerical investigation of nanofluid heat transfer in the wall cooling panels of an electric arc steelmaking furnace. SN Appl Sci 5. https://doi.org/10.1007/s42452-023-05327-6

55. Treppschuh A, Krueger K, Kuehn R, Schliephake H (2008) Thermal based power control of a DC-EAF. Arch Metall Mater 53:425–430

56. Utlu Z, Paralı U, Gültekin Ç (2018) Applicability of thermophotovoltaic technologies in the iron and steel sectors. Energy Technol 6:1039–1051. https://doi.org/10.1002/ente.201700607

57. Saboohi Y, Fathi A, Škrjanc I, Logar V (2018) EAF heat recovery from incident radiation on water-cooled panels using a thermophotovoltaic system: a conceptual study. Steel Res Int 89:1–8. https://doi.org/10.1002/srin.201700446

58. Pérez SL, López SH, Astigarraga EU, Arce IDH, Botas MG de A, Iñarga JI et al (2021) Design of a radiant heat capturing device for steel mills. J Sustain Dev Energy Water Environ Syst 9:1–15. https://doi.org/10.13044/j.sdewes.d8.0365

Chapter 13
Electric Power Quality

13.1 Basic Concepts Related to the Electric Power System

A metallurgical engineer in charge of the operation of the EAF should have the basic knowledge on electric engineering concepts in order to understand how the electric power parameters can be optimized. Optimization of the electric power parameters requires knowledge on both disciplines. This section will provide a review of the basic concepts.

Historical background: To develop the knowledge on the basic concepts of electric engineering has taken a long time. The first recorded reference was magnetism. The compass was invented in China about 4000 years ago using magnetite. Credit is given to Tcheou-Koung to teach how to use the compass in about 1100 BC [1, 2]. Thales of Miletus (624/623–548/545 BC) was a Greek engineer who first reported the properties of amber to attract light objects. The Greek word for amber is electron. The scientific approach proposed by Roger Bacon (1220–1292) was a disruptive way to study natural phenomena, away from the Aristotelian philosophy who promoted rational thinking. This method accelerated the production of knowledge.

William Gilbert (1599) proposed the term *electricus* to describe the attraction phenomena between objects. In 1646 Thomas Browne introduced the term electricity in the English language. Charles du Fay reported in 1733 that objects can be charged with two opposite charges called vitreous (glass) and resinous (amber). Benjamin Franklin in 1752 coined the terms positive for vitreous electricity and negative for resinous electricity indicating that both were part of the same fluid. He arbitrarily defined the convention for the flow of current, from the positive to the negative terminal. Actually, the flow of electrons is in the opposite direction, however the convention proposed by Franklin has remained. Charles Coulomb measured the electric force between two opposite charges in 1785. Alessandro Volta developed the first battery to produce electricity in 1800. Humphrey Davy studied in detail the chemical reactions in the battery developed by Volta founding the principles of electrochemistry and sometime between 1802 and 1807 produced an electric arc using

© The Author(s), under exclusive license to Springer Nature Singapore Pte Ltd. 2024
A. N. Conejo, *Electric Arc Furnace: Methods to Decrease Energy Consumption*,
https://doi.org/10.1007/978-981-97-4053-6_13

two electrodes, developing the arc lamp. In 1820 Hans Ørsted defined the existence of a magnetic field around a conductor and Andre Ampere measured the magnetic field. George Ohm reported in a simple equation the relationship between voltage, current and resistance in 1826. Michael Faraday in 1831 found that a magnet can induce an electric current, inventing the transformer. James Maxwell in 1860 developed the mathematical theory of electromagnetism. Heinrich Hertz in 1880 proved the existence of electromagnetic waves predicted by the Maxwell's equations. In the same year Thomas Edison invents the incandescent lamp using the vacuum pump invented previously by Otto von Guericke in 1650. Nikola Tesla in 1897 developed a new induction motor employed to produce and distribute electricity using AC, in opposition to Edison who defended the use of DC. Joseph Thomson estimated the mass of an electron in 1899. The charge of the electron was experimentally measured by Robert Millikan between 1909 and 1913.

Electric quantities. Dimensions and units: The definition of dimensions and units to describe a natural phenomenon is not old. The meter was defined in 1793 and the kilogram in 1795. Maxwell since 1871 suggested the definitions of units in terms of dimensionless constants [3]. Maxwell employed brackets to define a dimension, for example [M] for mass, [L] for length and [T] for time. He was part of a committee who reviewed various systems of units for electricity and magnetism. This committee had been founded by the British Association for the Advancement of Science (BAAS) in 1861. The committee was chaired by William Thompson. Gauss in 1850 defined electromagnetic quantities using two systems; electric system (ESU, electrostatic system of units) and magnetic system (emu, electromagnetic system of units). Both systems were officially recognized in the UK in 1873. The metric system was established in 1795 and once it was accepted by many countries in 1872 it became the International System of Units or Systeme Internationale (SI). The SI has seven basic units, as shown in Table 13.1. Derived units for some electromagnetic quantities are shown in Table 13.2. Today, the British system is no longer used by the UK and mainly the USA sticks to this old system.

What is static electricity? Every substance is made of atoms. At the center of the atom is the nucleus with protons and neutrons and around the nucleus are the electrons

Table 13.1 International system of units (SI)

Physical quantity	Dimension	Unit name	Symbol
Mass	M	Kilogram	kg
Length	L	Meter	m
Time	T	Second	s
Temperature	θ	Kelvin	K
Amount of substance	N	Mole	mol
Electric current intensity	I	Ampere	A
Luminous intensity	J	Candela	cd

Dimensions and base units

Table 13.2 Derived electromagnetic quantities

Physical quantity	Units	Unit name	Symbol
Electric charge	A s	Coulomb	C
Voltage	kg m^2/A s^3	Volt	U
Resistance, reactance, impedance	kg m^2/A^2 s^3	Ohm	Ω
Active power, reactive power	kg m^2/s^3	Watts	P, Q
Magnetic field	kg m^2/A^2 s^2	Henry	H
Magnetic induction	kg/A s^2	Tesla	T
Electrical conductance	A^2 s^3/kg m^2	Siemens	S
Magnetic flux	kg m^2/A s^2	Weber	Wb

occupying different orbits, with a maximum of 8 electrons per orbit. The number of electrons in the last orbit defines much of the properties of every substance. If the number of electrons is closer to 8 the atom will be more stable and will not release electrons. A triboelectric series classifies materials based on their capacity to release or accept electrons. In this series, amber is more stable than hair, if both materials are placed in contact and rubbed each other to increase its temperature to transfer electrons faster, hair will transfer electrons to the amber.

Number of electrons in an atom: The number of electrons in an atom is equal to its atomic number. The number of electrons is also equal to the number of protons. One coulomb was defined as 6.25×10^{18} electrons, from this definition, one elementary charge is the reciprocal value, 1.6×10^{-19} C.

Anode/Cathode: There is a definition based on the electrochemical cell. An electrochemical cell is a device with two electrodes that can generate electrical energy from the chemical reactions occurring in it. The electrode which gives electrons is called the anode and the electrode who accepts electrons is called the cathode. Since the anode loses electrons becomes the positive terminal and the cathode the negative terminal. Electrons flow from the anode to the cathode and positive charges flow in the opposite direction from cathode to anode. This scheme strictly applies to DC circuits where the polarity is constant because the flow of electrons is unidirectional but in AC the polarity switches. The current increases smoothly to a peak, decreases back to zero, then increases again in the opposite direction to a similar peak before returning to zero. This cycle repeats continuously. In EAF practice the electrode is assumed to be the cathode and the anode is the solid scrap or liquid metal. In AC circuits the leads do not alternate, only the current.

AC/DC currents: AC means alternate current and DC direct current. In alternate current the flow of electrons goes back and forth. The flow of electrons is in two directions. The voltage and current is not constant, it goes through a maximum and minimum values in the form of a sinusoidal wave. The time to complete one cycle is called period. Frequency is the number of cycles per second (Hertz). Typical frequencies are 50 or 60 Hz. The peak value in either side of the wave is called amplitude. If the values of voltage and current pass through zero at the same time, it is called the signals are in phase, otherwise is called out of phase.

Due to variable values of voltage and current, there are several definitions;

(i) Peak values or maximum values: Maximum values with respect to zero, V_P or i_P.

(ii) Instantaneous values: the small letters v and i are used to represent instantaneous values.

$$v = V_P \text{sen}\, \theta$$

$$i = C \frac{dv}{dt}$$

(iii) Average values: Due to symmetry the average value in an entire cycle is zero. The average values are taken over the half positive cycle, as follows:

$$V_{avg} = \left(\frac{2}{\pi}\right) \times V_p \cong 0.637 V_p$$

$$I_{avg} = \left(\frac{2}{\pi}\right) \times I_p \cong 0.637 I_p$$

(iv) Effective value (rms value): *In AC circuits, any values given for current or voltage are assumed to be effective values unless otherwise specified.*

$$V_{rms} = \sqrt{0.5} \times V_p$$

$$I_{rms} = \sqrt{0.5} \times I_p$$

At the point where the flow of electrons changes direction their values of voltage and current are zero and the flow of electrons is stopped, however the speed at which the electrons move makes this effect to be undetected. As long as there is flow of electrons there is flow of electricity.

In direct current the flow of electrons is unidirectional and the values of voltage and current are constant. Back in the days when household were being supplied by electricity there was a strong debate over which one to use. Edison was a strong defender of DC while Tesla was a defender of AC. The Edison model would work only if the power plant was immediate to the final user but that is not the general case and has to be transported over long distances. In order to transport electricity over long distances it is needed a large voltage (electric pressure) if the current is low and the thickness of the conductor is small. Transporting DC would require high currents and that would involve huge heat losses.

Ohm's law: Ohm's law is a very valuable equation due to its simplicity to describe the main electric parameters; voltage, current and resistance. Current (I) is measured in amperes. The ampere is the amount of electric charge per second (C/t). Voltage (V) or electromotive force (emf) is the analog of a water pump in a hydraulic circuit.

It represents the pressure drop between two points. It was called electric pressure. In an electric circuit voltage is represented by 4 vertical lines. The cathode side has a shorter line. When voltage is zero there is no flow of electrons.

A conductor offers some resistance (R) to the flow of electric current, as a consequence part of the electric energy is transformed into thermal energy, known as Joule effect. The experimental work carried out by Ohm concluded that keeping the voltage constant and decreasing the electric resistance, increases the electric current. In symbols it can be expressed as follows:

$$I = \frac{V}{R}$$

This is one form of Ohm's law. One way to remember this useful equation is to consider the analogy of marbles running on a table. If the table is horizontal (V = 0), marbles will not move (I = 0) but as the table is more inclined (voltage drop increases) the marbles move faster. For any inclination in the table, if the number of obstacles increases also the motion of the marbles decreases.

The electric resistance depends on 4 variables; type of material, length of conductor, cross-sectional area and temperature. The larger the cross-sectional area, the lower the resistance, also with a longer length, higher resistance. Temperature and material properties are defined by the electric resistivity.

$$R = \frac{\rho l}{A}$$

where ρ is the electric resistivity in Ω m, l is the conductor's length in m and A is the cross-sectional area in m^2. As temperature increases, the electric resistivity increases because with a higher temperature there are more collision among the electrons.

Electric circuits: A basic electric circuit consists of a source of electrons, electric conductors connecting the voltage source with the load and the load itself. There are two types of electric circuits, in series and in parallel. In an electric circuit in series the electric current flows only in one way (same value at any point) and in parallel there are several paths. The total resistance and total voltage for an electric circuit in series is the sum of all resistances and voltages, respectively:

$$I_t(\text{serie}) = I_1 = I_2 = \cdots = I_i$$
$$R_t(\text{series}) = R_1 + R_2 + \cdots + R_i = \sum R_i$$
$$V_t(\text{serie}) = V_1 + V_2 + \cdots + V_i = \sum V_i$$

Kirchoff reported in 1847 two laws which are useful to do calculations in electric circuits. They state that, for a closed circuit, that the sum of voltages is cero and also the sum of electric current entering into a point is equal to the sum of current leaving at that point.

An AC circuit includes a ground line. Its voltage is zero and is needed for safety reasons to carry electrical current only under short circuit or other conditions that would be potentially dangerous. The short-circuit occurs when the electric arc is extinguished, for example when the electrode gets attached to the scrap. Under those conditions there is the largest flow of electric current. This condition should be prevented but is useful to measure the electric parameters under those conditions because the design parameters are based on those boundaries.

Skin effect: Skin effect is related to one feature of the alternate current; the electric current increases towards the surface of the electrode. This effect reduces the surface area for the flow of electrons. The skin effect resembles conducing electricity through a hollow electrode. According to Lupi [4] the induced eddy currents will increase the total volume of the current near the surface and weaken it in the central part of the conductor producing an uneven distribution of the current density in the cross-section. Its depth is related to the frequency, as follows:

$$d = k \times 66.1\sqrt{f}$$

where d is the depth of the skin effect, f is the frequency of AC in Hz, k is a constant related to the resistivity of the conductor.

The skin effect is illustrated in Fig. 13.1, after Bowman and Krüger [5]. It is observed that current density is higher for a frequency of 60 Hz in comparison with 50 Hz.

Herland et al. [6] investigated the effect of the skin effect using Sodeberg electrodes for three conditions; (i) single electrode, (ii) three electrodes including proximity effects between the electrodes and (iii) three electrodes including induced currents in the furnace steel shell. The depth of the skin effect, defined as the distance where the current density is reduced by the factor 1/e from the surface was computed using the following expression:

Fig. 13.1 Skin effect. After [5]

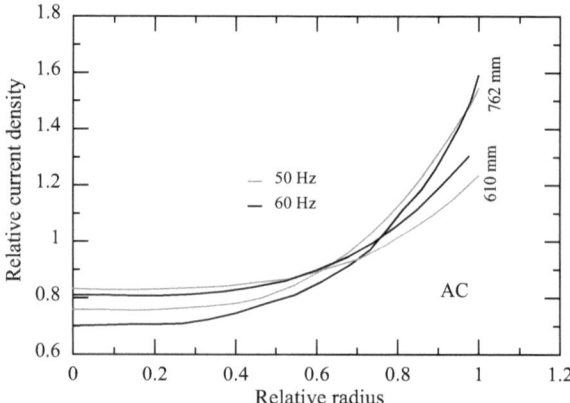

Fig. 13.2 Distortion of the
sinusoidal waveform due to
harmonics

$$\delta = \sqrt{\frac{1}{\pi f \mu \sigma}}$$

where δ is the depth of the skin effect, f is the frequency, μ is the magnetic permeability
and σ is the electric conductivity.

Middleton [7] investigated causes of failure analysis in electrodes and reported
that under tightening or overtightening may cause the current to flow through the
nipple itself.

Harmonics: The EAF is a non-linear load, which means the current and voltage are
not perfectly sinusoidal. The presence of non-linear loads disturbs the characteristics
of current and voltage waveforms from sinusoidal. The fundamental frequency of
the EAF can be 50 Hz but those non-linear loads generate voltage and current that
are the integer multiple of 50 Hz. Harmonics of third grade are 150 Hz. The higher
frequencies are called power system harmonics and degrade the power quality due to
lower power factor and voltage fluctuations. Figure 13.2 is an example of distortion
of the sinusoidal waveform due to harmonics.

Flicker: Flicker is a repetitive fluctuation in the voltage level. The sinusoidal waveform shows multiple spikes in a short period of time. It can be observed by the naked
eye by changes in the intensity of light emitted by lamps. There are many types of
voltage fluctuation, for example a single and momentary spike in the voltage is called
an impulse. There are several ways to quantify flicker. The following expression
define the flicker severity factor:

$$K_{st} = \frac{P_{st\,95\%}}{SCVD}$$

where K_{st} is the flicker severity factor, SCVD is the short circuit voltage depression
and $P_{st95\%}$ is the measure of short-term perception of flicker. K_{st} values range from
60 to 85.

Capacitors: A capacitor is formed by two conductive metal plates separated by an
insulator. Capacitors are used in power systems for voltage control, power-factor
correction, filtering, and reactive power compensation. Its misapplication can have
negative effects on the electric utility. A capacitor stores electric energy for a short
period of time, a property called capacitance (C). Energy is stored during the first

quarter cycle and returned in the second quarter. The final effect of a bank of capacitors is to decrease reactive power and increase the power factor.

Capacitance is measured in Farad. One farad represents the capacitance which stores the charge of one coulomb across a potential difference of one volt. A capacitor "absorbs" voltage but at the same time *has no influence on the current* [8]. A capacitor adds capacitive reactance, X_C, defined as follows:

$$X_C = \frac{1}{2\pi f C}$$

Inductors: A wire wound into a coil is called an inductor. The principle of electromagnetic induction discovered by Faraday indicates that a moving magnet inside a coil induces a voltage which is proportional to the rate of change of the magnetic field. The induced voltage is called inductance (L), measured in Henrys. One Henry (H) is the inductance when an electric current that is changing at one ampere per second results in an electromotive force of one volt across the inductor. Similar to the capacitor, an inductor absorbs current during the first half period of the sine wave and stored in its magnetic field but does not change the voltage. The inductive reactance (X_L) is defined as follows:

$$X_L = 2\pi f L$$

Reactance (X): Reactance is also a resistance. This term is used to differentiate resistance derived from capacitors and inductors. Resistance (R) to the flow of electrons is due to friction, reactance on the other hand is an inertial force to the movement of the electrons. When electrons flow through an inductor it is created an electromagnetic field which induces a voltage but of opposite sign to the voltage of the primary side in the transformer, as a consequence its final effect is a decrease in the voltage of the primary side. Inductors oppose a current change and capacitors oppose a voltage change.

Since the capacitor counteracts the induced reactance, the total reactance (X) is defined as follows:

$$X = X_L - X_C$$

Impedance (Z): The total resistance in an electric circuit which include capacitors and inductors is called impedance.

$$Z = R^2 + X^2 = R^2 + \left(X_L^2 - X_C^2\right)$$

Ohm's law for AC circuits becomes:

$$I = \frac{V}{Z}$$

Conversion from electric energy to thermal energy: The rate of production or consumption of electric energy to carry out electric work is known as electric power. In the SI system the unit of energy is Joule (J) and the rate of change of energy J/s is called Watt. Watt multiplied by time gives unites of energy, such as kWh.

Since voltage is the work required to move an electric charge (W/q) and current the rate of change of electric charge per unit time (q/t), electric power can be defined as the product of voltage times current, as follows:

$$P = \frac{\text{work}}{\text{time}} = \frac{W}{t} = \frac{W}{q}\frac{q}{t} = V \cdot I$$

From Ohm's law; $V = I \cdot R$ or $I = V/R$

$$P = V \cdot I = (I \cdot R) \cdot R = I^2 R$$

$$P = V \cdot I = V \cdot \frac{V}{R} = V^2 R$$

The previous equations were developed by Joule from 1840 to 1843 and are known are Joule's first law, Joule effect or Ohmic heating, is the process by which a conductor produces heat due to the passage of an electric current.

Electric power: There are two types of resistances and each one provides a totally different electric power. In a resistive circuit, voltage is consumed in overcoming the ohmic resistance of the circuit itself and then dissipates heat. This resistance defines the useful or active power (P), the only one which will contribute to actually melt the metallic charge in the EAF. Reactance on the other hand creates reactive power (Q). Reactive power can be seen as a waste since no reactive power at all would mean perfect power quality [8], why is then the reason to have reactance in an AC circuit? First, to visualize reactive power with an example consider beer and foam on the top. The liquid beer is the active power and foam the reactive power. Dutta [9] put it in this way: "…suppose you are said to fill a water tank with a bucket… the bucket filled with water is the real power… you carry the real power… but after a certain time you need to drink some water to do work… this energy drink do not contribute to fill the water tank but it gives you energy to do work… if you are tired due to lack of energy you cannot carry the bucket… you have a bucket full of water but you cannot transfer it to the tank… this is why reactive power is important in electrical engineering". Reactance is very important in AC circuits and plays a decisive role to improve the quality of electric power because when voltage and current reach zero the arc becomes extinguished and is the stored electric energy which contributes to restore the electric arc and provides a more smooth operation.

Capacitors store energy in an electric field and inductors also store energy but in a magnetic field [8], therefore, active power is the power consumed by resistors and reactive power is the power stored in the reactive components.

The following scheme illustrates the power triangle.

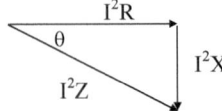

From this triangle the following relationships can be derived:

$$P = P_{ac}(MW) = I^2R = S \cos \theta$$

$$Q = P_r(MVAR) = I^2X = S \sin \theta$$

$$S = P_{app}(MVAR) = I^2Z = \sqrt{P_{ac}^2 + P_r^2}$$

$$MVA^2 = MW^2 + MVAR^2$$

where P is the active power, Q is the reactive power and S represents the apparent electric power.

The instantaneous active power is the product of the instantaneous voltage times the current over a period of time, as follows:

$$P = \frac{1}{T} \int_0^t vi \, dt$$

where T is the period (1/f).

Power factor: Power factor is an important variable because defines electric efficiency. It is defined as a ratio between the active electric power and the total or apparent power.

$$PF = \frac{P_{ac}}{P_{app}} = \frac{S \cos \theta}{S} = \cos \theta$$

$$\cos \theta = \frac{P_{ac}}{P_{app}} = \frac{I^2R}{I^2Z} = \frac{R}{Z} = \left[\frac{MW}{MVA} \right]$$

Then,

$$P_{ac} = P_{app} \times \cos \theta$$

$$P_{ac} = I^2Z \times \cos \theta$$

$$P_{ac} = VI \times \cos \theta$$

From the definition of the power factor, to increase the active power, the power factor should also be increased and the reactive power decreased. Increasing the power factor has its own limits because of arc instability, as will be discussed in a following section.

For a three phase AC series circuit, electric power is defined as follows:

$$P_{el} = I_1^2 R_{eq1} + I_2^2 R_{eq2} + I_3^2 R_{eq3}$$

$$P_a = \sum_{p=1}^{3} \frac{1}{T} \int_0^t v_{po} i_p \, dt$$

$$P_{arc} = P_a - P_{el}$$

where P_{el} are the power circuit losses, P_a is the active power and P_{arc} is the arc power, v_{po} is the instantaneous phase (p) to ground voltage, i_p is the instantaneous current per phase, I_i and is the rms voltage and per phase, R_{eqi} is the equivalent resistance per phase.

Transformer: A transformer is an equipment that transforms the voltage-current properties of alternate current. It basically consists of two coils close to each other, current passes through one of them in order to induce voltage on the other coil. The first coil is called primary coil or primary side and the second, secondary coil or secondary side. The voltage in the secondary side is controlled by the number of coils, as follows:

$$\frac{V_p}{V_s} = \frac{N_p}{N_s}$$

where V is voltage, N is the number of coils, p is primary and s is secondary. Controlling the number of coils in each side is possible to increase or decrease the voltage induced on the secondary side. A step-up transformer increases the voltage on the secondary side and is employed at the power plants to distribute electric energy over long distances. At the steel plant step down transformers decrease the voltage.

The electric power on both the primary and secondary side is the same because the electromagnetic field is the same. This condition allows to define the current on the secondary side, as follows:

$$V_p I_p = V_s I_s$$

$$I_s = I_p \frac{N_p}{N_s}$$

One of the main roles of the EAF is to melt as fast as possible. To achieve this, it is necessary to have transformers with higher capacity. An EAF transformer is one of the main components of the power system, exposed to critical conditions due to high secondary currents, heavy current fluctuations, harmonics, short circuits, mechanical stresses, frequent overloading conditions, vibrations, pollution and dust.

Calculation of transformer capacity: Its capacity is mainly defined on the amount steel to produce per year per furnace, the nominal capacity of the EAF and the energy consumed to produce one ton of steel. The following example illustrates a case using 100% DRI with an specific electric energy consumption of 650 kWh/ton.

- Assuming a steel shop capacity of 1 M ton/year, the working hours results from working 20 h during 331 days plus 24 h during 52 days, then the expected productivity is:

$$\text{Productivity} = 1{,}000{,}000 \ \frac{\text{ton}}{\text{year}} \times \frac{\text{year}}{(313 \times 20 + 52 \times 24) \ \text{h}} = 133.19 \ \text{ton/h}$$

- If the EAF has a nominal capacity of 220 ton/heat, the maximum tap-to-tap should be:

$$\text{tap-to-tap time} = \frac{200 \ \text{ton/heat}}{133.19 \ \text{ton/h}} = 1.5 \ \text{h} = 90 \ \text{min}$$

- Considering 15 min of delays per heat, *the power-on time* $= 90 - 15 = 80$ min/ heat $= 1.33$ h/heat
- Electric energy required per heat

$$\frac{\text{kWh}}{\text{heat}} = 650 \ \frac{\text{kWh}}{\text{ton}} \times 200 \ \frac{\text{ton}}{\text{heat}} = 130 \ \frac{\text{MWh}}{\text{heat}}$$

- Transformer capacity with a power factor of 1.0

$$\text{MW} = \frac{130 \ \text{MWh/heat}}{1.33 \ \text{h/heat}} \cong 98$$

- Transformer capacity with a power factor of 0.8

$$\text{transformer capacity} = \frac{98}{0.8} \cong 130 \ \text{MVA}$$

The capacity of transformers has grown from a fraction of MVA/ton in the early twentieth century to values higher than one for more recent EAF's, as shown in Table 13.3.

The term Ultra High Power-EAF (UHP-EAF) was applied to transformers with capacities above 0.6 MVA/ton in 1963 [10]. In 1980 the rated capacity of the transformer was divided into three categories; ordinary power furnace, high power furnace and ultra-high-power furnace. Higher capacities have been called Super UHP (SUHP). Modern transformers have capacities at least 1 MVA/ton in order to provide

Table 13.3 Evolution of EAF transformer capacity

Year	1906	1937	1963	1980	1998	2010
EAF, tons	1.5	30	150	100	100	250
Transformer, MVA	0.4	10	80	66	140	300
MVA/ton	0.2	0.3	0.5	0.6	1.4	1.2

an accelerated melting rate of the metallic charge and a shorter power-on time. This condition has a positive effect on energy consumption.

Looking back in time, in 1933, the installation of one of the largest EAF's in the world with a capacity of 100 ton was designed with two sets of three electrodes, each one powered by a transformer of 7.5 MVA, and an elliptical shell. It appears the reason for six electrodes was concerns on the limits of electric power that each electrode was able to transport [11]. It was speculated a limit of 3 MVA for each electrode.

Worrell et al. [12] indicated that transformer losses can be as high as 7% of the electrical inputs based on a report from the center for materials production from 1992. They estimated that increasing the transformer capacity to UHP the electric losses could be decreased 4%, suggesting energy savings in the order of 17 kWh/ ton.

Increasing the transformer capacity has a direct impact only on the power-on time, however if the power off-time is high the final effect on the tap-to-tap time can be drastically affected. It has been suggested that power-off times higher than 30% make UHP furnaces economically unsuitable [13].

The transformer is the most expensive part of the entire EAF, about 40% [14].

13.2 Components of the EAF Electric Power System

Electric energy is produced at power plants at about 11 kV. This voltage is raised to about 130 or 230 kV for transmission purposes. At the steel plant electric energy arrives at an electric sub-station and from there is delivered to different parts of the plant. The EAF is an electric circuit in series divided in two parts, the primary circuit from the substation to the primary circuit of the EAF transformer and the secondary circuit from the secondary circuit of the EAF transformer to the electric arc, as shown in Fig. 13.3.

13.2.1 Primary Circuit

The primary circuit involves the following components:

- Electric sub-station: In Mexico the electric current is supplied at 230 kV. The point of arrival is called Point of Common Coupling (PCC). Step-down transformers decrease the voltage to 69 kV.
- Circuit breaker: Circuit breakers operate under vacuum in the range from 15 to 115 kV. In north America the original supplier was Joslyn, in the west are commonly known by this name.
- Reactance devices: Reactors serially connected to control arc stability.

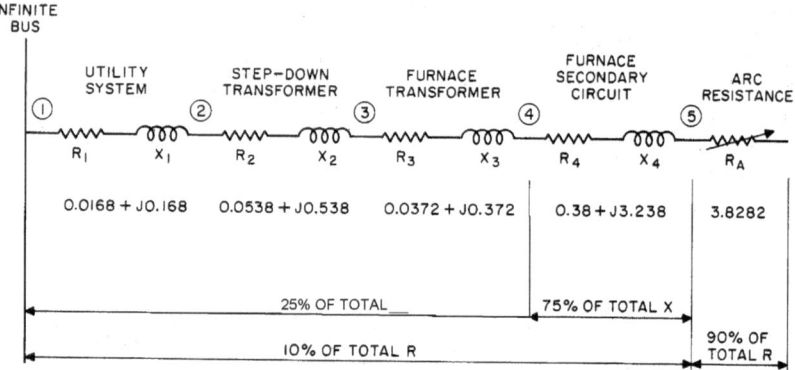

Fig. 13.3 EAF electric circuit

- Primary circuit of the EAF transformer: This circuit operates at high voltages and
  low currents, about 1–3 kA and 34.5 kV, respectively. The tap changer is on the
  primary circuit and its function is to control current and voltage on the secondary
  circuit by selecting the number of coils on the primary side. The tap changer
  works under load. The induced voltage and current on the secondary circuit are
  variable, depending on the melting stage at the EAF, usually in the range from
  100 to 1200 V and 20–80 kA.

13.2.2 Secondary Circuit

The secondary circuit involves the following components:

- Bus bars (delta closure): The secondary coil terminations from the furnace transformer carry out a huge electric current and require cooling. Cooling is carried
  out by forced oil circulation. Due to risks of fires, the transformer is enclosed in
  a separate room.
- Flexible cables: Flexible cables allow the movement of the current conducting
  arms. They are water cooled. There are several cables per phase.
- Current conducting arms: Arms can be made of steel coated with copper or
  aluminum and carry two functions, conduct electricity and support the electrodes.
  Aluminum is lighter and helps in the speed to move the electrodes but its corrosion
  is higher.
- Electrode holders: Finally, the electric current is transferred to the electrode using
  one or two clamps that hold each electrode.
- Electrodes: Electrodes are made of graphite and its purpose is to transfer electric
  current to melt the metallic charge.
- Electric arc.
- Metallic charge (load).

13.3 DC-EAF

The steel plant is supplied with AC. The conversion to DC is carried by rectification using thyristors. Due to differences in the characteristics of DC, the transformers are different to AC. The desired voltage in the secondary circuit is controlled by the amount of current. As the current increases there is a voltage drop, about 100 V per 100 kA [15]. Because thyristors handle a few kA, a bank of thyristors is required for each DC-EAF.

To make the electrical connection with DC, a positive metallic anode is placed at the bottom, within the rammed refractory. There are three types involving pins, sheets and billets. The top of the steel conductor melts during the heat but resolidifies during power-of time once scrap is charged.

Bowman [16] in 1993 compared both AC and DC furnaces and concluded that DC furnaces have lower flicker but both also need additional reactance. Electric energy consumption is similar. The main difference was a lower electrode consumption for DC, 1.2 kg/ton, in comparison with 1.6 kg/ton for the AC-EAF because employs only one electrode and electric operation is more stable. At the time Bowman presented his results the practice of slag foaming was not fully developed. Later on, with improvements on slag foaming it was possible to achieve similar values and is one of the reasons why the AC-EAF is still the dominant version.

Madias et al. [17] reported a survey of 190 EAF's in 2017, from this work they reported a relationship between EAF capacity and electrode diameter, as shown in Fig. 13.4. It is clear that for the same capacity a DC-EAF require a larger electrode diameter. The same report also compared energy consumption of both types of EAF, as shown in Fig. 13.5. In this figure it can be observed that there is not a clear difference in the specific electrode consumption, although DC-EAF are in the lower range.

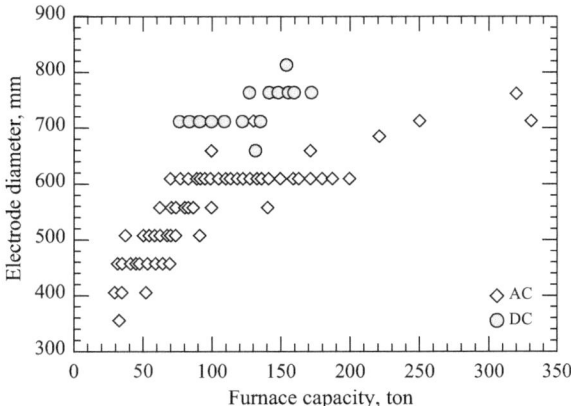

Fig. 13.4 Relationship between EAF capacity and electrode diameter for AC-EAF and DC-EAF. After [17]

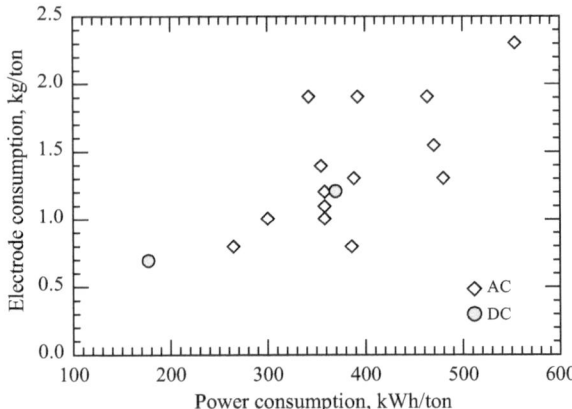

Fig. 13.5 Relationship between energy consumption and electrode diameter for AC-EAF and DC-EAF. After [17]

Jones et al. [18] reviewed the advantages and disadvantages of DC furnaces and points out that DC furnaces are well proven for smelting but not necessarily under the oxidizing conditions prevailing in steelmaking.

13.4   Electrodes

There are many materials with good electric current conducting capacity but graphite is superior in many ways [19], as shown in Table 13.4.

13.4.1   Production of Electrodes

H. Davy in 1810 was the first engineer to employ carbon to conduct electricity using an electrode made of charcoal. W. Lawrence founded the company National Carbon in 1886, which was famous for the production of the dry cell battery ever-ready, later merged as Union Carbide (UCAR) in 1917. Pritchard [20] was the first engineer describing in detail the production of graphite in 1890 who indicated that

Table 13.4 Properties of graphite and copper

	Graphite	Copper
Discharge speed	2–3 times faster than Cu	–
Softening point, °C	Evaporates at 3650	1000
Thermal expansion	1/30 Cu	
Density	1/5 Cu	
Price	30–60% lower Cu	Higher

Table 13.5 Evolution in diameter of electrodes

	1914	1924	1930	1937	1910	1980
D (mm)	356	406	457	508		813

one of the main properties of a good electrode was its electric conductivity which in turn depends on the firing process. The dominant technologies for the production of graphite electrodes were developed from 1893 to 1895. In 1893 H. Castner reported in the UK the Length Wise Graphitization (LWG) process where electric current passes longitudinally through the electrodes [21]. His patent was granted in the USA in 1896. E. G. Acheson invented a graphitization process in 1896 where the electrodes are aligned perpendicular to the pass of the current, called cross-wise graphitization of pre-baked carbon bodies. The materials employed were petroleum coke or pitch coke with coal-tar pitch as binder. The Acheson process is inferior because consumes more energy and takes longer time to complete graphitization, however it became more popular in the early years of its development, producing small size electrodes. Heroult in 1912 also invented a process similar to Castner due to the problems faced with the Acheson process to produce large electrodes.

Graphitization refers to the transformation by heat treatment of a disordered structure to a well-ordered structure. The heat treatment temperature is in the order of 2600 °C. Higher heating rates produce a more ordered structure. The Castner process allows higher heating and cooling rates in comparison with the Acheson process, therefore yields higher quality and shorter production times [22].

The electric current capacity in kA or the electrode current density in A/cm^2, both increase with the quality of the electrodes; high electric conductivity, low resistivity, low thermal expansion coefficient, good thermal shock resistance, density, high mechanical strength, optimum graphite shape and size, etc. however, also its price increases. Another way to increase that capacity is by increasing the electrode diameter. Table 13.5 summarizes the evolution in diameter of electrodes.

Figure 13.6 shows the relationship between electrode diameter and current conducting capacity. The electric current capacity increases from 50 to 100 kA by increasing the diameter from 500 to 700 mm, respectively. The current carrying capacity of DC electrodes is higher than AC electrodes because with DC the whole area of the electrode is available for current flow and current can be increased with overheating the electrode in comparison with AC.

13.4.2 Temperature Distribution

The temperature of the electrode increases due to the Joule effect and also due to heating from the hot gases inside the EAF, consequently there is some thermal expansion. The temperature gradient between the center and outer surface creates thermal stresses which induce the formation of cracks. In addition to thermal stresses

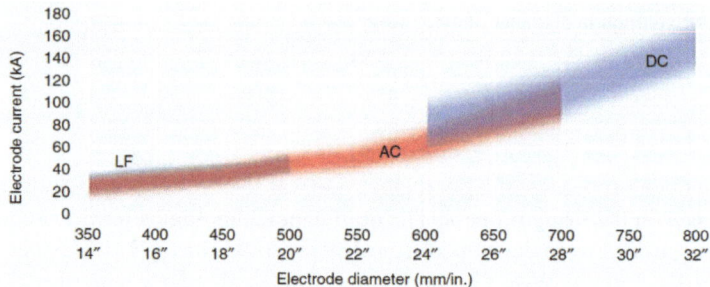

Fig. 13.6 Dependence of EAF current and required electrode diameter. After [23]

there is also vibration which results from the intense electromagnetic forces at high currents. Figure 13.7 shows the formation of cracks on the electrode and a typical temperature distribution.

Water added by a spray ring installed at the bottom of the electrode holder should evaporate once it enters the furnace. A typical water flow rate for a 600 mm electrode is from 7 to 15 l/min per electrode [15]. The water flow rate for electrode cooling should be optimized because an excess of water increases energy consumption due to formation of water vapor. Bowman et al. suggests a flow rate of 7–15 l/min per electrode of 600 mm.

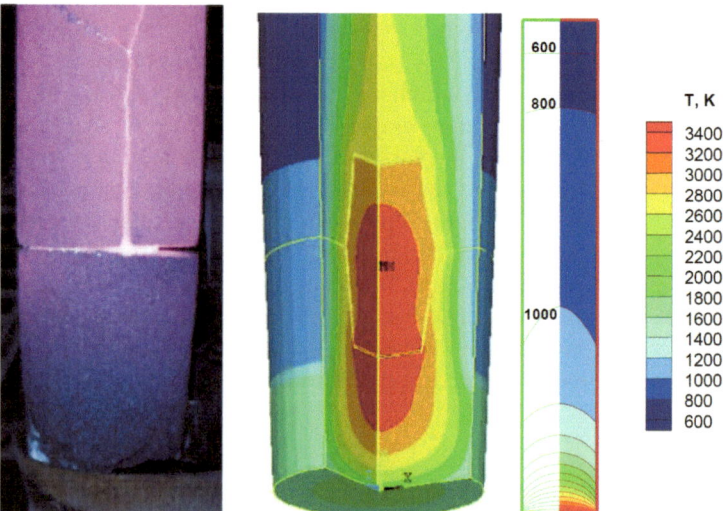

Fig. 13.7 Left: Graphite electrode with crack. After [23]. Right: Numerical simulation of temperature distribution. After [24]

13.4.3 Consumption Mechanisms and Methods to Enhance Its Life

Electrodes are quite expensive due the large amount of energy and time for graphitization and for this reason its specific consumption in kg/ton of liquid steel is an important process variable.

Figure 13.8 summarizes the major developments to improve the electrode quality starting with the use of needle coke in the 1960s [22] and the effect of improvements in both electrode quality and metallurgical practices. The specific consumption has decreased 7 times from the 1960s with more than 7 kg/ton to 2020 with an specific consumption below 1 kg/ton.

Electromagnetic forces induce a rotation movement. The direction of this rotation in AC furnaces is counterclockwise and help to tighten the joints and in the case of DC furnaces the direction is random, on both directions [15].

13.5 Physics of the Electric Arc

Bowman and Krüger [5] in his book "Arc Furnace Physics" summarize the fundamental aspects of electric arcs. They state that gas is drawn into the arc around the cathode and due to the extremely large temperatures in the electric arc, from 10,000 to 15,000 °C in the upper central part of the arc and about 2000 °C at the boundaries, a fraction of the gas atoms become ionized forming a plasma, electrons and ions are pumped by magnetic forces at high speed. The average velocity of the plasma

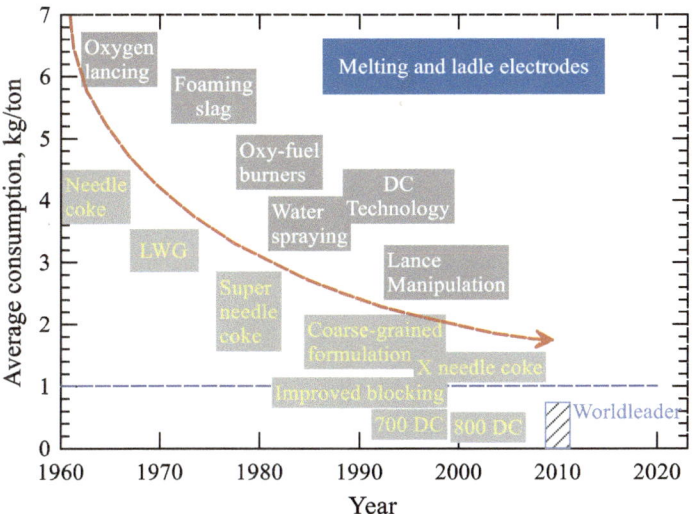

Fig. 13.8 Developments of the specific consumption of graphite electrodes [22]

is about 3500 m/s. The arc column expands away from the cathode. Ions recombine to atoms at lower temperatures. The arc plasma is electrically neutral because the amount of electrons is equal to amount of ions, electrons due to their lower mass have a higher mobility. The diameter of the electric arc is about 100–200 mm, lower than the diameter of the electrode [15]. Due to the high temperature the density of the plasma is low and can be easily displaced, modifying the arc length and then the arc voltage.

Similar to an oxygen jet, the electric arc due to the large electromagnetic forces also forms a depression on the molten bath. This depression depends on the arc current. The electric current in electric arcs vary from less than 0.1 kA for welding operations to more than 100 kA in electric and submerged arc furnaces. Bowman and Krüger [5] reported an arc depression from 8.2 to 33.2 cm by increasing the current from 40 to 80 kA, using an arc length of 400 mm. The vertical angle of electric arcs with AC is in the order of 25°, oscillates horizontally ± 40° [5]. This inclination angle promotes splashing.

A plasma having low density is equivalent with a low mass and then a low thermal capacity. This characteristic has several implications, on one hand when the arc gets momentarily extinguished it cools down fast and a second aspect it that is subject to deflection due to interaction with the other electrodes. Such deflection is observed in Fig. 13.9, as reported by Jordan in 1976 [5].

The electric arc causes ionization of liquid iron. Iron ions increase thermal radiation. In addition to this, immersing the electric arc in a foaming slag would change the composition of the plasma. Bowman and Krüger [5] simulated a case with 50% CO and 50% slag at 1600 °C and found ionization of Si and Ca ions in the order of 5% which would increase the electric conductivity.

The electromagnetic forces have an important contribution to the stirring conditions in the EAF, especially for DC furnaces. This aspect is reviewed in detail in Chap. 12.

Fig. 13.9 Deflection of the electric arc and heat exchange. After [5]

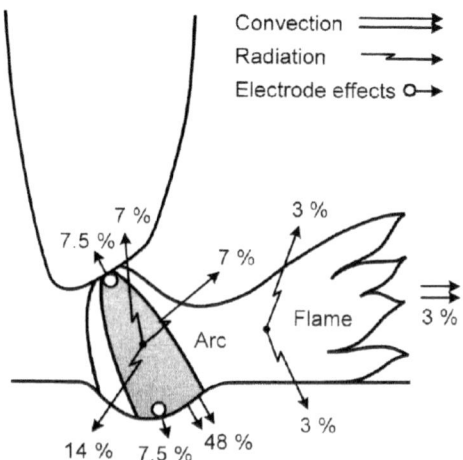

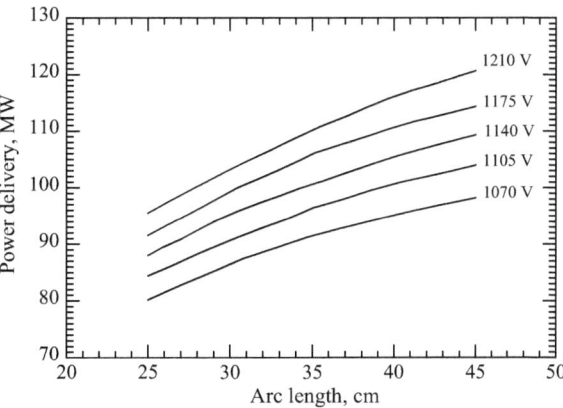

Fig. 13.10 Predicted electric power supply as a function of arc length and arc voltage. After [28]

13.5.1 Channel Arc Model

An arc model is extremely important to predict the instantaneous electric power supply. There are two types of models; Magnetohydrodynamic model (MHD) and the Channel Arc Model (CAM). MHD is more accurate but also more complex, on the contrary CAM is simple and essentially provides similar results to MHD. CAM assumes the electric arc is a cylinder, the inner part with a radius r has a constant temperature which depends on the energy input to the arc and energy output from the arc. The assumption of a constant temperature, independent of the radial and axial position, indicates an infinite thermal conductivity. The first modeling for AC electric arcs was reported by Larsen [25] and Bakken et al. [26, 27]. Sanchez et al. [28] described in detail the CAM, their results indicate heat transfer from the arc is primarily due to convection (72%), followed by radiation (24%) and then electron flow (3.5%). Figure 13.10 reports the predicted electric power supply as a function of arc length and arc voltage, which has a fairly good agreement with the actual values. The figure clearly illustrates that a long arc operation increases the active power. A short arc has low resistivity and supplies less energy in comparison with a long arc.

Fathi et al. [29] also reproduced the channel arc model and predicted heat transfer from the arc. They reported radiation as the primary heat transfer mechanism followed by convection and finally electron flow, with the following average values; 80%, 15% and 5%, respectively.

13.5.2 Arc Length

During the initial period of scrap melting or boredown, the boredown diameter increases as the arc length increases which is beneficial to increase the melting rate, as follows [5]:

$$D_b \cong D_t + 2L \sin(\theta/2)$$

where D_b is the boredown diameter, D_t is the diameter at the tip of the electrode, θ is the cone angle and L is the arc length.

The relationship between arc voltage and arc length has been investigated since 1902. Ayrton [30] reported an inverse relationship between arc voltage and arc current as well as a linear relationship between arc voltage and arc length, valid for arc lengths below 10 mm.

$$V_{arc} = V_{AC} + V_g l_{arc}$$

where V_{arc} is the arc voltage, V_{AC} is the voltage drop between the anode and cathode, l_{arc} is the arc length in mm and V_g is the arc coefficient or voltage gradient in V/mm.

Martell-Chávez et al. [31] assumes each term on the right side has a different contribution. The term that contributes to power dissipation is $V_g l_{arc}$ and the other term, V_{AC}, is independent of current and remains constant, contributes to the generation of electrons.

From the previous expression the arc length can be expressed as follows:

$$l_{arc} = \frac{V_{arc} - V_{AC}}{V_g} = \frac{V_{arc} - 40}{10}$$

where l_{arc} is the arc length in cm, V_{arc} is the arc voltage, V_{AC} is the voltage drop between the anode and cathode and V_g is the voltage gradient in the arc column in V/cm.

Pauna et al. [32] reviewed values reported for V_{AC}. The most common value for EAF conditions is 40 V but values from 30 to 80 have also been explored.

From experiments it has been found that V_g increases between 5 and 10 V per cm of arc length. The value of 10 V/cm has been suggested even for high currents [5]. Miller in 1982 reported a value of 11.5 V/cm and more recently, 14.49 V/cm by Pauna et al. [32].

Arc length can also be computed using arc voltage and arc current values. Warrington in 1931 proposed the following relationship [33]:

$$l_{arc} = \frac{V_{arc} I_{arc}^{0.4}}{8750} = \frac{V_{arc} \sqrt[2.5]{I_{arc}}}{8750}$$

where all units involve ft.

In the previous expression arc length is proportional with the arc current. Dittmer and Krüger, taken from ref [31], reported a relationship where arc length is inversely proportional with the arc current, and another relationship to estimate radiation from the arc:

$$l_{arc} \approx \frac{V_{arc}}{\sqrt{I_{arc}}}$$

$$P_{\text{rad}} \approx \frac{V_{\text{arc}} - 80}{\sqrt[8]{I_{\text{arc}}}}$$

Zhao et al. [34] derived the following expression to describe arc length:

$$l_{\text{arc}} = \frac{V_{\text{arc}} - V_{\text{AC}}}{V_g}$$

$$V_{\text{AC}}^2 = (V_{\text{arc}} + I_{\text{arc}}R)^2 + (I_{\text{arc}}X)^2$$

From the previous expression:

$$V_{\text{arc}} = \sqrt{V_{\text{AC}}^2 - (I_{\text{arc}}X)^2} - I_{\text{arc}}R$$

$$l_{\text{arc}} = \frac{1}{V_g}\left[\sqrt{V_{\text{AC}}^2 - (I_{\text{arc}}X)^2} - I_{\text{arc}}R - V_{\text{AC}}\right]$$

The following relationship has been employed in the industry (from unknown sources):

$$l_{\text{arc}} = \frac{1}{V_g}\left(\frac{1000\,P_a}{2I_{\text{sec}}} - I_{\text{sec}}R_{\text{sc}} - V_{\text{AC}}\right)$$

where P_a is the active power in MW, I_{sec} is the current in the secondary circuit in kA, R_{sc} is the short-circuit resistance in mΩ. Suggested values; $V_g = 10$ V/cm, $V_{\text{AC}} = 35$ V, $R_{\text{sc}} = 0.4$ mΩ. With this equation and using $P_a = 140$ MW, $I_{\text{sec}} = 86$ kA it is obtained an arc length of 47 cm.

Another expression from unknown sources:

$$l_{\text{arc}} = \frac{1}{V_g}\left(V_{\text{po}} - I_{\text{po}}R_{\text{tip}} - V_{\text{AC}}\right)$$

where V_{po} and I_{po} are the voltage and current ground to neutral, respectively. R_{tip} is the resistance of the electrode when the tip just contacts the hot bath surface.

UCAR has suggested the following expression to compute arc length:

$$l_{\text{arc}} = \frac{\cos\theta}{\cos 45°}\frac{0.85 \times P_{\text{arc}}}{36 \times I_{\text{p0}}} - 2.5 = \frac{\cos\theta}{\cos 45°}\frac{0.85 \times 3 \cdot V_{\text{arc}}I_{\text{arc}}}{36 \times I_{\text{p0}}} - 2.5$$

The effect of electric variables on arc length from the previous expressions can be summarized as follows:

- Arc length increases by increasing: (i) the secondary voltage (arc voltage increases), (ii) power factor, (iii) arc power.
- Arc length increases by decreasing: (i) voltage drop between the anode and cathode (ii) the voltage gradient in the arc column, (iii) arc current.

Since the active power increases by increasing the power factor it is important to work with a long arc operation. Resistance increases with a longer arc length decreasing the flow of current, beneficial for electrode consumption.

It is easy to visualize which arc length predominates by visual inspection of the electrode tip. In a long arc operation, the arc is deflected with an inclination angle from 20 to 30°. The electrode tip forms an angle of 45°. With a short arc operation, the tip of the electrode is more uniform at the tip.

13.6   Optimization of Electric Power Quality

13.6.1   Early Concepts on the Optimization of the Secondary Circuit

The ideal operation mode of an EAF from an electric power perspective keeps changing through time, this is because their contexts also change. An example is the initial changes to improve power quality when the UHP emerged in the 1960s. At that time the slag foaming practice was not developed and EAF's had to be operated under short arc length conditions. These conditions automatically involve lower voltages and lower power factors. The following is a detailed description given by Ciotti [35] in 1971 about the implementation of a electric balanced system using a new secondary circuit design: Before the UHP, the design was a co-planar or side-by-side arrangements of conductors, then replaced by an equilateral triangulation of those conductors. The co-planar array resulted in an unbalanced network that produced an apparent resistive component which impedes the flow of current in the leading phase. With the equilateral triangulation, the result was a lower impedance secondary circuit, as shown in Table 13.6. These data were obtained for an EAF of 150-ton, 6.7 m diameter, 400 V phase-to-phase, power factor 0.707 and AC current 60 Hz. As can be observed the triangular array yields a higher arc power, confirming that the electric power can be increased by decreasing the total circuit impedance. This author mentions that before the 1960s the power factor was high, from 0.8 to 0.85, which gave a higher arc length indicating that after the 1960s the practice was changed for lower power factors, from 0.65 to 0.7.

13.6.2   Refractory Wear Index (RWI) or Radiation Index

The refractory wear index is an important parameter that measures the radiation intensity from the electric arc. It was proposed by Schwabe in 1962 [36] to describe the radiation intensity to the refractory walls.

Table 13.6 Resulting electric power employing co-planar and triangulated arrangement of conductors

	Phase A, × 10^{-3} Ω	Phase B, × 10^{-3} Ω	Phase C, × 10^{-3} Ω	Power input, MW	I^2R losses. MW	Arc power, MW
Co-planar	0.81 + j2.92	0.200 + j2.34	− 0.070 + j2.70	30.27	3.49	26.78
Triangulated	0.39 + j2.53	0.35 + j2.35	0.26 + j2.56	32.05	4.35	27.70

After [35]

$$RWI = \frac{V_{arc}^2 \cdot I_{arc}}{d^2}$$

Hassan et al. [37] provided an alternate expression:

$$RWI(kWV/cm^2) = \frac{(arc\ power\ in\ MW) \times (arc\ voltage) \times 1000}{3d^2}$$

where V_{arc} and I_{arc} are the arc voltage and arc current, respectively, d is the smallest distance from the center of the electrode to the furnace walls, in cm. The arc voltage is equal to the arc length multiplied by the voltage drop (12 V/cm). According with Schwabe's expression increasing both the arc voltage and current voltage increases the radiation index and decreases by increasing the distance d. Due to the power function, voltage has the largest influence on the radiation index.

Gottardi [38] has reported an increment of the radiation index from 130 to 250 kWV/cm^2 due to enhanced chemical energy using oxy-fuel burners. Jansen et al. [39] reported a value of 342 kWV/cm^2 in an EAF with a nominal capacity of 80 ton and a transformer capacity of 100 MVA. Hassan et al. [37] reported a value of 166 KWV/cm^2 using a short arc length.

13.6.3 Power Quality

During normal operation, in particular during scrap-melting operations, the electric arc is subject to a large frequency of fluctuations in its length. Once the arc length changes, it changes the arc resistance and then the arc voltage and arc current. These changes affect the smooth sinusoidal waveform of AC and any distortion to that waveform is related to poor electric power quality.

The EAF has severe non-linear fluctuations in both voltage and current over time which affect the electric power quality, decreasing the electric efficiency then decreasing the arc power due to a lower active power. The final effect is a lower melting rate which increases the power-on time and energy consumption. The electric power quality is affected by the presence of the following disturbances:

- Short-circuit condition
- Low power factor
- Harmonics
- Flicker
- Unbalance between phases
- Arc instability.

To counteract these problems there are several solutions, such as:

- Harmonic filters
- Synchronous condenser (SC)
- Capacitor bank
- Static Var Compensators (SVC)
- Static Synchronous Compensator (STATCOM)
- Load balancing.

Harmonic filters are designed to absorb specific harmonic currents generated by non-linear loads (passive filters). Active filters are tuned to account for changes in the system impedance and loads.

The synchronous condenser (SC) is one of the earliest devices employed for reactive power compensation under no-load condition. Gives or absorbs reactive power.

The capacitor bank is the most common device employed to correct electric disturbances. One of the benefits is that increases the power factor. If the power factor is less than unity the current flow should be increased for a given load. Many utility companies impose financial penalties on large industrial consumers when the power factor is lower than a certain value, 0.90 or 0.95.

Static Var Compensators (SVC) are a parallel controlled compensation system based on thyristors, first developed in 1967 and later developed into two categories, using a Thyristor Controlled Reactor (TCR) and a Thyristor Switched Capacitor (TSC). SVC's reduces voltage fluctuations.

The Static Var Generator (SVG), more commonly known as Static Synchronous Compensator (*STATCOM*) was developed by EPRI and Westinghouse in 1986.

Čerňan et al. [40] argues that Static Var Compensators (SVC) are more common than STATCOM but in spite that is more effective to control flicker involves a higher investment cost. The advantages of SVC, in addition to lower investment costs are easier maintenance, capacity to add reactance in the order of hundreds of MVAr's and longer service life.

Qiao et al. [41] compared advantages and disadvantages of SC, Capacitors, SVC and SVG. They found the cheapest option was the bank of capacitors with additional advantages such as easy installation, lower investment, easier maintenance, however the response speed is low and also, if harmonics exist in the power system they are amplified. SVG has a faster response and doesn't generate harmonics but is more expensive. Due to the non-linearity of the bigger power systems, there are still engineering problems to be solved.

Deaconu et al. [42] made a techno-economical comparison of three reactive energy compensation methods; passive filter, SVC and STATCOM. They simulated conditions for 100-ton EAF with a transformer's capacity of 75 MVA and suggested a reactor with a capacity of 45 MVAr. The reactor's advantage is that can be switched off in the final stages of the melting process when the arc is stable.

Considering the high arc instability during the initial stages of scrap melting, Labar et al. [43, 44] suggested to replace electric energy by chemical energy during the boring phase.

13.6.4 Optimizing Reactance to Control Flicker and Harmonics

Bowman et al. [15] described the importance of adding reactance into the AC circuit: The total reactance of a common transformer, 2.5–4 mΩ per phase is low to operate with voltages in the range from 400 to 500 V. Additional reactance can reach up to 5–10 mΩ per phase. The ultimate benefit with the added reactance is that the active power can be increased because the secondary voltage waveform is much higher when the current passes through zero, then the risk of arc circuit interruption is reduced and the waveform approaches to a smooth sinusoidal wave.

The role of capacitors is to increase the power factor. One strategy is to add capacitors with a reactance equal to the inductive reactance from the furnace. Figure 13.11 shows the effect of added reactance [5]. The highest stability is obtained using both SVC and reactors.

13.6.5 Harmonics Control

Harmonics originate due to voltage or current replicas. It has been suggested that although a bank of capacitors is employed to solve electric disturbances, it should be the first step in harmonic analysis and check its performance due to fuse blowing and/or capacitor failure [45].

Furthermore, it has also been suggested that a star/delta arrangement (Y/Δ) minimizes harmonics.

Gandhare and Lulekar [46] measured the Total Harmonic Distortion (THD) in a 50 ton EAF equipped with a transformer of 40 MVA. The peak values of THD reached 20% during scrap melting. Martell et al. [47] reported THD values as a percentage of the fundamental arc voltage, shown in Table 13.7. It can be observed that during melting the third and fifth harmonics predominate, with 20% and 10% respectively.

Voltage THD can be calculated as follows:

Fig. 13.11 Added reactance
to improve arc stability,
a without reactance, **b** SVC,
c SVC and reactor. After
Kruger [5]

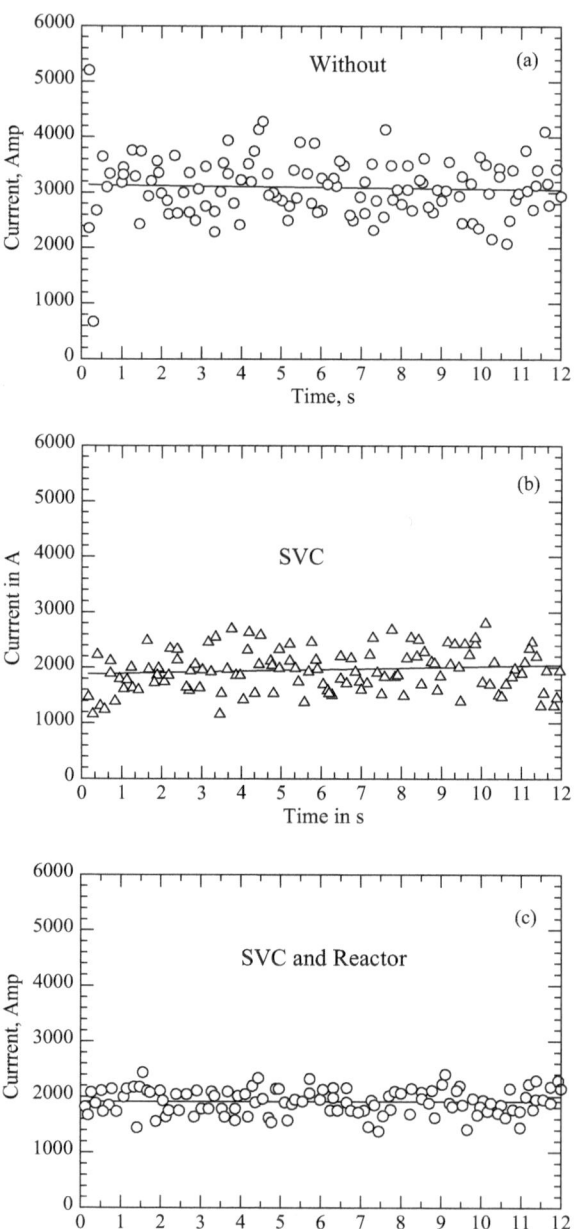

Table 13.7 Typical harmonic voltages according to [47], in %

	2nd	3rd	4th	5th	6th	7th	8th	9th	11th	13th
Melting	5	20	3	10	1.5	6	1	3	2	1
Refining	2	10	2	10	1.5	6	1	3	2	1

$$THD_V = \frac{\sqrt{V_{rms}^2 - V_1^2}}{V_1}$$

where V_1 fundamental voltage, V_{rms} is the total voltage harmonics.

Nikolaev et al. [48] reported THD (or K_I) as shown in Fig. 13.12, indicating a decrease from about 20 to 4% from scrap melting to flat bath operation. The lowest value, 4–5% occurs under slag foaming conditions. THD (or K_I) was defined as follows:

$$K_I = 100\left(\frac{\sqrt{I_{(0)}^2 + I_{(1)}^2 + I_{(2)}^2 + \cdots + I_{(n)}^2}}{I}\right)$$

where $I_{(n)}$ is the effective value of the harmonic.

13.6.6 Flicker Control

Flicker occurs due to fluctuations in arc voltage ($\Delta V/V$), however its measurement for different systems and its control are complicated. The time scale is very small, it can be deduced from the ratio L/R in the secondary side, for example if R = 0.4 mΩ and L = 12 µH, the time constant for transients is about 30 ms [5]. The variations in arc voltage mainly occur during the scrap melting stage. These variations occur independently of the other phases and a change in one phase does not affect the change in the other phase. In order to control flicker, the first step is to measure and/ or predict its value and the second step is its control. The earliest industrial flicker compensators were installed in the late 1950s.

Many algorithms have been developed to measure flicker over a long period of time. Zhang et al. [49] indicated that the first flicker meter was developed in the UK in 1972. The equivalent 10 Hz voltage flicker (ΔV_{10}) was developed in Japan in 1978 and the international standards in 1980 by the International Electrotechnical Commission (IEC) was accepted in Europe but not in North America, as of 2003 [50]. The basic measurements include short-term flicker severity (P_{st}) and long-term flicker severity (P_{lt}), Voltage flicker (V_{fg}), voltage depression (ΔV_f) and equivalent 10 Hz voltage flicker (ΔV_{10}). The maximum voltage depression $\Delta V_f = \Delta V/V$ in most countries is in the range 1.5–2.5% [49]. This value is a reference to define the other parameters; $V_{fg} = 0.12\Delta V_f$, $P_{st} = 0.5\Delta V_f$, $P_{lt} = 0.36\Delta V_f$.

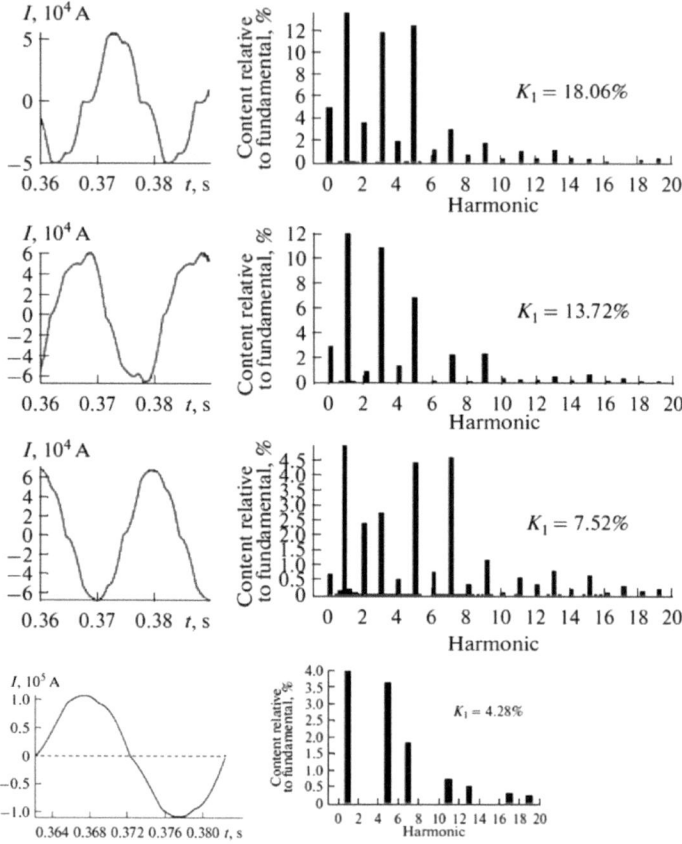

Fig. 13.12 Total harmonics distortion from scrap melting to flat bath operation. After [48]

Deaconu et al. [42] investigated these systems and found that introducing series reactance mitigate flicker but results in voltage reduction. SVC can improve power quality but have a slow response to fast-varying flicker. They suggested to switch off the series reactance under flat bath conditions when arc ignition is not an issue. Samet et al. [51] observed that adding series reactance mitigates voltage flicker but decreases voltage, then to compensate for this effect the tap voltage has to be increased but again the new change disrupts arc stability. The authors calculated optimum setpoints with and without SVC for a transformer with a capacity of 120 MVA, 21 taps ranging from 600 to 1300 V and series reactance with 7 taps. For this case they identified 6 optimum set points. For the final selection they suggested to consider the arc length, recommending short arc length. This final suggestion is wrong because their objective is to maximize the transfer of electric power and that is achieved with a long arc operation.

Morello et al. [52] describe in some detail the procedure to define the capacity of the SVC system and filter of harmonics as a consequence of revamping a steel shop with a new transformer. The analysis starts from an estimation of the amount of flicker, using an expression recommended by the UIE:

$$P_{st} = C \cdot \left(\frac{S_{ccf}}{S_{ccn}} \right)$$

where S_{ccf} is the short-circuit level of the arc furnace flicker at the point of common coupling (PCC), S_{ccn} is the fault level of the network at the PCC and C is the characteristic emission coefficient, its value is in the range from 40 to 85 with a mean value of 60.

13.6.7 Effect of the Slag to Improve Electric Parameters

Basic slag has a higher electric conductivity than acid slags, facilitating the passage of current through zero which decreases harmonics. Liu et al. [53] reported the electric conductivity of slags, as shown in Fig. 13.13 for two basicities 0.3 and 0.6. The electric conductivity increases 3–5 times, from about 0.06 to 0.12 Ω^{-1} cm^{-1} by increasing the basicity. Further studies also reported that increasing FeO in the slag increases the electric conductivity even further to values from 0.1 to 0.4 Ω^{-1} cm^{-1} [54]. In both cases, also increasing the temperature, increases the electric conductivity.

Slag foaming is one of the main methods to decrease electric energy consumption in the EAF. This method has been described in detail in Chap. 6. There is one aspect that is worth mentioning; the electric conductivity of foaming slags is small [5] due to the dominant role of the gas phase, about 50–66%, however, on the other hand, the presence of carbon particles increases the electric conductivity, however no results have been reported under these conditions.

Fig. 13.13 Effect of temperature and basicity on the electric conductivity of slags [53]

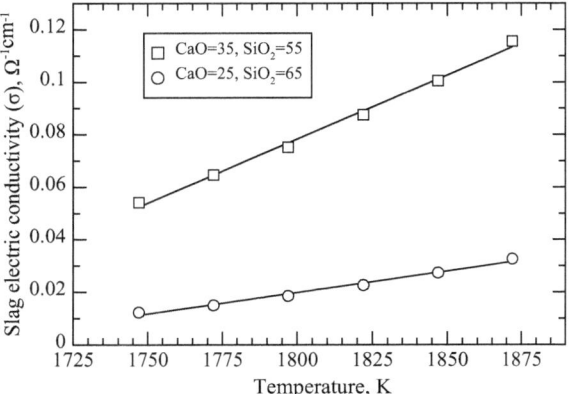

13.6.8 Arc Stability

Arc stability from a practical viewpoint means the electric arc doesn't extinguish and therefore the flow of current is continuous. Under these conditions electric disturbances are decreased. During the scrap melting process the electric arc is highly instable and the associated changes in the electric parameters affect power quality and decrease the active power. Arc instability can be detected and measured based on the noise or the arc distortion.

Sedivy and Krump compared noise and total harmonic distortion (THD) to evaluate arc stability and found a close agreement between the two signals. Martell et al. [47] reported that during scrap melting the third harmonic predominate, producing arc instability. Since the neutral to ground secondary voltage also changes due to arc instability, Martell et al. [47] proposed this variable as another stability index, defined as follows:

$$V_{Ng} = \sqrt{\frac{1}{T} \int_0^t v_{Ng}^2 dt}$$

$$3v_{Ng} = v_{Rg} + v_{Sg} + v_{Tg}$$

where V_{Ng} is the total virtual to ground voltage and v_{Ng} is the instantaneous virtual to ground voltage. v_{Ng} is calculated based on measured neutral to ground voltages. The results reported indicate a good relationship with the measurement of THD with the advantage that using V_{Ng} is simpler and doesn't require signal processing techniques (FFT-Fast Fourier Transform).

13.6.9 Optimum Electric Parameters Using Circle Diagrams

The ideal optimum electric parameters would correspond to a maximum arc power; however, this point is obtained for a maximum secondary current depending on the transformer rated capacity $\left(I_2^{rat}\right)$ which cannot be reached or exceeded for a long time due to risk of transformer overheat, in spite of this, this point defines the upper limit of current and arc power. In practice the arc power is defined for a current below that maximum, whose minimum value is called minimum secondary current $\left(I_2^{min}\right)$. Operation falls between I_2^{min} and I_2^{rat}. Nikolaev et al. [48] indicated the need to define an operation point that reaches maximum arc power with lower currents in order to minimize electric energy consumption. Since modern furnaces are provided with a variable-inductance reactor, Fig. 13.14 (left) illustrates how to keep the same arc power by increasing reactance and arc voltage and at the same time decreasing the arc current. In this example, voltage and reactance are increased from 1127 to 1345 V and 0 to 1.44 Ω, the arc current is decreased from 83.5 to 70 kA, keeping a constant arc power of 102.7 MW. Figure 13.4 (right) shows that increasing voltage also increases the arc length, from 424 to 505 mm. A long-arc operation requires a good slag foaming practice to shield the electric arcs.

A heat with steel scrap has two typical stages, scrap melting and flat bath operation. Each one of these stages require different sets of electrical parameters. During scrap melting is possible to operate with short arcs. The operation point for the lower current is defined from another curve defined by the Refractory Wear Index (RWI) called by Nikolaev et al. lining-wear coefficient (LWC), as shown in Fig. 13.15. In this figure is also included a heating rate coefficient ($HRC = P_{arc}I_{arc}$). The maximum LWC corresponds to the operation point.

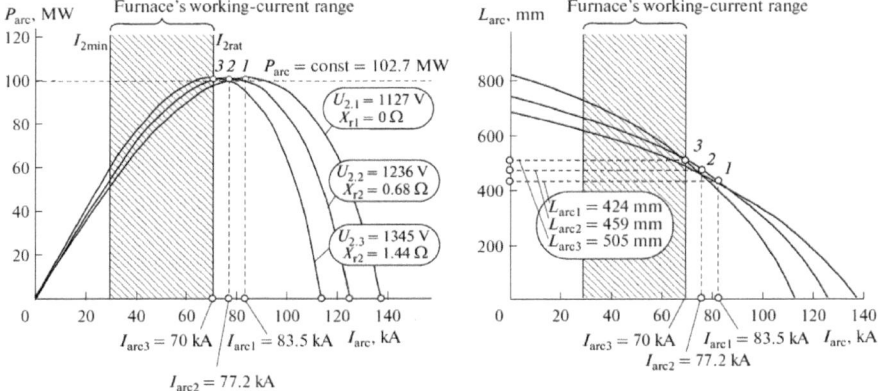

Fig. 13.14 Left: $P_{arc} - I_{arc}$ relationship changing reactance and voltage, (right) $l_{arc} - I_{arc}$ relationship for the three cases on the left figure. After [48]

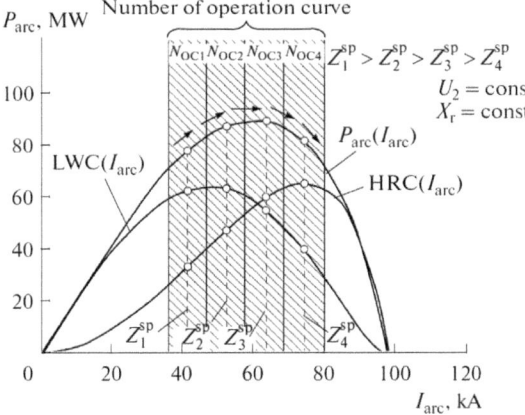

Fig. 13.15 Lining wear coefficient (LWC) and heating rate coefficient (HRC). After [48]

A circle diagram is a graphical representation of the performance of an electrical machine drawn in terms of the locus of the machine's input voltage and current. It was originally developed by Heyland in 1894 [55, 56]. To simplify the analysis with one phase, it is assumed the arcs are balanced. The basic information required is a no-load test to measure short-circuit parameters such as short circuit current (I_{sc}), short circuit reactance (X_V) and short circuit resistance (R_V). Under short-circuit conditions the ohmic resistance is zero (R_B). In an ideal short circuit, there is only pure reactive current with a maximum current. The real short circuit current during dipping of electrodes is slightly smaller than the ideal short circuit.

There are several ways to represent the circle diagrams:

- Voltage (V) versus Current (I): Defines the optimum current and power factor to get a maximum active power. The optimum power factor can be taken as half of the short circuit power factor and extrapolating this power factor as a vector to the V-I line, the intersection defines the optimum current.
- Active power (P) versus current (I): Called power diagram. Similar to the P-Q diagram.
- Active power (P) versus Reactive power (Q): this is the more general approach. A set of semi-circles are developed for each voltage in the secondary circuit. The maximum active power is obtained for a power factor of 45° ($\cos \theta = 0.7071$). For this condition the active power and the reactive power are equal.

Timm [57] has reported an example for a furnace transformer with a rated capacity of 75 MVA (S_N) and 10 taps from 240 to 600 V. The following calculations correspond to the tap with 600 V.

From short-circuit measurements on the primary side, the following reactance (X_{sc}) and resistance (R_{sc}) were obtained:

$$X_{sc} = 2.8 \text{ m}\Omega$$
$$R_{sc} = 0.45 \text{ m}\Omega$$

Short-circuit power factor:

$$\tan \theta_{sc} = \frac{X_{sc}}{R_{sc}} = \frac{2.8}{0.45} = 6.22$$
$$\theta_{sc} = 80.9°$$
$$(\cos \theta)_{sc} = 0.16$$

Maximum active and reactive power:

$$P_{max} = \frac{U^2}{2X_{sc}} = \frac{(600)^2}{2 \times (2.8 \times 10^{-3})} \frac{V^2}{\Omega} = 64.3 \text{ MW}$$
$$Q_{max} = \frac{U^2}{X_{sc}} = \frac{(600)^2}{(2.8 \times 10^{-3})} = 128.6 \text{ MW}$$

Maximum current based on the transformed rated capacity:

$$I_{max} = \frac{S_N}{\sqrt{3} \cdot U_{max}} = \frac{75}{\sqrt{3} \cdot 600} \frac{MVA}{V} = 72.2 \text{ kA}$$

The previous calculations can be graphically represented in a circle diagram, as shown in Fig. 13.16 for all the taps available. It can be observed that the current for the ideal maximum active power cannot be supplied by the transformer. Its value corresponds with a power factor of 0.70. The real maximum active power is defined based on the maximum current supplied by the transformer, for a power factor of 0.82, giving about 61 MW.

1. Maximum current from the transformer (line 1): 72.2 kA
2. Real maximum active power (point 2): 61 MW
3. Maximum voltage (line 3): 600 V
4. Maximum power factor (line 4): A very high-power factor makes the arc unstable. In the figure it has been chosen a limit at 0.86
5. Minimum secondary voltage (line 5): 240 V
6. Short circuit line (line 6): power factor equal to 0.16.

In summary, operating with a tap of 600 V, the maximum active power would be 64.3 MW but this condition is obtained for a current that exceeds the maximum

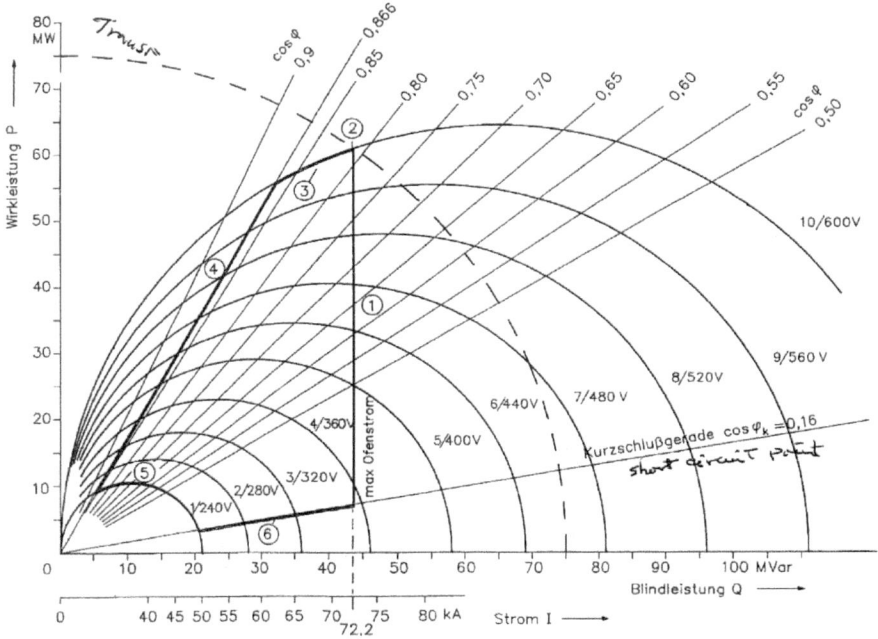

Fig. 13.16 Graphical representation of electric power parameters for a transformer of 75 MVA. After [57]

current capacity from the transformer, 72.2 kA. Instead, taking as a reference the real maximum current, point 2 in the figure, the real maximum active power corresponds to 61 MW and a power factor of 0.86.

In the circle diagrams is possible to add multiple axis, for example, the V-I diagram can also include the reactive power, for a constant reactance, in this case the reactive power is proportional to the square of the current.

$$Q = 3I^2X \approx I^2$$

Reactance is necessary to promote arc stability at the expense of active power. For DRI melting under slag foaming conditions is convenient to operate with maximum tap, the highest possible power factor and minimum or zero reactance. For scrap melting operations is convenient to increase reactance in the beginning but decrease it as melting progresses.

13.6.10   Optimization of the Electric Power Program

The active power can be increased in two ways:

- Increasing arc voltage (V_{arc}): it increases arc length but also the radiation intensity from the arc
- Increasing arc current (I_{arc}): This option is limited by the electric conductivity of the electrodes.

The power program or power profile depends on the type of metallic charge. Scrap-based operations typically include melting several scrap buckets followed by a flat bath operation. DRI-based operations charge metallic particles continuously and in the case of hot metal is already liquid, therefore the power profile can vary to a large extent.

The initial step on scrap melting is the boredown period. During this time is possible to supply the largest amount of energy and use a long arc operation because the arcs are covered by scrap. In this stage the electric arc is highly unstable. Once the scrap is melted there is no need to work with a short arc operation if there is foaming slag. During flat bath conditions the electric arc is stable. Changing to short arc operation decreases both current and radiation to the walls, however if there are good foaming conditions the long arc operation can be maintained. Martell et al. suggested a power profile that maximizes power delivery, deriving expressions for both the peak power and peak current using a simplified electric circuit. The currents calculated were employed to adjust the power profile but keeping the same tap voltages.

Jingshe et al. [58] improved the power program in a 50 ton EAF with a 25 MVA furnace transformer. The secondary voltage ranges from 269 to 519 V (taps 1–18). The objective is to achieve as much arc power as possible with the transformer's capacity. They computed a circular diagram for this EAF and defined a region of permissible operation defined by six points:

- Maximum rated current, $I < I_{max}$
- Power factor to achieve a stable electric arc, $\cos \theta < 0.866 < 0.866$
- Maximum transformer's apparent power, $S < 35$ MVA
- Maximum transformer's secondary voltage, $U_2 < 519$ V
- Minimum transformer's secondary voltage, $U_2 > 269$ V
- Avoid short circuit condition, $\cos \theta > 0.15$.

The metallic charge consisted of two charges of scrap and the original power profile maintained the same taps to melt each charge of scrap with only one tap change. The power-on time was 63.2 min. Boring was carried out with tap 16 for about 2 min and then switched to maximum power with tap 18. Tap changing takes about 20–30 s. With this power profile the maximum active power was about 27 MW. A new power profile was defined considering that under flat bath operation, radiation from the arc increases energy losses. The maximum tap was employed but probably due to the use of additional reactance the active power was increased to about 29 MW and during the final melting stage of the second scrap bucket the tap was decreased to 16. The new profile reduced the power-on time to 61.9 min.

It has been indicated previously that increasing the arc voltage also increases the arc length and then the active power also increases. However, the resulting longer electric arc can become unstable and with arc instability the amount of harmonics and flicker increases. This problem can be solved adding more reactance [59]. Adding more reactance is also called "high impedance operation". With a high impedance operation, the power factor increases and the stability of long arcs increases. The added reactor affects the voltage drop, to counteract this effect the tap voltage should also be increased simultaneously. Increasing the arc voltage will decrease the arc current. Table 13.8 summarize the results. It can be observed that with a long arc operation and higher voltage the refractory wear index (RWI) increases and would require improvements to the slag foaming practice to shield the arcs.

Table 13.8 High impedance operation

	Units	700 V	800 V	900 V	1000 V
Series reactor	mΩ	0	1.0	2.1	3.7
Secondary current	kA	64.4	55.5	48.8	42.4
Active power	MW	50.0	50.1	50.3	48.8
Arc length	mm	160.2	200	241	279.4
Arc length index	mm/kA	2.5	3.6	4.9	6.6
P/I	MW/kA	0.8	0.9	1.0	1.2
Electrode cons index		4148	3080	2385	1801
RWI	MW/cm	9487	12,338	15,155	17,301

After Martinez [59]

13.6.11   Optimizing the Electrode Regulation System

The electrode regulation system is a critical component in the EAF operation. An efficient system has multiple benefits; improvement to adjust the electric parameters which then improves the melting rate and then a lower power-on time.

The first electrode regulation system was designed by Thury, a Swiss engineer. In the 1940s it was developed the first system to make changes in the electrode position as a function of voltage and current.

Bowman et al. [15] points out that there are different ways to achieve optimum control but also many different opinions in regard to the best one. The electrode regulation system is primarily limited by the large tonnage of the electrode/arm/mast/cable assembly. This assembly has a large weight, in the order of 20 tons with a response time to reach 90% of full speed in the order from 0.2 to 0.5 s, compared with a frequency of 1/50th seconds for the changes in current and voltage [15], indicating that the mechanical system cannot respond fast enough to the changes needed by the electrical system. The control strategy is to choose a controlling variable, such as impedance, over-current, arc resistance, arc current or arc voltage. The movement of the assembly should respond to the set point of one of those variables.

Regulating valves interact with the electrode regulation system, converting the electric reference signal into a mechanical movement. Electronic controllers measure the electric parameters and compare them with a set point, then the regulating valves respond to eliminate the error.

The electrode regulation system also includes additional protections and electric measurements [60], such as the presence of non-conductive material and short-circuit conditions and the measurement of total harmonic distortion (THD). If the electrode reaches a non-conductive material it will try to push down with the risk of getting broken, also, if pieces of scrap fall or move in a way that touch the tip of the electrode it will create a short circuit and then creates a condition of overcurrent. In both cases the controller should detect the condition and react to raise the electrodes immediately.

Constant impedance: Is the most common method. This system measure both the arc current and arc voltage and compares those values with a pre-defined ratio. The difference or error is adjusted by movements of the electrodes. Voltage increases by raising the electrode. If voltage increases, impedance also increases:

$$Z = \frac{V}{I} = \text{Constant}$$

The main limitation is the capacity of the hydraulic system to move the electrodes. Modern furnaces can reach speeds in the order of 17 m/min. Since 1939 it appeared the remote system to hold the electrodes.

Constant current: Controls based on the arc properties are more complicated because its measurement is not simple [15]. The control by constant electrode current is employed in cases where the electrode consumption should be minimized.

13.6.12 Optimization of Electric Power by Mathematical Modelling

Many integral or comprehensive models describing the rate phenomena during EAF steelmaking have been reported [61, 62], however those models have focused on the description of rate phenomena related to changes in slag and steel chemistry and coupling with the instantaneous electric power delivery was not included. Logar et al. [63] in 2011 reported the first integral model to describe the rate of electric power supply and its coupling with chemical rate phenomena in the EAF. Their electric model computes electric power using computed values of voltages and currents using Kirchoff's current and voltage laws. To reach that point, all impedances in the circuit should be measured or computed. The arc resistance is calculated using the hybrid Cassie-Mayr model and the arc reactance using Köhl's model. The Mayr model's accuracy is low for currents higher than 1 kA. The hybrid model requires some parameters that are obtained empirically and is being modified continuously [64]. To provide for a more realistic representation of their model's results, Logar et al. added two randomness models. Their numerical predictions were in total agreement with experimental measurements. Opitz and Treffinger [65] introduced further improvements to Logar's model. Instead of the Cassie-Mayr model which depends on empirical parameters, they employed Nusselt correlations to define heat transfer coefficients.

13.6.13 Calculation of Electric Losses

Diaconu et al. [66] reported several expressions to calculate electric losses at the transformer, power cables, reactor and electrodes. Below is a detailed example of his calculations. The information required is indicated in Table 13.9. The total electric losses are defined by the following expression:

$$\Delta H_e (kWh/heat) = \Delta H_e^{tr} + \Delta H_e^{cable} + \Delta H_e^{reactor} + \Delta H_e^{electrodes}$$

(a) Transformer: electric losses from the transformer are calculated using the following expression:

$$\Delta H_e^{tr}(kWh/heat) = \left(P_{01} + \beta_1^2 P_{sc-1}\right)\tau_1 + \left(P_{02} + \beta_2^2 P_{sc-2}\right)\tau_2 + \left(P_{03} + \beta_3^2 P_{sc-3}\right)\tau_3$$

where P_0 are the power losses under no load, P_{sc} are the power losses under short circuit, τ is the time connected to the grid, β is load coefficient defined as follows:

Table 13.9 Transformer electric losses

Transformer capacity	6 kVA	50 Hz	
Reactive losses transformer	220 V	$P_{r1} = 5.7$ kW	
Reactive losses transformer	127 V	$P_{r2} = 2.2$ kW	
Power transformer Joule losses, short circuit	$P_{j1} = 29$ kW @ 5900 A	$P_{j2} = 23.2$ kW @ 220 V	
Operating loads	$S_1 = 2250$ kVA* @ 220 V	$S_2 = 1550$ kVA** @ 220 V	$S_3 = 1100$ kVA @ 127 V
Power-on time per heat	$\tau_{supply} = 3.5$ h		
Power profile	220 V	$\tau_{supply1} = 0.25$ h	*With reactor
	220 V	$\tau_{supply2} = 1.25$ h	**Without reactor
	127 V	$\tau_{supply3} = 2.0$ h	Without reactor
Cable resistance HV side	3.25×10^{-2} Ω/phase		
Electrode resistance	4.2×10^{-4} Ω/phase		
Reactor resistance	3.2×10^{-2} Ω/phase		
Electric energy from the grid	5903 kWh/heat		
Current	5900 A		

After [66]

$$\beta_i = \frac{S_i}{S_n}$$

$$S_n = \sqrt{3}U_j I_j$$

where S_i are the operating loads (2250, 1550 and 1100 kVA), S_n is the nominal power transformer at different voltages/currents (j).

$S_1 = 2250$ kVA U = 220 V	$\beta_1 = \frac{S_1}{S_n} = \frac{2250}{2250} = 1$	$S_n = \sqrt{3}U_j I_j = \sqrt{3} \times 220 \times 5900 = 2250$ kVA
$S_2 = 1550$ kVA U = 220 V	$\beta_1 = \frac{S_1}{S_n} = \frac{1550}{2250} = 0.69$	$S_n = \sqrt{3}U_j I_j = \sqrt{3} \times 220 \times 5900 = 2250$ kVA
$S_3 = 1100$ kVA U = 127 V	$\beta_1 = \frac{S_1}{S_n} = \frac{1100}{1298} = 0.85$	$S_n = \sqrt{3}U_j I_j = \sqrt{3} \times 127 \times 5900 = 1298$ kVA

$$\Delta H_e^{tr}(\text{kWh/heat}) = \left(5.7 + 1^2 \times 29\right) \times 0.25 + \left(5.7 + 0.69^2 \times 29\right)$$
$$\times 1.25 + \left(2.2 + 0.85^2 \times 29\right) \times 2$$
$$\Delta H_e^{tr}(\text{kWh/heat}) = 75.5 \text{ kWh/heat}$$

(b) Joule losses: Joule losses for the cable, reactor and electrodes are calculated by the following general expression:

$$\Delta H_e^{joule} = RI^2 \tau$$

Currents at HV and LV sides:

$$I = \frac{S}{\sqrt{3}U}$$

	HV ($U_{HV} = 6$ kV)	LV ($U_{HV} = 0.22$ kV)
$S_1 = 2250$ kVA	$I_{HV1} = \frac{2250}{\sqrt{3}\times 6} = 216$ A	$I_{LV1} = \frac{2250}{\sqrt{3}\times 0.22} = 5900$ A
$S_2 = 1550$ kVA	$I_{HV2} = \frac{1550}{\sqrt{3}\times 6} = 149$ A	$I_{LV2} = \frac{1550}{\sqrt{3}\times 0.22} = 4065$ A
$S_3 = 1100$ kVA	$I_{HV3} = \frac{1100}{\sqrt{3}\times 6} = 106$ A	$I_{LV3} = \frac{1100}{\sqrt{3}\times 0.22} = 5000$ A

Cable:

$$\Delta H_e^{cable} = 3R_{cable} \sum_{i=1}^{3} I_{HV,i}^2 \tau_i = 3$$
$$\times 0.0325 \times \left(216^2 \times 0.25 + 149^2 \times 1.25 + 106^2 \times 2\right) \times 10^{-3}$$
$$\Delta H_e^{cable} = 6.1 \text{ kWh/heat}$$

Reactor:

$$\Delta H_e^{reactor} = 3R_{reactor} I_{HV,i}^2 \tau_i = 3 \times 0.032 \times 216^2 \times 0.25 \times 10^{-3}$$
$$\Delta H_e^{reactor} = 1.1 \text{ kWh/heat}$$

Electrodes:

$$\Delta H_e^{electrodes} = 3R_{electrode} \sum_{i=1}^{3} I_{LV,i}^2 \tau_i = 3 \times 0.042$$
$$\times \left(5900^2 \times 0.25 + 4065^2 \times 1.25 + 5000^2 \times 2\right) \times 10^{-3}$$
$$\Delta H_e^{electrodes} = 101 \text{ kWh/heat}$$
$$\Delta H_e (\text{kWh/heat}) = \Delta H_e^{tr} + \Delta H_e^{cable} + \Delta H_e^{reactor} + \Delta H_e^{electrodes} = 183 \text{ kWh/heat}$$

This value, represented about 3% of the heat input. It is clear that due to their higher current, electric losses are the highest at the electrodes followed by electric losses at the transformer.

References

1. K. The early history of magnetism. Nature 523–524. https://doi.org/10.1038/013523a0
2. Meyer H (1972) A history of electricity and magnetism. Burndy Library, Connecticut, USA
3. Maxwell J-C (1869) Remarks on the mathematical classification of physical quantities. Proc London Math Soc 1–3:224–233
4. Lupi S (2017) Fundamentals of electroheat. Electrical technologies for proccess heating. Springer International Publishing, Switzerland. https://doi.org/10.1007/978-3-319-46015-4
5. Bowman B, Krüger K (2009) Arc furnace physics. StahlEisen
6. Herland EV, Sparta M, Halvorsen SA (2019) Skin and proximity effects in electrodes and furnace shells. Metall Mater Trans B Process Metall Mater Process Sci 50:2884–2897. https://doi.org/10.1007/s11663-019-01651-8
7. Middleton G (1985) The failure of graphite arc-furnace electrodes. University of Durham
8. Wild J (2022) Electrical engineering I step by step: basics, components & circuits explained for beginners M.Eng. Independent, Landau
9. Dutta S (2016) What is reactance in transmission lines? 1. https://www.quora.com/What-will-happen-if-the-transmission-reactance-becomes-zero. Accessed 18 May 2022
10. Toulouevski YN, Zinurov IY (2010) Innovation in electric arc furnaces. Scientific basis for selection. Springer, Berlin
11. Arnold S (1933) The 3-phase electric arc furnace. Electr Eng 52:839–843
12. Worrell E, Martin N, Price L (1999) Energy efficiency and carbon dioxide emissions reduction opportunities in the U.S. iron and steel sector. NBNL report
13. Toulouevski YN, Zinurov IY (2013) Innovation in electric arc furnaces. Scientific basis for selection, 2nd edn. Springer, Berlin
14. DanCarbon (2020) How much does an electric arc furnace cost? 1. https://www.dancarbon.com/q/eaf/electric-arc-furnace-cost-212.html. Accessed 19 May 2022
15. Jones JAT, Bowman B, Lefrank PA (1998) Electric arc furnace steelmaking. In: Fruehan RJ (ed) Making, shaping, and treating of steel, 11th edn. Steelmaking and refining volume, Pittsburgh, PA USA, pp 525–660
16. Bowman B (1993) A technical comparison between AC and DC furnaces. Rev Metall 90:809–816
17. Madias J, Bilancieri A, Hornby S (2017) The influence of metallics and EAF design. Steel Times Int 35–40
18. Jones RT, Reynolds QG, Curr TR, Sager D (2011) Some myths about DC arc furnaces. J South Afr Inst Min Metall 111:665–673
19. Carbon G (2020) What are the characteristics of graphite electrode materials 1. http://www.guotaitansu.com/en/news/87.html%0A. Accessed 19 May 2022
20. Pritchard OG (1890) The manufacture of electric light carbons. The Electrician, London
21. Castner H (1896) Anode for electrolytic processes. US patent 572472
22. Jager H, Frohs W (2021) Industrial carbon and graphite materials: raw materials, production and applications. Wiley-VCH, Weinheim
23. Steppich D (2021) Graphite electrodes for electric arc furnaces. In: Industrial carbon and graphite materials: raw materials, production and applications, chap 6.5.3. Wiley-VCH, Germany, p 39
24. Guo D, Irons GA (2003) Modeling of radiation intensity in an EAF. In: Proceedings of 3rd international conference on CFD minerals and process industries, Melbourne, Australia, 10–12 Dec 2003, pp 223–228
25. Larsen HL (1996) AC electric arc models for a laboratory set-up and a silicon metal furnace. PhD thesis, Norwegian University of Science and Technology
26. Larsen HL, Saevarsdottir GA, Bakken JA (1997) Simulation of AC arcs in the silicon metal furnace. In: 54th Electric furnace conference, Dallas, TX, USA, 9–12 Dec 1997, pp 157–168
27. Bakken JA, Gu L, Larsen HL, Sevastyanenko VG (1997) Numerical modeling of electric arcs. J Eng Phys Thermophys 70:532–544

28. Sanchez JLG, Ramírez-Argaez MA, Conejo AN (2009) Power delivery from the arc in AC electric arc furnaces with different gas atmospheres. Steel Res Int 80. https://doi.org/10.2374/SRI08SP079
29. Fathi A, Saboohi Y, Škrjanc I, Logar V (2015) Low computational-complexity model of EAF Arc-heat distribution. ISIJ Int 55:1353–1360. https://doi.org/10.2355/isijinternational.55.1353
30. Ayrton H (1902) The electric arc. The Electrician, London
31. Martell-Chávez F, Ramírez-Argáez M, Llamas-Terres A, Micheloud-Vernackt O (2013) Theoretical estimation of peak arc power to increase energy efficiency in electric arc furnaces. ISIJ Int 53:743–750. https://doi.org/10.2355/isijinternational.53.743
32. Pauna H, Willms T, Aula M, Echterhof T, Huttula M, Fabritius T (2020) Electric Arc length-voltage and conductivity characteristics in a pilot-scale AC electric arc furnace. Metall Mater Trans B Process Metall Mater Process Sci 51:1646–1655. https://doi.org/10.1007/s11663-020-01859-z
33. Terzija VV, Koglin HJ (2001) A new approach to arc resistance calculation. Electr Eng 83:187–192. https://doi.org/10.1007/s002020100074
34. Zhao H, Fazheng C, Zhao Z (2010) Study about the methods of electrodes motion control in the EAF based on intelligent control. In: 2010 International conference on computer, mechatronics, control electronic engineering, pp 68–71
35. Ciotti JA (1971) A new era in melting. JOM 23:30–35
36. Schwabe WE (1962) Arc heat transfer and refractory erosion in electric steel furnaces. In: EAF conference, pp 195–206
37. Hassan A, Abou-Ghazala A, Megahed A (2018) Field-verified integrated EAF-SVC-electrode positioning model simulation and a novel hybrid series compensation control for EAF. Turkish J Electr Eng Comput Sci 26:363–377. https://doi.org/10.3906/elk-1702-215
38. Gottardi R, Miani S, Partyka A (2006) Ultra high chemical power EAF. In: AISTech 2006 conference, Cleveland, OH, USA, 1–4 May, pp 1–10
39. Jansen H, Castillo J, Bruzal C, Schemmel T, Lecca M (2012) MgO-C bricks with catalytically activated resin for EAF's with DRI charge: industrial experiences at Corporacion Aceros Arequipa. In: 5th Congreso y exposición de la industria del acero, CONAC 2012, Monterrey, NL, Mexico, 8–11 Oct, pp 1–11
40. Čeřňan M, Müller Z, Tlustý J, Valouch V (2021) An improved SVC control for electric arc furnace voltage flicker mitigation. Int J Electr Power Energy Syst 129. https://doi.org/10.1016/j.ijepes.2021.106831
41. Qiao X, Bian J, Chen C, Li H (2019) Comparison and analysis of reactive power compensation strategy in power system. In: 2019 IEEE sustainable power and energy conference on grid modernizations for energy revolution, pp 689–692. https://doi.org/10.1109/iSPEC48194.2019.8975301
42. Deaconu I, Nicolae P, Latinovic T (2010) Comparative study for EAF's reactive energy compensation methods and power factor improvement. WSEAS Trans Syst 9:979–988
43. Labar H, Dgeghader Y, Kelaiaia MS, Bounaya K (2009) Closely parametrical model for an electrical arc furnace. World Acad Sci Eng Technol 40:96–100. https://doi.org/10.5281/zenodo.1055622
44. Labar H, Djeghader Y, Bounaya K, Mounia K (2009) Improvement of electrical arc furnace operation with an appropriate model. Energy 34:1207–1214. https://doi.org/10.1016/j.energy.2009.03.003
45. Masoum M, Fuchs E (2015) Power quality on power systems and electrical machines, 2nd edn. Elsevier; Academic Press
46. Gandhare WZ, Lulekar DD (2007) Analyzing electric power quality in arc furnaces. Renew Energy Power Qual J 1:286–290. https://doi.org/10.24084/repqj05.272
47. Martell F, Deschamps A, Mendoza R, Meléndez M, Llamas A, Micheloud O (2011) Virtual neutral to ground voltage as stability index for electric arc furnaces. ISIJ Int 51:1846–1851. https://doi.org/10.2355/isijinternational.51.1846
48. Nikolaev AA, Kornilov GP, Anufriev AV, Pekhterev SV, Povelitsa EV (2014) Electrical optimization of superpowerful arc furnaces. Steel Transl 44:289–297. https://doi.org/10.3103/S0967091214040135

49. Zhang Z, Fahmi NR, Norris WT (2001) Flicker analysis and methods for electric arc furnace flicker (EAF) mitigation (a survey). In: 2001 IEEE Porto power tech proceedings, vol 1, Porto, Portugal, pp 508–513. https://doi.org/10.1109/PTC.2001.964651
50. Halpin SM, Bergeron R, Blooming TM, Burch RF, Conrad LE, Key TS (2003) Voltage and lamp flicker issues: should the IEEE adopt the IEC approach? IEEE Trans Power Deliv 18:1088–1097. https://doi.org/10.1109/TPWRD.2003.814261
51. Samet H, Ghanbari T, Ghaisari J (2014) Maximizing the transferred power to electric arc furnace for having maximum production. Energy 72:752–759. https://doi.org/10.1016/j.energy.2014.05.105
52. Morello S, Dionise TJ, Mank TL (2015) Comprehensive analysis to specify a static var compensator for an electric arc furnace upgrade. IEEE Trans Ind Appl 51:4840–4852. https://doi.org/10.1109/TIA.2015.2451072
53. Liu JH, Zhang GH, Chou KC (2015) Study on electrical conductivities of CaO-SiO$_2$-Al$_2$O$_3$ slags. Can Metall Q 54:170–176. https://doi.org/10.1179/1879139514Y.0000000174
54. Liu JH, Zhang GH, Wu YD, Chou KC (2016) Study on electrical conductivity of FexO–CaO–SiO$_2$–Al$_2$O$_3$ slags. Can Metall Q 55:221–225. https://doi.org/10.1080/00084433.2016.1150545
55. Heyland A (1894) A graphical method for the prediction of power transformers and polyphase motors. Elektrotechnische Zeitschrift (ETZ) 15:561–564
56. Heyland A (1906) A graphical treatment of the induction motor. McGraw-Hill
57. Timm K (2006) Circle diagram of AC-furnaces. In: Electrical engineering of arc furnaces, Dusseldorf, Germany, 23–26 Oct. Steel Academy, pp 130–141
58. Li JS, Liu RZ, Li SQ, Wu J, Jiang GL, Wei QS et al (2004) Optimization of power input diagram of 50 t UHP-EAF. J Univ Sci Technol Beijing 11:10–12
59. Martinez F (1997) Experiences of high impedance AC EAF arc regulation using the new AMI automation's power input optimization. In: McCaster symposium on iron mills steel mills, Ontario, CA, pp 1–13
60. Origoni L, Kyle S, Memoli F (2010) The basic principles of an electrode regulation system. Ind Heat 77:63–67
61. Morales RD, Conejo AN, Rodríguez HH (2002) Process dynamics of electric arc furnace during direct reduced iron melting. Metall Mater Trans B Process Metall Mater Process Sci 33
62. Morales RD, Rodríguez-Hernández H, Conejo AN (2001) A mathematical simulator for the EAF steelmaking process using direct reduced iron. ISIJ Int 41
63. Logar V, Dovžan D, Škrjanc I (2011) Mathematical modeling and experimental validation of an electric arc furnace. ISIJ Int 51:382–391. https://doi.org/10.2355/isijinternational.51.382
64. Guardado JL, Maximov SG, Melgoza E, Naredo JL, Moreno P (2005) An improved arc model before current zero based on the combined Mayr and Cassie arc models. IEEE Trans Power Deliv 20:138–142. https://doi.org/10.1109/TPWRD.2004.837814
65. Opitz F, Treffinger P (2016) Physics-based modeling of electric operation, heat transfer, and scrap melting in an AC electric arc furnace. Metall Mater Trans B Process Metall Mater Process Sci 47:1489–1503. https://doi.org/10.1007/s11663-015-0573-x
66. Diaconu B, Anghelescu L, Cruceru M (2020) Analysis of energy balance for a steel electric arc furnace. WSEAS Trans Environ Dev 16:48–56. https://doi.org/10.37394/232015.2020.16.6

Chapter 14
EAF Design and Automation

14.1 Design

14.1.1 Inventors of the Original Designs

Carl Wilhelm Siemens (1823–1883) [1–3] *also known as Sir Charles William Siemens* was the inventor of the first electric arc furnace design in 1878. He was born in Lenthe, 10 km southwest of Hanover Germany. He was part of a family of inventors and the origin of the Siemens conglomerate today.

He attended secondary schools first in Lubeck and then in 1838 in Magdeburg until 1841. At age 19 he became an apprentice mechanical engineer. In 1839 his mother died and his father also died a year later. By 1844 he had settled in London and worked for a company in Birmingham. In 1847 he built the first prototype of his major invention, the regenerative furnace, at a time when the concept of heat was still under discussion.

His elder brother, Werner, built a Telegraph manufacturing company in 1847, based on his own invention and William became the company representative in

A. N. Conejo, *Electric Arc Furnace: Methods to Decrease Energy Consumption*, https://doi.org/10.1007/978-981-97-4053-6_14

London in 1850. In this year he got a financial success with his invention on a water flowmeter and also got an award for his invention on the regenerative furnace. He got married in 1859 and became a British citizen, changing his name from Carl Wilhelm to Charles William.

The idea of the regenerative furnace was to recover the sensible heat in the off-gas. This idea was proved to be a success in 1861 in glass manufacturing. In 1866 he carried out experiments to produce steel and got a British patent in 1866 and a US patent in 1867. In 1865 the French engineer Pierre Martin bought a license and was able to produce steel at a commercial scale. The process is known as Siemens-Martin or Open Hearth Furnace (OHF). Eventually, it replaced the Bessemer process. In the OHF the charge is composed of pig iron. Siemens employed iron ore to remove carbon and Martin employed steel scrap to reach the final carbon content. It was the main manufacturing process for more than a century.

At about 1877 he proposed the use of electric energy to melt steel scrap using direct current (DC-EAF). His first idea was to use indirect heating from the electric arc. Later on, in a new development, he placed one electrode on the top using direct current. To close the circuit and take the electric current away from the steel a bottom electrode was needed. He attached a block of iron at the bottom covered by concrete. The direct current was provided by 5 dynamos driven by a 12-horsepower steam engine. A patent was grated to Siemens for this idea in 1878. The cost of DC and control of the bottom electrode made this technology unsuccessful at that time. The modern dynamo was invented by his brother Werner in 1867. Figure 14.1 is a schematic representation of the DC-EAF invented by W. Siemens.

Few months before his death on November 1883, he was knighted by Queen Victoria. He is buried in Kensal green cemetery in London.

Ernesto Stassano (1859–1922) [5, 6], was a captain and engineer in the Italian army. He developed the first commercially successful EAF in 1898 using indirect heating from the electric arc, using AC. Figure 14.2 shows the electrodes placed on top of the metallic charge. This type of heating using arc radiation has the big limitation of large heat losses.

This design was initially commercially successful. Pring [5] reported results of a furnace built in 1901 indicating a consumption of electric energy of 4000 kWh/

Fig. 14.1 DC-EAF design
proposed by Siemens in
1878. After [4]

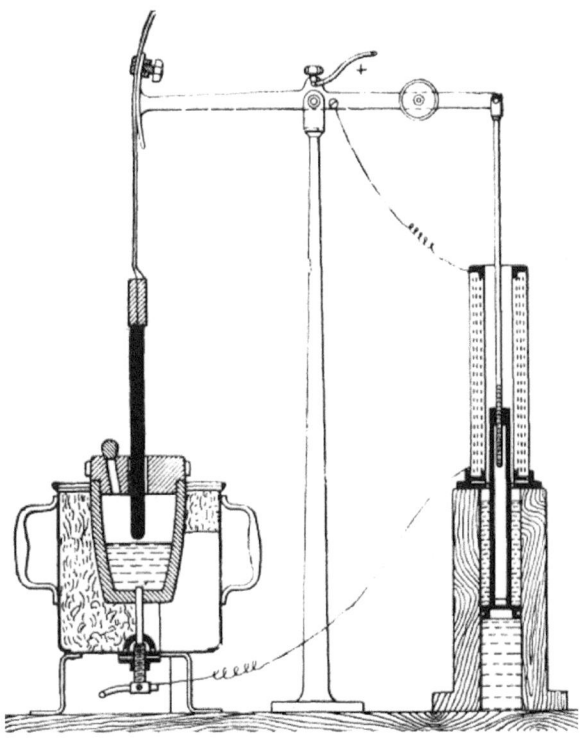

ton and a consumption of electrodes of 12 kg/ton, using an AC-EAF with a capacity
of 1 ton, with 2 kA and 170 V. Its dimensions were a diameter of 1.2 m and height
of 0.9 m. The furnace was equipped with three electrodes inclined slightly to the
horizontal pointing downwards towards the center of the furnace, their separation
angle was 120°.

A typical charge consisted of 450 kg scrap, 250 kg oxidized turnings and 50 kg
lime. After heating for three hours the first slag was removed and formed a second
one using 15 kg of calcium carbide. The second slag was removed after 30 min and
ferroalloys were then added. In 1921, Richards [1] reported a total of 40–45 Stassano
furnaces worldwide. He reported the prices of scrap in those days about 5 USD/ton.

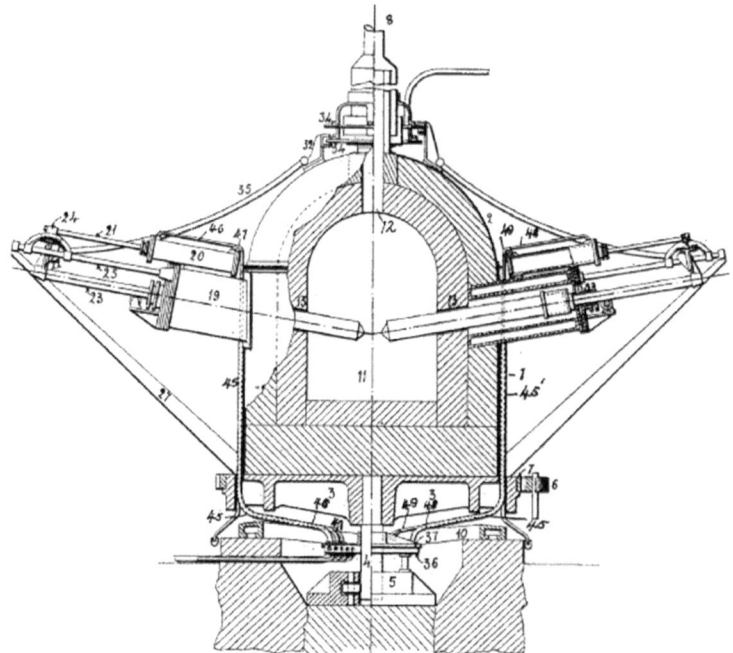

Fig. 14.2 Stassano furnace. After [6]

Paul Girod (1878–1951) [7] was a Swiss engineer. Graduated as a chemical engineer in 1897. In 1898 developed an electric furnace to extract vanadium. Throughout his life he worked as inventor but devoted more time as entrepreneur. In 1903 he founded Société Anonyme Electrometallurgique Processes in Ugine, France focused on the production of ferroalloys. His French patent 350250 from 1905 consisted of DC current and several bottom electrodes. The bottom electrodes were made of iron and was in contact with liquid steel. A proper cooling rate of these bottom electrodes avoids melting, which only occurs on the upper section. In 1909 he started producing steel. WWI allowed to enhance production capacity in his plant but at the same time

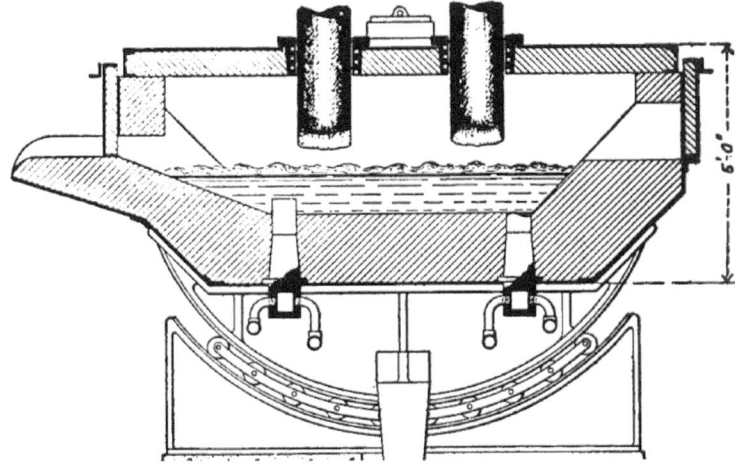

Fig. 14.3 Giroud furnace. After [8]

once war was over it left him with problems, the workers increased from 2500 in 1914 to 3800 in 1918. In 1921 he was forced to retire. Figure 14.3 is a schematic of his design.

Paul Louis Toussaint Héroult or *Paul Héroult* for short was the inventor of the modern EAF [9–12]. He was born in a small town, Thury-Harcourt in Normandy, in northwestern France on April 10, 1863. His father, Patrice, owned a small tannery. His mother, Elise was the daughter of a hotelier. In 1870 he spent three years in London with his maternal grandfather. He attended the Lyceum of Caen Normandy, then St. Barbe in Paris until 1882. At this time, he was 19 years old, entered the École des Mines in Paris. One of his teachers was Henry LeChatelier. For a short time, he served at the French army and then returned to the school for another short time before returning home due to the death of his father in 1883 who previously in 1875 had moved his tannery to Gentilly on the suburbs of Paris. His interest in electrochemistry emerged after reading a book, when he was 15 years old, on the production of Aluminum by Sainte-Claire Deville. During his short stay at the Ecole des Mines he wanted to study this field in more detail. Once he left the school in 1883, Héroult continued his investigations on the production of aluminum in Gentilly with the financial support from his mother and some friends, after a large number of experimental work Héroult finally was able to define the right conditions to produce aluminum. He was 23 years old when he patented his invention No. 175,711 on April 23, 1886. A rare coincidence is that the very same year, Charles Martin Hall, an American chemist not also patented the same idea to produce aluminum but also was born and died the same year as Héroult. Before they invented the so-called Hall-Héroult process, aluminum had the same price as silver but after that, its price was decreased by a factor of 200 and its production grew exponentially. Both inventors founded companies. Alcoa in the USA and the French electrochemical company in France.

Héroult married two times, first to Berthe Béliot in 1888. She passed away in 1898 leaving two children, Paul and Henriette. He married again in 1890 and had three children: Patrice, Anne-Marie and Elisabeth.

About 1892 he became general manager of a new aluminum plant built in La Praz. In 1900 he took a trip around the world with a couple of friends for eight months. His daughter's notes on his father indicate that he was a person who enjoyed the pleasures of life: good food, good wine, sailing and travelling but also an exceptional worker who could forget about time when tried to solve a problem. He hated walking because he was chubby and got tired but instead liked swimming.

His research work on the electric arc furnace in 1900 initially focused on the production of ferrochrome. On October 9th, 1900 he produced steel using an electric arc furnace he invented and on November 12th, 1900 he filed a French patent. On August 21, 1901 filed US Patent 707776, granted to him on August 26th, 1902 [13]. In this patent he disclosed the basic characteristics of the modern EAF, as shown in Fig. 14.4: Alternate current (AC) with a voltage from 1 to 100 V, electric current of 2–5 kA, two square electrodes, capacity of about 1.5 ton, tapping spout and tilting furnace, refractory lined using MgO, dolomite or silica, it could be partially tapped leaving a remanent liquid, it can be charged with hot metal and several tuyeres will assist the oxidation of the melt. In his patent from 1905 [14] he changed the shape of the crucible to a semi-sphere in order to decrease heat losses and increased the number of electrodes from 2 to 3. Operation was fully manual.

In 1904 Héroult came to the United States to attend the St. Louis World's Fair. The USA became the main market for his new invention. In 1905 he filed a new US patent [14]. The main change was an EAF with three square electrodes. In this patent he indicates the maximum EAF capacity was about 4.5 tons. He suggests his design can be scaled up to 300–400 ton.

The Sanderson family had been producing steel since 1776. In 1900 formed crucible steel and in 1902 built the Sanderson-Halcomb company in Syracuse New York. In 1906 Héroult installed the first EAF at the Sanderson-Halcomb company. The first heat was produced on April 5th, 1906. The furnace capacity was 3.4 ton with two square electrodes of 300 mm (12 in.) made of amorphous carbon. The crucible was of rectangular shape and manually charged through the door. The refractory

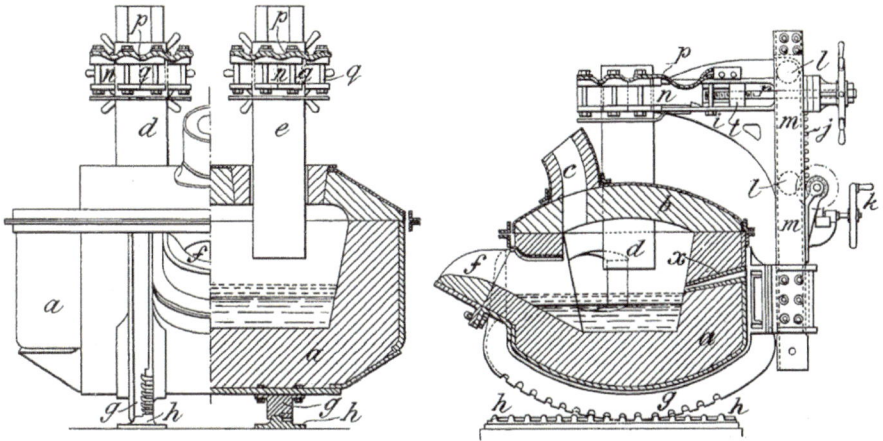

Fig. 14.4 EAF design by Héroult from 1902. Left: front-view, right: cross-view. After [13]

from the shell and roof were made of silica bricks. The sub-station was located on the back of the furnace. The transformer capacity was 0.5 MVA, single voltage at 90 V, and an estimated current of 4 kA [15]. This furnace was operated with short arc and large electrodes to sustain high currents. These conditions increased C content in the liquid steel. It was employed to produce tool steels. Not possible to produce low carbon steel < 0.25%C. Figure 14.5 shows the first EAF in America, located in the former plant in Syracuse NY.

Is interesting to notice that the EAF didn't start as a scrap melting reactor because of the limitations on electric power if cold scrap was employed, instead it started employing a partial charge of liquid steel from the OHF. The EAF was employed for the production of alloy steels and also opened the way for the study of alloying elements in steel.

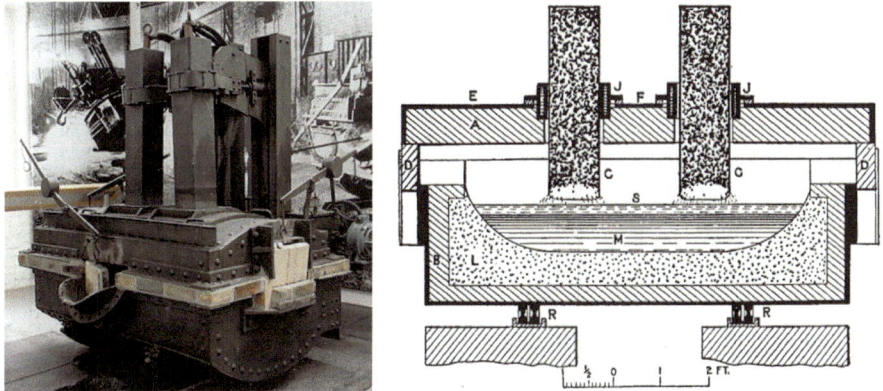

Fig. 14.5 First EAF in America. Left picture taken from [16] and right figure from [8]

In 1906 he was invited by the Canadian Bureau of Mines to help on the reduction of iron ores with the electric arc furnace. The United States Steel Corporation in 1910 acquired the rights to commercialize the EAF process. In 1902, the University of Aachen conferred on him the title of doctor-engineer honoris causa, in 1904 was awarded the Lavoisier medal and the Grand Prix at the Saint-Louis exhibition in the USA. His daughter mentions that in 1912 he was supposed to board the Titanic but missed it. Héroult died on May 9, 1914 of typhoid, in Antibes, during a cruise he was taking on his yacht in the Mediterranean.

14.1.2 Evolution of EAF Design

The advantages of AC current with lower heat losses when transferred from central power stations promoted the design developed by Héroult in 1900. An EAF of 7.5 ton capacity that shows its main characteristics is shown in Fig. 14.6.

From 1900 to 1920 many designs were developed and went temporarily into commercial operation, for example, Fig. 14.7 shows an EAF with 4 electrodes. Due to costs, the number of electrodes should be minimized with an optimum for a single electrode operation, however in terms of heat transfer rate is possible to achieve higher heating rates with a larger number of electrodes [17]. In 1912 Fitzgerald [18] estimated (incorrectly) that increasing the nominal capacity would also require a similar increment in the electrode diameter, for example for the largest EAF in those days, 15 ton, the electrode diameter was 508 mm (20 in.), doubling the steel capacity

Fig. 14.6 Early EAF, 7.5 tons capacity, from 1920. After Pring [5]

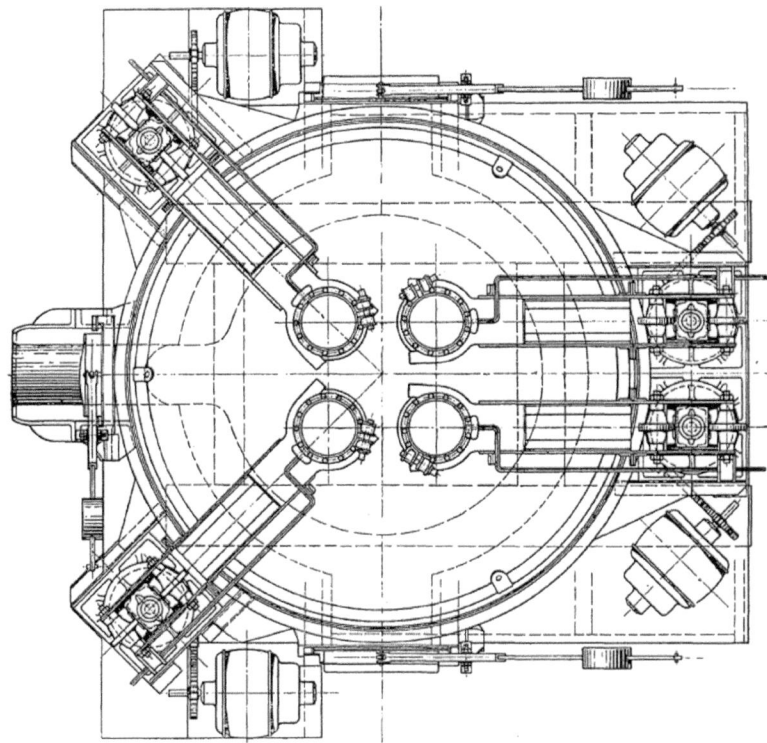

Fig. 14.7 Design of EAF with four electrodes. After [17]

would require electrodes of about 685 mm (28 in.), suggesting that a better option would be to increase the number of electrodes.

Stansfield [8] provides a large review on different EAF designs before 1914.

During the last 121 years the EAF has drastically evolved, however still preserves some of the basic features. It is this technological evolution that has allowed the EAF to decrease its specific consumption of electric energy.

Casey [19] in 1956 reviewed the evolution in EAF design for the first 50 years after the first heat of steel was produced in the EAF. Moore [20] made a technological update until 1930. Ciotti in 1971 described improvements in the secondary circuit to achieve higher power from UHP furnaces [21].

Below is a review in chronological order of some important developments from the EAF:

1907: The roof was raised, about 30 cm from the hearth surface.

1908: First Héroult furnace with three electrodes installed at the Illinois steel company in South Chicago with a capacity of 15 ton, the largest in the world at that time.

1908: Some small Girod and Stassano furnaces are built in the USA but went out of operation due to larger energy consumption in comparison with Héroult.

Westinghouse electric develops an electric regulator to handle the furnace electrodes.

1916: Lectromelt designs its first commercial furnace. This company becomes a top brand in the next 50 years, later on merged with CORE and finally became part of TENOVA. Lectromelt designed its first EAF with a cylindrical shell and spheroidal bottom crucible, adjustable transformer voltage taps, tilting in two directions for tapping and deslagging, included an electric sub-station located at right angles to the tilting plane. The original Héroult design had the substation in front of the door resulting in higher length of the flexible cables and higher impedance in the secondary circuit.

1920s: The furnace capacity increases to 15–20 ton with 0.3 MVA/ton transformers.

1920: Snyder designed an EAF with a tapping spout in the form of a coffee-pot. This tapping practice remained for more than 50 years until the EBT replaced this design.

1920s: The electrodes were originally made of amorphous carbon, mixing petroleum coke with tar, then baked and machined. Amorphous electrodes were very brittle, spoiling the heat when carbon felt into the liquid. Square electrodes also produces big losses due to the stub, about 40%, has to be thrown away when the electrode gets too short [22]. Amorphous electrodes for a 6-ton EAF required a bigger diameter in comparison with graphite, 431 mm (17 in.) and 254 mm (10 in.), respectively. Graphitization at high temperatures provided higher mechanical strength and higher electric conductivity. In 1919 Etchells [23] compared the electric conductivity for amorphous carbon and graphite, 10 and 40 A/cm^2, respectively. Round electrodes and the use of screw threads improved the strength of the electrode column.

1924: Swindell incorporated the first version of top charging using a gantry and a horizontal movement of the roof. This change proved to be very efficient. By 1935, using this method, the tap-to tap time was decreased increasing production 25% and 87% more steel per man hour in comparison with manual door charging.

1925: Variable voltage using a tap changer from 90 to 135 V.

1928: Transformer capacity increased to 7.5 MVA, secondary voltages increased up to 275 V, using Delta-Delta or Delta-Star switching modes. A limitation to keep increasing the voltage above 300 V was a long electric arc and intense heat radiation to the walls.

1929–1939: During the great depression in the USA the EAF proved its flexibility to adapt to difficult times. This experience stimulated its growth after WWII for some time.

1933: EAF with elliptical shell [24]. A 100-ton EAF with an elliptical shell was installed at Timken USA, with two sets of three electrodes, the major axis was 8.84 m and the minor axis 6.1 m, each set of electrodes employed a transformer with a capacity of 7.5 MVA, electrodes of 457 mm (18 in.) and two doors. 25 years later a new furnace was installed with a circular shell of 6.1 m, transformer of 20 MVA, three electrodes of 508 mm (20 in.). The elliptical shell reported higher melting rates. The author didn't state the reason but having more electric arcs and a bigger heated area, enhanced the melting rate.

1937: The shell height increased progressively, up to 70%. This change allowed to charge scrap of lower density, about 1.4 ton/m^3, that could not be charged before. Further increments in the shell height allowed the use of even lighter scrap, in the order of 0.48 ton/m^3.

1945: The air circuit breaker was developed.

1952: The transformer capacity increased from 3 MVA in 1930 to 35 MVA in 1952. The maximum furnace diameter increased from 4 to 7.3 m in the same period of time.

1960s: The EAF started as a specialized reactor to produce high alloy and stainless steels due to the cost of the electric energy. Ciotti [21] indicates this changed after WWII when carbon steels started to be produced in the EAF. After the 1960s the EAF was more popular to produce carbon steels.

1963: First UHP at Northwestern steel and wire in Sterling Illinois USA, 135 ton, 70–80 MVA, 0.5–0.6 MVA/ton, short arc operation. Increasing the power transformer from 25 to 80 MVA increased productivity from 27 to 80 ton/h, the largest in the world in those days. Due to a higher productivity, energy consumption was decreased in this plant from 550 to 480 kWh/ton. In order to operate with short arc length, the power factor was decreased from 0.8 to 0.64. The UHP required an electric balanced system. The co-planar secondary circuit was replaced by the equilateral triangulation of the conductors to improve stability and active power.

1965: Franzen [25] reports one of the earliest applications of water cooling in the EAF roof in USA.

1971: First mega EAF with a nominal capacity of 400 ton at Northwestern steel and wire (NWSW) in Sterling Illinois USA; 9.75 m diameter, electrodes of 711 mm (28 in.) diameter, 30° tilting angle, total weight of 1500 ton including 400 ton of liquid steel, 22 water cooled panels, transformer of 162 MVA, 840 V and 120 kA. The active power ranged from 95 to 105 MW. The productivity reported in 1972 was low, similar to an UHP of 135-ton, electrode consumption was 4.98 kg/ton and energy consumption 460 kWh/ton.

1985: Transformer capacity 1 MVA/ton.

2007: Siemens VAI (now Primetals) built a 365 ton AC-EAF (Ultimate), Colakoglu plant in Gebze Türkiye.

2010: Tenova/Danieli built at Tokyo Steel's Tahara Plant a Consteel EAF of 420 ton of nominal capacity, 256 MVA, twin electrode DC-EAF, energy consumption of 387 kWh/ton, productivity of 320 ton/h, annual production of 2.6 Mton.

2011: First plant with largest transformer capacity, 300 MVA, starts operations in Türkiye.

2012: Transformer capacity 1.5 MVA/ton.

Tables 14.1, 14.2, 14.3 summarize different aspects of the EAF evolution.

A full review of the changes in design is out of the scope of this book, however it should be mentioned some of the major achievements throughout time:

Table 14.1 Evolution in some EAF consumption factors

Reference	Year	EAF, ton	kWh/ton		Electrodes, kg/ton	Tap-to-tap, time
Ayres et al. [26]	1917		1550			
Pring [5, p. 198]	1920	3–3.5		700–800	13–18[a]	4–6 h
Pring [5, p. 198]	1920	6–7			6–7[b]	5–8 h
Moore [27]	1940	12		680		2 h (p-on)
Ayres et al. [26]	1947		1060			
Ayres et al. [26]	1960		800			
Ayres et al. [26]	1968			526		
Ayres et al. [26]	1971		700	500		
Ayres et al. [26]	1998			415		

[a]Amorphous carbon
[b]Graphite

Table 14.2 Evolution in EAF dimensions, capacity, electrode diameter and power transformer

Reference	Year	Shell diameter, m	EAF avg., tons	EAF max, tons	Electrode diameter, mm (in.)	MVA
King [28]	1915	3.3	6	15	431 (17)	
Moore [20]	1930			50		
Moore [20]	1950			100		
	1956			200		
Schwabe [29]	1963	6.7	100		609 (24)	70–80
King [28]	1971	9.7–11.6		400	711 (28)	162

Table 14.3 Examples of EAF pitch circle

Steel plant	EAF, ton	Diameter, mm	MVA	Electrode diameter, mm	Pitch circle, mm	Pitch circle, r/R
Bastug, Turkey [30]	165	8500	150	710	1400	0.16
Emirates steel [31]	159	7000	130	710	1400	0.20

- Source of oxygen for decarburization; replacement of iron ore by air, subsequently by oxygen using consumable lances, later supersonic jets and currently coherent jets, allowing higher melting and decarburization rates.
- Major replacement of refractory lining by water cooled panels which allowed the use of long arc operation with higher voltages, increasing the electric power and the melting rate.

- Slag foaming practice has allowed higher electric power transformers with long arc operation.
- Development of the DC-EAF design, although big differences in energy consumption have not been found between AC and DC EAF's [32].
- Installation of multiple burners for both scrap preheating and post-combustion.
- Minimization of slag-carry over slag using Eccentric Bottom Tapping (EBT).
- EAF stirring using bottom gas injection or electromagnetic stirring.
- Scrap-preheating in batches or continuous scrap preheating.
- Higher replacement of electric energy by chemical energy, which involves; higher supply rates of both oxygen and carbon, higher post-combustion of the CO produced and recycling of the sensible heat in the off-gas. Hot metal has become a common raw material in the EAF.
- Hot heel practice is now common but still remains as an empirical practice.
- Almost fully automated operation is now possible using Artificial Intelligence (AI) through Machine Learning (ML) algorithms.

Figure 14.8 was initially reported by Szekely and Trapaga in 1994 [33]. It became a classic plot to describe the evolution of the EAF and has been updated but the plot needs some comments. First of all, the dates for the technological developments are subjective, usually related to the time they were developed but in practice takes longer time to get accepted, second, there are technological improvements with more relevance than others which is not clear when they are mixed together and third, the final performance values describe the state of the art using all the latest technologies but not a world average. Among the technological achievements, the development of the ladle furnace is the most important in the last 45 years because it marked a technological revolution, the EAF was no longer a reactor to produce the final steel quality but to melt and decarburize as fast as possible a metallic charge. Before the ladle furnace was developed, the EAF was a melting and refining reactor, each of those stages require completely different chemical and thermal conditions, demanding a double slag practice. The figure assigns this development in 1970 but it was until the early 1980s that this technology was adopted worldwide, in the same way, slag foaming is shown to appear in 1980 but it was until 10 years later in the 1990s that became a popular practice. Currently, electric energy consumption can be zero using a high share of hot metal, about 70%, transforming the EAF into a BOF, however the EAF was not designed to use pure chemical energy and cannot reach the decarburization rates of the BOF due to limitations in oxygen injection and reactor geometry. Each technological innovation implied a change in the original design of the EAF, but a change in tapping using a spout for the EBT and scrap pre-heating technologies have made the largest changes in design.

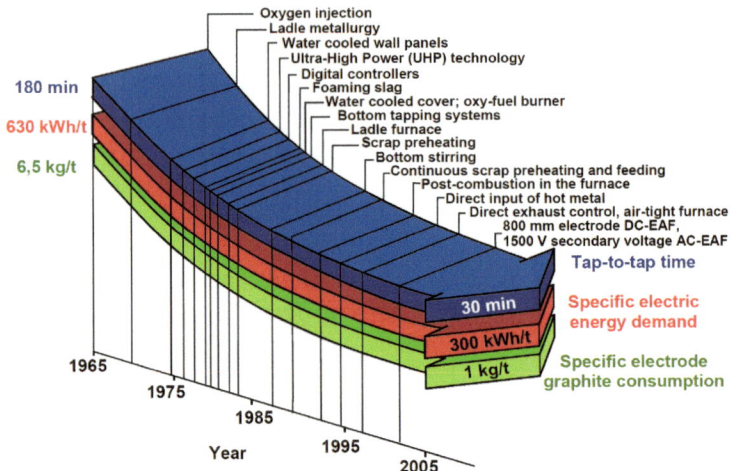

Fig. 14.8 Technological evolution of the EAF. After [33]

14.1.3   H/D Ratio and EAF Geometry

Old EAF's were characterized by large diameters to protect the refractory walls from hot spots radiation. Still today, the conventional EAF is characterized by a low ratio height of liquid steel to heart diameter, in the order of 0.14–0.25. This feature has remained almost unchanged in the entire history of the EAF. A group of experts forecasting the future of the EAF in 2000 anticipated the shape of the EAF would be modified by 2010 to be similar to that of the converter [34] but the idea in early 2024 has not materialized yet.

In 1963, Franzen [35] reported an EAF designed by Lectromelt which consisted of a tapered shell, as shown in Fig. 14.9. They needed to increase the furnace capacity from its original value of 30 tons. The diameter was increased from 3.26 to 3.87 m and its capacity to 56 ton of nominal capacity. The higher diameter resulted in a larger volume of steel scrap per charge, higher slag surface area, better electrode performance but also lower refractory life. The consumption of electrodes was decreased from 3.99 to 3.54 kg/ton, productivity increased from 8.45 to 9.74 ton/h. The refractory life was decreased from 185 to 143 heats, also, in spite of lower refractory life, due to a larger volume of steel produced per campaign the net cost of refractory per heat was decreased.

The shallow bath of the EAF is a factor that limits oxygen gas injection, when it is compared to the BOF. Timoshenko [36] has reported benefits by increasing the H/D ratio which decrease heat losses. He suggests for a 120 EAF ton, a shorter diameter from 5.5 to 4.35 m and an increased height of liquid steel from 1.0 to 1.5 m, representing a change in the H/D ratio from 0.18 to 0.33. Adams et al. [32] pointed out that conventional EAF's have a low H/D ratio in order to protect the refractory.

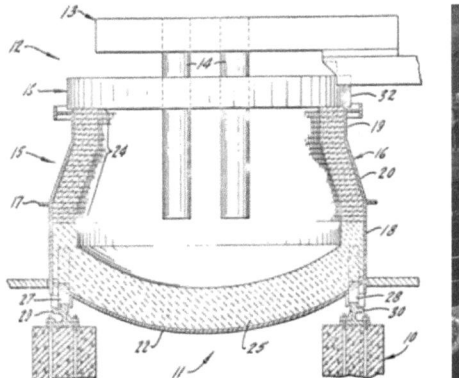

Fig. 14.9 EAF with tapered design. After [35]

By increasing the H/D ratio the roof area decreases, decreasing heat losses and the gas residence time also increases, improving post-combustion heat transfer.

Jones et al. [37] suggest a higher H/D ratio if the injection of solids is increased in the EAF. Currently, several new technologies are available that produce metallic particles (iron carbide and DRI particles from fluidized bed processes). A deeper bath allows a higher injection rate of solids.

The trend in increasing furnace capacity was consistent with an increment in furnace diameter. Casey [19] pointed out a limit to the growth in diameter because it results in cold metal around the banks. Danieli [38] reported a relationship between furnace diameter and furnace tapping weight capacity, shown in Fig. 14.10.

Carney [39] suggested a parabolic shape between the furnace shell and furnace roof would produce homogeneous heating and higher refractory life. He also suggested the need to define a relationship between the input power, pitch circle and furnace diameter to control heat radiation to the walls.

Fig. 14.10 Relationship between tapping weight and furnace diameter. After [38]

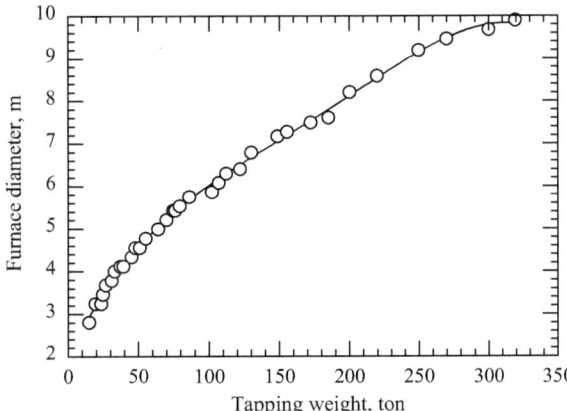

A typical AC-EAF with a capacity of 220 ton, from the 1980s is shown in detail in Fig. 14.11. The upper figure is a plane from the slag door to the EBT. The middle figure is a middle plane as seen from the slag door and the lower figure is a top view. It has a diameter of 6.69 m.

14.1.4 Single Scrap Charge EAF

Laurenti et al. [40] reported a large increment in the shell volume for an EAF with a nominal capacity of 82 ton, in order to achieve single-bucket charging. Both the diameter and height were increased, the diameter 17% and the height 29%, as shown in Table 14.4.

Nal et al. [30] indicated the shell height of a standard EAF with single charge varies from 3.3 to 4.0 m. The maximum value is limited by the distance to lift the electrode columns, liquid steel capacity and bath depth/area.

Daido steel [41] reported in 2017 a movable shell to allow for a uniform melting of the steel scrap. The shell is moved when the scrap at the hot spots is melted. This design allows a higher energy efficiency of 5%.

14.2 Automation

14.2.1 Machine Control Through Algorithms

Our society is moving into more automation because of its large benefits which include higher productivity, lower energy consumption, replacement of labor in dangerous activities, etc. At the same time is moving so fast that is replacing labor at

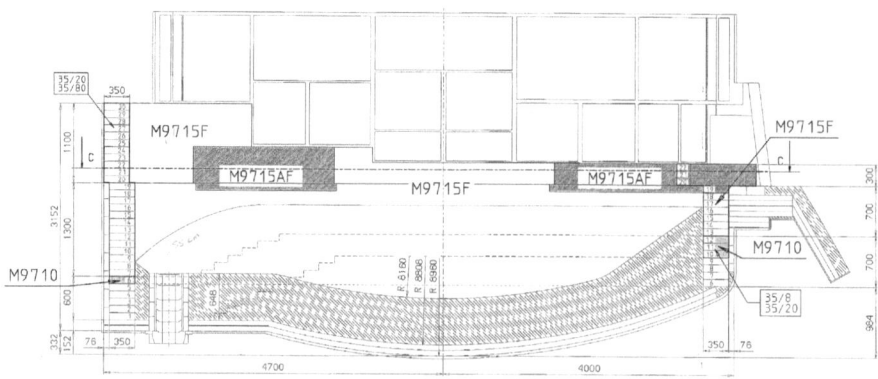

Fig. 14.11 Conventional AC-EAF with 220 ton of nominal capacity

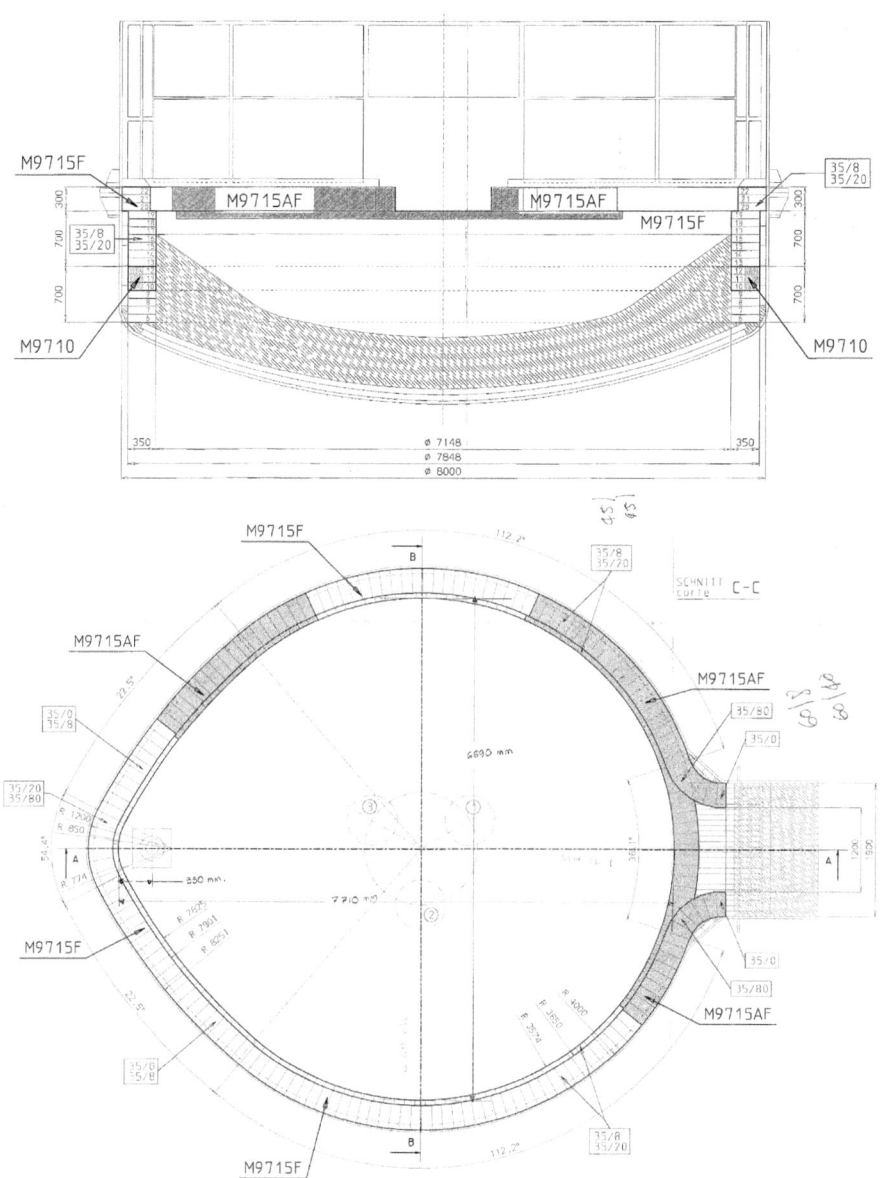

Fig. 14.11 (continued)

Table 14.4 EAF modifications to achieve single-bucket charging

	82 ton std.	82 ton large
Shell diameter, mm (D)	5700	6700
Height from top to sil line, mm (H)	2700	3500
Height WC panels	2300	3100
D/H, shell	2.47	1.91

After [40]

a massive scale. Many futuristic scenarios have been suggested. In her essay on the ironies of automation, Bainbridge [42] points out that one of the ironies is that "an automatic control system has been put in because it can do the job better than the operator, but yet the operator is being asked to monitor that it is working effectively" because a machine will eventually fail and will repeat the error if it is not stopped and corrected by a human. The philosopher H. Dreyfus argued that computers, who have no body, no childhood and no cultural practice, cannot acquire intelligence at all [43]. Harari has suggested that artificial intelligence (AI) should be controlled to avoid catastrophic inequalities [44].

The fact is that after the creation of computers scholars are producing more and more algorithms that control machinery and help produce steel with a higher productivity and minimum labor. The model based on automation is called intelligent manufacturing or industry 4.0. Full automation controlled by AI seems to be impossible if first it is recognized that intelligence is an ability to handle unknown events, as was first identified by the German philosopher Nicholas of Cusa in 1444 [45]. Cusa divided intelligence as a result of two components; ratio and intellect. Ratio is the ability to calculate based on existing knowledge and intellect the ability to handle the unknown and the uniqueness of unfolding situations.

One of the big challenges for automation is the processing of a huge amount of data, in the order of several terabytes per year due to the increased number of sensors installed in all the areas of a plant [46]. Automation was classified in 1980 by Sheridan [47] into 10 levels, from full human operation (level 1) to full machine operation (level 10).

Automation is one of the fields where the EAF has made great progress, in particular in the last 50 years. As an example, in Mexico in the early 1980s, all ferroalloys were charged manually and the operator relied on the empirical observation of the fumes to estimate FeO, the spark test was common to estimate the carbon content and observing the color of a steel rod immersed into liquid steel to estimate its temperature. The modern furnace today is equipped with multiple sensors to measure on-line different signals from the whole process. Tenova [48] has proposed an automation system based on three levels; first is the level of sensors and instrumentation, second is the level of dynamic process models and the third one corresponds to control and optimization modules. One of the main components is the computation of the energy balance using measurements of the off gas at different points in the exhaust system, the energy input and heat losses. The instantaneous energy balance allows for the computation of the melting rate and progress in scrap melting.

14.2.2   Industry 4.0

Mathematical models either based on machine learning, thermodynamics and rate phenomena, CFD, etc., are one key component of the current "Fourth industrial revolution" or Industry 4.0. The first industrial revolution from the eighteenth–nineteenth centuries mechanized manual labor. The 2nd industrial revolution was industrialization. The 3rd industrial revolution is marked by the use of computers and the internet. This applies for the industry in general. For the steelmaking industry, the first industrial revolution was the development of the Bessemer process in 1856 which represented the massification in steel production. The second industrial revolution occurred in 1981 with the introduction of the ladle furnace which transformed the main role of the BOF and EAF as melting reactors. Today, in the steel industry it would be Industry 3.0, characterized by the use of computers, algorithms, artificial intelligence and automation.

The general concept of Industry 4.0 is under development and is characterized by the industrial internet of things, artificial intelligence and automation. Machines work with little human intervention. According with the concept Industry 4.0 [49], manufacturing enterprises should be highly efficient in resource utilization, highly automated and quickly adapt to customer requirements. Liboni et al. [50] has addressed its implications on education, employment and human–machine interactions. Branca et al. [51] suggest that digitalization is a pre-condition for industry 4.0 but its implementation requires availability of skilled people in the labor market. Pellegrini et al. [46] have described several examples of automation in the steel industry, illustrating the complimentary roles between sensors and machine learning algorithms.

14.2.3   Delays (Power-Off Time) Due to Manual Operations

Many of the delays are due to manual operations. Delays have a high impact on productivity and energy consumption. In this area, there are many developments in automation to decrease human intervention.

Energy losses due to delays can range from 0.4 to 1.7 kWh/ton per minute, depending on slag foaming conditions and are below 0.5 kWh/ton per minute between heats because the hot heel is covered by a slag layer [32].

A benchmarking reported by Lugo [52] indicates an average power-off time of 15 min, similar value for AC and DC-EAF's. The shortest value of power-off time, two minutes, was obtained using twin shell EAF's and using the best metallurgical practices the range can be from 6 to 9 min.

The more typical delays are:

- *Tilting for tapping and de-slagging.* Furnace movements are made using hydraulic systems. In an old EAF with spout tapping the required angle was about 45° but with the EBT system the angle is shorter, from 15 to 20°. Modern EAF's have a tilting speed in the range from 3 to 4 degrees per second [37]. Tapping has a high manual intervention.

- *Scrap-charging*: The roof should be raised and swing to charge the scrap baskets. The swing speed is about 4–5 degrees per second and about 30–40 s to rise and swing the roof [37]. Scrap preparation to decrease the number of recharges and handling the scrap buckets involves high manual intervention.

- *Turnaround time*: is the time between the completion of tapping and the furnace is ready for the next heat. The length of electrodes can be adjusted or electrodes are added, the tap-hole is sealed with sand, the furnace is inspected and gunned if necessary. This a stage with high manual intervention.

- *Non-scheduled maintenance*: Maintenance affects productivity and cost of steel. Junger et al. [53] in 2009 compared BOF maintenance philosophies in USA and Europe, in the first case using maximum maintenance to produce an everlasting refractory campaign and the second with a zero-maintenance concept (barely the minimum), in other words repair versus replacement. Maintenance decreases productivity, 18–26 heats/day (tap-to-tap 50–70 min) for USA and 24–36 for Europe (tap-to-tap 40–60 min). With zero maintenance, productivity increases but the refractory campaigns are lower, 20,000–35,000 heats/campaign versus 2500–5000 with zero maintenance. The authors suggested an optimum of 5000–7000 heats/campaign to produce the lower operation costs. Life Cycle Cost Analysis (LCCA) has been suggested as a tool to make a more accurate decision on the optimum maintenance schemes [54].

- *Steel and slag samplings*. Mostly done manually.

  - Delayed response on chemical analysis,
  - Changes in steel grade,
  - Lack of materials.

The progress made on automation in the EAF has brought a faster response. For example, in the past, the turnaround time was higher because of the use of a tapping spout and the need to rotate the furnace using slow mechanical systems.

Further improvements on automation will keep decreasing the power-off time. In fact, EAF manufacturers claim to have developed enough software and hardware to fully automate steelmaking operations [55]: measurements on the off-gas system are used to control post-combustion, the electrode control system adjusts the arc length, computing the energy radiated can be used to adjusts the water flow rate for the WCP and optimize slag foaming, as well as the water flow rate for electrode cooling, use of robots to measure temperature and sampling without turning the power off, etc.

14.2.4 Furnace Productivity

The practical importance of productivity is because increasing productivity also decreases production costs. Productivity can be defined as a ratio between the outputs generated from a system to the inputs that are used to create those outputs. In EAF steelmaking, productivity can be defined in different ways, for example:

1. Ratio between the energy in liquid steel with respect the total energy supplied. This ratio is the *thermal efficiency*.
2. Amount of steel per unit time, for example, tons of liquid steel per hour (ton/h), heats per day, ton per hour per MW. This definition corresponds to *furnace productivity*. Time is related to the tap-to-tap time.
3. Tons of liquid steel per man-hour. This definition corresponds to **labor productivity**. New technologies with automation can replace human labor and increase productivity.
4. Mass of liquid steel with respect to the mass of metallic charge. This is the definition of *metallic yield*.
5. Selling price of steel with respect to the production cost *(profits)*. A similar ratio is the **profits per worker**.

From the previous definitions, productivity can be increased in many ways; by increasing the thermal efficiency which involves minimization of energy losses, increasing the furnace capacity, increasing the transformer capacity, enhancing automation, improving the reduction rate of FeO or the mixing conditions to approach equilibrium, using cheaper raw materials which do not significantly affect the tap-to tap time and energy consumption, etc.

Madias et al. [56] reported a survey of 190 EAF's and defined a relationship between furnace capacity and furnace productivity, as shown in Fig. 14.12. This figure shows that increasing the furnace capacity it also increases the furnace productivity.

There is no doubt that increasing furnace capacity increases furnace productivity, however there are limitations to this growth because of requirements which cannot be matched with other ancillary equipment, for example, higher electrode diameter and higher capacity of the furnace transformer. Toulouevski and Zinurov [57] argue that installing one EAF is better compared to several EAF's, in particular if the annual capacity is above 2 Mton, installing EAF's with a capacity from 300 to 400 ton is

Fig. 14.12 Technological evolution of the EAF. After [56]

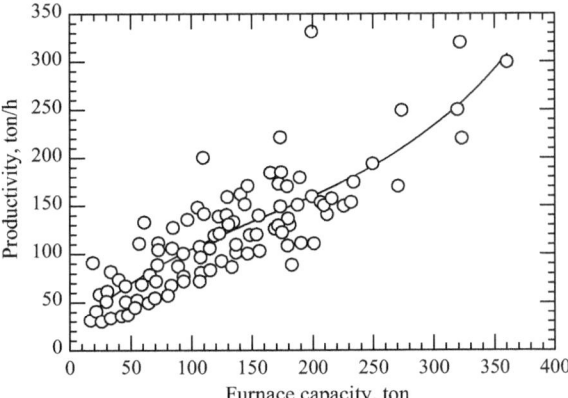

better, because in the case with several furnaces disruptions in one furnace affect the other furnaces. It seems that the number of EAF's above 300 ton is rare, three examples are the Colakoglu plant in Türkiye (2007), Tahara plant in Japan (2010) and Iskenderun plant in Turkiye (2011) with 365, 420 and 300 ton, respectively. Abel and Hein [58] reported design details for the AC-EAF at Colakoglu, summarized in Table 14.5. This EAF is an Ultimate model built by Siemens-VAI (currently Primetals), characterized by a high input of chemical energy.

The EAF installed at the Iskenderun plant in Türkiye was built by Danieli for MMK-Atakas, a joint venture between the Russian MMK Magnitogorsk Iron & Steel works and Turkish Atakas. The design parameters reported by Sellan and Fabbro [38] are given in Table 14.6. The EAF also has a considerable input of chemical energy, a high input of electric power with a transformer rated 1.2 MVA/ton. The oxygen jets and carbon injection lances are shown in Fig. 14.13.

For the same nominal capacity, the furnace productivity of an EAF can also be defined in terms of the number of heats. Dollé reported than in 30 years, from 1965 to 1995, the number of heats increased from 2500 to 8000 heats/year due to shorter tap-to-tap times. Still far from the productivity from the BOF but the gap is becoming closer.

Table 14.5 Details of large EAF at Colakoglu plant

Tapping weight, ton	320	Porous plugs	6
Max productivity, ton/h	356	Electrode diameter, mm (in.)	750 (30)
Av. productivity, heats/day	24	Pitch circle, mm	1450
Production, Mton/year	2.78	Electrode consumption, kg/ton	0.97
Scrap, %	100	Burner capacity, MW (each)	4
Transformer, MVA	240 + 20%	Number of oxy-fuel burners	8
Current, kA	85.7	Number post-combustion inj.	4
Voltage, V	800–1616	Fuel consumption, N m^3/ton	3.6
Av. active power, MW	148	O_2 flow rate (each), N m^3/h	2888
Hot heel, ton	45	O_2 flow rate (total), N m^3/h	19,878
Av. tap-to-tap time, min	60	O_2 consumption, N m^3/ton	32.4
Av. power-on, min	43	Number carbon injectors	6
Av. power-off, min	15	Carbon flow rate (each), N m^3/h	63
Volume, m^3	268	Carbon injected, kg/ton	13.4
Shell diameter (D), m	9.14	Avg. power input, MW	160
Bath depth (H), m	1.35	Scrap baskets/heat	2–3
H/D	0.147	Charged carbon, kg	1133
Panels height, m	3.15	Energy consumption, kWh/ton	326

After [58]

Table 14.6 Design details of large EAF at Iskenderun plant

Nominal capacity, ton	300	Porous plugs	Yes
Tapping weight, ton	250	Electrode diameter, mm (in.)	810
Production, Mton/year	2.4	Metallic charge 1: scrap-pig iron	80–20
Melting rate, ton/min	9.6	Metallic charge 2: scrap-DRI	65–35
Charge, scrap-pig iron, %	80–20	Number of oxy-fuel burners	8
Transformer, MVA	300	O_2 flow rate (total), N m^3/h	20,400
Transformer, MVA/ton	1.2	O_2 consumption, N m^3/ton	40–45
Max. current, kA	110	Number carbon injectors	5
Av. active power, MW	190	Carbon flow rate (total), N m^3/min	210
Tap-to-tap time, min	47	Scrap baskets/heat	2
Power-on, min	36	Energy consumption, kWh/ton	340–390
Volume, m^3	330	DeC rate, kg/h m^2	300

After [58]

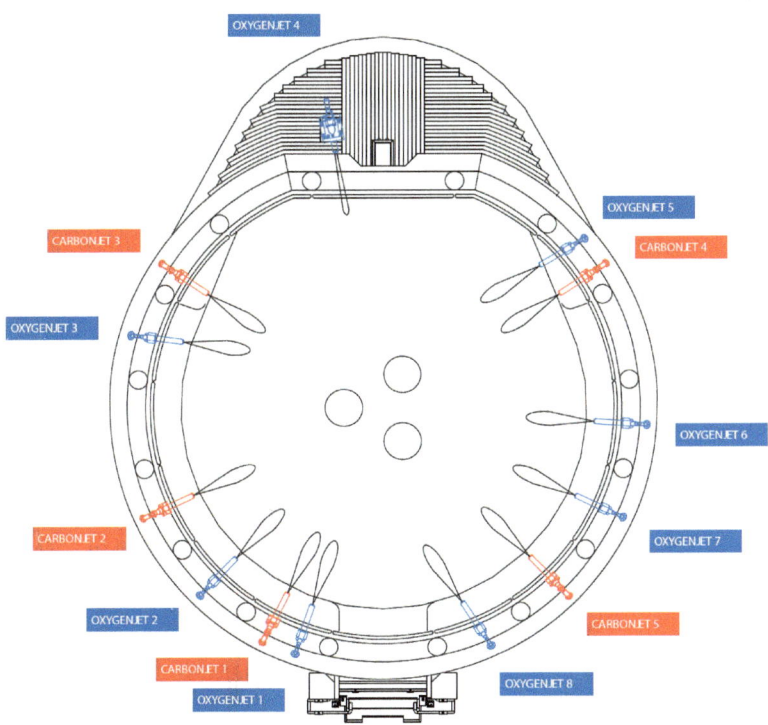

Fig. 14.13 AC-EAF 300 ton. After [38]

Pesamosca et al. [59] reported an example to increase furnace productivity in terms of ton/h and heats/day, by increasing both furnace capacity and process optimization. The furnace considered in their analysis was originally designed for a future revamping from 120 to 150 ton of nominal capacity and the main changes consisted of a decrease in the heart thickness refractory. The dimensions are shown in Table 14.7. The H/D ratio was increased from 0.146 to 0.158. The furnace transformer's capacity was 140 MVA. The increment in productivity from 190 to 212 ton/h and a maximum of 34 heats/day was due to a higher volume per heat, higher electric power and optimization of chemical energy. The active power was increased from 87 to 95 MW with a refractory wear index of 240 KVW/cm^2. The EAF employed 5 oxy-fuel burners, as shown in Fig. 14.14, in oxygen injection mode 2300 N m^3/h per injector and in burner mode a power of 4 MW per burner. The specific consumption of oxygen was 32 N m^3/ton, consumption of charged carbon 10 kg/ton, specific consumption of injected carbon 9.9 kg/ton, specific consumption of lime and dolomite, 43.6 and 6.7 kg/ton, respectively, tap-to-tap time 44 min, power-on time 34.6 min. The total energy consumption was 701 kWh/ton and the electric energy consumption was 365 kWh/ton, equivalent to 52% of the total input energy.

Increasing transformer capacity can increase productivity if the fraction of power-on time is at least 0.7–0.8. A high power off time off-sets the advantages of a high melting rate, similarly, by increasing the size of the EAF it increases the output of liquid steel but the size of the ladle limits this revamping strategy.

Memoli et al. [60] has reported that the productivity of the Consteel process can be increased by increasing the hot heel, from 1.8 to 2.3 ton/h per MW if the hot heel is increased from 36 to 56%. Further increments in the hot heel decrease the EAF productivity.

Laurenti et al. [40] in 2005 provided an example how to enhance productivity, considering two main aspects; *technological improvements and smart management*. On one side, a medium size furnace with a nominal capacity of 82 ton was provided with state-of-the-art equipment; expanded volume to reach a single-bucket charge practice, S-UHP transformer with a rated capacity of 110 MVA (813–1201 V and up to 70 kA). This high-capacity transformer is equivalent to 1.35 MVA /ton (0.9 MW/ton of effective active power, about 75 MW total). 5-step serial reactor with maximum reactance of 2.7 Ω. Tilting for tapping at 2° per second, roof swiveling at 5° per minute, roof lifting at 50 mm per second, standard electrode speed at 120 mm per second and 400 m per second for emergencies, four injectors, three mounted on the WCP and one on the EBT balcony, in burner mode each one with a power up to 5 MW with 500 N m^3/h of natural gas and 1100 N m^3/h of O$_2$, as post-combustion can inject

Table 14.7 Dimensions and characteristics of modified EAF

D$_h$, m	H$_h$, m	H$_h$/D$_h$	D$_b^s$, m	D$_t^s$, m	H$_{wcp}$, m	Electrodes diameter, mm	Pitch circle, mm	V, m^3
6.35	1.0	0.158	7.4	7.2	3	714	1350	140

After [59]

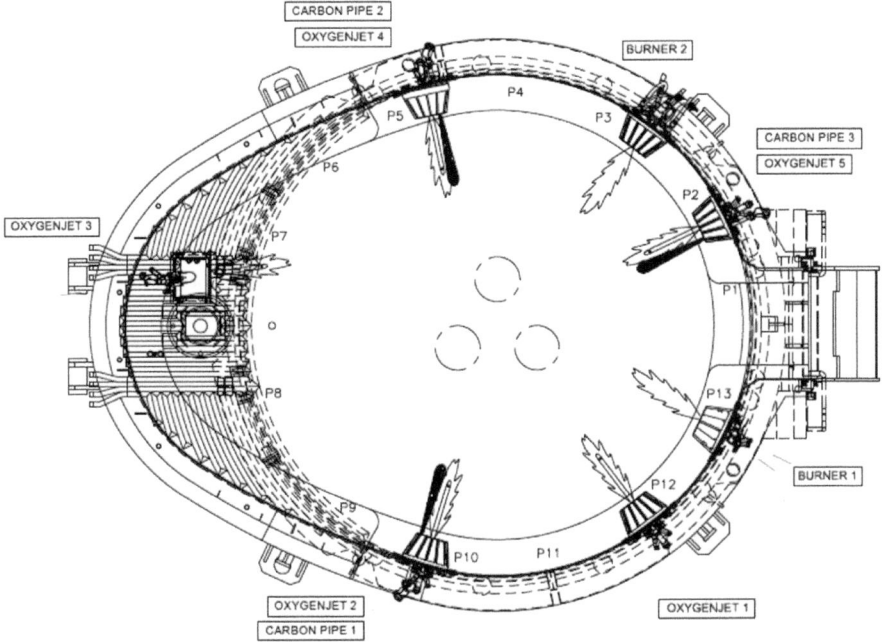

Fig. 14.14 Characteristics of EAF with enhanced productivity. After [59]

600–1000 N m³/h of O_2 and as oxygen lance from 1300 to 2500 N m³/h of O_2. This gives a total burner power of 20 MW and about 8000 N m³/h of lance oxygen to maximize post-combustion. Each injector can introduce up to 80 kg C/min for slag foaming. On the other hand, smart management which involved; outsourcing scrap management, the entire scrap storage area was served by movable scrap loaders, scrap was transferred to the bucket charging area by movable scrap loaders, reducing crane movements using a shuttle car to move the ladle from the EAF to the LMF, EAF control room located well above the roof level to get excellent visibility, from this control room the operator charges the scrap bucket taking from 60 to 80 s per heat including electrode movements, roof opening and closing. These changes were aimed at producing 1 Mton of steel per year and a productivity of 140 ton/h (or 1.86 ton/h per MW).

Kirschen et al. [61] has analyzed the effect on productivity by decreasing the total number of EAF's in Germany and replaced with more efficient EAF's. In the period from 1975 to 2000 the number of EAF's decreased from about 95 to 27 but the annual steel production increased from 5 to 13 Mton.

14.2.5 Labor Productivity

Labor productivity has been continuously increased due to many factors such as implementation of new technologies but decreasing the labor force has had a huge impact. Figure 14.15 shows the number of employees in the iron and steel industry in the USA from 1950 to 2021 [62–64]. It should be noticed that the workforce is divided in two parts; the total number of employees in the iron and steel industry involves steel manufacturing (steel industry or finished steel), and the second group is only the number of employees for the processes involved in the production of liquid iron and liquid steel. Labor productivity for the global steel industry involves the first group and labor productivity for steelmaking should only include the second group of employees. The workforce in the steel industry in the USA has been dramatically decreased, from 674,000 workers in 1950 to 131,400 workers in 2021. During this period of 70 years, the production of steel decreased from 96.8 to 85.8 Mton/year, in terms of labor productivity, it increased from 144 to 652 ton/worker per year.

If labor productivity is defined in terms of the number of employees to produce crude or liquid steel, the values are higher. Figure 14.16 reports USA labor steelmaking productivity from 1990 to 2022 [64]. With data from the world steel association for the liquid steel produced per year and the USA Bureau of labor statistics, the increment in USA labor steelmaking productivity from 1990 to 2003 increased from 527 to 911 ton/worker per year, about a two-fold increment. Similar results on productivity were reported for the EU and Japan [65]. USA labor productivity experienced a rapid growth from 1980 until 2004 but has remained almost stagnant for the last twenty years.

According to the American Iron and Steel Institute, USA labor productivity saw a five-fold increment since the early 1980s, going from an average of 10.1 man-hours

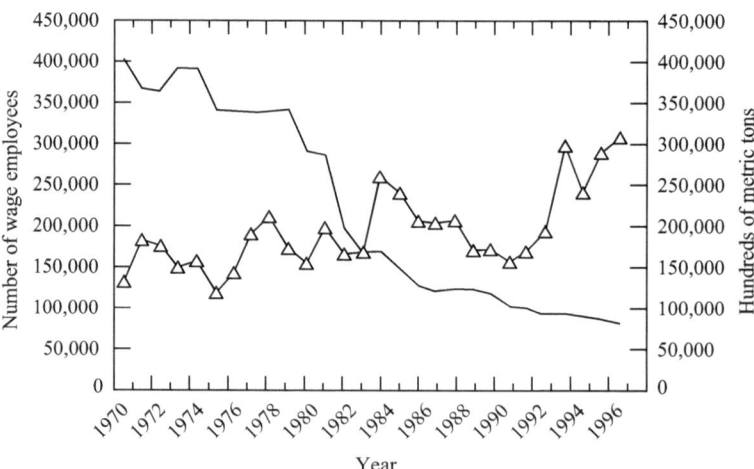

Fig. 14.15 Number of workers in the USA steel industry and steelmaking. After [62–64]

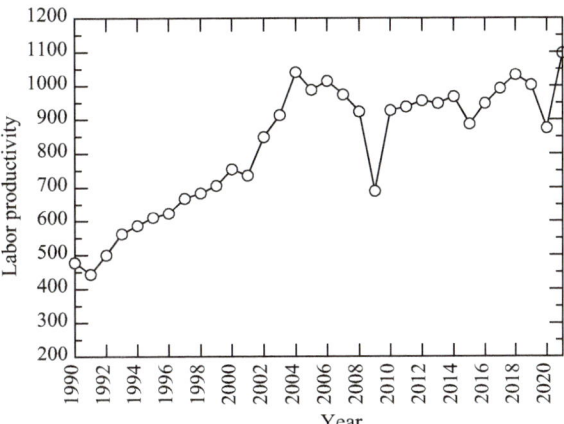

Fig. 14.16 Labor steelmaking productivity in the USA. After [64]

per finished ton of steel to an average of 1.9 man-hours per finished ton in 2015 [66], as shown in Fig. 14.17.

Labor productivity and wages: The relationship between labor productivity and wages has been investigated indicating a strong relationship [68, 69]. Kuznietsova [69] studied the relationship between productivity and wages in four countries (France, Indonesia, Kenya and China) and the result was a positive correlation, which leads to conclude that a raise in salaries can also be used to promote productivity.

One factor that contributes towards a policy of better wages is the progressive decline of the labor force on the cost of steel compared with the cost of raw materials. Table 14.8 summarizes the contribution of labor and materials to the production cost. Raw materials represent more than 63%.

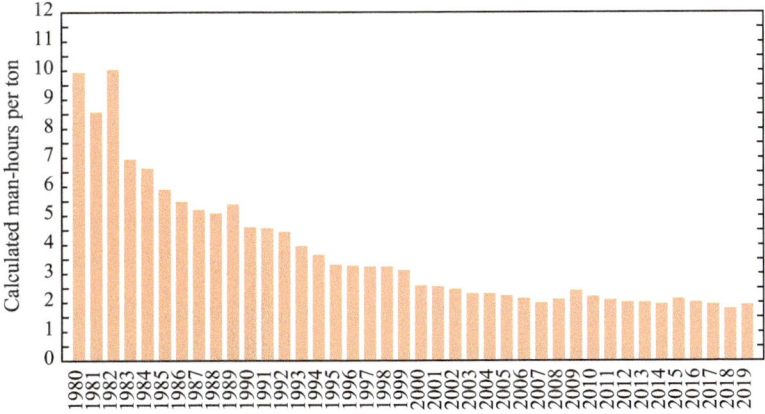

Fig. 14.17 Productivity of the USA steel industry in man-h/ton. After [67]

Table 14.8 Contribution of labor and materials to the production costs

		Labor, %	Scrap, %	Energy, %	Refractories, %	Electrodes, %	Waste magmt, %
[70]	IISI, 2006	3	73	7	3	5	2
[65]	French steelmakers assoc. 2002	6	65	10	3	3	
[71]	AM Mexico, 2000	*	63	15	4.4	3.56	

Labor productivity and non-unionized workforce: It would be extremely complex to generalize if a unionized or a non-unionized workforce is better in terms of productivity because productivity depends on many variables such us country, region of the country, industrial sector, specific characteristics of the employees and management in each company, etc. There are examples that exhibit benefits from each side. In countries where the management care more about profits for the shareholders, unions play a vital role, however, there is also the other extreme with powerful unions which care more about salaries irrespective of performance. The author has experienced both extremes and both extremes cause both sides losing in the end, leading to an unhealthy atmosphere with low productivity, typical but not limited to non-developed countries.

In the steelmaking industry there are two classic examples of non-unionized companies with high labor productivity due to both worker's performance and intensive application of new technologies, Dofasco in Canada and Nucor in the USA [72]. Non-unionized companies pay lower wages in comparison with unionized companies but in the end the final salary can be higher due to workers performance, high productivity and higher profits.

Dofasco (Dominion Foundries and Steel Company) was founded in 1912 in Hamilton, Ontario. Its owner C. W. Sherman was against unions. The way to consolidate Dofasco as non-unionized company was not easy. Storey [73] describes in detail a history that involved ruthless workers intimidation to reject a workers' union. Eventually, after the 1950s, the profit-sharing fund was the central part of benefits to accept a non-unionized workers organization. In the period from 1964 to 1980 the numbers of employees grew from about 8000 to 14,000 and then progressively decreased, during this time, in spite of a larger number of workers, labor productivity increased from about 200 to 300 ton/worker per year.

Nucor steel was made by K. Iverson who took a company on the verge of bankruptcy in 1965 and made significant decisions to make it the most successful in the USA steel industry, one of his main decisions was to produce steel using the EAF in 1969. Its high labor productivity also benefited from intensive application of the latest technologies for the EAF. Nucor expanded the minimill concept. In addition to this, Iverson also is credited with the final development of thin slab casting, a technology developed by SMS Schloemann-Siemag but implemented on a commercial scale by Nucor. Its labor productivity has been the highest in the USA since the early 1970s; from a very low labor productivity of 23 worker hours per ton in 1958, decreased to 10 in 1974 and in 1993 to less than 2 worker hours per ton, according to Lieberman [74] who compared labor productivity of the steel industry in Japan and USA from 1958 to 1993. Nucor has been taken as an example of those companies focused on the workers and the combination with efficient management and implementation of new technologies [62] which resulted in a successful and profitable company.

14.2.6 Effect of Technology on Productivity

In a large perspective, there are two general ways to increase productivity, improving the efficiency of existing processes or installing new technologies. Baran and Rokicki [75] compared the impact of technical efficiency and technological change in the productivity of the metal construction industry in the USA from 2006 to 2014 using the Malmquist Productivity Index (MPI). An index higher than 1 indicate an increment in productivity. The found a higher value due to technological progress in comparison with technical efficiency. Therefore, they concluded that the main driving force to improve the metal manufacturing industry was the implementation of new technologies, not improvements in technical efficiency. A similar study for the iron and steel industry in China from 2001 to 2008 indicated energy inefficiency and also the technological change as the main contributor to the growth rate of productivity. The study suggested the elimination of low energy efficiency plants. The average energy efficiency was 61% and the annual growth rate of productivity was 7.96%.

Collard-Wexler and De Loecker [76] studied the effect of technology on productivity in the steel industry in a period of 40 years, from 1963 to 2002. They reported a sharp increase in productivity and the cause was a new technology; the minimill. The implementation of the new technology didn't cause a replacement of the old technology, on the contrary also involved its own productivity resurgence. The displacement of the older technology was responsible for one third of the increment in productivity. The minimill is less labor intensive and was a factor that promoted a reduction in the workforce of 0.4 million people from 1962 to 2005. The result was that labor productivity increases by a factor of five and the total factor productivity increased 38%.

14.2.7 Effect of Business Model on Productivity

The business model, either state ownership or private ownership, has shown to influence productivity. Previous to the 1980s a large number of steel plants were state owned enterprises but that changed in the 1980s. Not only became private owned but also became giant steel groups. Closing a steel plant was easier for the owners because labor could be relocated to other plants. These changes in the business model are a central part of the neoliberal model. In Mexico, starting in 1988, the government sold all steel companies at prices much below its real value. The situation of the Mexican steel industry was bad due to many reasons, one of them was poor management. In private hands the steel plants increased its productivity because of massive layoffs and capital investments. China, currently is an example of state-owned enterprises with high productivity based on performance bonuses. To avoid a further discussion on this complex subject, I would say that there is a major factor that defines which business model is better; the one that results in higher productivity and better economic equality for the share-holders and workers.

14.2.8   Effect of Skilled Labor on Productivity

The most important asset in a company is the human capital and how it is managed in a highly competitive market. Ouvradou [65] has pointed out that productivity is entirely dependent on the company's organization. In addition to efficient technology a lean management with high level of staff training which grants them a significant level of responsibility makes the difference to reach high performance. Nda and Fard [77] state that "in order to maintain sustainability, organizations must see continuous employee training and development as invaluable. Training and development are very essential at all employee levels, due to the reason that skills erode and become obsolete over a period of time and has to be replenished". Training is the most effective way of motivating and retaining high quality in human resources within an organization.

Lee et al. evaluated six strategies related to Human Resource Management (HRM) and its effect on productivity in the steel industry in Taiwan. The HRM practices included; Training, teamwork, wages, HR planning, employee performance and employment security. Their study found that all six practices help improve firm performance, in line with other numerous investigations included in the same report.

Measuring the relationship between training and productivity is a complex problem, however it has been concluded that training has direct effect on three levels; learning new skills, motivation and loyalty to the firm [78]. Eventually better skills increase productivity and therefore the possibility of higher wages.

Today, training is even more essential for the steel industry because of the highly competitive market. Current success rest at the hands of decisions made by skilled people. Survival depends on the technical improvements and innovations which can be created by its workers.

References

1. Richards JW (1920) The electric melting furnace. J Am Inst Electr Eng 39:1034–1036. https://doi.org/10.1109/joaiee.1920.6594644
2. König W (2023) William Siemens: an engineer and industrialist in Germany and England. In: Springer International (ed) Professions and proficiency, pp 207–219
3. Jones RT (2016) DC arc furnaces—past, present, and future. In: Celebrating the megascale: David Robertson symposium, pp 129–139
4. Keller CA (1909) A contribution to the study of electric furnaces as applied to the manufacture of iron and steel. Trans Faraday Soc 5:113–134
5. Pring JN (1921) The electric furnace. Longmans, Green and Co., New York
6. Stassano E (1905) Revolving electric furnace. US patent 799105
7. Gavard-Perret F (2009) Paul Girod d'Ugine: la carrière d'un ingénieur entrepreneur de l'électrométallurgie. Cah d'histoire l'aluminium 1:108–129
8. Stansfield A (1914) The electric furnace: its construction, operation and uses. McGraw-Hill
9. Roeber E, Parmelee H (1914) Obituary of Paul Héroult. Metall Chem Eng 12:382
10. Bally J (1937) Biographies of Paul Héroult. Les Ann Des Mines. https://www.annales.org/archives/x/heroult.html. Accessed 4 Oct 2023
11. Héroult E (n.d.) Souvenirs sur Paul Héroult. https://paul-heroult.pagesperso-orange.fr/Souvenirs.htm
12. Habashi F (2010) A short history of electric furnaces in iron and steelmaking. Part 1 the pioneer days of the arc furnace. Steel Times Int 52
13. Heroult PLT (1902) Oscillating electric furnace. US patent 707776
14. Heroult PLT (1905) Steel mixing process. US patent 807027
15. Casey SB (1956) The first half-century of electric furnace steelmaking. J Met 1638–1640
16. Wikipedia (2010) Heroult furnace. https://en.wikipedia.org/wiki/Crucible_Industries#/media/File:Remscheid_-_Werkzeugmuseum_in_-_Lichtbogenofen_02_ies.jpg. Accessed 9 Oct 2023
17. Bibby J (1919) Developments in electric iron and steel furnaces. J Inst Electr Eng 57:231–246
18. Fitzgerald FAJ (1912) Thirty years' progress in the electric furnace. Proc Am Inst Electr Eng 31:875–888
19. Casey SB (1956) The first half century of electric furnace steel making. JOM 1638–1641
20. Moore WE (1931) Twenty year advance in electric arc furnaces for the production of iron and steel. Trans Electrochem Soc 60:165. https://doi.org/10.1149/1.3497861
21. Ciotti JA (1971) A new era in melting. JOM 23:30–35
22. Mathew JA (1932) The electric furnace and the alloy age. Trans Electrochem Soc 61:143–160
23. Etchells H (1919) Application of electric furnace methods to industrial processes. Trans Faraday Soc 14:71–78
24. Preston JK (1954) Elliptical electric furnace outperforms conventional circular type. JOM 6:18–20
25. Franzen EF. New furnace innovations-2. Internally water-cooled roof. JOM 17AD 4
26. Ayres R, Ayres L, Pokrovsky V (2004) On the efficiency of US electricity usage since 1900. IIASA report IR-04-027, Laxenburg, Austria
27. Moore WE (1940) Present status of the electric arc furnace in industry. Trans Electrochem Soc 77:63–71
28. King PE (1989) Electro-magneto-hydrodynamics: the processes of electric arc steelmaking. MSc thesis, Oregon State University
29. Schwabe WE (1962) Arc heat transfer and refractory erosion in electric steel furnaces. In: EAF conference, pp 195–206
30. Nal Ö, Dolapçioğlu S, Gottardi R, Partyka A, Miani S (2017) Efficiency of the EAF with telescope roof. In: AISTech 2017 conference, vol 1, pp 1149–1160
31. Razza P, Patrizio D (2010) Operating results with hot DRI charge at Emirates Steel Industries. Millenn Steel 39–44

32. Adams W, Alameddine S, Bowman B, Lugo N, Paege S, Stafford P (2001) Factors influencing the total energy consumption in arc furnaces. In: 59th Electric arc furnace conference, 11–14 Nov, Phoenix, AZ, USA, pp 691–702
33. Szekely J, Trapaga G (1994) Future prospects for new technologies in the steel industry. Stahl Eisen 114:43–55
34. Birat JP (2000) A futures study analysis of the technological evolution of the EAF by 2010. Rev Métallurgie 97:1347–1363. https://doi.org/10.1051/metal:2000114
35. Franzen EF (1965) New furnace innovations—1. Tapered shell furnace. JOM 17
36. Timoshenko SN (2016) Computer modeling bath geometry to improve energy efficiency of electric arc furnace, system technologies. Reg Collect Sci Work Dnipro 104:33–39
37. Jones JAT, Bowman B, Lefrank PA (1998) Electric arc furnace steelmaking. In: Fruehan RJ (ed) Making, shaping and treating of steel, 11th edn. Steelmaking and refining volume, Pittsburgh, PA, USA, pp 525–660
38. Sellan R, Fabbro M (2009) The 300-Ton "Jumbo-Size" FastArc EAF at MMK Iskenderun new Danieli Minimill Complex (Turkey). In: AISTech 2009 proceedings, pp 657–666
39. Carney DJ (1974) Electric furnace steelmaking in the next decade. JOM 41–47
40. Laurenti S, Gottardi R, Miani S, Partyka A (2005) High performance single-bucket charging EAF practice. Ironmak Steelmak 32:195–198. https://doi.org/10.1179/174328105X38099
41. Tanaka Y, Yamauchi T, Ogawa M (2017) Development of electric arc furnaces for uniform melting. In: 15th ISIJ-VDEh seminar, Stockholm, Sweden, 12–13 June, pp 202–209
42. Bainbridge L (1993) Ironies of automation. Automatica 19:775–779. https://doi.org/10.1016/j.ifacol.2020.12.2122
43. Fjelland R (2020) Why general artificial intelligence will not be realized. Humanit Soc Sci Commun 7:1–9. https://doi.org/10.1057/s41599-020-0494-4
44. Harari YN (2017) Reboot for the AI revolution. Nature 550:324–327. https://doi.org/10.1038/550324a
45. Svensson J (2023) Artificial intelligence is an oxymoron. Ai Soc 38:363–372. https://doi.org/10.1007/s00146-021-01311-z
46. Pellegrini G, Sandri M, Villagrossi E, Challapalli S, Cestari L, Polo A et al (2019) Successful use case applications of artificial intelligence in the steel industry. Iron Steel Technol 16:44–53
47. Sheridan TB (1980) Computer control and human alienation. Technol Rev 83:60–73
48. Scipolo V, Zuliani DJ (2018) Industry 4.0 leading to the evolution of intelligent EAF steelmaking. In: AISTech 2018 conference, Philadelphia, PA, 7–10 May, pp 1–11
49. Bahrin MAK, Othman MF, Azli NHN, Talib MF (2016) Industry 4.0: a review on industrial automation and robotic. J Teknol 78:137–143
50. Liboni LB, Cezarino LO, Jabbour CJC, Oliveira BG, Stefanelli NO (2019) Smart industry and the pathways to HRM 4.0: implications for SCM. Supply Chain Manag 24:124–146. https://doi.org/10.1108/SCM-03-2018-0150
51. Branca TA, Fornai B, Colla V, Murri MM, Streppa E, Schröder AJ (2020) The challenge of digitalization in the steel sector. Metals (Basel) 10:1–23. https://doi.org/10.3390/met10020288
52. Lugo N (2014) Electric arc furnace best operation practices. In: 45th ABM steelmaking seminar, Porto Alegre, Brazil, 25–28 May, pp 44–53. https://doi.org/10.5151/1982-9345-24178
53. Junger HJ, Jandl C, Cappel J, Zettl KM (2009) The link between BOF maintenance and productivity of steelmaking. Rev Metall Cah D'Informations Tech 106:168–174. https://doi.org/10.1051/metal/2009029
54. El-Akruti K, Zhang T, Dwight R (2016) Developing an optimum maintenance policy by life cycle cost analysis—a case study. Int J Prod Res 1–17. https://doi.org/10.1080/00207543.2016.1193244
55. Piazza M, Ometto M, Bianco F, Patrizio D (2016) EAF process optimization through a modular automation system and an adaptive control strategy. Metall Ital 108:21–30
56. Madias J, Bilancieri A, Hornby S (2017) The influence of metallics and EAF design. Steel Times Int 35–40
57. Toulouevski YN, Zinurov IY (2017) Fuel arc furnace (FAF) for effective scrap melting: from EAF to FAF. Springer Nature, Singapore. https://doi.org/10.1007/978-981-10-5885-1

58. Abel M, Hein M (2008) The Simetal ultimate at Colakoglu/Turkey. In: AISTech 2008 proceedings, Pittsburgh, PA, USA, 5–8 May, AISTech, p Paper 67
59. Pesamosca A, Kuran O, Olivieri L, Sellan R (2014) Heat size upgrade in Kroman Celik EAF. In: 45th ABM steelmaking seminar, Porto Alegre, Brazil, 25–28 May, ABM, pp 764–771
60. Memoli F, Jones JAT, Picciolo F (2013) How changes in scrap mix affect the operation of Consteel® EAF. In: AISTech 2013 conference proceedings, vol 1, pp 795–808
61. Kirschen M, Pfeifer H, Wahlers F-J, Mees H (2001) Off-gas measurements for mass and energy balances of stainless steel EAF. In: 59th Electric furnace conference, pp 737–745.
62. Rogers RP (2009) An economic history of the American steel industry. Routledge Taylor & Francis. https://doi.org/10.4324/9780203881033
63. Haller W (2005) Industrial restructuring and urban change in the Pittsburgh region: developmental, ecological, and socioeconomic trade-offs. Ecol Soc 10:Art 13
64. Watson C (2022) Domestic steel manufacturing: overview and prospects. CRC report R47107
65. Ouvradou C (2006) The electric furnace situation and European perspectives. Rev Métall 218–225
66. NMC (2018) A brief history of the American steel industry. Natl Mater Co. https://www.nationalmaterial.com/brief-history-american-steel-industry/. Accessed 1 Nov 2023
67. AISI (2021) Sustainability of the American steel industry, pp 1–11. https://www.steel.org/wp-content/uploads/2021/03/Sustainability-Key-Messages.pdf. Accessed 1 Nov 2023
68. Strain M (2019) The link between wages and productivity is strong. In: Kearney M, Ganz A (ed) Expanding economic opportunity for more Americans, Aspen Institute, pp 168–179. https://doi.org/10.1080/02692170701189151
69. Kuznietsova K (2018) How salaries influence on labour productivity? Int J Manag Humanit 3:1–5
70. Malfa E, Nyssen P, Filippini E, Dettmer B, Unamuno I, Gustafsson A et al (2013) Cost and energy effective management of EAF with flexible charge material mix. BHM Berg-Huettenmaenn Monatsh 158:3–12. https://doi.org/10.1007/s00501-012-0103-y
71. Torres R, Aguilar S, Conejo AN (2000) Evolution of refractory performance and metallurgical practices at IMEXSA. In: 58th Electric furnace conference, 12–15 Nov, Orlando, Florida, pp 415–423
72. Kawalec C (1991) The effects of technology on the labour force of DOFASCO INC. BSc thesis
73. Storey R (1983) Unionization versus corporate welfare: the "Dofasco Way." Labour/Le Trav 12:7–42
74. Lieberman MB, Johnson DR (1999) Comparative productivity of Japanese and U.S. steel producers, 1958–1993. Japan World Econ 11:1–27. https://doi.org/10.1016/s0922-1425(98)00032-2
75. Baran J, Rokicki T (2005) Productivity and efficiency of US metal industry in 2006–2014. In: Met. 2015, Brno, Czech Republic, 3–5 June, p 24
76. Collard-Wexler A, De Loecker J (2015) Reallocation and technology: evidence from the US steel industry. Am Econ Rev 105:131–171
77. Nda MM, Fard RY (2013) The impact of employee training and development on employee productivity. Glob J Commer Manag Perspect 2:91–93
78. De Grip A, Sauermann J (2013) The effect of training on productivity: the transfer of on-the-job training from the perspective of economics. Educ Res Rev 8:28–36. https://doi.org/10.1016/j.edurev.2012.05.005

Chapter 15
Effect of Energy Consumption on the Environment

15.1 The Problem of High Energy Consumption

15.1.1 CO_2 Was Left Unattended in the Past Centuries, then Became a Huge Problem

Before the industrial revolution, the problem of anthropogenic CO_2 didn't exist. If the beginning of the industrial revolution is taken from 1778, the year James Watt perfected the steam engine, that year marks the beginning of mass industrialization. Ritchie [1] from ourworldindata indicates that during the whole nineteenth century and until the middle of the twentieth century both the UK and the USA remained as the largest producers of CO_2. Their cumulative emissions in 1950, were 39.16 billion ton and 91.9 billion ton, respectively, with the European Union (EU) in second place producing 65.6 billion ton. In this year, China's cumulative emissions were only 2.27 billion ton. Up to 1980 both Europe and the USA produced almost 70% of the total CO_2 world emissions. The concept of sustainability was still not defined. Pollution was everywhere and out of control. The industrialization of China after the 1980s drastically increased its cumulative emissions of CO_2 from 23 billion ton to 237 billion ton in 2020, but still the USA and the EU had larger cumulative emissions, 416 and 290 billion ton, respectively. In 2020, USA and Europe still keep the record of highest emitters of CO_2, about 46%, compared to China and India combined with 17%. The exponential growth in cumulative emissions of CO_2 from both the USA and the EU started at the beginning of the twentieth century. The USA and Europe have the historical responsibility of those cumulative emissions. Since 2006, China is the largest producer of CO_2 per year, when it reached the levels of the USA, a total of about 6000 Mton [2]. The total production of CO_2 since 1850 has been estimated about 2.5 trillion ton and for every 1 trillion ton it has also been estimated that there is a rise in global warming of about 0.45 °C, resulting in a temperature rise of about 1.2 °C up to 2020 [3].

Worldwide efforts to face the problem of global warming are not new but the commitment to enforce policies on CO_2 had been weak by a large fraction of the international community until today. The Kyoto protocol was signed in 1997 but enforced until February 2005 when negotiations concluded with a minimum of 55 countries accepting its terms. The USA and Canada declined to ratify the Kyoto protocol. The Paris agreement, signed in 2015, is the successor of the Kyoto protocol. The main target is to keep a maximum rise in global warming, with respect to the pre-industrial era, of 2 °C and ideally below 1.5 °C.

In steel plants all over the world, reality on global warming has overcome the refusal to invest on environmental protection. This refusal is not particular of developing countries but also developed countries as reported by Shoji and Miyamoto for the case of Japan [4], who identified to companies behavioral principles; "First, pollution arises because companies focus on reducing environmental protection costs within a given production technology. Second, firms may employ more polluting technologies if they generate higher profits than cleaner alternatives". Kawabata [4] argued that after WWII when Japan increased steel production, the steel industry adopted pollution control measures, not voluntarily but in response to pressure from local governments and citizen movements. The refusal to adopt pollution controls is still a reality in less developed countries because the off-gas collection and cleaning system represent a large capital cost in the order of 25% of a new meltshop [5]. These systems are designed for peak operating conditions that usually exist during 10–20% of the heat cycle.

Table 15.1 summarizes the emission of CO_2 in the steel industry and compares with the production of other metals [6]. Comparing cement, steel and aluminum, it can be observed that aluminum produces sevenfold higher specific CO_2 emissions in comparison with steel but the steel industry produces more CO_2 per year because the production of steel is 27 times higher.

Wang et al. [7] in a very comprehensive analysis have reported the accumulated production of CO_2 by the steel industry from 1900 to 2015, reaching the conclusion that unless more severe measures are taken it will not be possible to reach the preferred climate target of 1.5 °C, which requires CO_2 emissions cut by half by 2030 and net zero emissions by 2050. These authors reported that about 45 Gton of steel were produced from 1900 to 2015 producing about 147 Gton of CO_2-eq. In spite of a 67% higher thermal efficiency CO_2 still increased 17-fold due to a 44-fold increase in steel production. They suggest six urgent measures; (1) Emerging steel producers like India

Table 15.1 Global CO_2 emissions for different materials

Material	Production in 2017, Mton/year	Mton CO_2/year	kg CO_2/ton	BAU demand in 2050, Mton/year
Cement	4050	2200	860	4682
Iron and steel	1736	3700	2000	2535
Aluminum	64.3	1000	14,400	110

After [6]

should install low carbon technologies, (2) Early retirement of blast furnaces. China should eliminate about 170 Mton capacity in the next 15 years and about 500 Mton by 2050, (3) higher use of steel scrap, in particular by developed nations, (4) material efficiency measures, (5) global cooperation because production of green steel is 20–40% more expensive, and (6) higher carbon taxation (100 USD/ton CO_2 + 4% per year since 2020).

15.1.2   Energy Consumption and CO_2

The steel industry is the largest producer of CO_2 among the industrial sector, as shown in Fig. 15.1. The US Energy Independent Administration (US-EIA) reported in 2022 that the steel industry alone produced 7% of the total emissions of CO_2 worldwide [8]. This result is in great part due to the fact that steel production is a very high energy intensive industry. In 2006 it consumed 24 EJ of energy [9]. The total energy consumption has been increased year by year in spite of increasing thermal efficiency represented by a decrease in the specific energy consumption of 50–75 GJ/ton steel in 1970 to 20–30 GJ/ton in 2010, as shown in Table 15.2, the total production of CO_2 has increased due to a higher production of steel/year. The specific energy consumption in 2019 has been further decreased to 10–20 GJ/ton [10].

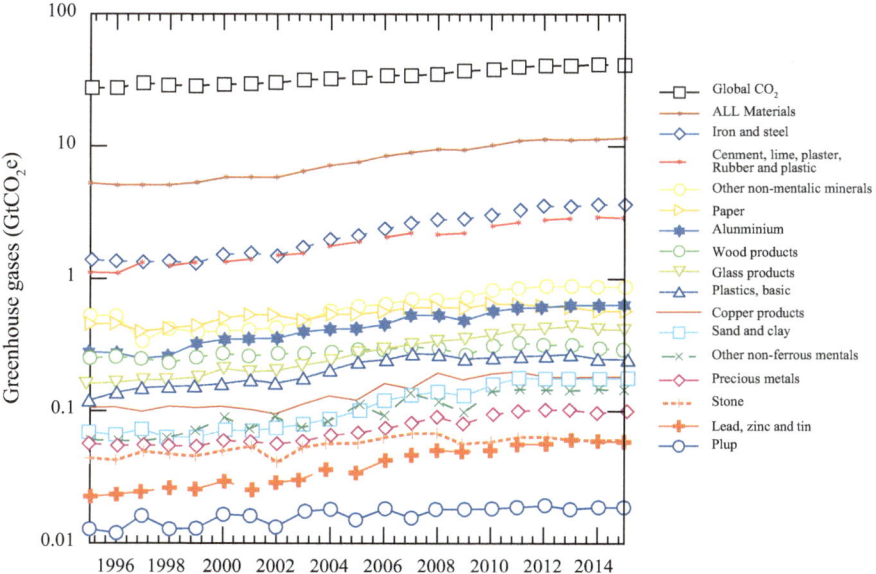

Fig. 15.1 The contribution of metal production to GHG. Taken from [11]

Table 15.2 Evolution on energy consumption in the steel industry

Reference	Total energy consumption, GJ/ton				EAF, kWh/ton	
	[12]	[12]	[13]	[14]	[15]	[15]
Year	1980	1991	2006	2018	2006	2017
China	52	43	26.3		651	488
Japan	22	22				
IEA (world)			19.8	18.6		

The specific energy consumption in the steel industry using BAT is close to its theoretical minimum defined by thermodynamics. It is now possible to use a small fraction of electric energy and even no electric energy at all using scrap preheating, hot metal, etc., however the means to achieve this goal are the use of fossil fuels that bring a drastic increment of GHG emissions contributing to global warming. The total energy consumption in steelmaking ranges from 510 to 880 kWh/ton [16], this value depends on multiple factors such as the type of raw materials. It can be decreased close to the theoretical values if the energy efficiency is increased by decreasing energy losses in all the effluent streams of the EAF, in particular the off-gas and the water-cooling system. The CO_2 emission intensity in the EAF is in the order of 0.15 kg CO_2/kWh [17]. Kirschen reported a range from 0.11 to 0.21 kg CO_2/kWh for Germany [16].

Today, China is the largest producer of CO_2 in the steel industry because is the largest producer of steel and also because their dominant route to produce steel is through the blast furnace, therefore highly dependent on the consumption of fossil fuels. In China, the integrated route is responsible for 70% of the energy consumption and the production of 85% of CO_2 emissions [15], however its steel industry has been making large improvements. EAF energy consumption decreased on average from 651 to 488 kWh/ton [15] as shown in Table 15.1. The decrease in specific energy consumption in the steel industry has decreased the total production of CO_2 from 2326 Mton in 2019 [18] to 2104 Mton in 2020 [10]. These values include direct and indirect CO_2 emissions. Indirect emissions represent about 10–15% and correspond to emissions from electricity generation and combustion of CO from the off-gas. The declining trend according to Lin et al. [19] started in 2015 with specific emissions of 2.0 ton CO_2/ton then declining to 1.6 ton CO_2/ton in 2019, indicating that the steel industry in China has already achieved the peak in CO_2 emissions. Figure 15.2 shows possible future scenarios on CO_2 emissions until 2060 [20]. The Business-as-usual (BAU) scenario only considers a reduction in steel production. The major contribution derives from a share of steel production by the EAF process which would decrease CO_2 emissions to less than 400 Mton CO_2 per year by 2050.

Kirschen et al. [21] summarized energy consumption and CO_2 emissions analyzing results from 70 EAF's. The total energy consumption ranged from 510 to 880 kWh/ton, as can be seen in Fig. 15.3 and Table 15.3. The reported thermal efficiency ranged from 45 to 60%, which depends on one side on the energy losses; off-gas, slag, water cooled panels and radiation heat losses but also depend on the

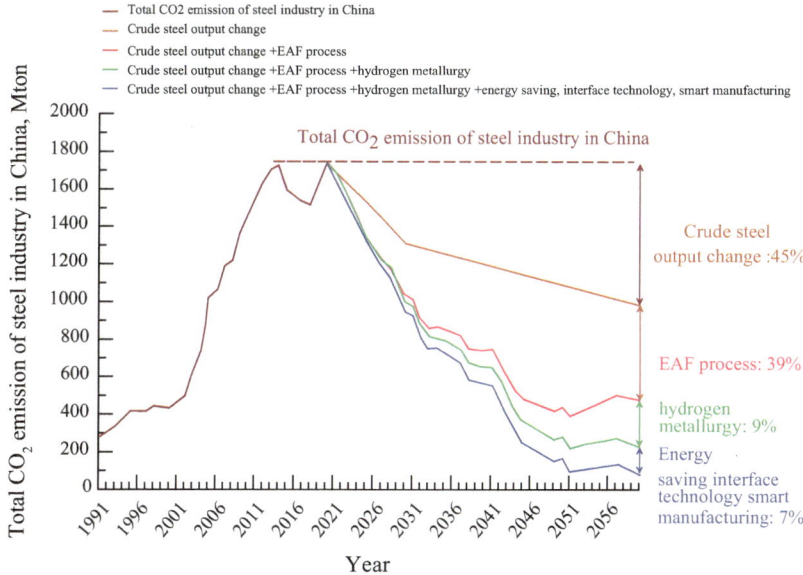

Fig. 15.2 Current and future scenarios on CO_2 generation in the steel industry. After [20]

total energy input. If the total energy requirements increases, for example due to lower quality raw materials, thermal efficiency also decreases.

Kirschen et al. [16] summarized several reports about the production of CO_2, as shown in Table 15.4. Using 100% scrap the value ranges from 350 to 450 kg CO_2/ton and considering the energy involved in the production of DRI, 700–1300 kg CO_2/ton using from 50 to 100% DRI, in this case, the largest value is still below in comparison with modern BF plants in Europe that produce about 1630 kg CO_2/ton. The EAF by

Fig. 15.3 Summary of energy balances of 70 EAF's. After Kirschen et al. [21]

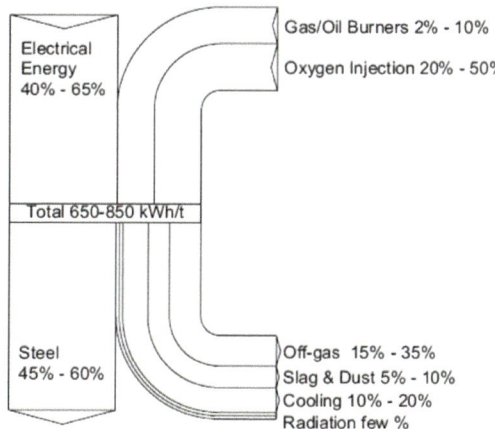

Table 15.3 Typical CO_2 emission with different types of natural gas

NG grade	kWh/m^3	ρ, kg/m^3	kg CO_2/MJ	kg CO_2/kWh	kg CO_2/m^3	kg C/kg NG
Combined L	9.28	0.8161	0.0562	0.2023	1.87	0.627
Holland L	9.33	0.8288	0.0563	0.2027	1.89	0.622
Russia H	9.97	0.7304	0.0549	0.1976	1.97	0.735
Mixed H	10.48	0.8128	0.0564	0.2030	2.12	0.714
North sea H	10.70	0.8147	0.0665	0.2034	2.17	0.729

After Kirschen et al. [21]
CO_2 emissions based on combustion with air

Table 15.4 CO_2 emissions by process

	kg CO_2/ton
Scrap-EAF	350–450
DRI-EAF	700–1300
BF-BOF	1630–1960

After [16]

itself produces about 70–200 kg CO_2/ton which results from the use of 20–55 kg C/ton.

The blast furnace traditionally employs sinter or pellets. Sintering increases CO_2 emissions and its production involves five times more dust in comparison with pellet production, 41–559 g/ton and 14–150 g/ton, respectively [22]. Ruukki Metal-Finland replaced sinter by pellets in the blast furnace and reported a decrease in CO_2 emissions in the order of 0.4 Mton/year [23].

Hornby-Anderson et al. [24] compared the blast furnace with both DRI and new ironmaking technologies in terms of energy consumption and carbon emissions. Their results indicate the best solution is using 100% scrap in the metallic charge, both in terms of energy consumption and carbon emissions. They reported an average production of 441 kg CO_2/ton using 100% scrap. Using cold DRI, energy consumption is higher compared with the BF/BOF, from 4624 to 5065 kWh/ton using 80% cold DRI, however carbon emissions are lower, decreasing from 1922 kg CO_2/ton for the BF/BOF to 1163 kg CO_2/ton with 80% cold DRI, due to the use of natural gas.

15.2 Effects of CO_2 Emissions on Human Health

15.2.1 Toxic Components in the Off-Gas

Dioxins (PCDD) and furans (PCDF) [25–27]

The most toxic dioxin is Tetrachloro-Dibenzo-p-Dioxin (TCDD), for short Dioxin. Similar to dioxins are the Polychlorinated-Dibenzo-furans (PCDF) for short Furans. Dioxins and Furans are grouped together simply as Dioxins. Dioxins are a persistent organic pollutant (POP).

Dioxins are persistent stable organic pollutants that can be formed when scrap containing paints, cutting oils, PVC or other organic substances impregnated with halogens (chlorine, bromine, etc.) are heated, especially at the beginning when the gas temperature is low (250–500 °C). It can also be formed during gas cooling by chemical reactions between chlorinated materials, oxygen and water during cooling of the off-gas, in the temperature range from 250 to 400 °C. It can be drastically decreased by rapid quenching the off-gas from 800 °C to less than 300 °C using spray water, to levels in the range 0.1–0.2 ng I-TEQ/N m^3 [26]. The unit I-TEQ means International Toxicity Equivalents. TEQ units provide a reference concentration with the most toxic Dioxin, TCDD. In Japan, from 2001, the concentration of dioxins should be lower than 0.5 ng TEQ/m^3 for new steel plants [27]. Different methods can be employed to capture dioxins, such as activated carbon. Sofilić et al. [28] compared dioxins and furans emissions between an old and an a new steel shop. With the new plant the total concentration of dioxins and furans was 0.022 ng I-TEQ/N m^3, and for the old plant 0.22–1.09 ng I-TEQ/N m^3.

NO$_x$ [26, 29]

NO$_x$ includes nitrous oxide (N$_2$O), nitrogen monoxide (NO), nitrogen dioxide (NO$_2$), nitrogen trioxide (NO$_3$), dinitrogen trioxide (N$_2$O$_3$), dinitrogen tetroxide (N$_2$O$_4$) and dinitrogen pentoxide (N$_2$O$_5$). NO$_2$ is the second most powerful global warming gas. NO$_x$ are formed at high temperatures. NO$_x$ can be formed in two regions, when nitrogen passes through the electric arc or the combustion temperature is high enough close to the burners or gases passing through the air gap. Typical levels are 36–90 g NO$_x$/ton steel [26]. The main solution is to avoid air ingress through the slag door and maintain a foamy slag. Once the arcs are covered by a foamy slag contact of air with the electric arc is almost eliminated.

Sulfur oxides (SO$_x$)

Sulfur oxides refer to SO$_2$ and SO$_3$. SO$_2$ combines with oxygen from the air to slowly form SO$_3$, then rapidly combines with water to form sulfuric acid. Released to the environment promotes acid rain.

Volatile organic compounds (VOC) [29]

VOC are formed by combustion of hydrocarbons coming from oil or plastics. As the name implies, they have a very high volatility at room temperature. VOC react with NO$_x$ to form ozone. The EU defines VOC as an organic compound of anthropogenic nature capable to produce photochemical oxidants in the presence of nitrogen oxides and sunlight.

Polycyclic aromatic hydrocarbons (PAH)

PAH is a persistent organic pollutant (POP), present in the scrap but can also be formed during EAF operation due to combustion of fossil fuels or organic matter. Lawal [30] reviewed in detail this subject. PAH only consist of carbon and hydrogen ring systems and different benzene ring arrangements. They are divided into light and heavy PAH. Heavy PAH are more stable and more toxic. Most PAH are associated with fine dust particles. Measurements from one steel shop in Poland reported a concentration of 95.5 $\mu g/m^3$ in $PM_{2.5}$ and 9.1 $\mu g/m^3$ in PM_{10}, indicating that finer dust contains a higher concentration of PAH [31]. PAH are carcinogenic.

Polychlorinated biphenyls (PCB)

PCB are a class of chlorinated semi-volatile organic compounds and a persistent organic pollutant (POP). From a total of 209, a group of 12 PCB exhibit "dioxin-like" behavior and are more toxic because are carcinogenic and affect the reproductive system. PCB were produced and employed in different applications, such as fluids in transformers but were prohibited in the 1970s. Shen et al. [32] measured the concentration of PCB in three steel shops that employ scrap pre-heating, the results had large differences, 1236.1, 81,664.4 and 669.8 pg/N m^3, due to different raw materials in the metallic charge.

In 2012, Ocio et al. [33] carried a comprehensive evaluation of EAF emissions in Europe and compared with 7 steel plants in Turkey. Table 15.5 summarizes the inputs in raw materials and Table 15.6 summarizes the outputs of emissions. The large ranges are due several factors such as differences in raw materials, types of steels produced, plant productivity, etc. The amount of by-products illustrate the EAF is an important source of dust and slags. Dust in the primary emissions is about 10 times higher in comparison with secondary emissions; 0.3–1.0 kg/ton during scrap charging, 0.5–2 kg/ton due to fume leakages during EAF operation, 0.2–0.3 kg/ton during tapping. The total amount of secondary dust emission is in the order of 1.4–3 kg/ton.

Table 15.7 compares the final emissions after the gas is processed by the powder emission collecting system (PECS) using bag filters and electrostatic precipitators.

There is an international agreement on POP, the Stockholm convention held in 2001, which defined the need to eliminate POP because they remain intact in the environment for a long time, bio-accumulate in the fatty tissue of living organisms, have a long-range transport and high level of toxicity. The objective of the Stockholm Convention is to protect human health and the environment from the risks posed by POP's. The agreement entered into force in 2004 with 151 countries signing the agreement but only 83 have ratified it [34].

Yang et al. [35] recently reported a review on the generation of persistent organic pollutants (POP). The EAF produces more POP than the BOF because it consumes a higher amount of scrap. Scrap pre-heating in particular enhances the formation of POP due to heating in the temperature range of formation of POP, 250–400 °C.

Table 15.5 Comparison of inputs to the EAF

			EU	Türkiye
Metallic charge	Scrap	kg/ton	39–1232	90–1176
	Pig iron	kg/ton	0–153	0–150
	Hot metal	kg/ton		
	DRI (HBI)	kg/ton	0–215	0–6.6
Lime/Dolomite		kg/ton	25–140	42–145
Carbon (coal, anthracite and coke)		kg/ton	3–28	8–25
Graphite electrodes		kg/ton	2–6	1.3–2.9
Refractories		kg/ton	4–60	4–10
Ferroalloys	Carbon steel	kg/ton	11–40	-
	High and S steels	kg/ton	23–363	-
Gases	Oxygen	N m^3/ton	5–65	38–55
	Argon	N m^3/ton	0.3–1.45	0.16–2.61
	Nitrogen	N m^3/ton	0.8–1.2	0.2–12
	Steam	kg/ton	33–360	0–40
Energy	Electricity	kWh/ton	404–748	435–592
		MJ/ton	1454–2693	469–1576
	Fuels (NG, liquid)	MJ/ton	50–1500	50–1500
Water		N m^3/ton	1–42.8	0.2–2.85

After [33]
Ton refers to ton of liquid steel

15.2.2 Effect of EAF-Dust Emissions on Human Health

The off-gas contains about 10–30 kg dust/ton for carbon steels and 10–18 kg dust/ton for high alloyed steels [33]. The BAT-associated emission level for dust is < 5 mg/ N m^3, determined as a daily mean value. EAF-dust emissions have been classified as carcinogenic for humans by the International Agency for Research on Cancer (IARC) because, depending on its concentration, it increases mortality. About 40% of EAF-dust is made of very fine particulate matter (PM$_{2.5}$) [36]. These particles can persist over a long period of time in the environment and body tissues. An extensive study carried out by Cappelletti et al. [36] from 1979 to 2009 for an Italian EAF shop without capture of secondary emissions from the EAF found a strong relationship with the causes of mortality due to higher frequency of cardiovascular diseases and lung cancer. In 1979 the workers were exposed to a higher load of EAF dust emissions, about 80 kg/h, decreased to 2–5 kg/h using systems to capture secondary emissions in 2011.

Dust is part of particulate matter (PM), defined as a suspension of solids, liquid or both in the air. PM$_{10}$ and PM$_{2.5}$ are defined as a fraction of particles in air with a size

Table 15.6 Primary EAF emissions

		EU	Türkiye
Off-gas flow rate	Million N m^3/h	1–2	1–3.55
	N m^3/ton	8000–10,000	11.6–17.7
Dust	g/ton	4–300	13–127
	mg/m^3	0.35–52	0.53–11
Hg	mg/ton	2–200	14
Pb	mg/ton	75–2850	56–97
Cr	mg/ton	12–2800	7.6–18
Ni	mg/ton	3–2000	14–44
Zn	mg/ton	200–24,000	0–137
Cd	mg/ton	1–148	2.3–18
Cu	mg/ton	11–510	13–34
HF	mg/ton	0.04–15,000	–
HCl	mg/ton	800–35,250	–
SO$_2$	g/ton	5–210	5–100
NO$_x$	g/ton	13–460	46–190
CO	g/ton	50–4500	42–360
CO$_2$	kg/ton	72–180	175
TOC	g C/ton	35–260	1.35
Benzene	mg/ton	30–4400	17
Chlorobenzenes	mg/ton	0.2–12	–
PAH	mg/ton	9–970	93–103
PCB	mg/ton	0.01–5	–
Dioxins (PCCD/F)	μg I-TEQ/ton	0.04–6	0.09–1.23
By-products			
Slag from EAF	kg/ton	60–270	60–178
Slag from LMF	kg/ton	10–80	15–25
Dust	kg/ton	10–30	10–21
Spent refractory	kg/ton	1.6–22.8	1.5–20
Noise	dB (A)	90–133	70–115

After [33]
Ton refers to ton of liquid steel

smaller than 10 and 2.5 μm, respectively. These particle sizes are more dangerous because can penetrate deep into the lungs.

Table 15.7 Comparison of final emissions to the atmosphere. After [33]

Parameter		Units	Bag filter	Electrostatic precipitator
Dust		mg/N m^3	0.35–3.4	1.8
CO		mg/N m^3	88–256	
NO$_x$		mg/N m^3	0.97–70	
SO$_x$		mg/N m^3	8–17	
Metals	Hg	mg/N m^3	0.016–0.019	< 0.0003
	Sb, Pb, Cr, CN, F, Cu, Mn, C, Se, Te, Ni, Co, Sn	mg/N m^3	0.006–0.022	0.01–0.07
	Cr (excepto Cr VI)	mg/N m^3	0.013	
	Mn	mg/N m^3	0.036	
	Ni	mg/N m^3	0.003	
PAH		mg/N m^3	< 0.00001	< 0.001
PCDD/F		mg/N m^3	0.0015–0.1	
HF		mg/N m^3	0.085–0.2	
HCl		mg/N m^3	3–5.4	
Cl$_2$		mg/N m^3	< 3	

15.3 Methods to Decrease CO_2 in EAF Steelmaking

15.3.1 Increase Thermal Efficiency of Existing Technologies

A hypothetical energy efficiency of 100% could be achieved if all the input energy corresponds to the sensible heat in liquid steel, the sensible heat in the off gas, liquid slag and water-cooled panels is fully recycled and there are no other heat losses. Since this is not possible, the options available to increase the thermal efficiency are a maximum recovery of the sensible heat and minimization of heat losses. In this book 17 methods have been described in detail to increase not only the thermal efficiency but also to decrease the consumption of electric energy. The following is a summarized analysis of additional methods to decrease CO_2 emissions.

15.3.2 Replace Obsolete Plants with Existing BAT

The IEA suggest that deployment of the Best Available Technologies (BAT) in the steel industry would involve extra energy savings of 20% of today's consumption [9]. Their analysis on the share of steel produced by different processes also suggests that applying BAT, the specific energy savings potential range from 1.4 to 8.7 GJ/ton, and on average 4.1 GJ/ton. Ukraine has the largest specific energy savings potential,

8.7 GJ/ton, and both South Korea and Japan the lowest, 1.4 GJ/ton but the major total contribution comes from China due to the mass of steel produced.

15.3.3 Scrap Recycling

Scrap recycling is one of the simplest ways to decrease CO_2. The International Energy Agency (IEA) predicted in 2012 a growth for the EAF for the year 2030 from 30 to 50% and a growth of scrap as raw material from 27 to 43% [37]. Since high quality steel scrap is almost pure iron, the energy requirements are the lowest among all raw materials. The production of CO_2 is the lowest, about 441 kg CO_2/ton [24]. In order to achieve the climate target of 1.5 °C, worldwide the specific production should be below 0.85-ton CO_2/ton steel [7]. The subject of scrap recycling is treated in detail in Chap. 8.

15.3.4 Green Electricity

The specific consumption of electric energy in the EAF, using steel scrap, is the lowest in comparison with other alternatives, however this entails a hidden aspect; most of the electric energy produced in the world is made using fossil fuels. This is called grey electricity. Green electricity is produced by renewable energy and its facilities including photovoltaic arrays, wind turbines, hydropower, nuclear and other renewable energy sources. The emissions of CO_2 depending on the source of energy and by country are summarized in Table 15.8. In China in 2022, coal-fired power generation was 69.8% followed by hydropower, wind power and solar power, with 14.3%, 8.2% and 2.7%, respectively. Green electricity represented 30.2% of China's total power generation. Green electricity generation is closely related to seasonal factors and weather. In China, green electricity has a price about 0.012 USD/kWh, which in spite to subsidies is at least twice more than that for grey electricity [38].

Table 15.8 Emission of CO_2 in the production of electric energy

Energy from	Coal	NG	Renewable	China	India	EU	World
Emissions, g CO_2/kWh	820	500	10–50	620	723	282	484

After [39]

15.3.5 Carbon Capture Sequestration-Utilization (CCS, CCSU)

CO$_2$ Capture and Sequestration (CCS) are mature technologies but not efficient due to the following reasons:

- Scale: The amount of CO$_2$ generated in 2022, according to Statista, was in the order of 40,000 Mton. CCS is expensive, unsafe and captures only a very small fraction of the CO$_2$ produced. The global CCS institute who monitors all facilities and projects worldwide reported that during 35 years from 1972 until 2018, CCS captured only 230 Mton [40]. From a total of 196 facilities only 30 are in operation with a current capacity of 40 Mton, with capacities from 0.1 to 7 Mton.
- Expensive: the only case involving a steel plant is United Arab Emirates in Abu Dhabi. The CCS plant was commissioned in 2016 to capture 0.8 Mton CO$_2$ for enhanced oil recovery (EOR) by the Abu Dhabi National Oil company with a cost of US\$ 122 millions [41].
- Majority of the projects have failed or are underperforming [41].
- Main successful application is the use of CO$_2$ to enhance oil recovery (EOR) which is not a way to reduce CO$_2$ [41].
- In addition to the previous problems by capture and underground sequestration, Greenpeace [42] added the following: (1) Risks of huge disasters due to leakages caused by the continuous movement of the earth crust, (2) Requires energy, about 10–40% of the energy consumed by a power plant, (3) makes no sense to capture CO$_2$ if there is no space to store it, (4) Transport of CO$_2$ over distances greater than 100 km is likely to be prohibitively expensive.

The release of CO$_2$ in the lake Nyos in Cameroon was due to a volcanic explosion but gives an idea of the disaster caused by the release of 1.6 Mton CO$_2$, that suffocated and killed 1746 people and 3500 livestock in 1986. CO$_2$ has 1.5 times the density of air.

In addition to underground sequestration, there is also the option of mineral sequestration. This method offers the advantages that fixes CO$_2$ and the product can be employed as a source of carbonates. A review paper by Zevenhoven et al. [43] describes in detail direct and indirect methods for mineral sequestration but the conclusion is that both methods have scale problems. The amount of minerals containing Mg or Ca is in the order of 1.8–3.0 ton mineral/ton CO$_2$.

CO$_2$ Capture and Sequestration and Utilization (CCSU) is more useful because obtains valuable products. Xie et al. [44] describes the use of slags to fix CO$_2$ and produce carbonates, such as CaCO$_3$. The largest majority of new projects from 2024 involve CCSU projects for ethanol or methanol production. Methanol production in particular is very promising. Biswal et al. [45] have reviewed some of the challenges, in particular the development of a proper catalyst. Depending on the ratio CO$_2$/H$_2$, pressure and temperature the reaction CO$_2$-H$_2$ can lead to different products, as shown below. Production yield of methanol increases at temperatures below 150 °C and pressures from 5 to 10 mPa. A large number of investigations are trying to develop suitable catalyst because the stability of CO_2 is high.

$$CO_2 + 3H_2 = CH_3OH + H_2O$$
$$CO_2 + H_2 = CO + H_2O$$
$$CO_2 + 2H_2 = CH_3OH$$

15.3.6 Change of Non-renewable Fossil Fuels by Renewable Fuels (H_2 and Biomass)

Steelmaking technology has evolved in the last 4000 years, based on fossil fuels. In a first stage the iron ore is reduced to metallic iron using carbon to remove oxygen, then in a second stage oxygen is injected to remove the excess carbon and oxidize alloying elements to a desired level by the final steel specification. Carbon is also added because it promotes stirring of the molten metal by the decarburization reaction, in turn the CO gas formed is essential to promote slag foaming. If carbon is employed to the minimum concentration, then stirring should be provided by bottom gas injection or EMS, but still is needed to reduce the iron oxide in the slag which is created due to oxygen injection. The reduction of iron oxide using hydrogen is possible but its role on foaming has never been investigated in the past. In order to decrease production costs, electric energy consumption is replaced by chemical energy which involves higher rates of both oxygen and carbon injection. From the previous discussion it is clear that carbon is employed both as a reducing element and fuel. Hydrogen can replace carbon to reduce liquid FeO [46] but not as a fuel because the reaction between FeO and hydrogen dissolved in liquid steel is slightly endothermic [47]. This analysis indicates that carbon as a fuel is required even if the input materials are carbon-free, in order to decrease the consumption of electric energy. One option to this problem is to employ biomass which is a renewable material, in addition to this, hydrogen can replace natural gas as a fuel in burners.

15.3.6.1 Bio-mass

One option to replace fossil fuels is the use of renewable energies, however, renewable energies are inefficient to produce electric energy [48] or available in a limited supply [49]. The shortest growth development time of a tree plantation is about 7–8 years. Echterhof [50] reviewed the use of alternative carbon sources (biomass, plastic and tires) in EAF steelmaking, indicating their potential for replacement as a carburizer or foaming agent. Hammerschmid et al., taken from Babich et al. [51] reported that biomass can be employed to replace natural gas in the production of direct reduced iron, reducing the emission of CO_2 by 80% compared with the BF. High quality biomass depends on the pyrolysis conditions and is possible to obtain low sulphur, high carbon and a highly reactive material. Using this type of biomass, it can replace coal injection and achieve a decrease from 35 to 45% CO$_2$. The use of alternative carbon sources in the EAF can decrease CO$_2$ in the EAF from 8 to 12% [49].

Once hydrogen replaces carbon as a reducing material and as a fuel in some applications, the carbon requirements will be minimized, but as pointed out before carbon is essential to provide for chemical energy and slag foaming.

Carbon from non-renewable sources can be replaced by biomass which is a renewable form for carbon. But… is it renewable? There is an academic debate about whether biomass is or is not a renewable. Different groups of experts have written to the EU and US Congress about the type of biomass which should be considered renewable. Burning biomass releases more CO$_2$ than coal and cutting trees eliminates a natural consumer of CO$_2$. New crops take decades or even a century to grow. The U.S. Environmental Protection Agency (EPA) found that "carbon neutrality cannot be assumed for all biomass energy a priori." It depends on the type of biomass, the combustion technology and what forest management techniques are employed in the areas where the biomass is harvested. Renewable or modern biomass has been defined as biomass produced in a sustainable way, if it is realistically replaced, avoiding deforestation [52]. Biomass is an organic material. Traditional biomass has been employed since antiquity as a fuel, for example cooking, especially in its form as charcoal. Charcoal is produced in the absence of oxygen to remove moisture and volatile matter. The calorific value of biomass is lower than that for charcoal and coal; 15, 30 and 35 MJ/kg, respectively [52], but on the other hand, biomass has low S, high reactivity, high specific surface area and stable pore structure. Due to its low density, humidity and volatile matter biomass requires pre-treatment before it is employed.

Using biomass is returning to the oldest fuel used by mankind until the middle of the eighteenth century but now in a sustainable way. Biomass is divided into plant biomass and animal biomass but the latter has a negligible share compared with the first one. Plants capture about 1% of the solar energy. Biomass as a fuel currently has a small contribution, about 1% of the total energy [53]. Biomass has huge potential as a reducing agent in both ironmaking and steelmaking. In BF ironmaking it can fully replace PCI. In steelmaking it can replace the addition of carbon. The consumption of coal in the BF-BOF route is much higher, about 800 kg C/ton compared to the EAF, 17 kg C/ton for scrap based operations and 23 kg C/ton, for DRI based operations, respectively [16].

15.3.6.2 Hydrogen Metallurgy

Burners Using Hydrogen

Burners are commonly employed in scrap-based operations in the EAF. The typical fuel is natural gas. This fuel can be replaced by green hydrogen. Chemical energy in an EAF is in the order of 30–50%, depending on the metallic charge. Burners contribute with approximately 40–80 kWh/ton, equivalent to 3.7–7.5 N m^3 NG/ton

[54]. In the modern EAF, the contribution of the chemical energy for scrap melting and refining is in the range of 25–45% of the total energy required. The Natural Gas (NG) burners provide chemical energy in the range of 40–80 kWh/t of energy. It means that the production of 100 tons of steel requires the combustion of 370–750 N m^3 of NG with CO_2 emissions of 0.75–1.5 tons. The substitution of just 10% of NG with hydrogen in the whole steel European production will bring a remarkable reduction of CO_2 emissions up to 0.1 Mtons/year.

The technology for H_2 burners has been recently developed by several companies; Linde in Sweden and Taiyo Nippon Sanso in Japan. von Scheele [55] reported completion of pilot tests in 2020, and Nippon Sanso [56] in 2021.

The Challenges of Hydrogen Metallurgy

Today, there is general agreement that fossil fuels have to be replaced to eliminate the problems of carbon foot print and the option is to use hydrogen. Hydrogen has been employed as a reducing gas in the blast furnace and DRI processes since the nineteenth century. In the early 1950s the H-iron process was developed and reached the pilot plant stage [57]. In this process iron ore fines were reduced with 100% hydrogen at 480 °C and a high pressure at 17 atm. Circored, developed by Lurgi is a commercial process which employs 100% hydrogen and fluidized beds, started operations in 1999. This technology has some limitations due to sticking problems associated with fluidized bed reactors and the high price of hydrogen, as described in detail by Sun et al. [58].

Most of the hydrogen that is currently produced employs fossil fuels. This is called grey hydrogen. Multiple colors are associated with the way hydrogen is produced. Green hydrogen is the result of using nonrenewable resources. Blue hydrogen is produced by steam reforming of natural gas. Before green hydrogen replaces fossil fuels many challenges have to be solved, such as the following:

- Scale: One of the main applications of hydrogen is in the reduction of iron ores to produce DRI. Elliot [59] has reported that in order to produce 1 ton of DRI it is required 650 N m^3 of green hydrogen (58 kg H_2/ton), therefore for a DRI plant producing 2 Mton/year the requirements would be 162,500 N m^3/h (14,500 kg H_2/h). Similar results were reported by Vogl and Ren. Vogl et al. [60] reported requirements of 5.22 TWh/year, about 595 MW, for a plant producing 1.5 Mton DRI/year. Ren et al. [61], considering a specific consumption of 55 kg H_2/ton DRI, estimated that for 2050 the production of DRI would be in the order to 243–300 Mton, requiring from 13 to 16 Mton of green H_2. To have a better idea on the current scale limitations to produce green electricity, the installed capacity of electrolisers is about 1.1 GW, its demand for 2030 is about 130 to 345 MW, more than one order of magnitude.
- Price: The price of green hydrogen is currently at least three times higher than blue hydrogen and forecasts indicate that even by 2040–2050 a similar situation will still remain. A forecast from 2020 suggest that using DRI produced with 100% H_2 would not compete with the price of hot metal unless the price of green electricity is below 35 USD/MWh [14]. In this scenario, different markets will

compete for hydrogen, not only the steel industry. Birat [48] has pointed out that the transport industry could use more hydrogen than the steel industry.

- Production scale: There are many technologies to produce green hydrogen but the most common is by water electrolysis. The capacity of the largest electrolysers is in the order of 20 MW, capable to produce 3000-ton H_2/year. Common electrolysers have a capacity of 6 MW. One single DRI plant would require 135 electrolysers.
- Storage and Transportation: H_2 storage and transportation have several problems [62]: The H_2 flammability limits range from 4 to 74% in comparison with methane from 5 to 15%. It has a very low ignition energy, 0.017 mJ. It has an ultra-low boiling point ($-250\,°C$), high thermal conductivity (~ 0.1 W/m/K) and high latent heat (448.7 kJ/kg), which combined lead to frostbite/cold burns. The low boiling point imposes severe insulating conditions to keep H_2 stored as a liquid. If liquid H_2 is left confined in a sealed vessel its expansion can cause significant pressure build up because it has a large liquid to vapor volume expansion ratio and occupies 860 times higher volume as a gas phase at ambient conditions compared to that as liquid phase. It is the smallest molecule/atom and can easily leak through most materials. It has the largest energy density on mass basis (~ 119.96 MJ/kg). It has the lowest energy density on volume basis (~ 8.7 MJ/lt) in liquid phase, therefore, requires a much larger vessel to store an equivalent amount of energy compared to other fuels. In addition to this, H_2 creates embrittlement of carbon and high strength steels, except for austenitic stainless steel (AISI 316). All these properties make storage and transportation a big problem. As an example, the transportation cost has been estimated for two locations, Cologne and Houston delivered from Egypt, was estimated at 9.4 and 8.6 €/ton, respectively, decreasing to 7.8 and 6.8 €/ton by 2050, respectively [63].
- Investments: The change to H_2 metallurgy will involve huge investments. Berger consultants [64] in 2020 have estimated that only for Germany with a production of steel in the order of 30 Mton per year, would require 30 billion euro of capital investments to shift from the blast furnace to H_2-DRI.

The concept of net zero carbon emissions is just a concept but it has to be understood that carbon-free steelmaking is an oxymoron. Some carbon in DRI will be needed to reduce the remaining iron oxide, in addition to this there are other sources of carbon emissions that usually are unavoidable and are not taken into account, for example the oxidation of electrodes and the production of metallurgical lime, combined represent about 53 kg CO_2/ton of steel [60].

15.3.7 Improve Materials Efficiency

According with the International Energy Agency (IEA), increasing material efficiency can contribute to decrease 40% of CO_2 [65]. Increasing materials efficiency is the result of a higher metallic yield during steel production, a decrease in the mass of steel for every application because of higher mechanical properties and a longer

Table 15.9 End of life of steel products

	Construction	Vehicles	Machinery	Consumer goods
Demand, %	50	15	20	15
End-of-life, years	25–35 (com.) 75–100 (resid.)	15–20 (cars)	20–30	10–12 (appliances) < 1 (packaging)

After [65]

end-of-life of steel products, all of which decrease the demand of steel. This efficiency depends on the type of steel products, as shown in Table 15.9. Since construction steel is the largest steel consumer, IEA proposes to improve buildings design and develop/use steels with a higher mechanical strength. Demolition of commercial buildings can be made after the real end-of-life of 75 years. The decrease in steel for vehicles can result from less mass of steel with a higher mechanical strength, policies oriented to the use of efficient public transport and home office. In addition to this, materials efficiency also involves recovery of iron units and recycling of iron and steel slags in different applications.

15.3.8   Commercial Application of Breakthrough (Ironmaking) Technologies

As pointed out before, the existing best available technologies are close to their theoretical limits and the only way to sustain in the long term the production of green steel is with the development of breakthrough technologies that are based on green hydrogen and green electricity, consequently its success is pending to their production problems, as previously discussed in this section. An important feature of the emerging technologies is that all of them are focused on ironmaking. There are many reviews that describe those emerging technologies [66–72]. What follows is a summarized version. Most of these technologies have resulted from multi-national research programs (European Union), national programs (Japan, USA) and from big steel companies (POSCO, BAOWU). The first program started in 2004, called ULCOS (Ultra Low CO_2 Steelmaking program), originated by the European steel industry, through the European Steel Technology Platform (ESTEP). The program was in operation 6 years until 2009 with huge achievements because new breakthrough technologies were proposed. Initial funding was €59 millions. ULCOS phase II continued with funding from the RFCS until 2012. The legacy of this program continues, the ideas proposed have been taken by different research groups worldwide. The majority of the proposed technologies employ Vacuum Pressure Swing Adsorption (VPSA), a CCS technology developed by Air liquide to separate CO_2 from the top gas. In addition to ULCOS BF-TGR and ULCOS Hisarna that have been reached pilot plant tests, ULCOS proposed two more processes, ULCORED and ULCOWIN. Ulcored involved CCS and use of alternative reducing gases, for example COG, to produce DRI. Those ideas have been already implemented on a commercial scale by the major technologies producing DRI. Ulcowin is about the production of iron by an electrochemical process at low temperatures using iron ore and iron as electrodes.

Japan, through NEDO (New Energy and Industrial Technology Development Organization) integrated steel companies in the program COURSE 50 (CO$_2$ Ultimate Reduction in Steelmaking process by innovative technology for cool Earth 2050) focused on the reduction of emissions CO$_2$ from the integrated plants using CCS and intensive use of H$_2$ in the blast furnace. Due to the endothermic reactions with H$_2$, the endothermic direct reduction with carbon has to be minimized and coke strength increased. COURSE50 has been in operation since 2008. This program was mainly focused on coke strength and the blast furnace. During 2020–2030 the program considers preliminary experimental work at industrial scale.

The American Iron and Steel Institute (AISI) and the USA-Department of energy (USA-DOE) have funded different research initiatives that have been coordinated by specific universities in the USA. The first project was funded in 1991, coordinated by Prof. Fruehan at CMU with a budget close to USD$70 million and focused on a bath smelting process, called AISI direct ironmaking [73, 74]. In this process iron ore pellets are pre-reduced and then smelted using coal injection in conjunction with a high degree of post-combustion. Pre-reduction is carried in a shaft reactor with a pre-reduction degree around 27%, then charged to the smelter at 800 °C. The gas from the smelter containing about 48%CO and 9%H$_2$ passes through a scrubber to remove the dust and then employed in the DRI reactor. It reached the pilot plant stage. Another important technology funded by AISI-DOE was the Flash Ironmaking Technology (FIT), developed by Prof. H.-Y. Sohn from the University of Utah [75], starting around 2010 and received about USD$ 16 million. The flash reactor has been widely employed in copper smelting but contrary to copper flash smelting, flash ironmaking is endothermic using hydrogen, also different to fluidized bed reactor that employ fines (+ 1 mm to − 10 mm), flash ironmaking employs iron ore concentrate < 100 μm and much higher reduction temperatures, from 1200 to 1550 °C. Contrary to fluidized bed reactors that have severe sticking problems, in the flash reactor this problem is avoided because reduction takes place in a matter of a few seconds. One version of this technology is to produce briquettes with the reduced particles. The process reached the pilot plant stage. Another promising emerging technology also funded by the AISI CO$_2$ breakthrough program is a high temperature electrochemical process to produce metal from metal oxides without carbon, called MOE (Molten Oxide Electrolysis). In the MOE cell, an inert anode is immersed in an electrolyte containing iron ore, and then it's electrified. When the cell heats to 1600 °C, the electrons split the bonds in the iron oxide to produce pure liquid metal. No carbon dioxide or other harmful by-products are generated in this process, only oxygen. Since the early twentieth century many metals have been produced by MOE. Prof. Allanore at MIT developed the process fundamentals for the production of molten iron [76] in 2013. The process is also known as Ulcolysis.

POSCO has kept the development of the original FIOR process, invented by ESSO in the early 1960s. The first commercial plant operated from 1976 to 2000 in

Venezuela. The successor was FINMET a development from FIOR Venezuela and VAI. Two Finmet plants were built in the late 1990's, one in Venezuela and another in Australia, producing briquettes. Finmet was a much-improved version of Fior, with specific consumptions of iron ore of 1.55 ton/ton DRI and energy of 12.4 GJ/ton DRI, in comparison with 1.98 ton/ton DRI and 20.6 GJ/ton DRI, respectively [77]. POSCO has been successful to connect the fluidized bed reactors with a smelter (FINEX process), therefore the reduction degree in the fluidized bed is lower in comparison to Finmet. The pilot plant stage was in 1996, the first demonstration plant started operations in 2003 and the first two commercial plants in 2007 and 2014. One of the major problems of Finex is that is a smelting technology and therefore produces large emissions of CO_2. CO_2 in Finex gas is higher than in the blast furnace, 40.3% and 24.8%, respectively [78].

China has lacked of national programs in the steel industry to develop new technologies, focused more on the acquisition of commercially available BAT to decrease its huge CO_2 emissions. The two largest steel producers in China, Baowu and HBIS, have pledged its commitment to achieve carbon neutrality by 2050. HBIS Xuanhua steelworks started operations of the first DRI plant in the world using coke oven gas in early 2023 and will progressively increase H_2.

In addition to the previous emerging technologies to produce hot metal, there is also a large number of programs to produce DRI using 100% hydrogen. The main technology suppliers, MIDREX and HyL have already developed experience in this field. In addition to them, several steel companies have also started research programs to produce DRI with green hydrogen (H-DRI). Millner et al. [79] concludes that although the technology is available, the main limitation is the high production costs using green hydrogen. Table 15.10 summarizes the emerging technologies to produce hot metal and H-DRI.

Hasanbeigi et al. [80] compared 12 emerging technologies on ironmaking that reduce CO_2 emissions. It was surprising that they suggested three options that produce a large amount of CO_2 emissions; COREX, FINEX and Coal-based HYL.

15.3.9 Carbon Trading Emission System (TES) and the Carbon Tax System (CTS)

The carbon trading emission system is a market strategy to decrease CO_2 emissions. Lohmann [81] describes the TES with an example, as follows; Two companies need to cut CO_2 emissions, company A has the BAT and is cheaper to do it in comparison with plant B which is obsolete, therefore plant B buy credits from plant A that effectively reduced its emissions. If this is scaled to the level of developed and undeveloped countries it seems that it can lead to brutal distortions of the original idea. The EU first applied TES in 2005, the USA in 2013 and China in 2021.

Table 15.10 Emerging technologies to produce hot metal and DRI with lower CO$_2$ emissions

Emerging technologies to produce hot metal			
2004	Europe	ULCOS BF-TGR	Top gas is recycled after CCS, heating and injection of recycled gas at both shaft and hearth tuyeres. TGR contains about 12%H$_2$ and 74%CO. Pure gas is injected at the hearth tuyere instead of hot blast air (eliminating N$_2$). Tested on experimental furnace of 1.7 ton/h. CO$_2$ emissions decreased in the range from 21 to 25% depending on the version. If successful it has the potential to decrease C consumption from 405 to 295 kg C/ton
2004	Europe	ULCOS Isarna or Hisarna	The name Isarna is Celtic (strong metal). Combines two existing technologies; CCF (cyclone converted furnace) and Hismelt. The reactor is formed by an upper part, the cyclone, the iron ore reduction degree is 20–25%. The lower part is the smelter chamber. Oxygen is injected at both the cyclone and smelter for post-combustion. Pilot plant results indicate a decrease in CO$_2$ emissions of 20%, relative to that of the BF
2008	Japan	COURSE 50	Decrease 30% CO$_2$ emissions from BF using CCS and H$_2$ from COG (57%) an RCOG (78%)
1991	USA	AISI-DOE	The AISI smelter reached the pilot plant stage. Not better than COREX, then discontinued
2010	USA	AISI-DOE	The flash ironmaking is an emerging technology that eliminates the problems of fluidized beds. It has reached the pilot plant stage
1996	Korea	POSCO FINEX	FINEX combined FINMET and COREX. It produces more CO$_2$ than the BF. Two commercial plants in operation
Hydrogen-DRI			
2016	Sweden	HYBRIT	Production of H-DRI using a shaft furnace. Pilot plant with support from TENOVA (HyL) completed in 2021 in Lulea. Partners LKAB, SSAB and Vattenfall. H$_2$ from water electrolysis
2017	Germany	SALCOS	Salzgitter started a research program to produce DRI using NG and H$_2$. H$_2$ from water electrolysis
2019	Germany	MIDREX-AM	H-DRI production at ArcelorMittal Hamburg using MIDREX reactor. 0.1 Mton annual capacity. Green H produced by steam methane reforming, 3 times more expensive than NG but H$_2$ from electrolysis is 1.5 more expensive. Production costs of H-DRI twice than conventional DRI [79]

There is a big cost to reduce CO$_2$ emissions. A report by the OECD/IEA in 2009 indicated a cost from 50 to 100 USD/ton CO$_2$ [9], suggesting investments in the order of 2–2.5 trillion USD between 2009 and 2050. Not all companies, out of free will, will be willing to cut CO$_2$ emissions and some companies actually prefer to pay taxes because it is cheaper.

Carbon tax is another strategy to reduce CO_2 emissions. In theory, the tax rate is equal to the prices to restore the damages of CO_2. Both, carbon taxes and TES reduce the demand of carbon intensive goods and services and provides incentives for individuals and companies to make investments in renewable energy and energy efficiency.

Wu et al. [82] have reviewed the effect of the TES in China with mixed results, both good and bad, but in general there was a positive effect. At present only the power sector is included in TES. Their study indicates the following findings; positive effects are obtained when TES has been implemented for a longer time, is a motivation for technological innovation but also reported a regional heterogeneity in results. Guandong had the largest share of TES, about 42%. Their results suggest the government should develop differentiated carbon trading policies in different regions and incorporate the iron and steel sector as soon as possible.

15.4 EAF Slags Recycling

Slags are valuable by-products from the iron and steel industry. The subject of energy recovery from EAF slags was discussed in Chap. 11. Slags were considered a waste for a very long time. In fact, the full potential of steelmaking slags recycling is still not fully developed in most countries. The amount of slag generated is huge [83]; Worldwide, the amount of iron and steelmaking slags produced is approximately 400 million/year, the Blast Furnace (BF) produces 250–300 kg/ton. The Basic Oxygen Furnace (BOF) produces 100–150 kg/ton and the electric Arc Furnace (EAF) produces approximately 100 kg/ton in scrap-based operations and 150–250 kg/ton in sponge iron-based operations. These numbers indicate that production of every ton of steel involves the production of 10–30% of slags at high temperatures.

Slag recycling has been a common practice in the past for blast furnace and secondary refining slags because those are basic slags and can be fully replace clinker in the production of cement, if the transportation distance is not far from the steel shop to the cement plant. The problem is in the recycling of steelmaking slags because they are of different chemical nature, they are oxidizing slags. There are many reviews on the progress made on slag recycling dealing with the recycling options [83–86]. Delbecq et al. have provided a critical assessment of the current status of slag valorization worldwide, pointing out the intrinsic value of all of its mineralogical components, for example the CaO content alone is worth 50 €/ton slag and the iron oxide 30 €/ton slag. A remarkable observation, in line with the trend in the prices of raw materials, was the suggestion to treat slags in a way similar to iron ores, involving fine crushing, selective separation and enrichment. Conejo [83] reviewed a large number of applications for steelmaking slags. The main problem is that except for road applications (it represents more than 50% of the commercial applications), the majority of the applications available employ a small fraction of the total amount of slags produced. Guo et al. [85] described the status of slag recycling in China up to 2013. China was one of the countries with the lowest recycling rate for steelmaking slags, below 20%. Matsuura et al. [86] in a more recent update reported an increment to 30% in 2017 and 2018. The large amount of steel steelmaking slag generated in China, about 250 Mton/year indicate the need to improve existing technologies, management and policies related to slag recycling. Conejo [83] concluded that one of the main problems for the recycling of steel slags is the need of better characterization, on a steelshop basis, of its physical and chemical properties due to their highly heterogeneous nature, in order to select the best alternative for recycling and the risks associated with harmful components to human health. Wang et al. [87] concluded that iron and steel slags investigated in China exceeded the limits of harmful elements and therefore should not be used for agricultural applications, contrary to a common practice in Japan in rice fields.

15.5 The EAF Steel Shop of the Future

The current decade will mark the peak in production for the conventional route through the BF-BOF. The main competitor for the EAF in the future will be the EAF itself by solving its own limitations in terms of energy efficiency, decarburization rates, metallic yield, etc. Worldwide the share of the EAF will keep increasing for a large part of this century, before a commercial continuous steelmaking process is developed. If new blast furnaces are built in the future, one option, suggested by Schmöle et al. [88], is the combination of multiple reactors that benefit from recycling the off-gases among them, for example, COREX supplies the off-gas to a DRI plant, the DRI plant feeds both the BF and EAF and COREX-BF both feed a BOF. This is an example of symbiotic mutualism and synergy.

Based on the changes in the last few decades on the EAF and the need to improve its current limitations, the following is a description of the EAF by 2050 along with the existing challenges:

1. *Energy consumption*:
    Electric energy consumption and renewable energies: By 2050 the EAF will consume close to 100% of green electricity in developed countries, produced by wind and solar energies. Currently there are examples of steel shops running with solar energy. Evraz in Pueblo CO USA has made the first steps to invest in a 240 MW solar facility to run a steel mill [89].
    Thermal efficiency: The thermal efficiency of the EAF will increase from current values around 50–55 to 70–80%. This will be possible by improving all the methods to decrease energy consumption described in this book, such as optimization of slag foaming using biomass or hydrogen, scrap pre-heating, energy recovery from liquid slags, hot DRI, etc. Electric energy consumption with steel scrap will be below 200 kWh/ton and below 400 kWh/ton with hot H_2-DRI. These values are not far from current values because of two opposite trends; automation decreasing energy consumption and higher gangue DRI increasing energy consumption.

2. **Raw materials**: The EAF has proved to be highly versatile in terms of raw materials, however it is also known that steel scrap provides the lowest energy consumption and lowest CO_2 emissions and therefore it should be kept as the primary raw material:

   Steel scrap: As explained in Chap. 1, in order to increase the share of scrap in the EAF there are three challenges; availability price and residual elements. It has been discussed in Chap. 9 the need to radically improve the way scrap is collected and the non-ferrous materials are also removed. Residual elements will become a bigger problem.

   DRI: DRI will play a bigger role than today in particular due to the presence of higher concentration of residual elements in steel scrap, however, the available DRI will bring two problems; a higher concentration of phosphorous and acid gangue, on one hand, and on the other hand it will have a minimum amount of carbon because it will be produced using hydrogen as a reducing gas. It is urgent to improve and invest on new technologies on iron ore beneficiation which separate as much as possible the impurities in the iron ore.

   In the last 6 years the consumption of steel scrap has remained 600 Mton/ year on average by contrast the consumption of DRI has increased from 86 to 135 Mton from 2017 to 2023.

   Hot metal: Hot metal will still be used, especially in China due to its large availability and lack of steel scrap, even though it could be a cheaper way to produce steel, the production of large amounts of CO_2 make this option unsuitable.

3. **Sensors and AI**: Automation will keep developing faster displacing direct human intervention. Technology producers have already developed sensors to monitor online a large number of process variables. The level of automation will require only one person per furnace. EAF engineers will develop dynamic control models that are interconnected with up-stream and down-stream processes.

4. **Stirring**: Bottom gas injection, electromagnetic stirring, oxygen injection jets, as well as changes in the current EAF geometry will increase the stirring conditions to provide similar decarburization rates to the BOF. Radical improvements to the current stirring conditions are essential to increase the decarburization rate and achieve a similar productivity to the BOF. Submerged oxygen injection is also important to enhance oxygen efficiency and reach higher DeC rates.

5. **EAF design and geometry**: The crucible geometry of the EAF has remained with minor changes with respect to the original designs. This area requires further research to define the optimum geometry in terms of crucible diameter and crucible height. It is expected an increase in the H/D ratio to values higher than 0.2.

6. **New EAF technologies**:

   DRI injection: An improvement in DRI is the elimination of pelletizing. Currently several technologies can produce DRI from iron ore concentrates but fluidization problems still remain. Once all these problems are solved DRI will be injected.

Submerged oxygen injection: Submerged gas injection is a common in the Q-BOF. This technology has been applied to the EAF and will expand further due to multiple benefits.

Hollow electrodes: Solid electrodes are highly expensive due to both the large energy consumption and time involved in its production. Hollow electrodes have been proposed since the early 1970s [90], which not only significantly decreases its mass but also can allow to introduce powders (such as DRI) through the center. Another alternative is the production of graphite by cheaper new methods such as graphite precipitation from cooling of hot metal [91].

Direct iron and steelmaking process: EAF ironmaking has been explored since the 1950s. In the past the proposals have employed fossil fuels. If hydrogen succeeds at reducing liquid iron oxides and promote slag foaming, steel will be directly produced in the EAF.

7. **AC-EAF or DC-EAF**: The AC will be the dominant type of electric energy. The DC-EAF has not exceeded in overall performance the AC-EAF.

8. **Environment**: Increasing the share of EAF to about 40–50% by 2050 will decrease CO_2 emissions. The EAF will employ a much higher share of H_2-DRI, from 30 to 40%. H_2-DRI, steel scrap and green electricity combined, will decrease the world average emissions from 1.85-ton CO_2/ton steel from today to values in the range from 0.4 to 0.6-ton CO_2/ton steel for the whole steel industry. This goal is possible but requires international synergies.

9. **Productivity**: Chaubal [92] CTO at ArcelorMittal recently described the future of the steel industry in terms of having steel plants with no dust, no noise, and with productivities that allow the supply of steel products in less than a week, rather than 3–6 months. The model of bonus based on workers performance should be fairly stimulated.

10. **Human resources**: In addition to major innovations for the conventional EAF, it is even more important to prepare the human resources which will operate the equipment. Pradip et al. [93] describe a large number of tools that the future metallurgical engineer will need, which seems so overwhelming. In my opinion, instead of more information, engineering students should receive an education based not only on critical thinking and how to make things in a better way, but also on a mindset open for collaboration and sharing knowledge without borders.

The future of the EAF will be largely influenced by China and India, in particular the changes they made in the next 10 years. More important, the future depends on our decisions from today to reach a symbiotic relationship with people, resources and technologies.

References

1. Ritchie H (2019) Who has contributed most to global CO_2 emissions? Ourworldindata. https://ourworldindata.org/contributed-most-global-co2. Accessed 7 Nov 2023

 2. Zhang X, Jiao K, Zhang J, Guo Z (2021) A review on low carbon emissions projects of steel industry in the world. J Clean Prod 1–11
 3. Evans S (2021) Analysis: which countries are historically responsible for climate change? CarbonBrief. https://www.carbonbrief.org/analysis-which-countries-are-historically-responsible-for-climate-change/. Accessed 7 Nov 2023
 4. Kawabata N (2023) Evaluating the technology path of Japanese steelmakers in green steel competition. Japan Polit Econ 49:231–252. https://doi.org/10.1080/2329194X.2023.2258162
 5. Jones J, Safe P, Wiggins B (1999) Optimization of EAF operations through offgas system analysis. In: Electric furnace conference, Pittsburgh, PA, USA, 14–16 Nov, pp 459–480
 6. Daehn K, Basuhi R, Gregory J, Berlinger M, Somjit V, Olivetti EA (2022) Innovations to decarbonize materials industries. Nat Rev Mater 7:275–294. https://doi.org/10.1038/s41578-021-00376-y
 7. Wang P, Ryberg M, Yang Y, Feng K, Kara S, Hauschild M et al (2021) Efficiency stagnation in global steel production urges joint supply- and demand-side mitigation efforts. Nat Commun 12:1–12. https://doi.org/10.1038/s41467-021-22245-6
 8. US-EIA (2022) IEO2021 issues in focus: energy implications of potential iron and steel-sector decarbonization pathways
 9. Tam C (2009) Energy technology transitions for industry. OECD/IEA. International Energy Agency, pp 1–326. https://www.iea.org/reports/energy-technology-transitions-for-industry
10. Hasanbeigi A (2022) Steel climate impact—an international benchmarking of energy and CO_2 intensities. Global Efficiency Intelligence. Global Efficiency Intelligence, Florida
11. Raabe D (2023) The materials science behind sustainable metals and alloys. Chem Rev 123:2436–2608. https://doi.org/10.1021/acs.chemrev.2c00799
12. Worrell E, Price L, Martin N, Farla J, Schaeffer R (1997) Energy intensity in the iron and steel industry: a comparison of physical and economic indicators. Energy Policy 25:727–744. https://doi.org/10.1016/s0301-4215(97)00064-5
13. Hasanbeigi A, Price L, Aden N, Chunxia Z, Xiuping L, Fangqin S (2011) A comparison of iron and steel production energy intensity in China and the U.S. Lawrence Berkeley National Laboratory, Res Rep LBNL 4836E, pp 1–77
14. https://www.iea.org/reports/iron-and-steel. IEA report on iron and steel 2020
15. He K, Wang L, Li X (2020) Review of the energy consumption and production structure of China's steel industry: current situation and future development. Metals (Basel) 10:18. https://doi.org/10.3390/met10030302
16. Kirschen M, Badr K, Pfeifer H (2011) Influence of direct reduced iron on the energy balance of the electric arc furnace in steel industry. Energy 36:6146–6155. https://doi.org/10.1016/j.energy.2011.07.050
17. Worrell E, Price L, Martin N (2001) Energy efficiency and carbon dioxide emissions reduction opportunities in the US iron and steel sector. Energy 26:513–536. https://doi.org/10.1016/S0360-5442(01)00017-2
18. Song X, Du S, Deng C, Shen P, Xie M, Zhao C et al (2023) Carbon emissions in China's steel industry from a life cycle perspective: carbon footprint insights. J Environ Sci (online). https://doi.org/10.1016/j.jes.2023.04.027
19. Lin Y, Yang H, Ma L, Li Z, Ni W (2021) Low-carbon development for the iron and steel industry in china and the world: status quo, future vision, and key actions. Sustainability 13:1–28. https://doi.org/10.3390/su132212548
20. Yin R, Liu Z, Shangguan F (2021) Thoughts on the implementation path to a carbon peak and carbon neutrality in China's steel industry. Engineering 7:1680–1683. https://doi.org/10.1016/j.eng.2021.10.008
21. Kirschen M, Risonarta V, Pfeifer H (2009) Energy efficiency and the influence of gas burners to the energy related carbon dioxide emissions of electric arc furnaces in steel industry. Energy 34:1065–1072. https://doi.org/10.1016/j.energy.2009.04.015
22. Mourao J, Cameron I, Huerta M, Patel N, Pereira R (2013) Comparison of sinter and pellet usage in an integrated steel plant. In: 1st Brazilian symposium on agglomeration of iron ore, Belo Horizonte, Brazil, 1–4 Sept, p 11

23. Arasto A, Tsupari E, Kärki J, Pisilä E, Sorsamäki L (2013) Post-combustion capture of CO_2 at an integrated steel mill—Part I: Technical concept analysis. Int J Greenh Gas Control 16:271–277. https://doi.org/10.1016/j.ijggc.2012.08.018

24. Hornby-Anderson S, Metius G, McClelland J (2002) Future green steelmaking technologies. In: 60th Electric furnace conference, San Antonio, TX, USA, 10–13 Nov, pp 175–191

25. Cavaliere P (2016) Dioxin emission reduction in electric arc furnaces for steel production. In: Ironmaking and steelmaking processes greenhouse emissions, control, and reduction. Springer, Berlin, pp 215–222. https://doi.org/10.1007/978-3-319-39529-6

26. Jones JAT, Bowman B, Lefrank PA (1998) Electric arc furnace steelmaking. In: Fruehan RJ (ed) Making, shaping and treating of steel, 11th edn. Steelmaking and refining volume, Pittsburgh, PA, USA, pp 525–660

27. Yamaguchi R, Mizukami H, Maki T, Ao N (2000) ECOARC technology. In: 58th Electric furnace conference, Orlando, FL, USA, 12–15 Nov, pp 325–336

28. Sofilić T, Jendričko J, Kovačevic Z, Ćosić M (2012) Measurement of polychlorinated dibenzo-p-dioxin and dibenzofuran emission from EAF steel making process. Arch Metall Mater 57:811–821. https://doi.org/10.2478/V10172-012-0089-1

29. Ilutiu-Varvara A (2016) Dangerous emissions during steelmaking in electric arc furnaces. In: Ironmaking and steelmaking processes greenhouse emissions, control, and reduction, chap 15. Springer, Berlin, pp 247–265

30. Lawal AT (2017) Polycyclic aromatic hydrocarbons. A review. Cogent Environ Sci 3:1–89. https://doi.org/10.1080/23311843.2017.1339841

31. Baraniecka J, Pyrzyńska K, Szewczyńska M, Pośniak M, Dobrzyńska E (2010) Emission of polycyclic aromatic hydrocarbons from selected processes in steelworks. J Hazard Mater 183:111–115. https://doi.org/10.1016/j.jhazmat.2010.06.120

32. Shen J, Yang L, Yang Q, Zhao X, Liu G, Zheng M (2021) Polychlorinated biphenyl emissions from steelmaking electric arc furnaces. Bull Environ Contam Toxicol 106:670–675. https://doi.org/10.1007/s00128-021-03105-x

33. Ocio JA, Bilbao H, Gayo JL, Garcia N, Seoanez C, Demir M et al (2012) BAT guide for electric arc furnace iron & steel installations, p 236. https://webdosya.csb.gov.tr/db/ippceng/icerikbelge/icerikbelge866.pdf

34. Hagen P, Walls M (2005) The Stockholm convention on persistent organic pollutants. Nat Resour Environ 19:49

35. Yang Q, Yang L, Shen X, Zheng M, Liu G (2021) Organic pollutants from electric arc furnaces in steelmaking: a review. Environ Chem Lett 19:1509–1523. https://doi.org/10.1007/s10311-020-01128-0

36. Cappelletti R, Ceppi M, Claudatus J, Gennaro V (2016) Health status of male steel workers at an electric arc furnace (EAF) in Trentino, Italy, pp 1–10. https://doi.org/10.1186/s12995-016-0095-8

37. International Energy Agency (2012) Energy technology perspectives 2012—how to secure a clean energy future. IEA

38. Li Y, Wang Z (2023) Green and grin. China Dly Glob 1. https://www.chinadaily.com.cn/a/202304/04/WS642b6baca31057c47ebb82e8.html. Accessed 7 Nov 2023

39. Holappa L (2020) A general vision for reduction of energy consumption and CO_2 emissions from the steel industry. Metals (Basel) 10:1–20. https://doi.org/10.3390/met10091117

40. Liu H, Consoli C, Zapantis A (2018) Overview of carbon capture and storage (CCS) facilities globally. In: 14th International conference on greenhouse gas control technologies, Melbourne, AU, 21–25 Oct, pp 1–10

41. Robertson B, Mousavian M (2022) The carbon capture crux; the lessons learned. Institute for Energy Economics and Financial Analysis. https://ieefa.org/resources/carbon-capture-crux-lessons-learned. Accessed 7 Nov 2023

42. Greenpeace (2008) False hope. Why carbon capture and storage won't save the climate

43. Zevenhoven R, Fagerlund J, Songok JK (2011) Review CO_2 mineral sequestration: developments toward large-scale application. Greenh Gases Sci Technol 57:48–57. https://doi.org/10.1002/ghg3

44. Xie H, Yue H, Zhu J, Liang B, Li C, Wang Y et al (2015) Scientific and engineering progress in CO_2 mineralization using industrial waste and natural minerals. Engineering 1:150–157. https://doi.org/10.15302/J-ENG-2015017
45. Biswal T, Shadangi KP, Sarangi PK, Srivastava RK (2022) Conversion of carbon dioxide to methanol: a comprehensive review. Chemosphere 298. https://doi.org/10.1016/j.chemosphere.2022.134299
46. Ban-ya S, Iguchi Y, Nagasaka T (1984) Rate of reduction of liquid wustite with hydrogen. Testu Hagane 70:689–696
47. Chen S, Zhang J, Wang Y, Wang T, Li Y, Liu Z (2023) Thermodynamic study of H_2-FeO based on the principle of minimum Gibbs free energy. Metals (Basel) 13:1–12
48. Birat J (2020) Society, materials, and the environment: the case of steel. Metals (Basel) 10:331
49. Jahanshahi S, Mathieson JG, Somerville MA, Haque N, Norgate TE, Deev A et al (2015) Development of low-emission integrated steelmaking process. J Sustain Metall 1:94–114. https://doi.org/10.1007/s40831-015-0008-6
50. Echterhof T (2021) Review on the use of alternative carbon sources in EAF steelmaking. Metals (Basel) 11:1–16. https://doi.org/10.3390/met11020222
51. Babich A, Senk D, Fernandez M (2010) Charcoal behaviour by its injection into the modern blast furnace. ISIJ Int 50:81–88. https://doi.org/10.2355/isijinternational.50.81
52. Goldemberg J, Coelho S (2004) Renewable energy—traditional biomass vs. modern biomass. Energy Policy 32:711–714
53. Abbasi T, Abbasi SA (2010) Biomass energy and the environmental impacts associated with its production and utilization. Renew Sustain Energy Rev 14:919–937. https://doi.org/10.1016/j.rser.2009.11.006
54. RINA (2021) Developing and enabling H_2 burner utilization to produce liquid steel in EAF (DevH2forEAF). RFCS
55. von Scheele J (2020) Embracing hydrogen flameless oxyfuel for CO_2-free heating. Iron Steel Today 12–14
56. TNS (2023) Development of hydrogen-oxygen burner for industrial furnaces. Taiyo Nippon Sanso, pp 1–4. https://www.tn-sanso.co.jp/LinkClick.aspx?fileticket=qpEb4LTAa08%3D&tabid=253&mid=994. Accessed 11 Nov 2023
57. Squires A, Johnson C (1957) The H-iron process. JOM 586–590
58. Sun M, Pang K, Jiang Z, Meng X, Gu Z (2023) Development and problems of fluidized bed ironmaking process: an overview. J Sustain Metall. https://doi.org/10.1007/s40831-023-00746-6
59. Elliot A (2022) La transicion a la economia de hidrogeno. IAS. https://www.siderurgia.org.ar/webinars.php. Accessed 3 May 2022
60. Vogl V, Åhman M, Nilsson LJ (2018) Assessment of hydrogen direct reduction for fossil-free steelmaking. J Clean Prod 203:736–745. https://doi.org/10.1016/j.jclepro.2018.08.279
61. Ren L, Zhou S, Peng T, Ou X (2021) A review of CO_2 emissions reduction technologies and low-carbon development in the iron and steel industry focusing on China. Renew Sustain Energy Rev 143. https://doi.org/10.1016/j.rser.2021.110846
62. Ratnakar RR, Gupta N, Zhang K, van Doorne C, Fesmire J, Dindoruk B et al (2021) Hydrogen supply chain and challenges in large-scale LH_2 storage and transportation. Int J Hydrogen Energy 46:24149–24168. https://doi.org/10.1016/j.ijhydene.2021.05.025
63. Collis J, Schomäcker R (2022) Determining the production and transport cost for H_2 on a global scale. Front Energy Res 10:1–24
64. Ito A, Langefeld B, Götz N (2020) The future of steelmaking—how the European steel industry can achieve carbon neutrality
65. IEA (2020) Iron and steel technology roadmap. Towards more sustainable steelmaking. https://doi.org/10.1787/3dcc2a1b-en
66. der Stel J, Hattink M, Zeilstra C, Louwerse G, Hirsch A, Janhsen U et al (2014) ULCOS top gas recycling blast furnace process (ULCOS TGRBF). RFCS 26414
67. Meijer K, Zeilstra C, Teerhuis C (2013) Developments in alternative ironmaking. Trans Indian Inst Met 66:475–481. https://doi.org/10.1007/s12666-013-0309-z

68. Meijer K, Denys M, Lasar J, Birat J, Still G, Overmaat B (2009) ULCOS: ultra-low CO_2 steelmaking. Ironmak Steelmak 36:249–252. https://doi.org/10.1179/174328109X439298

69. Post G (1973) The optimum method for converting sponge iron to steel in arc furnaces. In: Symposium on science and technology of sponge iron and its conversion to steel, Jamshedpur, India, 19–21 Feb, National Metallurgical Laboratory

70. Jahanshahi S, Mathieson JG, Reimink H (2016) Low emission steelmaking. J Sustain Metall 2:185–190. https://doi.org/10.1007/s40831-016-0065-5

71. Quader MA, Ahmed S, Dawal SZ, Nukman Y (2016) Present needs, recent progress and future trends of energy-efficient ultra-low carbon dioxide (CO_2) steelmaking (ULCOS) program. Renew Sustain Energy Rev 55:537–549. https://doi.org/10.1016/j.rser.2015.10.101

72. Junjie Y (2018) Progress and future of breakthrough low-carbon steelmaking technology (ULCOS) of EU technologies in global steel industry. Int J Miner Process Extr Metall 3:15–22. https://doi.org/10.11648/j.ijmpem.20180302.11

73. Sarma B, Fruehan J (1997) Bath smelting slag reactions. In: Fifth international conference on molten slags, Fluxes Salts, Sydney, Australia

74. Fruehan RJ (1999) Reduction smelting processes—technology and economics. In: Direct-reduced iron: technology and economics of production and use, chap 12. ISS, Warrendale, pp 163–171

75. Sohn HY (2023) Flash ironmaking. CRC Press

76. Allanore A (2015) Features and challenges of molten oxide electrolytes for metal extraction. J Electrochem Soc 162:E13-22. https://doi.org/10.1149/2.0451501jes

77. Zeller S, Deimek G, Milionis K, Gould L, Whipp R (1998) FINMET-fine ore reduction for HBI. Steel Times Int 18–19

78. Jeong W, Lee J, Ko C, Yi S, Lee JH (2023) Development and evaluation of FINEX off-gas capture and utilization processes for sustainable steelmaking industry. Int J Greenhouse Gas Control 127

79. Millner R, Ofner H, Boehm C, Ripke SJ, Metius M (2017) Future of direct reduction in Europe-medium and long-term perspective. Eur Steel Technol Appl Days-ESTAD 2017:26–29

80. Hasanbeigi A, Arens M, Price L (2014) Alternative emerging ironmaking technologies for energy-efficiency and carbon dioxide emissions reduction: a technical review. Renew Sustain Energy Rev 33:645–658. https://doi.org/10.1016/j.rser.2014.02.031

81. Lohmann L (2006) Carbon trading: a critical conversation on climate change, privatisation and power. Dev Dialogue 1–362

82. Wu R, Tan Z, Lin B (2023) Does carbon emission trading scheme really improve the CO_2 emission efficiency? Evidence from China's iron and steel industry. Energy 277:127743. https://doi.org/10.1016/j.energy.2023.127743

83. Conejo A (2014) Steel slags: characterization and alternatives of recycling. In: CONAC, Monterrey, NL, Mexico, 23–26 Mar. AIST Mexico, p 12

84. Delbecq J-M, Franceschini G, Fixaris M (2013) Blast-furnace and steelmaking slags: which future valorisation in the next 20 years? In: Third slag valorization symposium, Leuven, Belgium, 19–20 Mar, p 257

85. Guo J, Bao Y, Wang M (2018) Steel slag in China: treatment, recycling, and management. Waste Manag 78:318–330. https://doi.org/10.1016/j.wasman.2018.04.045

86. Matsuura H, Yang X, Li G, Yuan Z, Tsukihashi F (2022) Recycling of ironmaking and steel-making slags in Japan and China. Int J Miner Metall Mater 29:739–749. https://doi.org/10.1007/s12613-021-2400-5

87. Wang X, Li X, Yan X, Tu C, Yu Z (2021) Environmental risks for application of iron and steel slags in soils in China: a review. Pedosph An Int J 31:28–42. https://doi.org/10.1016/S1002-0160(20)60058-3

88. Schmöle P, Lüngen HB, Noldin JH (2012) New ironmaking technologies: will the dominance of the blast furnace be ever challenged? In: 6th International congress on science and technology of ironmaking 2012, ICSTI 2012—including Proceedings from 42nd ironmaking and raw materials seminar. 13th Brazilian symposium on iron ore 3, pp 2093–2104

89. Best A (2020) Making steel with solar energy. https://MountaintownnewsNet/2020/08/07/ Evraz-and-Solar-Project/
90. CMP (1992) Electric arc furnace efficiency. Center for Materials Production. Report 92-10, Pittsburgh, PA USA
91. Gu Y, Gong X, Lan X, Guo L, Guo Z (2024) Experimental study on precipitation behavior of Kish graphite during cooling of molten iron. Chin J Process Eng 24:71–78. https://doi.org/10. 12034/j.issn.1009-606X.223116
92. Chaubal P (2022) Transforming technology: decarbonising steel production through innovation. Futurising Epis 2. https://www.youtube.com/watch?v=8udByIWW92g
93. Pradip, Gautham BP, Reddy S, Runkana V (2019) Future of mining, mineral processing and metal extraction industry. Trans Indian Inst Met 72:2159–2177. https://doi.org/10.1007/s12 666-019-01790-1